Winnacker   **From Genes to Clones**

Prof. Dr. Ernst-Ludwig Winnacker
Chemisches Laboratorium
der Universität München
Institut für Biochemie
Karlstraße 23
8000 München 2
Federal Republic of Germany

Editorial Director: Dr. Michael G. Weller
Production Manager: Dipl.-Ing. (FH) Hans Jörg Maier

Library of Congress Card No. 86-24676

CIP-Kurztitelaufnahme der Deutschen Bibliothek

Winnacker, Ernst-Ludwig:
From genes to clones : introd. to gene technology /
Ernst-L. Winnacker. Transl. by Horst Ibelgaufts. –
Weinheim ; New York : VCH, 1987.
  Dt. Ausg. u.d.T.: Winnacker, Ernst-Ludwig:
  Gene und Klone
  ISBN 3-527-26644-5 (Weinheim)
  ISBN 0-89573-614-4 (New York)

Composition: hans richarz publications service, D-5205 St. Augustin 1
Printing: VOD Vereinigte Offsetdruckereien Mannheim Heidelberg, D-6904 Eppelheim
Bookbinding: Wilh. Osswald + Co · Großbuchbinderei, D-6730 Neustadt/Weinstraße
Printed in the Federal Republik of Germany

Preface and Words of Thanks

You see things, and say why?
But I dream things that never were, and I say, why not?

George Bernard Shaw

Gene technology has revolutionised biology in the last decade. Landmarks in this development were the discovery of the mosaic structure of genes, the expression of information stored in the gene by different organisms, the identification of control elements, and observations concerning the mobility of genes and variations in the genome.

"From Genes to Clones" gives a summary of the basic concepts and strategies behind the art of gene cloning. Thus, it is not a textbook of molecular biology (although it includes descriptions of a variety of phenomena relevant to the general subject) nor is it merely a book on laboratory recipes (for these the reader is referred to, e.g., the excellent "Molecular Cloning" by Maniatis, Fritsch and Sambrock). Rather, it gives an introduction to the field now called "Gene Technology" and its specific unit operations, i.e., the isolation and characterisation of DNA and of genes, the development of cloning vectors, and the characterisation of recombinant DNA molecules. Within this broad scope, "From Genes to Clones" covers vector developments in a variety of biological systems, from micro-organisms (E. coli, Streptomyces, B. subtilis, yeast) via plants to higher eukaryotic cells and organisms (including transgenic animals). In addition such a book would never be useful without chapters on the methodological prerequisites of cloning techni-ques, i.e. DNA sequencing, chemical synthesis of DNA, mapping techniques and in vitro mutagenesis.

Many of these steps have been treated in the literature, but, to my knowledge, a unified approach is still missing. Both from the point of view of a teacher and that of an experimental scientist I felt that a text which summarises all relevant aspects was long overdue. And even more so, since genetic engineering, apart from its importance in research, not only gains momentum in university courses but also is about to involve a variety of industries and may eventually create a novel industrial world.

Finally, an entire chapter of this book is devoted to the problems of safety in recombinant DNA work. It does not only contain a description of the historical developments leading to the recombinant DNA debate in the seventies and to the establishment of the NIH guidelines (the 1986 version of which is included); but it also deals with more recent concerns and potential risks including large-scale industrial production, deliberate release of micro-organisms and plants and the issues associated with genome analysis and gene therapy. The author's interest in these matters does not only stem from a two-and-a-half-year tenure as a member of a parliamentary commission established to assess chances and risks of this new technology. It also reflects a genuine concern. The author thus sincerely hopes that readers of this text will conduct their experiments with the utmost care and consideration for themselves and the environment.

There is no doubt that a book of this type will never be complete. I am fully aware, for example, that the reader will miss an extended discussion on adenovirus vectors although the author himself is working on this subject. Some readers may find that the references to "protein engineering" appear rather short. This field, however, is exploding and it was decided only to include the necessary methodology together with some early applications.

The photographs honour pioneers in the field of gene cloning. Their selection was an almost impossible task which I choose to undertake in order to add a personal touch to the otherwise rather impersonal and detached assembly of facts and data in such a textbook. I am grateful to all colleagues and friends who have contributed their photographs and apologise sincerely to all of those who have been forgotten.

Finally, there is a problem of topicality and updating inherent and associated with textbooks on any rapidly expanding field. The reader of this book will have to realise this constantly. The book, however, as a consolation will prepare him/her for the task of updating his/her knowledge in the primary literature. Students in the field, in addition, will have to, and should, exploit the chance of attending appropriate lectures.

It is customary for an author of such a text to assume responsibility for all omissions and mistakes. I certainly accept this responsibility and thus kindly ask everyone to inform me about possible and necessary corrections.

I owe an immense debt of gratitude to numerous friends who have helped me to write and to complete this book. The initial idea originated with Drs. Ebel and Giesler from VCH Publishers. Considering their initial proposals I was reminded of the comments of a cleaning lady in a professor's office filled with books up to the ceiling: "Why do you write a book, you already own so many". This argument, fortunately, could be dealt with and dispensed quite easily.

More critical was the problem of language. One's native speech, as Erwin Schrödinger has noted in his preface to "What is Life", is a closely fitting garment and one never feels quite at ease when it is not immediately available and has to be replaced by another one. This book was initially written in German and appeared in a German edition in the fall of 1984. The English translation was performed by Dr. Horst Ibelgaufts. Based on his critical and constructive contributions, an update of its scientific contents (up to the state of the art in early 1986) was undertaken by myself, while I enjoyed the hospitality of Dr. Howard Goodman in his department at the Massachusetts General Hospital. He and his colleagues, Fred Ausubel, Jack Szostak, Bob Kingston and Brian Seed were never hesitant to offer their comments and advise. To Brian Seed, in particular, I owe the appendix with its physical maps and DNA sequences of representative vectors. He even engulfed himself in the German edition of this book and thus contributed several critical corrections. I adopted most, if not all of them although Brian may still be unhappy with the chapter on cDNA cloning. cDNA cloning, indeed, is a difficult subject, central to gene cloning, but in practice more of an art than a real scientific endeavour. The scope of the book, unfortunately, did not warrant any lengthier and more detailed elaborations.

The English version would never have been completed without Leslie Taylor. In numerous discussions she has contributed novel perspectives and, if anyone, has understood the ideas behind my efforts. She has introduced me to the world of computer sciences but at the same time has prevented me from writing and inserting a chapter on computer applications into this book. The immediate applications in, *e.g.,* DNA sequencing or sequence comparisons, are readily available to everyone, while more ambitious projects, *e.g.,* molecular dynamics, are far beyond the scope of this book. Leslie also introduced me to Alison DeLong. Alison has given lots of her time to read our version of this book. Due to her efforts, her efficiency, her critical mind and her own scientific skills it gained the necessary final touch.

The collaboration with Horst Ibelgaufts turned out to be delightful. He was gracious enough to accept and integrate the various corrections made and proposed by myself and Alison. He also prepared the manuscript for typesetting and diligently and with endless patience screened the manuscript for typing and more substantial errors.

Of particular significance was the extraordinary skill displayed by those who typed various versions of the manuscript and provided the drawings. With greatest pleasure and gratitude I thus acknowledge the help of Gudrun Kausel and Kathy Dinwoodie as well as that of F. Wagner from Hoechst AG.

Dr. Weller and R. Maier of the VCH Publishers have coordinated all these various activities and thus are responsible for the preparation and presentation of the book in its printed form. Finally, there are my colleagues at the University of Munich, Drs. G. Hartmann and E. Fanning who have tolerated a considerable negligence on my part with respect to various administrative duties during my tenure as department chairman.

Last but not least, I like to mention my family. Since the contributions of Antonet, Vera and Thomas are of such an enormous extent, the book is dedicated to them.

München, August 1986

Ernst-Ludwig Winnacker

Contents

1 Introduction

The fundamental aspect of gene technology is the creation of new combinations of genetic material from DNA molecules of different origin. Such recombinant molecules are introduced into appropriate, albeit unnatural, host organisms, where they can be multiplied and selected for. This definition sets the scene for the basic steps of a recombinant DNA experiment (Fig. 1-1). First of all, the DNA to be cloned, *i.e.*, the "foreign" DNA, and a suitable vehicle for transmission, called vector DNA, are joined covalently. Vector DNAs are able to replicate in certain host cells due to the presence of a specific DNA structural element known as the "origin of replication" (*ori*). Since "foreign" DNA and vector DNA are covalently linked, the specific property of self-replication is also bestowed upon the "foreign" DNA molecule. The second step is to introduce the composite recombinant DNA molecule into a suitable host cell by way of transfection or transformation. Cells harbouring recombinant DNA molecules can be recognised and isolated by suitable selection procedures. Individual positive colonies, called clones, can then be propagated and grown in bulk, thus allowing the cloned DNA to be isolated from its new host in large amounts. Of course, it may also be desirable in certain cases that the "foreign" DNA be expressed in its new host, *i.e.*, translated into the protein for which it codes.

The chapters in this book will mirror these individual steps involved in a typical recombinant DNA experiment. Special techniques have been developed for each step. The selection of appropriate methods and experimental strategies pursued will naturally depend on individual biological problems; the properties of the desired host organisms, the size of the DNA to be cloned, and the desire to express the cloned DNA will necessarily influence any decisions about a suitable vector. Eventually the methods employed for the ultimate identification of the desired cloned DNA molecules will depend on the quality of this DNA and will have to be more elaborate, selective and specific, if the cloned DNA is not pure. It should be emphasised that cloning techniques *per se*, however elegantly designed, do not alter the heterogeneity of a DNA sample. If a certain mRNA can be purified easily, for example, it may be advantageous to invest some time and effort in the purification of mRNA to simplify the subsequent identification and selection of clones. Finally, selection of a vector also may have to take into account considerations of biological safety.

These and other problems will be discussed in detail in the following chapters. I hope that this will enable the reader not only to understand the underlying principles of experiments involving recombinant DNA, but that this book will also afford assistance in the design of experiments according to specific needs.

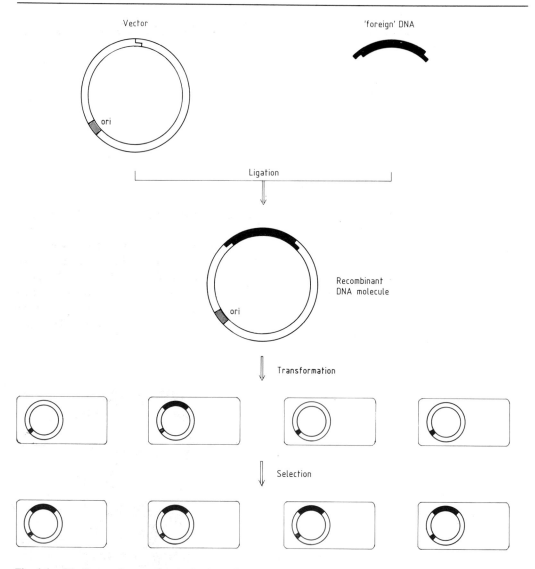

Fig. 1-1. Fundamental steps of a cloning experiment.
The experiment involves (1) the production and preparation of vector and foreign (passenger) DNA, (2) the ligation of vector and passenger DNA so as to construct a recombinant DNA molecule, (3) the transformation of suitable host cells, and (4) the selection of those cells which have acquired recombinant DNA molecules.

2 Isolation, Identification, and Characterisation of DNA Fragments

One of the prerequisites for an experiment with recombinant DNA is the availability of DNA fragments. Four different methods can be utilised to generate them:
- cleavage by restriction endonucleases;
- mechanical shearing;
- RNA-directed synthesis;
- chemical synthesis.

Historically speaking, it was the discovery of specific restriction endonucleases that actually gave birth to, and laid the foundations of, DNA manipulation as a science. This discovery has been a significant contribution towards the current position of DNA manipulation in the biological sciences.

2.1 Restriction Endonucleases

Techniques for the dissection of DNA molecules into discrete fragments by specific enzymes were virtually unknown until the late sixties. A solution to this fundamental problem eventually grew from long-standing research into the phenomenon of host-controlled restriction and modifica-

tion of a bacterial virus, phage λ (Arber and Linn, 1969). The terms "restriction" and "modification" describe the observation that the behaviour of bacteriophage may differ in different host bacteria. Bacteriophage λ, for example, grows quite well in *Escherichia coli* C, but its yields in *Escherichia coli* K may be decreased by as much as three to five orders of magnitude. Phage

Theodor Escherich (1857-1911)
pediatrician, discoverer of
Escherichia coli in Munich, 1886

particles that succeed in growing on *Escherichia coli* K will thrive in consecutive infections of the same strain; one growth cycle in *Escherichia coli* C, however, will result in poor growth if *Escherichia coli* K is re-infected. It appears that *Escherichia coli* K, known as the restricting host, possesses the ability to reduce the biological activity of phage DNA; however, the restricting host *Escherichia coli* K also appears to be able to effectively modify some of the infecting λ DNA molecules in such a way that they will now grow efficiently in this host in subsequent infections. The former process has been termed "restriction"; the latter is known as "modification". Elegant experiments by Arber and Linn (1968) and Meselson (Meselson and Yuan, 1968) revealed that restriction is caused by an enzymatic cleavage of infecting DNA by restriction endonucleases, while modification is the result of DNA methylation. Methylation of certain bases in the DNA sequences recognised by restriction endonucleases serves to protect an organism's own nucleic acids from being digested by its own nucleases: the organism is "immune". Unmethylated "foreign" DNA is recognised as soon as it enters the bacterial cell; it is immediately digested and rendered biologically inactive. This explains why infecting phage DNA performs so poorly in a restricting host, and why some DNA molecules which have survived restriction will grow efficiently in following infections. The small fraction of surviving DNA is replicated in the presence of modifying methylases and becomes methylated. In subsequent infections of the same host this DNA will, therefore, be protected from further attacks by the restriction endonucleases.

A restriction system always consists of two enzymatic elements, namely a nuclease and a methylase. These enzymes are encoded either by the bacterial chromosome or by phage or plasmid DNAs. It is not only the course of a phage infection which is influenced by such enzymes; other biological phenomena, such as conjugation, transduction, and transfection, all of which involve the transfer of DNA, are also affected. Three kinds of restriction endonucleases, designated types I, II, and III, are currently known (Yuan, 1981). These will be discussed in the following sections.

2.1.1 Type I Endonucleases

Type I endonucleases are complex nucleases, and comprise enzymes found in *Escherichia coli* B (*Eco* B), *Escherichia coli* K12 (*Eco* K), and the phage-encoded enzyme complex P$_1$. Each enzyme functions simultaneously as an endonuclease and a methylase and requires ATP, Mg^{2+}, and S-adenosylmethionine as cofactors. Restriction and modification activities are located on differ-

Werner Arber
Basel

ent subunits of these multifunctional enzyme complexes.

The DNA recognition sites of class I enzymes are stretches of DNA, 15 base pairs in length, which can be either methylated or cleaved at positions approximately 1 000 base pairs away from the 5′ end of the sequence "TCA" located in

```
                 *
Eco K    5'---A A C N N N N N N G T G C---3'
         3'---T T G N'N'N'N'N'N'C A C G---5'
                                 *

               *
Eco B    5'---T G A N N N N N N N N T G C T---3'
         3'---A C T N'N'N'N'N'N'N'A C G A---5
                                       *
```

Fig. 2.1-1. Recognition sequences for the *Eco* K and *Eco* B restriction and modification enzymes.
N represents an unspecified base, N′ its corresponding complementary base. The adenine residues in the *Eco* B sequence marked with an asterisk are methylated by the *Eco* B enzyme; adenine residues in the corresponding positions of the *Eco* K sequence are thought to represent the methylation target sites for the *Eco* K methylase.

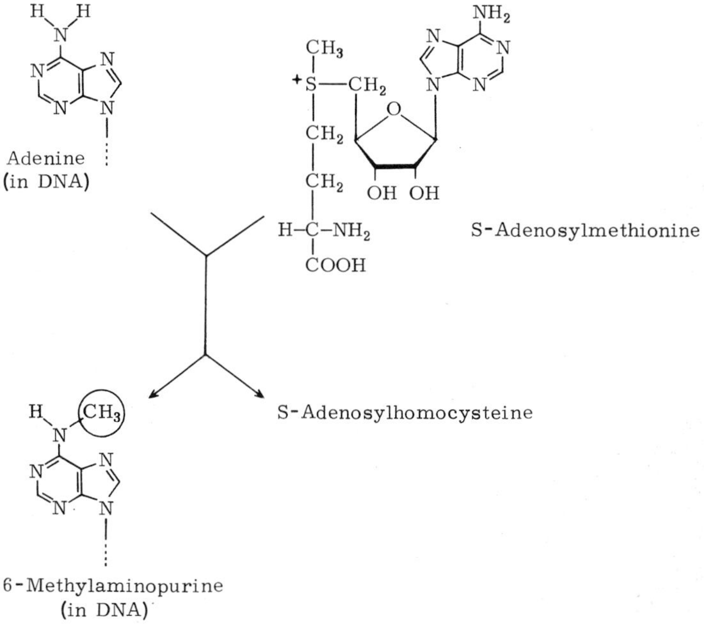

the 15 base pair recognition site (Fig. 2.1-1) (Lautenberger *et al.*, 1978). DNA molecules with recognition sequences methylated at both adenine residues on either strand (Fig. 2.1-1) are resistant to type I endonucleases. Heteroduplex molecules consisting of one modified and one unmodified strand are ideal substrates for the methylating activities of these enzymes; unmodified DNAs are restricted, *i.e.*, cleaved. The donor of the methyl group is S-adenosylmethionine (Fig.2.1-2). It is noteworthy that type I endonucleases exhibit a remarkably high ATPase activity: more than 10 000 molecules of ATP are hydrolysed to ADP and inorganic phosphate for each phosphodiester bond which is cleaved. However, interesting as these phenomena may be, apart from their methylating activities, type I endonucleases are of minor importance in gene technology, because only their recognition sequences, and not their cleavage sites on the DNA display specificity.

Fig. 2.1-2. Conversion of adenine to 6-methylaminopurine.

2.1.2 Type II Endonucleases

Unlike type I enzymes, the DNA binding sites of type II endonucleases are not only specific but also coincide with the DNA cleavage sites. These enzymes are remarkably stable and only require Mg^{2+} as a cofactor. Comparatively crude enzyme preparations can be used effectively, because a variety of very sensitive test systems are available. In a recent review Kessler *et al.* (1985) list 355 enzymes with over 100 different specificities (Appendix A; *cf.* also Roberts, 1983 and 1984). These enzymes were isolated from more than 200 bacterial strains representing almost all groups of prokaryotes. This diversity of enzymes required a suitable nomenclature and, following recommendations of Smith and Nathans (1973), each enzyme is now represented by a three-letter code derived from the genus name of the bacteria from which the enzyme was isolated. *Hae*, for example, stands for the restriction enzyme isolated from *Haemophilus aegypticus*; the enzyme *Sma* is

obtained from *Serratia marcescens*. Different serotypes of the same organism are identified by adding a fourth letter. *Hinf*, for example, stands for the enzyme isolated from serotype f of *Haemophilus influenza*. Occasionally, it is possible to isolate two or more enzymes from the same bacterial strain; these are then usually distinguished from each other by Roman numerals, such as *Hae* II, or *Hae* III, etc. Kessler *et al.* (1985) have arranged restriction enzymes by their recognition sequences. This inventory, in which each restriction enzyme is given a number, is very helpful for designing experiments based on site specificity (see Appendix A).

2.1.2.1 Recognition Sequences

Most of the type II restriction endonucleases recognise tetra-, penta-, or hexanucleotide sequences on the DNA. Such recognition sequences are conventionally written from left to right in 5' to 3' direction. GAATTC, for instance, represents the sequence 5'-GAATTC-3'.

The majority of the known recognition sequences have an axis of rotational symmetry, which means that the recognition sequence of *Eco* RI, for example, which is
5'-GAATTC-3'
3'-CTTAAG-5'
reads the same in either direction in opposite strands. Theoretically, there are 16 possible symmetrical tetranucleotides and 64 such hexanucleotides; however, only about 50% actually function as target sites for known restriction endonucleases (Fig. 2.1-3). GC-rich sequences predominate for unknown reasons. Since new restriction endonucleases are continuously being discovered, it can be expected that the entire spectrum of possible recognition sequences will be available sooner or later.

Richard J. Roberts
Cold Spring Harbor

	AATT	ACGT	AGCT	ATAT	CATG	CCGG	CGCG	CTAG	GATC	GCGC	GGCC	GTAC	TATA	TCGA	TGCA	TTAA
↓ _ _ _ _									Mbo I							
_ ↓ _ _ _		Mae II				Hpa II		Mae I		Sci NI				Taq I		
_ _ ↓ _ _			Alu I				Fnu DII		Dpn I		Hae III	Rsa I				
_ _ _ ↓ _										Hha I						
_ _ _ _ ↓																
A↓ _ _ _ _ T			Hind III				Mlu I		Bgl II / Xho II							
A _↓ _ _ _ T														Cla I		
A _ _↓ _ _ T											Stu I / Hae I	Sca I				
A _ _ _↓ _ T																
A _ _ _ _↓ T					Nsp CI					Hae II						
C↓ _ _ _ _ G					Nco I	Xma I / Ava I		Avr II			Xma III / Gdi II / Cfr I			Xho I / Ava I		Afl II
C _↓ _ _ _ G				Nde I												
C _ _↓ _ _ G			Pvu II			Sma I										
C _ _ _↓ _ G							Sac II		Pvu I							
C _ _ _ _↓ G															Pst I	
G↓ _ _ _ _ C	Eco RI						Bss HII		Bam HI / Xho II	Hgi CI		Asp 718		Sal I		
G _↓ _ _ _ C		Acy I								Nar I / Acy I		Acc I		Acc I		
G _ _↓ _ _ C				Eco RV		Nae I								Hind II		Hpa I / Hind II
G _ _ _↓ _ C		Aat II	Sac I													
G _ _ _ _↓ C		Hgi AI / Hgi JII			Nsp CI / Sph I					Bbe I / Hae II	Apa I / Hgi JII	Kpn I			Hgi AI	
T↓ _ _ _ _ A								Xba I	Bcl I		Cfr I					
T _↓ _ _ _ A														Asu II		
T _ _↓ _ _ A		Sna BI					Nru I			Mst I	Bal I / Hae I					Aha III
T _ _ _↓ _ A																
T _ _ _ _↓ A																

Fig. 2.1-3. Index of palindromic tetra- and hexanucleotide sequences and corresponding type II restriction endonucleases (courtesy of New England Biolabs, Inc.).

Hydrolytic cleavage of the two DNA strands by type II endonucleases occurs within their recognition sequence. This process generates single-stranded recessed, or protruding, 3' or 5' ends if the enzyme in question produces staggered cuts (Fig. 2.1-4). Other enzymes make even cuts so that the resulting DNA fragments possess flush or blunt termini. 3' ends always carry a hydroxyl group; 5' ends always bear a phosphate group.

Table 2.1-1 shows a list of several restriction enzymes grouped according to the types of termini they generate. The knowledge of such termini can be very useful for the construction of new recombinant DNA molecules from DNA fragments.

The symmetrical arrangement of cleavage sites has two important consequences: the generation of staggered cuts produces single-stranded termi-

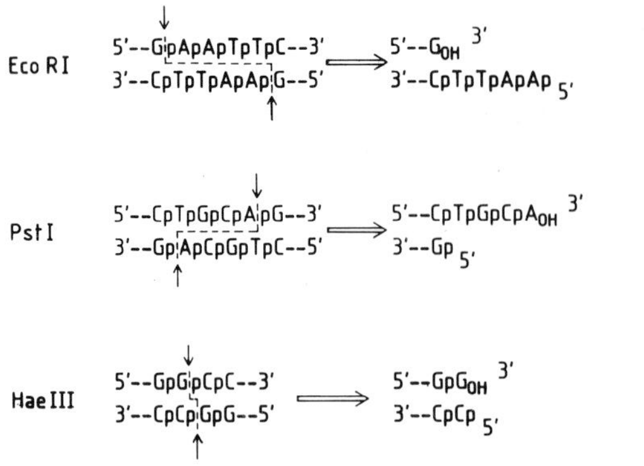

Eco RI
```
      ↓
5'--GpApApTpTpC--3'          5'--G_OH        3'
3'--CpTpTpApAp G--5'    ⟹    3'--CpTpTpApAp  5'
      ↑
```

Pst I
```
      ↓
5'--CpTpGpCpA pG--3'         5'--CpTpGpCpA_OH   3'
3'--Gp ApCpGpTpC--5'    ⟹    3'--Gp             5'
    ↑
```

Hae III
```
      ↓
5'--GpGpCpC--3'              5'--GpG_OH    3'
3'--CpCpGpG--5'        ⟹     3'--CpCp      5'
    ↑
```

Fig. 2.1-4. Types of type II restriction endonuclease cleavage patterns.

ni which are complementary in antiparallel, but identical in parallel configuration. The DNA ends of one molecule generated by a given enzyme with a given recognition sequence can, therefore, form complementary base pairs with any other DNA molecules provided they possess the same ends (Fig. 2.1-5a). This important observation was first made by Mertz and Davis (1972) while studying the enzyme Eco RI (G/AATTC). It must be

regarded as one of the most important milestones in the development of gene technology.

It should be noted, however, that not all recognition sequences are symmetrical. Due to ambiguities in their recognition sites, several

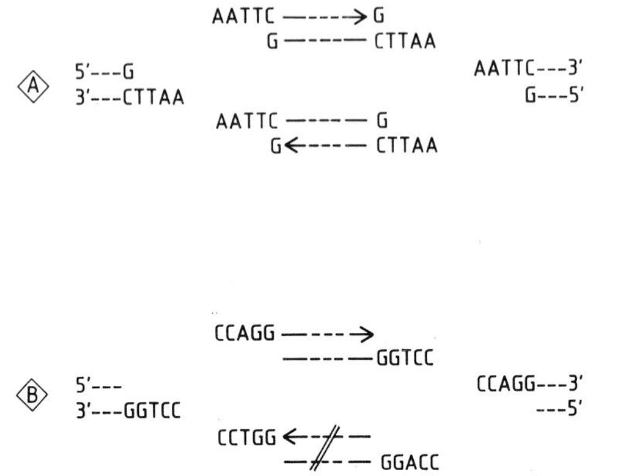

```
Ⓐ              AATTC ----→ G
                G ------ CTTAA
    5'---G                         AATTC---3'
    3'---CTTAA                     G---5'
                AATTC ----- G
                G ←---- CTTAA
```

```
Ⓑ              CCAGG ---- →
                ---- GGTCC
    5'---                          CCAGG---3'
    3'---GGTCC                     ---5'
                CCTGG ←-//-
                -//-- GGACC
```

Fig. 2.1-5. Possible arrangements of DNA fragments with either symmetrical or asymmetrical ends.
As shown for an Eco RI site (A), a DNA fragment can be inserted in either direction (arrows) while the insertion into an Eco RII site (B) occurs in one direction only since the recognition sequence is asymmetrical.

Table 2.1-1. Catalogue of some restriction endonucleases ordered according to the nature of their cleavage sites.

5'-overhanging ends		3'-overhanging ends		Blunt ends	
Enzyme	Sequence	Enzyme	Sequence	Enzyme	Sequence
TaqI	T/CGA	PstI	CTGCA/G	AluI	AG/CT
ClaI	AT/CGAT	SacI	GAGCT/C	FnuDII	CG/CG
MboI	/GATC	SphI	GCATG/C	DpnI	GA/TC
BglII	A/GATCT	BdeI	GGCGC/C	HaeIII	GG/CC
BamHI	G/GATCC	ApaI	GGGCC/C	PvuII	CAG/CTG
BclI	T/GATCA	KpnI	GGTAC/C	SmaI	CCC/GGG
HindIII	A/AGCTT			NaeI	GCC/GGC
NcoI	C/CATGG			HpaI	GTT/AAC
XmaI	C/CCGGG			NruI	TCG/CGA
XhoI	C/TCGAG			BalI	TGG/CCA
EcoRI	G/AATTC			MstI	TGC/GCA
SalI	G/TCGAC			AhaIII	TTT/AAA
XbaI	T/CTAGA			EcoRV	GAT/ATC

Slashes indicate cleavage sites.

Table 2.1–2.   Enzymes with multiple recognition sequences.

Enzyme	Recognition sequence	Number of sites on		
		pBR322 DNA	M13mp8 DNA	SV40 DNA
*Eco*RII	/CC $\boxed{A}$ GG /CC $\boxed{T}$ GG	6	7	17
*Acc*I	GT/ $\boxed{AG}$ AC GT/ $\boxed{AT}$ AC GT/ $\boxed{CG}$ AC GT/ $\boxed{CT}$ AC	2	1	1
*Ava*I	C/ $\boxed{C}$ CG $\boxed{A}$ G C/ $\boxed{C}$ CG $\boxed{G}$ G C/ $\boxed{T}$ CG $\boxed{A}$ G C/ $\boxed{T}$ CG $\boxed{G}$ G	1	2	0
*Hae*II	$\boxed{A}$ GCGC/ $\boxed{C}$ $\boxed{A}$ GCGC/ $\boxed{T}$ $\boxed{G}$ GCGC/ $\boxed{C}$ $\boxed{G}$ GCGC/ $\boxed{T}$	11	6	1
*Hind*II	GT $\boxed{C}$ / $\boxed{A}$ AC GT $\boxed{C}$ / $\boxed{G}$ AC GT $\boxed{T}$ / $\boxed{A}$ AC GT $\boxed{T}$ / $\boxed{G}$ AC	2	1	7

Variable positions are boxed.

enzymes recognise up to four different target sequences. Digestion of a DNA molecule with such enzymes may, therefore, yield non-identical termini which cannot be recombined at random. Table 2.1-2 lists some enzymes with multiple recognition sequences. As exemplified by a digestion with *Eco* RII (Fig. 2.1-5b), the generation of non-identical ends by asymmetrical cleavage favours certain orientations of the DNA insert during *in vitro* recombination.

Several enzymes, known as isoschizomers, possess identical recognition sites. *Hind* III and *Hsu* I (A/AGCTT), for example, share the same cleavage and recognition sequences. On the other hand, enzymes such as *Sma* I (CCC/GGG) and *Xma* I (C/CCGGG), possess the same recognition sequence but differ in their cleavage sites. Some

enzymes may also differ in their recognition sequences but still generate identical overlapping termini. Digestions with *Bam* HI (G/GATCC), *Bgl* II (A/GATCT), *Bcl* I (T/GATCA), and *Sau* 3A (/GATC), for example, generate cohesive 5'-GATC-3' ends. All DNA fragments generated by cleavage with any combination of these enzymes, for example, by *Bam* HI and *Bgl* II, can recombine with each other. In this case, recombinants will have the sequence 5'-GGATCT-3' and will be resistant to digestion with either enzyme because the flanking bases in the hexanucleotide recognition sequences of *Bam* HI and *Bgl* II differ (G/C and A/T, respectively) (Fig. 2.1-6); yet recombinant and parental molecules can still be cleaved by enzymes such as *Dpn* I, *Sau* 3A, and *Mbo* I, and are, therefore, easily distinguished.

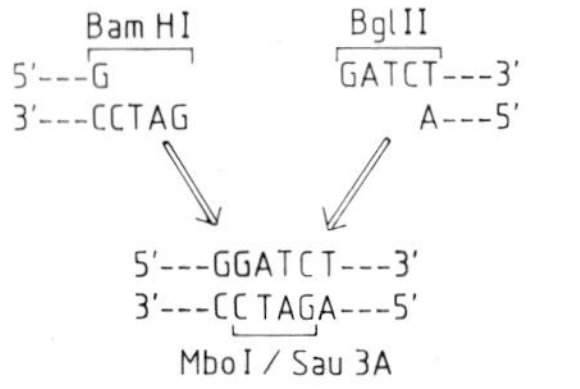

```
   Bam HI            Bgl II
  ┌──┐            ┌─────────┐
5'---G               GATCT---3'
3'---CCTAG               A---5'

         ╲          ╱
          ╲        ╱

        5'---GGATCT---3'
        3'---CCTAGA---5'
            └──┘
         Mbo I / Sau 3A
```

Fig. 2.1-6. Recombination of DNA fragments obtained by digestion with different enzymes (*Bam* HI, (G/GATCC); *Bgl* II, (A/GATCT)).
The resulting fragments have complementary ends. Recombinants are resistant to *Bam* HI and *Bgl* II but can be cleaved *e.g.* with *Mbo* I and *Sau* 3A.

These enzymes recognise the tetranucleotide sequence 5'-GATC-3' and are not influenced by flanking nucleotides. A similar situation exists for the 5'-TCGA terminus, which remains after cleavage by enzymes such as *Sal* I (G/TCGAC), *Xho* I (C/TCGAG), and *Ava* I (C/PyCGPuG). The 5'-CG end left by *Cla* I (AT/CGAT) and *Taq* I (T/CGA) is yet another example. These enzyme families are also listed in Table 2.1-1.

2.1.2.2 Detection and Purification of Restriction Endonucleases

Restriction endonucleases can be isolated by certain standard procedures which, above all, exploit the DNA binding properties of these enzymes. Enzymes are also purified by various methods using hydroxyapatite, phosphocellulose, heparin sepharose or hydrophobic chromatography (Greene *et al.*, 1978). Enzyme yields can be very variable. Five grams of *Haemophilus aegypticus* cells, for instance, contain sufficient amounts of the enzyme *Hae* III to cleave ten grams of DNA completely. The same mass of *Brevibacterium albidum* cells, on the other hand, just yields enough of the enzyme *Bal* I to cut 500 micrograms of DNA. Enzyme units for type II nucleases are usually defined as the amount of enzyme required for complete digestion of one microgram of DNA in 15 or 60 minutes at 37 °C. The above-mentioned amounts of *Hae* III and *Bal* I are, therefore, equivalent to 10^7 and 5×10^2 units, respectively.

The enzymatic activities of endonucleases can be detected in various ways. Initially, they were determined predominantly by measuring such parameters as the loss of biological activity, or changes in viscosity or sedimentation behaviour of DNA preparations. Today, only sedimentation analysis retains any practical value. In September 1971, D. Nathans surprised the scientific community by being able to separate electrophoretically SV40 DNA into eleven discrete fragments on agarose gels. This opened new vistas for the separation of DNA molecules by their molecular weights and eventually promoted the spread of gene cloning techniques. Agarose gels are prepared by boiling suspensions of solid agarose in suitable buffers until the solution clarifies, which can most effectively be accomplished in microwave ovens, if available. Clear agarose solutions are allowed to cool to 60 °C before the gel is cast. Horizontal gel electrophoresis systems should be used for agarose concentrations lower than 0.5%; higher agarose concentrations allow the use of conventional vertical slab gels cast between glass plates. Buffers that have proven particularly effective contain 50-100 mM Tris-acetate or Tris-borate, pH 7.4, and 1 mM EDTA. The electrophoresis of DNA is usually carried out at 5-10 V/cm. DNA fragments can be visualised under UV light by exploiting the fluorescent properties of DNA-ethidium bromide complexes (Fig. 2.1-7). Light of 254 nm wavelength gives maximum sensitivity. It should be noted, however, that this light will damage DNA; if intact DNA molecules

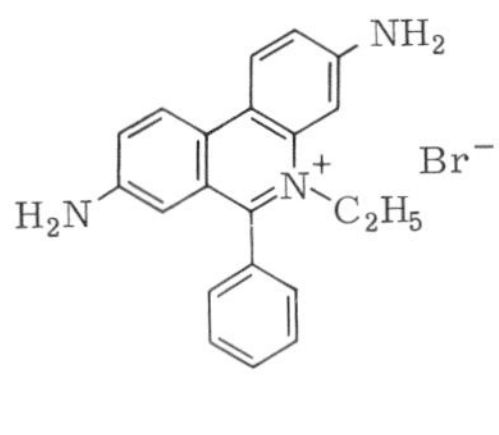

Fig. 2.1-7. Structure of ethidium bromide

are to be recovered from gels, it is advisable to use light of 366 nm wavelength, although this light is approximately ten times less sensitive than that of shorter wavelengths. In practice, long-wave red light is commonly filtered out by red interference filters, because photographic material normally used for documentation is very sensitive to this light. If radioactively labelled DNA fragments are separated by gel electrophoresis, DNA bands can also be visualised by covering the gel with Saran wrap and directly exposing it to photographic films. Special two-dimensional methods and cylindrical gels have been developed for the preparative separation of milligram amounts of DNA. For detailed protocols including appropriate equipment, the reader is referred to Southern (1980).

The mobility of a DNA fragment in agarose and polyacrylamide gels is proportional to the logarithm of the molecular weight of that fragment (*cf.* Maniatis *et al.* 1975). This allows molecular weights to be determined graphically, but this rule is applicable only to a limited range of molecular weights. It is, therefore, advisable to use DNA fragments of known molecular weights as standard molecular weight markers and run them in a separate slot of the same gel. The number of completely analysed DNA sequences is rising rapidly, and this facilitates the analysis considerably. Plasmid pBR322 DNA, for instance, can be easily prepared and provides molecular weight standards for DNA fragments up to 4 kb in length (Fig. 2.1-8A) (Sutcliffe 1978); phage λ DNA is also easily accessible and can be used as marker DNA for fragments up to 50,000 base pairs in length (Fig. 2.1-8B) (Table 2.1-3) (Parker *et al.*, 1977). DNA fragments larger than 50 kb cannot be resolved efficiently by agarose or polyacrylamide gel electrophoresis, because electrophoretic resolution diminishes above this

molecular weight and larger DNA fragments show anomalously high electrophoretic mobilities. It has been hypothesised that this behaviour may be due to condensation and orientation effects. Schwartz and Cantor (1984) therefore subjected large DNA molecules in agarose gels to alternating pulses of perpendicularly oriented electrical fields, one of which at least was non-uniform. Indeed, this technique of pulsed field gradient electrophoresis allowed DNAs from 30 to 2,000 kb in length to be separated, and was successfully employed for the separation of chromosome-sized DNA from yeast (Carle and Olson, 1985), *Trypanosoma brucei* (Guyaux *et al.*, 1985) and from the malaria parasite (Kemp *et al.*, 1985).

Certain difficulties may be experienced with preparative agarose or polyacrylamide gels from which the DNA is to be recovered quantitatively. A number of procedures have been described to

Daniel Nathans
Baltimore

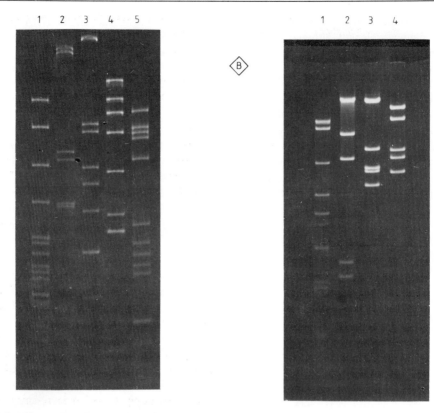

Fig. 2.1-8.   Electrophoretic separation of DNA fragments:
Part A: separation of *Hpa* II (1), *Taq* I (2), *Hinf* I (3), *Alu* I (4), and *Hae* III (5) DNA fragments of plasmid pBR322 on an 8% polyacrylamide gel. Fragment sizes are listed in Table 2.1-3.
Part B: separation of *Cla* I (1), *Hind* III (2), *Eco* RI (3), and *Bam* HI (4) fragments of λ DNA on a 0.8% agarose gel. Fragment sizes are listed in Table 2.1-3. Fragments were made visible by ethidium bromide staining. Larger fragments bind more of the dye and therefore light up brighter than smaller fragments.

recover DNA from such gels almost completely (Southern, 1979). Four of them, which appear to be the most effective ones, will be discussed below.

Method 1:

This method (Langridge *et al.*, 1980) relies on the conversion of DNA fragments into their hexadecyltrimethylammonium salts. These are soluble in organic solvents, such as butanol, and the DNA can be easily separated from agarose moieties which remain in the aqueous phase. Derivatised DNA fragments are subsequently transferred to the aqueous phase by raising the NaCl concentration and the quaternary ammonium salts are removed by chloroform extraction. It is important to use low-melting agarose for this procedure, because gel slices containing electrophoretically separated DNA fragments can then be heated to 70 °C and the DNA can be extracted in a homogeneous aqueous phase.

Method 2:

Slices of agarose containing DNA are placed on top of an agarose column in a cylindrical glass tube which is then sealed with agarose and receives a top layer of hydroxyapatite. The DNA can be transferred into the hydroxyapatite layer

Table 2.1–3. Sizes (in bp) of restriction fragments from bacteriophage λ and pBR322 DNA.

Lambda-DNA (48502 bp)			
ClaI	HindIII	EcoRI	BamHI
11 381	23 131	21 225	16 836
10 493	9 419	7 421	7 234
4 398	6 559	5 806	6 770
3 674	4 361	5 646	6 529
2 614	2 322	4 876	5 624
2 064	2 023	3 530	5 509
1 916	564		
1 806	125		
1 703			
1 112			
972			
657			
622			
537			

pBR322-DNA (4363 bp)				
HpaII	TaqI	Hinf	AluI	HaeIII
622	1 444	1 631	910	587
527	1 307	517	659	540
404	475	506	655	504
309	368	396	521	458
242	315	344	403	434
238	312	298	281	267
217	141	221	257	234
201		220	226	213
190		154	136	192
180		75	100	184
160			63	124
160			57	123
147			49	104
147			19	89
122			15	80
110			11	64
90				57
76				51
67				21
34				18
34				11
26				7
26				
15				
9				

by subsequent reversed current electrophoresis in a standard tube gel electrophoresis chamber. Hydroxyapatite containing DNA is then carefully layered on top of a sephadex G25 or biogel P30 column equilibrated with low salt buffers, and the DNA is eluted with 0.7 M sodium phosphate buffer pH 6.8. Most of the salt is retained by the column matrix during the elution step. If individual DNA fragments have been sufficiently resolved in the agarose gel, the hydroxyapatite can also be loaded into a slot cut directly in front of the DNA fragment to be isolated. DNA molecules can then be transferred directly into this hydroxyapatite layer by continuing electrophoresis (Tabak and Flavell, 1978).

Method 3:

DNA fragments are separated in a horizontal gel, and a slot is cut directly in front of the fragment to be eluted. This slot is filled with buffer, and electrophoresis is continued until the desired DNA fragment has completely entered the slot, which can be monitored by following the migration of ethidium bromide-stained DNA bands under UV light. The DNA is recovered with a Pasteur pipette before its re-entry into the gel. It may not always be possible to cut out a slot directly in front of a DNA fragment. In this case, a slice of the gel containing the desired DNA fragment is cut out from the gel and put into a dialysis bag. This bag is positioned in a slab gel apparatus and oriented perpendicular to the flow of electricity. Electroelution of the DNA from the gel slice can be monitored under UV light. The dialysis bag minimises losses of DNA caused by diffusion into the electrophoresis chamber (McDonnell et al. 1977).

Method 4:

DNA fragments which are smaller than 500 bp are conveniently extracted from agarose or polyacrylamide gel slices by a prolonged incubation in a salt solution (500 mM ammonium acetate, 10 mM magnesium acetate) (Maxam and Gilbert, 1980). Traces of the gel matrix can be removed by subsequent phenol extraction.

Other extraction methods rely on chaiotropic salts, such as NaI or NaClO$_4$, which dissolve agarose. The DNA can be recovered from the salt solutions by binding to glass filters (Vogelstein and Gillespie, 1979). The final decision about the method of choice will, of course, depend on the yield and quality of the eluted DNA; for example, this DNA should be amenable to subsequent enzymatic reactions with restriction endonucleases, DNA ligase, polynucleotide kinase, terminal transferase, etc.

The recognition sequences of restriction endonucleases can be determined by at least four different techniques:

Method 1:

The 5′ ends of a DNA fragment obtained by digestion with the enzyme under study are labelled with γ-^{32}P-ATP by using polynucleotide kinase. Labelled DNAs are then treated with pancreatic DNAse so as to obtain the terminal 5′ mononucleotides and also the corresponding di-, tri-, tetranucleotides, and so on, which can then be identified after electrophoretic separation. It is also possible to partially digest the 5′ phosphate-labelled DNA fragment with exonuclease I from the 3′ end of the DNA chain until 5′ dinucleotides are obtained. This procedure has been used by Kelly and Smith (1970) who determined for the first time the recognition sequence of the restriction endonuclease *Hind* II.

Method 2:

This procedure can be used for isoschizomeric enzymes and whole enzyme families. If identical patterns of DNA fragments are obtained after digesting a DNA with one known enzyme and another enzyme whose recognition sequence is unknown, this suggests that these two enzymes possess the same recognition sequence. A similar conclusion can be reached if the cleavage sites are related rather than identical. The recognition sequence of *Sau* 3A (5′-GATC-3′), for example, is contained within the hexanucleotide recognition sequence of *Bam* HI (5′-GGATCC-3′). A *Sau* 3A digest can be expected to yield a pattern of DNA fragments more complex than that produced by *Bam* HI. However, any *Bam* HI fragment which does not contain an internal *Sau* 3A site will be produced by digestion with either *Bam* HI or *Sau* 3A, and *Bam* HI and *Sau* 3A digests therefore may have some fragments in common.

Method 3:

This method relies on the availability of computer-stored nucleotide sequence data. A DNA fragment of known sequence is digested with the restriction enzyme of unknown recognition sequence. The number of resulting DNA fragments and their sizes are determined experimentally. A subsequent computer analysis of the sequence can then be used to deduce the cleavage site which explains the observed cleavage pattern.

Thomas J. Kelly, Jr.
Baltimore

Method 4:

This method is applicable to any enzyme which has a cleavage site not identical to its recognition sequence. In these cases, an answer may only be obtained if the DNA in the neighbourhood of the cleavage site is sequenced. This approach has been used for the determination of cleavage sites of type III enzymes.

2.1.2.3 Specificities of Recognition Sequences

The number of type II enzyme cleavage sites in a DNA molecule depends on the size of this DNA, its base composition, and the GC content of the recognition sequence. The probability P for a particular tetranucleotide consisting of all four bases to occur in a DNA molecule containing 50% AT and 50% GC base pairs is $1/4^4$. Similarly, the probability for a given hexanucleotide is $1/4^6$. On the average, then, tetranucleotide cleavage sites will occur every 256 base pairs, hexanucleotide sites every 4096 base pairs. A hexanucleotide containing only two of the four bases will occur less frequently; if it contains only G and C residues, it will only be found every 65 536 base pairs (*i.e.*, $1/(4^6 \times 4^2)$). In general, enzymes with hexanucleotide recognition sequences consisting of all four bases have more than one recognition

site in DNA molecules larger than 4 kb, such as plasmid pBR322 (4,363 bp) or SV40 DNA (5,243 bp).

Hexanucleotide recognition sites consisting of C and G residues only, for example, *Sma*I (CCC/GGG) or *Xma*I (C/GGCCG) sites (Table 2.1-4), are not generally observed in DNA molecules as small as pBR322 or SV40. Of course, DNA hardly ever contains exactly 50% AT and 50% GC; *Euglena gracilis* chloroplast DNA, for instance, is 130 kb in length and contains approximately 75% AT and 25% GC (Gray and Hallick, 1977). The probability that a hexanucleotide sequence, such as 5'-CCCGGG-3' (*Sma*I), which comprises only two different bases, will occur in this DNA is $1/(4^6 \times 8^2)$. This sequence should therefore only occur every 262 144 base pairs, and, indeed, *Sma*I sites have not been found in *Euglena gracilis* chloroplast DNA.

Moreover, it should be noted that the dinucleotide sequence CpG is extremely rare in eukaryotic DNAs. In a recent analysis of eukaryotic DNA comprising a total of 43 000 nucleotides, Nussinov (1980) found 956 CpG residues as opposed to 2 406 GpC residues; recognition sites containing the CpG dinucleotide should therefore be frequent in prokaryotic DNAs, such as plasmid pBR322 DNA, while they should be rare in eukaryotic DNAs such as SV40 DNA. The data

Table 2.1–4. Frequency of hexanucleotide recognition sites in pBR322 and SV40 DNA.

Enzyme	Recognition sequence	Number of recognition sequences on	
		pBR322-DNA 4363 bp; 54 % GC	SV40-DNA 5243 bp; 40,8 % GC
*Bam*HI	(GGATCC)	1	1
*Eco*RI	(GAATTC)	1	1
*Hind*III	(AAGCTT)	1	6
*Hpa*I	(GTTAAC)	0	4
*Pst*I	(CTGCAG)	1	2
*Pvu*II	(CAGCTG)	1	3
*Sma*I	(CCCGGG)	0	0
*Sac*II	(CCGCGG)	0	0
*Xma*III	(CGGCCG)	1	0

compiled in Table 2.1-5 show that this is indeed the case. For example, the enzyme *Fnu* DII (CG/CG) cleaves plasmid pBR322 DNA 23 times, while SV40 DNA, which is only negligibly larger in size, remains uncut; however, both DNAs are cut with approximately the same frequency by *Hae* III (GG/CC), the recognition sequence of which does not contain the CpG dinucleotide.

Of particular interest in this context are enzymes with hexanucleotide target sites containing two CpG dinucleotides, for example, *Sst* II (CCGC/GG) and *Pvu* I (CGAT/CG), which, on the average, cleave eukaryotic DNAs every 150 kb. The gene encoding the human blood coagulation factor VIII, for example, is 200 kb in length, and contains the expected 50 *Eco* RI sites (200 000/4 096), but only one *Sst* II and one *Pvu* I site. Such enzymes are, therefore, ideal for isolating very large DNA fragments which can be subfractionated by employing the technique of pulsed field gradient electrophoresis (Schwartz and Cantor, 1984). Another application for these enzymes lies in the construction of cosmid libraries. The *Bam* HI cloning site of cosmid pGcos4, for example, is flanked by two *Pvu* I sites (Gitschier *et al.*, 1984) (see also Section 4.2.3; Fig. 4.2-16). Insertions of eukaryotic DNA, which are at best 45 kb in length, can usually be cut out from this vector in an intact form because, for statistical reasons, *Pvu* I sites should not occur in fragments of this size.

An alternative strategy for the generation of large DNA fragments employs longer recognition sequences which occur less frequently on the genome. At present, only two suitable enzymes are known, namely, *Not* I (GC/GGCCGC), and *Sfi* I (GGCCNNNN/NGGCC), but the combined use of methylases and restriction endonucleases can also be used to generate longer cleavage sites. For this purpose DNA is first treated with the modifying methylases M.*Taq* I or M.*Cla* I, which convert the corresponding cleavage sites TCGA and ATCGAT to TCG^{m}A and ATCG^{m}AT, respectively. The methylated DNA is then cleaved with *Dpn* I which recognises and cleaves the sequence G^{m}ATC only if both strands are methylated. If M.*Taq* I has been used, only the octamer shown in Fig. 2.1-9A will be cleaved; in the case of M.*Cla* I it will be the decamer shown in Fig. 2.1-9C. The hexameric sequence TCGATC is not cleaved by *Dpn* I after M.*Taq* I methylation, although it contains the recognition sequence for *Dpn* I, (Fig. 2.1-9B) because it is methylated only in the A residue of one strand within the *Dpn* I recognition site, and thus remains refractory to *Dpn* I digestion (McClelland *et al.*, 1984). Similar procedures have been described for other combinations of enzymes (Nelson *et al.*, 1984).

Restriction endonuclease digestions are influenced heavily by reaction conditions. Deviations from optimal reaction conditions may even alter the cleavage specificity. In buffers containing 100 mM NaCl and 5 mM MgCl$_2$, pH 7.3, for example,

Table 2.1–5. Occurrence of tetranucleotide recognition sites containing CpG or GpC dinucleotides in various DNA molecules.

| Enzyme | Sequence | Number of sites on | | |
		pBR322 DNA	ΦX174 DNA	SV40 DNA
*Fnu*DII	(CGCG)	23	14	0
*Hha*I	(GCGC)	31	18	2
*Hpa*II	(CCGG)	26	5	1
*Taq*I	(TCGA)	7	10	1
*Hae*III	(GGCC)	22	10	19

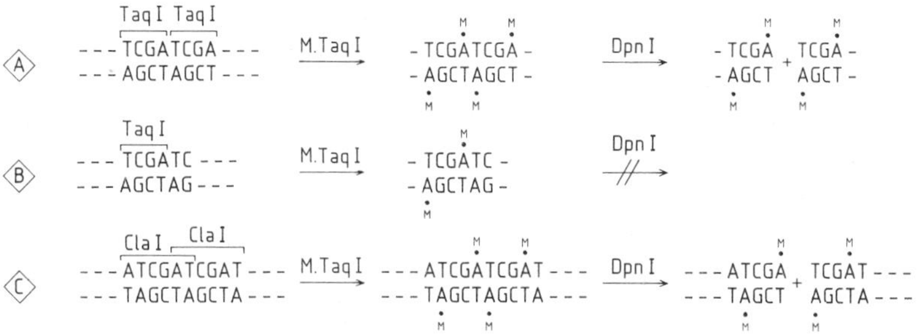

```
      Taq I  Taq I                        M   M                      M      M
 (A) ---TCGATCGA---   M.Taq I    - TCGATCGA -    Dpn I    - TCGA   TCGA -
     ---AGCTAGCT---   ──────►    - AGCTAGCT -    ──────►  - AGCT + AGCT -
                                   •   •                    •      •
                                   M   M                    M      M

      Taq I                           M
 (B) ---TCGATC---     M.Taq I    - TCGATC -      Dpn I
     ---AGCTAG---     ──────►    - AGCTAG -      ─╫─►
                                      •
                                      M

      Cla I  Cla I                     M   M                    M          M
 (C) ---ATCGATCGAT---  M.Taq I  ---ATCGATCGAT---  Dpn I  ---ATCGA   TCGAT---
     ---TAGCTAGCTA---  ──────►  ---TAGCTAGCTA---  ──────► ---TAGCT + AGCTA---
                                    •   •                    •          •
                                    M   M                    M          M
```

Fig. 2.1-9. Formation of long recognition sequences.
Recognition sequences can be extended by methylation and subsequent cleavage with *Dpn* I. In the case of *Taq* I, an extension to 8 bp is required to permit cleavage of this sequence after methylation with M.*Taq* I (A). The hexanucleotide recognition site (B) is not cleaved since one of the two adenine residues in the GATC portion of the lower strand is not methylated by M.*Taq* I. The smallest sequence which can be cleaved with *Dpn* I after M.*Cla* I methylation is a decamer (C).

Eco RI recognises and cleaves the sequence 5'-GAATTC-3'. Increasing the pH, lowering the concentration of NaCl, or replacing magnesium by manganese, will alter the specificity in such a way that *Eco* RI now recognises the tetranucleotide sequence 5'-AATT-3', which occurs much more frequently (Polisky *et al.*, 1975; Hsu and Berg, 1978; Woodhead *et al.*, 1981). This new enzyme specificity is known as *Eco* RI* (Eco star) activity. Alterations in the specificity have also been described for other enzymes, such as *Bsu* I (Heininger *et al.* 1977), *Bst* I (Clarke and Hartley, 1979), *Bam* HI (George *et al.*, 1980), and *Xba* I, *Hha* I, *Sal* I, *Pst* I, *Sst* I (Malyguine *et al.* 1980), although in most cases, the new specificities are still unknown. Their existence has been inferred merely from new DNA bands observed after electrophoretic separation of DNA digested with these enzymes.

Alterations in the specificity of endonucleases may also be due to changes in their nucleic acid substrates. It has been known for some time that the enzymes *Hae* III, *Hha* I, *Hinf* I, *Hpa* II, *Pst* I, and *Ava* I are also capable of hydrolysing single-stranded phage DNAs, such as M13 or ΦX174 DNA (Blakesley *et al.* 1977; Baralle *et al.*, 1980a), although much longer reaction times and much higher enzyme concentrations are required. Occasionally, these enzymes have been exploited for characterising single-stranded DNA intermediates occurring during the synthesis of cDNA (Ullrich *et al.*, 1977; Seeburg *et al.*, 1977; Baralle *et al.* 1980b). It remains to be shown whether these enzymes do indeed cleave single-stranded DNA; several observations, including DNA sequence analyses, suggest that single-stranded molecules may contain regions which are capable of forming intramolecular double strands, and that these double-stranded regions are attacked by the enzymes.

Several restriction enzymes do not only cleave double-stranded DNA but also DNA-RNA hybrids (Molley and Simons, 1980). This is an observation which may be of interest because, in principle, such enzymes should allow the physical mapping of RNA molecules by converting RNA molecules into RNA-DNA hybrids in a reaction catalysed by reverse transcriptase.

Reaction temperatures also may influence specificities of restriction endonucleases. It has been observed, for instance, that short oligonucleotides containing a known recognition sequence are not cleaved by the corresponding restriction enzyme. The octanucleotide 5'-TGAATTCA-3', for example, contains the *Eco* RI target site but is not cleaved by this enzyme; likewise, the decanucleotide 5'-CCAAGCTTGG-3' is not cleaved by *Hsu* I and *Hind* III, although it contains both sites (Miller *et al.* 1980). The melting temperatures of these linker molecules lie between 10 °C and 20 °C. Under normal reaction conditions, *i.e.*, at 37 °C, such linkers are single-stranded. DNA melting phenomena may also play an important role if closely neighbouring sites are to be cleaved on large DNA fragments. Polyoma virus, for instance, contains two *Mbo* I sites (/GATC), which are only five base pairs apart. Either site can be easily cleaved by the enzyme; however, cleavage of one site renders the remaining neighbouring site resistant to *Mbo* I cleavage. Similar problems have occasionally arisen with vectors such as M13mp7, 8, and 9, because these vectors contain chemically synthesised universal linkers with closely adjacent and even overlapping recognition sites (see Section 2.4.2).

In principle, reaction temperatures are also important for restriction enzymes isolated from thermophilic organisms. The optimal reaction temperature for enzymes, such as *Taq* I, *Tth* 111 I, and *Tth* 111 II, is 65-79 °C (Malcolm, 1981), at which DNA molecules are thought to undergo conformational changes. As yet however, DNA cleavage patterns have not been reported to be influenced by these elevated temperatures, and it seems unlikely that such conformational changes occur at all.

Alterations in the specificity of restriction enzymes may also be due to methylation of the DNA substrate. DNA can be methylated at position N^6 of adenine and/or position C^5 of cytosine (Fig. 2.1-2 or 8-25). Eukaryotic cells usually contain methylated cytosine residues, and up to 10% of all cytosines may be methylated. Since there is a close relationship between methylation and restriction, restriction endonucleases are excellent tools for studying methylation patterns. While some restriction enzymes recognise only unmethylated target sequences, their isoschizomers cleave both methylated and unmethylated sites. *Hpa* II (C/CGG), for instance, cleaves the unmethylated tetranucleotide, while its isoschizomer *Msp* I cleaves both the unmethylated and the methylated tetranucleotide (CC^mGG) (Waalwijk

and Flavell, 1978). By comparing DNA patterns obtained by *Hpa* II and *Msp* I digestion, it is possible to determine the numbers and locations of methylated cytosine residues in a DNA fragment. It is obvious, of course, that this type of analyses is restricted to cytosine residues in the recognition sequences of the enzymes. Nevertheless, such methylation analysis can play an important role in studying the control of gene expression. An inverse relationship between methylation and transcription has been shown to exist in several systems: genes which are expressed show a low level of methylation, while genes which are turned off appear to be highly methylated (Felsenfeld and McGhee, 1982; Doerfler, 1983). Apart from *Hpa* II and *Msp* I, there is another pair of isoschizomers which may be used to differentiate between methylated and unmethylated cytidine residues. *Fnu* DIII and *Hha* I both cleave the sequence 5'-GCGC-3', but unlike *Hha* I, *Fnu* DIII is also capable of cleaving the C-methylated recognition sequence (GC^mG/C^m).

In contrast to C methylation, another form of methylation, *i.e.*, methylation of adenine residues, appears to be of minor importance for eukaryotic DNAs but highly relevant in prokaryotic DNAs. Methylation of adenine residues can be tested by enzymes such as *Sau* 3A (/GATC), *Mbo* I (/GATC), and *Dpn* I (GA/TC). While *Sau* 3A cleaves both 5'-GATC-3' and 5'-GA^{m}TC-3', *Mbo* I cleaves only the unmethylated recognition sequence, and *Dpn* I cleaves 5'-GA^{m}TC-3' exclusively. A hemimethylated sequence, *i.e.*, a sequence methylated only in one strand, is resistant to *Dpn* I.

This property of *Dpn* I and the absence of adenine methylation in eukaryotic cells can be exploited for the development of a very sensitive assay allowing detection of vector DNA replication in higher cells. For this purpose plasmids are first isolated from *E. coli* cells expressing suitable methylating activities, *i.e.*, cells which are Dam$^+$ (DNA methylase-positive; see Section 12.4.2). Such vector molecules will be fully methylated and will be sensitive to *Dpn* I digestion. They are subsequently used for transfections of tissue culture cells or treated with suitable cellular extracts. Hemimethylated or unmethylated DNA molecules generated during semiconservative DNA replication become resistant to *Dpn* I digestion, while unreplicated DNA remains sensitive. This system has been used by Li and Kelly (1984) and Hay (1985) as an assay for the development of an *in vitro* replication system for SV40 and adenovirus DNA, respectively.

2.1.2.4 Modification of DNA Ends

Various methods are available for modifying the termini of DNA fragments generated by restriction enzymes. Such modifications are often mandatory if DNA fragments are to be manipulated further. 5′ phosphate groups can be removed by alkaline phosphatase treatment and can be re-attached to free 5′ hydroxylated termini in a reaction catalysed by T4 polynucleotide kinase (Fig. 2.1-10, Reactions a and b). Annealing of two incompatible ends requires overlapping ends to be removed (Fig. 2.1-10, Reaction c), or recessed ends to be filled-in (Fig. 2.1-10, Reaction d). The latter two processes yield blunt ends, which can then be joined in a reaction catalysed by T4 polynucleotide ligase (*cf.* Chapter 3).

It may also be necessary to convert blunt ends into protruding single-stranded ends. This can be accomplished most effectively by using λ exonuclease for 5′ end modification (Little *et al.*, 1967), and *E. coli* exonuclease III for 3′ end modification (Richardson *et al.*, 1964). The treatment of double-stranded DNAs with exonuclease *Bal*31 can be used for simultaneously degrading both strands of a DNA molecule; resulting DNA fragments are shortened and possess blunt ends at both termini (Legerski *et al.*, 1977).

The manipulations discussed so far play an important role in recombinant DNA technology and are frequently employed, for example in DNA sequencing, ligation of DNA fragments, or

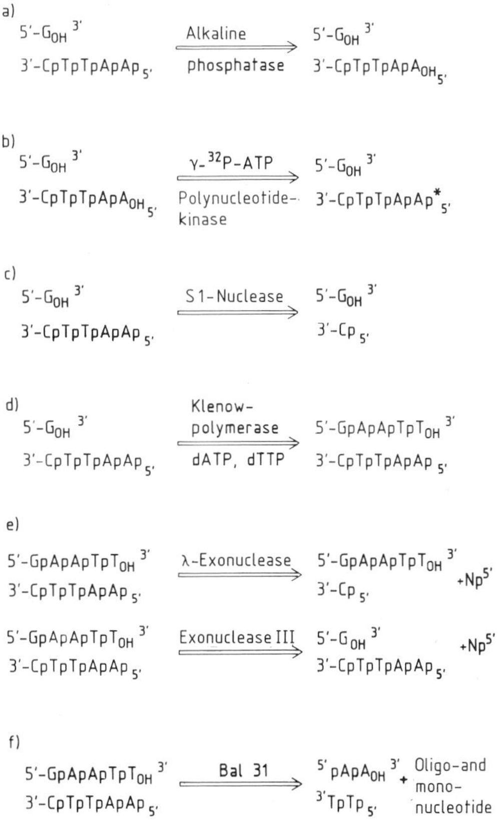

Fig. 2.1-10. Enzymatic methods for the modification of DNA ends.

the construction of expression vectors. Detailed discussions will be found in various chapters of this book, in which specific references are made to particular applications.

2.1.2.5 Physical Mapping

Type II restriction endonucleases are used mainly to provide detailed physical maps of genes and chromosomes. Functional sequences in a genome are mapped, for example, by taking positions of restriction cleavage sites as reference points. Only a few such reference points, including, for example, the ends of a linear DNA molecule, are required to establish a preliminary map. Circular molecules can be linearised easily using a nuclease that only cleaves once. Various strategies have been developed for obtaining the rough outline of a map, and for further fine structure mapping.

Mapping usually begins by employing enzymes which cleave the molecule in question only a few times. A preliminary map can be easily derived from an analysis of partial digests, or of the products yielded by simultaneous digestion with two or more enzymes. An analysis by electron microscope of partial denaturation patterns of DNA molecules may also considerably facilitate the construction of maps. Since a DNA molecule will display a characteristic partial denaturation pattern which depends on the distribution of GC and AT base pairs in the molecule, it is often possible to assign fragments to certain regions of the intact molecule by comparing the patterns of the whole DNA molecule with those of isolated restriction enzyme fragments.

A considerable asset is the specific end-labelling of DNA fragments. If DNA molecules are labelled before digestion, terminal fragments can be identified simply by electrophoretic separation of the resulting mixture and autoradiography of the gel. End-labelling after digestion facilitates the identification of smaller DNA fragments which may have escaped visual detection because of their small size. End-labelling is not influenced by the sizes of DNA fragments, because it is only the number of ends which counts.

In attempting to map DNA molecules as described above, one will almost inevitably observe discrepancies and inconsistencies in the maps, which are caused mainly by difficulties in determining the correct sizes of the DNA fragments. Although it is possible to determine the exact lengths of DNA fragments with the electron microscope (Thomas and Davis, 1975), it is considerably easier to make length measurements by comparing the electrophoretic mobilities of the DNA fragments in question with those of known standard DNA fragments (see Maniatis *et al.*, 1975, for example).

Frequently occurring restriction sites can also be mapped through the methods outlined above, but this may be a laborious and exacting task. There are, however, various other methods, which can be used instead.

One strategy (Smith and Birnstiel, 1976) is based on the same principles that apply to DNA sequencing. DNA fragments are labelled at one end only, and are then partially digested with the desired enzyme to introduce, on the average, only one cut per molecule. This procedure generates a nested set of fragments which differ from each other by the length of one restriction fragment. These fragments are separated by gel electrophoresis and visualised by autoradiography. The length of each fragment and its relative position then can be read directly from the X-ray film. Problems arise if fragments are similar in size or if there are large differences in the rates at which

Max L. Birnstiel
Zürich

individual recognition sites are cleaved. The linear DNA molecule in Fig. 2.1-11, which has three cleavage sites for a restriction endonuclease may serve to illustrate the technique. Partial digestion of the DNA labelled only at one end produces four DNA bands which correspond to the undigested molecules and three other fragments which differ from each other by the length of one of the possible fragments: the smallest fragment D, which is nearest to the 5' end, moves fastest; the other fragments are 10, 5 or 6 arbitrary units longer than fragment D. Their order is therefore DACB.

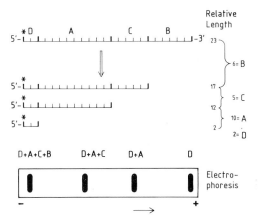

Fig. 2.1-11. Physical mapping of DNA fragments. A DNA fragment, 23 arbitrary units in length, is partially digested and yields sub-fragments of 17, 12, and 2 units. After electrophoresis and subsequent autoradiography of the gel only those bands containing fragments with common 5' ^{32}P-labelled ends are visible (Smith and Birnstiel, 1976).

An interesting adaptation of this method has been described by Rackwitz et al. (1984) for the mapping of phage λ clones. Right-hand or left-hand cohesive termini of λ DNA are selectively labelled by hybridisation with ^{32}P-labelled oligonucleotides complementary to the single-stranded cos ends. Since the oligonucleotides are specific for one end only, either the left-hand or the right-hand terminus, this procedure obviates the need for an additional cleavage step which is

required, for example, when a DNA fragment is end-labelled with T4 polynucleotide kinase. An adaptation for the mapping of cosmid clones has been described by Little and Cross (1985).

Another method for mapping frequently occurring restriction sites has been described by Parker et al. (1977). It relies on the fact that, under certain conditions of digestion, i.e., in the presence of ethidium bromide, circular DNA molecules are converted to linear molecules rather than being digested to completion. The circular molecule depicted in Fig. 2.1-12 possesses four cleavage sites for a type II restriction endonuclease, which we will designate X. After digestion such circular molecules will give rise to four different permuted linear DNA molecules with a length which equals that of the circular molecule. Let us consider now an enzyme Y which cuts each of these linear molecules only once at the position designated A. Each of the four linearised molecules will then give rise to two new molecules, the molecular weights of which will add up to the total molecular weight of the original linearised molecules. After gel electrophoresis one will obtain a pattern of DNA fragments with bands arranged in such a way that the sum of the molecular weights of the largest and the smallest fragments will equal the sum of the molecular weights of the second largest and second smallest DNA fragments, and so on. One end of each of these fragments will be the cleavage site of enzyme Y, which cuts only once (see arrows in Fig. 2.1-12); the other end of each fragment will be the cleavage site of the enzyme X which introduces multiple cuts in the starting molecule. Taking into account fragment sizes and the fact that fragments can be grouped pairwise, it is possible to build up the scheme depicted in Fig. 2.1-12. As a first step, the smallest fragment (No.8) is identified and drawn at the left-hand margin. The largest fragment (No.1) must then be positioned at the right-hand margin, because it carries the other part of the cleavage site (A''). The other fragments are also grouped in pairs, i.e., No.2 with No.7, and No.3 with No.6. They are aligned in

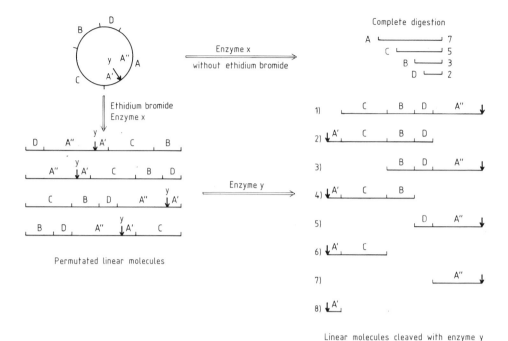

Fig. 2.1-12. Mapping of restriction endonuclease cleavage sites on circular molecules. The principle is explained in the text. (Parker *et al.*, 1977).

such a way that those fragments which differ in their molecular weights by a value corresponding to the molecular weight of a fragment occurring in the total digest, will come to lie on top of each other. In our example, fragments No.8 and No.7 differ by a value of 4.5, which does not occur among the molecular weights of the fragments observed in the total digest; these two fragments are positioned opposite to each other, *i.e.*, to the left and to the right, respectively. Their partners are fragments No.1 and No.2, which are positioned at appropriate positions at the right and left margin, respectively. The next largest fragment, No.6, differs from fragment No.8 by the size of a fragment (C) occurring in the complete digest and is positioned on top of fragment No.8 at the left hand margin. From this, the arrangement A'C can be deduced. Fragments No.5 and No.4 are treated similarly, and eventually the sequence A'CBDA'' can be derived. It is important to note

that the results obtained by this mapping procedure are only meaningful if fragment sizes are determined exactly and unambiguously (see also Maniatis *et al.* 1975; Parker, 1977). If this can be assured, in principle any circular DNA molecule, DNAs cloned in a plasmid, for example, can be mapped.

A completely different mapping protocol has been developed by Baralle *et al.* (1980a). Mapping of *Hinf* I fragments, for example, starts with a *Hinf* I digestion (G/ANTC) of a DNA molecule. This enzyme generates protruding ends, which are subsequently filled-in with the Klenow fragment of *E. coli* DNA polymerase I in the presence of α-^{32}P dATP (Fig. 2.1-10d). Each fragment is then isolated by gel electrophoresis and hybridised under suitable conditions with a single-stranded clone of M13 containing the entire original DNA molecule. After serving as a primer for a Klenow reaction with unlabelled nucleoside

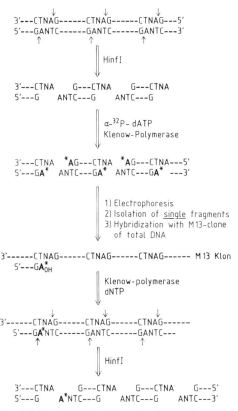

Fig. 2.1-13. Principle of mapping *Hinf* I cleavage sites (see text for details; Baralle *et al.*, 1980a).

triphosphates (Fig. 2.1-13), the filled-in DNA molecules are re-digested with *Hinf* I, and the fragments are again separated electrophoretically. After this second step only those DNA fragments which were adjacent to the DNA fragment initially labelled with α-^{32}P dATP will be labelled by dA. This experiment is repeated with all *Hinf* I fragments of the molecule to be mapped, and the analysis of labelled neighbouring fragments will then yield a physical *Hinf* I map of the original DNA molecule. In principle, this strategy can be applied to all enzymes which generate protruding 5' ends.

It should be noted here that difficulties are almost certain to arise with one or another

method; however, practical mapping should never rely exclusively on one mapping strategy; it should always be possible to obtain the desired results by combining several approaches.

2.1.2.6 Mapping of Large Genomes: Restriction Fragment Length Polymorphisms

The structural organisation of genes in large genomes, such as the human genome (3×10^9 bp), can be elucidated by making use of restriction fragment length polymorphisms (RFLP) (Botstein *et al.*, 1980). The phenomenon of polymorphism has been known for a long time at the level of proteins, where it is exemplified by proteins specifying blood groups or histocompatibility. Proteins are polymorphous if slight alterations in their primary structures are observed in different individuals of the same species, and if functions and chromosomal locations of such proteins are otherwise identical. Differences in the primary structures of proteins also manifest themselves at the level of DNA, and may occasionally lead to the generation of new restriction cleavage sites. Fig. 2.1-14 shows two different DNA strands, both of which encode a hypothetical gene *X* which is flanked by cleavage sites for enzymes A and B. The two genes (alleles) are identical with respect to the position of the cleavage sites for enzyme A, but they differ from each other by the position of the cleavage sites for B. The pattern of DNA fragments observed after hybridisation with a radioactively labelled probe of gene *X* yields one band if enzyme A is used for cleaving the corresponding cellular DNA, and two bands if enzyme B is used. In the latter case the two bands differ in size and correspond to the two different allelic forms. Gene *X* is said to be polymorphous with respect to the target site for enzyme B. In most cases such polymorphisms, also known as allelic variations, are caused by simple base substitutions destroying an existing or generating a new recognition site; occasionally, RFLPs may also be caused by more complex

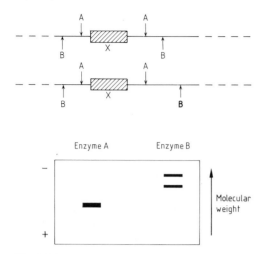

Fig. 2.1-14. Principles of restriction fragment length polymorphism.
The two alleles of a hypothetical gene X are flanked by cleavage sites for restriction enzymes A and B on both chromosomes. The positions of the two A sites are identical; however, the two alleles differ in the position of one of the B sites. The B site at the right-hand end of gene X is missing in one allele, but another B site (bold type) is found further to the right. If cellular DNA is cut either with enzyme A or B, a Southern blot analysis reveals only one band in DNA cut with enzyme A, but two bands in DNA cut with B after hybridisation with a radioactively labelled probe of gene X (Botstein *et al.*, 1980).

phenomena, such as deletions or insertions. A polymorphism which is due to the insertion or absence of a repetitive 5.8 kb DNA fragment is found, for example, in the gene specifying tyrosine biosynthesis (*sup4*) in yeast (Cameron *et al.*, 1979).

Unlike protein polymorphisms, DNA polymorphisms offer a number of advantages for mapping genomes; firstly, the number of known DNA markers already exceeds that of suitable protein markers, and secondly, a DNA sequence does not necessarily have to express a protein in order to be identified by polymorphous cleavage sites. DNA polymorphisms can, of course, occur in any DNA sequence, particularly in introns.

RFLPs are especially useful for identifying genetic defects in humans and can be exploited for diagnostic purposes as long as the DNA alterations involved do not occur several times, and are associated only with single genes. Most of the RFLPs known today appear to have occurred randomly and bear no relation to neighbouring genes.

RFLPs were first used for characterising mutant viruses (Grodzicker *et al.*, 1974). In humans, RFLPs were first identified in the vicinity of the globin gene and have been used for diagnosing sickle cell anaemia (Kan and Dozy, 1978). As shown in Fig. 2.1-15, the DNA polymorphism in this case is due to the presence of a *Hpa* I cleavage site 7.6 kb away from the β-globin gene itself. A certain percentage of individuals do not possess this site, and it is in 87% of patients carrying the sickle cell allele that an abnormal β-globin *Hpa* I fragment of 13 kb is observed. The detection of this fragment can be taken as a diagnostic indication of sickle cell anaemia. It is important to be aware of the fact that the polymorphous marker, *i.e.*, the mutated *Hpa* I site, and the β-globin gene have nothing in common and are only related in so far as one is located in the vicinity of the other. It is probably by pure chance that the mutation leading to sickle cell anaemia occurred a long time ago in a certain individual lacking the *Hpa* I site in question, and it is equally by pure chance that this RFLP has been conserved. Conservation, in this case, may be due to the close linkage of marker cleavage site and mutated gene.

The molecular defect of sickle cell anaemia is known today on the nucleotide level. It has been possible to develop other diagnostic techniques which guarantee that the genetic defect is identified with an accuracy of 100% rather than 87%. This approach, which is based on the use of chemically synthesised oligonucleotides, is described in Section 12.4.4.

Provided that polymorphous markers are closely linked with a gene, *i.e.*, both are inherited together, RFLPs can also be used for mapping

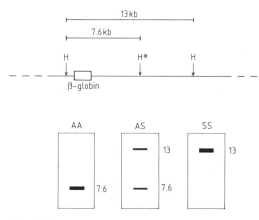

Fig. 2.1-15. *Hpa* I RFLP in the vicinity of the β-globin gene.

The three *Hpa* I sites are marked H; the variable cleavage site is marked H*. "A" stands for the normal, "S" for the sickle cell allele of the β-globin gene. The sickle cell allele lacks the H* site. A Southern blot analysis of healthy homozygous (AA) DNA donors shows a 7.6 kb band; sickle cell donors (SS) display a 13 kb band, and heterozygous donors (AS) two bands (Kan and Dozy, 1978).

genes and, hence, for characterising genetic defects even if the gene in question is completely unknown. Fig. 2.1-16 shows two polymorphous markers, *A* and *B*, which are located close to a gene. In the example, the normal allele is observed in the male, the mutant allele in the female; the mother is thus heterozygous, the father homozygous for the normal variant. According to Mendel's rules, these four alleles (two paternal and two maternal) lead to four possible combinations in the offspring. If marker *A* and the defective gene are closely linked, they are co-inherited, *i.e.*, all children with the genetic defect will also inherit the *A* allele of the marker. One might be tempted to draw the conclusion that the presence of the polymorphous marker *A* implies the presence of the genetic defect; strictly speaking, this correlation applies only to situations in which the gene in question and the polymorphous marker are closely linked. If marker and gene are widely separated, recombinational events can place the defective gene on to the

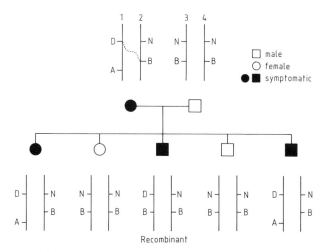

Fig. 2.1-16. Pedigree for two adjacent markers.

One of the two markers is polymorphous and gives rise to the two alleles A and B. The other marker has a normal allele (N) and a defective allele (D) the presence of which leads to clinical symptoms. In the parental DNA, alleles N and D are linked with the polymorphous markers B and A, respectively. In the first generation, the four alleles lead to four different combinations, with two normal and two diseased subjects. In the two latter cases, the defective marker is linked to the polymorphous marker A. If a recombinational event (dotted line) causes an exchange of genetic material between chromosomes 1 and 2 in the DNA of the mother, the defective gene will become associated with the polymorphous marker B (Gusella *et al.*, 1984).

chromosome bearing the polymorphous marker *B*. The probability that such a genetic exchange will take place increases with increasing distance between the marker and the gene in question. In the case of the *Hpa* I polymorphism in the vicinity of the β-globin gene it may be argued that the correlation factor of only 0.87 for sickle cell anaemia and a missing *Hpa* I site is the result of such recombinational events. The closer two markers are located together, the stronger are the observed correlations. Therefore, it should be possible to trace the inheritance of any polymorphous marker by pedigree analysis and to correlate its presence with the phenotypic expression of a genetic defect, if the particular marker maps sufficiently close to the defective gene.

In practice, the DNA of many individuals of one lineage is first cleaved with a restriction enzyme which exhibits a polymorphism in the probe DNA to be used subsequently. Typically, approximately one million different DNA fragments will be obtained; these can be separated by agarose gel electrophoresis and subjected to Southern blot analysis (*cf.* Chapter 11), that is the fragments are transferred to nitrocellulose filters and hybridised with the radioactively labelled probe DNA recognising the expected polymorphism. At present, approximately 200 different RFLP probes for a total of ten restriction enzymes have already been identified and employed for mapping purposes (Sparkes *et al.*, 1984).

It is worthwhile to consider how many polymorphous markers would be necessary to cover the entire human genome (Lange and Boehnke, 1982). The most important limiting factor is the frequency of recombination. Two markers should be close enough to be co-inherited; their linkage should not be abolished by recombinational events too often. Classical genetics measures the length of a genetic map in Morgan units, one Morgan being the length of DNA in which there will occur, on the average, only one recombinational event during meiosis, for example, during oogenesis or spermatogenesis. Pedigree analyses have shown that two markers which are 0.1

Morgan units apart are usually co-inherited with significant probabilities, and that these markers can still be identified as being closely linked even if the number of individuals available for pedigree analysis is relatively small. The human genome has a length of approximately 33 Morgan units (Renwick, 1971). If the distance between a gene and a marker must not exceed 0.1 Morgan units in order to guarantee that they segregate together, $0.1/33 = 165$ markers would be required to cover the entire human genome. Of course, this value must be regarded as a lower limit: for practical purposes one must assume a much higher value, possibly 1 000 to 2 000, because, *inter alia*, the above calculation assumes that the markers are evenly distributed along the genome; this can hardly be expected.

If the human genome with its length of 3×10^9 bp were to be covered by 1 000 DNA fragments, each fragment would be approximately 3 000 kb in length. Such fragments are still extremely long for a molecular characterisation, *e.g.*, by DNA sequencing. Subcloning, chromosome walking and chromosome hopping facilitate such analyses today and should eventually lead to the molecular characterisation of the fragment within a reasonable time.

An important application of restriction fragment length polymorphisms has been the detection of a marker for a disease known as Huntington's chorea (Gusella *et al.*, 1984), a severe neurodegenerative defect, which occurs with a frequency of ten afflicted individuals per 100 000. Its molecular and biochemical causes are still unknown. RFLP mapping, in this case, was made possible by the availability of two different pedigrees of disease carriers from Venezuela and the United States, and a collection of approximately 600 permanent lymphoblastoid cell lines derived from these patients from which sufficiently large amounts of DNA could be obtained for analysis. Initially, it was completely unknown whether any of the known polymorphous markers mapped in the vicinity of the gene in question and segregated together with the Huntington gene. It

Haplotype   Restriction fragments

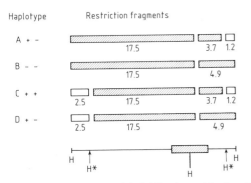

Fig. 2.1-17. Pattern of hybridization with the G8 probe.
The marker is characterised by two polymorphous *Hind* III sites (marked as H*). Absence (-) or presence (+) of this marker leads to four different patterns of hybridisation with the G8 probe which are designated as haplotypes A to D. Numbers below the bars indicate fragment lengths (in kb). Sequences represented by hatched bars can be recognised as bands in Southern blots; sequences shown as open bars cannot be identified by hybridisation since they are not overlapping with the probe (Gusella *et al.*, 1984).

was quickly observed that one of the probes, G8, yielded the desired results (Gusella *et al.*, 1983). As shown in Fig. 2.1-17, this probe, or a suitable subclone of it, recognises two RFLPs in genomic DNA cleaved with *Hind* III. This polymorphism is caused by two variable *Hind* III sites. The presence or absence of the first *Hind* III site leads to DNA fragments which are 15 or 17.6 kb in length respectively, while the presence or absence of the second *Hind* III site yields fragments of 3.7 and 4.9 kb. There are thus four possible fragment patterns, known as haplotypes *A*, *B*, *C*, and *D*. Since there are always two alleles of a given gene in a somatic cell, the four haplotypes give rise to ten different double combinations (*e.g.*, AA, AB, etc.; *cf.* also Fig. 2.1-18).

The frequency of these double combinations in the normal population is variable. A pedigree analysis of the Venezuelan family revealed that haplotype *C* (presence of both *Hind* III sites; Fig. 2.1-18) always segregated together with the

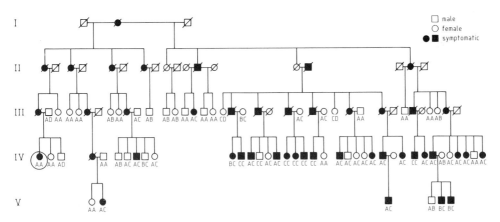

Fig. 2.1-18. A portion of a Venezuelan Huntington's Disease (HD) pedigree.
This subset of members of the Venezuelan family illustrates the inheritance of the HD gene on a chromosome bearing the C haplotype at the G8 locus. A single individual, representing a recombination event that transferred the HD gene to a chromosome containing the A haplotype at the G8 locus, is indicated by a circle. Deceased individuals are marked by symbols with diagonal slashes. In this family individuals with the haplotype C are at risk although they may not always be affected; individuals lacking haplotype C are never at risk. The odds in favour of this linkage are greater than one trillion to one (Gusella *et al.*, 1984).

disease. Since this haplotype also occurs in the normal population, the only plausible conclusion at this point is that a member of the family which did not inherit haplotype *C* from its parents would most likely not be afflicted with the disease; on the other hand, an individual with haplotype *C* could not be certain whether the disease would develop or not. Only approximately 25% of the normal population carry haplotype *C*. The chances of an individual from the afflicted Venezuelan family with haplotype *C* to develop the disease are, therefore, rather high.

Two conclusions can be drawn from the observations made so far: firstly, conclusive predictions can only be made by an analysis of pedigrees (*cf.* also White *et al.*, 1985). Since haplotype *C* also occurs in the normal population, its identification in an individual without family predisposition makes little or no sense at all. Secondly, the observed association of the *G8* marker with the disease does not imply that the gene responsible for Huntington's chorea has been identified. The authors observed a single recombinant in the Venezuelan family (Fig. 2.1-18) in which the Huntington gene had been translocated to a chromosome with haplotype *A*. From the occurrence of this single event they concluded that the distance between the two markers (Huntington gene and *Hind* III sites) was on the order of at least 2 000 kb. Even with currently available techniques it is not an easy task to characterise such a long stretch of DNA. A first approach would be to find additional polymorphisms within this DNA region. In addition, it is known that the Huntington gene and also the *G8* marker map in the short arm of chromosome 4. It should therefore be possible to construct a library containing only chromosome 4 DNA which would be of lower complexity and thus much easier to characterise than a full genomic library. This could be achieved, for example, by chromosome sorting (Davies *et al.*, 1981; Labo *et al.*, 1984) or by making use of human-hamster hybrid cell lines (Ruddle and McKusick, 1977; for a recent example involving the chromosomal assignment of the

human T-cell receptor α-chain gene, *cf.* Collins *et al.*, 1985).

In summary, RFLPs will not only allow mapping of genetic defects but also construction of a genetic map of human chromosomes within the next few years. The consequences are manifold. It can be envisaged, for example, that typing of organ donors for transplantation surgery by serological techniques will quickly be superseded by the more accurate and quicker determination of DNA polymorphisms. It will also be possible to detect correlations between the predisposition of individuals towards certain diseases and their genetic constitution. Such correlations are currently being discussed, for example, for an association between the susceptibility to certain viral infections and characteristic patterns of histocompatibility antigens. Of course, one may also envisage possible abuses; these are now being discussed by a concerned and distrustful public and will be analysed in Chapter 13.

2.1.3 Type III Endonucleases

These nucleases also cleave DNA at specific sequences, but cleavage takes place only in the immediate vicinity of the binding/recognition site. Fig. 2.1-19 shows three of the six cleavage sites on ΦX174 RF DNA of the enzyme *Hga* I, which cleaves DNA five to ten bases away from its recognition sequence 5'-GACGC-3' and generates protruding 5' ends (Brown and Smith, 1977). In contrast to staggered ends generated by type II endonucleases, such single-stranded ends always differ from each other and cannot be recombined at random; on the contrary, there is only one way in which the fourteen *Hga* I fragments obtained after digesting ΦX174 DNA can be recombined to regenerate infectious DNA molecules. It is apparent that this particular feature of type III nucleases greatly facilitates the isolation of certain DNA molecules from a complex mixture of eukaryotic DNA fragments. A whole set of such

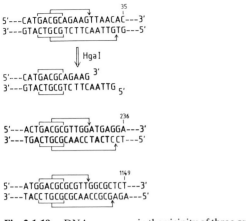

Fig. 2.1-19. DNA sequences in the vicinity of three out of a total of 14 *Hga*I cleavage sites on ΦX174 RF DNA.
The cleavage product for the first sequence is also shown. Recognition sequences (GACGC) are bracketed; cleavage sites are marked by arrows. The numbering is that of the original sequence described by Saenger *et al.* (1977).

enzymes possessing characteristic non-palindromic target sequences is now available (Table 2.1-6).

Recently, a number of enzymes related to type I nucleases have also been classified as type III endonucleases. Like type I enzymes they require ATP and Mg^{2+} (Yuan, 1981; Kanc and Piekarowicz, 1978); however, these enzymes lack both the ATPase activity of class I enzymes and their absolute requirement for S-adenosylmethionine. In addition, they produce homogeneous, rather than heterogeneous, populations of DNA fragments similar to those generated by type II enzymes since they cleave DNA at rather fixed positions, 25-27 bp downstream of their recognition sequences (Hadi *et al.*, 1983). As yet, the four known enzymes of this type have not been employed in gene cloning.

Table 2.1–6. Some restriction enzymes with non-palindromic recognition sequences.

| Enzyme | Recognition sequence | Number of sites on | | |
		pBR322 DNA	M13mp8 DNA	SV40 DNA
*Hga*I	GACGC(N)$_5$ CTGCG(N)$_{10}$	11	7	0
*Hph*I	GGTGA(N)$_8$ CCACT(N)$_7$	12	18	4
*Mbo*II	GAAGA(N)$_8$ CTTCT(N)$_7$	11	11	16
*Fok*I	GGATG(N)$_9$ CCTAC(N)$_{13}$	12	4	11
*Sfa*NI	GCATC(N)$_5$ CGTAG(N)$_9$	22	7	6
*Tth*111II	CAACA(N)$_{11}$ GTTGT(N)$_9$	5	7	11

pBR322-DNA: 4363 bp; M13mp8-DNA: 7229 bp; SV40-DNA: 5243 bp.

References

Arber, W., and Linn, S. (1969). DNA modification and restriction. Ann. Rev. Biochem. 38, 467-500.

Baralle, F.E., Shoulders, C.C., Goodbourn, S., Jeffrey, A., and Proudfoot, N.J. (1980a). The 5' flanking region of the human globin gene. Nucleic Acids Res. 8, 4393-4404.

Baralle, F.E., Shoulders, C.C., and Proudfoot, N.J. (1980b). The primary structure of the human globin gene. Cell 21, 621-626.

Blakesley, R.W., Dodgson, I.B., Nes, I.F., and Wells, R.D. (1977). Duplex regions in single-stranded ΦX174 DNA are cleaved by a restriction endonuclease from *Haemophilus aegyptius*. J. Biol. Chem. 252, 7300-7306.

Botstein, D., White, R.L., Skolnick, M., and Davis, R.W. (1980). Construction of a genetic linkage map in man using restriction fragment length polymorphism. Am. J. Hum. Genet. 32, 314-331.

Brown, N.L., and Smith, M. (1977). Cleavage specificity of the restriction endonuclease isolated from *Haemophilus gallinarum* (*Hga* I). Proc. Natl. Acad. Sci. USA 74, 3213-3216.

Cameron, J.R., Loh, E.Y., and Davis, R.W. (1979). Evidence for transposition of dispersed repetitive DNA families in yeast. Cell, 16, 739-751.

Carle, G.F., and Olson, M.V. (1985). An electrophoretic karyotype for yeast. Proc. Natl. Acad. Sci. USA 82, 3756-3760.

Clarke, C.M., and Hartley, B.S. (1979). Purification, properties and specificity of the restriction endonuclease from *Bacillus stearothermophilus*. Biochem. J. 177, 49-62.

Collins, M.K.L., Goodfellow, P.N., Spurr, N.K., Solomon, E., Tanigawa, G., Tonegawa, S., and Owen, M.J. (1985). The human T-cell receptor α-chain gene maps to chromosome 14. Nature 314, 273-274.

Doerfler, W. (1983). DNA methylation and gene activity. Ann. Rev. Biochem. 52, 93-124.

Felsenfeld, G., and McGhee, J.D. (1982). Methylation and gene control. Nature 296, 602-603.

George, J., Blakesley, R.W., and Chirikjian, J.G. (1980). Sequence specific endonuclease *Bam*H I. J. Biol. Chem. 255, 6521-6524.

Gitschier, J., Wood, W.I., Goralka, T.M., Wion, K.L., Chen, E.Y., Eaton, D.H., Vehar, G.A., Capon, D.J., and Lawn, R.M. (1984). Characterization of human factor VIII gene. Nature 312, 326-330.

Gray, P.W., and Hallick, R.B. (1977). Restriction endonuclease map of *Euglena gracilis* chloroplast DNA. Biochemistry 16, 1665-1671.

Greene, D.J., Heynecker, H.L., Bolivar, F., Rodriguez, R.L., Betlach, M.C., Covarrubias, A.A., Backman, K., Russel, D.J., Tait, R., and Boyer, H.W. (1978). A general method for the purification of restriction enzymes. Nucleic Acids Res. 5, 2373-2380.

Grodzicker, T., Williams, J., Sharp, P., and Sambrook, J. (1974). Physical mapping of temperature-sensitive mutations of adenoviruses. Cold Spring Harbour Symp. Quant. Biol. 39, 439-446.

Gusella, J.F., Wexler, N.S., Conneally, P.M., Naylor, S.L., Anderson, M.A., Tanzi, R.E., Watkins, P.C., Ottina, K., Wallace, M.R., Sakaguchi, A.Y., Young, A.B., Shoulson, I., Bonilla, E., and Martin, J.B. (1983). A polymorphic DNA marker genetically linked to Huntington's disease. Nature 306, 234-238.

Gusella, J.F., Tanzi, R.E., Anderson, M.A., Hobbs, W., Gibbons, K., Raschtchian, R., Gilliam, T.C., Wallace, M.R., Wexler, N.S., and Conneally, P.M. (1984). DNA markers for nervous system diseases. Science 225, 1320-1326.

Guyaux, M., Cornelissen, A.W.C.A., Pays, E., Steinert, M., and Borst, P. (1985). *Trypanosoma brucei*: a surface antigen mRNA is discontinuously transcribed from two distinct chromosomes. The EMBO J. 4, 995-998.

Hadi, S.M., Bächi, B., Iida, S., and Bickle, T.A. (1983). DNA restriction-modification enzymes of phage P1 and plasmid p15B: Subunit functions and structural homologies. J. Mol. Biol. 165, 19-34.

Hay, R.T. (1985). The origin of adenovirus DNA replication: minimal DNA sequence requirement *in vivo*. The EMBO J. 4, 421-426.

Heininger, K., Hort, W., and Zachau, H.G. (1977). Specificity of cleavage by restriction nucleases from *Bacillus subtilis*. Gene 1, 291-303.

Hsu, M., and Berg, P. (1978). Altering the specificity of restriction endonuclease: Effect of replacing Mg^{2+} with Mn^{2+}. Biochemistry 17, 131-138.

Kan, Y., and Dozy, A. (1978). Antenatal diagnosis of sickle-cell anaemia by DNA analysis of amniotic-fluid cells. Lancet 2, 910-912.

Kanc, L., and Piekaroicz, A. (1978). Purification and properties of a new endonuclease from *Haemophilus influenzae* Rf. Europ. J. Biochem. 92, 417-426.

Kelly, T. Jr., and Smith, H.O. (1970). A restriction enzyme from *Haemophilus influenzae*. II. Base sequence of the recognition site. J. Mol. Biol. 51, 393-409.

Kemp, D.J., Corcoran, L.M., Coppel, R.L., Stahl, H.D., Bianco, A.E., Brown, G.V., and Anders, R.F. (1985). Size variation in chromosomes from independent cultured isolates of *Plasmodium falciparum*. Nature 315, 347-350.

Kessler, C., Neumaier, P.S., and Wolf, W. (1985). Recognition sequences of restriction endonucleases and methylases – a review. Gene 33, 1-102.

Lange, K., and Boehke, M. (1982). How many polymorphic genes will it take to span the human genome? Am. J. Hum. Genet. 34, 842-845.

Langridge, I., Langridge, P., and Bergquist, P.L. (1980). Extraction of nucleic acids from agarose gels. Anal. Biochem. 103, 264-271.

Lautenberger, J.A., Kan, N.C., Lackey, D., Linn, S., Edgell, M.H., and Hutchinson, C.A. III. (1978). Recognition site of *Escherichia coli* B restriction enzyme on ΦX174 and simian virus 40 DNAs: An interrupted sequence. Proc. Natl. Acad. Sci. USA 75, 2271-2275.

Lebo, R.V., Gorin, F., Fletterick, R.J., Kao, F.T., Cheung, M.C., Bruce, B.D., and Kan, Y.W. (1984). High-resolution chromosome sorting and DNA spot-blot analysis assigning McArdle' syndrome to chromosome 11. Science 225, 57-59.

Legerski, R.J., Gray, H.B., and Robberson, D.L. (1977). A sensitive endonuclease probe for lesions in deoxyribonucleic acid helix structure produced by carcinogenic or mutagenic agents. J. Biol. Chem. 252, 8740-8746.

Li, J.L., and Kelly, T.J. (1984). Simian virus 40 DNA replication *in vitro*. Proc. Natl. Acad. Sci. USA 81, 6973-6977.

Little, J.W., Lehmann, I.R., and Kaiser, A.D. (1967). An exonuclease induced by bacteriophage λ. I. Preparation of the crystalline enzyme. J. Biol. Chem. 242, 672-678.

Little, P.F.R., and Cross, S.H. (1985). A cosmid vector that facilitates restriction enzyme mapping. Proc. Natl. Acad. Sci. USA. 82, 3159-3163.

Malcolm, A.D.B. (1981). The use of restriction enzymes in genetic engineering. In "Genetic Engineering", R. Williamson (ed.), Academic Press, New York, Vol. 2, pp.129-173.

Malyguine, E., Vannier, P., and Yot, P. (1980). Alterations of the specificity of restriction endonucleases in the presence of organic solvents. Gene 8, 163-177.

Maniatis, T., Jeffrey, A., and v.d. Sande, H. (1975). Chain length determination of small double- and single-stranded DNA molecules by polyacrylamide gel electrophoresis. Biochemistry 14, 3787-3794.

Maxam, A., and Gilbert, W. (1980). Sequencing end-labelled DNA with base-specific chemical cleavages. Methods in Enzymology 65, 499-580.

McClelland, M., Kessler, L.G., and Bittner, M. (1984). Site-specific cleavage of DNA at 8- and 10-base pair sequences. Proc. Natl. Acad. Sci. USA 81, 983-987.

McDonnell, M.W., Simon, M.N., and Studier, F.W. (1977). Analysis of restriction fragments of T7 DNA and determination of molecular weights by electrophoresis in neutral and alkaline gels. J. Mol. Biol. 110, 119-146.

Mertz, J.E., and Davis, R.W. (1972). Cleavage of DNA by RI restriction endonuclease generates cohesive ends. Proc. Natl. Acad. Sci. USA 69, 3370-3374.

Meselson, M., and Yuan, R. (1968). DNA restriction enzyme from *E. coli*. Nature 217, 1110-1114.

Miller, P.S., Cheung, D.M., Dreon, N., Jayamaram, K., Kan, L.S., Lentzinger, E.E., and Pulford, S.M. (1980). Preparation of a decadeoxyribonucleotide helix for studies by molecular magnetic resonance. Biochemistry 19, 4688-4698.

Molloy, P.L., and Symons, R.H. (1980). Cleavage of DNA/RNA hybrids with type II restriction endonucleases. Nucleic Acids Res. 8, 2939-2946.

Nelson, M., Christ, C., and Schildkraut, I. (1984). Alteration of apparent restriction endonuclease recognition specificities by DNA methylases. Nucleic Acids Res. 12, 4507-4517.

Nussinov, R. (1980). Some rules in the ordering of nucleotides in the DNA. Nucleic Acids Res. 8, 4545-4558.

Parker, R.C., Watson, R.M., and Vinograd, J. (1977). Mapping of closed circular DNAs by cleavage with restriction endonucleases and calibration by agarose gel electrophoresis. Proc. Natl. Acad. Sci. USA 74, 851-855.

Polisky, B., Greene, P., Garfin, D.E., McCarthy, B.J., Goodman, H.M., and Boyer, H.W. (1975). Specificity of substrate recognition by the *Eco* RI restriction endonuclease. Proc. Natl. Acad. Sci. USA 72, 3310-3314.

Rackwitz, H.R., Zehetner, G., Frischauf, A.M., and Lehrach, H. (1984). Rapid restriction mapping of DNA cloned in λ phage vectors. Gene 30, 195-200.

Renwick, J.H. (1971). The mapping of human chromosomes. Ann. Rev. Genet. 5, 81-120.

Richardson, C.C., Lehman, I.R., and Kornberg, A. (1964). A deoxyribonucleic acid phosphatase-exonuclease from *Escherichia coli*. J. Biol. Chem. 239, 251-258.

Roberts, R.J. (1983). Restriction and modification enzymes and their recognition sequences. Nucleic Acids Res. 11, r135-r167.

Roberts, R.J. (1984). Restriction and modification enzymes and their recognition sequences. Nucleic Acids Res. 12, r167-r204.

Ruddle, F., and McKusick, V. (1977). The status of the gene map of the human chromosomes. Science 196, 390-405.

Sanger, F., Air, G.M., Barrell, B.G., Brown, N.L., Coulson, A.R., Fiddes, J.C., Hutchinson III, C.A., Slocombe, P.M., and Smith, M. (1977). Nucleotide sequence of bacteriophage ΦX174 DNA. Nature 265, 687-695.

Schwartz, D.C., and Cantor, C.R. (1984). Separation of yeast chromosome-sized DNAs by pulse field gradient gel electrophoresis. Cell 37, 67-75.

Seeburg, P.H., Shine, J., Martial, J.A., Baxter, J.D., and Goodman, H.M. (1977). Nucleotide sequence and amplification in bacteria of structural gene for rat growth hormone. Nature 270, 468-494.

Smith, H.O., and Nathans, D. (1973). A suggested nomenclature for bacterial host modification and restriction systems and their enzymes. J. Mol. Biol. 81, 419-423.

Smith, H.O., and Birnstiel, M.L. (1976). A simple method for DNA restriction site mapping. Nucleic Acids Res. 3, 2387-2398.

Southern, E. (1979). Gel electrophoresis of restriction fragments. Methods in Enzymology 68, 152-191.

Sparkes, R.S., Berg, K., Evans, H.J., and Klinger, H.P., eds.; in "Human Gene Mapping"; Proceedings of the 1983 Los Angeles Conference, Karger, Basel, 1984.

Sutcliffe, J.G. (1978). pBR322 restriction map derived from the DNA sequence: accurate DNA size markers up to 4361 nucleotide pairs long. Nucleic Acids Res. 5, 2721-2728.

Tabak, H.F., and Flavell, R.A. (1978). A method for the recovery of DNA from agarose gels. Nucleic Acids Res. 7, 2321-2332.

Thomas, M., and Davis, R.W. (1975). Studies on the cleavage of bacteriophage λ DNA with *Eco* RI restriction endonuclease. J. Mol. Biol. 91, 315-328.

Ullrich, A., Shine, J., Chirgwin, J., Pictet, R., Tischer, E., Rutter, W.J., and Goodman, H.M. (1977). Rat insulin genes: Construction of plasmids containing the coding sequences. Science 196, 1313-1319.

Vogelstein, B., and Gillespie, D. (1979). Preparative and analytical purification of DNA from agarose. Proc. Natl. Acad. Sci. USA 76, 615-619.

Waalwijk, C., and Flavell, R.A. (1978). *Msp* I, an isoschiozomer of *Hpa* II which cleaves both unmethylated and methylated *Hpa*II sites. Nucleic Acids Res. 5, 3231-3236.

White, R., Leppert, M., Timothy Bishop, D., Barker, D., Berkowitz, J., Brown, C., Callahan, P., Holm, T., and Jerominski, L. (1985). Construction of linkage maps with DNA markers for human chromosomes. Nature 313, 101-105.

Woodhead, J.L., Bhave, N., and Malcolm, A.D.B. (1981). Cation dependence of restriction endonuclease *Eco* RI activity. Europ. J. Biochem. 115, 293-296.

Yuan, R. (1981). Structure and mechanism of multifunctional restriction endunucleases. Ann. Rev. Biochem. 50, 285-315.

Zabeau, M., and Roberts, R. (1979). The role of restriction endonucleases in molecular genetics. In: "Molecular Genetics", J.H. Taylor (ed.), Academic Press, New York, Part III, pp.1-63.

2.2 Preparation and Cloning of Eukaryotic cDNA

2.2.1 Definition and Structure of cDNA

Complementary DNA (cDNA) is a double-stranded DNA copy of the mRNA which is found in the cytoplasm of eukaryotic cells and serves as a template for protein biosynthesis. (It should be pointed out here that the entire chapter will exclusively be concerned with eukaryotic rather than prokaryotic mRNA.) As shown in Fig. 2.2-1, there is a linear correspondence between the information encoded in the mRNA and the protein that is read from it by the ribosomes (see also Breathnach and Chambon, 1981). mRNA molecules contain regions at their 5' and 3' ends which are not translated, and possess a poly(A) tail at the 3' end. Apart from these structures, eukaryotic mRNA molecules are entirely free of non-coding regions, called introns, which usually disrupt coding regions, or exons, of eukaryotic genes (Darnell, 1982). RNA molecules having just been transcribed from the DNA still contain introns. These non-coding regions are removed from the primary RNA transcripts by a process known as splicing, during their passage from the nucleus to the cytoplasm, where they serve as functional mRNAs (Fig. 2.2-1). Whenever eukaryotic split genes containing non-coding regions are dealt with, cDNA is a source of intron-free genetic information, because it is an exact copy of the mRNA. There are only a few known eukaryotic genes, α- and β-interferon genes, for instance, which do not contain introns.

The use of cDNA is an absolute prerequisite for the expression of eukaryotic proteins in bacteria. According to current knowledge, bacterial genes do not possess the characteristic exon-intron structure of eukaryotic genes, suggesting that bacteria are not endowed with mechanisms for processing genetic information containing such intervening sequences and thus cannot express complexly structured eukaryotic DNA. When gene cloning aims at the production of eukaryotic proteins in bacteria, the methods employed rely exclusively on cloning cDNA or chemically synthesised cDNA.

Successful cDNA cloning depends entirely on the purity of mRNA. The probability of finding a desired clone in a population of cloned molecules, or in other words, the amount of experimental work involved, stands and falls with the purity of the starting material. If, for instance, the percentage of the desired mRNA in the total poly(A)-containing RNA of a eukaryotic cell is taken to be on the order of 1%, only one out of 100 cDNA clones, on the average, may contain the desired genetic information. Depending on the procedures available for identifying this particular clone, even this comparatively favourable signal-to-noise ratio may give rise to considerable problems, because cloning *per se* does not improve the situation. It is, therefore of prime importance for our discussion and for any experimental approaches to consider in detail the nature of mRNA molecules employed in cloning experiments.

2.2.2 Isolation, Identification, and Purification of mRNA

The major aim of all methods devised for the isolation of mRNAs is a fast and effective inhibition of endogenous ribonuclease activities (Taylor, 1979). Of course, all solutions used for extractions must themselves be free of nucleases; this, in most cases, can be brought about by autoclaving. If solutions, for instance sucrose solutions, cannot be autoclaved, RNAse activities can be destroyed by diethylpyrocarbonate, which reacts quickly with proteins, and, due to its instability, is quickly decomposed to ethanol and carbon dioxide by Tris buffers. It should be kept in mind that this compound may also attack nucleic acids, albeit in a delayed manner. Certain ribonucleoside-vanadyl complexes (Puskas *et al.*,

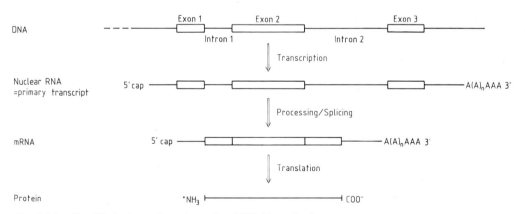

Fig. 2.2-1. Simplified scheme for eukaryotic mRNA biosynthesis. Boxes represent exons, *i.e.*, coding regions, lines are non-coding regions, or introns.

1982) are also potent inhibitors of ribonuclease activities. They inhibit many of the known ribonucleases and do not appear to interfere with most of the important reactions involving recombinant DNA molecules. These compounds are thought to act as analogues of cyclic $2',3'$ monophosphates which have been postulated to be intermediates of reactions catalysed by ribonucleases.

The disruption of tissue culture cells can be achieved effectively by treatment with hot phenol. Solid tissues must be frozen first, with either pulverised dry ice or liquid nitrogen, and can then be reduced to small pieces in a Waring blendor (Benveniste *et al.*, 1973). Phenolisation has been observed to be associated with the formation of aggregates between mRNA and other RNA species; such aggregates can be destroyed easily with formamide or by heating to 65 °C.

Tissues and tissue culture cells particularly abundant in nuclease activities have also been disrupted successfully by incubation in 4 M guanidine isothiocyanate and 1 M β-mercaptoethanol (Chirgwin *et al.*, 1979). The resulting extracts are usually deproteinised with chloroform/isoamylalcohol (24:1), and the RNA, which is heavy, is pelleted by centrifugation onto a CsCl cushion with a density of 1.75 g/ml (Ullrich *et al.* 1977). If nuclease activities are low, cell suspensions can be treated hypertonically or with detergents before they are separated into a nuclear and a cytoplasmic fraction by centrifugation; the cytoplasm is then phenolised, and the RNA can be obtained practically free of DNA. Residual DNA can be destroyed with DNAse, and this step is also appropriate for the other procedures mentioned above. It should be noted that commercially available DNAse preparations have frequently been repurified in order to ensure that

the enzyme is absolutely free of ribonuclease activities (see Maxwell *et al.*, 1977).

The crucial task is to isolate from RNA preparations mRNA which is essentially free of proteins and DNAs. Since most eukaryotic mRNAs possess a poly(A) tail of 50-200 bases at their $3'$ ends, one can enrich for the desired mRNAs by affinity chromatography on poly(U) sepharose (Lindberg and Persson, 1972) or oligo(dT) cellulose (Aviv and Leder, 1972). Experience has shown that the yield of mRNA bound specifically to the column can be significantly increased by recycling RNA solutions several times over the same column. Oligo(dT) cellulose columns have the advantage of being re-usable because they can be washed with alkali. If oligo(dT) columns are used, the total RNA is applied at high salt concentrations, and mRNA specifically retained is eluted with low salt buffers, ethanol-precipitated, and stored in ethanol at –20 °C. Unfortunately, yields are almost never quantitative, probably because poly(A) tails may

Philip Leder
Boston

be rather short and can be very heterogeneous in length.

Several methods are available to enrich for a particular mRNA from the total RNA preparation obtained from a tissue. The purification of certain mRNAs is facilitated considerably if they possess a characteristic base composition. In the case of silk fibroin and collagen, for instance, the known periodic amino acid sequences predict a special composition of the corresponding mRNAs. Indeed these could be isolated from total RNA preparations relatively easily by density centrifugation, because these mRNAs were guanine-rich, as expected, and were rather large (32S and 27S, respectively). Sometimes it is also possible to exploit the fact that certain cells contain relatively high amounts of a particular mRNA species. Examples are ovalbumin mRNA isolated from chick oviduct cells, β-globin mRNA isolated from reticulocytes, and insulin mRNA isolated from pancreatic β cells (Islets of Langerhans). Globin mRNA, for example, accounts for 98% of all poly(A)-containing mRNAs in reticulocytes of anaemic mice (Bastos et al., 1977).

Even under the best possible circumstances, mRNA enrichment hardly ever exceeds 1%. Occasionally it is advantageous to prepare mRNA from polysomes. Active polysomes are complexes of ribosomes and mRNA which contain a certain fraction of nascent polypeptide chains. Provided that these polypeptides are sufficiently long, they may already possess antigenic determinants and may therefore form specific antigen-antibody complexes with appropriate antisera. Yet binding of a single antibody is not sufficient to precipitate the polysomes. Several methods have been devised to precipitate such antigen-antibody complexes by letting them bind to a solid matrix or by reacting them with a second antibody directed against the first. In contrast to direct immunoprecipitation involving the original antigens, the latter process is called an indirect immunoprecipitation reaction, because it makes use of anti-antibodies. The complexes consisting of polyso-mes, antibodies, and anti-antibodies can be separated from other polysomes and antibodies in a discontinuous sucrose gradient. Detailed protocols for indirect immunoprecipitation have been described (Shapiro et al., 1974; Gough and Adams, 1978; Maurer, 1980).

An alternative to immunoprecipitation is the use of a solid phase. Staphylococcus aureus cells have proven very successful (Gough and Adams, 1978). The protein A component in the cell walls of these bacteria specifically binds F_c fragments of immunoglobulins. Antigen-antibody complexes and, hence, polysomes are thus bound to the surfaces of these bacteria, and the mRNA can be easily isolated from precipitates by phenol extraction. In many cases, a further enrichment of the purified mRNAs may be accomplished by separation in sucrose gradients according to size. Individual fractions of the gradient can be assayed for the presence of the desired mRNA by translation in vitro and subsequent detection of the protein in question. If this protein is precipitated, for example by specific antibodies, fractions which contain mRNA directing the synthesis of this protein can serve as the starting material for cDNA synthesis.

Three systems, prepared from reticulocytes, wheat germ, or Xenopus laevis oocytes are available for cell-free protein biosynthesis. All three systems are capable of translating exogenous, heterologous mRNA into proteins. Initially, reticulocyte lysates of anaemic rabbits had the disadvantage of being somewhat insensitive to exogenously added mRNA, due to their high content of endogenous globin mRNA. This endogenous background activity can be reduced substantially by treating the lysates with the calcium-dependent micrococcal ribonuclease, which then can be inactivated by EGTA (Pelham and Jackson, 1976). Today, rabbit reticulocyte lysates are used routinely and are even commercially available. The improved system can even be useful for estimating mRNA concentrations in different mRNA populations during the purification procedure by measuring the amount of ^{35}S-labelled proteins synthesised after the addition of constant amounts of different mRNA preparations; if all assays are carried out under non-saturating conditions, the amount of newly synthesised protein is proportional to the amount of exogenously added mRNA.

Wheat germ extracts are easily prepared and possess a low endogenous activity; however, they are very susceptible to slight alterations in reaction conditions, and the synthesis of larger proteins may create problems.

Strictly speaking, the oocytes of *Xenopus laevis* constitute whole cells rather than lysates, although the term "*in vitro* translation system" has come to stay. The oocyte system described by Gurdon (Laskey *et al.*, 1973; Gurdon and Melton, 1981) responds well to low concentrations of mRNA and is active for several days. It has been of special significance in gene technology and in particular for the isolation of mRNA and cDNA of various interferons (Gray *et al.*, 1982; Derynck *et al.*, 1980), for which only minute amounts of mRNA were recovered from density gradients. Only the oocyte system allowed such minute quantities of mRNA to be translated efficiently and to yield extracts or incubation fluids with sufficiently high antiviral activity; an advantage and prerequisite in this case was, of course, the very high specific activity of interferons (10^8 to 10^9 units/mg protein).

Another approach to determine amounts and yields of mRNA during purification is by hybridisation rather than by *in vitro* translation. One advantage of hybridisation is that it allows the size of mRNA molecules to be determined in a way which is completely independent of gel electrophoresis. There is a plethora of different hybridisation protocols (Flint, 1980), but hybridisation analysis of mRNA usually involves the reassociation of a large excess of RNA with double-stranded DNA labelled by nick translation or obtained by cDNA synthesis. These labelling procedures guarantee such high specific activities that the necessary excess of RNA can be easily obtained.

Strictly speaking, RNA-DNA hybridisation is a second order reaction, characterised by the equation

$$dD/dt = - k \cdot D \cdot R,$$

where t represents the time (in seconds), D the concentration of single-stranded DNA (in moles of nucleotides/l), R the concentration of RNA (moles of nucleotides/l), and k the velocity constant of the hybridisation reaction (mol/l · s). DNA self-reassociation can be avoided by either immobilising the DNA or by considerably lowering its concentration. Under conditions of RNA excess, the RNA concentration will then remain practically constant, and hybridisation reactions will, therefore, follow pseudo first-order kinetics. The velocity equation can then be simplified to

$$dD/dT = - k \cdot R_0 \cdot D,$$

where R_0 represents the RNA concentration at $t = 0$. By substituting $D = D_0$ at $t = 0$, the integrated equation will be

$$\ln D_0/D = k \cdot R_0 \cdot t.$$

When 50% of the DNA (cDNA) strands have been converted into DNA-RNA hybrids, D becomes $D_0/2$, and $R_0 t$ becomes $R_0 t_{1/2}$. This yields:

$$\ln \frac{D_0}{D_0 2} = k \cdot R_0 \cdot t_{1/2}$$

$$k = \frac{\ln 2}{R_0 \cdot t_{1/2}}$$

The velocity of the hybridisation reaction is, therefore, inversely proportional to the value of $R_0 t_{1/2}$, which is directly related to the genetic complexity, *i.e.*, the size of a pure and uniform DNA or RNA molecule. If the $R_0 t_{1/2}$ values of a pure mRNA (x) with known length and another mRNA (y) of unknown length are known, the complexity (size) of mRNA (y) can be determined by the following equation:

$$\frac{\text{complexity of mRNA } (y)}{\text{complexity of mRNA } (x)} =$$

$$\frac{R_0 \, t_{1/2} \, [\text{mRNA } (y)]}{R_0 \, t_{1/2} \, [\text{mRNA } (x)]}$$

For an mRNA of 2 000 nucleotides in length, such as ovalbumin mRNA, the $R_0 t_{1/2}$ value is approximately $4 \cdot 10^{-3}$ mol s/l. Accordingly, an mRNA of 900 nucleotides in length (prolactin mRNA) has a smaller $R_0 t_{1/2}$ value of approximately $1.8 \cdot 10^{-3}$ mol s/l (Maurer, 1980).

$R_0 t_{1/2}$ values can also be used to determine the fraction of a certain mRNA in a population of different mRNA molecules by hybridising "pure" cDNA with "pure" mRNA and also mRNA mixtures under conditions of RNA excess. Since the complexity of the RNA in question is constant, reaction rates depend solely on the concentration of the mRNA in the population: the slower the rate of the hybridisation reaction, the smaller is the proportion of a particular mRNA in the population, according to the equation:

$$\frac{R_0 \, t_{1/2} \, (\text{"pure" mRNA})}{R_0 \, t_{1/2} \, (\text{unknown sample})} =$$

$$\frac{\% \text{ mRNA in the unknown sample}}{\% \text{ mRNA in the "pure" sample}}$$

In practice, the course of a hybridisation reaction between labelled cDNA and unlabelled RNA is monitored by determining the number of double-stranded molecules, *i.e.*, the number of DNA-RNA hybrids, at various times during the reaction. This is usually accomplished by testing aliquots either by hydroxyapatite (HAP) chromatography, or by determining the amount of DNA resistant to S1 nuclease.

If HAP is used, the reaction mixture is applied to a HAP column in 0.05 M phosphate buffer of pH 6.8 at 60 °C, from which single-stranded DNA

can be eluted with 0.14 M phosphate buffers, and double-stranded hybrid molecules with 0.45 M phosphate buffers.

S1 nuclease digests single-stranded DNA, while RNA-DNA hybrids remain unaffected (Vogt, 1973). For the use of S1 nuclease, samples containing the reassociated DNAs are first adjusted to optimal salt concentrations and pH for S1 nuclease digestion. Undigested nucleic acids, *i.e.*, double-stranded hybrid DNA molecules, which are S1-resistant, are precipitated with trichloroacetic acid (TCA) after the reaction. The fraction of S1-resistant DNA is determined by comparing the amounts of radioactivity in these samples with the amounts of radioactivity observed in untreated samples. As shown in Fig. 2.2-2, the results

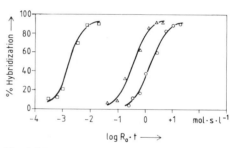

Fig. 2.2-2. Determination of mRNA probe purity. The purity and complexity of a prolactin cDNA was determined by hybridisation with the enriched preparation of mRNA from which the cDNA had initially been synthesised (squares), with total pituitary gland mRNA of untreated rats (circles), and total mRNA from pituitary glands of estrogen-treated rats (triangles). The different $R_0 t_{1/2}$ values are a measure of mRNA purity (Maurer, 1980; see text for further explanations).

are plotted as per cent hybridisation *versus* log $R_0 t$. This example shows the result of hybridisations of "pure" prolactin cDNA with its corresponding mRNA, with total mRNA from pituitary glands, and with total mRNA from the pituitary glands of rats treated with estradiol. The $R_0 t_{1/2}$ values are $1.8 \cdot 10^{-3}$ mol s/l for pure mRNA, 1.6 mol s/l for total pituitary mRNA, and 0.39

mol s/l for total pituitary mRNA from hormone-treated animals. By applying the formula given above, one obtains mRNA concentrations of 0.11% and 0.46% in the total pituitary mRNA and total pituitary mRNA of hormone-induced animals, respectively (Maurer, 1980).

If either the complete or a partial amino acid sequence of a protein is known, enrichment for the desired genetic information can be achieved by specific priming on the cDNA level or even after cloning. In such cases an oligonucleotide can be synthesised, which may then serve as a specific primer for cDNA synthesis or as a specific hybridisation probe, which allows, for example, the identification of the desired sequences in a cDNA library. In each case, the individual experimental approach will depend on the availability of starting material, such as suitable tissues or cells, the availability of antibodies, or a knowledge of structural information. As far as genomes with high complexities are concerned, cDNA cloning provides a means of eventually obtaining genomic DNA sequences which otherwise cannot be isolated due to their low concentration; on the other hand, genomes with low complexities, such as viral genomes, are frequently available in unlimited amounts. Here, the genomic DNA serves as starting material to obtain specific mRNA. In this case, even the minute amounts of viral mRNA observed, for example in the very early phases of a viral infection, can be identified and isolated by specific RNA-DNA hybridisation with genomic DNA. Specific DNA fragments of a viral genome can also be used for assigning specific mRNA species to them. Cloning techniques indeed have played a major role, for example, in the elucidation of transcription maps, because only cloned DNA fragments are absolutely pure and available in unlimited amounts.

2.2.3 Synthesis of cDNA

2.2.3.1 Reaction Conditions of Reverse Transcriptase and Synthesis of the First Strand

cDNA synthesis starts with mRNA and follows the scheme depicted in Fig. 2.2-3. The first step is the synthesis of an mRNA-DNA hybrid molecule with the help of reverse transcriptase from avian myeloblastosis virus (AMV). Being an RNA-dependent DNA polymerase, the enzyme requires an RNA template and a primer providing the necessary 3' hydroxyl group (Temin and Mizutani, 1970; Baltimore, 1970). Usually, oligo(dT)$_{12}$ chains which hybridise with the poly(A) tail of the mRNA are used. In principle, it should be possible to obtain a complete copy of the mRNA including the 5' and the 3' untranslated regions.

David Baltimore
Cambridge, Massachusetts

The chemical synthesis of specific primers gains more and more significance. If the amino acid sequence of the protein in question is known, the structure of suitable primers can be deduced. The problem of wobble bases in the genetic code, *i.e.*, the ambiguity observed in the third base of many codons, can be easily circumvented by employing suitable mixtures of derivatised nucleotide bases rather than individual bases at the positions in question during the chemical synthesis of the primer; the resulting oligonucleotide mixtures will then always contain the desired primer species in sufficient concentrations. As a rule, primers should be at least 12 bases long; usually primers of 15 bases are used. This ensures the formation of stable hybrids under conditions of

the reverse transcriptase reaction, and makes it more likely that only the desired mRNA molecules, but not other mRNA species, are primed. The probability of a sequence consisting of N bases to occur in a genome is 1 in 4^N; a sequence consisting of nine bases can be expected to occur every 262 144 base pairs, and for the human genome, even a dodecanucleotide primer ($N = 12$) would be too unspecific.

Examples for reverse transcription using defined oligonucleotides as primers are the cloning of the genes for gastrin (Mevarech *et al.*, 1979), human transplantation antigens (Sood *et al.*, 1981), and rat insulin (Chan *et al.*, 1979). It is obvious that the use of specific primers also allows specific regions in a mRNA to be copied. Long

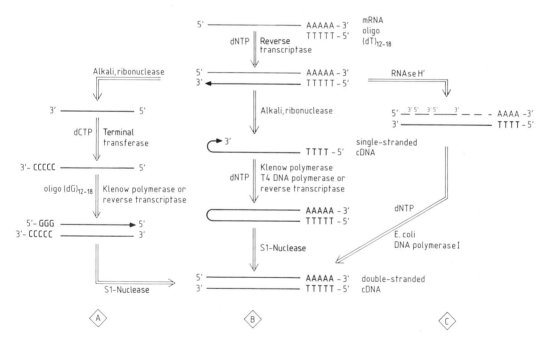

Fig. 2.2-3. Reaction steps for the preparation of double-stranded cDNA.
Following the synthesis of the first strand, three approaches may be followed for the synthesis of double-stranded cDNA. The "classical" approach (B) employs a self-priming mechanism followed by digestion of the hairpin with S1 nuclease. An alternative route (A) relies on the addition of a short homopolymeric tail to the single-stranded DNA (Land *et al.*, 1981), while strategy (C) involves the combined use of RNAse H and the Kornberg enzyme (Gubler and Hoffman, 1983; see text).

mRNA species, in particular, are frequently not copied by reverse transcriptase in one contiguous stretch and can thus be primed several times (Uhlén *et al.*, 1982).

Yet primers for cDNA synthesis do not always have to be specific. It is often quite useful, especially for cDNA synthesis from long mRNA species, to employ short tetra- to heptanucleotides, which can then prime at various positions of the genome. It is very difficult, for example, to prepare intact cDNA copies of the long RNA of tobacco mosaic virus (6393 nucleotides). The problem has been overcome by using short oligonucleotides for priming, which, though it did not yield clones containing the entire genome in an intact form, at least provided a small collection of clones which together represented the entire genome (Goelet *et al.*, 1982).

Sometimes mRNA molecules are heterogeneous with respect to the positions of their 5' ends.

By specifically priming with a defined tetradecamer, the 3' end of which corresponded to the translation start signal, it was possible to identify five different populations of mRNA for the fibre gene of adenovirus type 2. These mRNAs differed in their structure and the length of their 5' untranslated regions (Fig. 2.2-4). Oligonucleoti-

$$
\begin{array}{ccccc}
& 8 & 10 & 12 & 14 \\
& \downarrow & \downarrow & \downarrow & \downarrow \\
3' \leftarrow \text{TAC TTT GCG CGG TC 5'} & & & & \text{Primer} \\
5'\text{---GCAG ATG AAA CGC GCC AGA C---3'} & & & & \text{mRNA} \\
\text{met}
\end{array}
$$

Fig. 2.2-4. Defined oligonucleotides of different lengths as primers for reverse transcription.
This example shows the region of the adenovirus fiber protein mRNA with the translation initiation signal ATG for the fiber protein (Uhlén *et al.*, 1982). Primer-directed DNA synthesis is carried out to determine the structure of the 5' untranslated region of the mRNA. Numbers and arrows indicate different lengths of the primers.

des consisting of 8, 10, 12, and 14 bases were utilised in these experiments. While the octamer did not prime at all, the decamer primed moderately, and only the dodecamer had good priming activity. The method of specific priming with reverse transcriptase in order to identify the 5' termini of an mRNA is known as "primer extension".

Non-polyadenylated RNA molecules, such as ribosomal RNA or the fragmented RNA genomes of certain RNA viruses, have to be prepared for cDNA synthesis by addition of homopolymeric tails to their 3' ends (Emtage *et al.*, 1980). This can be accomplished with *E. coli* poly(A) polymerase, which, like terminal transferase, does not need a template for the addition of homopolymeric deoxyribonucleotides (Sippel,

Howard M. Temin
Madison

1973). It is also possible to exploit the specific property of T4 RNA ligase, which efficiently joins oligoribonucleotides to the 3′ ends of double-stranded RNA molecules. Both methods have been used for cloning the fragmented RNA genomes of reo- and rotaviruses (Cashdollar *et al.*, 1982; Imai *et al.*, 1983).

As far as the yields of cDNA and the lengths of the DNA products obtained from mRNA are concerned, optimising the reacting conditions for reverse transcription itself is of primary importance. Frequently the amount of mRNA available is limited, and very often the aim is to obtain cDNA copies as long as possible. The reaction conditions originally described for reverse transcription (see Verma *et al.*, 1972, for instance) have been improved considerably by using high concentrations of dNTPs, up to 1 mM (Efstradiatis *et al.*, 1975), high temperatures (46 °C) (Monahan *et al.*, 1976), and methylmercuryhydroxide (2-4 mM) (Wickens *et al.*, 1978). Only recently has it been demonstrated that the modifications mentioned above were necessary because the preparations of reverse transcriptase contained contaminating ribonuclease activities; highly purified reverse transcriptase does not cause any problems (Retsel *et al.*, 1980). Under optimal reaction conditions, the yields of cDNA obtained from the input poly(A^+) RNA should be in the range of 5 and 20% (microgram of cDNA/microgram of poly(A^+) RNA).

There are many examples from the field of virology which demonstrate that RNA-DNA hybrids can be cloned directly (Wood and Lee, 1976; Zain *et al.*, 1979); more frequently, however, a second strand is synthesised in order to obtain double-stranded cDNA molecules.

2.2.3.2 Synthesis of Double-Stranded cDNA

For the synthesis of the second strand of cDNA, reverse transcriptase, or the Klenow fragment of *E. coli* DNA polymerase I (Kornberg enzyme) or

T4 DNA polymerase can be used. The latter two enzymes lack the 5′ to 3′ exonuclease activity of DNA polymerase I, and do not degrade the newly synthesised double-stranded DNA. Both the Kornberg enzyme and T4 DNA polymerase require a 3′ hydroxyl primer. This primer can be generated spontaneously by hairpin formation within the single-stranded DNA, *i.e.*, by the 3′ end of the single-stranded DNA folding back on itself (Fig. 2.2-3B). It is also possible to extend the 3′ ends of the cDNA by homopolymeric tails (Fig. 2.2-3A), in which case the synthesis of the second strand can be initiated by using a complementary, homopolymeric primer (Rougeon and Mach, 1977). In the case of influenza virus RNA, which consists of eight RNA fragments, specific primers were used for the synthesis of the first and the second strand, because previous sequencing studies had shown that all eight RNA molecules possessed the same sequences of 12 and 13 bases at their 3′ and 5′ ends, respectively. One single primer could therefore be used for the synthesis of all eight first strands and another primer for the synthesis of all eight second strands (Winter *et al.*, 1981); however, apart from this and some other exceptions, "self-priming" is preferred because it leads to higher yields of cDNA.

The reactivity of reverse transcriptase differs from that of T4 DNA polymerase or the Klenow fragment. It has been observed, for example, that the use of AMV reverse transcriptase instead of polymerase I generally produces only full-length double-stranded cDNA copies and hardly ever any shorter reaction intermediates. The synthesis of the second strand directed by AMV reverse transcriptase, therefore, generally yields cDNA populations which are more homogeneous than those produced by *E. coli* polymerase I; on the other hand, Rougeon and Mach (1977) have demonstrated that certain cDNA molecules are good templates for *E. coli* polymerase I but not for AMV reverse transcriptase (Fig.2.2-5). These molecules do not possess base-paired 3′ ends, and it may be that in these cases the 3′ to 5′ exonuclease activities of the Klenow fragment and T4 DNA polymerase are required for the elimination of protruding single-stranded termini in order to generate active 3′ hydroxyl ends

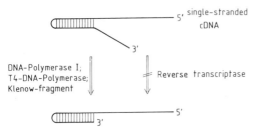

Fig. 2.2-5. Differences in the reactivities of *E. coli* DNA polymerase I, Klenow fragment, T4 DNA polymerase, and reverse transcriptase.
In each case the substrate is a single-stranded, but partially double-stranded, cDNA with a characteristic hairpin structure at its 3′ end.

(Englund, 1971). AMV reverse transcriptase would be inactive in this case because it does not possess the required 3′ to 5′ nuclease activity. In view of the experience reflected in various published reports, it is impossible to recommend solely the use of one or the other enzyme; instead, one should keep in mind that all these enzymes possess their characteristic individual properties.

The yields which have been reported in the literature for the synthesis of the second strand vary between 30% and 100%. The second reaction is usually carried out in two steps. In the first step, the RNA is removed by alkaline hydrolysis or digestion with ribonuclease, while the second step involves synthesis of the DNA strand (see Goodman and McDonald, 1979, for instance). Wickens *et al.* (1978) have devised a single-flask procedure, in which the RNA is not destroyed but separated from the DNA by heating.

In contrast to specifically primed double strands, hairpins of spontaneously (intramolecularly) primed strands must be treated with a single-strand specific nuclease, S1 nuclease (Fig. 2.2-3B). The cleavage of a phosphodiester bond within the hairpin loop produces double-stranded

ends, which make the molecule amenable to subsequent cloning. In order to avoid the digestion of the double-stranded parts of the cDNA which correspond to the 5′-terminal ends of the mRNA, this nuclease treatment has to be performed very carefully. The reaction, and also the course of cDNA synthesis, can be easily monitored if radioactively labelled precursors, for example, ^{3}H-dNTPs for the first reaction and ^{32}P-dNTPs for the second reaction, are used. These radioactive labels are important for two reasons: first of all, the yield of double strand synthesis can be determined by monitoring the extent to which the reaction products become S1-resistant during the reaction; secondly, labelling has a practical aspect, if the entire cDNA is not cloned directly but digested with restriction endonucleases before cloning. Inhomogeneous cDNA populations can be easily identified after gel electrophoresis, because very broad smears rather than discrete fragments are observed upon autoradiography; if, however, one or more mRNA species in the initial population have been particularly abundant, a pattern of specific bands may be observed, from which DNA can be isolated and used for cloning. cDNA can thus be amplified and purified, and even if a particular fragment represented only a particular part of the desired mRNA, clones obtained in this way can subsequently be used to isolate and identify intact mRNA molecules, which, in turn, can be employed for cloning full-length cDNAs.

The fractionation of cDNA by digestion with at least one enzyme prior to cloning also may be important for reasons of biological containment and safety. In many cases, the use of purified cDNA fragments instead of unpurified cDNA for shot-gun-type cloning experiments may reduce the required safety precautions by one or even several levels. In the early days of gene cloning, such considerations played an important role, for example, in the cloning of peptide hormones such as insulin and growth hormone.

A critical step in the synthesis of cDNA as described in Fig. 2.2-3B is the S1 nuclease treatment mentioned above. Gubler and Hoff-man (1983) have described a procedure which replaces this S1 nuclease digestion step (Fig. 2.2-3C). The RNA moiety of the RNA-DNA hybrid obtained after the synthesis of the first strand is not removed by a treatment with alkali, but rather by combined RNaseH/*E. coli* DNA polymerase I treatment. RNaseH (the H stands for hybrid), which specifically attacks the RNA portion of RNA-DNA hybrids, introduces a number of nicks, *i.e.*, 3' hydroxyl ends in the RNA strand which may serve as primers for the nick translation reaction catalysed by DNA poly-merase I. Although reverse transcriptase itself possesses an endogenous RNaseH activity, it cannot be used for this purpose because this activity is too weak and too variable. Yields of up to 10^6 clones per microgram of RNA have been described for instance in the cloning of cDNAs for the precursor of human growth hormone releas-ing factor (Gubler *et al.*, 1983) and of rat thymosin β4 (Wodner-Filipowicz *et al.*, 1984).

2.2.4 *Cloning of cDNA*

A double-stranded cDNA molecule must be ligated with an appropriate vector molecule for cloning. Often the DNA is inserted into the *Pst* I site of plasmid pBR322, either with linkers or *via* homopolymeric tails; the latter method may direct-ly produce expression vectors. In principle, cDNA molecules do not differ from other clona-ble DNA molecules; for a detailed discussion of the ligation of cDNA and vector molecules the reader therefore is referred to Chapter 3.

A fundamentally new method of cDNA cloning uses the vector itself for priming the synthesis of the first and the second strand (Okayama and Berg, 1982). The vector primer is a linear DNA molecule with an oligo(dT) extension at one end. It is derived from a pBR322-SV40 recombinant (Fig. 2.2-6) which contains a unique *Kpn* I cleav-age site. The protruding ends generated by cleav-

age at the *Kpn* I site are extended with poly(dT) tails in the presence of terminal transferase. One of these tails can be specifically removed by using the unique *Hpa* I site in the vicinity of one of the two poly(dT) ends. Hybridisation of the remain-ing poly(dT) tail with the poly(A) tail of the mRNA generates a substrate for reverse trans-cription (Fig. 2.2-7), in which the 3' hydroxyl group of the terminal (dT) residue at the poly(dT) tail of the vector serves as primer. Homopolymer-ic dC tails are added to both ends of the molecule when the first cDNA strand synthesis has been completed. The poly(dC) tail at one of the ends of the molecule can again be specifically removed, because the vector contains a single *Hind* III site near one of its ends. The molecule can then be circularised by using a short linker DNA fragment carrying a *Hind* III terminus at one end, and a poly(dG) tail at the other end (Fig. 2.2-6). The *Hind* III ends are covalently linked by DNA ligase, while the oligo(dC) and oligo(dG) tails are linked non-covalently by dG-dC base pairing.

The final step in the synthesis of cDNA is the replacement of the mRNA moiety by the corres-ponding DNA strand. Once strand breaks have been made in the mRNA moiety by *E. coli* RNAseH, the second cDNA strand is synthesised by *E. coli* DNA polymerase I in the presence of all four deoxyribonucleotide triphosphates. In an experiment carried out with β-globin mRNA, 10^5 ampicillin-resistant clones per microgram of mRNA were obtained upon transformation of *E. coli* strain HB101. 20% of these clones contained the entire mRNA sequence, and nearly 40% of all clones contained sequences lacking only a few nucleotides at the 5' end of the mRNA molecule (Okayama and Berg, 1982). A particular advan-tage of this cloning technique is the built-in selection, which favours the production of full-length reverse transcripts: the preferred substra-tes of terminal transferase are protruding 3' hydroxyl termini rather than termini covered by longer stretches of mRNA (Roychoudhury *et al.*, 1976). In the example given above, this selection becomes effective with the addition of homopolymeric dC tails after completion of the synthesis of the first strand. Of course, the technique outlined above also can employ a suitable pBR322-derived vector instead of one derived from the SV40 genome (Ko and Harter, 1984).

Okayama's and Berg's procedure has been effectively applied by Numa and coworkers (Kakidani *et al.*, 1982; Noda *et al.*, 1982; Noda *et al.*, 1984) for cloning cDNAs for porcine β-neo-endorphin/dynorphin precursors, the acetylcholi-

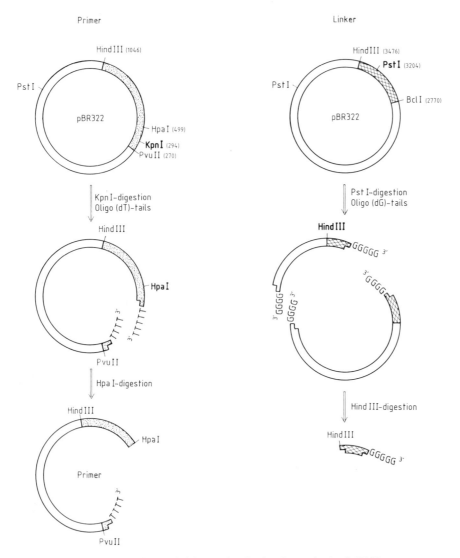

Fig. 2.2-6. Synthesis of primer and linker molecules for the synthesis of cDNA.
pBR322 sequences are represented by open, SV40 sequences by stippled or cross-hatched boxes. Numbers at cleavage sites are SV40 co-ordinates (Okayama and Berg, 1982; *cf*. also Fig. 8-10).

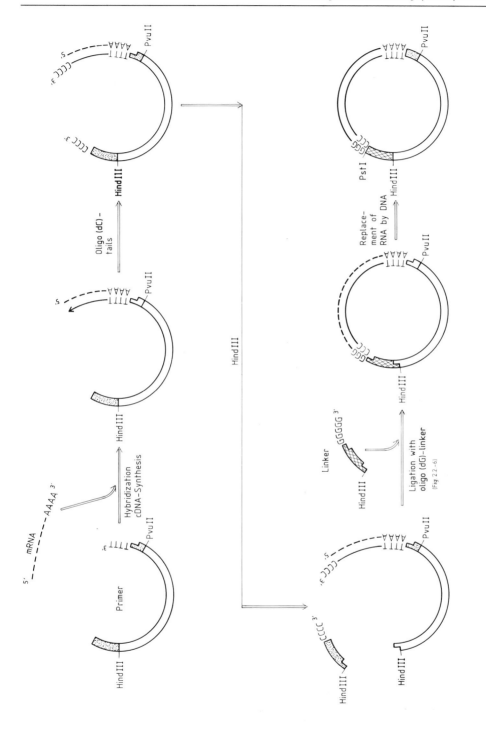

Fig. 2.2-7. Reaction steps for cDNA synthesis.
Broken lines represent mRNA. The synthesis of primer and linker DNA is described in Fig. 2.2-6 (Okayama and Berg, 1982).

ne receptors of the electric organs of *Torpedo californica*, and the sodium channel of *Electrophorus electricus*. If it proves effective in practice, this new technique should fundamentally alter cDNA cloning. The high yields of full-length cDNA might allow clones to be identified in suitable vectors by directly expressing the functional proteins; suitable expression vectors have already been described (Okayama and Berg, 1983): rather than circularising molecules with the linker described in Fig. 2.2-6, a new linker DNA fragment is applied instead. It contains the SV40 early promoter and a short intron derived from the SV40 late region, which carries the splice site for the 16S mRNA (see also Chapter 8, Fig. 8-16). This modification allows the production of recombinants with cDNA insertions which can be transcribed correctly in eukaryotic cells and can be translated into enzymatically active proteins. An interesting variation of this procedure which may be even simpler has been described by Heidecker and Messing (1983). It has proven quite effective for the preparation of maize endosperm cDNA libraries.

2.2.5 cDNA Libraries

A eukaryotic cell contains many thousand mRNA species. A cDNA gene library (or cDNA bank) is a population of bacterial transformants, in which each mRNA is represented as its cDNA insertion in a plasmid or a phage vector. The frequency with which individual inserts occur reflects the concentration of the corresponding mRNAs in the cell. There are several good reasons for making such gene banks:

– A cDNA library is much smaller than a library of genomic clones because a large proportion of eukaryotic DNA is never expressed, *i.e.*, it never appears in the form of mRNA. The RNA content of a human cell may be estimated to be 50 000 species, yet the cell may contain as much as ten times the amount of DNA required to

code for these RNA. If all mRNA species were to be represented in a library, a genomic library would have to be ten times larger than the corresponding cDNA library.

– Eukaryotic mRNAs are present in varying amounts. Their proportion among the total mRNA depends, for example, on specific growth phases, specific functions, or stages of development of a particular cell. A genomic library will contain all the genes of a species, while a cDNA library will reflect the specific state of cells or of particular cell types. Ninety per cent of all mRNA molecules in a reticulocyte, for instance, are globin mRNA molecules; this is reflected in the corresponding cDNA library, in which nine out of ten isolated clones will contain globin cDNA insertions. A genomic reticulocyte library, on the other hand, would contain only a relatively small number of globin clones, a number which would correspond to the number of globin genes in the total genomic DNA of the organism.

– Certain RNA sequences, such as the genomes of paramyxoviruses, myxoviruses or RNA phages, do not possess a true DNA counterpart. The genomes of these viruses can only be cloned after their conversion into cDNA copies.

– Apart from a few exceptions, the genes of eukaryotic organisms are interrupted by non-coding regions. cDNA clones prepared from corresponding mRNA molecules do not contain these non-coding regions, because they have been spliced out during mRNA maturation. Since bacteria are unable to remove non-coding regions from split genes, the use of cDNA copies is mandatory if cloned genes are also to be expressed as proteins in bacteria; of course, an alternative would be the chemical synthesis of genes.

The immediate question arising in the construction of a cDNA library concerns the number of clones that are required to make the library a representative one, because only in a complete library does the number of clones carrying a particular cDNA insertion reflect the relative fraction of the corresponding mRNA molecules in the cell. A good estimate can be obtained by determining the number of different mRNA sequences in an mRNA population. The method of choice is to hybridise highly labelled cDNA with an excess of the corresponding mRNA. As has been demonstrated already in Section 2.2.2, the amount of cDNA that exists in the form of DNA-RNA hybrids at any time point t is plotted as a function of R_0. The curve shown in Fig. 2.2-8 was obtained by plotting the data of an experiment in which HeLa mRNA was hybridised with its cDNA (Bishop et al., 1974). A computer analysis revealed the existence of three different kinetic classes of mRNA molecules, and Table 2.2-1 lists the corresponding proportions in which these classes occur in the total population, namely, 22%, 28%, and 50%, and the corresponding $R_0t_{1/2}$ values derived from the data in Fig. 2.2-8. The values have been corrected to obtain the $R_0t_{1/2}$ values which would have been observed had each of the three classes occurred on its own and represented the amount of mRNA with a concentration R_0. Moreover, the values are also corrected for the approximately 10% of poly(A) sequences which do not contribute towards hybridisation. The corrected $R_0t_{1/2}$ values can be used to determine the number of different sequences in each class by comparing them with $R_0t_{1/2}$ values of mRNAs of known complexity.

It has been shown in Section 2.2.2 that the $R_0t_{1/2}$ value of an mRNA with a molecular weight

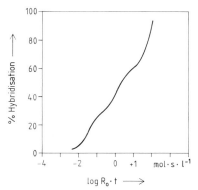

Fig. 2.2-8. Hybridisation of cDNA. Hybridisation of labelled cDNA representing the total mRNA of a cell with an excess of this mRNA. Results are plotted as a function of log R_0t (Bishop et al., 1974).

of $6 \cdot 10^5$ Da, the mean molecular weight of HeLa cell mRNA, is $9 \cdot 10^{-4}$ mol s/l. This value is divided by the corrected $R_0t_{1/2}$ values obtained for each of the different classes, and this yields the number of different mRNA species in each kinetic class. These numbers can be converted into copy numbers/cell (see Table 2.2-1), if it is assumed that each HeLa cell contains 0.625 pg of poly(A) mRNA (Flint, 1980). If the number of cDNA clones required for a representative cDNA library is to be calculated, it is important to focus the attention on the mRNA class containing the mRNA species with the lowest copy number which, in our example, is class 3. If mRNAs in this size class are to be represented at least once, the

Table 2.2–1. Composition and complexity of HeLa cell mRNA.

Class	%cDNA hybridised	$R_0t_{1/2}$ observed	$R_0t_{1/2}$ corrected	Number of different sequences	Copies/cell
1	22	5×10^{-2}	1.5×10^{-2}	17	8090
2	28	9×10^{-1}	3.35×10^{-1}	370	470
3	50	45	29.9	33000	9

Hybridisation experiments reveal three kinetic classes of HeLa cell mRNA. See text for details. $R_0t_{1/2}$ values can be used to calculate abundance and complexity of each RNA class (Flint, 1980).

other classes which occur 40 and 800 times more frequently will certainly be represented in the cDNA library.

In the example given in Fig. 2.2-8, 50% cDNA hybridisation corresponds to a complexity of 33 000 different mRNA sequences. The minimal number (n) of clones necessary to obtain a complete cDNA library representing all mRNAs will therefore be 33 000/0.50 = 66 000, although it should be noted that, for statistical reasons, this number is only a lower estimate of the number of clones required. The true number of clones, N, which is required to find a sequence of class 3 in a library with the probability P and the calculated minimal number of specific sequences (n), is:

$$N = \frac{\ln (1 - P)}{\ln (1 - 1/n)}$$

(Clark and Carbon, 1976). If a specific clone of this population were to be represented with a probability of 99%, 258 000 clones would have to be searched; 138 000 for a probability P of 90%. Considering that 10^6 Hela cells contain approximately 0.6 microgram poly(A) mRNA and that only 0.5% of this mRNA can be converted to double-stranded cDNA containing dC tails, approximately 3 ng of cDNA would be available for transforming bacteria. The mean molecular weight of Hela cDNA is 5.3×10^6 Da. To obtain equimolar conditions, these 3 ng of cDNA must be ligated to the approximately five-fold amount of pBR322 DNA (2.6×10^6 Da). Since at least 2×10^7 transformants per microgram of DNA can be expected, 18 ng of DNA would yield approximately 300 000 to 400 000 clones, and, according to the calculations outlined above, these clones would constitute a representative cDNA gene bank, in which a desired clone would occur with a probability of at least 99%. By starting with a single cell, the number of HeLa cells (10^6) mentioned above could be obtained after approximately 20 cell divisions, covering a petri dish with a diameter of 3 cm; these cells would correspond to approximately one milligram of tissue (wet weight), and this is on the order with which even biopsy specimens can be obtained.

The construction of a cDNA library from mRNA isolated from tissues does not differ very much from the methods described for cloning enriched mRNA preparations (Norgard et al., 1980). The most important question in each case is that of storage. It is obvious, for example, that several hundred thousand clones cannot be stored in individual test tubes. Moreover, there are some other essential considerations to keep in mind: do the inserts in the cDNA library really reflect the true mRNA composition? Can such a collection of clones be propagated without changing its initial composition? Does the library really contain a desired sequence with the expected probability?

The ways in which a library is stored depend on the number of bacterial clones to be dealt with. Comparatively small numbers, i.e., up to 5 000 clones can conveniently be stored on microtiter plates at –70 °C in 50% glycerol. Larger numbers of clones are grown on nitrocellulose filters placed on top of agar plates. The filters are then removed and stored frozen at –20 °C. This procedure allows the storage of up to 100 000 bacterial colonies on one filter with a diameter of 15 cm (Hanahan and Meselson, 1980); however, a cDNA library can also be stored as a mixture of clones. In this case, the transformants obtained in one transformation experiment are grown, for example, in large dishes (25 × 25 cm) which may contain as many as 100 000 to 200 000 clones. The agar plates are then overlayed with medium, bacterial colonies are scraped off, resuspended in medium, pelleted by centrifugation, and resuspended in a suitable buffer. Aliquots of this suspension kept frozen at –70 °C can be used as inocula for fresh cultures. A mixture of plasmids can later be isolated from these fresh cultures and may be used to isolate the desired insertions. cDNA insertions labelled, for example, by nick translation, can be tested in reassociation kinetic experiments, to gain information about their complexity. For the cDNA library of Hela mRNA discussed above, one should again observe three classes of molecules differing in their complexity. Such a pattern will usually be stable for many passages; yet, it should be noted that reassociation experiments hardly reveal any information about the fate of any one particular mRNA or cDNA species. It is advisable to avoid too many passages for cDNA amplification, because in the long run some cloned cDNA insertions might not be stable, due to their detrimental effects on the bacterial cell.

cDNA libraries can also be established in expression vectors. Clones can then be tested directly for the synthesis of a specific protein, for example, by immunological methods. This, of course, requires the cDNA to be positioned in proper orientation for reading, i.e., directly after suitable control elements of the vector. Helfman et al. (1983) have used vector pUC8, for example, which carries a polylinker directly after the control elements of the lac operon (Messing and Vieira, 1982). These authors have also devised a method with allows linkers with different specificities to be added to a cDNA. As demonstrated in Fig. 2.2-9, the first linker (SalI) is ligated to the ends of the DNA immediately after the synthesis of the second strand and before S1 digestion. This linker can be added selectively to one end of the molecule, because the formation of hairpin loops prevents its ligation at the other end. A second linker with a different specificity (Eco RI) can be added on this end of the molecule after S1 nuclease digestion. When this molecule is treated

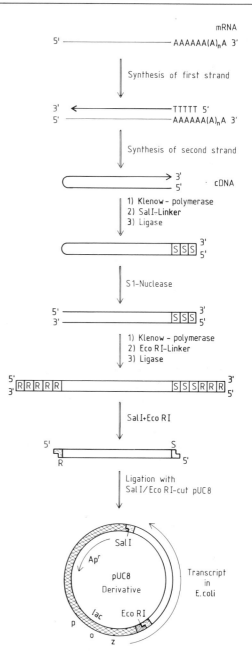

Fig. 2.2-9. Construction of a cDNA expression library.
A population of cDNA molecules is ligated in successive steps with linkers of different specificities in order to insert cDNA molecules with a defined orientation into an expression vector, e.g., pUC8. This strategy exploits the fact that only one end of a cDNA molecule is accessible for ligation after the synthesis of the second strand. Eco RI (R) and SalI (S) linkers (shown as boxes) are used in this example. The expression vector pUC8 is described in detail in Chapter 7 (Fig. 7-58). The polylinker insertion in the lacZ region of pUC8 is represented by a broken line; pUC8 sequences are shown as cross-hatched bars (Helfman et al., 1983).

with the two restriction enzymes *Eco* RI and *Sal* I, a cDNA molecule is obtained which can be inserted in the desired orientation into a vector which has also been cleaved with *Eco* RI and *Sal* I. In our example, a cDNA library consisting of approximately 9 000 clones was prepared from mRNA of chicken smooth muscle tissue. Two clones with tropomyosin gene insertions were identified by indirect immunofluorescence tests. The exact protocol is described in Chapter 11.

Of course, cDNA libraries can also be constructed in λ vectors, which are especially useful if representative libraries are required, for example, for the detection of rare mRNA species. The most suitable vectors are the immunity vectors such as λgt10 and NM1149 (*cf.* also Fig. 4.2-8.B). λgt10 and NM1149 are *Eco* RI-*Hind* III insertional vectors (Huynh *et al.*, 1985). cDNA to be cloned in these vectors must be furnished with suitable linkers. These vectors allow recombinant molecules to be identified by their clear plaques, which is due to the absence of the *cI* repressor. They also allow an efficient selection for recombinant phages in *hfl* indicator strains (see also Section 4.2.2.4). In both vectors, the repressor of phage *434* is substituted for the wild-type λ repressor. The advantage of the *434* repressor gene over the λ repressor gene is that the *434* gene possesses unique *Eco* RI and *Hind* III sites. Recent examples of cloning in λgt10 are cDNA libraries constructed from interferon-stimulated neuroblastoma cells (Friedman *et al.*, 1984), and libraries of cell lines producing human tumour necrosis factor (Pennica *et al.*, 1984).

The λ system also offers powerful expression vectors, such as vectors of the λgt11 type, which allow the efficient expression of individual cDNA sequences and their detection by specific antibodies (*cf.* also Fig. 4.2-8.B and Chapter 11).

In principle, both vectors, *i.e.*, λgt10 and λgt11, can be used for cDNA cloning. Both cloning vehicles are insertional vectors which accomodate DNA insertions of a similar size range. The main difference between these two vectors is that the immunity vector λgt10 allows not only easy identification of recombinant phages, but also selection for such recombinants. Since both vectors can also be packaged without DNA insertions, up to 90% of all phages of a cDNA library may be of the parental (vector) type without DNA insertions. If λgt10 is used, parental phage can be eliminated by a single passage on *E. coli hfl* strains because these strains allow only repressor-free phage, *i.e.*, those containing DNA insertions, to form plaques (*cf.* Section 4.2.2.4). This selection cannot be used with λgt11. The advantage of λgt11 is that it allows the expression of a cDNA insertion as a fusion protein with an *N*-terminal part of bacterial β-galactosidase, but this may prove a disadvantage upon amplification of the library, since small amounts of a toxic protein encoded by a cDNA insertion could be synthesised in spite of the presence of *lacI* repressor. Use of λgt11 then, selects against such cDNA molecules, and this vector should be employed only if possible screening methods depend on the use of antibodies *i.e.*, if no other structural information on the gene in question is available.

References

Aviv, H., and Leder, P. (1972). Purification of biological active globin mRNA by chromatography on oligothymidylic acid cellulose. Proc. Natl. Acad. Sci. USA 69, 1409-1412.

Baltimore, D. (1970). RNA-dependent DNA polymerases in virions of RNA tumor viruses. Nature 226, 1209-1210.

Bastos, R.N., Volloch, Z., and Aviv, H. (1977). Messenger RNA population analysis during erythroid differentiation: a kinetic approach. J. Mol. Biol. 110, 191-203.

Benveniste, K., Wilczek, I., and Stern, R. (1973). Translation of collagen mRNA from chick embryo calvaria in a cell-free system derived from Krebs II ascites cells. Nature 246, 303-305.

Bishop, J.O., Morton, J.G., Rosbash, M., and Richardson, M. (1974). Three abundance classes in HeLa cell messenger RNA. Nature 250, 199-204.

Breathnach, R., and Chambon, P. (1981). Organization and expression of eukaryotic split genes coding for proteins. Ann. Rev. Biochem. 50, 349-383.

Cashdollar, L.W., Esparza, J., Hudson, G.R., Chmelo, R., Lee, P.W.K., and Joklik, W.K. (1982). Cloning the double-stranded RNA genomes of reovirus: Sequence of the cloned *S2* gene. Proc. Natl. Acad. Sci. USA 79, 7644-7648.

Chan, S.J., Noyes, B.E., Agarwal., K.L., and Steiner, D.F. (1979). Construction and selection of recombinant plasmids containing full-length complementary DNAs corresponding to rat insulins I and II. Proc. Natl. Acad. Sci. USA 76, 5036-5040.

Chirgwin, J.M., Przybyla, A.E., MacDonald, R.J., and Rutter, W.J. (1979). Isolation of biological active ribonucleic acid from sources enriched in ribonuclease. Biochemistry 18, 5294-5299.

Clarke, L., and Carbon, J. (1976). A colony bank containing synthetic ColEI hybrid plasmids representative of the entire *E. coli* genome. Cell 9, 91-99.

Darnell, J.E. (1982). Variety in the level of gene control in eukaryotic cells. Nature 297, 365-371.

Derynck, R., Content, J., De Clercq, E., Volckaert, G., Tavernier, J., Devos, R., and Fiers, W. (1980). Isolation and structure of a human fibroblast interferon gene. Nature 285, 542-547.

Efstradiatis, A., Maniatis, T., Kafatos, F.C., Jeffrey, A., and Vournakis, J.N. (1975). Full-length and discrete partial reverse transcripts of globin and chorin mRNAs. Cell 4, 367-378.

Emtage, J.S., Tacon, W.C.A., Catlin, G.H., Jenkins, B., Porter, A.G., and Carey, N.H. (1980). Influenza antigenic determinants are expressed from haemagglutinin genes cloned in *Escherichia coli*. Nature 283, 171-174.

Englund, P.T. (1971). The initial step of *in vitro* synthesis of deoxyribonucleic acid by the T4 deoxyribonucleic acid polymerase. J. Biol. Chem. 346, 5684-5687.

Flint, S.J. (1980). Measurement of messenger RNA concentration. In "Genetic Engineering", Vol.2, pp.47-82. Setlow, J.H., and Hollaender, A. (eds.), Plenum Press, New York and London.

Friedman, R.L., Manly, S.P., McMakon, M., Kerr, I.M., and Stark, G.R. (1984). Transcriptional and post-transcriptional regulation of interferon-induced gene expression in human cells. Cell 38, 745-755.

Goelet, P., Lomonosoff, G.P., Butler, P.J.G., Akam, M.E., Gait, M.J., and Karn, J. (1982). Nucleotide sequence of tobacco mosaic virus RNA. Proc. Natl. Acad. Sci. USA 79, 5818-5822.

Goodman, H.M., and MacDonald, R. (1979). Cloning of hormone genes from a mixture of cDNA molecules. Methods in Enzymology 68, 75-90.

Gough, N.M., and Adams, J.M. (1978). Immunoprecipitation of specific polysomes using *Staphylococcus aureus* protein A. Purification of the immunoglobulin x chain messenger RNA from the mouse myeloma MPC11. Biochemistry 17, 5560-5566.

Gray, P.W., Leung, D.W., Pennica, D., Yelverton, E., Najarian, R., Simonsen, C.C., Derynck, R., Levinson, A.D., and Goeddel, D.V. (1982). Expression of human immune interferon cDNA in *E. coli* and monkey cells. Nature 295, 503-508.

Gubler, U., and Hoffman, B.J. (1983). A simple and very efficient method for generating cDNA libraries. Gene 25, 263-269.

Gubler, U., Monahan, J.J., Lomedico, P.T., Bhatt, R.S., Collier, K.J., Hoffman, B.J., Böhlen, P., Esch, F., Ling, N., Zeytin, F., Brazeau, P., Poonian, M.S., and Gage, L.P. (1983). Cloning and sequence analysis of cDNA for the precursor of human growth-hormone releasing factor, somatocrinin. Proc. Natl. Acad. Sci. USA 80, 4311-4314.

Gurdon, J.B., and Melton, D.A. (1981). Gene transfer in amphibian eggs and oocytes. Ann. Rev. Genet. 15, 189-218.

Hanahan, D., and Meselson, M. (1980). Plasmid screening at high colony density. Gene 10, 63-67.

Heidecker, G., and Messing, J. (1983). Sequence analysis of zein cDNAs obtained by an efficient mRNA cloning method. Nucleic Acids. Res. 11, 4891-5906.

Helfman, D.M., Feramisco, J.R., Fiddes, J.C., Thomas, G.P., and Hughes, S.H. (1983). Identification of clones that encode chicken tropomyosin by direct immunological screening of a cDNA expression library. Proc. Natl. Acad. Sci. USA 80, 31-35.

Huynh, T.V., Young, R.A., and Davis, R.W. (1985). Construction and screening cDNA libraries in λgt10 and λgt11. In "DNA Cloning Techniques: A practical approach", D. Glover (ed.), IRL Press, Oxford.

Imai, M., Richardson, M.A., Ikegami, N., Shatkin, A.J., and Furnichi, Y. (1983). Molecular cloning of double-stranded RNA virus genomes. Proc. Natl. Acad. Sci. USA 80, 373-377.

Kakidani, H., Furutani, Y., Takahishi, H., Noda, M., Morimoto, Y., Hirose, T., Asai, M., Inayama, S., Nakanishi, S., and Numa, S. (1982). Cloning and sequence analysis of cDNA for porcine β-neoendorphin/dynorphin precursor. Nature 298, 245-249.

Ko, J.L., and Harter, M.L. (1984). Plasmid-directed synthesis of genuine adenovirus 2 early-region 1A and 1B proteins in *Escherichia coli*. Mol. Cell. Biol. 4, 1427-1439.

Land, H., Grey, M., Hanser, H., Lindenmaier, W., and Schütz, G. (1981). 5'-terminal sequences of eukaryotic mRNA can be cloned with a high efficiency. Nucleic Acids Res. 9, 2251-2266.

Laskey, R.A., and Gurdon, J.B. (1973). Induction of polyoma DNA synthesis by injection into frog egg cytoplasma. Europ. J. Biochem. 37, 467-471.

Lindberg, U., and Persson, T. (1972). Isolation of mRNA from KB cells by affinity chromatography on polyuridylic acid covalently linked to sepharose. Europ. J. Biochem. 31, 246-154.

Maurer, R.A. (1980). Immunochemical isolation of prolactin mRNA. J. Biol. Chem. 255, 854-859.

Maxwell, I.H., Maxwell, F., and Hahn, W.E. (1977). Removal of RNase activity from DNase by affinity chromatography on agarose-coupled aminophenyl-phosphoryluridine 2'(3')-phosphate. Nucleic Acids Res. 4, 241-252.

Messing, J., and Vieira, J. (1982). A new pair of M13 vectors for selecting either DNA strand or double-digest restriction fragments. Gene 19, 269-267.

Mevarech, M., Noyes, B.E., and Agarwal, K.L. (1979). Detection of gastrin-specific mRNA using oligode-oxynucleotide probes of defined sequence. J. Biol. Chem. 254, 7472-7475.

Monahan, J.J., McReynolds, L.A., and O'Malley, B.W. (1976). The ovalbumin gene: In vitro enzymatic synthesis and characterization. J. Biol. Chem. 251, 7355-7362.

Noda, M., Shimizu, S., Tanabe, T., Takai, T., Kayano, T., Ikeda, T., Takahashi, H., Nakayama, H., Kana-oka, Y., Minamius, N., Kangawa, K., Matsuo, H., Raffery, M.A., Hirose, T., Inayama, S., Hayashida, H., Miyata, T., and Numa, S. (1984). Primary structure of Electrophorus electricus sodium channel deduc-ed from cDNA sequence. Nature 312, 121-127.

Noda, M., Takahishi, H., Tanabe, T., Toyosato, M., Furutani, Y., Hirose, T., Asai, M., Inayama, S., Miyata, T., and Numa, S. (1982). Primary structure of α-subunit precursor of Torpedo californica acetyl-choline receptor deduced from cDNA sequence. Nature 299, 793-797.

Norgard, M.V., Tocci, N.J., and Monahan, J.J. (1980). On the cloning of eukaryotic total Poly(A)-RNA populations in E. coli. J. Biol. Chem. 255, 7665-7672.

Okayama, H., and Berg, P. (1982). High-efficiency cloning of full-length cDNA. Mol. Cell. Biol. 2, 151-170.

Okayama, H., and Berg, P. (1983). A cDNA cloning vector that permits expression of cDNA inserts in mammalian cells. Mol. Cell. Biol. 3, 280-289.

Pelham, H.R.B., and Jackson, R.J. (1976). An efficient mRNA dependent translation system from reticulo-cyte lysates. Europ. J. Biochem. 67, 247-256.

Pennica, D., Nedwin, G.E., Hayflick, J.S., Seeburg, P.H., Derynck, R., Palladino, M.A., Kohr, W.J., Aggarwal, B.B., and Goeddel, D.V. (1984). Human tumour necrosis factor: precursor structure, expres-sion and homology to lymphotoxin. Nature 312, 724-729.

Puskas, R.S., Manley, N.R., Wallace, D.M., and Berger, S.L. (1982). Effects of ribonucleoside-vana-dyl complexes on enzyme catalyzed reactions central to recombinant DNA technology. Biochemistry 21, 4602-4608.

Retsel, E.F., Collett, M.S., and Faras, A.J. (1980). Enzymatic synthesis of deoxyribonucleic acid by avian retrovirus reverse transcriptase in vitro: Opti-mal conditions required for transcription of large ribonucleic acid templates. Biochemistry 19, 513-518.

Rougeon, F., and Mach, B. (1977). Cloning and amplification of rabbit α- and β-globin gene sequences into E. coli plasmids. J. Biol. Chem. 252, 2209-2217.

Roychoundhury, R., Jay, E., and Wu, R. (1976). Terminal labelling and addition of homopolymer tracts to duplex DNA fragments by terminal deoxy-nucleotidyl transferase. Nucleic Acids Res. 3, 863-877.

Taylor, M. (1979). The isolation of eukaryotic messen-ger RNA. Ann. Rev. Biochem. 48, 681-717.

Temin, H.M., and Mizutani, S. (1970). RNA-dependent DNA polymerase in virions of Rous sarcoma virus. Nature 226, 1211-1213.

Shapiro, D.J. Taylor, J.M., McKnight, G.S., Palacios, R. Gonzalez, C., Kiely, M.L., and Schimke, R.T. (1974). Isolation of hen oviduct ovalbumin and rat liver albumin polysomes by indirect immunoprecipi-tation. J. Biol. Chem. 249, 3665-3671.

Sippel, A.E. (1973). Purification and characterization of adenosine triphosphate: ribonucleic acid adenyl-transferase from Escherichia coli. Eur. J. Biochem. 37, 31-40.

Sood, A.K., Pereira, D., and Weissman, S.H. (1981). Isolation and partial nucleotide sequence of a cDNA clone for human histocompatibility antigen HLA-B by use of an oligodeoxynucleotide primer. Proc. Natl. Acad. Sci. USA 78, 616-620.

Uhlén, M., Svensson, C., Josephson, S., Aleström, P., Chattapadhyaya, J.B., Pettersson, U., and Philipson, L. (1982). Leader arrangement in the adenovirus fiber mRNA. The EMBO Journal 1, 249-254.

Ullrich, A., Shine, J., Chirgwin, J. Pectet,. R., Tischer, E., Rutter, W.J., and Goodman, H.M. (1977). Rat insulin genes: Construction of plasmids containing the coding sequences. Science 196, 1212-1219.

Verma, I., Temple, G.F., Fan, H., and Baltimore, D. (1972). In vitro synthesis of DNA complementary to rabbit reticulocyte 10S RNA. Nature (New Biol.) 235, 163-167.

Vogt, V.M. (1973). Purification and further properties

of a single-strand specific nuclease from *Aspergillus oryzae*. Europ. J. Biochem. 33, 192-200.

Wickens, M.P., Buell, G.N., and Schimke, R.T. (1978). Synthesis of double-stranded cDNA complementary to lysozyme, ovomucoid and ovalbumin mRNAs. J. Biol. Chem. 253, 2483-2495.

Winter, G., Fields, S., Gait, M.J., and Brownlee, G.G. (1981). The use of synthetic oligodeoxynucleotide primers in cloning and sequencing segment 8 of influenza virus (A/PR/8/34). Nucleic Acids Res. 9, 237-245.

Wodnar-Filipowicz, Gubler, U., Furinichi, Y., Richardson, M., Nowosniat, E.F., Poonian, M.S., and Horecker, B.L. (1984). Cloning and sequence analysis of cDNA for rat spleen thymosin β4. Proc. Natl. Acad. Sci. USA 81, 2295-2297.

Wood, K.O., and Lee, J.C. (1976). Integration of synthetic globin genes into E. coli plasmids. Nucleic Acid Res. 3, 1961-1971.

Zain, S., Sambrook, J., Roberts, R.J., Keller, W., Fried, M., and Dunn, A.R. (1979). Nucleotide sequence analysis of the leader segments in a cloned copy of adenovirus 2 fiber mRNA. Cell 16, 851-861.

2.3 Chemical Synthesis of DNA

For many years, the chemical synthesis of specific oligoribo- and oligodeoxyribonucleotides has been regarded as a somewhat esoteric domain of the chemistry of natural compounds. In spite of great successes, exemplified, for example, by the elucidation of the genetic code, chemists and molecular biologists alike have ignored this field and failed to realise its significance. This situation has changed dramatically with the advent of gene technology and the development of new chemical and analytical methods such as high pressure liquid chromatography (HPLC) and ^{31}P-NMR spectroscopy. Today, chemical synthesis is standardised to such an extent that the development of automated synthesis equipment has already led to commercially available "gene synthesising machines"; yet, it is obvious that this development has only just begun (Narang, 1983). The search

for optimal conditions continues and the final decisions about ideal solutions cannot yet be made.

The following fundamental prerequisites for a successful planning of oligonucleotide synthesis have been established:

- In order to be accessible to the repertoire of organic synthesis, reactants should be soluble in non-aqueous solvents. The amino and hydroxyl functions of nucleoside bases and sugar residues, respectively, must be suitably blocked.
- Protecting groups, which are introduced during synthesis, must remain stable under conditions of chain elongation, *i.e.*, during the formation of the internucleotide phosphodiester bonds; however, these groups should be labile enough to allow their removal at the end of the synthesis without damaging the reaction products. This is the only way in which the desired biologically active oligonucleotides can be obtained.
- Due to the costs of the starting material and the limitations presented by the various separation techniques, yields are an important consideration; the solid phase syntheses in particular require yields to be almost quantitative.

In principle, two strategies are currently known which comply with these limiting provisions, the phosphodiester and the phosphotriester approaches (Fig. 2.3-1). In both protocols the 3' and 5' hydroxyl groups of the deoxyribose are suitably protected (R_1 and R_2). In the phosphotriester method, a third protecting group, (R_3), is used for the hydroxyl group at the internucleotide bond. With respect to the strategies for using protecting groups, the approach known as the "phosphite" procedure (Section 2.3.4) for the synthesis of oligodeoxyribonucleotides, which employs compounds with trivalent phosphorus, should be regarded as a triester method; however, in the pertinent scientific literature the term phosphotriester method is reserved exclusively for methods utilising compounds of pentavalent phosphorus.

Historically speaking, the development of oligonucleotide synthesis started on the basis of the phosphotriester method (Michelson and Todd,

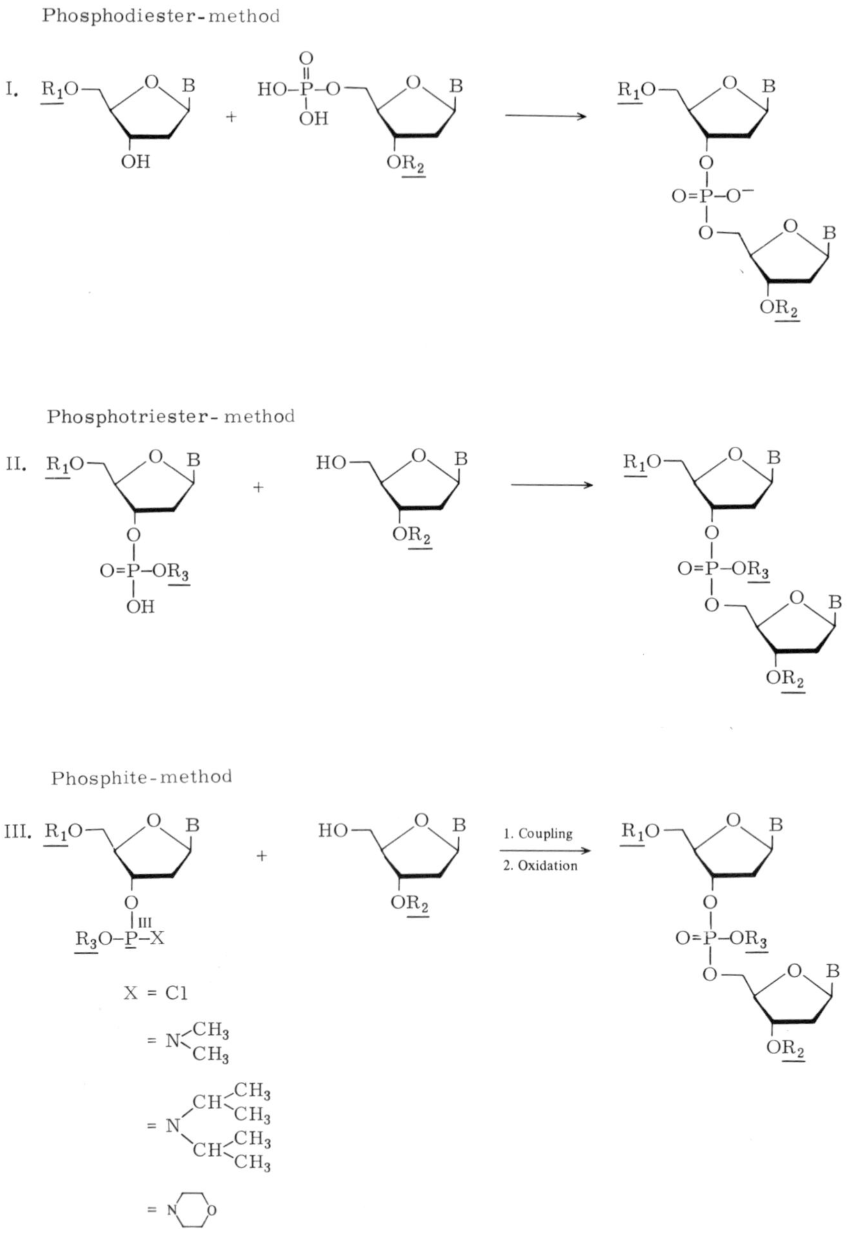

Phosphodiester-method

I.

Phosphotriester- method

II.

Phosphite-method

III.

X = Cl

= N⟨CH₃ / CH₃

= N⟨CH⟨CH₃ CH₃ / CH⟨CH₃ CH₃

= N⟨ ⟩O

Fig. 2.3-1. Strategies for the chemical synthesis of DNA.
In the phosphodiester approach, the phosphate group between the two nucleosides is unprotected. These compounds are therefore soluble in organic solvents only to a limited extent. "B" at the deoxyribose rings stands for one of the four bases A, T, G, or C.

1955); however, the first significant successes, such as the synthesis of the genes for alanine and tyrosine suppressor tRNAs of yeast and *E. coli*, respectively (Khorana *et al.*, 1972, 1976; Brown *et al.*, 1979), were gained with the phosphodiester method. Lately, this method, which presents a variety of solubility problems, has been superseded almost completely by the triester approach. This section will, therefore, be concerned exclusively with the applications of the triester method and its use for the synthesis of oligodeoxyribonucleotides. The synthesis of oligoribonucleotides requires effectively protected 2' hydroxyl groups and presents specific additional problems, which have been discussed in detail by Reese (1978); besides, to date, oligoribonucleotides have played a lesser role in gene technology.

The phosphotriester approach for the synthesis of oligodeoxyribonucleotides proceeds essentially in two steps: 1) preparation of suitably protected monomers and 2) coupling of the monomers in the desired sequence by an appropriate phosphorylation procedure.

fully acylated products, *i.e.*, also O-acylated compounds, which can be converted to the mono-N-acyl compounds by controlled alkali treatment (Fig. 2.3-2).

Fig. 2.3-2.   Structures of N-protected deoxyribonucleosides.
bzA = N-benzoyldeoxyadenosine; ibG = N-isobutyryldeoxyguanosine; anC = N-anisoyldeoxycytidine.

2.3.1 Preparation of 5'- and N-Protected 2'-Deoxyribonucleoside Compounds

When monomeric building blocks are synthesised, the bases themselves, as well as the 3' and 5' hydroxyl functions, must be protected. The groups protecting the bases should be more stable than those at the deoxyribose, because the former must be preserved during the entire synthesis. Since Khorana's pioneering work (*cf.* Schaller *et al.*, 1963; Brown *et al.*, 1979) N-acyl groups have proven particularly effective. They are introduced by a reaction of the bases with an excess of the corresponding acyl chlorides. This reaction yields

H. Gobind Khorana
Cambridge, Massachusetts

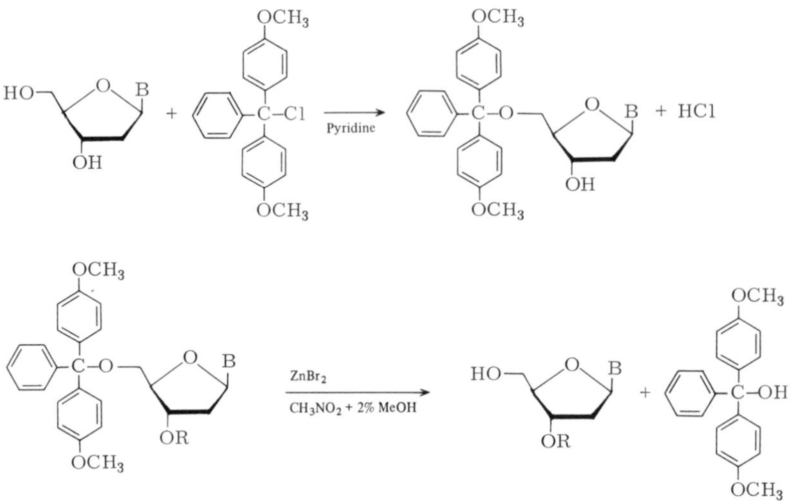

Fig. 2.3-3. Formation and removal of the 5'-O-dimethoxytrityl group.
R stands for a protecting group at the 3' position of the deoxyribose, and signifies that not only N-protected mononucleotides, but also oligonucleotides and oligonucleotides bound to a carrier, can be detritylated by $ZnBr_2$. Acid hydrolysis with 80% acetic acid is reserved for oligonucleotides which are only partially protected, and which are already soluble in the aqueous phase. Reaction yields can be determined by measuring the intensity of the yellow colour of the released triphenylmethyl alcohol.

A new concept has been introduced by Ti *et al.* (1982) who devised a one-flask procedure in which the 5'-O-dimethoxytrityl derivatives are treated simultaneously with trimethylchlorosilane and acylchloride. This procedure yields 3'-O-trimethylsilyl-N-acyl nucleosides, which are converted into the N-protected 3' hydroxyl nucleosides upon extraction with aqueous ammonia. With the exception of deoxyguanosine, this concept of a "transient" introduction of protecting groups guarantees high yields of phosphorylatable products.

Acid- and alkali-labile groups have been developed to protect 3' and 5' hydroxyl functions, which can therefore be manipulated selectively. The acid-labile 4,4'-dimethoxytrityl group serves to protect the 5' hydroxyl function (Fig. 2.3-3). It can be removed quantitatively by mild acid hydrolysis, with 80% acetic acid in particular, but also with Lewis acids, such as $ZnBr_2$ in nitromethane, without running the risk of depurination

(Kohli *et al.*, 1980; Matteuchi and Caruthers, 1980). The corresponding 3'-O-acyl derivatives usually are generated in a two-step procedure by starting with 5'-O-dimethoxytrityl derivatives, which first are acylated by suitable anhydrides and then hydrolysed with acids, which remove the 5'-O-trityl residue (Reese, 1978). Today, the established classical procedures for synthesising all these derivatives are complemented by time saving methods of purification such as HPLC.

2.3.2 Oligonucleotide Synthesis

The essential feature of this phase of the reaction is the formation of a genuine 3' to 5' internucleotide bond, which is accomplished in two phosphorylation steps. The first step is to synthesise the corresponding 3' and 5' monophosphate esters of the protected deoxyribonucleoside by employing

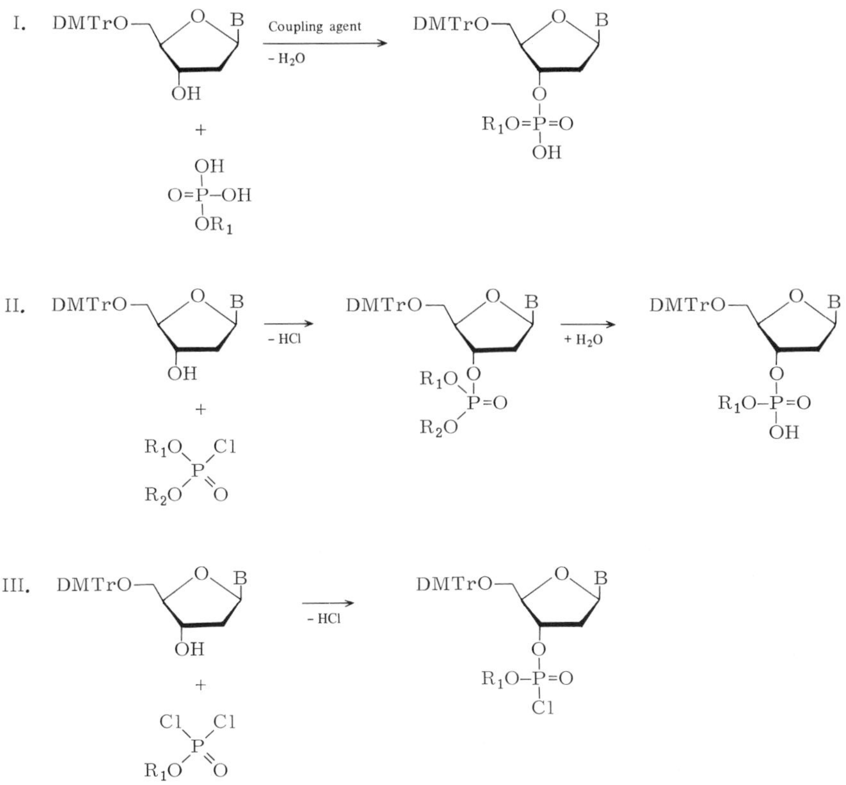

I. DMTrO—[sugar]—B $\xrightarrow[\text{- H}_2\text{O}]{\text{Coupling agent}}$ DMTrO—[sugar]—B

OH

+

OH
|
O=P—OH
|
OR$_1$

→ product with:

O
‖
R$_1$O=P=O
|
OH

II. DMTrO—[sugar]—B $\xrightarrow{\text{- HCl}}$ DMTrO—[sugar]—B $\xrightarrow{\text{+ H}_2\text{O}}$ DMTrO—[sugar]—B

OH

+

R$_1$O   Cl
   \ /
    P
   / \\
R$_2$O   O

R$_1$O   O
    \   ‖
     P=O
    /
R$_2$O

O
‖
R$_1$O—P=O
|
OH

III. DMTrO—[sugar]—B $\xrightarrow{\text{- HCl}}$ DMTrO—[sugar]—B

OH

+

Cl   Cl
  \ /
   P
  / \\
R$_1$O   O

O
‖
R$_1$O—P=O
|
Cl

Fig. 2.3-4. The first phosphorylation step.
I. Reaction with a non-functional reagent. In this case the reagent is a mono-ester of phosphoric acid (2-cyano-ethylphosphate (CNCH$_2$CH$_2$-O-PO(OH)$_2$).
II. Reaction with a monofunctional reagent. R$_1$ and R$_2$ are either 2-cyanoethyl, 2,2,2-trichloroethyl, benzyl, p-chlorophenyl, or o-chlorophenyl residues. In the more recent literature one of the two ester functions at the phosphate residue usually is the 2-cyanoethyl group.
III. Reaction with a bifunctional phosphorylating reagent such as o-chlorophenyl-diphosphochloridate. The o-chlorophenyl residue can be selectively removed from the nucleotide by treatment with an excess of triethylamine in absolute pyridine (see Fig. 2.3-8). The triethylammonium salts of the resulting phosphodiesters are soluble in absolute pyridine and can therefore be re-used for the second phosphorylation step.

suitable phosphorylating agents. Internucleotide bonds are then formed by reactions between the monophosphate ester and a second deoxyribonucleoside. The first phosphorylation step (Fig. 2.3-4) can be carried out with non-functional, mono- and bi-functional phosphorylating agents. Non-functional phosphoric acid esters, such as benzyl- or cyanoethylphosphate, react only with the 3' hydroxyl group of a fully protected 5'-O-derivatised mononucleoside in the presence of a coupling agent, which is required to split off water; the reaction products are the 3' phosphodiester compounds (Fig. 2.3-4.I). In the course of time, the classical dicyclohexylcarbodiimide has been replaced by arylsulfonylchlorides, such as TPS (Lohrmann and Khorana, 1966), or arylsulfonyltriazolides, for example MSNT (Jones et al., 1980) (Fig. 2.3-5).

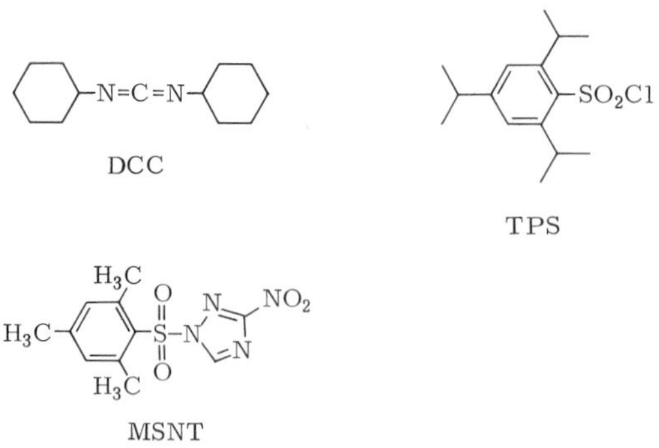

DCC

TPS

MSNT

Fig.2.3-5. Coupling or activating reagents.
DCC = Dicyclohexylcarbodiimide; TPS = Triisopropylsulfonylchloride; MSNT = 1-(mesithylene-2-sulfonyl)-3-nitro-1,2,4-triazole.

In contrast to non-functional reagents, mono- and bi-functional phosphorylating agents, such as mono- and dichlorides (Fig. 2.3-4.II and 2.3-4.III), react spontaneously with a 5'-O-derivatised mononucleoside, and coupling or activating agents can be dispensed with. As shown in Figs. 2.3-4.I and 2.3-4.II, phosphorylation reactions involving non-functional and mono-functional reagents yield identical products; in either case, an activating agent is required for coupling these products with a second nucleoside in a second phosphorylation reaction (Fig. 2.3-6).

Bi-functional agents have the advantage of yielding a mono-functional intermediate after the first phosphorylation step, which can then react with the 5' hydroxyl group of a second nucleotide without requiring a coupling agent (Fig. 2.3-6.II). The application of arylphosphodichlorides and the much more selective arylphosphotri- and tetrazolides has been described in the relevant scientific literature (Stawinski et al., 1977). Simple as they may seem, bi-functional reagents have the disadvantage of leading to the occasional synthesis of symmetrical 3' to 3' or 5' to 5' dimers, which are the result of a reaction between two identical 3' or 5' hydroxyl components and the

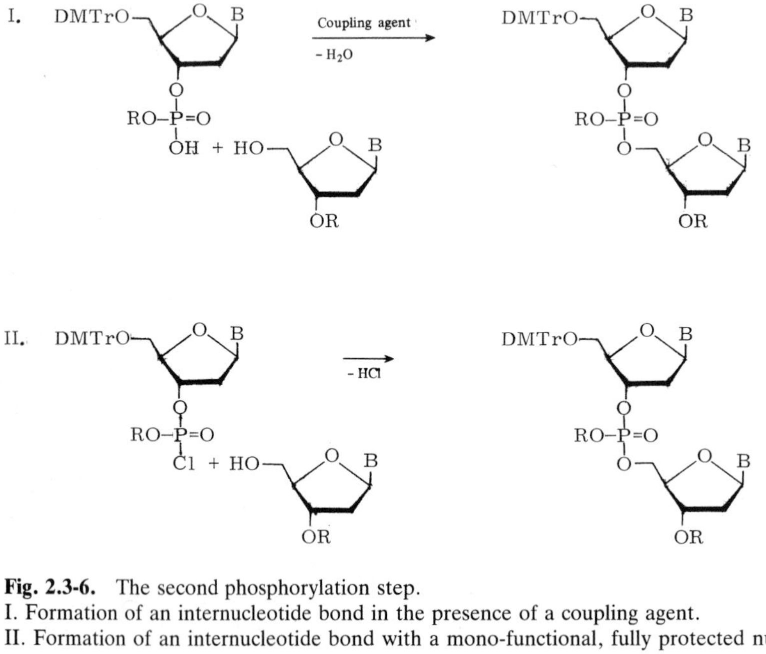

Fig. 2.3-6. The second phosphorylation step.
I. Formation of an internucleotide bond in the presence of a coupling agent.
II. Formation of an internucleotide bond with a mono-functional, fully protected nucleotide as phosphorylating agent.

Fig. 2.3-7. Synthesis of a fully protected mononucleotide phosphotriester. The bifunctional phosphorylating reagent p-chlorophenylphosphoryl-(1,2,4-triazolide) reacts sequentially with two alcohols, *i.e.*, with the 5' protected mononucleotide, and with 2-cyanoethanol (Stawinsky *et al.*, 1977). A coupling agent is not required. The triazole which is released in each step has been omitted.

bi-functional agent; a skillful selection of reaction conditions, however, and the use of arylphosphotriazolides have reduced these problems. If used at all, bi-functional agents are now predominantly applied during the first phosphorylating step. The mono-functional reagent generated in this reaction is used for the esterification with a suitable alcohol rather than with a second nucleoside (Broka *et al.* 1980), and in this way, a fully protected triester is easily obtained (Fig. 2.3-7).

A widely used and important strategy, uniting several of the methods outlined above, has been worked out by Catlin and Cramer (1973). Its key feature is a fully protected phosphotriester (Fig. 2.3-7), which can be partially unblocked at the 3' or 5' ends by treatment with alkali or acid, respectively (Fig. 2.3-8). This reaction yields two intermediates which can be joined in the presence of a coupling agent. Fig. 2.3-8 shows an example based on procedures described by Duckworth *et al.* (1981), Narang *et al.* (1979), and Itakura (1980). The reported yields of the coupling reaction *per se* are greater than 90%. The corresponding barium salts have been employed

successfully (Gough *et al.* 1979) as well as the usual triethylammonium salts. The fully protected dinucleotide generated in this reaction can be re-functionalised by treatment with acid or alkali, and can be re-used for a further coupling reaction; of course, the reactants in this coupling reaction can be other suitably activated mono- or oligonucleotides. Today, the use of a set of 16 di- or 64 trinucleotides allows entire genes to be

Fritz Cramer
Göttingen

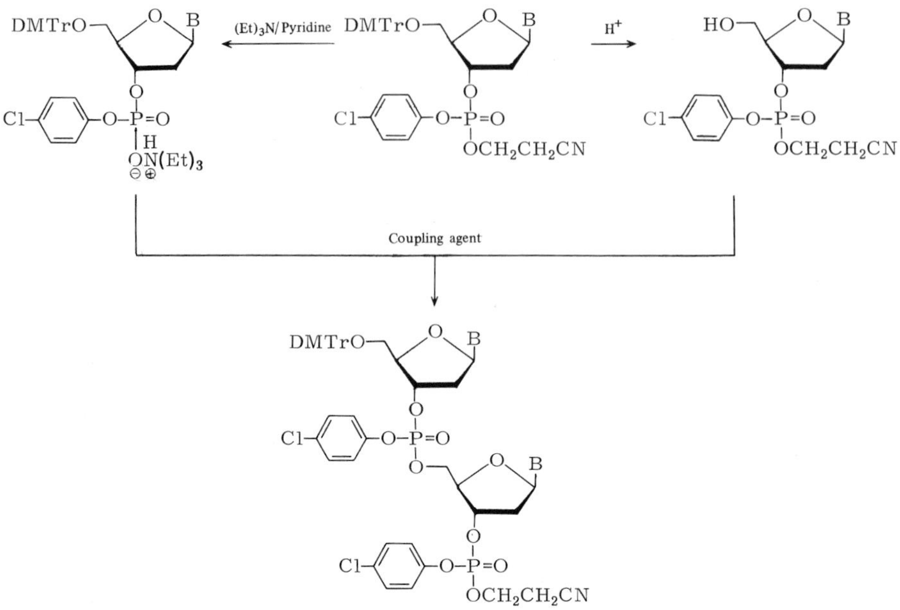

Fig. 2.3-8. Dinucleotide synthesis starting with a fully protected mononucleotide phosphotriester (Catlin and Cramer, 1973).

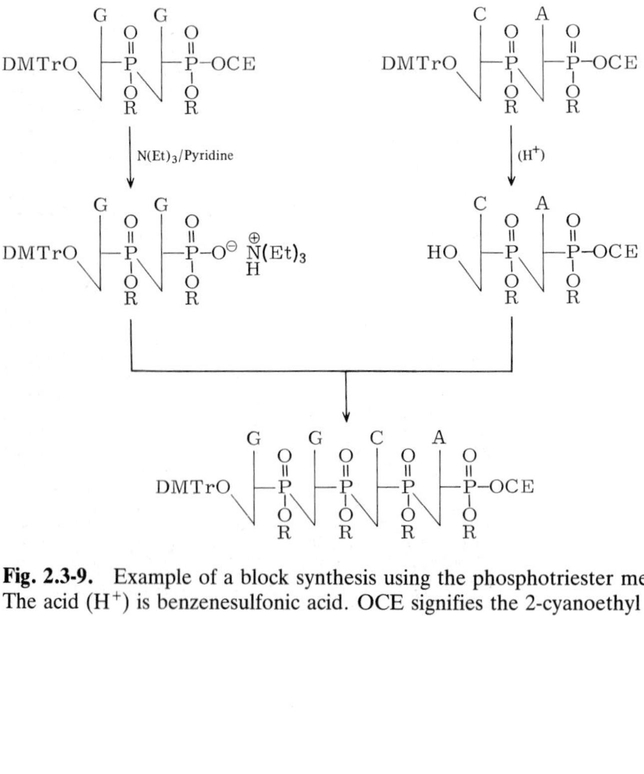

Fig. 2.3-9. Example of a block synthesis using the phosphotriester method.
The acid (H$^+$) is benzenesulfonic acid. OCE signifies the 2-cyanoethyl residue (see also Fig. 2.3-8).

pieced together in a relatively short time. This process is known as block synthesis (Itakura 1980) (Fig. 2.3-9). The stepwise increase in the size of the synthesised oligonucleotides, a key feature of the block synthesis procedure, simplifies the separation of the products from the starting materials. The strategy is explained in Fig. 2.3-9 for the synthesis of a tetranucleotide (Narang *et al.* 1979).

2.3.3 Solid Phase Synthesis

The multi-step synthesis of biopolymers on solid polymer carriers has been introduced by Merrifield (1963) for the synthesis of peptides and Letsinger and Kornet (1963) for the synthesis of polynucleotides. As compared with the synthesis in solution, solid phase systems have a number of advantages:
– the solid phase allows growing biopolymer chains to be easily separated from other components of the reaction mixture;
– chromatographic separation is replaced by simple washing and filtration procedures;
– the different solubilities of the components can be neglected because the reactions take place in a heterogeneous phase;
– individual reactions can be repeated as often as desired in order to increase yields;
– the simple washing and filtration procedures allow the process to be mechanised.

Several conditions have to be met, if these advantages are to be exploited, and of course, there are also certain limitations to be aware of:
– yields must be quantitative; if, for example, the yield in each addition step is assumed to be on the order of only 80%, the total yield of a decanucle-

otide will be decreased to $0.8^{10} \cdot 100$, *i.e.*, to 10%;
– the purity of the products and the yield of the end product cannot be determined during the synthesis but only after its completion;
– several oligonucleotide chains are usually attached to one polymer particle. A decreasing yield in each step will result in the formation of products which will be heterogeneous with respect to composition and length. The application of the solid phase method is, therefore, extremely limited by techniques available for subsequent separation;
– the chemical reactions take place in a heterogeneous phase and are therefore usually slower than in solution. Since the reaction half life, $t_{1/2}$, increases at least by a factor of 2 to 3, reaction conditions of the liquid phase can only be applied to solid phase synthesis if reaction times are short, *i.e.*, on the order of seconds or minutes;
– a number of additional and new reaction steps are required, for example, coupling of the monomers to the polymer carrier, and the removal of the end product from this solid support.

Robert L. Letsinger
Evanston

Fig. 2.3-10. Coupling of the first nucleoside to a polymer carrier. Starting material for the derivatisation is commercially available chloromethylpolystyrene.

Since its introduction two decades ago, peptide synthesis on solid carriers has presented satisfactory results; the real break-through for the solid phase synthesis of oligonucleotides has occured only during the past four years.

Currently used polymeric carriers for solid phase synthesis consist of polydimethylacrylamides (Duckworth *et al.*, 1981) or modified polystyrenes (Miyoshi *et al.*, 1980), to which the mononucleosides are attached by suitable "arms". The preferred direction of synthesis is 3' to 5', because the 3' phosphodiesters can then be activated in solution, *i.e.*, at the nucleoside which is added rather than at the polymer carrier. The first mononucleoside therefore is coupled to the carrier with its 3' end. The experimental strategy involving a polystyrene carrier is exemplified in Fig. 2.3-10. Starting with the mononucleoside bound to the carrier, an oligonucleotide is synthesised by the stepwise addition of suitably activated

Keiichi Itakura
Duarte

B$_1$

DMTrO —O—C(=O)—CH$_2$—CH$_2$—C(=O)—N(H)—CH$_2$—⟨ ⟩—(P)

↓ ZnBr$_2$

B$_1$

HO —O—C(=O)—CH$_2$—CH$_2$—C(=O)—N(H)—CH$_2$—⟨ ⟩—(P)

B$_2$

DMTrO —O—P(=O)(—O⁻)—O H N(Et)$_3$ O Cl

MSNT

B$_2$ B$_1$

DMTrO —O—P(=O)—O —O—C(=O)—CH$_2$—CH$_2$—C(=O)—N(H)—CH$_2$—⟨ ⟩—(P) Cl

Fig. 2.3-11. Schematic representation of a coupling reaction during solid phase phosphotriester synthesis. The structure of the coupling agent MSNT is shown in Fig. 2.3-6.

mono-, di-, and trimers. By using mesitylensulfonyl-3-nitrotriazole (MSNT) as a coupling agent, and 1-methyl-imidazole as a catalyst (Sproat and Bannwarth, 1983) (Fig. 2.3-7), coupling times of 15 minutes, and yields on the order of 95% can be obtained (Fig. 2.3-11). This procedure also can be mechanised, and the protocol shown in Table 2.3-1 demonstrates that a complete reaction cycle requires approximately 30 minutes. The full reaction cycle includes the coupling reaction itself, several washing procedures, the unblocking of the 5′ trityl group (detritylation), and the reaction known as "capping". The latter reaction is an acetylation step which prevents further reactions by blocking unreacted 5′ hydroxyl products. After the cycle has been completed, the newly synthesised oligonucleotides can be removed from the carrier by oximate cleavage (Reese and Zard, 1981). The chlorophenyl protecting groups are removed at the same time. Subsequent hydrolysis in concentrated ammonia removes the

Michael J. Gait
Cambridge

Table 2.3–1.   Chemical steps for a single synthesis cycle with the phosphotriester method.

Step	Reagent or solvent	Volume (ml)	Time (min)	Repetitions
1 wash	CH_2Cl_2/IPA	1	0.1	3
2 detritylation	$ZnBr_2$	1	0.1	2
		1	5	1
3 wash	CH_2Cl_2	1	0.1	2
4 detritylation	$ZnBr_2$	1	0.1	2
5 wash	CH_2Cl_2	1	0.1	3
6 wash	TEA-Ac/DMF	1	0.2	3
7 wash	THF	1	0.1	3
8 drying	N_2	–	5	
9 condensation	mono- or dinucle- otide; plus 3 equiv. MSNT; 1-methyl- imidazole	0.3	15	1
10 wash	pyridine	1	0,1	2
11 capping	THF/pyridine/DMAP/ Ac_2O	1	5	1
12 wash	pyridine	1	0,1	1

TEA = Triethylamine; IPA = Isopropanol; TEA-Ac = Triethylammonium-acetate; DMF = Dimethylformanide; FHF = Tetrahydrofuran; DMAP = Dimethylaminopyridine.

N-acyl protecting groups at the nucleotide bases, while the 5′ trityl groups are split off by treatment with 80% acetic acid. The oligonucleotide can then be purified by HPLC, ion exchange, and/or exclusion chromatography. Another procedure which has proven effective is the separation of the reaction mixture on 20% polyacrylamide/8M urea gels, which are also used for DNA sequencing

Helmut Blöcker, Ronald Frank, Braunschweig-Stöckheim, FRG

(Maxam and Gilbert, 1980; *cf.* also Section 2.4.1.4).

A recent and ingenious adaptation of the solid phase phosphotriester method by Frank *et al.* (1983) employs cellulose support papers, which allow the simultaneous synthesis of several hundred oligonucleotides with a length of up to twenty bases. The set-up consists of an array of cellulose disks, each of which carries the respective first nucleotide of the oligonucleotides to be synthesised. The filter disks are divided into four groups depending on whether the second nucleotide to be added is to be A, C, G or T. All filter disks of the same group are treated simultaneously with solvents and reagents. After the completion of one elongation cycle, the filter disks are re-grouped, again according to which base is to be added in the next step. The method is almost independent of the number of oligonucleotides to be synthesised, and this allows the number of cycles required for the synthesis of a gene to be reduced by up to 90%. The technique has recently been adapted to a 50-150 nanomole scale as compared to the 1-5 micromoles normally required, and this, of course, reduces the amount of labour involved and the cost of reagents considerably (Matthes *et al.*, 1984). It is to be hoped that this concept will not only simplify the synthesis of genes but also the sequencing of DNA in M13 vectors (see also Section 2.4.2; Brenner and Shaw, 1985).

2.3.4 Phosphite Triester Method

An important alternative to the solid phase phosphotriester method has been developed in the past few years. It is based on the use of phosphite, rather than phosphate, intermediates,

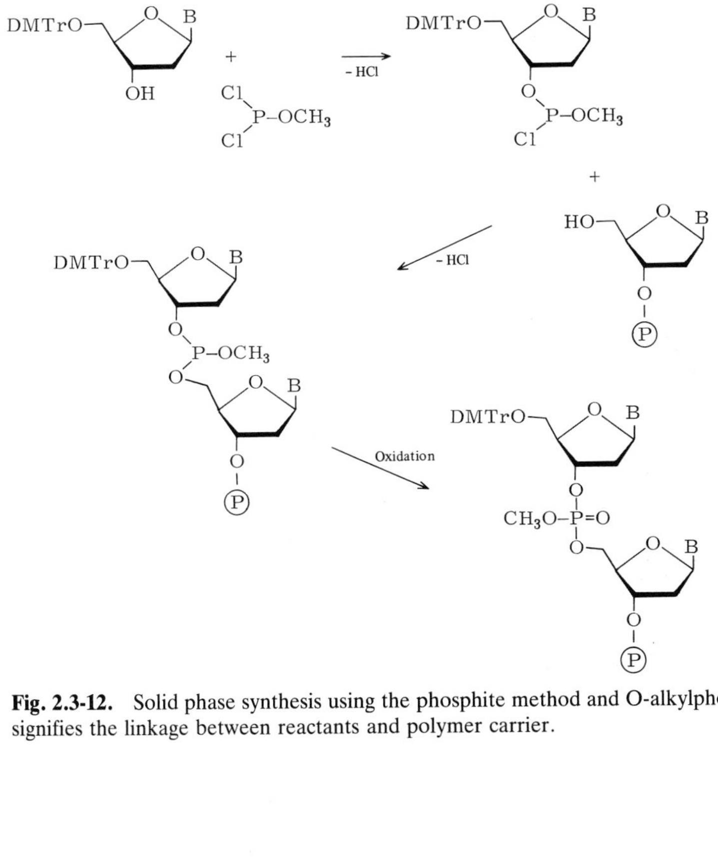

Fig. 2.3-12. Solid phase synthesis using the phosphite method and O-alkylphosphodichloridites. The encircled P signifies the linkage between reactants and polymer carrier.

in which bi-functional derivatives of trivalent phosphorus of the O-alkylphosphochloridite type react quickly and quantitatively with the 3' hydroxyl groups of suitably base-protected mononucleotides (Fig. 2.3-12; Letsinger and Lunsford, 1976). The second phosphorylation step, involving, for example, a carrier-bound 5'-unprotected mononucleotide, is also more or less quantitative. The phosphite approach is therefore well suited for mechanised solid phase synthesis and was, indeed, being used as such (Chow *et al.*, 1981; Matteucci and Caruthers, 1981).

A disadvantage of bi-functional phosphochloridites in their bi-functional as well as monofunctional form is the lability of the nucleoside phosphomonochloridites, because these compounds usually must be stored in hexane at –80 °C. Caruthers (Beaucage and Caruthers, 1981) has developed nucleoside phosphoramidites which are much more stable at elevated temperatures and less sensitive to oxidation. These new compounds allowed a new approach to synthesis, in which the phosphorylating agent is generated in a two-step procedure (Fig. 2.3-13).

First, phosphotrichloride and methanol react to yield dichloromethoxyphosphine (I), which is then converted to chloro-N,N-dimethylaminomethoxyphosphine (II) through a reaction with trimethylsilyldimethylamine. The reaction with 5'-O-dimethoxytritylnucleosides generates the corresponding nucleoside phosphoramidites (III). Apart from the dimethylamino compounds, a number of other secondary amines have also been used lately. Compounds derived from diisopropylamine and morpholine have been shown to be particularly stable (Dörper and Winnacker, 1983; McBride and Caruthers, 1983). The synthesis of the phosphorylating agents must be carried out with utmost care; purity controls should not only involve boiling point measurements, but especially a measurement of ^{31}P- NMR spectra, because only the latter method ensures that impurities, caused by phosphotrichloride or hydrolysis and oxidation products, which would interfere with the reaction, will be recognised with sufficient sensitivity.

In the presence of tetrazole (as a weak acid), phosphoramidites react almost quantitatively with the free 5' hydroxyl group of a nucleoside bound to a silica gel carrier *via* its 3' hydroxyl group (Matteucci and Caruthers, 1981) Fig. 2.3-14. Reactions involving pyrimidine bases are quantitative, while the yields obtained with purine bases are on the order of 90-95%, which allows longer oligonucleotide of up to 40 bases to be synthesised. Oligonucleotides with a length up to or even greater than 100 bases can be synthesised if CPG glass is used as the carrier material (Adams *et al.*, 1983). In the course of each addition step, which is shown in Fig. 2.3-15, the phosphite triester formed must be oxidised to phosphate by using iodine, and non-reacted 5' hydroxyl nucleosides must be inactivated by a capping reaction. The cycle is completed by a

Marvin H. Caruthers
Boulder, Colorado

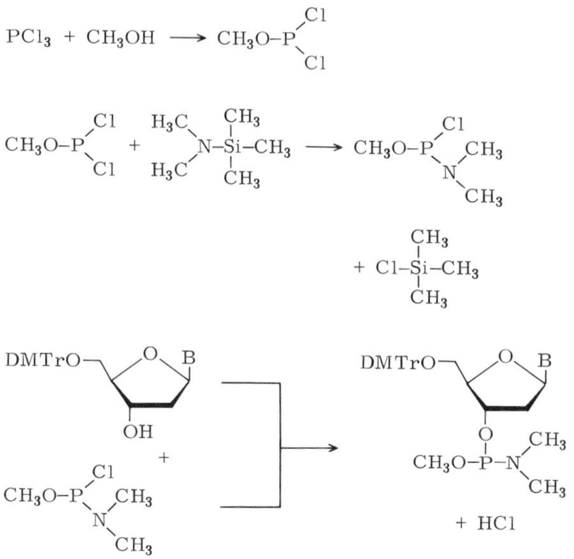

Fig. 2.3-13. Synthesis of deoxyribonucleoside phosphoramidites.
A dichloridite (I) can be converted to an amidite (II) by using either dimethylamine, or its derivative trimethylsilyl-dimethylamine. The reaction with the free amine yields its hydrochloride which is difficult to filtrate. Trimethylsilyl-dimethylamine which is also commercially available has the advantage that the reaction requires equimolar amounts rather than a two-fold excess.

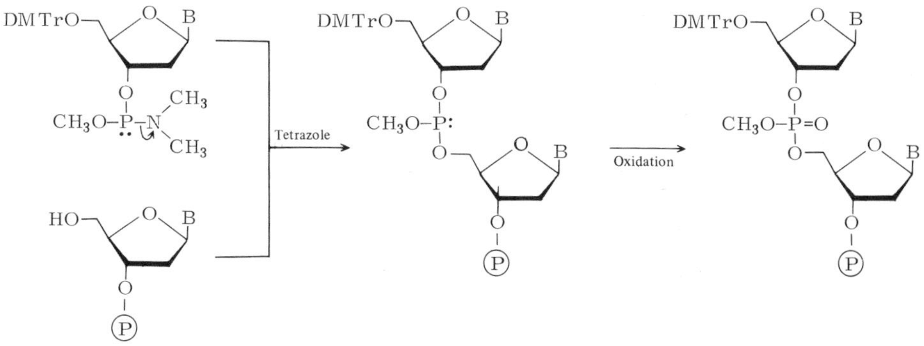

Fig. 2.3-14. Solid phase synthesis of oligonucleotides using the amidite method.
The polymeric carrier (encircled P) is either a silica gel or controlled pore glass (CPG). It is functionalised in a way similar to that described in Fig. 2.3-10. The oxidation is carried out with diluted I_2 solutions. (Caruthers, 1982).

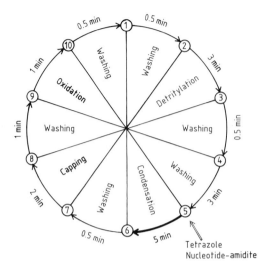

Fig. 2.3-15. Reaction cycle (A) and chemical steps (B) of solid phase synthesis using the phosphite amidite approach.
In each step the volume handled in this manual procedure is one millilitre except where otherwise indicated. Total time per cycle is about 15 to 17 minutes. Through optimisation of various parameters, synthesis machines have cycle times of only 5 to 8 minutes. In step 5, for each µmole of deoxynucleoside attached covalently to silica gel, 0,4 M tetrazole (0.2 ml) and 0.1 M deoxynucleoside phosphoramidite (0.2 ml) are premixed in acetonitrile. I_2 solution is THF:lutidine:H_2O (2:2:1, by volume) containing 0,2 M iodine. Abbrevations: DMAP, dimethylaminopyridine; THF, tetrahydrofuran. (from Caruthers, 1985, and Doerper and Winnacker, 1982).

subsequent detritylation. The iodine can be replaced by sulphur/lutidine, which yields the corresponding phosphorothioate analogues (Stec et al., 1984); nevertheless, a complete cycle takes only approximately 15 minutes rather than the 30 to 60 minutes required for the phosphotriester method.

The synthesis just described yields a fully protected polynucleotide bound to a carrier. In order to obtain the desired final product, the polynucleotide must be subjected to a number of relatively time consuming procedures. The methoxy group at the internucleotide bond is removed by hydrolysis with the nucleophilic thiophenolate anion (Daub and van Tamelen, 1977); this reaction is followed by the hydrolysis of the N-benzoylated bases and the removal of the polynucleotide from the silica gel carrier by concentrated ammonia. The remaining 5'-O-dimethoxytrityl group is usually removed by acid hydrolysis (80% acetic acid) after the oligonucleotide has been purified by HPLC. The unprotected oligonucleotide can then be obtained in microgram amounts after electrophoresis on denaturing polyacrylamide gels such as those used for DNA sequencing.

Caruthers' method (Caruthers, 1985), including the removal of protective groups and cleavage of the oligonucleotide from its carrier, has already been mechanised (Hunkapillar et al., 1984; Shep-

B)

Step	Reagent or solvent	Purpose	Time (min)
1	dichloroacetic acid in CH_2CL_2 (2:100, by volume)	detritylation	3
2	CH_2Cl_2	wash	0.5
3	acetonitrile	wash	1.5
4	dry acetonitrile	wash	1.5
5	activated nucleotide in acetonitrile	add one nucleotide	5
6	acetonitrile	wash	0.5
7	DMAP:THF:lutidine (6:90:10, weight/volume/volume) 0.1 ml acetic anhydride	cap	2
8	THF:lutidine H_2O (2:1:1, by volume)	wash	1
9	I_2 solution	oxidation	1
10	acetonitrile	wash	0.5
11	CH_2Cl_2	wash	0.5

pard, 1983). Since the starting materials, *i.e.*, the nucleoside 3' phosphoramidites, are also commercially available, there should be no limitations for further applications of this technique.

When the synthesis of an oligonucleotide has been completed, its sequence must be determined. Today, this is usually done by the chemical procedure described by Maxam and Gilbert (1980) with modifications introduced by Jay *et al.* (1982). Occasionally, problems may be experienced, and in this case, another method, which is based on the differential mobilities of oligonucleotides with different sequences during electrophoresis and homochromatography, can be used (Tu and Wu, 1980). Another more recent approach allows the sequence of molecules consisting of up to 15 nucleotides to be determined by mass spectroscopy (Rinehart, 1982; Grotjahn *et al.*, 1982).

2.3.5 Application of Oligonucleotide Syntheses

Synthetic oligonucleotides have found an extraordinary number of applications in gene technology, some of which are listed in Table 2.3-2. It can be assumed that this list will already have grown considerably by the time this book is printed. One of the most obvious applications is, of course, the synthesis of genes, for which the synthesis of the somatostatin gene (Itakura *et al.*, 1977), the gene for the A and B chains of human insulin (Crea *et al.*, 1978), the gene of human leukocyte interfe-

Table 2.3–2. Application of synthetic oligonucleotides.

–   Partial or total gene synthesis
–   Primers for DNA- and RNA sequencing
–   Hybridization probes for the screening of RNA, DNA and cDNA or genomic libraries
–   Adapters and mutagenesis
–   Site-directed mutagenesis
–   RNA/DNA structure determinations

ron (Edge *et al.*, 1980), and the human proinsulin gene (Brousseau *et al.*, 1982) may serve as examples.

The strategy introduced by Khorana employs the synthesis of overlapping oligonucleotides and the ligation of suitable mixtures of these oligonucleotides with polynucleotide ligase. A computer analysis is required to ascertain that there is only one way in which the synthesised oligonucleotides can reassociate. The overlaps have to be designed accordingly. In planning the chemical synthesis, attention can also be paid to special codon frequencies, *e.g.*, those of *E. coli* (*cf.* Table 7-4), and it is also possible to introduce suitable restriction cleavage sites. The human interleukin 2 gene, for example, has been constructed from four different fragments, each of which was pieced together by 6 to 12 oligonucleotides, 12-30 bases in length (Fig. 2.3-16). Codon usage was optimised by altering 102 bp in the human gene in 86 out of 133 codons, such that the synthetic gene showed only 74% homology with the genuine interleukin gene (Mertz *et al.*, 1986).

A modified and alternative strategy developed by Rossi *et al.* (1982) makes use of longer, only partially overlapping synthetic oligonucleotides. Gaps remaining after the reassociation are then closed enzymatically by DNA polymerase I. This approach may reduce the burden of chemical synthesis considerably, although today this factor is of lesser importance.

It is apparent that a completely new development in molecular genetics is about to take place, *i.e.*, the development of *synthetic biology*. Wetzel *et al.* (1981), for instance, have synthesised a new insulin gene variant which consists of intact A and B chains, but which contains a shortened C peptide comprising the sequence Arg-Arg-Gly-Ser-Lys-Arg instead of the entire 35 amino acids. Of course, the use of a shortened C peptide facilitates the synthesis considerably; on the other hand, the new protein possesses physical and biochemical properties which differ from those of natural proinsulin. These differences are not yet known in detail, but it is conceivable that the

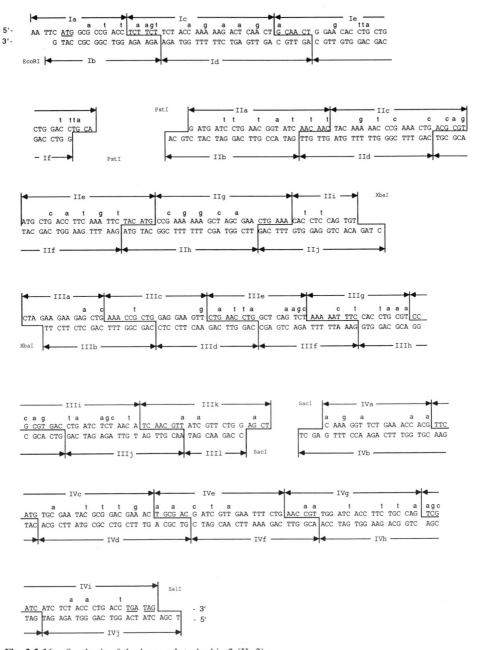

Fig. 2.3-16. Synthesis of the human Interleukin-2 (IL-2) gene.
The gene was assembled from four DNA fragments (I-IV) which in turn were pieced together by several oligonucleotides (a, b, c, etc.), indicated by lower case letters added to the fragment numbers I to IV. Lower case letters above the sequence indicate the bases in the original IL-2 sequence (Taniguchi *et al.*, 1983). Start and Stop codons are underlined.

chemical synthesis of genes and their direct manipulation may result in insulin gene variants with clinically interesting features. Other examples of this newly emerging field of "protein engineering" are mentioned in Section 12.4.4.

Many of the applications listed in Table 2.3-2 will be discussed in detail in the chapters about DNA sequencing, gene libraries, cDNA synthesis and ligation of DNA fragments. Site-directed mutagenesis will be treated in a chapter of its own (Chapter 12). The applications of synthetic oligonucleotides in the field of X-ray analysis involve the crystallisation of certain oligonucleotides and the study of their crystal structures. These studies already have led to the detection of a completely unexpected species of DNA helix, *i.e.*, that of the left-handed Z helix, which differs completely from the known two right-handed helix families, A and B (Drew *et al.* 1980; Conner *et al.* 1982).

As far as the use of oligonucleotides as hybridisation probes is concerned (item 3 in Table 2.3-2), it is essential to see the automated chemical synthesis of DNA molecules also in the context of other technical developments. Hunkapillar *et al.* (1984) have developed an integrated system especially for the isolation of weakly expressed genes, which does not only make use of chemical DNA synthesis, but also employs protein microsequencing, peptide synthesis, and the use of computers for the analysis of sequence data. The classical techniques of protein sequencing by mechanised Edman degradation employs a liquid phase sequenator, in which the proteins to be sequenced are covalently bound to a solid support. Amino acids are then sequentially removed, one residue at a time, from the amino terminus of the peptide, by Edman degradation. This procedure yields phenylthiohydantoines, which can be characterised by HPLC or mass spectroscopy in order to obtain the protein sequence. The application of the liquid phase sequenator is still limited, because the process requires between 5 and 10 nM of protein; in other words, approximately 250 to 500 microgram would be required for determining the sequence

of a protein with a molecular weight of 50 000 Da. Very often, it is difficult, if not impossible, to obtain such quantities. A new technique, known as gas phase sequencing, allows the sensitivity to be increased by three orders of magnitude into the picomole range. The chemistry of this process is analogous to conventional Edman degradation (Fig. 2.3-17), but in this case the reagents are

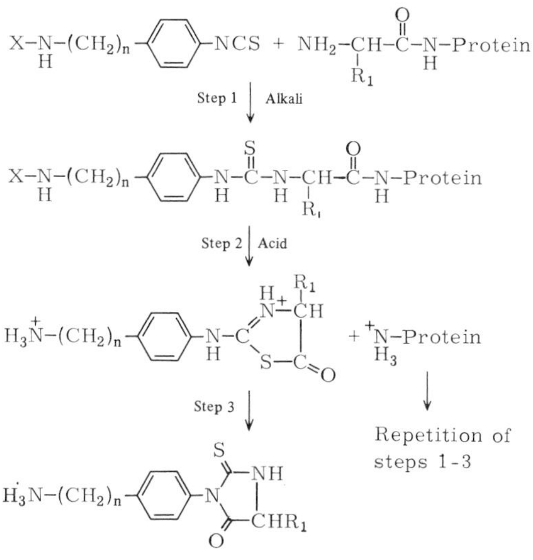

Fig. 2.3-17. Reaction steps for Edman degradation in a gas phase sequenator.
The α-amino group of a protein reacts with a modified phenylisothiocyanate in alcoholic solution. The reaction product allows the amino acid to be cleaved off by acid in the form of an anilinothiazolinone. An additional acid treatment results in the formation of phenylthiohydantoin derivatives which can subsequently be identified by HPLC. The three reactions are then repeated with the protein having been shortened by one amino acid. X signifies an alkali stable amino-protective group such as the tertiary butyloxy residue (t-Boc).

carried in the gas phase, by a stream of argon, to the protein which is bound to a small polybren film (Hewick *et al.*, 1981). This reduces the amounts of solvents, reagents, and byproducts, which, in turn, brings about the observed high degree of sensitivity (Hunkapillar and Hood, 1983).

Leroy E. Hood; Pasadena (left);
Brigitte Wittmann-Liebold; Berlin;
Michael W. Hunkapillar; Foster City

These new developments allow partial sequencing of microgram amounts of protein, which may be sufficient, for example, for the determination of the first 5-7 amino-terminal amino acids of a protein. Such a sequence can then be translated into all conceivable DNA sequences according to the genetic code. The corresponding oligonucleotides are synthesised by automated DNA synthesis and are used as hybridisation probes for the identification of cDNA clones. Such cDNA clones can be used to isolate the corresponding genes from gene libraries. For a successful analysis, by hybridisation, of cDNA libraries obtained with oligo(dT) as primers, the oligonucleotide probes should be specific for the 3' end of the mRNA, *i.e.*, for the *C*-terminal end of the corresponding protein. This would require not only a sequence

from the amino terminus of a protein but also from more internal regions, and thus would ask for the isolation of fragments of a given protein. It will be essential in the future that the sensitivity of protein sequencing should reach picomole and even lower ranges.

In this context, the synthesis of peptides becomes as important as protein sequencing (Schlesinger, 1983). Sequencing of viral genomes or large chromosomal DNA fragments frequently leads to the identification of genes for unknown and possibly unexpressed products. DNA sequence analysis can be used to identify hydrophilic peptide regions, which can be synthesised and employed in immunisation (Lerner, 1982; Lerner, 1984). The antibodies elicited by these antigens could be used as probes for the identification of the putative protein products. It is especially this experimental approach which illustrates the necessity of computers, for example, for the analysis of relationships between DNA and protein sequences, for the comparison of related DNA and protein sequences in a search of homologies, and for the prediction of hydrophobic properties and secondary structures. The necessary software is, in part, available and under constant development. Special issues of the journal "Nucleic Acids Research" are devoted regularly to this subject.

References

Adams, S.P., Karka, K.S., Wykes, E.J., Holder, S.B., and Galluppi, G.R. (1983). Hindered dialkylaminonucleoside phosphite reagents in the synthesis of two DNA 51-mers. J. Am.Chem. Soc. 105, 661-663.

Beaucage, S.L., and Caruthers, M. H. (1981). Deoxynucleoside phosphoramidites: a new class of key intermediates for deoxypolynucleotide synthesis. Tetrahedron Letters 22, 1859-1862.

Brenner, D.G., and Shaw, W.V. (1985). The use of synthetic oligonucleotide with universal templates for rapid DNA sequencing: results with staphylococcal replicon pC221. The EMBO J. 4, 561-568.

Broka, C., Hozumi, T., Areutzen, R., and Itakura, K. (1980). Simplifications in the synthesis of short oligonucleotide blocks. Nucleic Acids Res. 8, 5461-5471.

Brousseau, R., Scarpulla, R., Sung, W., Hsiung, H.M., Narang, S.A., and Wu, R. (1982). Synthesis of a human insulin gene. V. Enzymatic assembly, cloning and characterization of the human proinsulin DNA. Gene 17, 279-289.

Brown, E.L., Ramamoorthy, B., Ryan, M.J., and Khorana, H.G. (1979). Chemical synthesis and cloning of a tyrosine tRNA gene. Methods in Enzymology 68. 109-151.

Caruthers, M.H. (1982). In "Chemical and enzymatic synthesis of gene fragments"; Gassen, H.G., and Lang, A. (eds.), Verlag Chemie, Weinheim and New York; pp. 71-80.

Caruthers, M.H. (1985). Gene synthesis machines: DNA chemistry and its uses. Science 230, 281-285.

Catlin, J.C., and Cramer, F. (1973). Deoxyoligonucleotide synthesis via the triester method. J. Org. Chem. 38, 245-250.

Chow, F., Kempe, T., and Palm, G. (1981). Synthesis of oligodeoxyribonucleotides on silica gel support. Nucleic Acids Res. 9, 2807-2817.

Conner, B.N., Takano, T., Tanaka, S., Itakura, K., and Dickerson, R.E. (1982). The molecular structure of d(CpCpGpG), a fragment of right-handed double helical A-DNA. Nature 295, 294-299.

Crea, R., Kraszewski, A., Hirose, T., and Itakura, K. (1978). Chemical synthesis of genes for human insulin. Proc. Natl. Acad. Sci. USA 75, 5765-5769.

Daub, G.W., and van Tamelen, E.E. (1977). Synthesis of oligoribonucleotides based on the facile cleavage of methyl-phosphotriester intermediates. J. Amer. Chem. Soc. 99, 3526-3528.

Dörper, T., and Winnacker, E.-L. (1982). In: "Chemical and enzymatic synthesis of gene fragments"; Gassen, H.G., and Lang, A., (eds.), Verlag Chemie, Weinheim und New York; pp. 97-102.

Dörper, T., and Winnacker, E.-L. (1983). Improvements in the phosphoramidite procedure for the synthesis of oligodeoxyribonucleotides. Nucleic Acids Res. 11, 2575-2584.

Drew, H., Takano, T., Tanaka, S., Itakura, K., and Dickerson, R.E. (1980). High-salt d(CpGpCpGp), a left-handed Z-DNA double helix. Nature 286, 567-573.

Duckworth, M.L., Gait, M.J., Goelet, P., Hong, G.F., Singh, M., and Titmas, R.C. (1981). Rapid synthesis of oligodeoxyribonucleotides. VI. Efficient, mechanized synthesis of heptadecadeoxyribonucleotides by an improved solid phase phosphotriester route. Nucleic Acids Res. 9, 1691-1706.

Edge, M.D., Greene, S.R., Heathcliffe, G.R., Meacock, P.A., Schuch, W., Scanlon, D.B., Atkinson, T.C., Newton, C.R., and Markham, S.F. (1981). Total synthesis of a human leukocyte interferon gene. Nature 292, 756-762.

Frank, R., Heikens, W., Heisterberg-Montris, G., and Blöcker, H. (1983). A new general approach for the simultaneous chemical synthesis of large number of oligonucleotides: segmented solid support. Nucleic Acids Res. 11, 4365-4377.

Gough, G.R., Collier, K.J., Weith, H.L., and Gilham, P.T. (1979). The use of barium salts of protected deoxyribonucleoside-3'-p-chlorophenyl phosphates for construction of oligonucleotides by the phosphotriester method: high-yield synthesis of dinucleotide blocks. Nucleic Acids Res. 7, 1955-1964.

Grotjahn, L., Frank, R., and Blöcker, H. (1982). Ultrafast sequencing of oligodeoxyribonucleotides by FAB-mass spectrometry. Nucleic Acids Res. 10, 4671-4678.

Hewick, R.M., Hunkapillar, M.W., Hood, L.E., and Dreyer, W.J. (1981). A gas-liquid solid phase peptide and protein sequenator. J. Biol. Chem. 256, 7990-7997.

Hunkapillar, M.W., and Hood, L.E. (1983). Protein sequence analysis: Automated microsequencing. Science 219, 656-659.

Hunkapillar, M., Kent, S., Caruthers, M., Dreyer, W., Firea, J., Giffin, C., Horrath, S., Hunkapillar, T., Tempsito, P., and Hood, L. (1984). A microchemical facility for the analysis and synthesis of genes and proteins. Nature 310, 105-111.

Itakura, K., Hirose, T., Crea, R., Riggs, A.D.,

Heyneker, H.L., Bolivar, F., and Boyer, H.W. (1977). Expression in *Escherichia coli* of a chemically synthesized gene for the hormone somatostatin. Science 198, 1056-1063.

Itakura, K. (1980). Synthesis of genes. Trends Biochem. Sci. 5, 114-116.

Jay, E., Seth, A.H., Rommens, J., Sood, A., and Jay, G. (1982). Gene expression: Chemical synthesis of *Escherichia coli* ribosome binding sites and their use in directing the expression of mammalian proteins in bacteria. Nucleic Acids Res. 10, 6319-6329.

Jones, S.S., Rayner, B., Reese, C.B., Ubasawa, A., and Ubasawa, M. (1980). Synthesis of the 3'-terminal decaribonucleoside nonaphosphate of yeast alanine transfer ribonucleic acid. Tetrahedron 36, 3075-3085.

Khorana, H.G., Agarwal, K.L., Büchi, H., Caruthers, M.H., Gupta, N.K., Kleppe, K., Kumar, A., Ohtsuka, E., RayBhandary, U.L., v.d. Sande, J.H., Sgaramella, V., Terao, T., Weber, H., and Yamada, T. (1972). Studies on polynucleotides. CIII. Total synthesis of the structural gene for an alanine transfer ribonucleic acid from yeast. J. Mol. Biol. 72, 209-217.

Khorana, H.G., Agarwal, K.L., Besmer, P., Büchi, H., Caruthers, M.H., Cashion, P.J., Fridkin, M., Jay, E., Kleppe, K., Kleppe, R., Kumar, A., Loewen, P.C., Miller, R.C., Minamoto, K., Panet, A., RajBhandary, U.L., Ramamoorthy, B., Sekiya, T., Takeya, T., and v.d. Sande, J.H. (1976). Total synthesis of the structural gene for the precursor of a tyrosine suppressor transfer RNA from *Escherichia coli*, 1. General Introduction. J. Biol. Chem. 251, 565-570.

Kohli, V., Blöcker, H., and Köster, H. (1980). The triphenylmethyl (trityl) group and its uses in nucleotide chemistry. Tetrahedron Letters 21, 2683-2686.

Lerner, R.A. (1982). Tapping the immunological repertoire to produce antibodies of predetermined specificity, Nature 299, 592-596.

Lerner, R.A. (1984). Antibodies of predetermined specificity in Biology and Medicine. Adv. Immunol. 36, 1-44.

Letsinger, R.L., and Kornet, M.J. (1963). Popcorn polymer as a support in multistep syntheses. J. Am. Chem. Soc. 85, 3045-3046.

Letsinger, R.L., and Lunsford, W.B. (1976). Synthesis of thymidine oligonucleotides by phosphite triester intermediates. J. Am. Chem. Soc. 98, 3655-3661.

Lohrmann, R., and Khorana, H.G. (1966). Studies on polynucleotides. LII. The use of 2,4,6-triisopropylbenzenesulfonyl chloride for the synthesis of internucleotide bonds. J. Am. chem. Soc. 88, 829-833.

Matteucci, M.D., and Caruthers, M.H. (1980). The use of zinc bromide for removal of dimethoxytritylethers from deoxynucleosides. Tetrahedron Letters 21, 3243-3246.

Matteucci, M.D., and Caruthers, M.H. (1981). Synthesis of deoxyoligonucleotides on a polymer support. J. Am. Chem. Soc. 103, 3185-3191.

Matthes, H.W.D., Zenke, W.M., Grunström, T., Staub, A., Wintzerith, M., and Chambon, P. (1984). Simultaneous rapid chemical synthesis of over one hundred oligonucleotides on a microscale. The EMBO J. 3, 801-805.

Maxam, A.M., and Gilbert, W. (1980). Sequencing end-labelled DNA with base-specific chemical cleavages. Methods in Enzymology 65, 499-560.

McBridge, L.J., and Caruthers, M.H. (1983). An investigation of several deoxynucleoside phosphoramidites. Tetrahedron Letters 24, 245-248.

Merrifield, R.B. (1963). Solid phase peptide synthesis. I. The synthesis of a tetrapeptide. J. Am. Chem. Soc. 85, 2149-2154.

Michelson, A.M., and Todd, A.R. (1955). Nucleotides Part XXXII. Synthesis of a dithymidine dinucleotide containing a 3':5'-internucleotide linkage. J. Chem. Soc. 2632-2538.

Miyoshi, K., Miyake, T., Hozumi, T., and Itakura, K. (1980). Solid phase synthesis of polynucleotides. II. Synthesis of polythymidylic acids by the block coupling phosphotriester method. Nucleic Acids Res. 8, 5473-5489.

Narang, S.A. (1983). DNA synthesis. Tetrahedron 34, 3-22.

Narang, S.A., Hsiung, H.M., and Brousseau, R. (1979). Improved phosphotriester method for the synthesis of gene fragments. Methods in Enzymology 68, 90-98, s.a. 65, 610-620.

Reese, C.B. (1978). The chemical synthesis of oligo- and polynucleotides by the phosphotriester approach. Tetrahedron 34, 3143-3179.

Reese, C.B., and Zard, L. (1981). Some observations relating to the oximate ion promoted unblocking of oligonucleotide aryl esters. Nucleic Acids Res. 9, 4611-4626.

Rinehart, K.L. (1982). Fast atom bombardment mass spectrometry. Science 218, 254-260.

Rossi, J.J., Kierzek, R., Huang, T., Walker, P.A., and Itakura, K. (1982). An alternate method for synthesis of double-stranded DNA segments. J. Biol. Chem. 257, 9226-9229.

Schaller, H., Weimann, G., Lerch, B., and Khorana, H.G. (1963). Studies on polynucleotide XXIV. The stepwise synthesis of specific deoxyribopolynucleotides. Protected derivatives of deoxyribonucleosides and new synthesis of deoxyribonucleoside-3'-phosphates. J. Am. Chem. Soc. 85, 3821-3827.

Schlesinger, I.H. (1983). High-Performance Liquid Chromatography of side chain-protected phenylthiohydantions: Application to solid-phase peptide synthesis. Methods in Enzymology. 91, 494-502.

Sheppard, R.C. (1983). Continuous flow methods in organic synthesis. Chemistry in Britain, p.402-413.

Sproat, B.S., and Bannwarth, W. (1983). Improved synthesis of oligodeoxynucleotides on controlled pore glass using phosphotriester chemistry and a flow system. Tetrahedron Letters 24, 5771-5774.

Stawinski, J., Hozumi, T., Narang, S.A., Bahl, C.P., and Wu, R. (1977). Arylsulfonyltetrazoles, new coupling reagents and further improvements in the triester method for the synthesis of deoxyribooligonucleotides. Nucleic Acids Res. 4, 353-372.

Stec, W.J., Zon, G., Egan, W., and Stec, B. (1984). Automated solid-phase synthesis, separation and stereochemistry of phosphorothioate analogues of oligodeoxyribonucleotides. J. Am. Chem. Soc. 106, 6077-6079.

Taniguchi, T., Matsui, H., Iugita, T., Takaoka, C., Kashima, N., Yoshimoto, R., and Hamuro, J. (1983). Structure and expression of a cloned cDNA for human interleukin-2. Nature 302, 305-310.

Ti, G.S., Gaffney, B.L., and Jones, R.A. (1982). Transient protection. Efficient one-flask synthesis of protected deoxyribonucleosides. J. Am. Chem. Soc. 104, 1316-1319.

Tu, C.D., and Wu, R. (1980). Sequence analysis of short DNA fragments. Methods in Enzymology 65, 620-638.

Wetzel, R., Kleid, D.G., Crea, R., Heyneker, H.L., Yansura, D.G., Hirose, T., Kraszewski, A., Riggs, A.D., Itakura, K., and Goeddel, D.V. (1981). Expression in *Escherichia coli* of a chemically synthesized gene for a mini-C analog of human proinsulin. Gene 16, 63-71.

2.4 DNA Sequencing

Several methods for rapid DNA sequencing have been devised in the past few years. Two of them have gained a reputation of being particularly reliable and simple: sequencing of end-labelled DNA by base-specific chemical cleavage (Maxam and Gilbert, 1977), and sequence analysis by primed enzymatic synthesis (Sanger *et al.*, 1977; Smith 1980). The latter approach requires single-stranded DNA, and cloning of DNA in the single-stranded DNA phages M13 and fd is therefore becoming increasingly important. Neither of the fast sequencing methods is suitable for sequencing short stretches of DNA. For this purpose special methods were devised, methods which exploit the differential mobility of partially hydrolysed oligonucleotides in two-dimensional electrophoresis systems. These two-dimensional methods now can be complemented by mass spectroscopy. Finally, the problems associated with the analysis of sequence data have reached such dimensions that very often they can be solved only with the aid of computers. These problems, and the fundamental elements of the two approaches towards DNA sequencing alluded to above, will be the topic of the following sections.

Walter Gilbert
Genève

2.4.1 Sequencing of End-labelled DNA by Base-Specific Chemical Cleavage

The availability of single- or double-stranded DNAs specifically radiolabelled at one end is a prerequisite for sequencing by base-specific chemical cleavage (Maxam and Gilbert, 1980), in which the DNA is partially cleaved at each of the four bases in four different reactions. The cleavage products are separated in denaturing polyacrylamide gels according to size; subsequently the sequence can be read directly from an autoradiogram by determining which of the base-specific reagents cleaved adjacent nucleotides (Fig. 2.4-1). The key features of this technique are the three-step base-specific cleavage reactions, which include the modification of the base, the removal of the modified base from the deoxyribose moiety, and a strand break at the position of the modified base (Fig. 2.4-2). The modification reaction must be specific but limited; it should only occur with low frequencies at one, or, simultaneously, at two bases (G and A, or T and C, for example). Subsequent cleavage reactions must be quantitative. The reagents for the first and second reactions are dimethylsulfate and piperidine in the case of purines, and hydrazine and piperidine for pyrimidines. The steps which are generally carried out for Maxam-Gilbert sequencing are described in the following sections.

2.4.1.1 Isolation of Defined DNA Fragments

The DNA fragments used for sequencing are primarily generated by digestion with a restriction endonuclease. Occasionally the isolation of electrophoretically separated individual fragments from the gel matrix may prove problematic. Four different procedures for the preparative isolation of such DNA fragments already have been described in Section 2.1.1.2. The direct elution of fragments smaller than 500 bp with high salt buffers is most convenient and preferable. It is

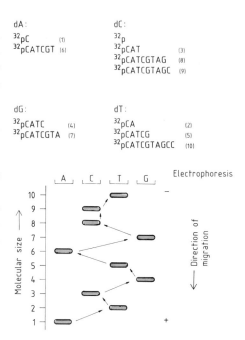

Fig. 2.4-1. Principle of DNA sequencing by chemical cleavage of DNA.
The upper panel shows the sequence of a DNA molecule which is labelled with ^{32}P-phosphate at its 5′ end, and the cleavage products obtained by partial, but specific, cleavage at dA, dC, dT, and dG residues. Numbers in brackets indicate the different lengths of the cleavage products. After completion of the chemical reactions, the individual reaction mixtures are applied to adjacent slots of a gel and are separated by electrophoresis under denaturing conditions. The reaction mixture obtained by dA-specific cleavage, for example, has been separated in the lane marked A. Bands are visualised by autoradiography. In practice, cleavage reactions are either specific for dG and dC, or for dG+dA and T+dC (*cf.* also Fig. 2.4-7).

frequently quite useful, and also advisable, to end-label the DNA before electrophoresis (see below) in order to circumvent any difficulties that might arise with the enzymatic treatment of DNA fragments eluted from the gel.

The amount of DNA that must be used for sequencing is determined by a variety of factors,

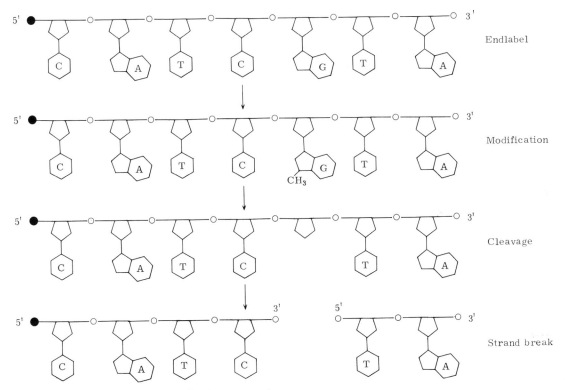

Fig. 2.4-2. The four separate steps of a cleavage reaction.
Shown is a dG-specific cleavage which yields an N^7-methyl dG residue by methylation.

such as the size of the DNA fragment, or the number of reactions involved. Last but not least, this amount also is influenced greatly by the experimenter's skill. As a rule of thumb, at least one picomole of DNA ends is required for end-labelling. In the case of linearised DNA of plasmid pBR322, which has a molecular weight of 2.6×10^6 Da, two picomoles of DNA ends correspond to 2.6 µg of DNA. If the specific activity of the ^{32}P-dATP used for labelling is 1 000 Ci/mmol, up to 1 µCi (approximately 10^6 cpm) theoretically could be incorporated into one DNA end. The radioactively labelled DNA is generally subjected to four or five cleavage reactions involving several precipitation steps, which are usually not quantitative. Under the most favorable conditions, then, 10^5 cpm will be available for the subsequent analysis by gel electrophoresis. Thus, one picomole of DNA ends constitutes the lower limit of DNA ends required for an experiment.

2.4.1.2 End Labelling of DNA Fragments

An essential feature of the Maxam-Gilbert sequencing approach is the specific labelling of one of the two ends of the DNA to be sequenced. Labelling generally involves the 5' phosphate and 3' hydroxyl residues at the ends of the DNA fragments. Of course, these fragments should possess defined termini, *i.e.*, they should either have been generated by cleavage with restriction endonucleases, or constitute the ends of intact chromosomes. End-labelling at the 5' end usually is achieved by transferring the γ-phosphate residue of γ-^{32}P-ATP in a reaction mediated by the enzyme polynucleotide kinase, which is encoded by bacteriophage T4. Since restriction endonucleases always generate 5' phosphate ends, these ends must be dephosphorylated by alkaline phosphatase, because the substrate for the kinase is a 5' hydroxyl terminus (Fig. 2.4-3). The phosphatase activity, which would later reverse γ-^{32}P-ATP labelling, can be removed by phenol extraction; however, it is also possible to carry out the phosphatase and kinase reactions in the same solution in two consecutive steps by exploiting the different sensitivities of the two enzymes towards inorganic phosphate. At potassium phosphate concentrations as low as 5 mM, the phosphatase is inhibited sufficiently to allow the subsequent phosphorylation to occur with yields in the range of 30-50 per cent. In this context, it is important to note that many restriction enzymes are stored in phosphate buffers. Use of such enzymes prior to phosphatase treatment therefore may give unsatisfactory yields because the phosphatase reaction may be inhibited. It can be assumed that, apart from contaminations of polynucleotide kinase preparations with nuclease activities, this is one of the most prevalent causes of unsatisfactory yields in end-labelling reactions.

Terminal 3' hydroxyl groups can be labelled with terminal deoxynucleotidyl transferase (Fig. 2.4-4). In the presence of Co^{2+} rather than Mg^{2+} ions, this enzyme catalyses the limited addition of the four ribonucleoside triphosphates to the 3' hydroxyl termini of double-stranded or single-stranded DNAs. The product obtained from a reaction with α-^{32}P-labelled ribonucleoside triphosphates is subsequently treated with alkali, which removes all but one of the added ribonucleotides. The terminus is now labelled by the two ^{32}P phosphate residues in the single remaining ribonucleotide. Only a single nucleotide is added, if α-cordycepine triphosphate (2',3' dideoxyadenosine triphosphate) is used as a substrate for the terminal deoxynucleotidyl transferase. In this

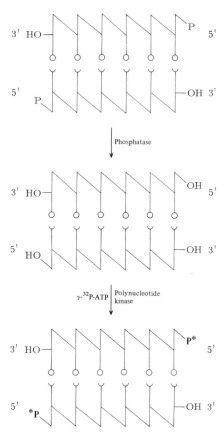

Fig. 2.4-3. Scheme for ^{32}P- labelling of 5' ends. Shown is a double-stranded blunt-ended DNA molecule, obtained for example, by cleavage with restriction endonuclease *Hae* III. Labelling is effected by incorporation of the γ-phosphate residue of γ-^{32}P-labelled ATP in a reaction catalysed by T4 polynucleotide kinase.

case, alkaline hydrolysis is not required. In principle, the enzyme also reacts with internal 3' hydroxyl groups, and it may be important to ensure that only intact double-stranded DNA is used.

Another way of obtaining end-labelled DNA is provided by T4 DNA polymerase, which displays both a 5' to 3' polymerase and a 3' to 5' exonuclease activity. Under optimal reaction conditions, the polymerase reaction is considerably faster than hydrolysis, and these kinetic

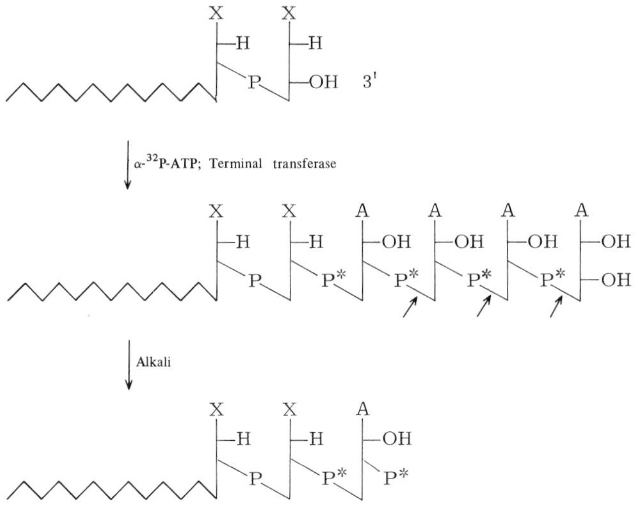

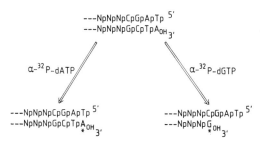

Fig. 2.4-4. Labelling of 3' hydroxyl termini by terminal transferase.
Shown is one 3' hydroxyl end with two terminal bases (X). In the presence of α-^{32}P-rATP and Co^{2+} ions terminal transferase adds ribo-adenosine residues. After alkali treatment (arrows), two labelled phosphate residues at the 5' and 3' positions of the terminal ribonucleotide are retained.

Fig. 2.4-5. End-labelling by T4 DNA polymerase.
The 3' to 5' exonuclease activity of the bacteriophage T4 encoded enzyme degrades double-stranded DNA from the 3' end. In the presence of suitable substrates (α-^{32}P-deoxyribonucleoside triphosphates) for the polymerisation reaction, the incorporation of triphosphates proceeds much faster than the excision reaction. This treatment therefore yields labelled 3' termini (see also text for details).

properties can be exploited for removing the terminal deoxyadenylate in the double-stranded DNA shown in Fig. 2.4-5 and for replacing it by ^{32}P-deoxyadenylate in the presence of ^{32}P-dATP. If α-^{32}P-dGTP is used, the terminal four nucleo-

tides from the 3' end are first removed by hydrolysis, and the missing G residue is immediately replaced by α-^{32}P-dGMP. The 3' terminus can therefore not only be labelled, but also can be sequenced to a limited extent.

2.4.1.3 Separation of the Two Labelled Ends of a DNA Fragment

Before sequencing a DNA fragment, it is necessary to separate from each other the two ends of the DNA molecule labelled by the procedures described above. One way to achieve this is by digestion of the DNA fragment with a restriction endonuclease which produces two double-stranded fragments of different size. The subfragments, each now labelled at only one end, can be separated from each other in non-denaturing gels (Fig. 2.4-6A). It is also possible to melt the double-stranded DNA fragment, or denature it by treatment with alkali in order to obtain the two complementary single strands (Fig. 2.4-6B). These

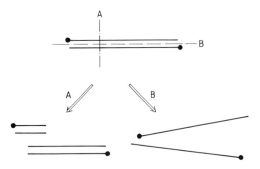

Fig. 2.4-6. Separation of the labelled ends of a double-stranded DNA molecule.
End-labels are shown as solid circles. Method A involves digestion with a restriction endonuclease, and yields two restriction fragments which differ in length. Method B involves denaturation of the labelled DNA, and yields two complementary single strands of equal lengths. In each case the different labelled ends are separated by electrophoresis.

strands can again be separated from each other by gel electrophoresis. This procedure, originally developed by Hayward (1972) for intact DNA of bacteriophage λ, has also been applied to shorter DNA fragments. It is based on the observation that, for reasons still unknown, complementary single strands often migrate with different velocities during agarose or polyacrylamide gel electrophoresis. Although this method may fail occasionally, it generally offers the advantage that sequencing from either of the two DNA termini usually ends in a region of overlap, the sequence of which, therefore, can be checked by comparing the results obtained from each strand.

2.4.1.4 Base-specific Chemical Cleavage

The experimental procedures for the four specific reactions (*i.e.*, G-, G+A-, C+T-, and C-specific) have been described in detail by Maxam and Gilbert (1980). The exact nature of these chemical reactions is not known. Essentially, the modification reaction is designed to alter the structure of the bases so as to weaken the

N-glycosidic bond between base and sugar. Cleavage of the base is usually mediated by piperidine, which also catalyses the subsequent β-elimination of the phosphate residues from the free sugar moiety (see also Fig. 2.4-2).

The number of bases cleaved in each reaction does not depend on the length of the DNA fragment but only on the concentration of the reagents, reaction temperature, and incubation time. Reaction conditions should be chosen in a way which guarantees that the bases in the desired region of each end-labelled molecule are cleaved. The resolution, *i.e.*, the ability to separate molecules with a chain length of *N* from those with a length *N+1* by gel electrophoresis, depends entirely on the nature of the gel, for example, its length and composition. A conventional gel composed of 8% polyacrylamide/8M urea, which is 0.3 mm thick and 400 mm long, for example, allows a satisfactory resolution of at least 200 nucleotides. Much longer gels have also been described, and the appropriate equipment is commercially available. It should be pointed out that the above-mentioned 200 nucleotides cannot be separated on a 40 cm gel in one single run (Fig. 2.4-7). This problem is usually circumvented by loading the gel with aliquots of the four different cleavage reaction mixtures at different times. The shortest molecules with a chain length of up to 300

Fig. 2.4-7. Autoradiograph of a sequencing gel. →
The 12% polyacrylamide/7 M urea gel (0.3 mm thick) was run at 1 350 V and 35 mA. The letters above the lanes indicate the individual cleavage mixtures (G-specific, (G+A)-specific, (C+T)-specific, and C-specific). The same mixtures that were loaded in lanes marked I were applied to the gel two hours later (II) to allow reading of a longer DNA sequence. The gel is read by starting at the bottom part of the lanes marked II (right). These lanes contain the mixtures having run for a shorter time. When a region of overlapping sequences is reached, reading is continued in lanes marked I (left). The first four bases marked in the autoradiograph are 5'-CGTC-3'. The switch from the right to the left side of the autoradiograph is indicated by an arrow. (Autoradiograph courtesy of W. Wetekam, Hoechst AG).

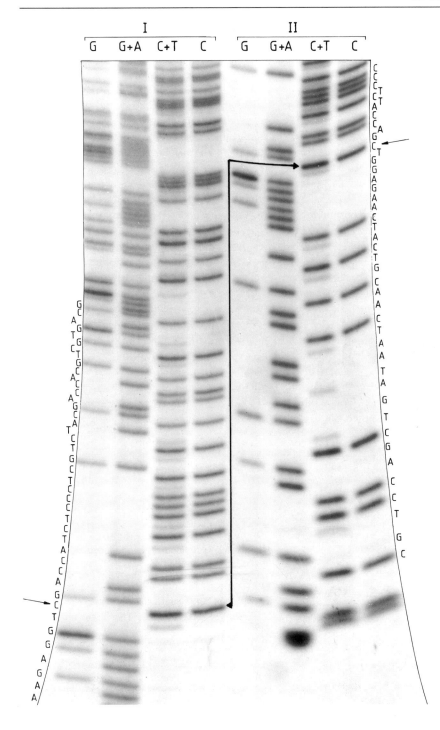

bases, for example, are resolved in the slot containing the last aliquot, which has only migrated for a short time. Reading the sequence from the autoradiograms, therefore, begins in the lane which contains this material. The lower part of the gel contains cleavage products which come from regions closest to the labelled end of the DNA molecule. In the case of 5' end-labelling, the lowest band, which has run fastest, will be closest to the 5' end. The sequence ladder is then read from bottom to top as far as possible, *i.e.*, as far as individual bands can be discriminated. Reading of the sequence ladder, always from bottom to top, then continues in lanes containing aliquots of the reaction mixtures having been applied to the gel at earlier times.

If the times at which aliquots were loaded have been chosen correctly, there will be two regions with identical sequences, one of which will lie in the top part of the sequence ladder read from material having been electrophoresed for the shortest time, *i.e.*, having been loaded last; the other region will lie in the bottom part of the sequence ladder obtained from the aliquot having been loaded before the first, *i.e.*, having run for the longest time. The sequence then can be read in one continuous stretch. Of course, the practice of using aliquots, and the desire to have sensible exposure times, again influence the input amount of DNA. Although the resolution can be increased, *i.e.*, by the use of gradient gels and ^{35}S-labelled dNTPs (Biggin *et al.*, 1983), it is apparent that the picomole of labelled DNA ends mentioned above constitutes, indeed, the lower limit of DNA to be used in a single experiment.

However elegant this approach may be, the amount of experimental work involved even today is justified only if correct results are obtained. Each attempt to determine a DNA sequence must, therefore, be accompanied by considerations of suitable strategies and critera which guarantee the desired correct result. The best strategy is, of course, to sequence the two complementary strands of a DNA fragment; yet, difficulties may be experienced if the two comple-

mentary strands cannot be separated from each other. This problem can usually be circumvented by beginning sequencing from the 3' end, or by using fragments generated with different restriction enzymes. Ten different restriction enzymes (out of sixteen possible) with tetranucleotide recognition sequences are now available. It should therefore be possible to divide the DNA region to be sequenced into different fragments.

Several strategies have been described for sequencing larger stretches of DNA such as whole viral chromosomes. It is relatively easy to produce a physical gene map from cleavage sites of several restriction enzymes and to sequence the individual DNA fragments. An alternative approach, which has been recommended and used, is shot-

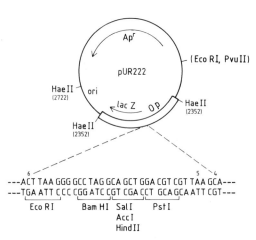

Fig. 2.4-8. Structure of plasmid vector pUR222. This vector contains pBR322 sequences spanning the region between the *Eco* RI site (4363) and the *Pvu* II site (2066) and therefore covers the β-lactamase gene, and the origin of DNA replication (*ori*). The tetracycline resistance region is lacking, and selection can therefore only be carried out for ampicillin resistance. The polylinker contains sites for *Eco* RI, *Bam* HI, *Sal* I, *Acc* I, and *Pst* I and is located between the codons of amino acids five and six of β-galactosidase. The *lacZ* region codes for 59 *N*-terminal amino acids of the β-galactosidase. In spite of the polylinker insertion which however conserves the correct reading frame, the sub-fragment of β-galactosidase shows α-complementation. (Rüther *et al.*, 1981).

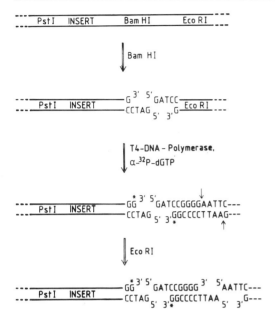

Fig. 2.4-9. DNA sequencing using vector pUR222. The vector contains a DNA insertion in the *Sal* I or *Acc* I site within the polylinker region. After *Bam* HI cleavage a ^{32}P-labelled dG residue derived from α-^{32}P dGTP is incorporated in a reaction catalysed by T4 DNA polymerase. Subsequent *Eco* RI cleavage yields two fragments. The mixture can be sequenced as such without separation of the fragments since one of the fragments is only 10 bases long. Only the first 10 bases will therefore be contaminated by bases derived from both sequences. (Rüther *et al.*, 1981).

gun sequencing, where the DNA in question is first cleaved with restriction enzymes and the resulting DNA fragments are then sequenced without knowing their relative order. Sequence data can be accumulated rapidly in this way, and overlapping sequences eventually can be aligned by means of appropriate computer programs. This procedure may be intellectually unsatisfactory and will sooner or later lead to redundant sequences; eventually, the construction of a detailed restriction map will be unavoidable. In spite of these reservations, shot-gun cloning and sequencing is gaining significance.

A useful vector, plasmid pUR222, has been described for shot-gun sequencing using the

Maxam-Gilbert technique (Rüther *et al.* 1981). In addition to markers which are typical for plasmids vector pUR222 contains a polylinker with six restriction enzyme recognition sequences which do not occur anywhere else on the vector. The polylinker lies between the codons for the amino acids 5 and 6 of the *lacZ* gene on the vector (Fig. 2.4-8). Its position in the amino-terminal region of the *lacZ* gene offers an advantage, because it allows recombinant DNA molecules to be identified by the loss of α-complementation (*cf.* Section 2.4.2.3). The polylinker also permits cloning of DNA fragments with different ends, for example, *Pst* I and *Bam* HI, or cloning of overlapping *Hpa* II (C/CGG) or *Taq* I (T/CGA) fragments in one and the same recognition site, for example, *Acc* I. The chief advantage of using this plasmid, however, is that it simplifies DNA sequencing, because the separation of both labelled ends of a DNA fragment can be omitted under certain conditions. Fig. 2.4-9 shows a chimaeric pUR222 plasmid, which carries an insertion in the *Acc* I site. This plasmid is cleaved with *Bam* HI and is then subjected to a T4 DNA polymerase treatment, which catalyses G-labelling in the presence of α-^{32}P dGTP (see Fig. 2.4-5); labelled molecules are subsequently digested with *Eco* RI. One labelled G will be in the immediate vicinity of the insert, while the other labelled G will be in a decameric DNA fragment. When this mixture of DNA fragments is analysed, the whole sequence of the insert will be available apart from the first 10 bases which, by necessity, are obscured by the presence of the pieces from the small decameric fragment. This, however, will be of no account, especially if overlapping fragments, for instance, *Hpa* II and *Taq* I fragments, are shot-gun sequenced. The pUC vectors described in Section 2.4.2.5 can be used in an analogous way.

2.4.1.5 Genomic Sequencing

Genomic sequencing allows sequence determinations of specific and selected DNA regions in mixtures of genomic DNA sequences (Church and Gilbert, 1984) by combining techniques of chemical DNA sequencing, and methods for the specific recognition of DNA sequences by Southern blotting (Southern, 1975; *cf.* also Section 11.2.2.2). The first step (Fig. 2.4-10) is the complete digestion of genomic DNA with a particular restriction enzyme, for example, with *Eco* RI. The DNA is then partially chemically cleaved, for example, at C residues, denatured, and the resulting fragments are separated on polyacrylamide gels and subsequently transferred to nitrocellulose filters. Specific DNA sequences are identified on such filters by hybridisation with labelled DNA probes. If only the lower strand of the DNA molecule shown in panel A of Fig. 2.4-10 is considered, probe 1 recognises the fragments shown in panel B, provided that the probe itself is without internal C residues; other DNA fragments, such as those shown in panel C, would not be recognised by the probe. An autoradiograph would, therefore, reveal a set of fragments with one common end and the respective other ends possessing a C residue. The new feature of this technique is the generation of a specific pattern of DNA fragments from a mixture which contains more than one million fragments (*cf.* also Fig. 2.4-7). Difficulties can arise if the probe itself covers one of the C residues, for example, the left-most C residue in Fig. 2.4-10A. In this case the first four fragments shown in panel C of Fig. 2.4-10 would also be detected and, hence, obscure the pattern of fragments visualised by hybridisation. This problem can be reduced if the frequency of C cleavage, (P), is kept relatively small. The probability of the occurrence of undesired internal DNA fragments with C residues at both ends would then be only P^2. The pattern of internal fragments would then be much fainter than that generated by the desired fragments possessing one C terminus and their other ends defined by the *Eco* RI site. It is, nevertheless, advisable not to underestimate the difficulties, because a low frequency of C cleavage produces only minute amounts of the desired fragments.

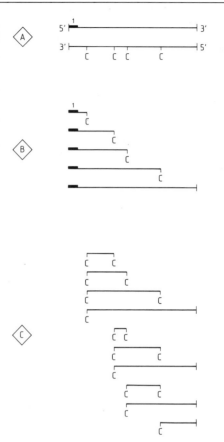

Fig. 2.4-10. Principle of genomic sequencing. Probe 1 is derived from the 5′ terminal region of the upper strand of a DNA fragment shown in (A). It is complementary to the 3′ terminal region of the lower strand. (B) shows a set of single-stranded DNA fragments obtained after partial C cleavage of the lower strand. These single-stranded fragments are identified by probe 1. Other fragments obtained by cleavage at C (shown in (C)) cannot be identified with probe 1.

This in turn renders hybridisation (which is concentration-dependent), and hence the detection of fragments, more difficult, because extremely long exposures of the X-ray films would be required. The technique has already been used successfully for the identification of methylated C residues. This application is based on the fact that 5-methylcytosine is hardly cleaved by hydrazine and that the corresponding DNA fragments therefore do not occur. In contrast to the technique employing restriction enzymes for the identification of methylated C residues (see also Section 2.1.2.3), this new technique allows the identification of all 5-methyl C residues, *i.e.*, also those which are not contained within the recognition sequences of *Hpa* II or *Msp* I. The new technique is also suitable for footprint analyses, *i.e.*, the identification of binding sites of DNA-binding proteins on DNA (Ephrussi *et al.*, 1985).

2.4.2 DNA Sequencing by Enzymatic Synthesis

2.4.2.1 Essential Features

This sequencing approach relies on the enzymatic synthesis of a radioactively labelled complementary copy of the single-stranded template strand to be sequenced. There are several variations of this procedure, including the classical plus-minus method and the recently developed chain termination technique (Sanger *et al.* 1977; Smith, 1980). As shown in Fig. 2.4-11A, sequencing requires a hybrid molecule consisting of a section of single-stranded DNA and a short complementary primer molecule. The 3' hydroxyl end of this primer is positioned in the vicinity of the stretch of DNA to be sequenced. In the presence of suitable substrates, the Klenow fragment of DNA polymerase I starts at this primer and synthesises a complementary copy of this particular DNA region. The primer is usually a short oligonucleotide or a restriction endonuclease fragment with a defined length. All synthesised DNA molecules which initiate at the 3' hydroxyl end of the primer

will therefore possess identical 5' ends, *i.e.*, the 5' end of the primer. The Klenow fragment is the larger (76 kDa) of two enzymatically active peptide fragments obtained by proteolytic digestion of DNA polymerase I (109 kDa) (Kornberg, 1980). It still retains the complete polymerase and 3' to 5' exonuclease activity of the native enzyme but lacks the 5' to 3' exonuclease activity. This is important, because the subsequent analysis of the reaction products separated by gel electrophoresis requires that all labelled oligonucleotides possess a common 5' end. Intact DNA polymerase I cannot be used, because, due to its 5' to 3' exonuclease activity, it would remove these 5' ends of the primer. Reverse transcriptase can replace the Klenow fragment of DNA polymerase I, since it lacks a 5' to 3' exonuclease activity in its native form.

DNA sequencing by enzymatic synthesis usually employs four different reaction mixtures, each of which contains all four deoxyribonucleoside triphosphates. At least one of these dNTPs (usually dATP) is α-^{32}P-labelled. Most recently, α-^{35}S-labelled thiophosphates have also been used (see Fig. 12-7 for chemical structures). They

Frederick Sanger
Cambridge

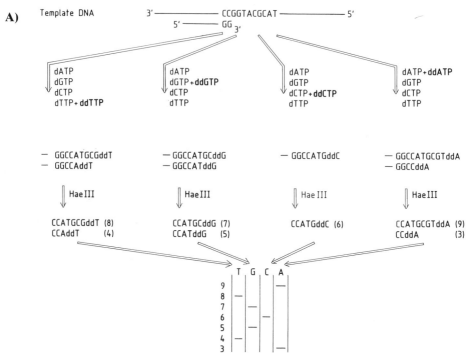

Fig. 2.4-11.   Sequencing by primed enzymatic synthesis. Part A.
The principle of this approach is explained in *part A*. Shown are the template strand, and a primer with the dinucleotide GG at its 3′ hydroxyl end. The four reaction mixtures contain all four α-^{32}P-labelled deoxyribonucleoside triphosphates, and each mixture contains, in addition, a different dideoxytriphosphate. This procedure yields nested sets of molecules each ending with the particular dideoxynucleotide contained in the reaction mixture. After cleavage with *Hae* III (GG/CC) these molecules are separated on denaturing gels. Numbers in brackets refer to the number of nucleotides in individual molecules.

have the advantage of producing much sharper bands and thus much higher resolution in the upper parts of gels (Biggin *et al.*, 1983). In addition to the four dNTPs, each of the four mixtures contains one of the four dideoxyribonucleoside triphosphates (ddNTP, Fig. 2.4-11). ddNTPs lack both the 2′ and the 3′ hydroxyl group, and, because the 3′ hydroxyl group is required for the formation of phosphodiester bonds, the presence of ddNTPs causes chain termination. Each reaction tube will therefore contain a population of radioactively labelled DNA molecules which possess common 5′ ends but differ in length. The lengths of these fragments are specified by the particular dideoxy analogue at

their 3′ ends. A mixture containing ddATP, for example, will contain a population of oligonucleotides all ending with ddA, which has become incorporated instead of a dA opposite a T residue in the template DNA. The ratio between dideoxy- and deoxyribonucleoside triphosphates must be chosen in a way which ensures the formation of long chains; on the other hand, a sufficient number of terminations must occur in the desired sequence. The base composition of the templates and the K_m values of polymerase for the different substrates and analogues play an important role. The concentration of each of the four dNTPs in the reaction mixtures is usually 2 μM, while the ratio between deoxy- and dideoxynucleoside tri-

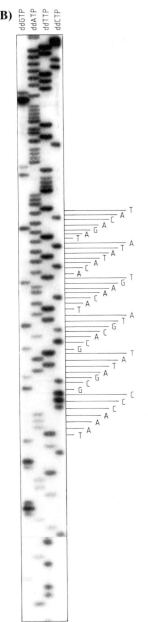

B)

phosphates may vary between 1/200 in A or T mixtures, and 1/100 in G and C mixtures. For practical reasons, the whole experimental procedure often includes a fifth mixture containing araCTP as a substrate analogue. In this case, the ratio between dCTP and araCTP is even lower, *i.e.*, 1/4 000, which is due to the extremely high K_m value of DNA polymerase for araCTP.

If a long primer is used, it must be separated from the newly synthesised oligonucleotides prior to electrophoresis, because it would unduly increase the electrophoresis times or even reduce resolution. In many cases, the primer is a restriction endonuclease fragment. Since the DNA polymerase reaction restores the recognition sequence which initially facilitated primer formation, the primer can again be removed using the same restriction endonuclease. In the example shown in Fig. 2.4-11, the enzyme is *Hae* III (GG/CC).

Problems may arise if a second recognition sequence with the same specificity is located in the vicinity of the primer, and in the direction of chain elongation. In this case, one exploits the fact that

Fig. 2.4-11. Part B. Autoradiograph of an authentic sequencing gel. (Courtesy of E. Penhoet, Chiron Corp.).

Hans Klenow
Copenhagen

DNA polymerase I is capable of incorporating ribonucleoside triphosphates (rNTPs) into DNA in the presence of Mn^{2+}. The example shown in Fig. 2.4-12 contains two neighbouring *Alu* I recognition sequences (AG/CT). The DNA is first subjected to a limited polymerase reaction in the presence of rCTP, and this is followed by the usual chain elongation reactions. The resulting oligonucleotides can be cleaved with ribonuclease A or alkali at the rC position (Brown, 1978). A second cleavage with *Alu* I would yield two families of fragments rather than one, and these could hardly be resolved.

Once all reactions are completed, the mixtures are loaded into neighbouring slots of a denaturing gel. The reaction products are separated according to size and the sequence can be read directly from the autoradiograph of the gel (Fig. 2.4-11, part B).

This sequencing method is extremely simple, accurate, and reproducible; however, it requires suitable primers and single-stranded DNA as templates. As has already been mentioned, primers are frequently obtained from double-stranded restriction endonuclease fragments. This is achieved by heating a mixture of the single-stranded template strand and the double-stranded restriction enzyme fragment to 100 °C, and letting the mixture reassociate at 67 °C in 1M NaCl. Heating dissociates the strands, while hybridisation at 67 °C results in the formation of double strands between the template DNA and that strand of the restriction endonuclease fragment which is complementary to the template. The recent advances in chemical DNA synthesis also allow suitable oligonucleotide primers to be synthesised. As sequencing proceeds, either new oligonucleotides are synthesised, or special oligonucleotides are used as universal primers in combination with special vectors (see Section 2.4.2.4).

Originally, the necessity of using single-stranded DNA as a template was regarded as a drawback, because double-stranded DNAs appeared to be more readily available and could be generated by restriction enzyme digestions; sequencing also required new primers for every 200 to 300 nucleotides. The decisive breakthrough has come with the use of filamentous single-stranded DNA phages. Before going into procedural details, the next section will first describe the growth cycle of these phages.

2.4.2.2 Growth Cycle of Single-stranded DNA Phages

Filamentous *E. coli* phages, such as M13, fd, and f1, have a single-stranded circular DNA genome. The sequences of fd (6 408 nucleotides), f1 (6 407 nucleotides), and M13 (6 407 nucleotides) are almost identical, showing 97% homology (Beck *et al.*, 1978; Van Wezenbeek *et al.*, 1980; Beck and Zink, 1981). The relative order of their ten genes (Fig. 2.4-13) is also almost identical, and these phages therefore often are referred to as the M13 class of filamentous bacteriophage. Incidentally,

Arthur Kornberg
Stanford

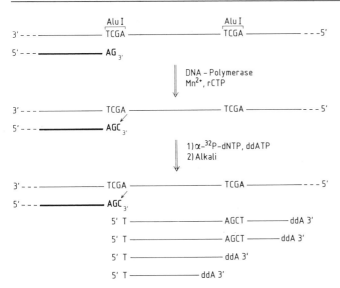

Fig. 2.4-12. Sequencing by primed enzymatic synthesis in the presence of ribonucleoside triphosphates. Synthesis initiates at the primer (marked by a thick line). In the presence of rCTP only rC is incorporated (arrow). Subsequent reactions are carried out as usual (*cf.* Fig. 2.4-11). After treatment with alkali which cleaves at the position of the ribonucleotide residue, a nested set of labelled molecules is obtained. They differ in lengths, but possess common 5' ends.

the "M" in M13 stands for the city of Munich, where M13 was discovered (Hofschneider, 1963). M13 genes comprise three functional units: the genes for replication (*II* and *V*), the virus coat genes (*IX*, *VIII*, *III*, *VI*), and genes for morphogenesis (*I* and *IV*). The known properties and functions of these genes are listed in Table 2.4-1. Two non-coding regions are found between genes *VIII* and *III*, and *IV* and *II*, respectively; the former region contains a *rho*-independent central terminator of transcription (*cf.* Fig. 7-33), and the latter an origin of DNA replication (position 5782). Apart from a short region comprising approximately 20 nucleotides shared by genes *I* and *IV*, the filamentous phages, in contrast to icosahedral phages, do not possess overlapping genes. M13 phages grow only on "male" strains of *E. coli*, *i.e.*, those strains which harbour an F episome. This conjugative plasmid (and derivatives thereof) carries a gene which codes for the sex pilus, a special structure on the surface of the bacterial cell, which serves as phage receptor, and which is the site of initiation of infection.

After adsorption, the single-stranded phage DNA, the designated (+) strand, enters the cell and is rapidly converted via the early intermediates (RI$_e$ forms) into the double-stranded replicative forms RF I and II (RF I and II; consisting of "+" genomic strand and "–" complementary strand; Fig. 2.4-14). Up to 100 RF I molecules per cell are produced within a few minutes, and can be isolated from infected cells like plasmids. Concomitantly with the synthesis and accumulation of gene *V* protein, which is a single-strand-specific DNA binding protein, replication of the RF DNA proceeds more and more asymmetrically. The (+) strand formed by rolling circle replication is complexed with gene *V* protein molecules, blocking the synthesis of the complementary (–) strand. The corresponding intermediates have been designated RI$_l$ (l = late). The only structures synthesised in the late phases of infection are

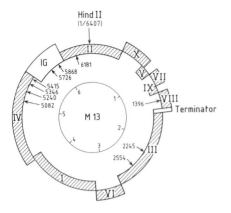

Fig. 2.4-13. Genetic map of bacteriophage M13. The single *Hind* II site (GTT/AAC) is used as a reference point. M13 genes are marked by Roman numerals. Transcription proceeds clockwise. The box marked IG signifies the 507 base pairs of the intergenic region between positions 5501 and 6007 (see Fig. 2.4-15). The numbers in the inner circle are the co-ordinates of the ten *Hae* III sites.

nucleoprotein complexes, which consist of single-stranded (+) strand DNA and gene *V* protein. The core protein, which is the product of gene *VIII*, is finally attached to this complex, and virus particles are then released from the infected bacteria. In contrast to lytic infections with other phages, host cells survive M13 infections, since M13 infection merely slows down the growth rate of the cells by a factor of approximately three. This slowed growth is also the mechanism underlying plaque formation. M13 plaques constitute regions of reduced bacterial growth rather than areas of virus-induced bacterial lysis, and therefore appear to be more translucent than neighbouring regions. Approximately 1 000 phage particles are produced per cell, which corresponds to 5 to 10 micrograms of single-stranded DNA per ml of bacteria with an OD_{600} of 1. The length of the phage particles depends on the length of the phage DNA. In principle, the length of the DNA packagable in this system should be unlimited;

Table 2.4–1. Genes of bacteriophage M13.

Gene	Start (position)	Number of amino acids	Function	Copy number per virion or per cell
II	6 007	410	initiation of DNA-replication; cleaves (+) RF-strand between positions 5781 and 5782	–
X	496	111	unknown	–
V	843	87	ss-DNA-specific protein	100 000
VII	1 108	33	unknown	
IX	1 206	32	capsid protein	<10
VIII	1 301	50 (precursor = 73)	major capsid protein	2 700
III	1 579	406 (precursor = 424)	pilot protein	5
VI	2 856	112	capsid protein	5
I	3 197	348	morphogenesis	–
IV	4 221	426	morphogenesis	–
IG	5 501–6 007	–	initiation of DNA replication; intergenic region	–

(from Beck and Zink, 1981; ss-DNA indicates single-stranded DNA.)

however, practical experience shows that insertions of more than 2 000 nucleotides result in a markedly reduced growth rate in infected cells. This is reflected in an altered plaque morphology, particularly in the production of smaller plaques. Cloning of small DNA fragments does not influence plaque formation. For this reason filamentous phages are exceedingly well-suited for the production of single-stranded recombinant DNA molecules carrying inserts not longer than 2 000 bp. Larger insertions can be obtained with single-stranded plasmid vectors derived from M13 structural elements (*cf*. Section 4.1.5).

2.4.2.3 Filamentous Phages as Vectors

Non-essential regions in a virus genome are an absolute prerequisite for the development of vector systems. It is only in these regions that the viral genome can be manipulated without influencing the viability of viral particles. A single non-essential region has been located in the M13 genome. It is located within the intergenic region between genes *II* and *IV* which comprises 507 nucleotides (Fig. 2.4-15). This observation is the basis for the development of cloning vectors in the M13 system. This work provides an object lesson in experimental approaches to such problems (Messing *et al.* 1977); similar systems have also been described for fd DNA (Herrmann *et al.*, 1980).

For the development of M13 vectors, RF I molecules were partially digested with *Hae* III (GG/CC) to yield singly cut full-length molecules. There are ten cleavage sites for this enzyme on M13 DNA, therefore, the ends of these linear

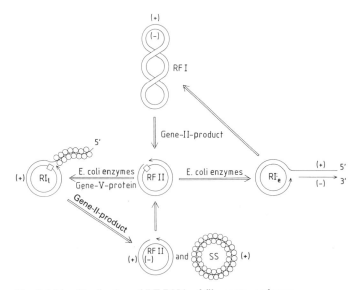

Fig. 2.4-14. Replication of RF DNA of filamentous phages.
Double-stranded RF I DNA is specifically nicked by the phage gene *II* product between position 5781 and 5782 in the intergenic region. The resulting RF II DNA contains one molecule of the gene *II* product bound to the (−) strand. Ensuing rolling circle replication is catalysed by three host-specific enzymes and results in the formation of RI$_e$ and RI$_l$ forms in the early and late phases of replication, respectively. (−) strands which are used as templates for mRNA synthesis are synthesised from the early intermediate RI$_e$. During the late phase, the synthesis of (−) strands on RI$_l$ forms is prevented by binding of gene *V* product to the overhanging (+) strand. During replication one molecule of the gene *II* product remains attached to the (−) strand. After the first round of replication it serves as a signal for a second molecule of the gene *II* product to cleave and circularise the displaced (+) strand (*cf*. also Meyer and Geider, 1982).

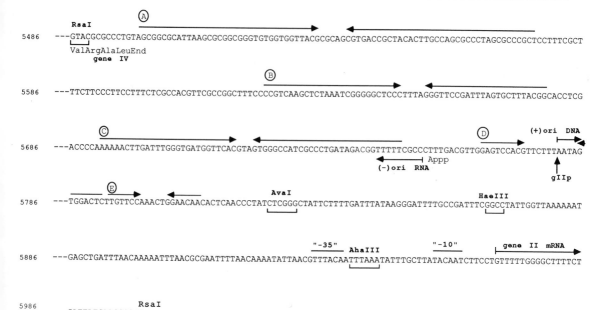

```
        RsaI                  Ⓐ
5486  ---GTACGCGCCCTGTAGCGGCGCATTAAGCGCGGCGGGTGTGGTGGTTACGCGCAGCGTGACCGCTACACTTGCCAGCGCCCTAGCGCCCGCTCCTTTCGCT
        ValArgAlaLeuEnd
           gene IV

                                              Ⓑ
5586  ---TTCTTCCCTTCCTTTCTCGCCACGTTCGCCGGCTTTCCCCGTCAAGCTCTAAATCGGGGGCTCCCTTTAGGGTTCCGATTTAGTGCTTTACGGCACCTCG

              Ⓒ                                                              Ⓓ            (+)ori  DNA
5686  ---ACCCCAAAAAACTTGATTTGGGTGATGGTTCACGTAGTGGGCCATCGCCCTGATAGACGGTTTTTCGCCCTTTGACGTTGGAGTCCACGTTCTTTAATAG
                                                        (-)ori  RNA    Appp
                                                                                           gIIp
              Ⓔ                         AvaI                                   HaeIII
5786  ---TGGACTCTTGTTCCAAACTGGAACAACACTCAACCCTATCTCGGGCTATTCTTTTGATTTATAAGGGATTTTGCCGATTTCGGCCTATTGGTTAAAAAAT

                                "-35"   AhaIII        "-10"      gene II mRNA
5886  ---GAGCTGATTTAACAAAAATTTAACGCGAATTTTTAACAAAATATTAACGTTTACAATTTAAATATTTGCTTATACAATCTTCCTGTTTTTGGGGCTTTTCT

              RsaI
5986  ---GATTATCAACCGGGGTACATATGATT---
                        MetIle
```

Fig. 2.4-15. Sequence of the intergenic region of M13 DNA.
The sequence is that of the viral (+) strand in the 5' to 3' direction. Possible stem and loop structures are indicated by arrows and are marked "A" through "E". The cleavage site in the RF I (+) strand for the gene *II* protein is marked as gIIp. Initiation sites for minus (–) strand DNA synthesis ((–) *ori* RNA) and gene *II* mRNA are marked by wavy lines. The *Hae* III site at position 5868 is exploited for the *lac* DNA insertions in M13mp vectors (*cf.* Fig. 2.4-16) (Zagursky and Berman, 1984).

molecules could occur in any one of ten different positions. Foreign DNA was ligated to the molecular ends of those molecules. Only those recombinants in which the original *Hae* III cut had occurred in a non-essential region could be replicated as hybrid phages. Recombinant DNA molecules with inserts in essential regions of the genome could not be replicated and were lost from ligated DNA preparations. In our example, the linearised M13 RF molecules were ligated with a 789 bp *Hind* II fragment (GTPy/PuAC), containing parts of the *lacI* gene, the complete regulatory region of the *lac* operon, and the first 146 codons of the *lacZ* gene (Fig. 2.4-16). The

Joachim Messing
Piscataway

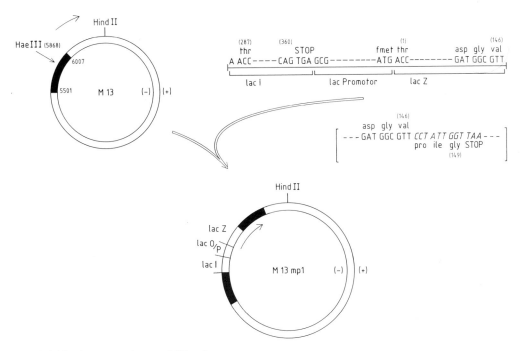

Fig. 2.4-16. Structure of vector M13mp1.
This vector was generated by joining an M13 RF DNA molecule partially cut with *Hae* III, and a 789 bp *Hind* II DNA fragment of the *lac* operon. The *Hae* III site is at position 5868 in the intergenic region (black bar). The *lac-Hind* II fragment contains the *C*-terminal region of the repressor (amino acids 287-360; 225 bp; Farabough, 1978), the entire *lac* regulatory (*OP*) region (125 bp), and amino acids 1-146 (438 bp) of the *N*-terminal region of the *lacZ* gene. These regions are not drawn to scale. The reading frame of the *lac* α-peptide extends into the M13 part (in italics) until it meets an ochre (UAA) codon at position 5879 still within the intergenic region (*cf*. Fig. 2.4-15). The actual α-peptide thus is 149 amino acids long. In other M13mp vectors this value increases by the contribution from the respective polylinkers, *e.g.*, by 14 amino acids in M13mp7. Only the (+) strand of the sequences of the *lac* insert is shown. In the recombinant vector M13mp1, the *lac* (+) strand and the (+) strand of M13 are co-linear, *i.e.*, transcription of the *lac* insert proceeds clockwise (arrows). Transducing particles, therefore package the *lac* coding (+) strand.

protein encoded by this fragment is known as the α-peptide. It is not a functional enzyme, but it displays α-donor activity (see below), and therefore can be detected easily after transformation of suitable host cells (Langley *et al.* 1975).

The host strain which allows the α-donor activity to be detected lacks the entire *lac-pro* region and is therefore Lac⁻ and Pro⁻. However, these bacteria harbour a conjugative F episome carrying a *pro* region complementing the Pro⁻ deficiency of the host, and simultaneously provid-

ing a selection against loss of the episome. This is essential, because the episome codes for the sex pilus required for M13 infection. In addition, the F episome also carries a *lac* operon with deletion *M15* in the *lacZ* gene. The *E. coli* strain containing this F plasmid therefore produces a plasmid-encoded enzymatically inactive β-galactosidase lacking the N-terminal amino acids 11-41 (of 1 021). This defect can be complemented by the protein fragment provided by the hybrid phage. This peptide, consisting of 146 amino-terminal amino acids from β-galactosidase and three amino acids derived from read-through into the M13 portion, possesses α-complementing activity. Alone, the protein encoded by either the phage or the bacterium is enzymatically inactive, yet together these proteins form an active β-galactosidase enzyme. Complementation can be detected by a suitable colour test. In practice, this is done by growing phage-infected bacteria on agar plates containing the colourless compound 5-bromo-4-chloro-indolyl-β-D-galactoside (Xgal), which is hydrolysed by β-galactosidase to yield a dark blue dye, 5-bromo-4-chloro-indigo (see also Fig. 7-6). Hybrid phages carrying the *Hind* II-*lac* fragment thus can be identified in mixtures with wild-type phage, because they form blue plaques on Xgal agar containing IPTG, which is an inducer of the episomal β-galactosidase.

The first vector isolated in the experiment described here was M13mp1. It carries the *lac* insertion in one of two *Hae* III sites in the intergenic region (position 5868; Fig. 2.4-13, 2.4-15 and 2.4-16). This demonstrates that the sequence around this *Hae* III site is, indeed, not essential for M13 replication.

M13mp1 itself is not a suitable cloning vector for the insertion of foreign DNA into the *lac* region, because it lacks suitable restriction endonuclease recognition sequences. Sequence analysis of the *lac* insert, however, then revealed that a base substitution of G to A within codon 5 of β-galactosidase (Fig. 2.4-17) would generate an *Eco* RI site (G/AATTC). Since it was known that O^6 methylation of G leads to base pairing of G with U, single-stranded M13mp1 (+) strand DNA was treated with the methylating agent N-methyl-N-nitrosourea (Gronenborn and Messing, 1978). The mutagenised DNA was used to transfect bacteria, and RF DNA isolated from these bacteria was subsequently linearised with *Eco* RI. Linearisation of DNA can be easily monitored in agarose gels and occurs only in adequately mutagenised RF molecules (Fig. 2.4-17). The linear molecules were isolated, recircularised using their *Eco* RI ends, and transfected again. Three clones were isolated, which had unique *Eco* RI sites at different positions in the hybrid phage genome. They were designated M13mp2, M13mp3, and M13mp4. The sequence of clone M13mp2, which is shown in Fig. 2.4-18, contains a new *Eco* RI site near the codons for amino acids 5 and 6 of β-galactosidase. The other two clones contained similar changes at the position of amino acid 119 of β-galactosidase and in gene *X* of the M13 region of the hybrid phages, respectively.

The M13mp2 vector can be used only for *Eco* RI cloning. In the course of further experiments, additional restriction sites have been introduced into the *lac* region by incorporating synthetic DNA fragments. Vector M13mp7, for example, contains a chemically synthesised 44 bp fragment flanked by *Eco* RI sites (Fig. 2.4-18). This new insert codes for two restriction sites each of *Eco* RI, *Bam* HI, *Sal* I, *Acc* I, and *Hinc* II.

```
                              thr met ile thr asp ser
                                1                10
M 13 mp 1              5'- - -ATG ACC ATG ATT ACG GAT TCA CT - - -3'
(+Strand)                                        CH₃

Complementary (-) strand   3'- - -TAC TGG TAC TAA TGC TTA AGT GA- - -5'

                                              Eco RI
M 13 mp 2             5'- - -ATG ACC ATG ATT ACG AAT TCA CA- - -3'
(+Strand)                     thr met ile thr asn ser
α-Peptide                      1   2   3   4   5   6
```

Fig. 2.4-17.   Introduction of an *Eco* RI site into vector M13mp1.
The (+) strand of M13mp1 contains a potential *Eco* RI site between codons 4 and 6 of the *lacZ* gene (bracketed). A GT pair can be formed during the synthesis of the complementary (−) strand by G methylation. Newly replicated (+) strands contain the desired *Eco* RI site. The substitution of aspartic acid by asparagine at position 5 of β-galactosidase does not influence α-complementation.

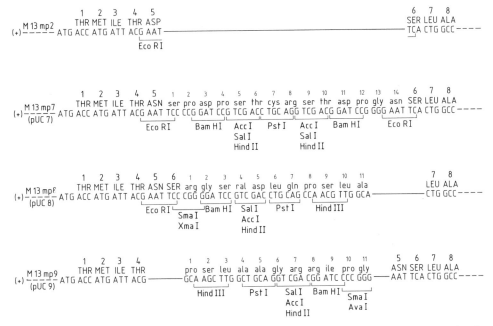

Fig. 2.4-18. Structure of polylinker insertions in vectors M13mp2, M13mp7, M13mp8, and M13mp9. Shown is the strand which is co-linear with the (+) strand of M13 DNA. Amino acids of β-galactosidase are written in upper case, and those of insertions in lower case letters. The insertions of derivatives M13mp12, mp13, mp18 and mp19 are shown in Fig. 7-58 as pUC plasmid insertions.

Since these sites do not occur anywhere else in the vector molecule, M13mp7 is ideal for cloning foreign DNA (Messing, 1981). As shown in Fig. 2.4-18, the arrangement of cleavage sites in M13mp7 is symmetrical, and all sites, with the exception of the *Pst*I site, occur twice. New improved vectors, such as M13mp8 and M13mp9, contain a synthetic DNA fragment, in which all restriction sites are unique (Messing *et al.*, 1981; Messing and Vieira, 1982) (Fig. 2.4-18). Fragments possessing different ends can therefore be cloned and forced into a certain orientation with respect to the M13 genome. The relative order of the recognition sites on the M13mp8 polylinker with respect to the M13 genome is reversed in M13mp9. This allows one to obtain either strand of a cloned fragment linked to the (+) strand of M13 DNA, and hence as single-stranded DNA packaged into phage particles. The polylinkers in vectors M13mp7-9, which are as long as 57 bp, do not interfere with α-complementation. However, the insertion of long stretches of foreign DNA will abolish α-complementation. Recombinants therefore can be detected as phages which no longer form blue plaques when plated in the presence of Xgal and IPTG. It should be noted that there are also several restrictions which have to be kept in mind. The N-terminal region of β-galactosidase is not essential and can therefore be modified considerably without impairing either enzymatic activity or α-complementation. Insertions of foreign DNA will thus only lead to the formation of white plaques if the reading frame of the α-peptide is destroyed, or if the insert contains stop codons in the correct reading frame. If the length of a DNA insert (in bp) is a multiple of three and if the insert does not contain any stop codons blue plaques will still be formed. Such insertions may be as long as several hundred base pairs. This property of M13 vectors can be exploited to detect and select for mutations creating new stop codons or altering the reading frame, *i.e.*, leading to frameshift mutations (Traboni *et al.* 1982) (see also Chapter 12).

2.4.2.4 Filamentous Phages in DNA Sequencing

The M13 vectors described above can also be used for sequencing DNA if the DNA fragment in question is first cloned in one of these vectors, for example, in M13mp8. These clones are then used to transform suitable bacteria by the usual calcium chloride technique (see also Section 4.1.7). White plaques are isolated and used to re-infect other bacteria. Single-stranded DNA isolated from these phages is sequenced by the chain termination method. A universal primer, the 3' hydroxyl end of which is positioned in the immediate vicinity of the insertion, has been devised for this purpose (Fig. 2.4-19). Sequencing from this primer permits approximately 300 to 400 nucleotides of the insert to be sequenced directly (Messing *et al.* 1981). The primer can be used for sequencing any DNA insertion, because it is located very close to, but outside, the polylinker cloning sites. The synthetic universal primer, which is also known as the master primer, is now commercially available. Its sequence is 5'-CCCAGTCACGACGTT-3'.

As shown in Fig. 2.4-19, this sequence is complementary to the coding sequences for amino acids 12-17 of β-galactosidase. DNA polymerase I-directed synthesis beginning at the 3' hydroxyl end of the terminal T residue will reach the recombinant DNA insert after only 16 nucleotides. It is the position of the primer in the immediate vicinity of the insertion which makes it truly universal.

Another notable feature of the M13 vector systems is the fact that the *lac* sequence in the single-stranded phage DNA always represents the (+) strand, *i.e.*, it corresponds with that of the mRNA. It is always the (+) strand which is packaged into mature phage particles; however, when M13mp1 was constructed (see Fig. 2.4-16), the *lac* region could have been inserted in either orientation. When the first Lac$^+$ isolates were studied it was necessary to distinguish between these two orientations, and this was readily achieved by restriction enzyme analysis of the RF forms. In all M13mp vectors, the (+) strand carries the same sequence information which appears as *lac* mRNA. If fragments with identical ends are cloned in vectors M13mp1-9, inserts will always be found in both orientations. Isolated phages thus always contain single-stranded DNA with the (+) strand of the *lac* region linked to either one of the two complementary strands of an insert. This is quite useful because it allows both strands to be sequenced, and of course, this increases accuracy.

Vectors such as M13mp8 and M13mp9 (Fig. 2.4-18), also allow fragments with different ends to be inserted in only one orientation. In this case, phage clones always will contain only the one strand of the insert which is associated with the (+) strand of the phage DNA in the RF form. It has been mentioned already that M13mp8 and M13mp9 contain the same cleavage sites, but in reversed order. It is therefore possible to choose which of the two complementary strands of the DNA insert is cloned in single-stranded form, by simply employing the appropriate vector.

If insertions are larger than 300-400 base pairs, they cannot be sequenced from a single primer;

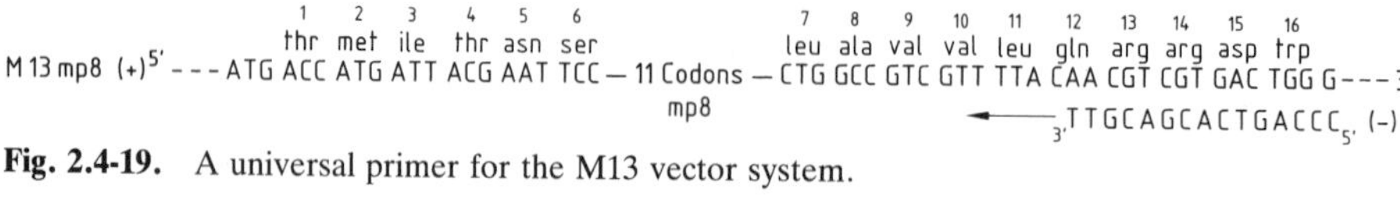

```
               1   2   3   4   5   6              7   8   9  10   11   12   13  14  15  16
              thr met ile thr asn ser            leu ala val val leu gln arg arg asp trp
M13 mp8 (+)5'---ATG ACC ATG ATT ACG AAT TCC—11 Codons—CTG GCC GTC GTT TTA CAA CGT CGT GAC TGG G---3'
                              mp8                          ⟵————3'TTGCAGCACTGACCC5' (-)
```

Fig. 2.4-19. A universal primer for the M13 vector system.

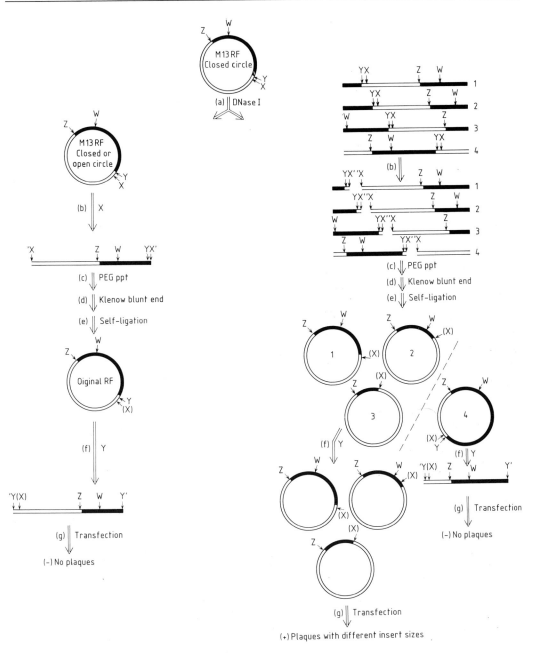

Fig. 2.4-20. A strategy for systematic subcloning.
Solid lines indicate DNA inserts into the polylinker. Z, W, Y, Z represent recognition sites for different restriction endonucleases. Sites X, Y and Z are part of the polylinker, while W is located within the insert. Open lines indicate M13 DNA. (X) denotes a restriction site altered by treatment with the Klenow fragment of DNA polymerase I (Lin et al., 1985). See text for details.

however, the currently available methods for the chemical synthesis of oligonucleotides allow the rapid synthesis of new primers from which another 300-400 nucleotides can be sequenced (see also Section 2.3); in addition, it is also possible to do shot-gun sequencing by randomly cloning restriction enzyme fragments in M13 vectors without having any information about their arrangement on a genome. The cloned DNA fragments are then sequenced from the master primer, and sequence data are later processed by suitable computer programs in order to identify overlapping sequences. The 8 031 base pairs of cauliflower mosaic virus, for example, have been determined by shot-gun sequencing and computer-aided sequence analysis (Gardner *et al.* 1981). While 80% of the entire sequence was, indeed, determined by random sequencing, the remaining 20% required a direct strategy for the identification of the missing, non-overlapping regions.

In order to overcome the difficulties associated with shot-gun sequencing, especially the filling-in of small gaps within otherwise contiguous sequences, several non-random subcloning strategies were developed. One such strategy for systematic subcloning is shown in Fig. 2.4-20 (Hong, 1982; Lin *et al.*, 1985). A DNA fragment to be subcloned for subsequent sequencing is cloned into an M13mp vector. The RF DNA, with the insert in either orientation, is treated with DNAseI under conditions which lead to linearisation at random positions. In order to obtain a nested set of fragments containing the intact M13 portion (required for plaque formation), but carrying insertions of variable length extending from site Z to site Y at the other end of the insert, the linearised mixture of fragments is treated with enzyme X. After treatment with polyethylene glycol, which precipitates only larger DNA fragments, these larger fragments are resuspended and religated to form circular DNA molecules. Among these molecules are those (1, 2 and 3) which have been cut by DNAse I within the insert and which thus do no longer contain site Y. Those which have not been cut with DNAse I in the first

place (reaction pathway to the left) or which have been cleaved within the M13 portion (molecule 4 in the pathway to the right) still contain site Y. Upon treatment of the whole reaction mixture with enzyme Y, the latter molecules will be linearised while the former ones with the nested set of inserts extending from site Z will remain circular. Transfection efficiencies with linear DNA are low or nil as compared to circular DNA, thus only the desired molecules will form plaques after transformation. Since all inserts have a common terminus (Z), sequencing of different clones by primed sythesis starting from site X must eventually cover the whole insert.

In the past, the application of single-stranded phage vectors was limited by reservations concerning their biological safety. It has been pointed out already that M13 and its derivatives grow only on *E. coli* strains harbouring F episomes. Conjugative plasmids are not permitted in EK1 host vector systems (see Chapter 13). The conjugative properties of these plasmids are dramatically reduced by introducing one or more mutations into their transfer genes. Tra⁻ mutants, such as F'*traD36*, still synthesise the sex pili required for phage absorption, but they can neither conjugate nor can they transfer DNA. Such M13 host/vector systems are, therefore, acceptable as EK1 vectors.

2.4.2.5 Plasmids in DNA Sequencing

The concept of the universal primer can also be applied to sequencing of double-stranded DNA molecules. There are, in fact, instances where this approach may be advantageous or even necessary. Time can be saved, for example, if a series of plasmid constructions containing mutated insertions can be analysed directly. Moreover, there are cases in which certain DNA sequences cannot be stably maintained and replicated in M13.

A variety of synthetic primers have become available today which are complementary to DNA sequences in the neighbourhood of unique restriction enzyme cleavage sites of plasmid

P$_2$
          3600   Pst I   3620 3'←—TGCGAGCAGCAAA 5'
5'–––TGCGCAACGTTGTTGCCATTGCTGCAGGCATCGTGGTGTCACGCTCGTCGTTTGG–––3'
3'–––ACGCGTTGCAACAACGGTAACGACGTCCGTAGCACCACAGTGCGAGCAGCAAACC–––5'
     5' CGCAACGTTGTT —→3'
       P$_1$

Fig. 2.4-21. Universal primers in plasmid pBR322.
Shown is a portion of the sequence of the β-lactamase gene around the *Pst*I site (CTGCA/G) at position 3609.
Primer sequences are marked P$_1$ and P$_2$, respectively.

pBR322 (Wallace *et al.* 1981). Fig. 2.4-21 shows the sequence around the *Pst*I site of plasmid pBR322 and the two primers P1 and P2, which permit an insertion into the *Pst*I site to be sequenced from both ends. In practice, the recombinant molecules are first linearised by cleavage at a site which does not influence the insert. The DNA strands are then separated by heating and subsequently reassociated in the presence of the primers. The mixture of primers and templates can then be used for sequencing by the chain termination technique. A major drawback of this method is that linear DNA of low complexity renatures rapidly, thereby leaving too little single-stranded template for the sequencing reactions. The problem has been solved by denaturing closed circular rather than linear DNA completely with alkali at pH above 12. Apparently, this material does not easily renature to its original structure, but rather to a network of hybrid molecules which expose sufficient amounts of single-stranded DNA (Chen and Seeburg, 1985).

A useful set of rectors for double-stranded DNA sequencing is a series of pBR322 derivatives containing M13mp insertions. Each plasmid in the pUC family of vectors (Fig. 2.4-22; Vieira and Messing, 1982; Yanisch-Perron *et al.*, 1985), for example, contains the larger of two DNA fragments obtained by digesting pBR322 with *Eco* RI

(4363) and *Pvu* II (2066). This DNA fragment is 2,297 base pairs in length and contains a *Hae* II site at position 2349. A 433 bp *Hae* II DNA fragment of the *lac* operon has been inserted into this site after a partial digestion of the pBR322-derived fragment with *Hae* II. The *Hae* II *lac* operon fragment codes for only 59 amino acids of β-galactosidase as compared with the 146 amino acids of β-galactosidase encoded by the fragment inserted into M13mp2. The *lac* insert in the pBR322 *Hae* II site at position 2349 allows read-through from the *lacZ* reading frame into pBR322, resulting in the formation of an α-peptide consisting of the 59 *lacZ*-derived amino acids and an additional 29 amino acids encoded by pBR322 DNA. This α-peptide still shows α-complementation, and thus allows direct screening, by loss of α-complementation, for inserts into vectors containing this fragment. The *Hae* II *lac* fragments of pUC7, pUC8, and pUC9 have been obtained from the replicative forms of vectors M13mp7, M13mp8, and M13mp9, respectively. These shortened β-galactosidase fragments, therefore, contain the corresponding poly-linkers at the amino terminus of the β-galactosidase (see Fig. 2.4-18). The polylinkers serve the same functions as those in the M13mp vectors, and, because of their structures, it is possible to use the same primers initially developed for the M13 vectors.

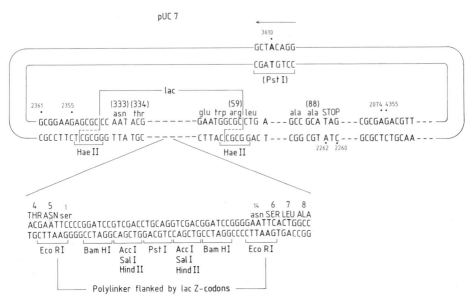

Fig. 2.4-22.   Structure of vector pUC7.

The pUC vectors are composed of the large *Eco* RI-*Pvu* II fragment from pBR322 (positions 4363-2066), and of a 433 bp *Hae* II fragment of the *lac* operon cloned into the *Hae* II site of the pBR322 sequence at position 2349 in the same orientation as the β-lactamase gene. The construction results in small deletions at the termini of the pBR322 portion, such that the fusion occurs between nucleotides 4355 and 2074. The *Acc* I, *Pst* I and *Hinc* II sites contained within the pBR322 sequence, which also occur within the linker, have been removed by mutagenesis. The GC pair at position 3610, for example, was replaced through EMS mutagenesis by an AT pair, thus eliminating the *Pst* I site. The *Hae* II *lac* fragment carries the coding sequences for amino acids 333-360 of the *lacI* gene (lower case letters), the regulatory (*OP*) region and coding sequences for 59 *N*-terminal amino acids of β-galactosidase (lower case letters). The polylinker insertion in pUC7 (blown-up insert; amino acids are indicated by lower case letters; flanking β-galactosidase amino acids by upper case letters) is derived from M13mp7. In other pUC derivatives the polylinkers derive from the corresponding M13mpX plasmids (*cf.* also Fig. 7-58). The α-peptide is 103 amino acids long; in addition to the 59 *N*-terminal amino acids of β-galactosidase there are 14 amino acids encoded by the polylinker region (19 amino acids in the polylinker regions of pUC18 and 19) and 29 amino acids obtained by read-through into the adjacent pBR322 portion (lower case letters). Numbers are pBR322 co-ordinates. (Vieira and Messing, 1982; Yanisch-Perron *et al.*, 1985).

References

Beck, E., Sommer, R., Auerswald, E.A., Kurz, Ch., Zink, B., Osterburg, G., Schaller, H., Sugimoto, K., Sugisaki, H., Okamoto, T., and Takanami, M. (1978). Nucleotide sequence of bacteriophage fd DNA. Nucleic Acids Res. 5, 4495-4504.

Beck, E., and Zink, B. (1981). Nucleotide sequence and genome organisation of filamentous bacteriophage f1 and fd. Gene 16, 35-58.

Biggin, M.D., Gibson, T.J., and Hong, G.F. (1983). Buffer gradient gels and ^{35}S label as an aid to rapid DNA sequence determination. Proc. Natl. Acad. Sci. USA 80, 3963-3965.

Brown, N.L. (1978). A primed-synthesis method for ribosubstitution of DNA at a single site. FEBS Letters. 93, 10-15.

Chen, E.Y., and Seeburg, P.H. (1985). Supercoil sequencing: A fast and simple method for sequencing plasmid DNA. DNA 4, 165-170.

Church, G., and Gilbert, W. (1984). Genomic sequencing. Proc. Natl. Acad. Sci. USA 81, 1991-1995.

Ephrussi, A., Chruch, G.M., Tonegawa, S., and Gilbert, W. (1985). B lineage-specific interactions of an immunoglobulin enhancer with cellular factors *in vivo*. Science 227, 134-140.

Farabaugh, P.J. (1978). Sequence of the *lacI* gene. Nature 274, 765-769.

Gardner, R.C., Howarth, A.J., Hahn, P., Brown-Buedi, M., Shepherd, R.J., and Messing, J. (1981). The complete nucleotide sequence of an infectious clone of cauliflower mosaic virus by M13mp7 shotgun sequencing. Nucleic Acids Res. 9, 2871-2888.

Gronenborn, B., and Messing, J. (1978). Methylation of single-stranded DNA *in vitro* introduces new restriction endonuclease cleavage sites. Nature 272, 375-372.

Hayward, G.S. (1972). Gel electrophoretic separation of the complementary strands of bacteriophage DNA. Virology 49, 342-344.

Herrmann, P., Neugebauer, K., Pirkl, E., Zentgraf, H., and Schaller, H. (1980). Conversion of bacteriophage fd into an efficient single-stranded DNA vector system. Mol. Gen. Genet. 177, 231-242.

Hofschneider, P.H. (1963). Untersuchungen über "kleine" *E. coli* K12 Bakteriophagen. Z. Naturforschung 18b, 203-210.

Hong, G.F. (1982). A systematic DNA sequencing strategy. J. Mol. Biol. 158, 539-549.

Kornberg, A. (1980). In: "DNA Replication", pp. 139-142. W.H. Freeman and Company, San Francisco.

Langley, K.E., Villarejo, M.R., Fowler, A.V., Zamenhof, P.J., and Zabin, I. (1975). Molecular Basis of β-Galactosidase α-Complementation. Proc. Natl. Acad. Sci. USA 72, 1254-1257.

Lin, H.C., Lei, S.P., and Wilcox, G. (1985). An improved DNA sequencing strategy. Anal. Biochem. 147, 114-119.

Maxam, A.M., and Gilbert, W. (1977). A new method for sequencing DNA. Proc. Natl. Acad. Sci. USA 74, 560-564.

Maxam, A.M., and Gilbert, W. (1980). Sequencing end-labelled DNA with base-specific chemical cleavages. Methods in Enzymology 65, 499-560.

Messing, J., Gronenborn, B., Müller-Hill, B., and Hofschneider, P.H. (1977). Filamentous coliphage M13 as a cloning vehicle. Insertion of a *Hinc* II fragment of the *lac* regulatory region in M13 replicative form *in vitro*. Proc. Natl. Acad. Sci. USA 74, 3642-3646.

Messing, J., Crea, R., and Seeburg, P.H. (1981). A system for shotgun DNA sequencing. Nucleic Acids Res. 9, 309-321.

Messing, J. (1981). M13mp2 and derivatives: A molecular cloning system for DNA sequencing, strand specific hybridization, and *in vitro* mutagenesis. In:

"Recombinant DNA", Proceeding of the Third Cleveland Symposium on Macromolecules. Cleveland, Ohio, 22-26, June 1981. A.G. Walter (ed.). Elsevier Scientific Publishing Company, Amsterdam, The Netherlands.

Messing, J., and Vieira, J. (1982). A new pair of M13 vectors for selecting either DNA strand of double-digest restriction fragments. Gene 19, 269-276.

Meyer, T.F., and Geider, H. (1982). Enzymatic synthesis of bacteriophage fd viral DNA. Nature 296, 828-832.

Rüther, U., Koenen, M., Otto, K., and Müller-Hill, B. (1981). pUR222, a vector for cloning and rapid chemical sequencing of DNA. Nucleic Acids Res. 9, 4087-4098.

Sanger, F., Nickler, S., and Coulson, A.R. (1977). DNA sequencing with chain-terminating inhibitors. Proc. Natl. Acad. Sci. USA 74, 5463-5467.

Smith, A.J.H. (1980). DNA sequence analysis by primed synthesis. Methods in Enzymology 65, 560-580.

Southern, E. (1975). Detection of specific sequences among DNA fragments separated by gel electrophoresis. J. Mol. Biol. 98, 503-517.

Traboni, C., Cilibert, G., and Cortese, R. (1982). A novel method for site-directed mutagenesis: its application to an eukaryotic tRNA^{pro} gene promoter. The EMBO Journal 1, 415-420.

Van Wezenbeek, P.M.G.F., Hulsebos, T.J.M., and Schoenmakers, J.G.G. (1980). Nucleotide sequence of the filamentous bacteriophage M13 genome: Comparison with phage fd. Gene 11, 129-148.

Vieira, J., and Messing, J. (1982). The pUC plasmid, an M13mp7-derived system for insertion mutagenesis and sequencing with synthetic universal primers. Gene 19, 259-268.

Wallace, R.B., Jokuson, M.J., Suggs, S.V., Miyoski, K., Bhatt, R., and Itakura, K. (1981). A set of synthetic oligodeoxyribonucleotide primers for DNA sequencing in the plasmid vector pBR322. Gene 16, 21-26.

Yanisch-Perron, C., Vieira, J., and Messing, J. (1985). Improved M13 Phage cloning vectors and host strains: nucleotide sequences of the M13mp18 and pUC19 vectors. Gene 33, 103-119.

Zagursky, R.J., and Berman, M.L. (1984). Cloning vectors that yield high levels of single-stranded DNA for rapid DNA sequencing. Gene 27, 183-191.

3 Ligation of DNA Fragments

Another fundamental step in recombinant DNA technology is the ligation of foreign DNA molecules to vector DNA. This process involves the formation of four phosphodiester bonds, *i.e.*, two at each end of the molecule. In principle, such bonds can be formed either *in vitro* or *in vivo*. While the *in vitro* reaction is catalysed by enzymes and is carried out under carefully controlled reaction conditions, *in vivo* ligation takes place in the organism that has been chosen to harbour and replicate the recombinant DNA. Various cloning strategies have been developed for each of these two processes and will be discussed in the following sections.

3.1 Ligation with DNA Ligases

3.1.1 Properties of DNA Ligases

3.1.1.1 Isolation and Reaction Mechanism

The generation of phosphodiester bonds between neighbouring 3' hydroxyl ends and 5' phosphate ends of double-stranded DNA chains is catalysed by DNA ligases (Fig. 3-1). The enzyme isolated from *E. coli* is a polypeptide chain with a molecular weight of 75 kDa, which requires NAD^+ as cofactor. In contrast, the ligase encoded by phage T4 (68 kDa), as well as enzymes

isolated from mammalian and plant cells, require ATP as a cofactor and as an energy source. The initial step of the ligase reaction is the formation of an adenylate-enzyme complex involving either NAD^+ or ATP. Complex formation is accompanied by the release of NMN, or pyrophosphate, while an AMP residue is bound covalently to the ε-amino group of a lysine residue in the enzyme *via* a phosphoric acid amide bond (Fig. 3-2). The 5' phosphate residue of the DNA is subsequently activated by transfer of this adenylate residue. Formation of the phosphodiester bond is finally accomplished by a nucleophilic substitution of the 3' hydroxyl substituent at the activated 5' phosphate residue. AMP is released as a reaction product (Fig. 3-3). DNA ligases from *E. coli* and phage T4 can be obtained in highly purified form, because suitable overproducers have been developed. Overproduction of ligase has been achieved using either T4 mutants defective in DNA replication, or recombinant λ phages containing a ligase gene. The T4 mutants, due to their genetic defect, overproduce large quantities of "early" T4 enzymes, including ligase. The recombinant λ phages carry the genes for *E. coli* ligase, or T4 ligase, under the control of phage promoters (Cameron *et al.* 1975; Wilson and Murray, 1979). One recombinant, NM989, in which the T4 DNA ligase gene (gene *30*) is controlled by the late λ promoter P_R', is a particularly efficient producer of T4 DNA ligase (Fig. 3-4) (Murray *et al.*, 1979). In this case, the T4 insertion is approximately 3.2 kb in length. It comprises the ligase gene with its own

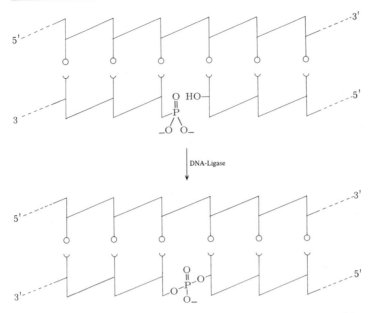

Fig. 3-1. Formation of a phosphodiester bond between a 5′ phosphate residue and a 3′ hydroxyl residue in a nicked double-stranded DNA molecule.

promoter and contains four *Eco* RI sites. In the phage genome itself, the ligase gene is transcribed from a 0.4 kb *Eco* RI fragment towards another 2.2 kb *Eco* RI fragment. The T4 gene *30* promoter is not very efficient and is far excelled by λ promoter P'_R, which predominantly drives synthesis in the λ NM989 recombinant (see Section 4.2). In this construction, up to 10% of the total protein in infected bacteria may be T4 ligase. This corresponds to an overproduction by a factor of 500 to 2 000. Only a few steps are therefore required for the purification of ligase (Tait *et al.* 1980). Similar protocols have been worked out for *E. coli* DNA ligase (Panasenko *et al.* 1978).

3.1.1.2 Substrate Specificity

The DNA ligases isolated from *E. coli* and T4 differ not only in their coenzyme requirements but also in their substrate specificities. The physiological substrate for both enzymes is thought to be the breakage point at a phosphodiester bond between neighbouring 3′ hydroxyl and 5′ phosphate ends still held together by an intact complementary strand (Fig. 3-5; type A). Another substrate for either enzyme contains the open and staggered phosphodiester bonds formed through reassociation of the protruding termini of different DNA molecules generated by digestion

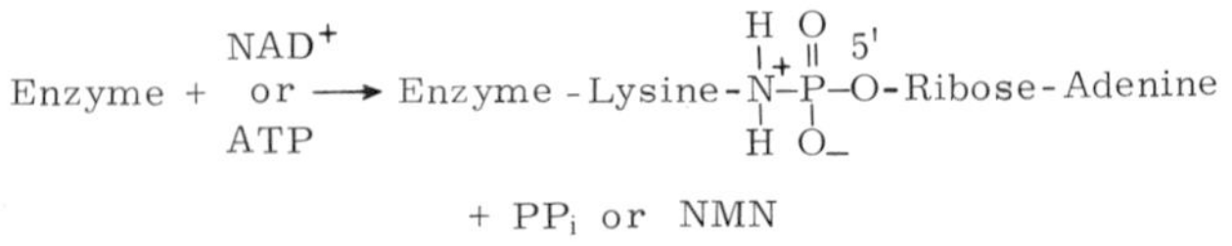

$$\text{Enzyme} + \begin{array}{c} \text{NAD}^+ \\ \text{or} \\ \text{ATP} \end{array} \longrightarrow \text{Enzyme - Lysine-} \overset{\overset{\text{H}}{|}}{\underset{\overset{|}{\text{H}}}{\text{N}}} \overset{\overset{\text{O}}{\|}}{\underset{\overset{|}{\text{O}^-}}{\text{P}}} \overset{5'}{-\text{O}-} \text{Ribose- Adenine}$$

$$+ \text{ PP}_i \text{ or NMN}$$

Fig. 3-2. Formation of an enzyme-adenylate complex as an intermediate of the DNA ligase reaction. The phage-encoded T4 DNA ligase requires ATP, and the *E. coli* DNA ligase NAD$^+$ as co-factor.

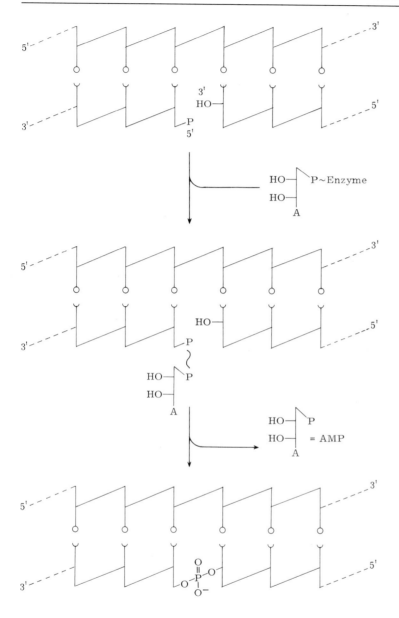

Fig. 3-3. Reaction mechanism of DNA ligases (see text for details).

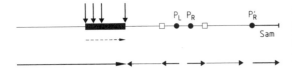

Fig. 3-4. Structure of phage expression vector NM989.
The T4 DNA ligase gene, shown as a thick bar, is inserted into the wild-type *Eco* RI site of bacteriophage λ DNA at position 43.8, immediately behind the late genes (*cf.* also Fig. 4.2-2). The four *Eco* RI sites of the insert (indicated by arrows) generate three T4 DNA fragments of 0.4, 0.5, and 2.2 kb. Horizontal arrows indicate the positions and directions of transcripts beginning at promoters P_L, P_R, and P'_R. Since the functional λ genome is circular, the transcript initiating from P'_R to the right also reaches the T4 DNA region. The broken line indicates the transcript initiating at the promoter of the T4 DNA ligase gene. Transcription initiating from promoters P_L and P_R terminates at positions marked by open squares and extends into the directions indicated by arrows in the presence of gene *N* product (*cf.* also Fig. 4.2-3; Murray *et al.*, 1979).

with certain type II restriction endonucleases. Here, the strands may be held together by base pairing between 2, 3, or 4 protruding nucleotides (Fig. 3-5; type B).

In addition, the virus-encoded T4 DNA ligase catalyses a number of other ligation reactions: it is capable, for example, of joining nicks in the RNA chains of double-stranded RNA-DNA hybrids, and it also anneals RNA termini with DNA strands (Fig. 3-5, type D). Such reactions may play an important biological role in the RNA-primed enzymatic synthesis of DNA. Neither of the two enzymes is capable of ligating single-stranded polynucleotides; however, this reaction can be carried out by another enzyme, RNA ligase, also induced and encoded by T4 DNA, which does not require a complementary strand for its activity (Silber *et al.* 1972).

One of the most remarkable properties of T4 DNA ligase, which distinguishes it from the bacterial DNA ligase, is its ability to accomplish *blunt end ligation* of double-stranded DNA molecules (Fig. 3-5; type C) (Sgaramella and Khorana, 1972). Little is known about the mechanism of

this reaction, apart from the fact that the reaction requires high concentrations of ligase, and that it is stimulated by T4 RNA ligase (Sugino *et al.* 1977).

The 68 kDa polypeptide chain of T4 DNA ligase is, however, capable of catalysing the reactions involving substrates of types A, B, and C (Fig. 3-5) by itself. This has been demonstrated by cloning of the T4 DNA ligase gene in λ phages and by the subsequent identification of the 68 kDa gene product in thermally induced *E. coli* lysates which did not contain any phage-encoded proteins other than T4 DNA ligase (Ferreti and Sgaramella, 1981a). The specificity of T4 DNA ligase for type A, B and C molecules can be switched to type A and B substrates simply by increasing ATP concentrations from 0.5 mM to 5 mM. At the elevated ATP concentration, the specificity of the enzyme for blunt ends is reversibly inhibited. An increase of the ATP concentration to 7.5 mM reversibly inhibits both activities. Therefore, DNA fragments with protruding and blunt ends easily can be ligated sequentially in the same assay mixture by altering the ATP concentrations appropriately (Ferreti and Sgaramella, 1981b).

3.1.1.3 Temperature Dependence

Reaction temperature is another important parameter which influences ligase reactions. The optimal ligation temperature is 37 °C. At this temperature, however, base pairing between complementary protruding ends of DNA fragments generated by restriction enzymes is very unstable, because such paired structures involve, at most, only four base pairs. As a compromise, a much more favourable temperature will therefore be one between the temperature optimum of the enzyme and the temperature which still guarantees that base pairing between protruding ends will be sufficiently stable. This optimal temperature increases with increasing G+C content at the site of ligation. At a DNA concentration of 100

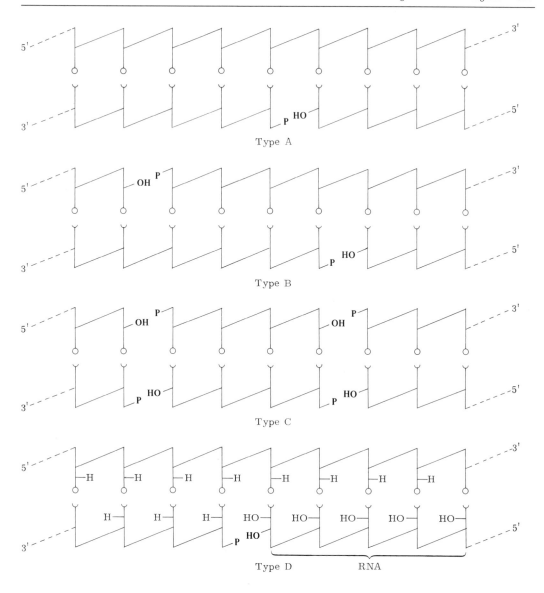

Fig. 3-5. Four substrates for DNA ligases.
E. coli DNA ligase only catalyses reactions with substrates of types A and B, while T4 DNA ligase is capable of using all four substrates.

micrograms per ml, the temperature optimum for circularisation of SV40 DNA, linearised by *Eco* RI (pAATT ends), was shown to be 12.5 °C (Dugaiczyk *et al.* 1975); at 25 °C, the yield of circularised SV40 DNA molecules still exceeds 50%, although base-paired AATT tetranucleotide termini melt between 5-6 °C (Ferreti and Sgaramella, 1981a).

3.1.1.4 Concentration Dependence

Several important parameters determine whether a ligase reaction will yield linear concatemers or circular structures. At constant ionic strength (salt concentration), preferences for intra- or intermolecular reactions depend on the length of the DNA fragments and on DNA concentrations. The smaller a DNA fragment at a given DNA concentration, the more intramolecular reactions, leading to circularisation, will be favoured. At constant lengths, the probability of circularisation increases with decreasing DNA concentrations. At a concentration of 50 µg/ml, a DNA fragment with a molecular weight of 10^6 Da (1.7 kb) will predominantly be converted to linear

concatemers, while the formation of circular monomers will be favoured at a concentration of only 10 µg/ml.

Of course it is usually the bimolecular reactions between vector and foreign DNA which are desired and preferred in gene cloning experiments. Since such reactions follow second-order kinetics, the fraction of recombinants will increase with increasing DNA concentrations. However, if vector and foreign DNA fragment differ in size, care must be taken to guarantee that bimolecular reactions are favoured, *i.e.*, that one end of a molecule will react more frequently with one end of another molecule than with its own other end.

The theory of intramolecular polycondensation reactions (Jacobson and Stockmayer, 1950) allows calculation of the concentration *j* of a DNA fragment, at which the initial velocity of the bimolecular reaction equals that of the monomolecular circularisation reaction. Equation 1 (Dugaiczyk *et al.* 1975) describes this relationship between the concentration parameter *j* and the size of a DNA fragment:

$$j \text{ (in g/l)} = 51.1 \times M_r^{-1/2} \tag{1}$$

DNA concentrations smaller than *j* drive the reaction towards circularisation; linear oligomerisation will predominate at concentrations higher than *j*. The *j* value of λ DNA (molecular weight 31 × 10^6 Da) is 10 µg/ml; for SV40 DNA, which is nine times smaller (molecular weight 3.3 × 10^6 Da), *j* is 28.4 µg/ml. For plasmid pBR322 DNA (molecular weight 2.6 × 10^6 Da), which is twelve times smaller than λ DNA, the value of *j* is 32 µg/ml, and a cDNA which is 1 000 bp in length (molecular weight 6 × 10^5 Da) has a *j* value of 65.5 µg/ml.

According to the theory of polycondensation, bimolecular reactions between monomers of dif-

Vittorio Sgaramella
Pavia, Italy

ferent length work best when the two reaction partners are both present in equimolar amounts. Hence, it follows that the length of the shortest DNA fragment will always determine the concentration of vector ends required to compete with the monomolecular circularisation of the shortest fragment. The relationship between the concentration of vector DNA and a foreign DNA to be cloned, is expressed by equation 2 (Williams and Blattner, 1980):

$$\text{"Foreign DNA" (in µg/ml)} = \qquad (2)$$
$$\frac{M_r \text{ (foreign DNA)}}{M_r \text{ (vector DNA)}} \times \text{vector DNA (in µg/ml)}$$

The following example from a cDNA cloning experiment in plasmid pBR322 will clarify the point. The minimal vector concentration j for pBR322 is 32 µg/ml (see above). An equivalent molar concentration of a cDNA which is 1 000 bp in length, would be $(0.6 \times 10^6)/(2.6 \times 10^6) \times 32$ = 7.38 µg/ml. This amount, however, is considerably less than the j value obtained by applying equation 1 (65.6 µg/ml). Note that at concentrations lower than 65.6 µg/ml, the undesired circularisation of this cDNA would be favoured, so its concentration should therefore be raised to at least 65.5 µg/ml. In order to obtain equimolar conditions again, the concentration of vector DNA would accordingly have to be raised to 65.5/7.38 × 32 = 284 µg/ml.

Table 3-1 lists the concentrations of λ and pBR322 DNAs which are required for an efficient competition with the free ends of foreign DNA fragments if reaction conditions are such that the rates of the monomolecular circularisation reactions and the rates of the bimolecular reactions between vector and foreign DNA molecules are equal. The larger the vector, the higher the DNA concentration must be in order to obtain the necessary concentration of DNA ends, because it is solely the concentration of such ends which drives the reaction. The vector concentrations obtained through these calculations are usually

Table 3–1. Minimal vector DNA concentrations for the cloning of foreign DNA.

Insert size (kb)	Vector DNA concentration (µg/ml)	
	Lambda-DNA	pBR322-DNA
0.5	9 750	813
1	3 447	286
2	1 218	102
3	663	55
4	430	36
5	308	26
10	109	9
15	59	5
20	38	3.2
30	21	1.8

Values were obtained from equations 1 and 2 analogous to the examples given in the text.

much higher than the j values (for example, 284 µg/ml as compared to 32 µg/ml for pBR322 DNA) and hence the bimolecular reaction is favoured.

By definition, the j values for foreign DNA obtained by applying Equation 1 specify DNA concentrations at which bimolecular and monomolecular reactions occur with the same rate. Experimental data demonstrate that the calculated j values should be at least two times higher to ensure that an efficient bimolecular reaction between vector DNA and foreign DNA will be accomplished (Dugaiczyk et al. (1975). In practice, however, such high concentrations often are unnecessary because a number of procedures have been developed which allow competing monomolecular circularisation reactions to be eliminated. DNA fragments smaller than 200 bp, for example, circularise very inefficiently. It is therefore quite unnecessary to use extremely high vector concentrations in order to compensate for the self-circularisation of the foreign DNA when this inserted DNA is small. An alternative approach is to prevent self-circularisation of the vector itself. Since ligase reactions require opposite 5' phosphate and 3' hydroxyl ends, the method of choice is to remove the 5' phosphate

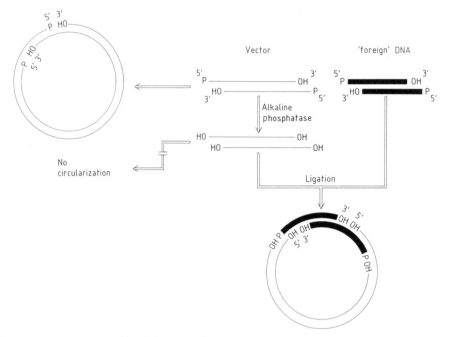

Fig. 3-6.   Treatment with alkaline phosphatase.
Alkaline phosphatase removes 5′ phosphate groups, and hence prevents re-circularisation of vector molecules. In the presence of foreign DNA a recombinant DNA molecule is generated in which two of four phosphodiester bonds can be closed by DNA ligase because the foreign DNA provides two 5′ phosphate residues. The two remaining open phosphodiester bonds are closed *in vivo* after transformation into the bacterial cell.

termini of linearised vector DNAs by treatment with alkaline phosphatase. Such DNA molecules can neither perform intermolecular nor intramolecular reactions with themselves. The lacking 5′ phosphate residues can only be provided by the foreign DNA fragment, and hence, circularisation can only be achieved by means of a bimolecular recombination event (Fig. 3-6). Although this leaves two open phosphodiester bonds in each molecule, these molecules still transform bacteria quite efficiently because cellular enzymes subsequently complete the double strand *in vivo*.

3.1.2 Ligation Strategies

3.1.2.1 Reconstitution of Cleavage Sites

There are several reasons why it is important to be able to cut out a cloned DNA fragment from a recombinant DNA molecule. If both the foreign DNA to be cloned and the vector DNA possess the same molecular ends because they have been generated by cleavage with the same enzyme (or an isoschizomer), cleavage sites are, by definition, regenerated during ligation. Very often, however, it is necessary to ligate DNA fragments with different and non-compatible ends, or blunt ends with either staggered 3′ or 5′ ends. Changing

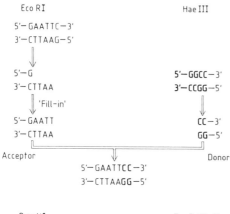

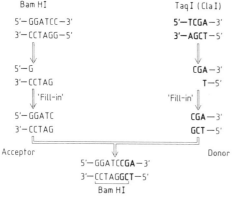

Fig. 3-7. Reconstruction of recognition sites. Ligase reactions between filled-in *Eco* RI or *Bam* HI acceptor sites and *Hae* III or filled-in *Taq* I donor sites regenerate the complete recognition sites for *Eco* RI and *Bam* HI, respectively.

a specific restriction site at will also is sometimes advantageous.

Incompatible DNA fragments with recessed ends can be ligated if the termini are first converted into flush ends. This can be achieved either by digesting protruding strands, or by filling-in protruding tails with complementary nucleotides. Single-stranded DNA regions are most conveniently digested with the single strand-specific endonuclease S1 from *Aspergillus oryzae*, which is capable of digesting 3′ as well as 5′ protruding ends, and which generates 3′ hydroxyl and 5′ phosphate termini, respectively (*cf.* Fig.

2.1-10c). The resulting flush ends are suitable substrates for ligases. Instead of removing nucleotides from the ends of a DNA molecule, it is also possible to fill-in single-stranded 5′ ends with T4 DNA polymerase (*cf.* also Figs. 2.1-10d and 2.4-5). Although this enzyme possesses a 3′ to 5′ exonuclease activity, which is active under these circumstances, there will be a turnover of only the terminal nucleotide in the presence of suitable (complementary) nucleoside triphosphates. T4 DNA polymerase is preferred to S1 nuclease, because, due to contamination with double strand-specific activities, the S1 reaction often leads to ill-defined reaction products. Alternatively the Klenow fragment of *E. coli* polymerase I or even AMV reverse transcriptase can be used.

If the four protruding bases of a hexameric cleavage site are filled-in, five of the six base pairs are restored. The recognition sequence can be fully restored by carefully selecting the terminal base pair of the DNA used for ligation in such a way that it supplies the missing sixth base pair. As shown in Fig. 3-7, a filled-in *Eco* RI site (G/AATTC) yields a 5′-GAATT-3′ sequence which, if annealed with a double-stranded molecule carrying a 5′-terminal C, is converted into a completely restored *Eco* RI site, 5′-GAATTC-3′. Such a 5′-terminal C can be provided, for example, by DNA fragments which have been digested with *Hae* III (GG/CC). The corollary is that the fusion of filled-in *Eco* RI ends with *Hae* III ends will restore *Eco* RI sites. There are many other combinations of such donors and acceptors which allow cleavage sites to be reconstituted; only two are shown in Fig. 3-7, and as can be seen here, flush ends required for ligation can also be obtained by filling-in donor DNAs.

Donoghue and Hunter (1982) have constructed the pBR322 derivative pDD52 which contains unique sites for the enzymes *Xba* I, *Eco* RI, *Xho* I, and *Hind* III. After digestion, the filled-in sites accept the four possible terminal base pairs A/T, C/G, G/C, and T/A (Table 3-2). In order to clone a DNA fragment possessing one *Fnu* DII

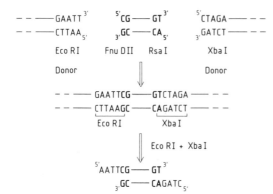

Fig. 3-8. Cloning of a DNA fragment with blunt *Fnu*DII-*Rsa*I ends in a donor plasmid with filled-in *Eco*RI or *Xba*I termini.
After cloning, the inserted DNA fragment can be cut out from the plasmid with *Eco*RI or *Xba*I. This approach does not alter the structure of the 3' ends of the inserted DNA fragment.

(CG/CG) and one *Rsa*I end (GT/AC), for example, pDD52 is cut with *Eco*RI and *Xba*I. These ends are filled-in, and the vector DNA is then ligated with the *Fnu*DII-*Rsa*I fragment. The desired recognition sequences (*Eco*RI and *Xba*I) are restored if the DNA fragment is inserted in the correct orientation (Fig. 3-8); the cloned fragment can then be cut out from the vector DNA using these two enzymes. It should be

noted, however, that the two 5' ends of the cloned DNA have now been structurally altered and carry the extensions derived from the *Eco*RI/*Xba*I cleavage. The 3' termini remain unchanged; single strands of fragments cloned in the way just described can thus be used as primers for DNA sequencing without any reservations (see following section).

3.1.2.2 Linker Technology

The synthesis of complementary DNA (cDNA) produces populations of DNA molecules with undefined and variable termini. Cloning of such molecules does not necessarily restore cleavage sites. This difficulty, as well as a few other problems, can be overcome using adaptor and/or linker molecules.

Linkers are short, chemically synthesised, self-complementary oligomers, which contain one or more restriction site(s). As exemplified by the decamer pCCGAATTCGG, which contains an *Eco*RI site, such molecules can be fused to any blunt-ended DNA molecule. Subsequent cleavage with a suitable restriction enzyme (*Eco*RI in this case) yields the corresponding protruding

Table 3–2.   Regeneration of restriction sites by ligation of filled-in termini with appropriate donor sites.

Enzyme	Acceptor sites in pDD52	Following digestion	After fill-in	Donor sites
*Xba*I	5'--TCTAGA-- 3'--AGATCT--	5'--T 3'--AGATC	5'--TCTAG 3'--AGATC	A--3' T--5'
*Eco*RI	5'--GAATTC-- 3'--CTTAAG--	5'--G 3'--CTTAA	5'--GAATT 3'--CTTAA	C--3' G--5'
*Xho*I	5'--CTCGAG-- 3'--GAGCTC--	5'--C 3'--GAGCT	5'--CTCGA 3'--GAGCT	G--3' C--5'
*Hind*III	5'--AAGCTT-- 3'--TTCGAA–	5'--A 3'--TTCGA	5'--AAGCT 3'--TTCGA	T--3' A--5'

The table displays four recognition sites of Type II restriction enzymes which have been modified by cleavage and fill-in reactions in such a way as to restore only five of the six base pairs. The four sites on vector pDD52 have been chosen in order to provide all four possible combinations of termini (A/T, T/A, C/G and G/C). Upon ligation to an appropriate donor site the complete recognition sites can thus be regenerated (Donoghue and Hunter, 1982).

ends required for cloning (Fig. 3-9). Oligomeric linkers are poor substrates for T4 DNA ligase. High concentrations of the enzyme, and a large excess of linker DNA are required, because the enzyme shows an increased apparent K_m value for the reaction with blunt ends. In practice, a 50 to 200-fold excess of linker is generally used; 200 micrograms (= 0.7 pmol ends) of a cDNA with a molecular weight of 600 000 (approximately 1 000 bp), for example, would then require approximately 500 micrograms (= 160 pmol ends) of an Eco RI decamer. The addition of linkers to cDNA molecules can be improved approximately tenfold by treating the cDNA with DNA polymerase I prior to the ligase reaction. Presumably, a cDNA which has been treated with nuclease S1 (Fig. 2.2-3) still contains single-stranded, recessed termini which are filled-in and converted to blunt ends by polymerase I treatment. Experimental details for these procedures have been described by Goodman and McDonald (1979), and Rothstein et al. (1979).

Difficulties may arise with the addition of linkers if the DNA to be cloned contains cleavage sites with the same specificity as the linker. The enzymatic cleavage of the linker termini following the ligation step (Fig. 3-9) would, of course, also affect these internal cleavage sites. One solution of this problem would be the specific methylation of internal cleavage sites by suitable modifying methylases prior to addition of the linker. In the case of the Eco RI linker described above, Eco RI methylase would have to be employed (cf. also Maniatis et al. 1978, for example). A variety of other methylases are commercially available today and permit a high degree of flexibility.

The problem of internal cleavage sites can also be overcome by using non-complementary single-stranded oligonucleotides which are known as adaptors (Bahl et al. 1978). These molecules can be used for cloning DNA molecules with different ends in conjunction with linkers or other adaptor molecules. The reassociation of a Bam HI adaptor (Fig. 3-10) with a Hpa II linker, for example, yields a short double-stranded DNA with one

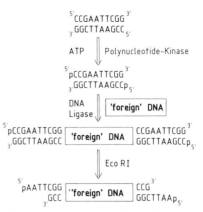

Fig. 3-9. Introduction of Eco RI recognition sites by Eco RI linkers.

blunt and one protruding end, which can be annealed with foreign DNA by blunt-end ligation so that the foreign DNA receives a protruding Bam HI end. In contrast to the example shown in Fig. 3-9, in which a linker was used, the DNA molecule which has been modified by such a linker-adaptor DNA does not have to be re-cut with the restriction enzyme; cloning of DNA molecules possessing internal Bam HI sites would therefore create no problems.

Cloning of DNA fragments obtained by digestion with a restriction enzyme X into a vector with another specificity, Y, can be achieved using

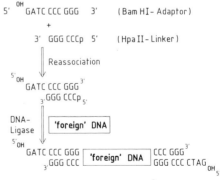

Fig. 3-10. Use of Bam HI adaptors and Hpa II linkers for cloning into a Bam HI-cut vector.
Bam HI adaptors carry a free 3' hydroxyl end which prevents self-polymerisation of the adaptor-linker molecules.

conversion adaptors. These adaptors are prepared from two different single-stranded oligodeoxynucleotides which carry the recognition sequence of different endonucleases at their 5′ ends, and a self-complementary DNA sequence at their 3′ ends. The example shown in Fig. 3-11 demonstrates that the reassociation of two such conversion adaptors yields a short DNA molecule with a *Bam* HI site at one end and an *Eco* RI site at the other end. The self-polymerisation of such an adaptor during the subsequent ligation reaction, *i.e.*, the generation of multiple tandem structures, can be prevented by phosphorylating only one of the two 5′ ends. The use of oligonucleotides with a self-complementary sequence at their 3′ ends also yields a novel cleavage site; in the example shown in Fig. 3-11, it is a new *Xho* I site. If the cloned DNA contains internal *Bam* HI or *Eco* RI sites, it can be cut out from the vector molecule with *Xho* I. Before an adaptor oligonucleotide is synthesised, it is advisable to test the DNA to be cloned for the absence of the recognition site in question.

Cloning with conversion adaptors requires high concentrations of adaptor molecules (10-20 μM)

and foreign DNA since double-stranded DNA molecules held together by only six base pairs are thermally labile. Such concentrations are not always practical. However, the use of such adaptors is often simplified by using them stepwise (Rothstein *et al.* 1979). As shown in Fig. 3-12, the first step is to mix and anneal two phosphorylated conversion adaptors; it is essential that these do not contain self-complementary regions. The reassociation products are ligated with DNA ligase and digested with *Bam* HI, which yields DNA fragments with internal *Hind* III and overlapping *Bam* HI ends. The ligated and *Bam* HI-cut molecules contain sixteen, instead of six, base pairs in the short duplex. The modified adaptors can be ligated with foreign DNA much more efficiently than the shorter pieces; following *Hind* III digestion, an adapted DNA molecule is obtained, which can be inserted into a suitably prepared vector cut with *Hind* III.

All linkers and adaptors described so far ligate 5′ protruding ends; however, some restriction enzymes, such as *Kpn* I, *Pst* I, or *Hha* I, generate 3′ protruding ends. These ends can be combined with the 5′ protruding ends of a vector, if

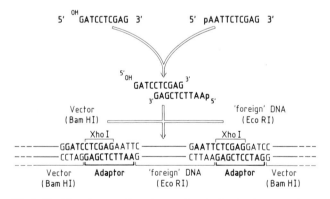

Fig. 3-11.   Use of conversion adaptors for cloning foreign DNA fragments with *Eco* RI termini into vectors cut with *Bam* HI.

The positions of the individual reaction partners are indicated by brackets. In contrast to the *Bam* HI adaptor shown in Fig. 3-10, conversion adaptors are self-complementary at their 3′ ends; therefore, they can reassociate with the respective other partner, as shown in the figure, but also with an identical partner. The latter reaction which leads to molecules with two identical termini is abortive since these molecules are not able to link the vector with the insert. In either case, the reassociation reaction generates a new restriction site, in this example an *Xho* I site (indicated in brackets).

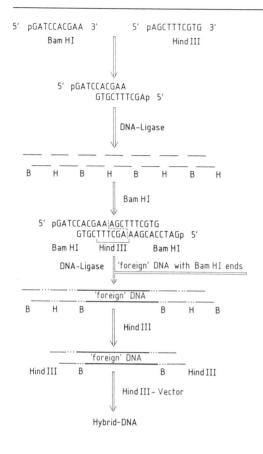

5' pGATCCACGAA 3'   5' pAGCTTTCGTG 3'
        Bam HI   Hind III

5' pGATCCACGAA
        GTGCTTTCGAp 5'

                DNA-Ligase

B   H   B   H   B   H   B   H

                Bam HI

5' pGATCCACGAAAGCTTTCGTG
        GTGCTTTCGAAAGCACCTAGp 5'
Bam HI   Hind III   Bam HI

DNA-Ligase   'foreign' DNA with Bam HI ends

                'foreign' DNA

B   H   B   B   H   B

                Hind III

                'foreign' DNA

Hind III   B   B   Hind III

                Hind III - Vector

                Hybrid-DNA

Fig. 3-12. Use of *Bam* HI and *Hind* III adaptors for cloning of *Bam* HI-cut foreign DNA into *Hind* III-cut vectors.

In contrast to the procedure described in Fig. 3-11, ligation and subsequent *Bam* HI cleavage of the adaptor first yields a double strand with 16 (6+10) complementary bases. Unlike adaptors used in Fig. 3-11 the adaptors of this example do not carry self-complementary, but only complementary sequences at their 3' ends; thus they can reassociate only with another complementary partner, but not with themselves. Due to the absence of self-complementary sequences the reassociation reaction only leads to a new internal restriction site after a tetrameric structure is formed. The adaptor described in Fig. 3-11 could also be employed in this strategy; however, upon reassociation and redigestion one would not only obtain double-stranded regions with 16 complementary bases, as above, but also those with $16+(10)_n$ bases. The final result, however, would be identical, since it would always create a new terminus for the insert to be compatible with the vector DNA.

single-stranded conversion adaptors are used which contain the complementary central parts of the cleavage sites in question. The decanucleotide 5'-AATTCCTGCA-3', for example, can be used to join a vector carrying *Pst* I termini with a DNA fragment generated by cleavage with *Eco* RI (Fig. 3-13). Although the two additional C residues in the adaptor introduce a gap in the resulting hybrid molecule, the advantage is that both recognition sequences are restored when the gap is closed *in vivo* after transformation.

Linkers and adaptors are widely used, and their applications cover, for example, such diverse

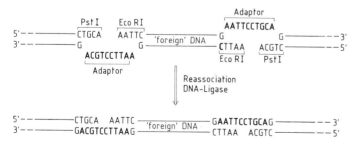

Fig. 3-13. Application of conversion adaptors in the ligation of DNA fragments with 3' and/or 5' protruding ends (see text for details).

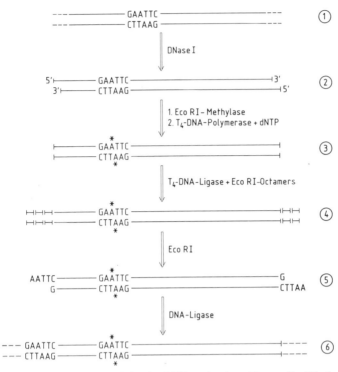

Fig. 3-14.   Construction of circular DNA molecules with new *Eco* RI sites.
Broken lines at the ends of DNA strands indicate circular molecules. The star(*) at the *Eco* RI site indicates A^6 methylation by *Eco* RI methylase. The *Eco* RI linker (short bars) in molecule 4 is added in multiple copies. The partial DNAse I treatment in the first step yields DNA molecules with staggered ends. These can be converted to blunt ends by treatment with T4 DNA polymerase. (Heffron *et al.*, 1978).

fields as cDNA cloning, specific mutagenesis, and the introduction of new cleavage sites into cloning vectors. Cloning of cDNA is certainly an outstanding example, because a cDNA library constitutes a mixture of DNA molecules with heterogenous ends. The other applications are becoming increasingly important. Heffron *et al.* (1978), for example, linearised circular DNA molecules unspecifically using pancreatic DNAse I, added *Eco* RI octamers to the linearised molecules, and cleaved with *Eco* RI. Upon recircularisation, new circular DNA molecules were obtained, each of which carried one new (*Eco* RI) cleavage site at a random site on the DNA (Fig. 3-14). Using these newly acquired *Eco* RI sites, a physical map was established and correlated to the observed bio-

logical phenotypes of the different clones. The DNA used by Heffron *et al.* was plasmid RSF1050 which is approximately 8 000 bp in length and carries transposon Tn3, coding for a penicillinase. This new method of mutagenesis made it relatively easy to identify those regions responsible for transposition, expression of the penicillinase, and initiation of DNA replication (Fig. 3-15).

Another choice example of the introduction of new restriction sites into vector molecules is the M13 vector system (Fig. 2.4-18). New multiple-purpose cloning sites comprising several cleavage sites were introduced into these vectors by using synthetic oligonucleotides. Vector M13mp7, for instance, contains such a polylinker, a 42 bp insertion with five adjacent cleavage sites for the

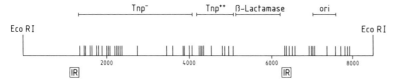

Fig. 3-15. Genetic map of plasmid RSF1050 containing transposon Tn3. The map was established by linker mutagenesis as described in Fig. 3-14. Short vertical lines indicate linker insertions. Mutations mapping in the region marked Tnp⁻ transpose at least 100 times less frequently than wild-type. Mutations in region Tnp⁺⁺ transpose more frequently than wild-type. Region Tnp⁻ must therefore be considered essential for transposition. Region Tnp⁺⁺ has a function in regulating transposition. *Ori* designates the origin of DNA replication of the plasmid. IR designates the presence of short inverted repetitions flanking the transposon. (Heffron *et al.*, 1978).

enzymes *Bam* HI, *Sal* I, *Acc* I, *Hinc* II, and *Pst* I. The application of this and other vectors has been described in Section 2.4.2.

An interesting problem, which is also important with respect to linker technology, is the cloning of several identical DNA copies within the same vector. It has been shown that clones with multiple insertions are only stable if all inserted DNA fragments show the same orientation. Since a DNA fragment generated by restriction enzymes has a rotational symmetry (Fig. 2.1-5), the probability of producing such polymers is drastically reduced with increasing length of the insertion. Hartley and Gregori (1981) have devised a technique which allows head-to-tail oligomerisation to be favoured by using one of the non-palindromic recognition sites of *Ava* I (C/PyCGPuG), for example, (C/TCGGG). As shown in Fig. 3-16, only head-to-tail ligation yields structures which are completely base-paired, while head-to-head and tail-to-tail pairings show two mismatches. This procedure may be of some significance, because it could be used for obtaining short DNA pieces, such as linkers or repetitive DNA, in large amounts. It is also conceivable to use this approach to increase the expression of a given gene by increasing the number of its copies in a vector molecule. Another approach to this problem makes use of DNA fragments with termini obtained by digestion with isoschizomeric enzymes. When such fragments are ligated in the presence of the pair of restriction enzymes which generated them, *e.g.* *Bam* HI/*Bgl* II or *Sal* I/*Xho* I, only head-to-tail polymers will be formed, since only these are resistant to digestion with either enzyme (*cf.* Fig. 2.1-6).

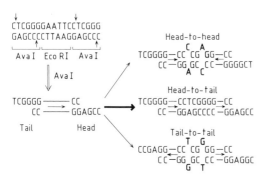

Fig. 3-16. Cloning of multiple copies of a DNA fragment by using *Ava* I cleavage sites. The plasmid shown in this example contains an *Eco* RI site flanked by two identical *Ava* I sites. *Ava* I digestion yields protruding non-symmetrical *Ava* I sites. Head-to-head and tail-to-tail ligations result in two mismatched base pairs, while head-to-tail polymerisation produces perfect base pairs. (Hartley and Gregori, 1981).

3.2 *In vivo* Ligation

In principle, *in vitro* ligation of DNA fragments could be dispensed with and delegated to the transformed cell itself, which, of course, contains the necessary DNA ligase to perform this task. An important parameter in this case is the *in vivo* stability, *i.e.*, the stability at 37 °C, of base pairs formed between the protruding ends of the two DNA fragments. It has been pointed out (*cf.* Section 3.1.1.3) that this prerequisite, *i.e.*, the necessary stability, is hardly fulfilled if DNA ends have been generated by restriction endonucleases, even if these termini should comprise four GC pairs. Such molecules should, therefore, be ligated prior to transfection.

Another solution for problems anociated with the instability of overlapping, protruding termini is the use of homopolymeric tails. This procedure is based on the addition of complementary homopolymeric tails to the respective 3' ends of the DNA fragments to be joined: one of the DNA fragments in question receives a dA tail, the other fragment a dT tail (Fig. 3-17). The desired recombinant DNA molecules are formed *in vitro* by base pairing between these complementary tails. Since such tails usually differ in length the resulting recombinant DNA molecules generally possess single-stranded gaps which are closed *in vivo* by bacterial DNA polymerases and DNA ligases. This procedure is of historical interest because the very early experiments in recombinant DNA technology involved the use of homo-

polymeric tails (Jackson *et al.* 1972; Lobban and Kaiser, 1973). Homopolymer tailing also permits the ligation of DNA fragments if one of the fragments in question does not possess termini capable of specific base pairing, which is always the case when cDNA molecules are cloned. Moreover, it is considered an advantage that homopolymer tailing ensures that the two reaction partners, vector and foreign DNA, will always possess complementary tails, *i.e.*, dA and dT, or dC and dG, respectively. This, in turn, guarantees that only different partners, and never molecules of the same kind, will recombine with each other; any clone obtained after transfection will, therefore, contain chimaeric DNA molecules. The latter requirement of different, but complementary, homopolymeric tails in vector and foreign DNA clearly eliminates the competition between monomolecular and bimolecular reactions. The relative and absolute DNA concentrations in the ligase assay are thus of lesser importance.

Homopolymeric tails are added to DNA fragments by the enzyme terminal deoxynucleotidyl transferase (Nakamura *et al.* 1981). This enzyme, which can be isolated from calf, rat, or mouse thymus, consists of a single polypeptide chain with a molecular weight of 60 kDa. It does not

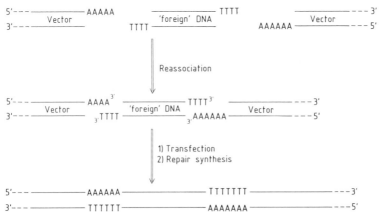

Fig. 3-17. Cloning by homopolymeric tails.
In this example the DNA to be cloned carries homopolymeric dT tails, and the vector DNA homopolymeric dA tails.

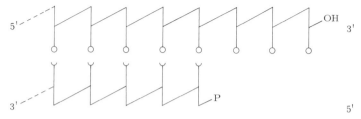

Fig. 3-18. A preferred substrate for terminal deoxynucleotidyl transferase.

require a template, nor, in cacodylate buffers, a primer; nevertheless, the preferred substrate of terminal transferase is the 3′ hydroxyl end of a protruding DNA end at least 3 bases in length (Fig. 3-18). DNA molecules possessing blunt ends or 5′ protruding ends are extremely poor substrates. In the past, this problem was solved by a short treatment with λ exonuclease, which digests DNA molecules from the 5′ end. Improved incubation conditions have been established and now even blunt ends and 3′ recessed ends can be tailed (Deng and Wu, 1981). A detailed analysis of the reaction conditions has revealed that the addition of dA and dT tails is very efficient if the buffers

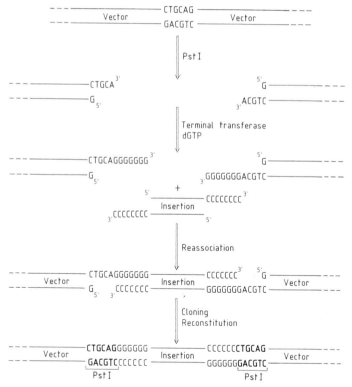

Fig. 3-19. Use of poly(dC)-poly(dG) tails for cloning into a *Pst* I site (Otsuka, 1981) (see text for details).

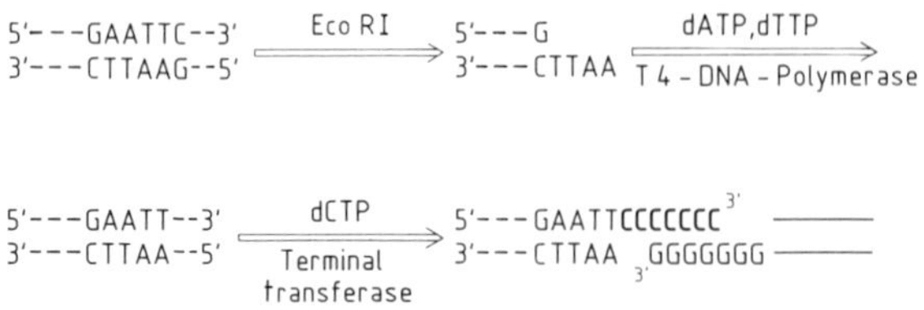

```
5'---GAATTC--3'    Eco RI      5'---G        dATP,dTTP
3'---CTTAAG--5'  ───────────►  3'---CTTAA  ──────────────►
                                            T 4 - DNA - Polymerase


5'---GAATT--3'     dCTP        5'---GAATTCCCCCCC 3'  ─────────
3'---CTTAA--5'  ───────────►   3'---CTTAA GGGGGGG ───────────
                 Terminal                    3'
                 transferase
```

Fig. 3-20.   Use of homopolymeric poly(dC) tails for the reconstruction of an *Eco* RI site.

contain cobalt ions; the buffers of choice for the addition of dG and dC tails, however, should contain manganese chloride. It is not only the fraction of molecules having received a tail which is increased under these particular conditions, but also the length of the tails. The lengths of the tails, which should be on the order of 10-30 bases, as well as total yield can be easily monitored by analysing the reaction products by gel electrophoresis (Deng and Wu, 1981); this is the method of

choice, although acid precipitation of radioactively labelled tailed DNA is still widely used.

It should be noted that homopolymeric tailing also has certain disadvantages. Firstly, the vector DNA must be completely intact, because terminal transferase also attacks free internal 3' hydroxyl groups. This leads to branched stuctures and inactivates the vector. In addition, tailing frequently destroys the restriction enzyme cleavage site on the vector DNA so that the inserted DNA

Table 3–3.   Regeneration of restriction sites by addition of homopolymeric tails.

Enzyme	Cleavage sites	Termini	fill-in	Nature of tail
*Pst*I	5'--CTGCAG–3' 3'--GACGTC–5'	5'--CTGCA 3'--G	–	dG
*Kpn*I	5'--GGTACC–3' 3'--CCATGG–5'	5'--GGTAC 3'--G	–	dC
*Sst*I	5'--GAGCTC–3' 3'--CTCGAG–5'	5'--GAGCT 3'--C	–	dC
*Hae*III	5'--GGCC–3' 3'--CCGG–5'	5'--GG 3'--CC	–	dC
*Eco*RI	5'--GAATTC–3' 3'--CTTAAG–5'	5'--G 3'--CTTAA	+	dC
*Bgl*II	5'--AGATCT–3' 3'--TCTAGA–5'	5'--A 3'--TCTAG	+	dT

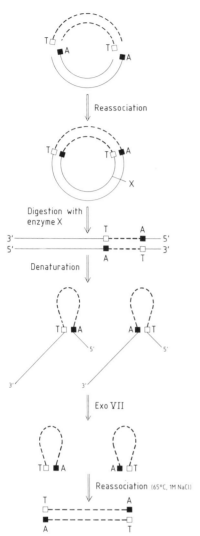

Fig. 3-21. Recovery of insertions cloned *via* poly(dA)-poly(dT) tails.
X stands for a restriction enzyme which cuts once outside of the cloned DNA fragment. The insertion is shown as a broken line. Tails are shown as solid (A) or open (T) squares (*cf.* text; Goff and Berg, 1978).

cannot be cut out in an intact form; however, several methods have been devised which allow the restriction site to be reconstituted after the addition of tails and cloning of the DNA (Otsuka,

1981). The digestion of a suitable vector molecule with *Pst* I, for instance, generates protruding 3' ends, which are ideal substrates for terminal transferase. Homopolymeric dG tails can be added to such ends by a reaction with dGTP; the DNA to be cloned receives homopolymeric dC tails (Fig. 3-19). After reassociation and cloning, the *Pst* I sites at either side of the insertion are regenerated, and the cloned DNA can be cut out from the vector together with its homopolymeric tails. Of course, the enzyme used for the tailing reaction should be free of any 3' exonuclease activities, because the removal of a single nucleotide would prevent the reconstitution of the cleavage site. This procedure has proven particularly effective for cDNA cloning in vector pBR322, which contains a single *Pst* I site in the β-lactamase gene. Insertions can therefore be detected by marker inactivation (*cf.* Chapter 11).

In principle, this strategy can also be used with all other restriction sites which, like *Pst* I, generate 3' protruding ends, for example, *Kpn* I and *Sst* I (Table 3-3). The technique can also be employed if the enzymes used generate blunt ends, such as *Hae* III. If the cleavage sites possess 5' protruding ends, which is the case, for example, with ends generated by *Hind* III, *Eco* RI, and *Sal* I, it is advisable to fill-in these ends with T4 DNA polymerase I in the presence of dNTPs. Restriction sites then can be reconstituted by adding appropriate tails (Fig. 3-20). If, for whatever reason, the reconstruction of a recognition sequence should prove difficult, there are at least two other methods which can be used for cutting out DNA molecules cloned by homopolymer tailing from recombinant DNA. One technique is based on the thermal instability of poly(dA)-poly(dT) regions. Such regions can be cleaved with S1 nuclease after partial denaturation in buffers containing 45-50% formamide (Hofstetter *et al.* 1976). Due to their thermal stability, this procedure cannot be used for poly(dG)-poly(dC) tails. The strategy described by Goff and Berg (1978) exploits the opposite polarity of homopoly-

meric tails for obtaining intramolecularly reassociated molecules. As shown in Fig. 3-21, the
recombinant DNA is first linearised with an
enzyme cleaving outside the DNA insert. After
denaturation and subsequent renaturation, intramolecular reactions lead to the formation of snap-
back molecules, from which the single-stranded
vector DNA can be removed by exonuclease VII
digestion. A second reassociation step then yields
double-stranded DNA molecules in which the
inserted DNA is flanked by homopolymeric tails.
After a brief treatment with λ exonuclease the
DNA fragment can be re-inserted into a vector
with suitable homopolymeric tails at its 3' ends.

References

Bahl, C.P., Wu, R., Brousseau, R., Sood, A.K.,
Hsiung, H.M., and Narang, S.A. (1978). Chemical
synthesis of versatile adaptors for molecular cloning.
Biochem. Biophys. Res. Comm. 81, 695-703.

Cameron, J.R., Panasenko, S.M., Lehman, I.R., and
Davis, R.W. (1975). *In vitro* construction of bacteriophage carrying segments of the *E. coli* chromosome: selection of hybrids containing the gene for
DNA ligase. Proc. Natl. Acad. Sci. USA 72, 3416-
3420.

Deng, G., and Wu, R. (1981). An improved procedure
for utilizing terminal transferase to add homopoly-
mers to the 3'-termini of DNA. Nucleic Acids Res. 9,
4173-4188.

Donoghue, D.J., and Hunter, T. (1982). A generalized
method of subcloning DNA fragments by restriction
site reconstruction: application to sequencing the
amino-terminal coding region of the transforming
genes of Gazdar murine sarcoma virus. Nucleic Acids
Res. 10, 2549-2564.

Dugaiczyk, A., Boyer, H.W., and Goodman, H.M.
(1975). Ligation of *Eco* RI endonuclease generated
DNA fragments into linear and circular structures. J.
Mol. Biol. 96, 171-184.

Ferretti, L., and Sgaramella, V. (1981a). Temperature
dependence of the joining by T4 DNA ligase of
termini produced by type II restriction endonucleases. Nucleic Acids. Res. 9, 85-93.

Ferretti, L., and Sgaramella, V. (1981b). Specific and
reversible inhibition of blunt end joining activity of

the T4 DNA ligase. Nucleic Acids Res. 9, 3695-
3705.

Goff, S., and Berg, P. (1978). Excision of DNA
segments introduced into cloning vectors by the
poly(dA/dT) joining method. Proc. Natl. Acad. Sci.
USA 75, 1763-1767.

Goodman, H.M., and MacDonald, R.J. (1979). Cloning of hormone genes from a mixture of cDNA
molecules. Methods in Enzymology 68, 75-90.

Hartley, J.L., and Gregori, T.J. (1981). Cloning multiple copies of a DNA segment. Gene 13, 347-353.

Heffron, F., So, M., and McCarthy, B.J. (1978). *In vitro*
mutagenesis of a circular DNA molecule by using
synthetic restriction sites. Proc. Natl. Acad. Sci. USA
75, 6012-6016.

Hofstetter, H., Schambök, A., van den Berg, J., and
Weissmann, C. (1976). Specific excision of the
inserted DNA segment from hybrid plasmids constructed by the poly(dA)poly(dT) method. Biochim.
Biophys. Acta 454, 587-591.

Jackson, D.A., Symons, R.H., and Berg, P. (1972).
Biochemical method for inserting new genetic infor-
mation into DNA of simian virus 40: Circular SV40
DNA molecules containing λ phage genes and the
galactose operon of *Escherichia coli*. Proc. Natl.
Acad. Sci. USA 69, 2904-2909.

Jacobson, H., and Stockmayer, W.H. (1950). Intramolecular reaction in polycondensations. I. The theory
of linear systems. J. Chem. Phys. 18, 1600-1606.

Lobban, P.E., and Kaiser, A.D. (1973). Enzymatic
end-to-end joining of DNA molecules. J. Mol. Biol.
78, 453-471.

Maniatis, T., Hardison, R.C., Lacey, E., Laner, J.,
O'Connell, C., Quon, D., Sim, G.K., and Efstradiatis, A. (1978). The isolation of structural genes from
libraries of eucaryotic DNA. Cell 15, 687-701.

Murray, N.E., Bruce, S.A., and Murray, K. (1979).
Molecular Cloning of the DNA Ligase Gene from
Bacteriophage T4. II. Amplification and preparation
of the gene product. J. Mol. Biol. 132, 493-505.

Nakamura, H., Tanabe, K., Yoshida, S., and Morita,
T. (1981). Terminal Deoxynucleotidyltransferase of
60 000 Daltons from Mouse, Rat and Calf Thymus. J.
Biol. Chem. 256, 8745-8751.

Otsuka, A. (1981). Recovery of DNA fragments inserted by the "tailing" method: regeneration of *Pst* I
restriction sites. Gene 13, 339-346.

Panasenko, S.M., Cameron, J.R., Davis, R.W., and
Lehman, I.R. (1977). Five hundredfold overproduction of DNA ligase after induction of a hybrid λ
lysogen constructed *in vitro*. Science 196, 188-189.

Rothstein, R.J., Lau, L.F., Bahl, C.P., Narang, S.A.,
and Wu, R. (1979). Synthetic adaptors for cloning
DNA. Methods in Enzymology 68, 98-109.

Sgaramella, V., and Khorana, H.G. (1972). Studies on

polynucleotides. CXVI. A further study of the T4 ligase-catalyzed joining of DNA at base-paired ends. J. Mol. Biol. 72, 493-502.

Silber, R., Malathi, V.G., and Hurwitz, J. (1972). Purification and properties of bacteriophage T4-induced RNA ligase. Proc. Natl. Acad. Sci. USA 69, 3009-3013.

Sugino, A., Goodman, H.M., Heynecker, H.L., Shine, I., Boyer, H., and Cozzarelli, N.R. (1977). Interaction of bacteriophage T4 RNA and DNA ligases in joining of duplex DNA at base-paired ends. J. Biol. Chem. 252, 3982-3994.

Tait, R.C., Rodriguez, R.L., and West, R.W., Jr. (1980). The rapid purification of T4 DNA ligase from a λ T4 lig lysogen. J. Biol. Chem. 255, 813-815.

Williams, G.B., and Blattner, F.R. (1980). Bacteriophage λ Vectors for DNA cloning. In: "Genetic Engineering", Setlow, J.K., and Hollaender, A. (eds.). Vol. 2, pp.201-281, Plenum Press, New York and London.

Wilson, G.G., and Murray, N.E. (1979). Molecular Cloning of the DNA Ligase Gene from Bacteriophage T4. I. Characterization of the Recombinants. J. Mol. Biol. 132, 471-491.

4 E. coli Vectors

One of the most important elements in gene cloning is the vector, which, in conjunction with the DNA molecule to be cloned, forms the recombinant DNA molecule, which can be propagated in suitable host cells. In order to perform its function, a vector must possess the following properties:

- Vector DNA molecules must contain a replicon, *i.e.*, a stretch of DNA which enables them to replicate in host cells. These specific DNA sequences usually span approximately 500 base pairs and interact with enzymes engaged in the initiation of DNA replication. All prokaryotic genomes and chromosomes studied so far, and also the genomes of eukaryotic viruses, contain such specific regions. It is still an open question whether eukaryotic chromosomes may not contain superordinate structural elements as start signals.
- Vectors should code for genetic information, the presence or absence of which can be used for the development of a selection system allowing to discriminate between, and select for, vector and recombinant DNA. Such genetic markers may be resistance to antibiotics, or certain auxotrophic markers such as the inability to metabolise certains amino acids and sugars. In this case, suitably mutated bacterial host strains must be available. Vector molecules which carry several of these markers are very useful because one of the markers then can be inactivated by cloning; marker inactivation easily allows differentiation between plasmid-free colonies, colonies containing vector DNA, and colonies containing recombinant DNA molecules.
- Vectors should contain unique cleavage sites for as many restriction enzymes as possible into which foreign DNA can be inserted. Such sites must lie outside of the origin of replication and other essential genes, but they should lie within one of the marker genes, so that the insertion of foreign DNA leads to marker inactivation.
- For the expression of cloned DNA, the vector DNA should contain suitable control elements, such as promoters, ribosome binding sites, etc. The structure of such expression vectors will be discussed in Chapters 7 and 8.
- It is very helpful to know the nucleic acid sequence of the vector. In this case the exact positions of the marker genes, cleavage sites for restriction endonucleases, and other vector elements of interest can be identified and analysed by suitable computer programs.

A variety of different cloning vectors have been developed by using the items mentioned above as a guideline. The various plasmid and phage vectors, or combinations of both, the choice of which depends, of course, on individual requirements, will be discussed in detail in the following sections. Eukaryotic vectors will be dealt with in separate chapters.

4.1 Plasmids

4.1.1 Properties of Plasmids

Plasmids are circular extrachromosomal DNA molecules found in most Gram-positive and Gram-negative organisms, and also in some yeasts, but not in higher eukaryotes (Helinski, 1979). A first indication of the existence of such genetic elements was gleaned in the early fifties from the appearance and rapid spread of multiple drug resistances. Indeed, plasmids confer upon their host cells a variety of phenotypic traits which are not required for bacterial reproduction and are usually dispensable; plasmids are, however, responsible for certain growth advantages, e.g, when such bacteria grow in the presence of antibiotics.

There are two groups of naturally occurring plasmids known as conjugative and non-conjuga-tive. Conjugation is a biological phenomenon observed in bacterial genetics: the directed trans-fer of DNA from a donor into a recipient cell (Williams and Skurray, 1980). This transfer is mediated by two groups of plasmid-coded genes, transfer genes (*tra*) and mobilising functions (*mob*). The transfer functions comprise a cluster of at least twelve different genes, which, among other things, are responsible for the synthesis of pili and other surface components. These pili allow physical contact – a prerequisite for conju-gation – between donor and recipient cells. The *mob* functions are defined by at least two regions on the plasmid DNA. One of these regions codes for a mobilising protein which binds to the other *mob* region on the plasmid DNA. The formation of this relaxation complex is associated with a strand-break in this region, which frequently is referred to as the *nic/bom* region (*nic* = nick; *bom* = basis of mobility).

Plasmids with functional transfer and mobility regions (Tra$^+$, Mob$^+$) are conjugative and mobil-isable. Examples are the sex factor F, and other plasmids belonging to incompatibility classes Inc FI, II, III, IV (see below). Plasmids lacking the transfer functions are non-conjugative but can be mobilised by another conjugative plasmid present in the same cell, if their phenotype is Tra$^-$, Mob$^+$, *i.e.*, if they possess an intact *mob* region. Plasmids ColE1 and cloacin DF13 (Clo DF13), which can be mobilised effectively by F factor and F-like plasmids, belong to this group.

Plasmids which have lost all *tra* and *mob* functions are neither conjugative, nor can they be mobilised; yet, if these plasmids have simply lost only that part of the *mob* region which codes for the mobilising protein, they can still be mobilised by another plasmid, if the latter codes for a functional Mob protein, which can act in *trans* on the Mob$^-$ plasmid. The *nic/bom* site, however, is a

Donald R. Helinski
San Diego

cis-acting element which cannot be provided by another plasmid. A number of specially designed non-conjugative and transfer-deficient plasmids have gained particular significance as cloning vehicles due to biological safety considerations.

Plasmids then are classified by their conjugative properties, but they are also discriminated according to the products they encode. Plasmids carrying genes which are responsible for antibiotic resistance, for example, are known as resistance transfer factors, or R plasmids. Genes for antibiotic resistance frequently reside in, or are derived from, transposons (Kleckner, 1981), and are flanked by 0.8-1.4 kb repetitions. These flanking regions are insertion sequences (*IS1*, *IS2*, *IS4* etc.) which can integrate into various sites of the host or plasmid chromosome. Spread of resistance is, therefore, due not only to mobilisation of plasmids *per se*, but also to insertion sequences flanking resistance genes which render such genes particularly mobile and confer on them the ability to "jump" to other chromosomal locations (*cf.* Jida *et al.*, 1983).

Apart from the functions described above, all plasmids contain specific regions responsible for their autonomous replication (*rep*), which include an origin of DNA replication (*ori*). Since all genes and DNA sequences in a plasmid are usually organised in functional clusters, it is possible to draw a general structure of plasmids, which is shown in Fig. 4.1-1.

Mechanisms which regulate the copy numbers of plasmids are of major importance for the maintenance of plasmids in bacteria. Copy numbers are either *strictly* controlled by, and correlated with, the number of chromosomal DNA molecules (*i.e.*, they are hardly ever greater than one), or *relaxed*, (*i.e.*, a bacterial cell may contain multiple copies of a plasmid). Plasmids can be categorised as stringent or relaxed, depending on whether they are maintained as a limited number of copies per cell, or as multiple copies, respectively. With some exceptions plasmids with low copy numbers usually have a relatively high molecular weight and are conjugative, while high

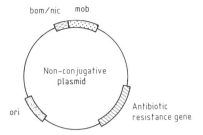

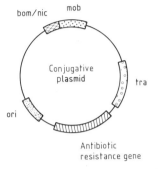

Fig. 4.1-1.   General structure of plasmids. Non-conjugative differ from conjugative plasmids by the absence of transfer (*tra*) functions.

copy number plasmids are normally non-conjugative and have low molecular weights (Table 4.1-1).

Plasmids replicating under stringent control require active protein biosynthesis but no DNA polymerase I for their replication; relaxed plasmids, however, need a functional DNA polymerase I (product of the *polA* gene), but they replicate without *de novo* protein biosynthesis. Many of the multicopy plasmids show quite unusual behaviour, because they are capable of continuing with replication even in the presence of an inhibitor of protein biosynthesis. Upon addition of chloramphenicol, for example, chromosomal DNA synthesis ceases after a few minutes, while plasmid replication continues unabated. This results in what has been termed amplification of a plasmid, *i.e.*, the synthesis of up to several thousand plasmid molecules per cell. Since amplification simultaneously affects recom-

Table 4.1–1. Properties of some naturally occurring plasmids.

Plasmid	Size (Dalton)	(kb)	Conjugative	Copy-number	Amplifiable	Selectable markers	References
ColE1	4.2×10^6	7	–	10–15	+	E1[imm*]	1
RSF1030	5.6×10^6	9.3	–	20–40	+	Ap[r]	2,3
CloDF13	6.0×10^6	10	–	10	+	DF13[imm*]	3
pSC101	5.8×10^6	9.7	–	1– 2	–	Tc[r]	3
R6K	25×10^6	42	+	10–40	–	Ap[r]Sm[r]	1
F	62×10^6	103	+	1– 2	–	–	1
R1	65×10^6	108	+	1– 2	–	Ap[r]Cm[r]Sn[r] Sn[r]Sm[r]Km[r]	3
RK2 (RP4; RP1)	38×10^6	56.4	+	3– 5	–	Ap[r]Km[r]Tc[r]	1,2

* Cells containing plasmids ColE1 or CloDF13 produce so-called immunity proteins (imm) which protect against lethal effects of the homologous but not the heterologous colicin. Plasmids pSC101 and R6K represent exceptions from the rule according to which plasmids with high copy numbers are non-conjugative and of low molecular weight while low copy number plasmids are conjugative and of high molecular weight. Abbreviations: Ap = ampicillin; Sm = streptomycin; Tc = tetracyclin; Cm = chloramphenicol; Sn = sulfonamide; Km = kanamycin. References: 1 = Kahn *et al.*, 1979; 2 = Thomas, 1981; 3 = Helinski, 1979.

binant DNA inserts on a plasmid molecule, relaxed multicopy plasmids play a central role in gene technology.

Copy numbers of plasmids are genetically controlled and can be altered by mutagenesis in such a way that plasmid replication becomes temperature-sensitive and either ceases after the temperature has been raised (Meacock and Cohen, 1979), or quite the reverse, becomes relaxed. Certain temperature-sensitive mutants of plasmid R1 (Tab. 4.1-1), for example, have a low copy number (1-2) at 30 °C, as expected. At 42 °C, however, replication control is lost and the plasmids begin to replicate much faster than the chromosome of the host cell. This phenomenon, known as runaway replication, is responsible for the fact that up to 50% of the total DNA in a bacterial cell may consist of plasmid molecules after only 4-5 cell divisions (Uhlin *et al.*, 1979). In contrast to the chloramphenicol-induced amplification of plasmid DNA, which requires inhibition of protein biosynthesis, runaway replication can proceed in the presence of normal protein biosynthesis. Since this leads to a considerable overproduction of plasmid-encoded gene products (Uhlin

and Clark, 1981), runaway plasmids play a certain role as expression vectors.

Suinsky *et al.* (1981) have developed a system which consists of two plasmids with temperature-sensitive replicons, one of which replicates at 30 °C, but not at 42 °C, and carries a gene coding for *lac* repressor. The second plasmid carries a runaway replicon which allows high copy numbers at the elevated temperature (42 °C), and a gene coding for the protein to be expressed in this system. The expression of this gene is controlled by the *lac* operator/promoter region. At low temperatures, this gene will not be expressed, because the first plasmid provides large amounts of *lac* repressor which blocks transcription by binding to the *lac* operator. When the temperature is raised, replication of the first plasmid ceases and, at the same time, the second plasmid which carries the *lac* control region is amplified. Since the cell soon runs out of repressor, the expression of the desired gene is derepressed. Such selective expression systems always play a particular role if the heterologous protein is toxic for the host cell, because production of the protein can be repressed in actively growing bacteria and switched on at will by a simple temperature shift, *e.g.*, when the cell density has reached the desired level.

The stability of low copy number plasmids, *i.e.*, their transmission to daughter cells during cell division, constitutes a particular problem (see also Nordström *et al.*, 1984). If we assume that there is no particular mechanism which controls the segregation of plasmids, and if a parent cell contains *n* molecules of a given plasmid, the probability *P* that a daughter cell will not obtain a copy of this plasmid after cell division can be calculated by the following equation:

$$P = 2 \times (1/2)^n$$

As shown in Fig. 4.1-2, the fraction of plasmid-free cells is directly correlated with copy numbers. Low copy number plasmids would, therefore, be lost extremely rapidly, if a special segregation mechanism did not exist. Such control mechanisms have indeed been defined genetically for the *E. coli* plasmids pSC101 and R1 (Meacock and Cohen, 1980; Nordström *et al.*, 1980). Plasmid pSC101 contains the *par* region (*par* =

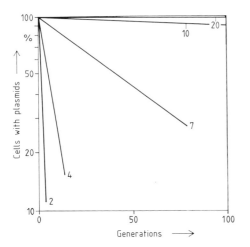

Fig. 4.1-2. Loss of plasmids as a function of cell generations.
Starting with different copy numbers (n = 2, 4, 7, 10, or 20), the proportion of cells containing plasmids is calculated by the formula given in the text, and plotted as a function of cell generations.

partition) which is responsible for the inheritance of stable copy numbers. The *par* function has been mapped to a short DNA region of 370 bp in the immediate vicinity of the origin of DNA replication, but the mechanism (*via* membrane interactions, for example) by which this sequence effects the correct partition of plasmids is still unknown (Tucker *et al.*, 1984). Plasmids in which this region has been deleted will be found only in less than 1% of the cells after 60 cell generations. F factors contain an additional control element, the *ccd* system (control of cell division), which prevents cell division until there are at least two copies of the plasmid per cell (Ogura and Hiraga, 1983). It may be important to consider such control functions when cells are grown in large fermenters, where it may not be possible, for practical reasons, to select for the presence of plasmid-encoded genes, for example, by adding suitable antibiotics.

It has already been pointed out that a bacterial cell has no difficulties in harbouring many hundred molecules of one and the same plasmid. This is not necessarily true for the coexistence of two different plasmids. Certain combinations of plasmids are incompatible, and more than 30 incompatibility (Inc) groups have now been defined on the basis of mutual incompatibility. By definition, two plasmids belong to the same incompatibility group if they are unable to coexist in the same cell. It appears that plasmid incompatibility and the exact control of copy numbers or of plasmid DNA replication are intimately linked. The corresponding underlying mechanisms have been particularly well studied for plasmid ColE1 and its derivatives (reviewed by Cesarini and Banner, 1985). In these plasmids, the region containing the origin of replication (*ori*) spans several hundred base pairs (Fig. 4.1-3) and codes for an RNA molecule, transcription of which initiates 555 bases upstream from, and proceeds in the direction of, the origin of DNA replication (Itoh and Tomizawa, 1980). Transcription of this RNA, which has been designated RNA II, frequently proceeds beyond this site. Since this RNA

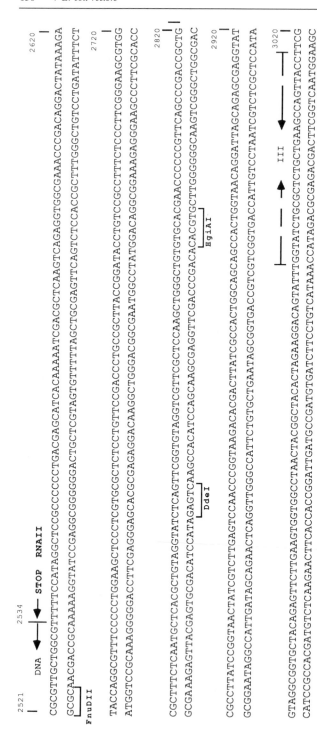
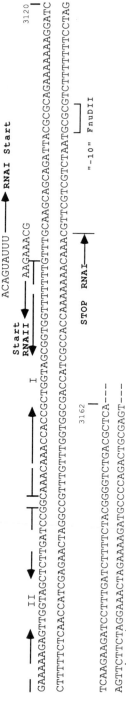

serves as a primer for DNA replication, it must be trimmed by a special ribonuclease, RNAse H, so as to provide a 3′ end, which lies exactly at the initiation site of DNA replication. In addition to this primer RNA II, the initiation region codes for a second RNA, which is 108 nucleotides in length, and is known as RNA I. RNA I transcription proceeds in the opposite direction with respect to that of RNA II and terminates close to the RNA II transcription initiation site. RNA I can assume a secondary stem-loop structure which resembles that of a tRNA molecule. The coding regions for RNA I and RNA II overlap and are, therefore, complementary. As shown in Fig. 4.1-4, the interaction of RNA I with nascent RNA II chains is similar to the interaction between anticodon and codon (Tomizawa, 1984). RNA I, when complexed to the nascent (but not to the mature) primer transcript RNA II, behaves like a repressor of primer formation, presumably by preventing RNA II from adopting the secondary structure required for primer activity (Wong and Polisky, 1985). This, in turn, prevents the initiation of DNA replication (Lacatena and Cesarini, 1981). This model which has been termed the "primer processing model" also explains the phenomenon of plasmid incompatibility: primer formation by RNA II from plasmid X is inhibited by RNA I produced by plasmid Y, if

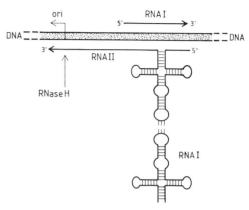

Fig. 4.1-4. DNA synthesis and incompatibility of ColE1 plasmids.
Shown is a region of ColE1 DNA, approximately 600 bp in length, which codes for two RNA molecules, RNA I and RNA II. RNA II serves as a primer of DNA synthesis at the origin of DNA replication. It possesses a special secondary structure near its 5′ end, which interacts with the much smaller RNA I (Lacatena and Cesarini, 1981).

RNA I$_X$ and RNA I$_Y$ are identical at the points of RNA I-RNA II interaction and can assume similar tRNA-like conformations. Mutations which alter the copy numbers of plasmids almost exclusively map in the region of "anticodon/codon" interaction between RNA I and RNA II. If the interaction between wild-type RNA II and mutant RNA I (occasionally referred to as the "kissing" interaction; Tomizawa, 1984) is less efficient, mutant plasmids grow better in the presence of wild-type plasmids than wild-type plasmids *per se*. Such mutants define new incompatibility groups (Watson, 1982).

The host-range of most of the naturally occurring plasmids is rather restricted; ColE1 and its derivatives, for example, grow only in *E. coli* and related organisms. The incompatibility group known as P1, however, is especially interesting for gene manipulation, because members of this group are capable of growing in a wide range of Gram-negative organisms (*cf.* also Section 4.1.2). This class of promiscuous plasmids will become more important as gene manipulation is applied

◄—— **Fig. 4.1-3.** The origin of pBR322 DNA replication. The relevant nucleotide sequence section from plasmid pBR322 extends from an *Fnu*DII site at position 2523 to position 3162 (*cf.* Appendix B-1). Although the 580 bp *Fnu*DII fragment (positions 2523-3103) has been shown to provide DNA replication functions, more recent constructions which increase plasmid yields include the "-35" prokaryotic promoter sequence upstream of the *Fnu*DII site (*cf.* Fig. 9-6). RNA primer (RNA II) and RNA I synthesis initiates at positions 3088 and 2978/2980, respectively. RNA I terminates at position 3087, while the transition from RNA primer synthesis to DNA synthesis, *i.e.*, the start of replication, takes place at position 2534. Some relevant restriction sites are also indicated. Horizontal arrows indicate three palindromes (I to III) which specify possible stem and loop structures for RNA I and II. Dots in the arrows indicate unpaired nucleotides.

to organisms other than *E. coli*, because such plasmids can replicate, for example, in *E. coli*, *Agrobacterium*, *Pseudomonads* (Bagdasarian *et al.* 1981), and *Rhizobium* species (Kahn *et al.* 1979). Moreover, these plasmids are compatible with plasmids of the pSC101 type and with ColE1 and its derivatives. They can therefore be used for conjugation experiments designed to bring several different plasmids into the same host cell.

Quite a different series of plasmids, which will be discussed in Chapter 6, has been developed for Gram-positive organisms.

4.1.2 Plasmids in Gene Technology

Ideally, plasmids designed for recombinant DNA technology should possess the following properties:
- They must be able to replicate autonomously in bacterial cells.
- They must possess at least one, but preferably several markers, such as resistance to antibiotics, which confer upon the cells a new selectable phenotype.
- Plasmid molecules should contain a whole spectrum of unique restriction endonuclease cleavage sites for the construction of recombinant DNA molecules. At least some of these sites should reside in a marker gene so that this gene can be inactivated by a recombinational event. A second marker on the same plasmid molecule should remain intact in order to provide a selectable phenotype.
- In view of biological safety considerations, certain plasmids should neither be transmissible nor mobilisable.
- The molecular weights of plasmids should be as low as possible, because the efficiency of transformation decreases with increasing molecular weights.
- Very often it will be desirable to maintain plasmids as multiple copies per cell. Such plasmids must then carry a replicon which is under relaxed control.

Naturally occurring plasmids usually do not fulfill all the requirements listed above. Artificial plasmid vectors combining several elements of various naturally occurring plasmids were therefore devised very early in the development of cloning techniques.

4.1.3 Plasmid pBR322 and its Derivatives

The prototype of an artificial vector is plasmid pBR322 (Bolivar *et al.*, 1977), the origin and structure of which will be discussed in detail in the following sections. As can be seen from the biological map in Fig. 4.1-5 this plasmid is composed of three parts, *i.e.*, a region of Tn3 and a region of pSC101 which specify ampicillin and tetracycline resistance, respectively, and a region of pMB1 which carries the origin of DNA replication (*ori*) (Sutcliffe, 1979).

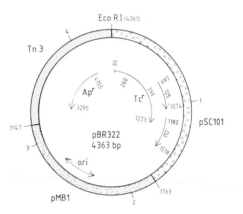

Fig. 4.1-5. Structure of pBR322.
The plasmid is composed of three sections, which are derived from Transposon Tn3 (Tn3), plasmid pMB1, and plasmid pSC101. The reading frame for β-lactamase (Ap^r = ampicillin resistance), and three reading frames of proteins within the section derived from pSC101 are shown. The positions of the first base of the initiation codon, the last base of the stop codon of these proteins, and their lengths in amino acids are also indicated. DNA replication initiates at position 2534 and proceeds anticlockwise (Sutcliffe, 1979).

The origin of the ampicillin resistance gene can be traced back to resistance factor R1 originally isolated in London in 1963 (Fig. 4.1-6). Ampicillin resistance, which is encoded by transposon Tn3, was transferred from R1drd19, a variant of R1, to the multicopy plasmid ColE1 and yielded the recombinant pSF2124. Plasmid pMB9, which codes for tetracycline resistance as well as colicin immunity (Col^{imm}), is a chimaeric molecule and was constructed by mixing *Eco* RI fragments of pSC101 with plasmid pMB8 which had been linearised at its unique *Eco* RI site. A new plasmid, pBR312, was obtained subsequently by transferring the ampicillin resistance of transposon Tn3 in plasmid pSF2124 to pMB9. Rearrangement of appropriate *Eco* RI fragments in pBR312 yielded pBR313, the immediate progenitor of pBR322. Although it has not been mentioned explicitly, it should be apparent that these constructions were guided by two principles. Firstly, transposability of transposon Tn3 was exploited, and the ampicillin gene was allowed to jump unspecifically between DNA molecules. This transposability is exemplified by the ampicillin gene jumping from plasmid ColE1 to a new DNA molecule to form pSF2124 (Fig. 4.1-6). Secondly, these constructions demonstrate that DNA rearrangements can be accomplished effectively by random or direct reassociation of mixtures of certain restriction fragments. The generation of pBR322 from pBR313 may serve to illustrate this point (Fig. 4.1-7).

The phenotype of plasmid pBR313 is Ap^{r}Tc^{r}Col^{imm}. It contains three *Pst* I sites, one of which is located in the ampicillin resistance gene. In order to eliminate the two other *Pst* I sites, and to maintain a functional ampicillin resistance gene, two new plasmids had to be constructed. Plasmid pBR318 was obtained by selecting for tetracycline resistance after transformation of *E.*

coli RRI with a mixture of the three pBR313 *Pst* I fragments. Two small *Pst* I fragments were lost during this procedure, and a new molecule was generated, which was tetracycline-resistant but ampicillin-sensitive (Fig. 4.1-7). In order to reconstruct the ampicillin resistance gene in pBR318, a second plasmid, designated pBR320, had to be constructed. Since it was known that the ampicillin resistance gene of pBR313 did not contain an *Eco* RII site, bacteria were transformed with unligated *Eco* RII fragments of pBR313. Ampicillin-resistant cells were selected and screened for tetracycline and colicin sensitivity. Plasmid pBR320 was one of sixteen clones with the desired phenotype, containing a unique *Pst* I site in the ampicillin resistance gene. This plasmid, in conjunction with pBR318, was used to obtain plasmid pBR322. For this purpose, plasmid pBR318 was simultaneously digested with *Pst* I and *Hpa* I, yielding two fragments with molecular weights of 1.95 and 2.2 × 10^6 Da, respectively; the smaller fragment contained the

Francisco Bolivar
Mexico City

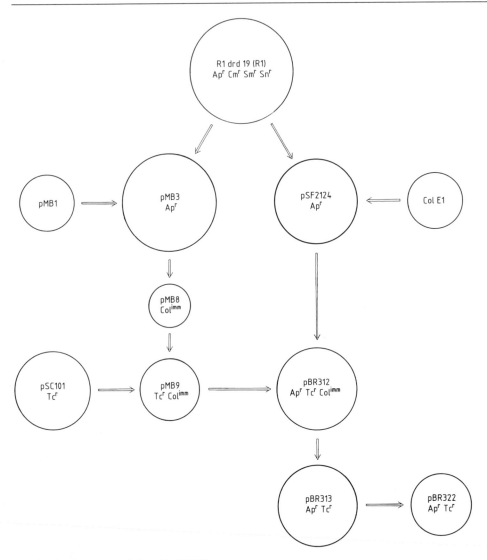

Fig. 4.1-6. Genealogy of plasmid pBR322.
This scheme shows the origin of the three elements employed in the construction of pBR322. Shown are replicon pMB1, the ampicillin transposon Tn3, obtained from plasmid R1drd19, and the tetracycline resistance region of pSC101. Abbreviations are explained in the legend of Tab. 4.1-1 (*cf.* also text; Sutcliffe, 1979).

entire tetracycline resistance gene and a part of the ampicillin resistance gene. Plasmid pBR320 was digested with *Pst* I and *Hinc* II, which yielded three fragments, the largest of which (open bar in Fig. 4.1-7) contained the origin of DNA replication and that particular part of the ampicillin

resistance gene lacking in pBR318. After transformation of *E. coli* strain RRI with a mixture of the DNA fragments obtained from pBR320 and pBR318, an ampicillin- and tetracycline-resistant, colicin-sensitive phenotype was selected. This selection regenerated a new functional ampicillin

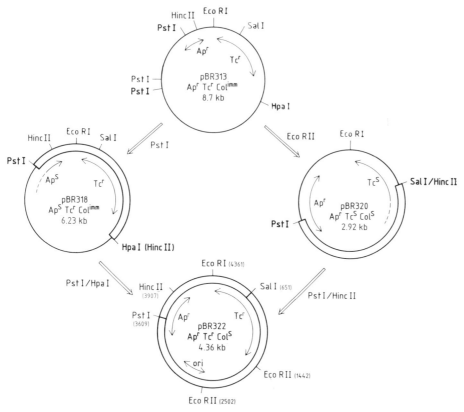

Fig. 4.1-7. Generation of plasmid pBR322 from pBR313.
Open bars in pBR318 and pBR320 indicate those regions which are incorporated into pBR322. The enzyme *Hinc* II (GTPy/PuAC) cleaves *Hpa* I (GTT/AAC) and *Sal* I (G/TCGAC) sites. The procedure is described in detail in the text (Bolivar *et al.*, 1977). Arrows represent functional regions such as ampicillin resistance (Ap^r), or the origin of DNA replication (*ori*). Relevant restriction sites are in bold-face (*cf.* also text).

resistance gene in the new plasmid and, at the same time, blocked the parent plasmids and their fragments, which could not be replicated, because they were either Tc^{r}Ap^S (pBR318) or Ap^{r}Tc^S (pBR320). This experiment did not only yield the expected plasmid with a molecluar weight of 3.1 × 10^6 Da, but also pBR322, which is 0.5 × 10^6 Da smaller. For unknown reasons, the loss of DNA, or more generally, the occurrence of DNA rearrangements is nothing unusual in such experiments.

In the meantime, several derivatives which are smaller than pBR322 have also been described.

pBR322 possesses six *Eco* RII sites and contains a long non-essential region, which is flanked by two *Eco* RII cleavage sites at positions 1442 and 2502, between the origin of DNA replication and the tetracycline resistance gene (*cf.* Fig. 4.1-8). This particular DNA fragment was removed by partially digesting pBR322 with *Eco* RII. Since the two *Eco* RII sites at positions 1442 (CCTGG) and 2502 (CCAGG) differ from each other, the resulting non-complementary ends had to be trimmed with the single-strand specific nuclease S1. The mixture of DNA fragments was subsequently subjected to gel electrophoresis, the

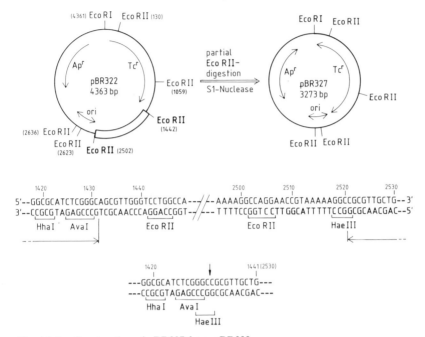

Fig. 4.1-8. Construction of pBR327 from pBR322.
Shown are the positions of the six *Eco* RII sites (or *Bst* NI; (CC/ATGG)) in pBR322 DNA. The two *Eco* RII sites at positions 1442 and 2502, and the location of the desired deletion (indicated by an open bar) can be seen in the portion of the sequence shown in the lower panels. The two horizontal arrows below the sequence mark the region deleted from pBR322 DNA. This region is larger than expected because S1 nuclease also hydrolysed part of the double strand beyond the two *Eco* RII sites. The vertical arrow indicates the site of ligation.

desired 3.3 kb DNA fragment isolated, circularised with T4 DNA ligase, and used for transformation. Plasmid pBR327 was isolated from Ap^r Tc^r transformants (Soberon *et al.*, 1980). The sequence analysis of the regions around the *Eco* RII sites of pBR327 and the comparison with the original sequences in pBR322 provide a choice example of what may frequently be observed when S1 nuclease is used: the enzyme does not only digest single-stranded regions, but also removes double-stranded stretches of DNA, up to 20 bp in our example (Fig. 4.1-8). This can be prevented only by using specially purified enzyme preparations.

pBR327 has two advantages over pBR322. Firstly, it is 1 089 bp smaller, and secondly, it lacks one mobilisation function. Initially, pBR322 had been certified as an EK2 vector, because it could not be mobilised by the conjugative plasmid

R64drd11; however, it has been shown in the meantime that pBR322 can be mobilised by another plasmid, ColK (Young and Poulis, 1978). Since the special relaxation (*bom*) site between positions 2207 and 2263 of pBR322 is missing from pBR327 (Clark and Warren, 1979), pBR327 offers additional safety and has, indeed, been certified as an EK2 vector. As far as biological safety is concerned, pBR327 even exceeds the X1776/pBR322 system.

A similar vector is pAT153 which lacks two *Hae* II fragments of pBR322 (positions 1644-2349) (Twigg and Sherratt, 1980). This deletion which is only 705 bp in length also removes sequences responsible for conjugational transfer, and hence pAT153 is no longer mobilisable (*cf.* also Fig. 8-27). In comparison to pBR322, pBR327 and pAT153 show a 2- to 4-fold increase

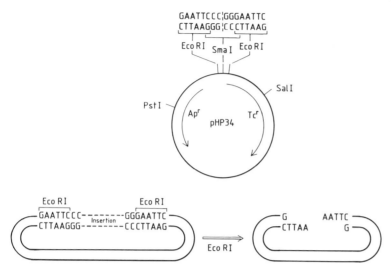

Fig. 4.1-9. Vector pHP34.
This plasmid is derived from pBR322 and contains in its *Eco* RI site a linker with an additional *Sma* I site. This site may be used for cloning of blunt-ended DNA molecules. The *Sma* I site is destroyed by the insertion but the insert can be recovered from the plasmid by *Eco* RI digestion (Prentki and Krisch, 1982).

in copy numbers; presumably, the deleted DNA codes for a repressor of replication which controls copy numbers in relaxed plasmids.

New improved pBR322 and pBR327 derivatives are constantly being developed. Plasmid pBR329, for example, carries an additional chloramphenicol resistance marker (Covarrubias and Bolivar, 1982). Another vector, pHP34, allows cloning of DNA fragments with flush ends (Prentki and Krisch, 1982). Although pBR322 already contains sites, *e.g.*, *Bal* I (1443; TTG/CCA) and *Pvu* II (2066; CAG/CTG) which would allow cloning of blunt-ended DNA fragments, DNA inserts which have not been prepared by cleavage with the same enzymes cannot be excised from the vector after ligation; in contrast, plasmid pHP34 contains a *Sma* I site (CCC/GGG) which is flanked by two *Eco* RI sites (Fig. 4.1-9). The inserted DNAs can therefore be excised from the vector by *Eco* RI digestion.

4.1.3.1 The β-Lactamase Gene of pBR322

pBR322 has been completely sequenced and comprises 4 363 bp (Sutcliffe, 1979; *cf.* also Appendix B.1). The β-lactamase gene, which is responsible for ampicillin resistance, plays an important role in pBR322 (Sutcliffe, 1978). The 5′ ends of two mRNA species coding for this enzyme have been mapped to positions 4189 and 37 (Fig. 4.1-10) (Brosius *et al.*, 1982). The first base of the methionine initiation codon AUG (ATG) is at position 4155, 209 bp away from the *Eco* RI site at position 4362. A Shine-Dalgarno box, which, in this case, shows a pronounced homology with the 3′ end of 16S ribosomal RNA, is found at the usual distance from the AUG codon. Counterclockwise transcription of a peptide 286 amino acids in length initiates at the AUG codon at position 4155. The 23 amino-terminal amino acids are regarded as a hydrophobic secretion signal

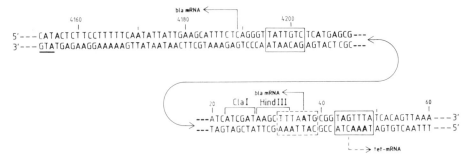

Fig. 4.1-10. Promoters for β-lactamase and tetracycline resistance genes in pBR322.
The initiation codon for β-lactamase at position 4155 (left) is underlined. Also shown are the two initiation sites of transcription for β-lactamase (*bla*) and the corresponding Pribnow-Schaller signal sequences (boxed). Tetracycline mRNA is transcribed clockwise and initiates at position 45. The corresponding Pribnow-Schaller region is boxed with dashed lines. *Cla*I and *Hind*III sites in the vicinity are also indicated.

and are, indeed, not observed in the mature periplasmic enzyme. pBR322 DNA contains a unique *Pst*I site at position 3609. Cleavage of the plasmid by *Pst*I therefore interrupts the β-lactamase gene in front of the codon for amino acid 182 (ala) (Fig. 4.1-11). This site can be reconstituted in a recombinant plasmid by poly(dG)-poly(dC) tailing, and hence also allows the insertion and subsequent excision of any desired DNA fragment (*cf.* also Chapter 3, Fig. 3-19). If the reading frame is conserved in spite of the homopolymeric tails, suitably transformed cells express a fusion protein which consists of

β-lactamase and the protein coded by the inserted DNA. This technique has allowed, for the first time, detection of antigenic determinants of a rat insulin fusion protein in the periplasmic space of transformed cells (Villa-Komaroff *et al.*, 1978). In practice, it is important to be aware of the fact that cells transformed with plasmids expressing such fusion proteins may show a certain degree of penicillin resistance despite the insertion of DNA and the resulting inactivation of the marker gene (*cf.* also Section 11.1.2). Occasionally, cloning into the β-lactamase gene also utilises the unique *Pvu*I site (CGAT/CG) at position 3734.

4.1.3.2 The Tetracycline Resistance Genes

The tetracycline resistance cassette of pBR322 contains three open reading frames which are partially overlapping and span approximately 1 200 bp (Fig. 4.1-4). The 386 amino acid peptide from the major reading frame (position 86-1273) is associated with the tetracycline-resistant phenotype because the synthesis of this protein ceases, and tetracycline resistance disappears, if the DNA is interrupted either by small deletion mutations (Backman and Boyer, 1983), or by the insertion of DNA fragments in the unique *Sal*I site (G/TCGAC; position 650) of pBR322. Other useful positions for marker inactivation are the

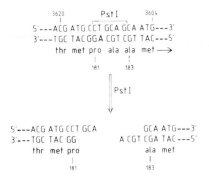

Fig. 4.1-11. The *Pst*I site of pBR322.
After *Pst*I digestion, the β-lactamase gene is interrupted immediately before the codon for amino acid 182 (ala).

Bam HI site (G/GATCC) at position 375, and the *Sph*I site (GCATG/C) at position 561. The neighbouring *Hind* III (A/AGCTT; position 29) and *Cla* I (AT/CGAT; position 23) sites are only of limited use for marker inactivation since they lie in the immediate vicinity of the promoter region for the *tet* resistance gene. This promoter is defined by a Pribnow box between positions 33 and 39 and an initiation site for mRNA synthesis at position 45 (Fig. 4.1-10). An insertion of foreign DNA at positions 23 or 29 would not necessarily influence this region.

Four open reading frames for *tet* had originally been observed in the published pBR322 sequence; the situation has been clarified in a revision of the DNA sequence by the insertion of a CG pair at position 526 (Peden, 1983).

4.1.4 Plasmids With Broad Host Range

The use of *E. coli* vector plasmids, for example pBR322 and its derivatives, is restricted to a narrow spectrum of Gram-negative organisms comprising, *e.g. Salmonella* and *Serratia*. These vectors cannot be used in soil bacteria, such as the *Pseudomonas* species, which include a wide range of industrially important organisms. Soil bacteria can utilise a wide range of different carbon and nitrogen sources, *e.g.* methanol, terpenes, alkaloids, and above all, aromatic compounds. They fix nitrogen (*Rhizobia*), possess photosynthetic activities, and are capable of symbiosis with plant cells (*Agrobacteria*). As will be seen in Chapter 10, such symbiotic relationships are of practical importance for the transfer of genes into plant cells. Methylotrophic bacteria already play an important role in biotechnology and are used for the production of tons of single cell proteins (SCP). In one of these strains, *Methylophilus methylotrophus*, the utilisation of methanol, and hence the energy balance, has been improved by approximately 5% by the introduction of the glutamate dehydrogenase gene from *E. coli* (Win-

dass *et al.*, 1980). In view of the large scale of industrial productions, this improvement is also of considerable economic interest.

The example discussed above demonstrates that *E. coli* genes can be expressed satisfactorily in other Gram-negative organisms; very often the reverse is not true. Genes from *Pseudomonads*, for example, are only poorly expressed in *E. coli* (Jacoby *et al.*, 1978) and have to be analysed in their original hosts.

Such analyses are facilitated by a range of broad host-range plasmids, which belong to two incompatibility groups, P and Q/P4. P group plasmids are usually derived from *Pseudomonads*. The prototype of this group is RP4 (also known as RP1 or RK2) (Barth and Grinter, 1977) (Fig. 4.1-12). RP4 is 56 kb in length and carries tetracycline, neomycin, kanamycin, and ampicillin resistance genes. Plasmids of this kind are characterised by a paucity of restriction cleavage sites and scattered replication and transfer functions. It would be expected for a plasmid of RP4 size, for example, to have 10-15 sites for *Eco* RI, *Bgl* II, *Bam* HI, or *Hind* III; instead, these sites occur only once and are usually located within regions coding for antibiotic resistance. One reason for this relative paucity of restriction sites may be the continuous transfer of these plasmids to a wide range of different hosts with as many different restriction systems. The presence of restriction sites within resistance genes suggests that these genes may have been acquired only fairly recently, for example, by transposition.

Another property of P plasmids, which distinguishes them from other plasmids, is that replication and transfer functions are spread over the entire DNA. This particular feature makes it extremely difficult to construct lower molecular weight derivatives which could be used as cloning vehicles, although two such cloning vectors have been described. Vector pRK2501, which is 11 kb in length, lacks all transfer and mobilisation functions (Kahn *et al.*, 1979). Since this vector is transfer-deficient and non-mobilisable, it satisfies important requirements for the biological safety

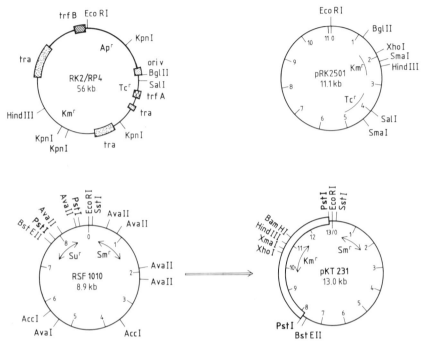

Fig. 4.1-12. Structure of plasmids with broad host specificity.
Plasmids RK2 and pRK2501 belong to incompatibility group IncP; RSF1010 and pKT231 to group IncQ/P4. In pKT231, the small *Pst*I fragment of RSF1010 has been replaced by a 3.8 kb *Pst*I fragment from transposon Tn601/903 with a kanamycin resistance gene (open bar). Its nucleotide sequence is not homologous to the Tn5-derived kanamycin resistance gene (*cf.* Fig. 8-7). Su = sulfonamide; Sm = streptomycin; Km = kanamycin.

of vectors with broad host-range. The usefulness of pRK2501 is limited in those cases where the host bacteria, *e.g.*, *Rhizobia*, cannot easily be transformed. In such instances, the transfer functions would not necessarily be desired, but it would be of advantage to use the mobilisation functions, because the vector could be mobilised by a second plasmid and transferred to the desired recipient cells. Such a system which employs two plasmids has been developed for cloning in *Rhizobia* by Ditta *et al.* (1980) (Fig. 4.1-13). The first plasmid, pRK290, is the cloning vehicle which is transfer-deficient but still possesses all mobilisation functions (Tra⁻ Mob⁺) and can replicate equally well in *E. coli* and *Rhizobium* because it contains an RK2 replicon. This vector

can be mobilised by the second plasmid residing in the donor cells, and hence can be transferred to suitable recipient cells, *e.g.*, *Rhizobium*. The mobilising plasmid in the donor is pRK2013, which possesses all transfer functions and also contains the neomycin resistance region of RK2 and the ColE1 replicon, which allows its replication in *E. coli* but not in *Rhizobium* (Figurski and Helinski, 1979). pRK2013 can therefore be transferred to other *E. coli* recipients but cannot direct its own transfer to *Rhizobium*. When this plasmid enters *E. coli* cells, the transfer functions of pRK2013 are expressed and mobilise plasmid pRK290, which can replicate in *Rhizobium* cells. The physical separation of transfer and mobilisation functions, *i.e.*, their localisation on two

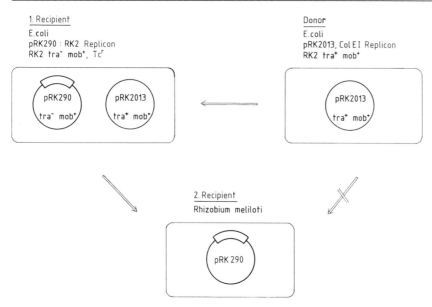

Fig. 4.1-13. Tripel system for cloning with plasmids of broad host specificity. The donor contains plasmid pRK2013 with RK2 transfer functions and a ColE1 replicon (Figurski and Helinski, 1979). The ColE1 replicon is responsible for a narrow host specificity, and the plasmid can therefore be transferred only to the first recipient which is an *E. coli* strain. This transfer mobilises plasmid pRK290 which carries the RK2 replicon and therefore possesses broad host specificity. pRK290 can exist in the second recipient, *Rhizobium meliloti*. pRK290 also carries the recombinant foreign DNA (open bar) which is thus transferred from *E. coli* to *Rhizobium* cells (Ditta *et al.*, 1981).

different replicons, allows a degree of biological safety which, in general, is observed only with plasmids of a very narrow host specificity.

The second incompatibility group, Q/P4, is defined by plasmid RSF1010 and its derivatives. RSF1010 itself is comparatively small (8.9 kb) and is non-conjugative, but can be mobilised by conjugative plasmids such as RP4 or the F factor. Since this plasmid contains only a few useful restriction sites in its sulfonamide and streptomycin resistance genes, Bagdasarian *et al.* (1981) have constructed a set of derivatives containing other resistance genes. Vector pKT231 for example contains an additional 3.8 kb DNA fragment, which codes for kanamycin resistance and was derived from transposon Tn601/Tn903. This DNA fragment provides *Hind* III, *Xma* I, *Xho* I, *Eco* RI, and *Sst* I sites for cloning by marker inactivation. For recipients which are poorly, or not at all, transformable, and into which plasmids can be transferred only by conjugation, safety requirements can be met by using conjugative plasmids with a narrow host-range, such as R64drd11, in order to mobilise pKT231 (Willetts and Crowthers, 1981).

An alternative for cloning in Gram-negative organisms would be provided by plasmids with a narrow host-range which exclude *E. coli*. At present, this solution seems remote, because the characterisation of such plasmids has only now begun (for *Methylomonas clara* plasmids *cf.* Marquardt and Winnacker, 1984).

4.1.5 The P15A Replicon

Plasmids carrying the P15A replicon rather than the ColE1 replicon gain increasing interest in gene cloning. Plasmid P15A from *E. coli* 15 is a cryptic plasmid which does not confer any detectable phenotype. It has been isolated, however, in pure form and its replicon has been be linked to a variety of selectable markers to yield a series of interesting cloning vehicles (Chang and Cohen, 1978). The prototype vector is plasmid pACYC177 with a size of 3.45 kb. It is composed of three parts, the replicon from plasmid P15A, a kanamycin resistance gene originally derived from plasmid R6-5, and a β-lactamase gene from transposon Tn3 (Fig. 4.1-14). The properties of pACYC177 and related plasmids are very similar to those of ColE1-type plasmids: they require DNA polymerase I for replication, they exist in multiple copies per cell, and they continue to replicate in the presence of protein synthesis

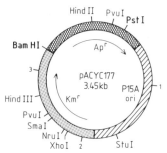

Fig. 4.1-14.   Structure of plasmid pACYC177. This vector is derived from a much larger 10.8 kb plasmid, pACYC142, which contains the P15A replicon (hatched bar), the β-lactamase gene from Tn3 (cross-hatched bar), and a kanamycin resistance gene (stippled bar). The size reduction to 3.42 kb was achieved by a simple "scrambling" procedure whereby pACYC142 was digested with *Hae* II into ten blunt-ended fragments which reassociated to the Km^r Ap^r vector pACYC177. Bold face letters indicate the *Bam* HI/*Pst* I sites required for the formation of a 2.7 kb *Bam* HI-*Pst* I fragment. The P15A replicon resides between position 0.6 and 1.9 although its exact extension is not known (Chang and Cohen, 1978).

inhibitors, for example chloramphenicol. Moreover, these two replicons seem to be closely related as judged by heteroduplex analysis. Nevertheless they are compatible with ColE1-type plasmids such as pBR322, and can thus be used in the development of two-plasmid systems.

Two-plasmid systems offer the considerable advantage that they separate the regulation functions required for the controlled expression of heterologous genes from the cloning of these genes itself. The two examples described in Fig. 7-36 and in Section 7.4.2.3 attest to these attractive possibilities. In both cases, these constructions employed a 2.7 kb *Bam* HI-*Pst* I fragment from pACYC177 which contains the P15A origin of replication and the kanamycin resistance gene.

Stanley N. Cohen
Stanford

Plasmid pcI857, for example, contains this pACYC177-derived DNA fragment in addition to the *cI857* allele of the λ repressor present on a 1.36 kb λ*cI857*-derived DNA fragment. Cells containing this vector are easily recognised since, at 28 °C, they are resistant to wild-type λ infections but sensitive to λ*vir* infections. λ*vir* mutants can grow in the presence of λ repressor due to mutations in the two operators O_L and O_R which affect repressor binding. At 42 °C, pcI857-containing cells are sensitive to both wild-type λ and λ*vir* since, at this temperature, they synthesise an inactive repressor. Plaques are thus formed by both viruses at this temperature. The copy number of pcI857 is almost independent of the temperature and of the copy number of a second, ColE1-derived vector residing in the same cell, with copy numbers varying between 30 and 40 copies per cell (3 to 4% of the total DNA). At 28 °C, the vector thus produces considerable amounts of an active λ repressor which can be used rather efficiently to limit the expression of a heterologous gene present on a second vector, and which is driven by the λ P_L promotor, while this control is released by a simple temperature shift. The P15A replicon, which is the basis of these and other constructions, will no doubt be of increasing value in various applications.

4.1.6 Single-stranded Plasmids

The availability of single-stranded DNA is of major importance for a variety of applications in gene technology, for example, DNA sequencing, site-directed mutagenesis *via* oligonucleotides, construction of strand-specific cDNA libraries, and the generation of highly radioactive probes. The development of M13 vectors (*cf.* Section 2.4.2.3) has been a major breakthrough although the length of DNA which can be inserted into these vectors is limited. It was therefore necessary to develop other single-stranded plasmid vectors

which allow larger DNA fragments (up to 10 kb) to be inserted. In these vectors, all important *cis* elements for viral DNA synthesis and packaging of the filamentous phages (M13, f1) are located in a small region of DNA known as the intergenic region. In M13, this region lies between nucleotides 5498 and 6005 (*cf.* also Fig. 2.4-12 and Table 2.4-1). Any plasmids into which this particular intergenic region of M13 is introduced, for example, by the insertion of the 514 bp M13 *Rsa* I fragment, should therefore be replicated efficiently after infection with wild-type M13. Such plasmids should also be excreted into the growth medium in a single-stranded and packaged form, and the supernatant of such cultures should contain M13 phage particles derived from the M13 helper phage and packaged single-stranded transducing plasmid particles. Indeed, the single-stranded DNA isolated from such particles is completely free of the complementary strand, and can therefore be used for the applications mentioned above.

Two classes of such vectors are available for practical applications. pEMBL vectors (Dente *et al.*, 1983) contain an *Eco* RI fragment of 1 300 bp carrying the f1 intergenic region, which is inserted into the *Nar* I site of vector pUC8 (Fig. 7-54). These constructions contain all pUC8-specific functions such as ampicillin resistance and α-complementation. As shown in Fig. 4.1-15, the *Eco* RI fragment, and hence the f1 intergenic region, can be inserted in two different orientations, determining which of the two strands will be incorporated into the transducing particles. In pEMBL8(+), the β-galactosidase gene is excreted as the complementary (–) strand, in pEMBL8(–) as the coding (+) strand (Fig. 4.1-16).

The second class of such vectors, which comprises the pSD plasmids, employs a DNA fragment from the intergenic region of M13 between position 5372 and 5943 (Levinson *et al.*, 1984). For unknown reasons, this fragment, which also includes 128 bp of the coding sequence of gene *IV* (*cf.* also Fig. 2.4-12), increases the yield of

Fig. 4.1-15. Structure of pEMBL8 vectors.
The pEMBL8 vectors contain a 1 247 bp *Eco* RI fragment with the intergenic region (*IG*) of phage f1, which extends from an *Hga* I site at position 5159 in the f1 genome to the *Hind* II site at position 6406 (hatched bar). Apart from the centrally located *IG* region it thus contains 276 *C*-terminal amino acids (out of 426) from gene *IV* and 279 *N*-terminal codons (out of 410) from gene *II* (see also Fig. 2.4-13). It is inserted in two orientations within the *Nar* I site of pUC8 (open bar), immediately at the transition between the *lacZ* insert and the pBR322 portion. Arrows indicate the direction of transcription of the β-lactamase gene and the *lacZ* insert; fat arrows designate the direction of *ori* RNA ((-) strand) synthesis initiated from the f1 insert (Dente *et al.*, 1983).

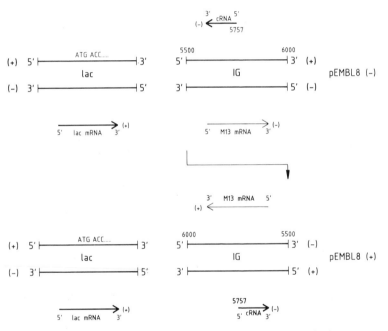

Fig. 4.1-16. Relative orientation of the *lac* and intergenic (*IG*) regions in vectors pEMBL8(+) and pEMBL8(-).
In pEMBL8(-), the directions of transcription of *lac* mRNA and of the viral (-) strand synthesis (cRNA) point in opposite directions. In this situation, the coding (+) strand of the *lac* insertion is packaged into transducing particles because it is colinear with the M13-*IG* (+) strand. In pEMBL8(+) the respective orientations are reversed, *i.e.*, *lac* mRNA and M13 (-) strand synthesis proceed in the same directions. Numbers at the termini of the *IG* region refer to the sequence published by Beck and Zink (1981) (*cf.* chapter 2.4). The section of the sequence shows the two initial codons of the *lacZ* gene. In the M13mp vector system, the arrangement of the *lac* insertion and the *IG* region is identical with that in pEMBL(-), *i.e.*, the (+) strand of M13 is colinear with the (+) strand (the sense strand) of the *lac* insert.

obtained after insertion of intergenic region sequences only. Plasmids pSDL12 and pSDL13 (Fig. 4.1-17) are composed of the M13 origin, the *lacZ* α-complementation fragment of pUC12 or pUC13 (Vieira and Messing, 1982), and the suppressor miniplasmid ΠAN7 (*cf.* Section 9.4.5). This mini-plasmid is an improved version of ΠVX (*cf.* Fig. 9-6). In contrast to ΠVX, ΠAN7 contains the entire promoter of the replication primer RNA of pBR322, *i.e.*, the "-10" as well as the "-35" region, and this increases plasmid yields.

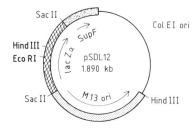

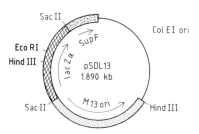

Fig. 4.1-17. Structure of vectors pSDL12 and pSDL13.
Apart from the *supF* and M13 regions mentioned in the text, vectors pSDL12 and pSDL13 contain the *lacZ* α-complementation region of pUC12 and pUC13, respectively. Transcription of the M13 and *lacZ* sequences in the pSDL vectors proceeds as in pEMBL(+) (Fig. 4.1-16), *i.e.*, in opposite direction to that observed in the M13mp series of vectors (Levinson *et al.*, 1984). In pSDL vectors it is, thus, the *lacZ*(-), or anti-sense, strand which is packaged into transducing particles. *Hind* III and *Eco* RI mark the respective flanking restriction sites of the polylinker insertions (*cf.* Fig. 2.4-18).

The propagation and identification of *supF*-bearing mini-plasmids requires special bacterial strains which must harbour the Tra⁻ derivative p3 of plasmid RP1. The resistance markers of the latter plasmid have amber mutations which can be suppressed by the suppressor on the mini-plasmid. Special strains which harbour plasmid p3 in conjunction with the F factor (JM103 for example) must therefore be used for M13/mini-plasmid derivatives. In comparison with pEMBL vectors and, above all, the M13mp vectors, the major advantage of pSD plasmids is that they are extremely small (1.9 kb) and therefore allow the incorporation of up of 10 kb of foreign DNA, which M13mp vectors would not accomodate.

4.1.7 Transformation Techniques

The introduction of biologically active recombinant plasmid DNA into bacterial cells holds a key position in all cloning experiments. The transfer of DNA can be effected by transformation, *i.e.*, by the immediate uptake of naked plasmid DNA by competent cells. An alternative approach is *in vitro* packaging of recombinant DNA into empty heads of bacteriophage λ, which then can be used for subsequent infections. *In vitro* packaging requires a certain minimal length of DNA and the presence of the *cos* sites (*cf.* also Section 4.2.2.5). These two features have been realised in λ vectors and cosmid cloning vehicles. If these vectors cannot be used, for whatever reasons, the technique of choice is bacterial transformation. It has been perfected to such an extent that it can be applied satisfactorily in most cases, although the technique is not as effective as a phage infection.

Most transformation techniques are based on the observation originally made by Mandel and Higa (1970), who noticed, for the first time, that the uptake of λ DNA by bacteria can be increased considerably in the presence of calcium chloride. Cohen *et al.* (1972) were the first to employ this technique for the transfection of plasmid DNA. Since then, this method has been improved by many modifications. The standard procedure described by Mandel and Higa (1970) yields approximately 10^6-10^7 transformants per microgram of DNA. The starting material is a population of growing bacteria with a cell density of approximately 5×10^7 cells/ml ($OD_{550} = 0.2$-0.4), which is harvested and resuspended in 50 mM calcium chloride at a concentration of 10^8 cells/ml. Dagert and Ehrlich (1979) have observed that the efficiency of transformation can be increased by a factor of 4 to 10 if the cells are resuspended at 5×10^{10} cells/ml and incubated in calcium chloride for 12-24 h. After the addition of plasmid DNA the competent bacteria are plated on suitable selective agars.

It may seem that transformation efficiencies of 10^7 transformants per microgram of DNA are extremely high, but it should be noted that this corresponds to only one per 10 000 plasmid molecules actually entering the cell. This ratio can be increased further by using the modified procedure described by Kushner (1978). Apart from special E. coli strains, such as SK1590 and SK1592, the use of rubidium chloride and dimethylsulfoxide plays a decisive role. A detailed analysis by Hanahan (1983) describes conditions for transformation under which only one in 400 plasmid molecules is required for cell transformation (5×10^8 transformants/ microgram of DNA). Apart from reaction conditions, such as elevated Mg^{2+} concentrations, or the presence of hexamine cobalt(III)chloride, the use of suitable E. coli strains is also an essential factor influencing transformation efficiencies. Such strains, the isogenic pair DH1 and MM294 in particular, are also discussed in Appendix E. Excellent transformation frequencies have also been obtained with strain X1776, which was originally developed as an EK2 host (Curtiss et al., 1977). This strain is auxotrophic for thymidine and diaminopimelic acid, and is also very sensitive to detergents; nevertheless, a special transformation technique allows one to obtain approximately 10^8 transfor-

mants per microgram of DNA (Hanahan and Meselson, 1980). Because of its excellent transformability it appears worthwhile, even today, to use this strain for pilot transformation experiments although its use is no longer necessitated by safety considerations. Plasmid DNA obtained from X1776 can subsequently be transferred to cells which are easier and more economic to grow.

References

Backman, K., and Boyer, H.W. (1983). Tetracycline resistance determined by pBR322 is mediated by one polypeptide. Gene 26, 197-203.

Bagdasarian, M., Lurz, R., Rückert, B., Franklin, F.C.H., Bagdasarian, M.M., Frey, J., and Timmis, K.N. (1981). Specific-purpose plasmid cloning vectors II. Broad host range, high copy number, RSF1010 derived vectors, and a host-vector system for gene cloning in Pseudomonas. Gene 16, 237-247.

Barth, P.T., and Grinter, N.J. (1977). Map of plasmid RP4 derived by insertion of transposon C. J. Mol. Biol. 113, 455-474.

Bolivar, F., Rodriguez, R.L., Greene, P.J., Betlach, M.C., Heyneker, H.L., Boyer, H.W. (1977). Construction and characterization of new cloning vehicles. II. A multipurpose cloning system. Gene 2, 95-113.

Brosius, J., Cate, R.L., and Perlmutter, A.P. (1982). Precise location of two promotors for the β-lactamase gene of pBR322. J. Biol. Chem. 257, 9205-9210.

Chang, A.C.Y., and Cohen, S.N. (1978). Construction and characterization of amplifiable multicopy DNA cloning vehicles derived from the P15A cryptic miniplasmid. J. Bacteriol. 134, 1141-1156.

Cesarini, G., and Banner, D.W. (1985). Regulation of plasmid copy number by complementary RNAs. Trends in Biochem. Sci. 10, 303-306.

Clark, A., and Warren, G.J. (1979). Conjugal transmission of plasmids. Ann. Rev. Genet. 13, 99-125.

Cohen, S.N., Chang, A.C.Y., and Hsu, L. (1972). Nonchromosomal antibiotic resistance in bacteria: Genetic transformation of E. coli by R-factor DNA. Proc. Natl. Acad. Sci. USA 69, 2110-2114.

Covarrubias, L., and Bolivar, F. (1982). Construction

and characterization of new cloning vehicles. VI. Plasmid pBR329, a new derivative of pBR328 lacking the 428 base pair inserted duplication. Gene 17, 79-89.

Curtiss, R.III., Pereira, D.A., Hsu, J.C., Hull, S.C., Clark, J.E., Maturin, L.F., Goldschmidt, R., Moody, R., Inoue, M., and Alexander, L. (1977). Biological Containment. The subordination of *Escherichia coli* K12. In: "Recombinant Molecules: Impact of Science and Society", pp.45-56. Beers R.R., and Basset, E.G. (eds.), Raven Press, New York.

Dagert, M., and Ehrlich, S.D. (1979). Prolonged incubation in calcium chloride improves the competence of *Escherichia coli* cells. Gene 6, 23-28.

Dente, L., Cesarini, G., and Cortese, R. (1983). pEMBL: a new family of single-stranded plasmids. Nucleic Acids Res. 11, 1645-1655.

Ditta, G., Stanfield, S., Corbin, D., and Helinski, D.R. (1980). Broad host range cloning system for Gram-negative bacteria: Construction of a gene bank of *Rhizobium meliloti*. Proc. Natl. Acad. Sci. USA 77, 7347-7351.

Figurski, D.H., and Helinski, D.R. (1979). Replication of an origin-containing derivative of plasmid RK2 dependent on a plasmid function provided *in trans*. Proc. Natl. Acad. Sci. USA 76, 1648-1652.

Hanahan, D., and Meselson, M. (1980). A protocol for high density screening of plasmids in ×1776. Gene 10, 63-67.

Hanahan, D. (1983). Studies on transformation of *Escherichia coli* with plasmids. J. Mol. Biol. 166, 557-580.

Helinski, D.R. (1979). Bacterial Plasmids: Autonomous replication and vehicles for gene cloning. CRC Critical Reviews in Biochemistry 7, 83-101.

Jida, S., Meyer, J., and Arber, W. (1983). Prokaryotic IS Elements. In: "Mobile genetic elements", pp.159-221. Shapiro, I.A. (ed.), Academic Press, New York and London.

Itoh, T., and Tomizawa, J. (1980). Formation of an RNA primer for initiation of ColE1 DNA by ribonuclase H. Proc. Natl. Acad. Sci. USA 77, 2450-2454.

Jakoby, G.A., Rogers, J.E., Jacob, A.E., and Hedges, R.W. (1978). Transposition of *Pseudomonas* toluene-degrading genes and expression in *Escherichia coli*. Nature 274, 179-180.

Kahn, M., Kolter, R., Thomas, C., Figurski, D., Meyer, R., Remaut, E., and Helinski, D.R. (1979). Plasmid cloning vehicles derived from Plasmids ColE1, F, R6K and RK2. Methods in Enzymology 68, 268-280.

Kleckner, N. (1981). Transposable Elements in Prokaryotes. Ann. Rev. Genet. 15, 341-404.

Kushner, S.R. (1978). An improved method for transformation of *Escherichia coli* with ColE1-derived plasmids. In "Genetic Engineering", pp.17-23. Boyer, H.W., and Nicoria, S. (eds.), Elsevier/North Holland, Amsterdam.

Lacatena, R.M., and Cesarini, G. (1981). Base pairing of RNA I with its complementary sequence in the primer precursosr inhibits ColE1 replication. Nature 294, 623-626.

Levinson, A., Silver, D., and Seed, B. (1984). Minimal size plasmids containing an M13 origin for production of single-stranded transducing particles. J. Mol. Appl. Gen. 2, 507-517.

Mandel., M., and Higa, A. (1970). Calcium dependent bacteriophage DNA infection. J. Mol. Biol. 53, 154-162.

Marquardt, R., and Winnacker, E.L. (1984). Characterization of plasmids from *Methylomonas clara*. J. Biotechnology 1, 317-330.

Meacock, P.A., and Cohen, S.N. (1979). Genetic analysis of the interrelationship between plasmid replication and incompatibility. Mol. Gen. Genet. 174, 135-147.

Meacock, P.A., and Cohen, S.N. (1980). Partitioning of bacterial plasmids during cell division: a *cis*-acting locus that accomplishes stable plasmid inheritance. Cell 20, 529-542.

Nordström, K., Molin, S., and Agaard-Hausen, H. (1980). Partitioning of plasmid R1 in *Escherichia coli*: I. Kinetics of lose of plasmid derivatives deleted of the *par* region. Plasmid 4, 215-227.

Nordström, K., Molin, S., and Light, J. (1984). Control of replication of bacterial plasmids: Genetics, Molecular Biology and physiology of the plasmid R1 system. Plasmid 12, 71-90.

Ogura, T., and Hiraga, S. (1983). Partition mechanism of F plasmid: two plasmid gene-encoded products and a *cis*-acting region are involved in partition. Cell 32, 351-360.

Peden, K.W.C. (1983). Revised sequence of the tetracycline-resistance gene of pBR322. Gene 22, 277-280.

Prentki, P., and Krisch, H.M. (1980). A modified pBR322 vector with improved properties for the cloning, recovery and sequencing of blunt-ended DNA fragments. Gene 17, 189-196.

Soberon, X., Covarrubias, L., and Bolivar, F. (1980). Construction and characterization of new cloning vehicles. IV. Deletion derivatives of pBR322 and pBR325. Gene 9, 287-305.

Suinsky, J.J., Uhlin, B.E., Gustafsson, P., and Cohen, S.N. (1981). Construction and characterization of a novel two-plasmid system for accomplishing temperature-regulated, amplified expression of cloned adventitious genes in *Escherichia coli*. Gene 16, 275-286.

Sutcliffe, J.G. (1978). Nucleotide sequence of the ampicillin resistance gene of *Escherichia coli* plasmid pBR322. Proc. Natl. Acad. Sci. USA 75, 3737-3741.

Sutcliffe, G. (1979). Complete nucleotide sequence of the *Escherichia coli* plasmid pBR322. Cold Spring Harbor Symp. Quant. Biol. 43, 77-90.

Thomas, C.M. (1981). Molecular genetics of broad host range plasmid RK2. Plasmid 5, 10-19.

Tomizawa, J. (1984). Control of ColE1 plasmid replication: the process of binding of RNA I to the primer transcript. Cell 38, 861-870.

Tucker, W.T., Miller, C.A., and Cohen, S.N. (1984). Structural and functional analysis of the *par* region of the pSC101 plasmid. Cell 38, 191-201.

Twigg, A., and Sherrat, D. (1980). Trans-complementable copy-number mutants of plasmid ColE1. Nature 283, 216-218.

Uhlin, B.E., Molin, S., Gustafsson, P., and Nordström, K. (1979). Plasmids with temperature-dependent copy number for amplification of cloned genes and their products. Gene 6, 91- 106.

Uhlin, B.E., and Clark, A.J. (1981). Overproduction of the *E. coli recA* protein without stimulation of its proteolytic activity. J. Bacteriol 148, 386-390.

Vieira, J., and Messing, J. (1982). The pUC plasmids, an M13mp7-derived system for insertion mutagenesis and sequencing with synthetic universal primers. Gene 19, 259-268.

Villa-Komaroff, L., Efstratiadis, A., Broome, S.W.L., and Gilbert, W. (1978). A bacterial clone synthesising proinsulin. Proc. Natl. Acad. Sci. USA 75, 3727-3731.

Watson, M. (1982). Replication control and incompatibility in bacterial plasmids. Trends in Biochem. Sciences 7, 198.

Willetts, N., and Crowthers, C. (1981). Mobilization of the non-conjugative IncQ plasmid RSF1010. Genet. Res. Camb. 37, 311-316.

Williams, N., and Skurray, R. (1980). The conjugation system of F-like plasmids. Ann. Rev. Genet. 14, 41-76.

Windass, J.D., Worsey, M.J., Pioli, E.M., Barth, P.T., Atherton, K.T., Dart, E.C., Byrom, D., Powell, K., and Senior, P.J. (1980). Improved conversion of methanol to single cell protein by *Methylophilus methylotrophus*. Nature 287, 396-401.

Wong, E.M., and Polisky, B. (1985). Alternative conformations of the ColE1 replication primer modulate its interaction with RNA I. Cell 42, 959-966.

Young, J.G., and Poulis, M.I. (1978). Conjugal transfer of cloning vectors derived from ColE1. Gene 4, 175-179.

4.2 Bacteriophage λ as Cloning Vehicle

Derivatives of bacteriophage λ have been developed as cloning vectors since the early days of gene technology (Murray and Murray, 1974; Rambach and Tiollais, 1974). Even today, these derivatives are considered to be the most suitable cloning vehicles for cloning genomic eukaryotic DNA, because they have a number of advantages over plasmids. Firstly, several thousand phage plaques can be screened and characterised, *e.g.*, by DNA-DNA hybridisation, on a single Petri dish; in addition, an *in vitro* packaging system is available, which allows DNA to be packaged into empty phage heads. This increases the infectivity of recombinant DNA by several orders of magnitude as compared to pure DNA preparations. The very same *in vitro* system also allows a size selection of the packaged DNA such that under suitable conditions only recombinant DNA molecules will be packaged. Finally, millions of independently cloned virus particles constituting a gene library which represents the entire genetic information of a complex organism can be stored in a few milliliters of broth. In order to understand these particular features, the structure of the phage genome and its interactions with a host cell have to be considered in detail.

Phage λ is representative of a number of related temperate *E. coli* phages, such as Φ80, 82, 21, 434, P2, and 299. An infection with these phages leads to either lytic growth or entry into the lysogenic pathway. The lytic or productive pathway ultimately causes cell death and leads to the production of approximately 100 progeny phage particles per infected cell. The lysogenic pathway allows cells to survive infection by enabling the phage genome to become covalently inserted (integrated) into the host genome, which then perpetuates the viral genome as a prophage. A number of essential genes and control circuits must be taken into consideration for the development of phage λ as a cloning vehicle. These elements will be discussed below.

4.2.1 Molecular Biology of the λ Genome

λ DNA has a molecular weight of 31×10^6 Da and is 48 502 bp in length (Sanger *et al.*, 1982). The DNA isolated from virus particles is a double-stranded linear molecule with short complementary single-stranded projections of 12 nucleotides at its 5′ ends. These cohesive termini, also referred to as *cos* sites, allow the DNA to be circularised after infection of the host cell (Fig. 4.2-1). The *cos* sites were the first biologically relevant DNA sequences to be structurally analysed (Wu and Taylor, 1971).

The genetic map of phage λ comprises approximately 40 genes which are organised in functional clusters (Fig. 4.2-2) (*cf.* also Appendix D; Szybalski and Szybalski, 1979). Genes coding for head and tail proteins (genes *A-J*) are on the left of the conventional linear map. The central region contains genes, such as *int*, *xis*, *exo* etc.,

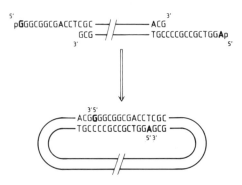

Fig. 4.2-1. Structures of linear and circular forms of λ DNA.
Intra-molecular circularisation takes place by base pairing between overlapping protruding complementary termini. Concatemeric forms are generated by inter-molecular reactions between different DNA molecules.

Noreen E. Murray
Ken Murray
Edinburgh

which are responsible for lysogenisation, *i.e.*, the process leading to the integration of viral DNA and other recombination events. Much of this central region is not essential for lytic growth and can be deleted for the construction of suitable vectors. Genes to the right of the central region comprise six regulatory genes (*cI*, *cII*, *cIII*, *cro*, *N*, and *Q*), two other genes, *O* and *P*, which are essential for DNA replication during lytic growth, and genes *S* and *R* which are required for the lysis of cellular membranes.

The lytic cycle can be divided into three stages which are known as early, delayed early, and late (Fig. 4.2-3). Immediately after infection, bacterial RNA polymerase binds to the four λ promoters P_L, P_R, P_0, and P'_R. Leftward transcription starting at promoter P_0 yields a transcript of 77 nucleotides which has been postulated as a primer of DNA synthesis, although this so-called *oop* RNA is 350 bases away from the origin of DNA replication (*ori*). Since this particular RNA species is irrelevant for a discussion of transcriptional control mechanisms, it shall not be considered further. Promoter P'_R regulates the synthesis of an RNA which is 194 nucleotides in length

when its transcription is terminated. In the late phase of infection (phase 3 in Fig. 4.2-3), this transcript is extended by the intervention of the gene *Q* protein (see below) such that it spans the entire late region. The extended transcript serves as mRNA for viral structural proteins in the late phase of infection.

Promoters P_L and P_R are particularly important for the early phase of infection. Due to the action of the cellular terminator protein, ϱ, leftward and rightward transcription initiated at P_L and P_R terminate after 850 (P_L) and 310 (P_R) nucleotides at the terminator sites t_{L1} and t_{R1}; nevertheless, these short transcripts lead to the synthesis of two proteins, N and Cro. Cro protein, which is coded for by the rightward transcript initiated at P_R, binds to the operator O_R and blocks the neighbouring promoter P_{RM}, but not its own promoter P_R (Fig. 4.2-5). This prevents initiation of leftward transcription at promoter P_{RM}, which would otherwise produce the mRNA coding for the cI repressor; this repressor could bind to operators O_L and O_R, and hence block transcription initiating at P_L and P_R, which would, in turn, prevent the entry into the lytic pathway (Ptashne *et al.*, 1976, 1980).

The product of the transcript initiating at promoter P_L is the N protein, which serves as an anti-terminator by releasing the transcriptional block at the terminator sites t_{L1}, t_{R1}, and t_{R2}. The mechanism of anti-termination is not completely understood; in principle, N protein acts in *trans*, but it also requires the presence of a certain short DNA sequence, which has been designated *nut* (N utilisation), and the co-operation of certain host proteins (*cf.* Ward and Gottesmann, 1982, for example). N protein allows transcription, which has been initiated at P_L and P_R, to proceed in leftward and rightward directions, and eventually, this causes the synthesis of proteins cII, cIII, O, and P.

Ray Wu
Ithaca

The infected cell now contains the proteins N, Cro, cIII, cII, O, P, and small amounts of Q protein and must decide between the lysogenic or lytic pathway. This decision is influenced by the relative amounts of cII, cIII, and a host protein known as Hfl. Hfl (high frequency of lysogenisation) is a cellular protease whose activity is influenced by the physiological state of cells and also by the amounts of cIII. At high multiplicities of infection, the cell contains large amounts of cIII, and Hfl protein is therefore inhibited. This inhibition, which can also be achieved by a mutation in the *hfl* locus, in turn increases the activity of protein cII, which would otherwise have been cleaved proteolytically by Hfl. Protein cII stimulates the transcription of the genes for cI and Int proteins. cI is a repressor of the lytic genes and blocks the synthesis of proteins N, O, P, and Q, by binding to operators O_L and O_R, thus preventing the lytic cycle.

At low concentrations of cIII, protein Hfl is not inhibited and reduces the activity, or the amounts, of protein cII. This prevents the synthesis of the cI protein, but the synthesis of N, O, P, and Q remains unaffected. The latter sequence of events precipitates the lytic infection. The particular role and overriding significance of the *hfl* helper function becomes apparent in mutants which do not express Hfl protein, because, in the absence of Hfl protein, suitable phages are quantitatively forced into the lysogenic pathway (Hoyt *et al.*, 1982). Cell lysis and plaque formation in Hfl⁻ strains is only observed when the cI repressor is inactivated, for example, by *ts* mutations or the insertion of recombinant DNA (*cf.* also Section 4.2.2.4).

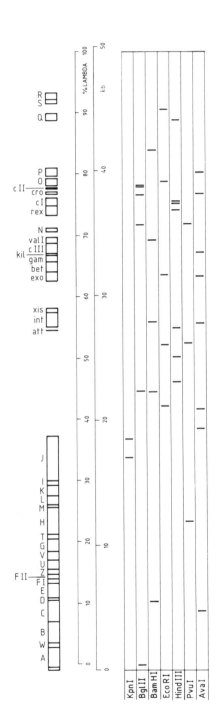

Fig. 4.2-2. Physical and genetic structure of the λ genome.

Shown are the positions of 38 genes and the recognition sites of seven restriction endonucleases. Map positions are based on the sequence published by Sanger *et al.* (1982). Many of the genes shown are mentioned in the text and have been described in detail by Szybalski and Szybalski (1979) (*cf.* also Appendix D).

The synthesis of large amounts of Q protein marks the beginning of the late phase of transcription (phase 3 in Fig. 4.2-3) and the synthesis of head and tail proteins. Like N, protein Q acts as an anti-terminator; both proteins are therefore genuine positive regulators of transcription. It is only in their presence that transcripts which have been prematurely terminated are extended into the delayed early genes (mediated by N gene product), or into the late genes (mediated by Q gene product).

λ DNA replication requires two virus-encoded *trans* functions, *i.e.*, the products of genes O and P, and the two *cis* functions *ori* and *ice*. DNA replication is initiated at *ori*, which is located within gene O. Due to its particular secondary structure, the *ice* (inceptor) site in gene *cII* promotes the switch from RNA synthesis (RNA serves as a primer) to DNA synthesis (Hobom *et al.*, 1979) in replication initiation. The gene O product is required for the recognition of *ori*. Its interaction with protein P mediates the association of *ori* with certain host proteins, *e.g.* the DnaB protein, which is also required for DNA replication (Klein *et al.*, 1978). Initially, DNA replication is bidirectional and involves Θ-type molecules; at a later stage of the lytic cycle, DNA replication proceeds via rolling circles (Fig. 4.2-4). Rolling circle replication yields DNA concatemers, which are cleaved exactly at the *cos* site by the protein coded by λ gene *A* when the viral DNA is packaged into phage heads. Virus particles contain monomeric linear DNA molecules with protruding ends. The lysis of the host cell and the subsequent release of phage particles is mediated by the products of genes S and R. Mutations in gene S can be very useful because they prevent the premature lysis of infected cells and allow accumulation of large amounts of viral particles.

A small fraction of cells infected with wild-type λ enters the lysogenic pathway, which leads to the integration of viral DNA into the host chromosome. Only the phage repressor cI is synthesised from the prophage in lysogenic bacteria. cI interacts with the left and right promoter regions and blocks early transcription at P_L and P_R; in addition, cI repressor also controls its own transcription. This feature can be understood if the interaction of cI with operator O_R is analysed in detail. As shown in Fig. 4.2-5, the operator O_R comprises three regions, each of which is 17 bp in length and is separated from the others by spacers of 6 or 7 nucleotides. The affinity of the repressor for region O_R1 is highest, and hence even marginal concentrations of cI block the promoter site P_R, and therefore transcription of *cro*, which is absolutely required for the entry into the lytic cycle. Binding of cI to O_R1 does not affect the cI promoter P_{RM}, and transcription controlled by cI

Waclaw Szybalski
Madison

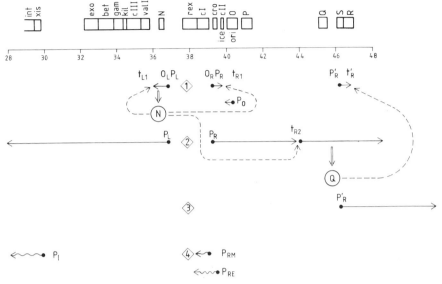

Fig. 4.2-3.   Gene regulation of phage λ.
Shown is a region of the λ genome between positions 28-48 kb encoding regulatory genes. 1, 2, and 3 mark the three phases of the lytic cycle, *i.e.*, the early (1), middle (2), and late phase (3). Products of genes *N* and *Q* act as positive regulators for middle and late transcription. The transcript initiating from P'_R (3) extends into the late genes since the genome is circular rather than linear in a lytically infected cell. In the lysogenic state (4) the only phage encoded products are repressors cI and cII initiating at promoters P_{RE} and P_{RM}, respectively, and the integration protein *Int* which is transcribed from promoter P_I.

remains unaffected. With increasing concentrations of cI, the repressor also binds to regions O_R2 and O_R3, so blocking the transcription initiating at P_{RM}, and hence its own transcription. Protein Cro also binds to operator regions O_R1, O_R2, and O_R3, but in reverse order; at low concentrations it binds predominantly to O_R3, which immediately blocks transcription initiating at P_{RM}, and hence also the synthesis of cI protein.

While the lysogenic state is being established, cI repressor formation is not controlled from promoter P_{RM} (promoter repressor maintenance) but from promoter P_{RE} (promoter repressor establishment), which lies several hundred base pairs upstream from P_{RM}. Promoter P_{RE} is activated by repressors cII and cIII, concentrations of which are very much influenced by the physiological state of the cells. Normally, cII and cIII have no effects at all. Under certain condi-

tions, such as a high multiplicity of infection, they may become active even during a lytic infection and activate the transcription of the *int* (integration) gene, which is initiated at promoter P_I (Fig. 4.2-3). Int protein recognises the *att* (attachment) site, which is 15 bp in length, and is present on both λ and host DNA. The *attP* sequence on phage λ is the site of integration of phage DNA into the homologous host sequence *attB*, on the host chromosome (Fig. 4.2-4). Since the locations of the *cos* sites at the extreme ends of the linear phage DNA and of the *att* site are not identical, the genetic maps of the integrated prophage and free phage DNA become cyclically permutated.

The prophage remains integrated as long as it synthesises active cI repressor, but all events which inactivate the repressor induce the lytic growth cycle. If the repressor is temperature-sensitive, for example, a temperature shift to the

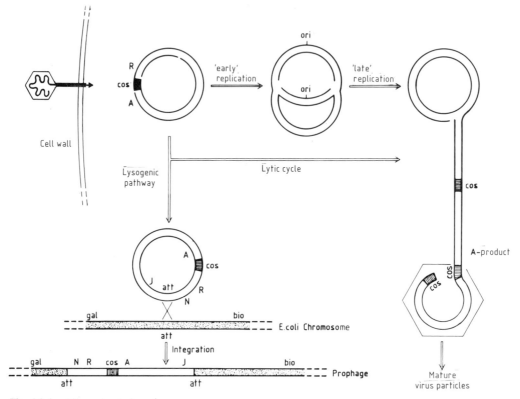

Fig. 4.2-4. Life cycle of phage λ.
As explained in the text, a phage infection may either result in a lytic or a lysogenic interaction between phage and bacterial cells.

non-permissive temperature leads to the excision of the prophage from the host chromosome and initiates the lytic growth phase. Lytic growth also ensues from the activation of the cellular RecA protease, synthesis of which is always induced in bacterial cells when the DNA is damaged, *e.g.* by UV light. The protease inactivates the cI repressor and transcription can then begin at promoter sites P_L and P_R. Among other things, this leads to the synthesis of the phage gene product Xis, which excises the prophage at the *att* site.

A thorough knowledge of λ physiology, the central role of promoters P_L and P_R, and the function of control proteins, in particular, allows the construction of suitable λ cloning vectors which replicate efficiently, and also allow the

DNA cloned in such vectors to be efficiently expressed under the control of λ promoters. The whole system has been extensively reviewed in the book Lambda II (see also Appendix D).

4.2.2 λ Vectors

4.2.2.1 The General Structure of λ Vectors

The region between genes *J* and *N* in the λ genome which spans map positions 38 to 68%, is not essential for lytic growth. In principle, a vector lacking this region could accomodate

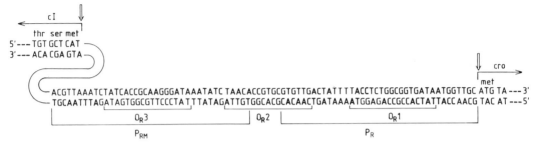

Fig. 4.2-5. Structure of the right λ operator.
Shown are the two promoter regions, P_{RM} and P_R, and the three regions O_R1, O_R2, and O_R3 (bracketed) which interact with cI repressor and cro protein. The promoter regions include the corresponding "-10" and "-35" regions. Vertical arrows mark the start of transcription for genes *cI* and *cro*. The affinities of proteins cI and cro for the two operator regions is as follows. *cI*: $O_R1 > O_R2 > O_R3$; *cro*: $O_R3 > O_R2 = O_R1$ (Ptashne *et al.*, 1980). The sequence shown is that published by Ptashne *et al.* (1976).

30 x 485 = 14 500 bp of foreign DNA, which would then restore the original length of the λ genome. However, it is known that up to 105% of the normal complement of λ DNA can be packaged into phage heads, and that λ arms contain some other non-essential regions which can be removed. By taking these observations into account, a maximum value of 24.6 kb of foreign DNA, which can be inserted into λ DNA, is obtained. While 105% of DNA (52 kb) constitutes the upper limit for packagable DNA, there is also a lower limit below which DNA cannot be packaged, which, in this case, is 75% (38 kb). To be more precise, DNA molecules shorter than 38 kb are not packaged efficiently and plaques are formed only if additional DNA is incorporated. A vector DNA which is smaller than 38 kb, therefore, offers a strong selection for those molecules which contain inserted foreign DNA.

Two kinds of λ vectors are known today (Fig. 4.2-6). Vectors which contain a unique site (x) for the insertion of foreign DNA have been designated *insertional* vectors (type A in Fig. 4.2-6); vectors with two cleavage sites, which allow foreign DNA to be substituted for the DNA

sequences between these sites, are known as *replacement* vectors (type B in Fig. 4.2-6). If the central portion of a replacement vector is removed, the residual genome, which is usually obtained by ligating the two λ arms, cannot be packaged and therefore allows the positive selection for recombinant phage genomes.

Gerd Hobom
Gießen, FRG

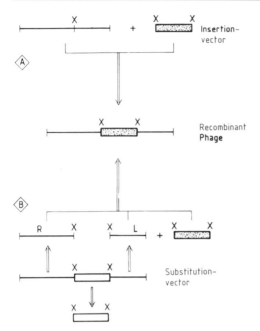

Fig. 4.2-6. Structure of insertional and replacement vectors.
The upper panel (A) shows the use of an insertional vector (solid line) for cloning a DNA fragment (stippled bar) obtained by digestion with endonuclease X for which there is a single site on the vector DNA. The lower panel (B) shows the use of a replacement vector. The replaceable fragment (open bar) is removed, and the space is filled by inserting foreign DNA obtained by digestion with endonuclease X between the left (L) and right (R) arms of the vector.

4.2.2.2 Rationale for Vector Constructions

Wild-type λ DNA itself cannot be used as a vector, since it contains too many restriction sites. Furthermore, these sites are often located within the essential regions described above. The distribution of sites can be altered by introducing point mutations, substitutions, and deletions in a two-step procedure. The first step is to look for λ mutants which do not contain a target site for a particular restriction enzyme; in the second step, the desired cleavage sites are introduced by

genetic crosses. It is comparatively easy to eliminate restriction cleavage sites completely by growing the phages on restricting hosts. Only phage DNA devoid of susceptible *Eco* RI sites will be able to survive in host cells containing, for example, the *Eco* RI restriction and modification system. Such restriction-resistant phages can be crossed subsequently with suitable susceptible phages in order to obtain the desired combination of cleavage sites.

The following example may illustrate this technique (Fig. 4.2-7) (Rambach and Tiollais, 1974). The starting material for the experiment was λ strain λ*b221*, which possesses only three, instead of five, *Eco* RI sites due to the presence of the *b221* deletion. By a fortunate coincidence, these remaining three sites lie within non-essential regions. An *Eco* RI-resistant mutant was obtained by 14 alternate passages on restriction-positive hosts carrying an *Eco* RI-producing plasmid, and restriction-negative strains of *E. coli*. The resistant mutant was crossed with a λ strain which still contained *Eco* RI sites in order to restore the *Eco* RI site at map position 65.5%. One parent strain was the *Eco* RI-resistant phage λ*b221*, the other parent strain contained a long *lac* insertion and two additional *Eco* RI sites in the region which is deleted in λ*b221*. As indicated in Fig. 4.2-7, these two phages can only recombine either before the *Eco* RI site at position 65.5% or after it, because recombination requires homologous DNA sequences. The genetic cross was made even more selective by taking λ*plac5imm*^{21} as the parent strain containing the *lac* insertion, and employing a Lac⁻ host bacterium lysogenic for λ^{imm21}. The two parent phages cannot replicate in this host when growing on lactose medium since λ*b221* is Lac⁻ and incapable of complementing the *lac* deficiency of the host; λ*plac5imm*^{21} could complement this deficiency since it is Lac⁺, but the host is an *imm21* lysogen and therefore produces an *imm21* repressor which interacts with the early control region of λ*plac5imm*^{21}, and thus prevents its replication. Replication of the desired recombinants, however, cannot be prevented,

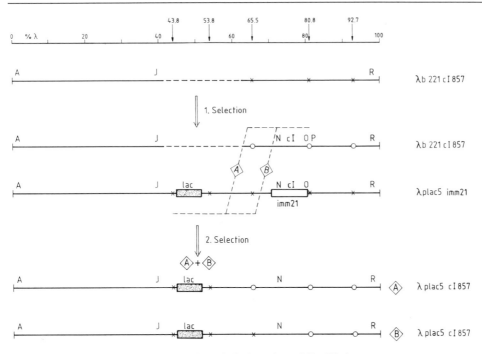

Fig. 4.2-7. Construction of λ vectors with a limited number of *Eco* RI sites.
The wild-type genome with its five *Eco* RI sites and their positions, expressed in per cent of λ DNA length, is shown at the top. Strain λ*b221cI857* carries deletion *b221* which removes two of the five *Eco* RI sites. After the first selection step on a restricting host, one obtains a λ derivative without any *Eco* RI sites. Recombination between this strain and λ*plac5imm²¹* which contains five *Eco* RI sites can occur only in a very limited region near the third *Eco* RI site from the right, because other regions in the vicinity are not homologous. Cross-overs may either occur before (A) or after (B) the *Eco* RI site. Intact *Eco* RI sites are marked "x", missing sites are marked by open circles.

because the recombinant genotype λ*plac5cI857* complements the Lac⁻ mutation of the host, and because its control regions P_L and P_R cannot interact with the imm21 repressor protein synthesised by the host cell. Depending upon whether the crossover took place before or after the *Eco* RI site in question, these crosses will yield two different λplac mutants with either two or three *Eco* RI sites in their non-essential regions.

Other λ derivatives with different restriction sites have been constructed in a similar way. The *Bam* HI site plays a particular role, because *Bam* HI ends can also be used for the insertion of DNA fragments obtained with a number of isoschizomeric enzymes (*cf.* Section 2.1.2.1). In

this case the strategy described above causes some problems, because one of the five *Bam* HI sites occurs in an essential region (position 11.4) and cannot easily be eliminated by passages on a restricting host. Klein and Murray have eliminated this particular *Bam* HI site by first mutagenising a phage lysate with nitrous acid. The mutagenised phage population was amplified, progeny phages harvested, their DNA isolated and cleaved with *Bam* HI. The small fraction of *Bam* HI-resistant DNA was isolated, packaged *in vitro*, and propagated in suitable hosts. The desired phage derivatives containing only one or two *Bam* HI sites were then obtained by genetic crosses similar to those described above.

4.2.2.3 Structure of Selected λ Vectors

Apart from containing useful target sites for restriction enzymes, λ vectors should also satisfy some other requirements. Firstly, they should allow cloning of DNA molecules of a broad size range. Secondly, it should be possible to distinguish recombinant and parent phages by plaque morphology or marker inactivation. Thirdly, recombinant phages should be obtainable with high yields, and finally, such vectors should guarantee a sufficient level of biological safety.

A number of λ vectors which fulfill these requirements have been constructed and some will now be discussed in detail. Their essential properties are also summarised in Table 4.2-1.

The set of Charon phages meets almost all of the requirements mentioned above (Fig. 4.2-8A) (Blattner *et al.*, 1977; Rimm *et al.*, 1980). These phages are named for the ferryman of Greek mythology who conveyed the spirits of the dead across the river Styx which separated the realm of

the living from Hades, the underworld. The *Eco* RI replacement vector Charon 4A, for example, contains three *Eco* RI sites in its nonessential region. Cleavage with *Eco* RI therefore yields four DNA fragments. The central fragments can be easily purified away from the two other fragments at the ends of the molecules by centrifugation. Since the left and the right arms are 19.9 and 11.04 kb in length, this vector can accomodate insertions between 7 and 20 kb. Charon 4A is therefore used for cloning large *Eco* RI fragments, such as those employed for the construction of genomic libraries (*cf.* Chapter 9). The separation of the internal fragments also eliminates the two markers *lac5* and *bio256*. The *lac5* region carries *lac* regulatory sequences and the *lacZ* gene which codes for β-galactosidase. In the presence of Xgal and suitable inducers, the intact vector forms blue plaques, while recombinants containing both vector arms and inserted foreign DNA yield colourless plaques (*cf.* also Fig. 7-6). However, this selection is only of minor importance, because the size of the inserted DNA itself exerts a much more powerful positive selection pressure for recombinant DNA molecules. The discrimination by plaque colour is quite useful and becomes important if insertional vectors of the Charon type are used, because, in this case, the minimal length of the inserted DNA is of no consequence. Vector Charon 16A, for example, possesses only one *Eco* RI site for the insertion of foreign DNA. The vector itself can be packaged because its size is never smaller than the critical 38 kb which are essential for packaging.

It should be noted that the use of Charon phages is not restricted to *Eco* RI target sites. A second generation of vectors has been developed to allow cloning at a number of other restriction sites. Charon 30, for example, is a *Bam* HI replacement vector (Rimm *et al.*, 1980), and λL47.1

Pierre Tiollais
Paris

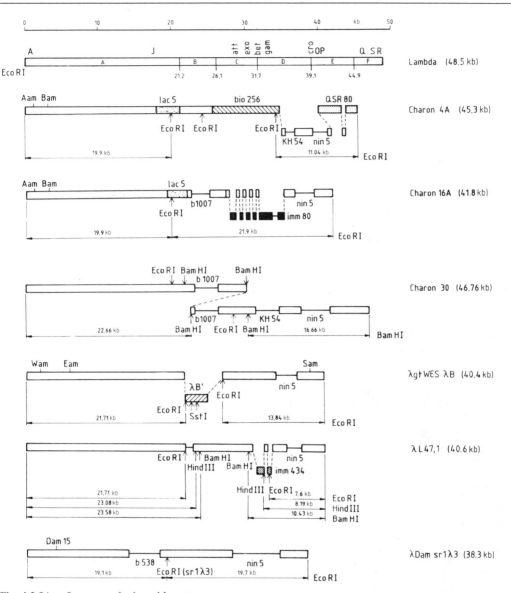

Fig. 4.2-8A. Structure of selected λ vectors.

A simplified genetic map with the positions (in kb) of the six *Eco* RI sites is shown at the top. For historical reasons the corresponding *Eco* RI fragments are labelled by letters A through F from left to right, and not according to size.

Wild-type λ sequences are shown as open bars, deletions such as *b189*, *b527*, *b1007*, *b558*, or *nin5* by solid lines. *Lac5* (stippled bar) and *bio256* (hatched bar) are substitutions with sequences derived from the *E. coli lac* and *bio* regions, respectively. *QSR80*, *imm80*, *int29*, and *imm434* are substitutions with sequences derived from λ-related phages Φ80, Φ29, and 434. B′ in λgtWESλB indicates that the *Eco* RI-B fragment is inserted in opposite orientation. KH 54 represents a deletion in the *rex-cI* region which prevents lysogenisation. Literature references are given in the text (in part taken from Maniatis *et al.*, 1982).

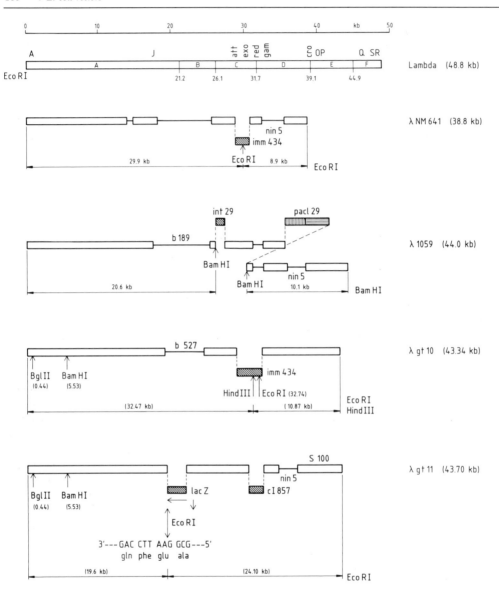

Fig. 4.2-8B. Structure of selected λ vectors. (Continuation of Fig. 4.2-8A).
For explanations see legend to Fig. 4.2-8A. pacl29 is the ColE1-derived sequence in λ1059. The hatched bar in λgt10 indicates the insertion of the repressor gene derived from phage 434 with its unique *Hind* III and *Eco* RI sites (Huynh *et al.*, 1985). λgt11 carries an insertion of the *E. coli lacZ* gene. The DNA sequence shows codons 1003 (ala) to 1006 (gln) of β-galactosidase with the relevant *Eco* RI site (Huynh *et al.*, 1985).

Table 4.2–1. Cloning capacity of λ vectors

Vector	Size (in kb)	Size (in % λ)	Enzyme	Length of DNA between arms (in kb)	Capacity (in kb)	Selection
Charon 4A	45.3	(93.4)	EcoRI	14.38	7–20	lac⁻ bio⁻
			XbaI		0–5.63	red⁻ gam⁻
Charon 16A	41.8	(86.2)	EcoRI	–	0–10	lac⁻
Charon 30	46.8	(96.5)	BamHI	14.86	6.1–19.1	size
			EcoRI	13.2	4.5–17.5	
λ L47.1	40.62	(83.6)	EcoRI	11.31	8.6–24	spi⁻
			HindIII	9.35	7.1–21.6	spi⁻
			BamHI	6.61	4.7–19.6	spi⁻
λ gtWESλB	40.4	(83.4)	EcoRI	4.85	2.2–15.1	–
λ Dam sr1λ3	38.3	(78.9)	EcoRI	–	0–13	D⁻ red⁻
λ 1059	44	(90.7)	BamHI	9.7	6.3–24.4	spi⁻
NM641	38.8	(80)	EcoRI	–	0–13	clear plaques

Vector size is not only given in kb but also in % λ DNA (100 % λ = 48 502 bp). Capacities are determined from minimal and maximal DNA sizes which can be packaged into λ heads. Possibilities for selection are explained in the text. All substitution vectors are Red⁻ and have to be grown on RecA⁺ hosts.

can be used as a replacement vector for DNA fragments with Eco RI, Bam HI, or Hind III ends (Loenen and Brammar, 1980).

Most λ vectors contain genetic markers which are important for their biological safety. The guidelines for EK2 vectors require that only one phage particle in 10^8 should survive under natural conditions if phages are employed as cloning vehicles. For this reason phage vectors frequently carry suppressor mutations in genes coding for capsid proteins A and B. Such mutants grow only on suppressor-positive host strains which do not occur in the natural environment. The EK2 vector λgtWESλB contains three such amber mutations, one of which, Sam1000, lies in gene S, the product of which is responsible for the lysis of bacteria. This S amber mutation thus renders the burst of mature phage particles more difficult (Tiemeier et al., 1976). This vector also lacks the Eco RI-C fragment which contains the genes int, xis, part of exo, and part of the attachment site, and hence all important regions required for lysogenisation. The att site is also missing from vectors Charon 16A and Charon 30, from which it has been removed by deletion b1007. A frequently used

deletion is nin5 which constitutes 5.75% (2.8 kb) of the normal complement of λ DNA. This deletion does not only increase the capacity of vectors for the insertion of foreign DNA but also removes the strong early terminator site t_{R2} (cf. Fig. 4.2-3). In wild-type phages this terminator is overcome by the N gene product, so the nin deletion makes phages N-independent. The nin deletion increases the synthesis of the Q protein immediately after infection. Since Q protein is an important regulator of the late phase of infection, nin deletions force phages into the lytic growth cycle, and hence provide an additional safety factor.

4.2.2.4 Selection Techniques

Two important principles for the selection of recombinant λ phages have already been mentioned. One is based on a size selection of DNA molecules and is realised in replacement vectors, such as Charon 4A and others listed in Table 4.2-1; the other is lac inactivation which is used, for example, in Charon 16A and other insertional vectors. There is, however, a wide range of other useful markers. The insertional vector λDamsr1λ3 (*cf.* Fig. 4.2-8A) carries two deletions, *nin5* and *b538*, which render its genome 11.4 kb smaller than wild-type λ (48.8 kb) (Enquist and Sternberg, 1980). This vector can accomodate up to 13 kb of foreign DNA. A unique *Eco* RI site is located within the *exo* gene (also known as *redα*) which codes for an exonuclease. This gene is non-essential and can be inactivated by cloning of *Eco* RI fragments. Phages with a genetic defect in the *exo* gene and/or in the neighbouring gene *redβ* (abbreviated *bet*, *cf.* also Appendix D) have the phenotype Red⁻. They plaque on wild-type *E. coli*, albeit with some reduced efficiency as compared to Red⁺ phages, but not at all on strains which are either defective in DNA polymerase I (*polA*) or DNA ligase. Vector and hybrid phages are therefore easily distinguishable. The *Dam* (*D* amber) mutation of this vector prevents the synthesis of capsid protein D and is as useful as the Red⁻ phenotype. λ phages can grow in the absence of capsid protein D as long as the length of their DNA does not exceed 82% (40.6 kb) of that of wild-type λ DNA. Since the vector itself contains deletions which shorten the molecule by 22.6%, the length of the vector is only 77.4% of that of wild-type λ DNA and it grows equally well on wild-type strains and strains containing amber suppressors; however, recombinants containing more than 4-5% of the normal complement of λ DNA (1.9-2.2 kb) require a functional gene *D* product and therefore grow on amber suppressor strains only.

In practice, the detection of Red⁻ phages described for the λDamsr1λ3 insertional vector is extremely laborious since it requires screening on *polA* or ligase-deficient mutants. It would therefore be advantageous to inactivate a phage gene by an insertion in a way which renders its loss of activity much more easily detectable. A suitable gene would be the λ repressor gene *cI*, which is responsible for the formation of lysogens. Plaques of lambdoid phages are normally not clear but slightly turbid, because a few cells will always survive infection. These cells are lysogenic and grow within the area of the plaque. The inactivation of the cI repressor is easily recognised since it results in clear plaques, and as a matter of fact, the repressor gene derives its name from the clear plaque morphology of *cI* mutants (*c* = clear). Murray *et al.* (1977) have developed a number of insertional vectors on the basis of cI repressor inactivation. In this case use was made of the repressor gene of a λ-related phage 434 since it contains unique *Eco* RI and *Hind* III sites. Fig. 4.2-8 shows the structure of a prototype vector designated λNM641 which has been successfully used by Scalenghe *et al.* (1981) for cloning of a particular DNA segment from the X chromosome of *Drosophila melanogaster*. This cloning experiment is quite remarkable since 80 clones were obtained from approximately 10 pg of DNA in a volume of only 10 nl.

The same strategy used for NM641 has also been applied for the insertional vectors λgt10 (Huynh *et al.*, 1985) and NM1149 (Murray, 1983) (Fig. 4.2-8B). These vectors, however, are also Red⁺, *i.e.*, they carry the phage-encoded recombination system consisting of genes *redα* and *redβ*, and therefore give much higher titers than the Red-negative immunity vector NM641. The presence of *red* genes may thus be of advantage especially for the construction of cDNA libraries, when high phage yields are required.

Immunity vectors which are based on the inactivation of the *cI* gene do not only allow the recognition of recombinants by their clear plaque morphology. If a suitable indicator strain is used, which can be lysogenised efficiently by unrecombined (vector) phages only, the system also offers a selection for recombinant DNA molecules. A mutation in the host protein Hfl, a protease, forces the phage to enter almost quantitatively the lysogenic pathway by stabilising the cII gene product (*cf.* Section 4.2.1) (Hoyt *et al.*, 1982). Only recombinant phages which cannot lysogenise due to the absence of the cI gene product will therefore give plaques on such an indicator strain; on the other hand, vector phages lysogenise quantitatively since they contain an intact cI repressor. An example taken from the literature (Friedman *et al.*, 1984) demonstrates the efficieny of this selection. In this particular example 10^8 phages were obtained from 20 ng of cDNA, but only 9% of these phages were recombinant. One passage in an Hfl-negative host (BNN150 carrying the *hflA150* mutation) increased the proportion of recombinants to 99.8%.

If replacement vectors of the Charon 4A type are used, the central or "stuffer" fragment must usually be removed before cloning by physical techniques, such as gel electrophoresis or gradient centrifugation, in order to prevent reformation of the vector itself. This rather laborious step can be avoided, and DNA fragments can be cloned in the presence of the central fragment, if the latter is selectively destroyed or eliminated by a biological selection directed against the multiplication of the vector. The central *Eco* RI-B fragment of vector λgtWESλB (Fig. 4.2-8A), for example, contains two *Sst* I sites which occur on this fragment only and not on the vector arms (Tiemeyer *et al.*, 1976). An additional *Sst* I digestion selectively inactivates the central fragment by providing *Sst* I termini which are not compatible with *Eco* RI sites.

An important biological selection for recombinant λ phages is based on the fact that wild-type λ does not grow on *E. coli* strains lysogenic for P2. This phenotype, which is known as Spi $^+$ (sensitive to P2 interference), is due to the products of λ genes *red* (*exo* and *bet*) and *gam*. λ mutants with defects in both genes are Spi $^-$ and grow on P2 lysogens. The *gam* gene product plays an important role in the transition from early to late phase of DNA replication and prevents concatemeric DNA forms which are the natural substrates for packaging (Fig. 4.2-4) from being digested by the *recBC* product, a host-specific exonuclease. In the absence of *gam* concatemeric DNA is digested. In this situation viable phage particles are only obtained if the monomeric circular DNA is converted into multimeric circular forms from which linear DNA molecules can be cut out and packaged subsequently. Such multimeric forms

are obtained from circular monomers by recombinational events catalysed by the host-specific recombination enzyme, RecA, or the phage-specific *red* system. *Red* genes are located in the vicinity of *gam* on the central fragment and are removed together with the stuffer fragment. Only the host *recA* system then remains to be used for recombination. Red $^-$/Gam $^-$ phages, which are also known as Fec $^-$ (feckless) phages, therefore, require a RecA $^+$ host for growth, while Red $^+$/Gam $^+$ phages also multiply in RecA $^-$ hosts.

Vector and recombinant phages, which are characterised by the presence or absence of *red* and *gam* genes, can be distinguished in two ways. Either the vector is Red $^+$/Gam $^+$, and hence possesses the phenotype Spi $^+$, and so does not grow on P2 lysogens while it does multiply in RecA $^-$ hosts; or the phage contain recombinant DNA, lack *red* and *gam* on the central fragment, and are Spi $^-$. Such molecules can be replicated in a P2 lysogen, but they require a RecA $^+$ host. A number of vectors, such as λL47.1 (Loenen and Brammer, 1980), or λ1059 (Karn *et al.*, 1980) are based on this principle. This selection fails in some other vectors although the *red* and *gam* genes are absent. It is known that *RecA*-dependent recombination, which leads to multimeric circular structures, requires the presence of *chi* sequences (*chi* = c̲ross-over h̲ot-spot i̲nstigator). These sequences are reponsible for efficient recombination (Smith *et al.*, 1981), but they do not normally occur on λ DNA (Chattoraj *et al.*, 1979); however, they can be generated and introduced into the vector arms by point mutations. In λ1059 and λL47.1 they are present on the right and left arm, respectively. Eukaryotic DNA may also carry such *chi* sequences (Kenter and Birshtein, 1981). If a vector without *chi* sequences is used, recombinants with insertions containing such *chi* sequences would be preferentially replicated and would be overrepresented in a library. It is therefore important that vector arms themselves contain these *chi* elements.

In spite of such possibilities it has proven most effective in practice to purify the vector arms before recombinant DNA molecules are con- structed. This is even more advisable if λ1059 is used since this vector contains a replicon derived from ColE1 (Fig. 4.2-8B), which enables this vector to replicate as a plasmid in λ-lysogenic hosts. Cloning vehicles which combine phage and plasmid properties are referred to as phasmids (Brenner et al., 1982). The presence of pBR322 sequences makes it rather difficult to screen genomic libraries with cDNA probes cloned in pBR322 even if these sequences are located

within the central vector fragment which is usually removed before cloning.

A series of vectors, the EMBL series derived from λ1059, does not have this disadvantage. They are very versatile since they also carry a polylinker (Frischauf et al., 1983). The polylinker insertions in EMBL3 and EMBL4 (Fig. 4.2-9) are derived from vector pUC7 (cf. Fig. 2.4-22). For this purpose two pUC7 derivatives were construct- ed (Fig. 4.2-10). The first was obtained by ligation of a mixture of Eco RI and Bgl II linkers,

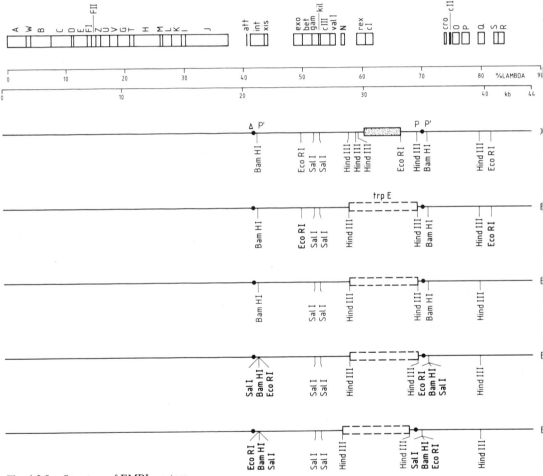

Fig. 4.2-9.   Structure of EMBL vectors.
For details of the constructions see text. The pBR322 (stippled bar) portion in the central fragment of λ1059 has been replaced by a *trp E* insertion (with broken lines) in EMBL vectors. Black dots represent the duplicated λ *att* sites (△ P′ and PP′).

subsequent cleavage of the reaction product with *Eco* RI, and insertion of the resulting fragments into pUC7 linearised with *Eco* RI. The new vector, designated pUC(*Eco* RI-*Bgl* II), contains the polylinker sequence *Eco* RI-(*Bgl* II)$_n$-*Eco* RI. The second pUC7 derivative was obtained in a similar fashion by the insertion of DNA fragments obtained from ligated *Sal* I and *Bgl* II linkers cut with *Sal* I into a *Sal* I-linearised pUC7. The new polylinker sequence in pUC(*Sal* I-*Bgl* II) is *Eco* RI-*Bam* HI-*Sal* I-*Bgl* II-*Sal* I-*Bam* HI-*Eco* RI. These different linkers were subsequently

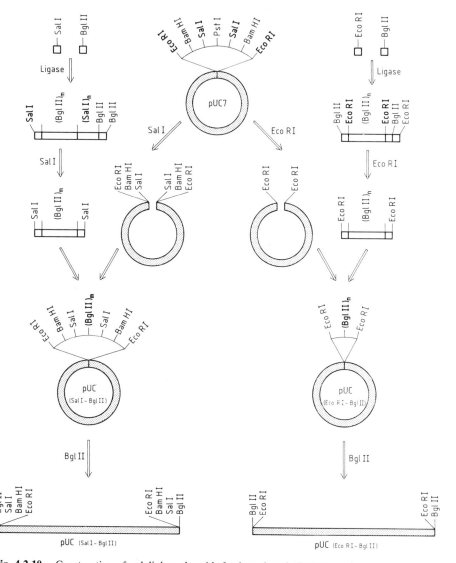

Fig. 4.2-10. Construction of polylinker plasmids for insertions in EMBL2.
Subscripts "m" and "n" denote multiple linker insertions. Linker sequences are shown as open bars, sequences derived from pUC7 by stippled bars. See text for details.

incorporated into EMBL2. EMBL2 is a derivative of λ1059 which carries a *Hind* III insertion with the *E. coli trpE* gene in its middle fragment. The *Eco* RI sites of λ1059 were removed by alternating growth cycles on restricting and non-modifying hosts. The construction of EMBL2 is not describ-ed in detail here since this strategy for the removal of *Eco* RI sites has been described earlier (Fig. 4.2-7). In order to introduce the linkers of pUC(*Sal* I-*Bgl* II) and pUC(*Eco* RI-*Bgl* II) into

EMBL2, these vectors were linearised with *Bgl* II and ligated with a *Bam* HI-cut EMBL2 phage (Fig. 4.2-11). Recombinant phages (EMBL Sal and EMBL RI) with flanking pUC insertions between the middle fragment and the vector arms were identified by selecting for blue plaques on Lac+ hosts in the presence of Xgal. These phages make blue plaques because endogenous *lac* re-pressor is titrated out by the many *lac* operator regions on the two pUC2 insertions on the

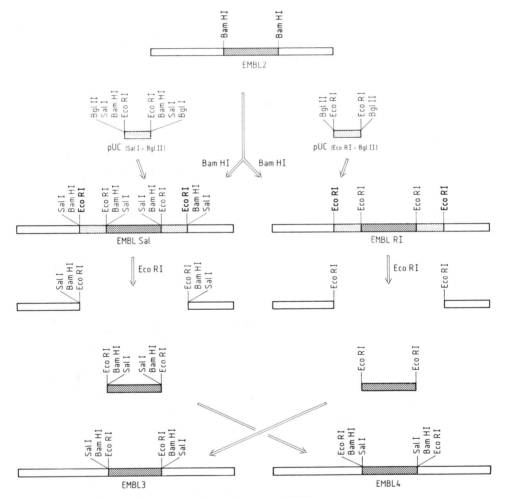

Fig. 4.2-11. Construction of EMBL3 and EMBL4 from EMBL2.
See text for details. Hatched bars represent the central λ DNA stuffer fragment, stippled bars are sequences derived from pUC7 (Frischauf *et al.*, 1983).

recombinant phages, which leads to an induction of the chromosomal *lac* operon. The presence of four *Eco* RI sites, and hence of two plasmid insertions, was tested by the sensitivity on an *Eco* RI-restricting host and suitable *Eco* RI digestions of the DNA.

The central fragments of vectors EMBL Sal and EMBL RI without the pUC insertions were then replaced by each other, *i.e.*, EMBL Sal received the central fragment of EMBL RI and *vice versa*. This replacement reaction was not carried out in one step. In an intermediate reaction (not shown in Fig. 4.2-11) the central regions were replaced by *Eco* RI fragments of *E. coli*. These intermediate vectors with *E. coli* DNA insertions could be isolated by a Spi⁻ selection since they lack the central λ DNA fragment still present in EMBL Sal and EMBL RI. *Eco* RI digests of the Spi⁻ derivatives were then ligated with the highly purified central λ DNA fragments of the appropriate opposite partner. The desired recombinants were identified by a Spi⁺ selection on a RecA⁻ host.

EMBL3A as opposed to EMBL3 carries the *Aam32Bam1* mutations of Charon 4A; in contrast to Charon 4A, however, EMBL3A also carries a mutation destroying the *Bam* HI site at 11.4%. For unknown reasons, EMBL3A grows poorly on *sup*E hosts and thus should be propagated on *sup*F hosts. EMBL3A also shows extremely low recombination frequencies in the *in vivo* recombination assay developed by Seed (1983) and cannot be used in this system (for a possible explanation *cf.* also Section 9.4.5).

EMBL vectors do not only permit a genetic selection for recombinant DNA by employing the Spi⁻ selection, but they also allow a simple biochemical selection by simultaneously digesting the vectors with *Eco* RI and *Bam* HI. After precipitation with isopropanol, the short linker fragments remain in the supernatant. The arms

and the central fragment carry different ends and cannot be re-ligated, which makes it unnecessary to isolate the vector arms. Because of these advantages EMBL vectors are currently in wide use.

The detection of recombinant DNA in λ phages by hybridisation will be described in detail in Chapter 11.

4.2.2.5 *In vitro* Packaging of λ DNA

An obstacle in gene technology is the reintroduction of cloned DNA into host bacteria. Although as many as 2×10^{10} plaques may be obtained from λ phage particles containing the equivalent of one μg of λ DNA, the standard $CaCl_2$ technique utilising unpackaged λ DNA yields at most 10^4 to

Barbara Hohn
Basel

10^6 plaques per μg (*cf.* also Section 4.1.5). The study of phage λ morphogenesis has led to an alternative, namely *in vitro* packaging of naked phage DNA into empty phage heads (Hohn, 1979). In the presence of suitable maturation factors *in vitro* packaging yields intact phage particles of an infectivity hardly distinguishable from that of normal phage particles.

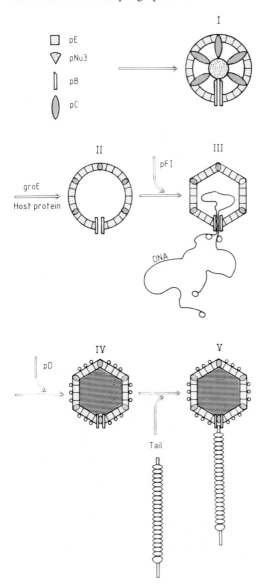

Phage λ consists of an icosahedral head structure and a flexible tail (Fig. 4.2-12). A complete capsid is composed of structural proteins encoded by genes *E* and *D*. One precursor in the complex biosynthesis of a capsid is the prehead (structure II, Fig. 4.2-12), which still lacks protein D but takes up phage DNA. The substrate for this packaging process is a concatemeric DNA molecule consisting of several λ DNA units. While entering prehead structures these concatemers are converted to monomeric forms by endonucleolytic cleavage by the product of gene *A*. This process yields strand breaks 12 base pairs apart at each *cos* site, generating linear λ DNA molecules with protruding single-stranded ends such as those isolated from mature phage particles (*cf.* Fig. 4.2-1). Linear monomeric phage DNA cannot be packaged *in vivo*, and neither are monomeric circular DNA molecules with a *cos* site suitable substrates for packaging. Packaging only occurs if a second *cos* site has been introduced into the DNA either *in vitro* or by a recombinational event *in vivo*. Thus the presence of at least two *cos* sites on DNA to be packaged is required. In addition, the distance between *cos* sites is important for packaging. Only those *cos* sites which are separated by 38-52 kb (75-105% of λ DNA) are correctly cleaved during *in vivo* morphogenesis. The nature of the DNA between these *cos* sites is of little or no consequence.

Empty phage heads, packaging factors, and phage tails required for packaging can be obtain-

Fig. 4.2-12. Morphogenesis of phage λ. Only those steps of the morphogenetic sequence of events are shown which are important for *in vitro* packaging of DNA. Mutants in gene *E* accumulate all structural proteins apart from E protein in a soluble form. Such mutants are incapable of forming structure I. Mutants in gene *D* are blocked at the level of structure II. Lysates obtained from the two mutants can complement each other. In the presence of suitable DNA molecules, head maturation (structures III and IV), and tail acquisition proceed normally so that a mature phage (structure V) is formed. pE, pB, etc. signify the products of genes *E*, *B*, etc. (Hohn, 1979).

ed from two lysates, each of which is derived from one phage strain with a genetic defect affecting a different step of morphogenesis. Both strains are used as temperature-sensitive lysogens carrying the temperature-sensitive repressor imm434cIts. While the lysogenic state is stably maintained at 32 °C, lytic growth, and hence synthesis of the desired structural proteins, ensues after a temperature shift to 40 °C. In order to accumulate these proteins and to prevent their premature release the two prophages also carry the amber mutation *Sam7*. The product of gene *S* is required for the lysis of bacterial cell walls. The first of the two lysogens, strain BHB2690, contains a λ prophage with an amber mutation in gene *D* (*Dam*). The D protein, the decoration protein, is located on the outside of mature phage particles and participates in the maturation of head structures and the threading of DNA into phage heads. Heat induction of a *Dam* lysogen leads to the accumulation of empty prehead particles (structure II, Fig. 4.2-12). The lacking D protein is provided by the second strain, BHB2688, which is mutated in gene *E*. Gene E protein is the major structural protein of phage heads and is required in an early phase of head assembly. Mutations in gene *E* accumulate all components of head structures, including protein D, in a free form. When the two lysates which complement each other are mixed, DNA can be packaged *in vitro*, and mature phage heads can be generated if exogenous ATP and biogenic amines are provided (structure V, Fig. 4.2-12).

The efficiency with which hybrid clones are generated is on the order of $1-5 \times 10^5$ plaques per microgram of inserted DNA. Untreated λ DNA may yield 10^7 to 10^8 plaques per microgram of DNA. A principal difficulty of this *in vitro* packaging system is the endogenous DNA from the phage lysates themselves, which competes with exogenously added recombinant DNA and gives a background of 10^2 to 10^3 plaques per packaging experiment. This background problem can be avoided completely by pretreating packaging lysates with UV light; the enzymatic packaging activities are fully retained if the lysates are irradiated at a dose which corresponds to 40 lethal hits per phage DNA molecule.

Phage particles obtained by packaging of endogenous DNA may require containment procedures to ensure their biological safety, because of the generation of wild-type (Am$^+$) recombinants from endogenous and exogenous (recombinant) DNA. Such recombinants may arise because mutations existing in exogenous and endogenous DNA may not map at exactly the same site within the same gene. In this case recombination between the two loci may lead to Am$^+$ revertants, albeit with a very low frequency. Such recombinational events can be prevented only by using RecA$^-$ host bacteria. Under such circumstances there should not be any restrictions for using these λ vectors as EK2 vectors.

4.2.3 Cosmids

The yield of phage particles in *in vitro* packaging systems described in Section 4.2.2.5 is independent of the genetic constitution of phage DNA. The only factor of importance is the presence of several λ *cos* sites, which must be separated by certain minimal distances on concatemeric DNA molecules. This system, therefore, allows packaging of any DNA molecule, even a plasmid, *in vitro*. Plasmids containing λ *cos* sites are known as cosmids (Collins and Hohn, 1979) and their development has widened the scope of plasmid cloning in several ways:

– The infectivity of plasmid DNA packaged in phage heads is at least three orders of magnitude higher than that of pure plasmid DNA;
– the process almost exclusively yields hybrid clones so that a subsequent selection for recombinant DNA becomes unnecessary;
– in contrast to normal plasmid transformations the system strongly selects for clones containing large DNA inserts. It is, therefore, particularly

Table 4.2–2. Structural features of some cosmids.

Cosmid	Size	Cleavage sites	Size of insertion	Reference
MUA3	4.76 kb	EcoRI/PstI/PvuII/PvuI	40–48 kb	1
pJB8	5.4 kb	BamHI	32–45 kb	2
Homer I	5.4 kb	EcoRI/ClaI	30–47 kb	3
Homer II	6.38 kb	SstI	32–44 kb	3
pJC79	6.4 kb	EcoRI/ClaI/BamHI	32–44 kb	4

References: 1 = Meyerowitz *et al.*, 1981; 2 = Ish-Horowicz and Burke, 1981; 3 = Chia *et al.*, 1982; 4 = Hohn and Collins, 1980.

well suited for generating genomic libraries (*cf.* Chapter 9).

The individual steps of cosmid cloning are described in Fig. 4.2-13. Cosmid vector DNA is first linearised by digestion with a suitable restriction enzyme. The foreign DNA to be inserted is partially cleaved with an enzyme generating compatible protruding ends. DNA fragments to be inserted may be as long as 45 kb if the cosmids employed are small (4-6 kb). Ligation of mixed DNA preparations with DNA ligase yields a wide range of recombinant DNA molecules, and the use of high DNA concentrations favours the generation of concatemeric forms. *In vitro* packaging of suitable DNA molecules with *cos* sites

positioned approximately 38-51 kb apart is accomplished, after cleavage of the DNA at the *cos* sites by gene *A* product, using the λ lysates described above. DNA positioned between two *cos* sites is incorporated into phage particles and used to infect suitable λ-sensitive *E. coli* strains. The plasmid, which does not carry any λ functions, is circularised *in vivo* and replicates without producing plaques. Under these conditions only plasmid-specific functions, such as the origin of DNA replication and the selective biological markers are biologically active. These markers, which normally are antibiotic resistance markers, can be used for the selection of recombinant plasmids.

The first generation cosmids required several improvements because they usually carried only a few restriction sites suitable for cloning, only a single antibiotic resistance marker, and, due to their sizes, did not allow one to exploit the full cloning capacity of cosmids. Today, a number of sophisticated cosmid cloning vehicles are available in which the initial disadvantages have been overcome. Some properties of these new cosmid vectors are listed in Table 4.2-2. Cosmid MUA-3, for example, carries a *cos* site in a 403 bp λ *Hinc* II fragment cloned into the *Pst* I site of pBR322 by the method of homopolymeric tailing (Meyerowitz *et al.*, 1980) (Fig. 4.2-14). The construction of this cosmid should have yielded predominantly

John Collins
Braunschweig, FRG

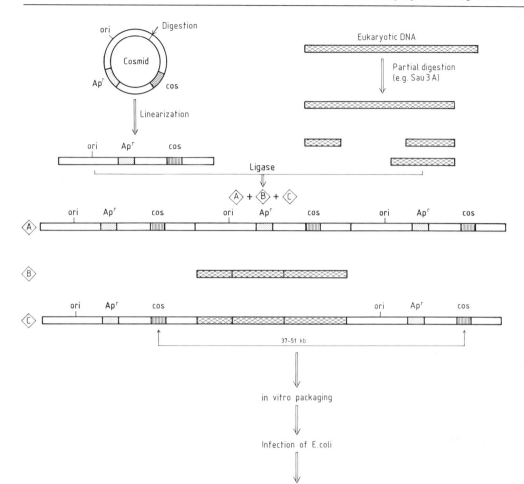

Fig. 4.2-13.   Principle of cosmid cloning.
Reactions A and B lead to the formation of oligomeric cosmid molecules, or to the association of fragments of foreign DNA. Reaction C yields recombinants between cosmids and foreign DNA, which can be packaged *in vitro* if their sizes are between 38-51 kb long as indicated.

molecules containing two rather than one *Pst* I sites. Indeed, such a plasmid (MUA-10) was also observed; however, the loss of restriction sites after cloning *via* homopolymeric tails is not uncommon. MUA-3 and MUA-5 are very useful vectors for cloning since they possess only a single *Pst* I site. Meyerowitz *et al.* (1980) have constructed a *Drosophila* genomic library with inserts of 40.3-48.1 kb in length and a mean insert length of 45.5 kb. The length of these inserts is approxi-

mately twice the maximum length which could have been obtained in λ DNA (24.6 kb).

Another cosmid, Homer I, is derived from plasmid pAT153 (Chia *et al.*, 1982). It contains a 1.78 kb *Bgl* II fragment with a *cos* site derived from circularised Charon 4A DNA inserted into the unique *Bam* HI site of the tetracycline resistance region (Fig. 4.2-15). Homer I has been widely used, in particular for cloning chromosomal eukaryotic DNAs obtained by partial *Eco* RI[*]

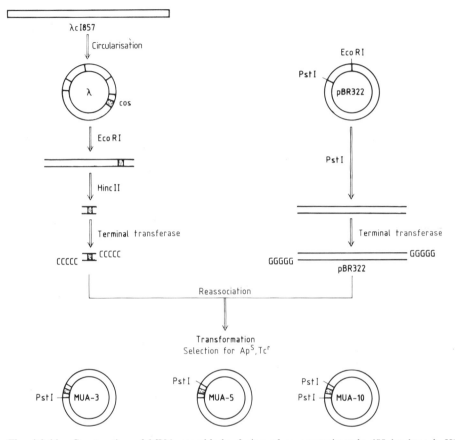

Fig. 4.2-14. Construction of MUA cosmids by fusion of an approximately 400 bp long λ *Hinc* II fragment containing a *cos* site with *Pst* I-linearised pBR322 DNA.
Cosmids MUA-3 lack one of the two expected *Pst* I sites. Although the *Pst* I sites should have been regenerated by employing homopolymeric dG-dC tails (*cf.* also Fig. 3-19) practical experience frequently shows that such sites are lost. Loss of a single nucleotide from *Pst* I-linearised vector molecules renders the reconstruction of sites impossible (Meyerowitz *et al.*, 1980).

cleavage. *Eco* RI* is an activity of *Eco* RI which usually is observed under special reaction conditions, predominantly in the presence of dimethylsulfoxide or at low salt concentrations. This so-called star activity recognises the innermost tetranucleotide sequence N/AATTN instead of the usual *Eco* RI recognition site G/AATTC. It is important to methylate eukaryotic DNA with *Eco* RI methylase prior to *Eco* RI* cleavage because *Eco* RI sites will otherwise be cleaved before *Eco* RI* sites: the advantage gained by

using an enzyme with a tetrameric recognition sequence would be lost. The *Drosophila* library generated by Chia *et al.* (1982) appears to be representative and contains DNA insertions with a mean length of 37 kb.

If a cosmid gene library of eukaryotic DNA is to be used directly in higher cells, for example, for the characterisation of an oncogene, cosmids with suitable dominant selectable markers are recommended. Cosmids Homer V and Homer VI, for example, carry the bacterial aminoglycoside

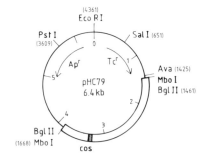

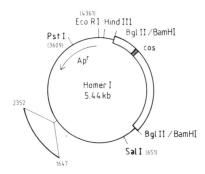

Fig. 4.2-15. Structure of cosmid vectors pHC79 and Homer I.

In cosmid pHC79 a *Sau* 3A fragment of pBR322 between positions 1459 and 1666 has been replaced by a 1.78 kb *Bgl* II fragment of Charon 4A which contains a *cos* site. Other pBR322 sequences such as antibiotic resistance genes have not been altered. With respect to the *cos* site the left end of λ DNA is located in clockwise direction (Hohn and Collins, 1980).

Cosmid Homer I contains the 1.78 kb fragment mentioned above cloned into the *Bam* HI site of pAT153. pAT153 is a pBR322 derivative which lacks sequences between positions 1647-2352 as indicated (*cf.* also Fig. 8-27; Twigg and Sherratt, 1980). Cloning into the *Bam* HI site of pAT153 destroyed tetracycline resistance (Chia *et al.*, 1982).

Plasmid DNA sequences are represented by single lines, λ DNA by open bars. *Cos* sites are shown as hatched boxes. Numbers are pBR322 co-ordinates.

phosphotransferase gene *APH(3')II* encoded by Tn5 (*cf.* Fig. 8-7). The presence of this gene in higher cells can be selected for by the gentamycin derivative G418 (Wolfe *et al.*, 1984). In Homer V (Fig. 4.2-16) the *APH(3')* gene is under transcrip-

tional control of the Herpes Simplex Virus thymidine kinase gene promoter while in Homer VI its 5' control region is replaced by the corresponding control region of the *LTR* region of Moloney Murine Sarcoma Virus (MoMSV). Both vectors also contain a *Hpa* II-*Hind* III fragment of SV40 DNA carrying the SV40 origin of DNA replication and transcriptional enhancer sequences. Due to these enhancers it was possible to use these vectors for the expression of the human T24 bladder carcinoma oncogene in primary hamster and rat cells (Spandidos and Wilkie, 1984). The selectable marker in cosmid pGcos4 is a methotrexate-resistant dihydrofolate reductase (Gitschier *et al.*, 1984) (Fig. 4.2-16). This vector can therefore also be used for selections in higher cells (*cf.* also Section 8.1.4).

A useful derivative of Homer I has been described by Ish-Horowicz and Burke (1981) who have introduced a synthetic *Bam* HI linker into the *Eco* RI site of Homer I, and thus obtained cosmid pIB8 (Fig. 4.2-17). Another cosmid, pHC79, appears to be even more versatile since it contains several useful cloning sites (Fig. 4.2-15), although its length of 6.4 kb is much greater than that of other cosmids (Hohn and Collins, 1980).

At first sight cosmid cloning may appear to be a very efficient method; nevertheless, it presented considerable problems. One difficulty was the self-ligation of vector molecules which are packaged without containing foreign DNA insertions (Fig. 4.2-13, reaction A). It has also been observed that cosmids can take up two or more unlinked DNA fragments. Other problems also arose with the necessity of screening large numbers of bacterial colonies, and the instability of certain cosmids. A strategy described by Ish-Horowicz and Burke (1981) appears to solve at least the first problem, *i.e.*, self-ligation of vector molecules. Two samples of vector pIB8 are treated with either *Hind* III/alkaline phosphatase or *Sal* I/alkaline phosphatase (Fig. 4.2-17). A subsequent *Bam* HI digestion yields two mixtures of molecules with different ends. Even as a mixture these molecules are unable to reassociate and form

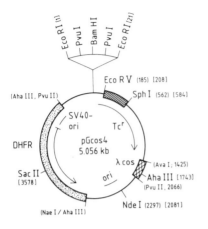

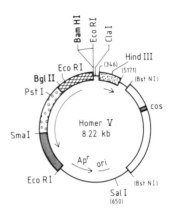

Fig. 4.2-16. Structure of cosmid vectors pGcos4 and Homer V.

Vector pGcos4 (*cf.* Appendix B-5 for the DNA sequence) is a composite molecule consisting of (1) a 1 497 bp *Sph*I-*Nde*I fragment derived from pBR322 with the 403 bp *Hinc*II fragment from λ*cI857* encoding the *cos* site inserted between the *Ava*I and *Pvu*II sites of pBR322. The deletion of the 641 bp *Ava*I-*Pvu*II fragment increases the copy number and removes the sequences interfering with replication in eukaryotic cells; (2) a 3 163 bp *Nde*I-*Eco*RV fragment with the SV40 origin of DNA replication, the early SV40 promoter, a methotrex-ate-resistant *dhfr* gene, and the polyadenylation site of hepatitis B virus surface antigen (Simonsen and Levinson, 1983); (3) a 376 bp *Eco*RV-*Sph*I fragment (hatched bar) comprising a section of the tetracycline resistance region of pBR322 lacking the *Bam*HI site, and (4) a polylinker sequence cloned into the *Eco*RI site of pBR322, which contains a *Bam*HI site flanked by *Pvu*I sites (*cf.* also Section 2.1.2.3). This *Bam*HI site can be used for constructing genomic libraries with inserts of up to 45 kb in length (Gitschier *et al.*, 1984). Thin lines represent sequences derived from pBR322; numbers in parantheses are pBR322 co-ordinates, numbers in brackets pGcos4 co-ordinates.

Cosmid Homer V is derived from Homer I (*cf.* Fig. 4.2-15) by insertion into its *Eco*RI site of a 2.78 kb DNA fragment composed of the HSV *tk* promoter (cross-hatched bar), the Tn5-derived *APH(3')II* gene (coarsely stippled bar), and the polyadenylation site from the HSV *tk* gene (hatched bar). The *Bgl*II site is located downstream from the cap site of the *tk* gene (*cf.* Fig. 8-2) and immediately upstream of the initiation site of the *APH* gene (*cf.* Fig. 8-7). The expression of the *APH* gene is thus driven by the *tk* gene promoter. A 415 bp insert containing the SV40 origin of replication and the 72 bp repeats required for efficient transcription of the SV40 early genes (positions 346 to 5141 of the SV40 genome) is represented as a stippled bar. (Wolfe *et al.*, 1984).

molecules long enough to be packaged. Package-able molecules are formed only in the presence of sufficiently long insertions.

This strategy also solves the second problem, namely cloning of DNA sequences originally not adjacent to each other. This is achieved by a careful size selection for molecules with a length of about 45 kb, which subsequently are treated with alkaline phosphatase. This technique has been employed for constructing not only genomic libraries of *Drosophila*, but also of representative libraries of murine and human DNA (Grosveld *et*

al., 1981; Groffen *et al.*, 1982; Cattaneo *et al.*, 1981).

Cos sites have also been introduced into plas-mids with broad host specificities. Friedman *et al.* (1982) have constructed a new cosmid, which is 21.6 kb in length, by inserting a 1.78 kb *Bgl*II fragment of pHC79 with its *cos* site (Fig. 7.2-15) into the single *Bgl*II site of vector pRK290 (Fig. 4.1-13). With suitable DNA insertions, the new vector, pLAFR1, can be packaged *in vitro* and can be mobilised from *E. coli* into a variety of other Gram-negative organisms, in which it is

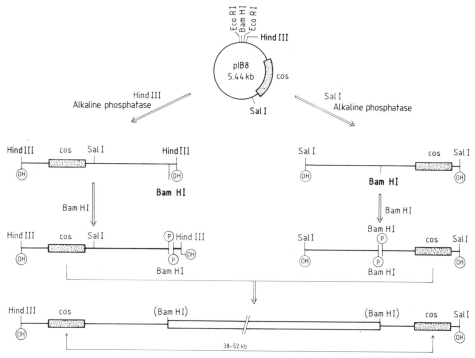

Fig. 4.2-17. Cosmid cloning: selection for recombinant DNA.
The exact procedure is described in the text. Open bars (bottom line) represent DNA to be cloned, which is ligated with the two vector elements (Ish-Horowicz and Burke, 1981).

stably replicated as a plasmid. This vector is also suitable for the selection of mutants by complementation analysis. A library of clones obtained from *Rhizobium meliloti* DNA in pLAFR1, for example, can be used to complement auxotrophic *Rhizobium* mutants. The nature of the complementing DNA can be studied immediately by analysing the corresponding plasmids.

In principle, cosmids should also be packageable *in vivo*. Indeed, this can be accomplished by superinfecting bacteria harbouring such cosmids with λ helper viruses. The helper does not only provide the necessary structural proteins but also those λ gene products which convert monomeric cosmid molecules into a packageable multimeric concatemer. The λ *gam* gene product mentioned above plays an important role since it is essential for late DNA replication (Feiss *et al.*, 1982).

References

Blattner, F.R., Williams, B.G., Blechl, A.E., Denniston-Thompson, K., Faber, H.E., Furlong, S.A., Grunwald, D.J., Kiefer, D.O., Moore, D.D., Sheldon, E.L., and Smithies, O. (1977). Charon Phages: Safer derivatives of bacteriophage λ for DNA cloning. Science 196, 161-169.

Brenner, S., Cesarini, G., and Karn, J. (1982). Phasmids: Hybrids between ColE1 plasmids and *E. coli* bacteriophage λ. Gene 17, 27-44.

Cattaneo, R., Gorski, I., and Mach, B. (1981). Cloning of multiple copies of immunoglobulin variable kappa genes in cosmid vectors. Nucleic Acids Res. 9, 2777-2790.

Chattoraj, D.K., Craseman, J.M., Dower, N., Faulds, D., Faulds, P., Malone, R.E., Stahl, F.W., and Stahl, M.M. (1979). Chi. Cold Spring Harbor Symp. Quant. Biol. 43, 1063-1066.

Chia, W., Scott, M.R.D., and Rigby, P.W.J. (1982).

Construction of cosmid libraries of eukaryotic DNA using the Homer series of vectors. Nucleic Acids Res. 10, 2503-2520.

Collins, J., and Hohn, B. (1979). Cosmids: a type of plasmid gene-cloning vector that is packageable *in vitro* in bacteriophage λ heads. Proc. Natl. Acad. Sci. USA 75, 4242-4246.

Enquist, L., and Sternberg, N. (1980). *In vitro* packaging of λ *Dam* Vectors and their use in cloning DNA fragments. Methods in Enzymology 68, 281-298.

Feiss, M., Siegele, D.A., Rudolph, C.F., and Frackman, S. (1982). Cosmid DNA packaging *in vivo*. Gene 17, 123-130.

Friedman, A.M., Long, S.R., Brown, S.E., Buikema, W.J., and Ausubel, F.M. (1982). Construction of a broad host range cosmid cloning vector and its use in the genetic analysis of *Rhizobium mutants*. Gene 18, 289-296.

Friedman, R.L., Manly, S.P., McMahon, M., Kerr, I.M., and Stark, G.R. (1984). Transcriptional and post-transcriptional regulation of interferon-induced gene expression in human cells. Cell 38, 745-755.

Frischauf, A.M., Lehrach, H., Poustka, A., and Murray, N. (1983). λ Replacement vectors carrying polylinker sequences. J. Mol. Biol. 170, 827-842.

Gitschier, J., Wood, W.I., Goralka, T.M., Wion, K.L., Chen, E.Y., Eaton, D.H., Vehar, G.A., Capon, D.J., and Lawn, R.M. (1984). Characterization of the human factor VIII gene. Nature 312, 326-330.

Groffen, J., Heisterkamp, N., Grosveld, F., Van den Ven, W., and Stephenson, I.R. (1982). Isolation of human oncogene sequences (*v-fes* homologue) from a cosmid library. Science 216, 1136-1138.

Grosveld, F.G., Dahl, H.H.M., de Boer, E., and Flavell, R.A. (1981). Isolation of β-globin-related genes from a human cosmid library. Gene 13, 227-237.

Hobom, G., Grosschedl, R., Lusky, M., Scherer, G., Schwarz, E., and Kössel, H. (1979). Functional analysis of the replicator structure of lambdoid bacteriophage DNAs. Cold Spring Harbor Symp. Quant. Biol. 43, 165-178.

Hohn, B. (1979). *In vitro* packaging of λ and Cosmid DNA. Methods in Enzymology 68, 299-309.

Hohn, B., and Collins, B. (1980). Small cosmid for efficient cloning of large DNA fragments. Gene 11, 291-298.

Hoyt, M.A., Knight, D.M., Das, A., Müller, H.I., and Echols, H. (1982). Control of phage λ development by stability and synthesis of cII Protein: Role of the viral *cIII* and host *hflA*, *himA* and *himD* genes. Cell 31, 565-573.

Huynh, T.V., Young, R.A., and Davis, R.W. (1985). Construction and screening cDNA libraries in λgt10 and λgt11. In "DNA Cloning Techniques: A practical approach"; Glover, D., ed; IRL Press, Oxford.

Ish-Horowicz, D., and Burke, J.F. (1981). Rapid and efficient cosmid cloning. Nucleic Acids Res. 9, 2989-2998.

Karn, J., Brenner, S., Barnett, L, and Cesarini, G. (1980). Novel bacteriophage λ cloning vector. Proc. Natl. Acad. Sci. USA 77, 5172-5176.

Kenter, A.L., and Birshtein, B.K. (1981). Chi, a promotor of generalized recombination in λ phage, is present in immunoglobulin genes. Nature 293, 402-404.

Klein, A., Bremer, B., Kluding, H., and Symmons, D. (1978). Initiation of λ DNA replication in vitro promoted by isolated *P* gene product. Europ. J. Biochem. 83, 59-66.

Klein, B., and Murray, K. (1979). Phage λ receptor chromosomes for DNA fragments made with restriction endonuclease I of *Bacillus amyloliquefaciens* H. J. Mol. Biol. 133, 289-294.

Loenen, W.A.M., and Brammar, W.J. (1980). A bacteriophage λ vector for cloning large DNA fragments made with several restriction enzymes. Gene 20, 249-259.

Maniatis, T., Fritsch, E.F., and Sambrook, J. (1982). In: "Molecular Cloning", Cold Spring Harbor Laboratory, Cold Spring Harbor, New York, 11724; pp.342-343.

Meyerowitz, E.M., Guild, G.M., Prestidge, L.S., and Hogness, D.S. (1980). A new higher capacity cosmid vector and its use. Gene 11, 271-282.

Murray, N.E., and Murray, K. (1974). Manipulations of restriction targets in phage λ to form receptor chromosomes for DNA fragments. Nature 251, 476-481.

Murray, N.E., Brammar, W.J., and Murray, K. (1977). Lambdoid phages that simplify the recovery of *in vitro* recombinants. Molec. Gen. Genet. 150, 53-61.

Murray, N.E. (1983). Phage λ and Molecular Cloning. In "Lambda II"; Hendrix, R.W. et al., eds; Cold Spring Harbor Laboratory, Cold Spring Harbor, New York.

Ptashne, M., Backman, K., Humayun, M.Z., Jeffrey, A., Maurer, R., Meyer, B., and Sauer, R.T. (1976). Autoregulation and function of a repressor in bacteriophage λ. Science 194, 156-161.

Ptashne, M., Jeffrey, A., Johnson, A.D., Maurer, R. Meyer, B.J., Pabo, C.O., Roberts, T.M., and Sauer, R.T. (1980). How the λ repressor and cro work. Cell 19, 1-11.

Rambach, A., and Tiollais, P. (1974). Bacteriophage λ having *Eco* RI endonuclease sites only in the non-essential region of the genome. Proc. Natl. Acad. Sci. USA 71, 3927-3930.

Rimm, D.L., Horners, D., Kucera, J., and Blattner,

F.R. (1980). Construction of coliphage λ Charon vectors with *Bam* HI cloning sites. Gene 12, 301-309.

Sanger, F., Coulson, A.R., Hong, G.F., Hill, D.F., and Petersen, G.B. (1982). Nucleotide sequence of bacteriophage λ DNA. J. Mol. Biol. 162, 729-773.

Scalenghe, F., Turco, E., Edström, J.E., Pirrotta, V., and Melli, M. (1981). Microdissection and cloning of DNA from a specific region of *Drosophila melanogaster* polytene chromosomes. Chromosoma 82, 205-216.

Seed, B. (1983). Purification of genomic sequences from bacteriophage libraries by recombination and selection *in vivo*. Nucleic Acids Res. 11, 2427-2445.

Simonsen, C.C., and Levinson, A.D. (1983). Isolation and expression of an altered mouse dihydrofolate reductase cDNA. Proc. Natl. Acad. Sci. USA 80, 2495-2499.

Smith, G.R., Kunes, S.M., Schultz, D.W., Taylor, A., and Triman, K.L. (1981). Structure of chi spots of generalized recombination. Cell 24, 429-436.

Spandidos, D.A., and Wilkie, N.M. (1984). Malignant transformation of early passage rodent cells by a single mutated human oncogene. Nature 310, 469-475.

Szybalski, E.H., and Szybalski. W. (1979). A comprehensive molecular map of bacteriophage λ. Gene 7, 217-270.

Tiemeyer, D., Enquist, L, and Leder, P. (1976). Improved derivatives of a phage λ EK2 vector for cloning recombinant DNA. Nature 263, 526-527.

Twigg, A.J., and Sherratt, D. (1980). Transcomplementable copy-number mutants of plasmid ColE1. Nature 283, 216-218.

Ward, D.F., and Gottesman, M.E. (1982). Suppression of transcription termination by phage λ. Science 216, 946-951.

Wolfe, J., Erickson, R.P., Rigby, P.W.J., and Goodfellow, P.N. (1984). Cosmid clones derived from both euchromatic and heterochromatic regions of the Y chromosome. The EMBO J. 3, 1997-2003.

Wu, R., and Taylor, E. (1971). Nucleotide sequence analysis of DNA. II. Complete nucleotide sequence of the cohesive ends of bacteriophage λ DNA. J. Mol. Biol. 57, 491-511.

5 Cloning in Yeasts

Genetic manipulations in yeasts are of interest for several reasons:

- Yeasts, like bacteria, are single-celled, but unlike bacteria they are eukaryotic and therefore may be the preferred organisms for the functional expression of eukaryotic genes. Important factors in this context are post-transcriptional and post-translational mechanisms for which there are profound differences between prokaryotes and eukaryotes.
- Yeast cells are not only much easier to grow in tissue culture than cells from higher eukaryotes, but it is also possible to draw knowledge from long-standing expertise in yeast biotechnology.
- Yeast strains are genetically well-characterised; detailed genetic maps are available for *Saccharomyces cerevisiae* and *Schizosaccharomyces pombe* (Petes, 1980a). Yeasts offer the general advantages that they can exist in haploid and diploid states. This particular feature is an essential prerequisite for detailed genetic analyses, and indeed, many auxotrophic and other markers of yeasts are known today.
- Since powerful techniques for the transformation of yeast cells with naked DNA are available it is possible to introduce individual genes into mutated yeast strains and to select transformed clones by complementation.
- Many yeast genes are functionally expressed in *E. coli*. This allows the isolation of a number of yeast genes for metabolically important enzymes by their ability to complement cognate

genetic defects in *E. coli* mutants. As far as gene manipulation is concerned it is important that bacterial genes can also be expressed in yeasts. Yeast cells, therefore, constitute a remarkably flexible system for the molecular biological analysis of eukaryotic cells.

5.1 The Life Cycle of Saccharomyces Cerevisiae

One of the main characteristics of diploid and haploid yeast cells is their method of vegetative (asexual) reproduction which occurs during a mitotic cell cycle as outlined in Fig. 5-1. Like cell cycles of other eukaryotic organisms, the yeast cell cycle can be divided into a G1 phase preceeding the initiation of chromosomal DNA synthesis, an S phase of chromosomal DNA replication, a G2 phase, and finally mitosis, with division of the nucleus and formation of daughter cells. More than 51 biochemical functions of this cell cycle have now been defined by cell cycle mutants (*cdc*) (for reviews see Strathern, Jones and Broach, 1981).

A particular feature of this cell cycle is the asymmetrical division of cytoplasm, which is known as budding and which is the prevailing asexual reproductive process in yeasts. In this

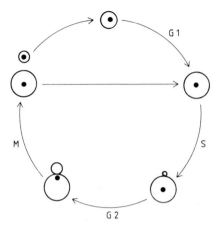

Fig. 5-1. Assymmetric cell division cycle of *S. cerevisiae*.
When the M (mitotic) phase has been completed and cells have divided, parent and daughter cells are unequal in size. Daughter cells grow during the G1 phase until they reach the size of parent cells. The S (synthetic) phase is characterised by chromosomal DNA synthesis and the emergence of buds. During the G2 phase the nucleus (closed circles) migrates to the periphery of the parent cell, and a spindle is formed in the daughter cell.

process, a portion of the protoplasm of a parental cell forms a lateral outgrowth (bud) which finally may split off to form a daughter cell (Fig. 5-1). The different sizes of parent and daughter cells are useful morphological markers since they allow identification of the stage individual cells occupy within the cell cycle.

As long as growth conditions are optimal cells may repeatedly go through mitosis; however, if nutrients become scarce in a diploid culture the mitotic cycle is switched to a meiotic cycle (Fig. 5-2), which is characterised by the generation of four haploid cells from the original diploid cell. These cells remain together in a sac (ascus) composed of the former cell wall. Two types of cells are observed in this tetrad stage, two α and two *a* cells. If the individual cells of an ascus are separated, by micromanipulation for example, each individual haploid cell can be propagated by vegetative reproduction without losing its former

cell type, *i.e.*, α or *a*. In order to obtain a diploid cell for sporulation, the two different cell types must be mixed. Conjugation occurs only between cells belonging to different cell types, never between cells of the same cell type. Differences have been observed in the stability of these cell types; heterothallic types are stable, homothallic

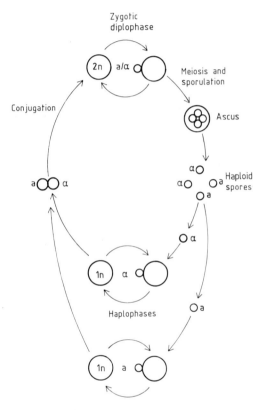

Fig. 5-2. Life cycle of *S. cerevisiae*.
Two types of haploid cells, *a* and α, are known. Each cell type has 16 chromosomes (1n). In a heterothallic strain these two types multiply vegetatively as stable entities for many generations by way of mitotic cell cycles. The same applies to the diploid zygote (2n = 32 chromosomes). Under certain growth conditions, however, a zygote may also sporulate and give rise to an ascus (4n), containing four haploid spores (2 a, 2 α). In homothallic strains the haploid cells are unstable and spontaneously revert to the opposite cell type. These cell strains are therefore capable of conjugation and can be kept and propagated only as diploid cells.

unstable. Haploid cells of heterothallic species always remain haploid in culture and always express the same cell type. In haploid cells of homothallic species the cell type is altered with a certain frequency and switched between α and *a* and *vice versa*. Conjugation between these two cell types yields diploid cells and cultures of homothallic strains therefore always contain (α/*a*) diploid cells.

Cell type is governed by a single gene locus known as *MAT* (mating type). Its allele α is responsible for the α cell type, *a* for cell type *a*. The rapid switch between cell types in homothallic species such as *S. cerevisiae* is caused by a translocation of genes, which is regarded as the prototype of similar translocations observed, for example, in the development of immunoglobulin genes (*cf.* Shapiro, 1983).

5.2 Genetics of Saccharomyces cerevisiae

A haploid cell of *S. cerevisiae* contains approximately 14 000 kb of DNA, which exceeds the DNA content of *E. coli* by a factor of 4. Apart from chromosomal DNA, which constitutes approximately 90% of *S. cerevisiae* DNA, there are at least two independent genetic elements; mitochondrial DNA and the 2μ plasmid. Some strains contain a third independent replicon, known as the killer plasmid; these plasmids are double-stranded RNA molecules coding for a toxin which kills other susceptible yeast strains (Wickner, 1981). The genetic analysis of the nuclear DNA has been facilitated by the isolation of suitable mutants and the establishment of genetic maps by recombination. In this context it is very important that there are stable haploid as well as diploid stages in the development of yeast cells. Haploid forms do not only allow the isolation of dominant but also of recessive mutants which can be classified subsequently in

diploid cells through complementation (Mortimer and Schild, 1981).

The genetic map of yeast comprises sixteen metacentric chromosomes with a total of 312 genes already mapped (Mortimer and Schild, 1985). It is striking that, with some exceptions, functionally related genes are generally not grouped together in yeasts whereas such genes are often linked in prokaryotes. The genes *GAL1*, *7*, and *10*, which code for galactokinase (E.C.2.7.1.6), galactose transferase (E.C.2.7.7.10), and galactose epimerase (E.C.5.1.3.2) are located on chomosome II. The five genes of the *ARO1* complex, which code for enzymes involved in the biosynthesis of aromatic amino acids, are located on chromosome IV. These two gene complexes are atypical of the genomic organisation of *S. cerevisiae*.

Genetic crosses have established that the genome of *S. cerevisiae* has a total length of 4 600 cM (centimorgan). Since the size of the genome is 14 000 kb, one centimorgan corresponds to approximately 3 kb. This value has been confirmed, for example, by the sequence analysis of cloned yeast DNA comprising the region between the *LEU2* and *CDC10* loci on the left arm of chromosome III (Clarke and Carbon, 1980a).

5.3 Identification of Yeast Genes

In principle, the same techniques used for the analysis of other eukaryotic genomes can also be employed for the identification of yeast genes in cDNA or genomic libraries. However, common *in vitro* cloning techniques, such as the use of enriched cDNA or RNA probes, the application of synthetic oligonucleotides, and also immunological detection systems are of lesser importance since powerful genetic tools are available for yeasts.

An important technique is the analysis of yeast genes by complementation cloning. This approach is based on the selection of yeast DNA

Table 5–1.   Yeast genes which can complement mutations in *E. coli*.

Yeast gene	Enzymes	*E. coli* mutations
HIS3	Imidazole glycerol phosphate (IGP) dehydratase	*his*B
TRP5	Tryptophan synthase	*trp*AB
LEU2	β-Isopropylmalate dehydrogenase	*leu*B
URA3	Orotidine-5'-phosphate decarboxylase	*pyr*F
ARG4	Argininosuccinate lyase	*arg*H

sequences in bacteria, and exploits the remarkable fact that approximately 20% of all yeast genes can complement mutations of *E. coli*. The functional expression of yeast genes in *E. coli* has been demonstrated in particular with auxotrophic mutants of the biosynthetic pathways for amino acids, but also with genes involved in the biosynthesis of pyrimidines.

The starting point for such experiments is usually the transformation of suitable *E. coli* mutants with plasmid mixtures containing yeast DNA. Desired transformants are obtained by plating on suitable selective media. Table 5-1 lists some yeast genes and the corresponding *E. coli* mutants which already have been used for this type of complementation analysis. The first example was the gene locus *HIS3* for imidazole glycerol phosphate dehydratase (IGP dehydratase) which is involved in the biosynthesis of histidine and complements the *E. coli* mutation *hisB* (Struhl *et al.*, 1976). A pertinent question was, of course, whether the complementing yeast sequence really contained the structural gene for IGP dehydratase and not another DNA sequence. In order to answer this, the sequence in question was used to isolate a homologous DNA sequence from a yeast strain carrying an amber mutation in the *HIS3* locus by DNA-DNA hybridisation. The mutated DNA was able to complement the genetic defect in *hisB* strains of *E. coli* carrying an amber suppressor, but not in suppressor-free mutants, proving beyond doubt the identity of the complementing DNA and of the structural gene for IGP dehydratase.

When these experiments were carried out in 1976-77 they created quite a sensation, since they showed for the first time that eukaryotic genes could also be expressed in prokaryotic cells. These experiments probably were successful because comparatively few yeast genes possess introns, for instance, the actin gene (Gallwitz and Sures, 1980), genes coding for some ribosomal proteins (Langford *et al.*, 1984) and the *MATal* gene (Miller, 1983).

Although such experiments were very helpful in the initial stages of gene manipulation in yeasts, today it is preferable to characterise yeast genes by complementation in yeast itself. A prerequisite for such experiments involving the complementation of yeast mutants by cloned yeast DNA sequences is a method which allows DNA to be introduced into yeast cells in the first place. The successful introduction of DNA into yeast cells was described for the first time by Hinnen *et al.* (1978) who succeeded in transferring and stably expressing the gene for β-isopropylmalate dehydrogenase (*LEU2*) into suitable *leu2* mutants of *S. cerevisiae*. A large proportion of the LEU-positive transformants, however, expressed the gene from chromosomal DNA sequences. The recombinant plasmid containing ColE1 and yeast sequences had become integrated into the *LEU2* region on chromosome III in more than 70% of the transformants (*cf.* also Section 5.5.1). The yield of transformants was on the order of one transformant per 10^7 cells, which makes it difficult to distinguish genuine transformants from revertants. The problem was solved at that time by using extremely stable double mutants carrying frame-shift mutations in *LEU2* as recipients. Transformation frequencies have now been improved by at least four orders of magnitude by using vectors which do not integrate into the yeast chromosomal DNA but replicate autonomously (see Section 5.5.2).

5.4 Genetic Markers and Selection Systems

In principle, every yeast gene for which there is a mutation available can also be used as a selective marker in a transformation experiment.

The selective pressure depends on the ratio of transformation frequency to reversion rate. It has already been mentioned in Section 5.3 that transformation frequencies below 10^{-6} require the use of double mutants since the reversion rates of many yeast mutations are on the order of 10^{-6} to 10^{-7}. The construction of such double mutations is comparatively laborious. An alternative would be to employ suitable dominant markers which could also be used in normal cells.

Although yeast cells are generally not very sensitive to common antibiotics, a number of selection systems based on *E. coli* resistance genes have been developed in the past several years. The first promising observations were made by Hollenberg (1979) who demonstrated the expression of a bacterial β-lactamase gene in yeast cells. This system could not be exploited, however, since *S. cerevisiae* cells are not sensitive to penicillin. Cohen *et al.* (1980) have successfully transferred and expressed the gene for chloramphenicol acetyltransferase, which inactivates the antibiotic chloramphenicol by acetylation (*cf.* Fig. 8-7). Since aerobically grown cells of *S. cerevisiae* are sensitive to chloramphenicol this enzyme can be used for a direct selection. Resistance to aminoglycoside antibiotics is only rarely exploited, and only in yeast strains for which suitable

mutants are not available. In bacteria this resistance is mediated by bacterial transposons, *e.g.*, Tn601 or Tn5, coding for an aminoglycoside phosphotransferase which inactivates antibiotics such as kanamycin or neomycin by phosphorylation of the 3′ hydroxyl group in the 2-deoxystreptamin component. If integrating yeast vectors are used which contain the transposon in conjunction with suitable yeast markers (*LEU2*, for example), the phosphotransferase gene is also expressed in yeast cells. The gentamycin derivative G418 has proven particularly useful since it is much more toxic for yeast cells than is kanamycin (Jiminez and Davies, 1980). Yeast transformants which carry the Tn601 resistance gene on a 2μ plasmid are resistant to G418 at concentrations above 1 mg/ml. If glycerol is used as the sole carbon source, the reversion rate, *i.e.*, the spontaneous occurrence of G418-resistant colonies, is lower than 10^{-7}, and the use of G418 therefore allows for a powerful selection system.

Nevertheless, in *S. cerevisiae* with its extensive collection of mutants, selections based on complementation of mutant strains are preferred over dominant markers of the kind described above and used in higher eukaryotes (*cf.* Section 8.1).

5.5 Vectors

The four different types of yeast vectors are distinguished by their interaction with recipient cells: they are either integrating vectors, designated YI$_p$ (yeast integrating plasmids), replicating vectors (YR$_p$) (yeast replicating plasmids), episomal vectors (YE$_p$) (yeast episomal plasmids), or artificial chromosomes.

5.5.1 Integrating Vectors

YI$_p$ vectors consist of bacterial vector components and a yeast gene with a selectable phenotype (Fig. 5-3). Site-specific integration of vector

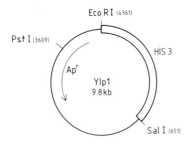 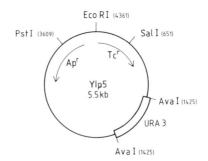

Fig. 5-3. Structures of two integrating yeast plasmids.
YI$_p$1 contains a 5.5 kb yeast DNA fragment with the coding region of *HIS3* cloned into pBR322 DNA linearised with *Eco* RI and *Sal* I. YI$_p$5 was constructed by cleaving pBR322 with *Ava* I (C/TCGGG), adding dC tails, and inserting *via* dG tails a 1.1 kb yeast *Hind* III DNA fragment carrying the *URA3* gene. The *URA3* gene codes for orotidine-5'-phosphate decarboxylase. The way in which YI$_p$5 was constructed conserved both the tetracycline and the ampicillin resistance markers of pBR322. Yeast sequences are represented as open bars.

DNA into chromosomal DNA may be mediated by homologous recombination between chromosomal and vector DNA. As shown in Fig. 5-4 for the example of the *HIS3* gene region on chromosome XV, homologous recombination results in a duplication of the cloned yeast gene (Struhl *et al.*, 1979). With approximately one transformant per microgram of DNA per 10^7-10^8 cells, transformation yields obtained with these vectors are comparatively low. Moreover, transformed cells often contain only one copy (sometimes a few copies) of the vector, which replicates under the control of, and together with, chromosomal DNA. In practice such properties of integrating vectors would be problematic were it not for the fact that these plasmids can be propagated in *E. coli via* their prokaryotic DNA sequences. It is therefore comparatively easy to obtain enough DNA for studies of their interaction with yeast cells.

It should be obvious that the recovery of integrating vectors requires special techniques. Chromosomal yeast DNA must be cleaved with a suitable restriction endonuclease which does not cleave within the cloned sequence or the prokaryotic vector sequences. The mixture of yeast DNA fragments obtained after digestion is ligated at low DNA concentrations since this favours intramolecular reactions, and desired plasmids are selected in *E. coli* cells. In the example shown in Fig. 5-4 the inserted *his3* allele would be recovered after *Eco* RI digestion, the *HIS3* allele after *Sal* I treatment and subsequent selection for ampicillin resistance.

Integrating vectors offer interesting applications. Not only do they allow the introduction of genes into yeast cells but also their integration at exactly their normal chromosomal locations (Scherer and Davies, 1979). This particular feature permits, for example, the introduction of specific mutations into the yeast genome through substitution of normal alleles for mutated genes.

The integration event shown in Fig. 5-4 suggests that the integration by general recombination of a mutated gene in the vicinity of a wild-type gene could be reversed. This process would lead to two different genotypes, a chromosome XV carrying the mutated (*his3*) gene and one carrying the functional (*HIS3*) gene. In both cases pBR322 sequences will be lost. Unfortunately, it is very difficult to identify the reversion of the transformation event as shown in Fig. 5-4, since the desired recombinational events cannot easily be selected for. This can be accomplished, however, by introducing another marker into the YI$_p$ plasmid. The exact procedure is explained

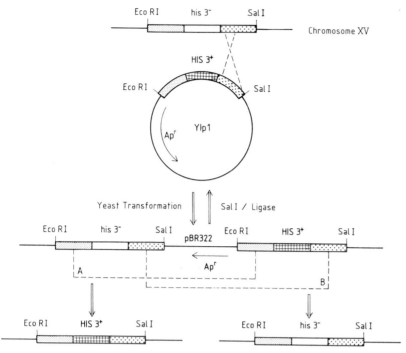

Fig. 5-4. Integration of a YI$_p$ vector.
Plasmid YI$_p$1 with its functional *HIS3* locus (HIS3$^+$) integrates into a homologous mutated his3$^-$ region on chromosome XV by homologous genetic recombination. The resulting duplicated structures are unstable. After 15 generations in a non-selective medium, approximately 1% of the colonies are his$^-$. As shown in reaction B, an exact reversal of the initial transformation process leads to a complete loss of the transforming HIS3$^+$ DNA and restores the initial (his3$^-$) state. Pathway A also yields a stable wild-type (HIS$^+$) configuration (see text).

schematically in Fig. 5-5. The starting point in this case is a YI$_p$ plasmid with two different yeast DNA sequences, one coding for a functional *URA3* locus (orotidine 5′ phosphate decarboxyl-ase), the other coding for a mutated *his3* gene which is to be integrated into its normal chromo-somal site. In principle, the vector contains two regions homologous to the yeast genome, *i.e.*, *HIS3* and *URA3*, which would allow its integra-tion into chromosomal DNA; however, the inte-gration event can be forced to take place at the *HIS3* locus if the recipient carries a *URA3* deletion which is large enough to prevent integra-tion at this site. After the integration event, the chromosomal DNA will contain two different *HIS3* regions, one being an unmutated *HIS3*, the other a mutated *his3* allele. As shown in Fig. 5-5, the homologous ends of these two regions can now recombine with each other. This allows the reverse reaction to take place, *i.e.*, the substitu-tion of the wild-type gene for the mutated gene. Since the *URA3* wild-type gene lies between the two *HIS* markers a recombinational event would delete these sequences like the pBR322 sequences shown in Fig. 5-4. A selection for uracil auxotro phy, *e.g.* by selection for 5-fluoro-orotic acid resistance (Boeke *et al.*, 1984), would select for the desired recombinants, 50% of which should be HIS3$^+$, and 50% his3$^-$. The two *HIS* regions cannot be lost from the chromosome since the sequences to the left of his3$^-$ and to the right of HIS3$^+$ are not homologous. The actual experi-

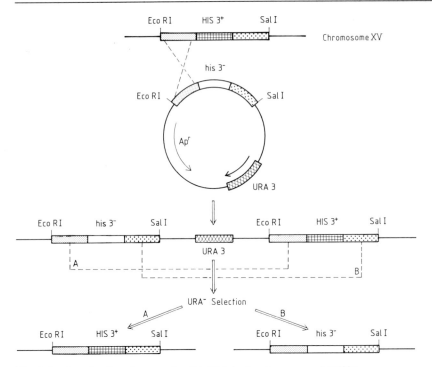

Fig. 5-5.   Introduction of a mutation (his3⁻) into the chromosomal *HIS3* locus.
After 20 generations on rich medium, approximately 1% of cells lose one of the duplicated *HIS* alleles. Such cells can be selected for by ura⁻ selection (Scherer and Davis, 1979). See text for details.

ment yielded approximately 1% ura3⁻ cells with only one of the two *HIS* alleles after twenty to thirty cell generations.

In principle, this technique should be applicable to all clonable yeast genes. Unfortunately, however, it is laborious since it it requires two selection steps to identify recessive mutations. The integration event results in a gene duplication, one copy being wild-type, and the other mutant. A recessive mutation thus would not be detected until the plasmid vector is excised in such a way that the mutation is retained on the chromosome (Pathway B in Fig. 5-5). Shortle *et al.* (1984) developed a new strategy (Fig. 5-6), known as "integration replacement/disruption strategy", which avoids the step of screening for excision. This approach employs a vector which does not contain a full-length (albeit mutated) copy of a gene; instead, it carries a fragment of this gene with a deletion at one end. The single cross-over event that leads to the integration of the vector thus yields two possible constructions. Depending upon the site where the cross-over takes place, *i.e.*, its position relative to the point mutation, the gene carrying the point mutation can be recombined into an intact copy of this gene on the chromosome, thus disrupting the wild-type gene, or *vice versa*. The first integration event therefore allows the mutated gene to be expressed even if it is recessive. Shortle *et al.* (1984) were able to recover temperature-sensitive mutations of an active gene (*URA3* in this case) by simply selecting for URA3⁺ transformants and screening them for temperature sensitivity. More recently,

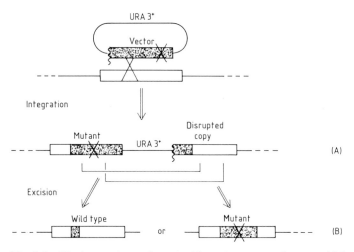

URA 3⁺

Vector

Integration

Mutant
URA 3⁺
Disrupted copy

(A)

Excision

Wild type   or   Mutant

(B)

Fig. 5-6. The integrative replacement/disruption strategy for recombining mutations constructed *in vitro* into a cellular gene (open bar).
A vector molecule carries a selectable marker (*e.g.*, URA3⁺) and a copy of a mutated gene (*X*) with a 5' deletion (shaded bar). Upon transfection into yeast cells, integration (A) results in transformants bearing an intact mutant gene and a disrupted wild-type gene. The other possibility, namely that the cross-over occurs on the other site of the mutation, placing the mutation into the disrupted copy of the gene, is not shown. Excision (B) of the plasmid leads either to a wild-type or a mutant strain (Holm *et al.*, 1985).

Holm *et al.* (1985) used this procedure to isolate a variety of temperature-sensitive DNA topoisomerase II mutations.

A different one-step gene disruption strategy, based on homologous recombination, was described by Rothstein (1983). It takes advantage of the observation that, during yeast transformation, the free ends of linear DNA molecules can recombine (Orr-Weaver *et al.* 1981). A selectable yeast gene, *e.g.*, *HIS3*, is inserted into a cloned copy of another gene, *GENE X*, so that this gene is disrupted (Fig. 5-7). The disrupted gene with its *HIS3* insertion is liberated from the bacterial plasmid sequences by restriction endonuclease digestion and transformed into a his3⁻, GENE X⁺ yeast cell. Provided that the linear fragment contains sequences homologous to the chromosomal *GENEX* region on either site of the inserted selectable yeast gene (*HIS3*) it is possible to isolate HIS3⁺ transformants which are simultane-

ously HIS3⁺ and gene X⁻. *GENE X* thus has been replaced by a disrupted *gene x* in a single step. This strategy can be used, for example, to determine whether a cloned yeast gene is essential, and also to alter or even delete specific chromosomal regions.

It should be mentioned that integration of the cloned yeast DNA also occurs at unexpected sites with low frequency. This is usually observed, for example, if the gene in question is flanked by repetitive sequences, which occur not only in the vicinity of the desired integration site but also in other regions on the chromosome. Homologous recombination will then take place between these repetitive flanking regions (Petes, 1980b). One practical aspect of this phenomenon is that it allows the expression of a gene to be studied in different surroundings, and hence demonstration of the influence of genetic context on gene expression.

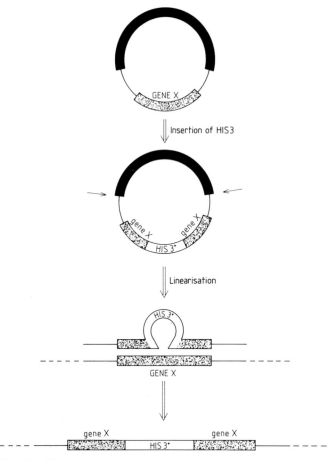

Fig. 5-7. The one-step gene disruption procedure.
A DNA fragment carrying the selectable marker gene (*e.g.*, HIS3$^+$) is inserted into a cloned copy of *GENE X*. The linear fragment with the disrupted *gene X* is liberated from plasmid sequences (black bar) by restriction enzyme digestion (arrows) and transferred into yeast cells. Integration *via* the flanking *gene X* sequences results in substitution of the linear disrupted *gene X* for the intact chromosomal *GENE X* (Rothstein, 1983).

5.5.2 Replicating Vectors

YR$_P$ vectors contain prokaryotic plasmid DNA sequences and parts of yeast DNA derived from YI$_p$ vectors, and additional chromosomal origins of DNA replication. The latter allow their replication as extrachromosomal elements presumably in the nucleoplasm of yeast cells. Origins of chromosomal DNA replication were initially identified in derivatives of YI$_p$ vectors which

transformed approximately two to three orders of magnitude better than expected (Stinchcomb *et al.*, 1979). This high frequency of transformation is thought to be due to the presence of certain DNA sequences, namely origins of DNA replication which allow these vectors to replicate autonomously. Such vectors, are often referred to as *ARS* vectors (*ARS* = autonomously replicating sequences) (Stinchcomb *et al.*, 1980). Their *ARS* portions can be replaced by origins of DNA replication of other lower eukaryotes and higher plant cells.

The prototype of an YR$_p$ vector is vector YR$_p$7 which is 5.7 kb in length (Fig. 5-8) (Stinchcomb *et al.*, 1979). It contains the entire pBR322 sequence and a 1.4 kb *Eco* RI fragment of yeast DNA coding for the *TRP1* gene (N-(5'-phosphoribosyl)-anthranilate isomerase) with a neighbouring *ARS* region (Fig. 5-8). Since the yeast components of this plasmid have been cloned into the *Eco* RI site of pBR322, YR$_p$7 contains two *Eco* RI sites, rendering *Eco* RI cloning more difficult. In order to eliminate the *Eco* RI site between the *ars* region and the β-lactamase gene, a vector lacking both *Eco* RI sites was constructed. For this purpose YR$_p$7 was first cleaved with *Eco* RI, the protruding ends were filled-in, ligated, and the molecules were then selected on trp$^-$ mutants of *E. coli*. The desired *Eco* RI site was reintroduced by substituting a *Hind* III-*Bam* HI fragment of YR$_p$7 (containing this *Eco* RI site) for the corresponding *Hind* III-*Bam* HI fragment in the new vector. The new plasmid, pFRL4, can also be used as an expression vector (Hitzemann *et al.*, 1980).

On the average, vectors of the YR$_p$7 type exist only in a few copies per cell and are quite unstable. In the absence of tryptophan, *i.e.*, if transformants containing YR$_p$7 are selected for, the plasmid is found in only 10-20% of cell after a the few cell generations. In the presence of tryptophan, *i.e.*, under conditions which also allow the growth of plasmid-free transformants, less than one per cent of the cells retain the Trp$^+$ phenotype (Kingsman *et al.*, 1979). The rapid loss of these plasmids in a cell population is due to preferential segregation into the parent cells which frequently accumulate as many as 50 copies of a given plasmid (Murray and Szostak, 1983a). This problem can be solved by providing YR$_p$ plasmids with functional centromeric DNA sequences. Such centromeric DNA regions, designated *CEN* DNAs, which already have been cloned from twelve of the sixteen yeast chromosomes, contain a central AT-rich sequence of

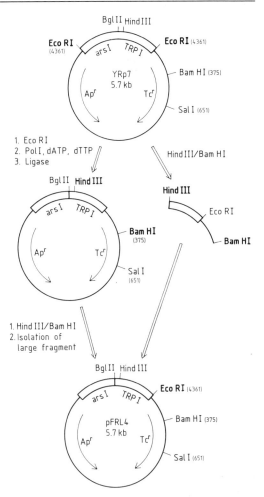

Fig. 5-8. Construction of vector YR$_p$7 and its derivative, pFRL4.
The derivative pFRL14 contains only one of the two *Eco* RI sites of YR$_p$7. Its construction is explained in the text. Yeast DNA sequences are shown as open bars (Hitzeman *et al.*, 1981).

approximately 90 bp flanked by highly conserved regions of 11 and 14 bp, respectively (Hieter *et al.*, 1985b). If such *CEN* DNA sequences are incorporated into plasmids they become true minichromosomes (Szostak and Blackburn, 1984), *i.e.*, they replicate only once per cell cycle, occur in a copy number of one, and are distributed evenly to daughter cells (*cf.* Section 5.5.4). It should be noted, however, that even these constructs do not possess the same stability as intact functional chromosomes. Colour tests which allow the copy numbers of such plasmids to be determined in individual cells have revealed that the remaining residual instability decreases with increasing length of such minichromosomes (Hieter *et al.*, 1985a; Koshland *et al.*, 1985).

5.5.3 Episomal Vectors

YE$_p$ vectors contain prokaryotic sequences, a selectable yeast gene, and the entire yeast 2μ plasmid (Hollenberg, 1982). The latter, also known as the scp plasmid (*S. cerevisiae* plasmid), is found in almost all strains of *S. cerevisiae* in 50-100 copies per haploid cell, which corresponds to approximately 2-3% of the entire chromosomal DNA. It is assumed that this plasmid replicates in the nucleus although the analysis of tetrads obtained from a cross of plasmid-containing strains (cir$^+$) with plasmid-free strains (cir^0) reveals that all haploid cells in an ascus contain 2μ plasmids (Livingston, 1977); in addition, the same experiment demonstrated that 2μ plasmids can be easily transferred into strains of *S. carlsbergensis* and other strains of *S. cerevisiae* which normally do not contain endogenous 2μ sequences. The primary structure of this plasmid, which is 6 318 bp in length, has been determined by Hartley and Donelson (1980). The most striking structural features are two inverted repetitions of 599 bp which divide the plasmid molecule into two non-identical regions of 2 774 and 2 364 bp (Fig. 5-9). These two identical repetitions can be exchanged by intra- and intermolecular site-specific genetic recombination which is very efficient in yeast. These processes yield a population of plasmids consisting of equal amounts of two types of molecules which differ only in the orientation of their unique sequences (Fig. 5-10). These two forms could be physically separated by cloning in *E. coli*.

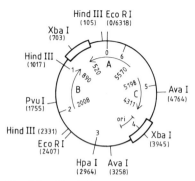

Fig. 5-9. Physical map of form A of the 2μ plasmid from *S. cerevisiae*.
Open bars represent the 589 bp inverted repetitions. Arrows marked A, B, and C indicate the positions of the three open reading frames, with arrow heads marking the 3' ends of the corresponding mRNAs and the C-terminal ends of the three peptides with molecular weights of 48 625, 43 245, and 33 199 Da. Numbers are the co-ordinates as described by Hartley and Donelson (1980).

The biological function of this autonomously replicating element is still unknown. Although this plasmid shows a variety of biological activities, cir$^+$ and cir^0 strains cannot be differentiated by their phenotypes. The primary structure of the 2μ plasmid reveals three open reading frames for larger proteins (Fig. 5-9). One of these proteins from coding region A, is involved in recombination processes generating the two different molecular species of this plasmid. Flp$^-$ strains which have been transformed with suitably deleted plasmids, contain only one of these molecular species.

The 2μ plasmid replicates autonomously and presumably possesses one origin of DNA replication. This region has been mapped by deletion mapping and comprises approximately 350 bp which are located predominantly on one of the inverted repetitions, although the origin also extends into the neighbouring unique DNA sequences (Fig. 5-9). The presence of this *cis* function is sufficient to allow replication, and chimaeric plasmids containing this region can replicate in cir^0 strains, albeit with low copy numbers. High copy numbers are observed only

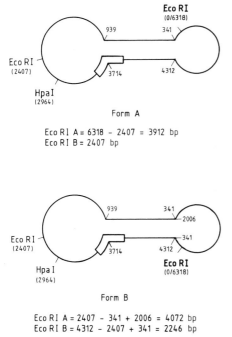

Eco RI A = 6318 - 2407 = 3912 bp
Eco RI B = 2407 bp

Eco RI A = 2407 - 341 + 2006 = 4072 bp
Eco RI B = 4312 - 2407 + 341 = 2246 bp

Fig. 5-10. Forms A and B of the 2μ plasmid. Parallel lines indicate the inverted repetitions. Open bars represent the location of the origin of DNA replication. The repetitions are separated by unique regions (shown as circles) of different lengths. The lengths of the *Eco* RI fragments of each form are listed below the schematic drawings. An equimolar mixture of the two forms yields four different *Eco* RI fragments.

in the presence of gene products encoded by the other two regions, *B* and *C*, often also referred to as *REP1* and *REP2*. These proteins are thought to be involved in, and required for proper and random segregation of the plasmid into the daughter cells.

Clearly then, the 2μ plasmid is a suitable vector for cloning in yeast cells. Prototype plasmids contain either the entire 2μ plasmid or selected *Eco* RI and *Hind* III fragments (see Fig. 5-19); in addition, these cloning vehicles contain a bacterial replicon (pBR322 or pMB9) and a yeast marker gene. Vector pJDB219, for example, contains the B form of the entire 2μ plasmid cloned into the *Eco* RI site of pMB9 (Fig. 5-11) (Beggs, 1978).

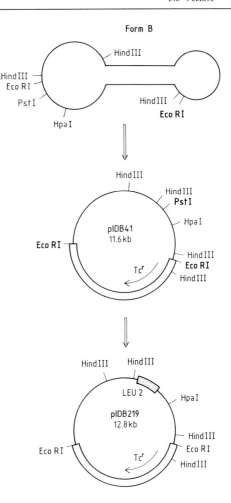

Fig. 5-11. Construction of YE_p vector pJDB219. The starting materials for this construction are form B of the 2μ plasmid, partially cleaved with *Eco* RI, and *Eco* RI-linearised plasmid pMB9 (5.3 kb). The relevant *Eco* RI site in form B of the 2μ plasmid is indicated in boldface type (top). pJDB219 is derived from pJDB41 and contains a 1.2 kb yeast DNA fragment with the *LEU2* gene inserted into the unique *Pst* I site by poly(dA)-poly(dT) tails. This insertion destroys the *Pst* I site. The *LEU2* gene (stippled bar) codes for β-isopropylmalate dehydrogenase. pMB9 sequences with the tetracycline resistance region are represented as open bars. (Beggs, 1978).

The B form contains two *Eco* RI sites (Fig 5-11, top). Partial cleavage with *Eco* RI, therefore, yields two different molecules linearised at different positions but equal in size. The same applies to the A forms of the 2μ plasmid. Since all four linear molecules can also be inserted with two different orientations, ligation of *Eco* RI-linearised 2μ plasmids with *Eco* RI-linearised pMB9 should lead to eight different molecules. Plasmid pJDB41 is one of these eight molecular forms. It can be opened at the *Pst* I site and annealed with suitable yeast DNA by using poly(dA)-poly(dT) tails. Vector pJDB219 contains a 1.2 kb insert with the coding region for the LEU2 gene of *S. cerevisiae*. These and similar constructions containing the entire 2μ DNA transform yeast strains with an efficiency which is appoximately one order of magnitude higher than that of YR_p vectors. Provided that cir^+ strains with endogenous 2μ plasmids are transformed, the same high transformation frequencies are also observed for plasmids containing only parts of the 2μ plasmid. If plasmid-free cir^0 strains are transformed, stable transformants are only obtained if the vectors contain the origin of DNA replication of the 2μ plasmid. Plasmid-containing cir^+ strains can still be transformed efficiently with vectors lacking the origin of DNA replication of the 2μ plasmid since the presence of the inverted repetitions allows these vectors to acquire an origin through recombination with endogenous plasmids.

Although the transformation frequencies obtainable with vectors containing 2μ DNA are high, their use is not unproblematic since they show a certain degree of instability. Firstly, recombinational events between exogenous and endogenous vector sequences may alter the structure of the cloning vehicle. Secondly, hybrid plasmids may be lost under non-selective conditions in the presence of endogenous 2μ plasmids with a frequency of 5-40% per cell division. This phenomenon, which is presumably due to the incompatibility of identical replicons competing for the same enzymes, is also observed in prokaryotic systems (Hollenberg, 1982). It is still not known why hybrid rather than endogenous vector plasmids are lost predominantly. The problem can be reduced, however, by using strains which do not contain endogenous plasmids.

5.5.4 Artificial chromosomes

Normal eukaryotic chromosomes are always linear. In yeast, the sequence elements required for circular plasmid vectors to be replicated and stably maintained in an extrachromosomal state have been identified as *ARS* and *CEN* sequences. Linear plasmids, however, must carry additional functional structures at their ends known as telomers (Murray and Szostak, 1983b). These

Jack Szostak
Boston

consist of a tandem array of a simple satellite-like repetitive sequence, which is $5'$-$(C_{1-3}A)_{30-100}$-$3'$ in yeast and $5'$-$C_{1-8}(T/A)_{1-4}$-$3'$ in other lower eukaryotes including slime moulds, ciliated protozoa and haemoflagellates (Blackburn, 1985; Blackburn and Szostak, 1984). Linear plasmids containing the *LEU2* gene as a selectable marker, and the elements *CEN3* and *ARS1* flanked by the terminal 0.7 kb *Tetrahymena* rDNA termini as telomers, are less stable than any circular plasmids shorter than 10-16 kb. With increasing length, however, mitotic stability of these constructs increases considerably, which indicates that there is a minimum size requirement for a linear centromeric plasmid to function as an artificial chromosome. The estimated lengths of genuine yeast chromosomes range fom 300 to 3,000 kb suggesting that, indeed, sheer size affects the segregation behavior of a chromosome.

Artificial chromosomes may not be useful vectors for expression of foreign proteins in yeasts; however they may permit a systematic exploration of a variety of chromosomal features and functions of yeasts and other eukaryotes as well. These phenomena could include structural requirements for homologous recombination, chromosomal position effects on gene expression, and gene dosage compensation (Blackburn, 1985).

5.6 Gene Expression in Eukaryotes

Yeasts are eukaryotic micro-organisms and therefore offer a number of advantages over bacteria for the expression of cloned eukaryotic genes; they do not only possess the eukaryotic machinery required for the functional expression of genes, but they can be grown in fermenters on a large scale, which again has many advantages over tissue culture of eukaryotic cells. Criteria similar to those that are important for the construction of expression vectors in other systems should also be taken into account for the construction of yeast expression vectors (*cf.* Chapters 7 and 8).

Some of these criteria are the nature and function of regulatory sequences for transcription and translation, codon selection, transport mechanisms of the synthesised gene products, control of gene expression, and the stability of synthesised proteins. Much is known about the structure of eukaryotic transcriptional and translational regulatory signals. Although very little work has addressed the rest of these problems, it has become apparent that eukaryotic systems differ from their prokaryotic counterparts in these important aspects.

5.6.1 The Eukaryotic Transcription Unit

A typical eukaryotic gene consists of a $5'$ untranslated region, a coding sequence which may be interrupted by introns, and a $3'$ untranslated region (Fig. 5-12). Transcription is catalysed by RNA polymerase II initiating in the $5'$ untranslated region. The primary transcript is colinear with the DNA and includes coding and non-coding regions. Transcription is finally terminated in the $3'$ untranslated region. Primary transcripts are characterised by a cap structure at the $5'$ end and a poly(A) tail at the $3'$ end. Non-coding regions (introns) are removed from the primary transcript by a splicing process which eventually yields translatable mRNA. The analysis of the primary structures of many eukaryotic genes has revealed the existence of specific signals for all steps leading from the primary transcript to functional mRNA. The sequence region at the initiation site of transcription is characterised by the canonical TATA box and the CAAT box (Breathnach and Chambon, 1981). The sequence TATA(^A_T)A(^A_T) is usually positioned 25-32 bp, and the sequence GCCTC*AA*TCT approximately 80 bp upstream from the initiation site (Grosveld *et al.*, 1982). Some genes, for instance the late SV40 genes and the early genes of region

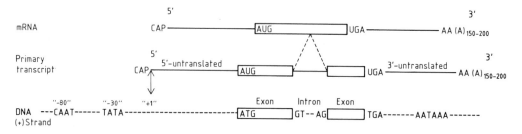

Fig. 5-12. Structure of a eukaryotic gene and its transcripts.
The transcription unit comprises the entire DNA region represented as a primary transcript plus several signal sequences (see text for details).

2 of type C adenoviruses, lack these sequences. Since it has not yet been proven that RNA polymerase II binds to these sequences, the exact role of these consensus sequences remains to be established. What is known is that the space between the TATA box and the initiation site is fixed. If the TATA box is moved, for example, by the introduction of deletions, the system reacts by utilising new transcription initiation sites at positions which retain the original distance from the TATA sequence.

Apart from TATA- and CAAT boxes there is now convincing evidence for the existence of additional sequences which are much farther away from the transcriptional start site and which are responsible for the efficiency of transcriptional initiation *in vivo* (summarized for the case of yeast genes by Brent, 1985). In yeast those stretches of DNA responsible for transcriptional activation are called upstream activation sites (or UAS), in higher enkaryots they are designated enhancers (*cf.* Section 8.4).

As far as the splicing signals and the corresponding splicing enzymes are concerned, the situation is extremely complex. A comparison of intron sequences has revealed that introns are always flanked by a 5′ GT and a 3′ AG dinucleotide (Mount, 1982; Fig. 5-12). It can hardly be assumed, of course, that splice sites are completely defined by these dinucleotides. The analysis of deletion mutations near exon/intron junctions has shown that such junctions are most probably characterised by 20-30 nucleotides around the actual splice site (*cf.* Wieringa et al., 1983, for example). The development of cell-free splicing systems has allowed possible reaction sequences of mRNA splicing to be identified. As shown in Fig. 5-13, the intron of the RNA forms a lariat structure with one branch point which is usually located at an A residue approximately 20-40 nucleotides upstream of the 3′ splice site. For β-globin, for example, the structure of this branch point is

$$A \begin{cases} 2'p5'G \\ 3'p5'U \end{cases}$$

This structure contains not only the expected normal 3′ to 5′, but also a novel 2′ to 5′ linkage (Keller, 1984). Little is known about the enzymology of the splicing reaction. Both in yeast and in human cells, the pre-mRNA was shown to be associated with a 40S to 60S particle termed the "spliceosome" (Brody and Abelson, 1985; Grabowski et al., 1985; Frendewey and Keller, 1985). The protein components of these complexes, however, have not been identified.

The 3′ ends of transcripts are also marked with characteristic sequences. The polyadenylated terminus is usually found approximately 20 bp downstream from an AATAAA sequence, although there are some indications that RNA polymerase II continues approximately 1 kb beyond this target site and that the 3′ end is eventually created by endonucleolytic attack and the addition of a poly(A) tail (reviewed by Birnstiel et al., 1985).

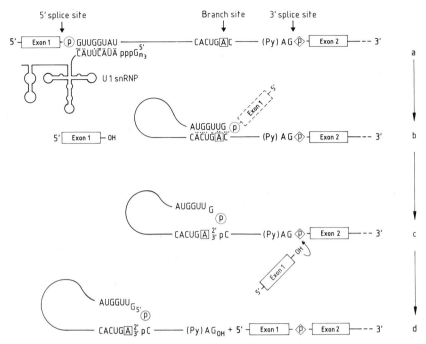

Fig. 5-13. Reactions involved in splicing of pre mRNA molecules.
Shown is a section of a pre-mRNA molecule with an intron and two flanking exons, 1 and 2. Splicing is initiated by the association of a sequence region immediately adjacent to the 5′ splice site with a section from the 5′ end of the U1 RNA (a). This RNA molecule is part of a ribonucleoprotein particle known as the U1 snRNP. Further steps are the formation of a lariat structure (b), the creation of a 2′ to 5′ phosphodiester bond (c), and eventually the ligation of the two exons, which is accompanied by the release of the lariat structure (d). In yeast DNA, the branch point is located within a unique sequence region, the TACTAAC box. The whole sequence of events is presumed to take place as a concerted reaction within a large protein/RNA complex such that exon 1, for example, does not have to exist in a free form (dashed box) (Keller, 1984).

It appears that the situation in *S. cerevisiae* is somewhat special. All yeast genes studied so far contain the TATA consensus sequence. Efficiently expressed genes, for example, the genes coding for enzymes of the glycolytic pathway, also contain the CAAT sequence (Dobson *et al.*, 1982); nevertheless, heterologous eukaryotic genes controlled by their own promoter signals generally are not expressed at all or are expressed inefficiently or aberrantly in yeast cells. Transcription of the rabbit β-globin gene in yeast cells, for example, initiates at a site differing from the normal initiation site and is prematurely terminated within the second intron (Beggs *et al.*, 1980); in addition, the first intron is not removed from the primary transcript. Similar difficulties have been described for the transcription of the herpes virus thymidine kinase gene if it is under the control of its own promoter sequences.

Such differences are also observed with termination signals. Only four of approximately fifteen yeast genes that have been analysed contain AATAAA as a termination signal. The consensus sequence TATGT, on the other hand, is found in all yeast genes studied so far. It is located approximately 30 to 40 bp downstream from the polyadenylation site. If this consensus sequence is deleted in the gene for iso-1-cytochrome c, the amounts of the protein and of its corresponding mRNA are decreased by more than 90%; instead, a biologically inactive mRNA is formed which is more than 1 000 nucleotides longer at the 3' end than is normal mRNA (Zaret and Sherman, 1982).

Splicing of primary transcripts in yeast cells appears to be identical to splicing in other higher eukaryotes, at least as far as lariat structures are concerned (Domdey *et al.*, 1984). In this case, however, the branch point is characterised by a 2' to 5' linkage between the G residue at the 5' terminus of the intron and the third A residue of a special sequence known as TACTAAC box. This sequence region, which is positioned near the 3' splice site, is essential for splicing and has been conserved in all nuclear genes, *i.e.*, in the corresponding introns, of *S. cerevisiae* (Langford *et al.*, 1984; Newman *et al.*, 1985).

Further differences between yeast genes and genes of higher eukaryotes will certainly be uncovered in the future. As long as the full details remain to be elucidated, it is prudent to employ homologous transcription signals for the construction of yeast expression vectors.

5.6.2 Translation Signals

In prokaryotes, the translation initiation site is defined by the ribosomal binding site on the mRNA. This site contains the AUG start codon and a neighbouring purine-rich sequence known as the Shine-Dalgarno box, which shows a certain degree of homology with the 3' end of 16S ribosomal RNA (*cf.* Fig. 7-38). This kind of homology in the vicinity of the AUG start codon is barely or not at all observed in eukaryotes. Eukaryotic initiation sites must therefore be defined otherwise. According to the scanning model proposed by Kozak (1981) for higher eukaryotes, the 40S ribosomal subunit binds to the 5' end of a mRNA, migrates along this mRNA molecule as far as the first AUG codon, and uses this AUG as the start codon. Indeed, it has been observed that in 90% of all eukaryotic mRNAs (142 of 153 studied) translation initiates at the AUG codon which is nearest to the 5' end of the mRNA. Exceptions are usually explained by the observation that a functional AUG codon should be flanked by a neighbouring purine (usually A) at position "−3" and/or a G at position "+4". If this is not the case, some 40S subunits will ignore this start signal and initiate at the following AUG (Kozak, 1982). The yeast mRNA species analysed so far also conform with Kozak's rules. In frequently expressed mRNAs, however, an additional U has been conserved at position "+6" (Dobson *et al.*, 1982). It is quite remarkable that the consensus sequence $CC(^A_G)CCAUGG$ within the region of the eukaryotic translation start

Dieter Gallwitz
Göttingen

signal derived from the analysis of more than 250 cellular and viral mRNAs (Kozak, 1984a,b) contains an *Nco*I site (C/CATGG). A number of eukaryotic genes (approximately 33/200) can therefore be isolated directly at their start codon from genomic DNA or cDNA libraries by *Nco* I digestion and insertion into suitable *Nco* I expression vectors.

Marked differences in the selection of codons have been observed between yeast and other higher eukaryotes and also between yeast and bacteria. In frequently transcribed yeast genes, such as the genes coding for alcohol dehydrogenase isoenzyme I and glycerol-3-phosphate dehy-

drogenase, 96% of the 1004 amino acids are coded by only 25 of the 61 possible codons (Bennetzen and Hall, 1982b; see also Table 7-4 and Table 11-2 for codon preferences in *E. coli* and human genes, respectively).

As expected, these codons are complementary with the anticodons of the most frequently used tRNA species in yeast. As shown in Table 5-2 for the degenerate codons of Leu, Arg, and Val, the special preference for certain codons is more marked, the more frequently a gene is transcribed; moreover, the data in Table 5-2 also demonstrate that there are also codon preferences in frequently transcribed *E. coli* genes and that

Table 5–2.   Codon usage for amino acids leucin, arginine and valine for some *S. cerevisiae* genes, for the human interferon αD gene and for strongly expressed genes of *E. coli*.

Amino acid	Codons	Genes									
		gap 491	gap 63	ADH	B1	B2	C1	C2	sum	IFN αD	E. coli
Leu	UUA	0	0	2	1	1	1	2	7	1	
	UUG	21	20	19	5	5	5	3	78	4	
	CUU	0	0	0	0	0	1	0	1	1	
	CUC	0	0	0	0	0	0	0	0	8	
	CUA	0	1	3	0	0	1	0	5	1	
	CUG	0	0	0	0	0	0	0	0	10	56 (58)
Arg	AGA	11	11	8	6	5	3	2	46	5	
	AGG	0	0	0	0	1	0	1	2	4	
	CGA	0	0	0	0	0	0	0	0	0	
	CGG	0	0	0	0	0	0	0	0	0	
	CGU	0	0	0	0	0	0	0	0	0	33 (40)
	CGC	0	0	0	0	0	0	0	0	0	
Val	GUU	22	23	19	4	5	1	1	75	1	
	GUC	15	12	17	1	2	0	1	48	2	62 (70)
	GUA	0	0	0	0	0	0	0	0	0	
	GUG	0	0	0	0	0	2	0	2	7	

Abbreviations for *S. cerevisiae* genes are: gap 491 and gap 63 = glyceraldehydphosphate dehydrogenase, clones 491 and 63; ADH = alcohol dehydrogenase; B1 = histone H2B, gene H2B1; B2 = histone H2B, gene H2B2; C1 = iso-1-cytochrome c; C2 = iso-2-cytochrome c; IFN αD = leukocyte interferon alpha-D (the intron-free gene codes for a total of 189 amino acids including a 23 amino acid long signal peptide). With respect to arginine codons AGA and AGG, codon usage in the human IFN-gene is similar to that of strongly expressed *S. cerevisiae* genes. *E. coli* stands for some strongly expressed *E. coli* genes which include the genes for the outer membrane protein (*omp*A), the lipoprotein (*lpp*), and the translational elongation factors *tuf*A and *tuf*B (from Bennetzen and Hall, 1982b). A more recent summary of codon usage in *S. cerevisiae* (as well as in other organisms) can be found in Maruyama *et al.* (1986), Nucleic Acids Res. 14, r151–197.

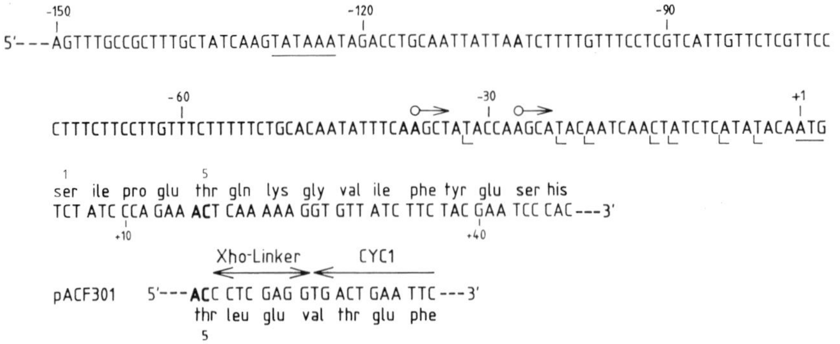

```
  -150                              -120                                      -90
   |                                 |                                        |
5'----AGTTTGCCGCTTTGCTATCAAGTATAAATAGACCTGCAATTATTAATCTTTTGTTTCCTCGTCATTGTTCTCGTTCC
```

```
      -60                                -30                                      +1
       |                                  |                                        |
CTTTCTTCCTTGTTTCTTTTTCTGCACAATATTTCAAGCTATACCAAGCATACAATCAACTATCTCATATACAATG
```

```
  1              5
ser ile pro glu thr gln lys gly val ile  phe tyr glu ser his
TCT ATC CCA GAA ACT CAA AAA GGT GTT ATC TTC TAC GAA TCC CAC---3'
            |                                   |
          +10                                 +40
```

```
                    Xho-Linker        CYC1
pACF301    5'---ACC CTC GAG GTG ACT GAA TTC ---3'
                thr leu glu  val  thr  glu  phe
                5
```

Fig. 5-14. Promoter region of the *ADHI* gene of *S. cerevisiae*.
The initiation points for transcription at positions "-27" and "-37" are marked by open circles with arrows. The TATA box and the ATG start codon are underlined. A portion of the structure of vector pACF301 (Hitzeman *et al.*, 1981) is shown below the coding sequence of the *ADHI* gene. In this vector the first five *N*-terminal sequences (only residue 5, thr, is shown) of the *ADHI* gene are linked with coding sequences for iso-1-cytochrome c (*CYC1*) *via* an *Xho* I linker (Smith *et al.*, 1979). This vector which also contains the *LEU2* gene and the origin of DNA replication of the 2µ plasmid, expresses iso-1-cytochrome c under the control of the *ADHI* promoter. The 5' ends of the transcripts are therefore identical with those of the normal chromosomal *ADHI* gene and not with those of the *CYC1* gene. The deletions in the 5' untranslated region of the *ADHI* gene mentioned in the text are marked by brackets. These deletions are generated by *Bal*31 treatment of *Xho* I-linearised pACF301 vector DNA.

codon usage of Arg and Val codons in strongly expressed genes of *E. coli*, for example, is not compatible with yeast codon usage. The same applies to some other amino acids not shown in Table 5-2.

The example of the human leukocyte interferon gene, LeIF-D, demonstrates how complex the situation is for the expression in yeast of genes derived from higher eukaryotes. As far as the leucine codons are concerned, yeast appears to be quite incompatible since UUG is only a minor codon in LeIF-D; on the other hand, *E. coli* appears to be quite suitable because the codon CUG is used in both human and *E. coli* genes. The reverse is true for the arginine codon. In practice, the situation will have to be analysed for each individual case before the decision between yeast and *E. coli* expression vectors is made. The chemical synthesis of genes allows, of course, such differences in codon usage to be taken into account. Codons can be chosen to suit the frequency with which isoaccepting tRNAs occur in the desired host organism.

5.7 Yeast Expression Vectors

A corollary of what has been said above is that eukaryotic genes to be expressed in yeast should not contain introns. In most cases it is best to use a cDNA with coding sequences under the control of yeast promoters. Four yeast promoters have been used extensively in expression vectors to date, namely, the promoter of the gene coding for alcohol dehydrogenase isoenzyme I (ADHI), the phosphoglycerol kinase (PGK) promoter, the repressible acid phophatase (PHO5) promoter and the α factor promoter (plus the signal sequence).

ADHI is a cytoplasmic enzyme responsible for the generation of ethanol from acetaldehyde and NADH. When cells are grown in the presence of glucose this enzyme accounts for at least 1% of the total cellular protein. The *ADHI* promoter is therefore quite attractive for the construction of expression vectors. As shown in Fig. 5-14, transcription initiates at positions "-27" and "-37" of the *ADH* promoter sequence (Bennetzen and

Hall, 1982a). The starting material for the construction of expression vectors was a series of plasmids with deletions extending into the *ADHI* promoter region as far as positions "-4", "-7", "-12", "-14", "-20", "-23", and "-32". All deletions remove the ATG codon but not the promoter-specific sequences TATA and CAAT. These deletions were generated in vector pACF301 which contains the *CYCI* gene (iso-I-cytochrome c) joined by *Xho* I linkers to the codon for the fifth amino acid (thr) of alcohol dehydrogenase (Fig. 5-14). Expression of the *CYCI* gene in this vector is, therefore, controlled by the *ADHI* promoter. A series of deletions extending into the 5′ untranslated region can be generated in pACF301 by limited *Bal*31 digestion starting at the single *Xho* I site. These deletions are marked by square brackets in Fig. 5-14. Addition of *Eco* RI linkers and subsequent simultaneous digestion with *Bam* HI/*Eco* RI yields a population of DNA fragments which are approximately 1 500 bp in length. These fragments are cloned into vector pFRL4 (Figs. 5-8; 5-15). The single *Eco* RI site of these expression vectors is used for cloning cDNA sequences which are then expressed under the control of the *ADHI* promoter. One published example is the cloning of leukocyte interferon D cDNA (Hitzeman *et al.*, 1981). As shown schematically in Fig. 5-16 for one of these cDNA clones, an adaptor with an ATG start codon and one protruding *Eco* RI end was substituted for the prepeptide sequence (Goeddel *et al.*, 1980 and 1981). This yielded a DNA fragment of 568 bp

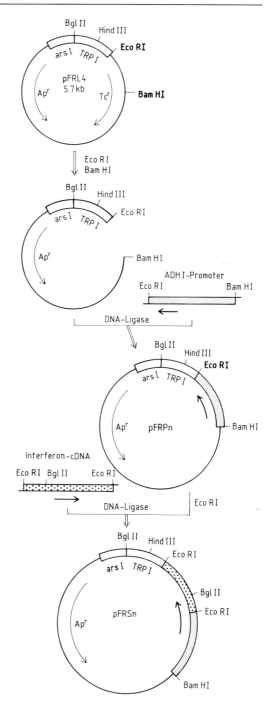

Fig. 5-15. Construction of an expression vector with the *ADHI* promoter.
The *ADHI* promoter (stippled bar; see text) is furnished with *Eco* RI-*Bam* HI ends and is inserted into vector pFRL4 (Fig. 5-8) which has been linearised with *Eco* RI and *Bam* HI. The unique *Eco* RI site can be used for cloning of LeIF-D cDNA. Neither of the seven *ADHI* promoter deletions in pFRPn contains an *ADHI*-derived ATG start codon (Fig. 5-14), and hence a met-leukocyte interferon D protein, rather than a fusion peptide, is generated. The arrows in pFRS indicate the directions of transcription. The interferon DNA fragment is represented by a dotted bar (coarse dots).

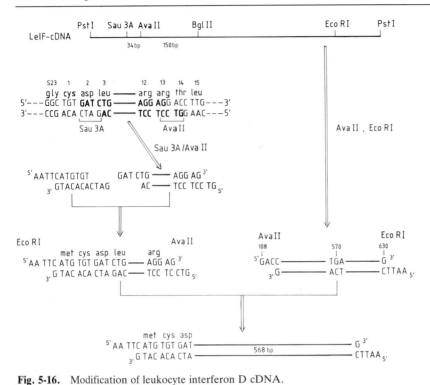

Fig. 5-16. Modification of leukocyte interferon D cDNA.
LeIF-D cDNA encodes a protein of 189 amino acid, comprising a signal sequence of 23 amino acids, and 166 amino acids of the mature interferon protein. The terminal amino acid of the signal peptide is marked S23. Regions coding for the signal peptide were removed by cutting the cDNA with *Sau* 3A and *Ava* II. The *Sau* 3A-*Ava* II fragment which is 34 bp in length was isolated and ligated with a chemically synthesised adaptor carrying *Eco* RI-*Sau* 3A ends. The adaptor was tailored so that a methionine residue is placed immediately in front of the first cysteine residue of the interferon peptide. The entire coding sequence was recovered by ligating the *Eco* RI-*Ava* II fragment with a 522 bp fragment directly obtained by *Eco* RI-*Ava* II treatment of the original cDNA. The stop codon, TGA, of LeIF-D is located 60 bp ahead of the *Eco* RI site; the new *Eco* RI fragment therefore covers the entire coding region. (Hitzeman *et al.*, 1981).

which could be placed downstream from the various *ADHI* promoter deletions (Fig. 5-15). As long as the interferon gene was correctly oriented, all promoter variants produced interferon in amounts accounting for 1-2% of the total cellular protein. Kozak's model for the initiation of translation is supported by the finding that none of the deletions in the 5' untranslated region (some of which were as long as 28 bp) blocked expression of the gene. The specific sequences of the 5' regions were therefore of no consequence in the efficiency of translation. Kozak's scanning model requires that the first AUG codon in an

mRNA is recognised as a signal and that it is not an extensive homology between mRNA and ribosomal RNA which plays a role in initiation.

Another series of expression vectors employs the phosphoglycerol kinase promoter which is attractive for two reasons. Firstly, the *PGK* gene, like other genes coding for glycolytic enzymes, belongs to a class of genes which are frequently transcribed; mRNA and protein produced from these genes may account for 1-5% of the total mRNA or protein of a cell. Secondly, the expression of this gene can be controlled by the presence of fermentable carbon sources. In the presence of glucose, for example, the level of glycolytic enzymes is approximately 100 times higher than in the presence of non-fermentable carbon sources (Holland and Holland, 1978).

The starting point for the construction of an expression vector with a *PGK* promoter was the *E. coli*/yeast shuttle vector pMA3-PGK (Fig. 5-17). It contains pBR322 sequences with an *Eco* RI fragment comprising the coding region for the *LEU2* gene and the 2μ plasmid origin of DNA replication cloned into its *Eco* RI site, and a 2.9 kb *Hind* III fragment containing the *PGK* gene (Tuite *et al.*, 1982). The *PKG* gene does not possess any useful restriction sites within its promoter region. pMA3-PGK was therefore digested with *Sal* I and subsequently treated with *Bal*31 which removed approximately 500 bp in the direction of the start codon (Fig. 5-15). Ligation in the presence of an excess of *Bam* HI linkers yielded a series of plasmids possessing a *Bam* HI site in the *PGK* promoter region. In one of these plasmids, pMA230, the *Bam* HI site is located immediately after the codon for amino acid 13. Insertions of coding sequences into this site therefore lead to the production of hybrid fusion proteins. One particular clone containing an interferon α2 cDNA, for example, produced a PGK-interferon fusion peptide which accounted for up to 2% of the total cellular protein (15 mg

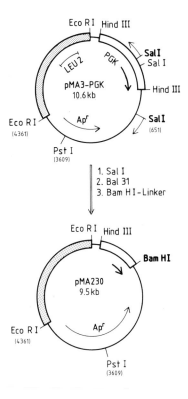

```
-160                                              -120
 |                                                 |
ACATTTACATATATATAAACTTGCATAAATTGGTCAATGCAAGAAA
```

```
                                   -80
                                    |
TACATATTTGTCTTTTCTAATTCGTAGTTTTTCAAGTTCTTAGATGCTT
```

```
                        -40
                         |
TCTTTTTCTCTTTTTTACAGATCATCAAGAAGTAATTATCTACTTTTT
```

```
                -1 +1
                 | |
ACAACAAATATAAAACA ATG TCT TTA TCT TCA AAG TTG TCT
                  met ser leu ser ser lys leu ser
                                              +5
```

```
GTC CAA CAT TTG GCC GGA TCC
val gln his leu ala Bam HI
 +10
```

Fig. 5-17. Construction of an expression vector on the basis of the phosphoglycerol kinase (*PGK*) gene. The direction of transcription of the *PGK* gene in vector pMA3-PGK is indicated by an arrow heading clockwise. The *PGK* gene (open bar) is located on a *Hind* III fragment. Another yeast DNA fragment (stippled bar) which is flanked by *Eco* RI sites codes for the *LEU2* gene and contains the origin of DNA replication of the 2μ plasmid. The promoter-proximal *Sal* I site in the *PGK* gene and the start codon are approximately 580 bp apart. These bases were removed by opening the plasmid with *Sal* I and subsequent digestion with *Bal*31 (see arrows). This results in deletions both in the *PGK* gene (counterclockwise) and the pBR322 portion (clockwise) which is removed up to position 1150. Ligation in the presence of the *Bam* HI linker, GGGATCCC, yields a set of plasmids, one of which, pMA230, is shown below pMA3-PGK. The sequence around the promoter (bottom) shows that the linker is located immediately adjacent to amino acid 13. The postulated TATA and CAAT regions are underlined. Numbers in brackets refer to the position of cleavage sites in pBR322 DNA.

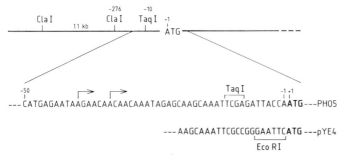

Fig. 5-18. Map of the 5′ flanking regions of the *PHO5* gene.
Shown on top are the start codon and the DNA fragments, *Cla* I-*Taq* I and *Cla* I-*Cla* I, which are important for the construction of vector pYE4. The expanded region shows the sequence between the ATG start codons (bold-face) and the 5′ flanking region, with the the 5′ ends of the two transcripts at positions "−40" and "−34" (arrows). Also shown is the *Taq* I site which has been replaced by an *Eco* RI site in pYE4 (Kramer *et al.*, 1984). Apart from a short GC-rich region immediately upstream of the *Eco* RI site, there are no differences between the 5′ flanking regions in pYE4 and *PHO5* (Thill *et al.*, 1983).

interferon/l of culture). In addition, interferon production also depended on the presence of glucose. Cells utilising glucose produced approximately 20-30 times more interferon than cells growing on acetate (Tuite *et al.*, 1982).

Although these vectors can be very useful, future work will have to be directed towards modifications such that genuine proteins rather than fusion products will be obtained; such advances already have been made in the case of interferon clones with *ADHI* promoters.

Another promising candidate for use in an expression vector is an acid phosphatase (APase) gene. Two such genes, *PHO3* and *PHO5*, are closely linked and lie on the right arm of chromosome II. The expression of the *PHO3* gene is constitutive; *PHO5* can be repressed by phosphate and is expressed only in the absence of inorganic phosphate. Its mRNA is expressed at the same level as those for the glycolytic enzymes 3-phosphoglycerate kinase and glyceraldehyde-3-phosphate dehydrogenase (GAPDH). Since GAPDH amounts to approximately 5% of the total yeast cell protein, it should be possible to develop an interesting vector system based on *PHO5*. A corresponding vector, pYE4, has been developed from plasmid pFRL4 (Fig. 5-19). It

contains approximately 1.3 kb of the 5′ flanking region of the *PHO5* gene and an *Eco* RI site into which suitable genes can be cloned. The unique *Eco* RI site was created from the original *Taq* I site by several linker additions (Fig. 5-18). In this construction the region upstream of position "-10" corresponds exactly to the *PHO5* gene, and there are some alterations only in the immediate vicinity of the start codon. These sequence alterations are thought to explain the low level expression of the γIFN-αD gene in this vector; the corresponding protein amounts to only 0.2% rather than the expected 5% of the total cellular protein. Nevertheless, the observed induction after a shift to phosphate-free growth medium was quite remarkable and resulted in a 200-fold increased interferon activity. This finding clearly demonstrates that the interferon gene is, indeed, controlled by the *PHO5* promoter.

The regulation of *PHO5* expression is mediated by a regulatory complex consisting of the gene products of *PHO2*, *PHO4*, *PHO80*, and *PHO85* (Bostian *et al.*, 1980). Several temperature-sensitive mutants are known (in *PHO4* and *PHO80*) which grow well at the non-permissive temperature (35 °C) although they do not synthesise any APase even at low phosphate concentrations;

however, at 23 °C, APase is produced in high amounts in medium with and without phosphate. Such mutant strains will allow replacement of the switch from phosphate-containing medium to phosphate-free conditions by a simple temperature shift, which may be particularly helpful if cells are to be grown in large quantities. When the vector described above was used, a temperature-dependent induction of interferon synthesis was indeed observed upon a shift to the lower temperature.

Another possibility for the expression of heterologous proteins is offered by fusion with the α factor. α factor is an effector similar to certain peptide hormones in higher eukaryotes, and is excreted by α haploid cells. This factor arrests "MATa" cells in the G1 phase of the cell cycle

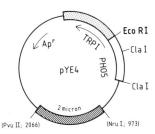

Fig. 5-19. Structure of expression vector pYE4. This vector is derived from pFRL4 (*cf.* Fig. 5-8) which contains the *TRP1* gene (dotted bar) on a 1.4 kb *Eco* RI fragment of pBR322. A 0.27 kb *Cla* I-*Eco* RI fragment with the initiation site of transcription for *PHO5* (*cf.* Fig. 5-18) has been inserted between the *Eco* RI and the *Cla* I site in the promoter region of the tetracycline resistance gene. The vector also contains the 1.1 kb *Cla* I-*Cla* I fragment (*cf.* Fig. 5-18) with additional flanking regions of the *PHO5* gene, and a 1.88 kb *Nru* I-*Hinc* II fragment derived from the 2μ plasmid (B form) with the origin of DNA replication (hatched bar). The 2μ sequences have been inserted between the *Pvu* II and the *Nru* I sites of the pBR322 region (indicated in brackets). The transcription termination site for the *TRP1* gene is approximately 0.8 kb downstream from the *Eco* RI site, which allows transcription of a gene inserted into the *Eco* RI site to be terminated either in its own 3' untranslated region, or by the *TPR1* termination signal. *PHO5* regions are shown as open bars (Kramer *et al.*, 1984).

and prepares these cells for conjugation with α cells. α factor, which is 13 amino acids in length is synthesised as a preproprotein precursor of 165 amino acids containing four copies of the α factor (Kurjan and Herskowitz, 1982) (Fig. 5-20). The precursor also includes a hydrophobic signal peptide of 22 amino acids, and a prosegment of unknown function which is 61 amino acids in length. The four α factor regions are separated by spacer peptides and are cleaved from the precursor by proteolytic activities (Julius *et al.*, 1983). α factor is synthesised in amounts similar to those observed for glycolytic enzymes and is excreted into the growth medium. It has been possible to exploit the mechanisms which allow the efficient expression and excretion of α factor for the excretion of foreign proteins.

Most applications described in the literature employ a *Hind* III site near the alanine codon 89 at the junction between the spacer peptide and mature α factor. Fusion peptides therefore contain the signal peptide, propeptide, first spacer, and the protein to be expressed. Apart from necessary leader sequences these fusion proteins also contain the lys-arg residues which are required for endoproteolytic cleavage of the precursor, and the glu-ala sequence required for exoproteolytic processing by a dipeptidylaminopeptidase. When the α factor gene was fused with almost the entire coding region of yeast invertase (*SUC2*; 528 of 532 amino acids, 513 amino acids corresponding to the mature enzyme) 90% of the invertase activity was detected in the extracellular periplasmic fraction (Emr *et al.*, 1983).

In the second example the gene fusions involved α factor and an analogue of human β endorphin of 31 amino acids, and a portion of α-interferon of 166 amino acids, respectively. Efficient expression and excretion of the two proteins in their mature forms, *i.e.*, without the prepro-α factor component, was also observed in these cases (Bitter *et al.*, 1984). Finally, Gardell *et al.* (1985) have expressed the cDNA for rat procarboxypeptidase A in a fusion with the α factor gene. The excretion of a protein with an

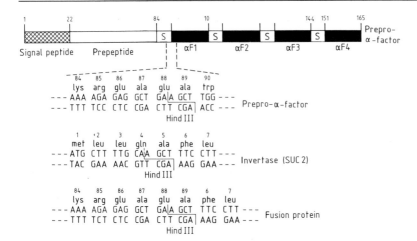

Fig. 5-20. Structure of an α-factor-invertase gene fusion.
The structure of the prepro-α-factor (165 amino acids) was derived from the sequence published by Kurjan and Herskowitz (1982). The top shows the signal peptide (cross-hatched bar), the prepeptide (open bar), the four α-factor regions (black bars; αF1-αF4), and the spacer regions (S). The expanded sequence is that of the junction between the first spacer region and the first α-factor region. The HindIII fragment with the *SUC2* gene is obtained from plasmid pRB58 (Carlson and Botstein, 1982).

activity and an electrophoretic mobility similar to rat carboxypeptidase A confirmed the correct processing of the fusion protein.

In summary, it can be stated that expression vectors in *S. cerevisiae* have come of age. They will be of increasing significance for the expression of eukaryotic proteins.

References

Beggs, J.D. (1978). Transformation of yeast by a replicating hybrid plasmid. Nature 175, 104-109.

Beggs, J.D., van den Berg, J., van Ooyen, A., and Weissmann, C. (1980). Abnormal expression of chromosomal rabbit β-globin gene in *S. cerevisiae.* Nature 283, 835-840.

Bennetzen, J.L., and Hall, B.D. (1982a). The primary structure of the *S. cerevisiae* gene for alcohol dehydrogenase I. J. Biol. Chemistry 257, 3018-3025.

Bennetzen, J.L, and Hall, B.D. (1982b). Codon selection in yeast. J. Biol. Chem. 257, 3026-3031.

Birnstiel, M.L., Busslinger, M., and Strub, K. (1985). Transcription termination and 3′ processing: The end is in sight. Cell 41, 349-359.

Bitter, G.A., Chen, K.K., Banks, A.R., and Lai, P.H. (1984). Secretion of foreign proteins from *S. cerevisiae* directed by α factor gene fusions. Proc. Natl. Acad. Sci. USA 81, 5330-5334.

Blackburn, E. (1985). Artificial chromosomes in yeast. Trends in Genetics 1, 8-12.

Blackburn, E., and Szostak, J.W. (1984). The molecular structure of centromers and telomers. Ann. Rev. Biochem. 53, 163-194.

Boeke, J.D., Lacroute, F., and Fink, G.R. (1984). A positive selection for mutants lacking orotidine-5′-phosphate decarboxylate activity in yeast: 5-fluoro-orotic acid resistance. Mol. Gen. Genet. 197, 345-347.

Bostian, K.A., Lemive, J.M., Cannon, L.E., and Halvorson, H.O. (1980). *In vitro* synthesis of repressible yeast acid phosphatase: identification of multiple mRNAs and products. Proc. Natl. Acad. Sci. USA 77, 6541-6545.

Breathnach, R., and Chambon, P. (1981). Organization and expression of eukaryotic split genes coding for proteins. Ann. Rev. Biochem. 50, 349-383.

Brent, R. (1985). Repression of transcription in yeast. Cell 42, 3-4.

Brody, E., and Abelson, J. (1985). The "spliceosome": Yeast pre-messenger RNA associates with a 40S complex in a splicing-dependent reaction. Science 228, 963-967.

Carlson, M., and Botstein, D. (1982). Two differentially regulated mRNAs with different 5'-ends encode secreted and intracellular forms of yeast invertase. Cell 28, 145-154.

Clarke, L., and Carbon, J. (1980a). Isolation of the centromer-linked *CDC10* gene by complementation in yeast. Proc. Natl. Acad. Sci. USA 77, 2173-2177.

Clarke, L., and Carbon, J. (1980b). Isolation of a yeast centromer and construction of a functional small circular chromosome. Nature 287, 504-509.

Cohen, J.D., Eccleshall, T.R., Needleman, R.B., Federoff, H., Buchferer, B.A., and Marmur, J. (1980). Functional expression in yeast of the *Escherichia coli* plasmid gene coding for chloramphenicol acetyltransferase. Proc. Natl. Acad. Sci. USA 77, 1078-1082.

Dobson, M.J., Tuite, M.F., Roberts, N.A., Kingsman, A.J., Kingsman, S.M., Perkins, R.E., Conroy,, S.C., Dunbar, B., and Fothergill, L.A. (1982). Conservation of high efficiency promotor sequences in *S. cerevisiae*. Nucleic Acids Res. 10, 2625-2637.

Domdey, H., Apostol, B., Lin, R.J., Newman, A., Brody, E., and Abelson, J. (1984). Lariat structures are *in vivo* intermediates in yeast pre-mRNA Splicing. Cell 39, 611-621.

Emr, S.D., Schekman, R., Flessel, M.C., and Thorner, J. (1983). An MFα1-SUC2 (α factor-invertase) gene fusion for study of protein localization and gene expression in yeast. Proc. Natl. Acad. Sci. USA 80, 7080-7084.

Frendewey, D., and Keller, W. (1985). Stepwise assembly of a pre-mRNA splicing complex requires U-snRNPs and specific intron sequences. Cell 42, 355-367.

Gallwitz, D., and Sures, I. (1980). Structure of a split yeast gene: complete nucleotide sequence of the actin gene in *Saccharomyces cerevisiae*. Proc. Natl. Acad. Sci. USA 77, 2546-2550.

Goeddel, D.V., Yelverton, E., Ullrich, a., Heynecker, H.L., Miozzari, G., Holmes, W., Seeburg, P.H., Dull, T., May, L., Stebbing, N., Crea, R., Maeda, S., McCandliss, R., Sloma, A., Tabor, J.M., Gross, M., Familletti, P.C., and Pestka, S. (1980). Human leukocyte interferon produced by *E. coli* is biologically active. Nature 287, 411-416.

Goeddel, D.V., Leung, D.W., Dull, J.J., Gross, M., Lawn, R.M., McCardliss, R., Seeburg, P.H., Ullrich, A., Yelverton, E., and Gray, P. (1981). The structure of eight distinct cloned human leukocyte interferon cDNAs. Nature 290, 20-26.

Grabowski, P.J., Seiler, S.R., and Sharp, P.A. (1985). A multicomponent complex is involved in the splicing of messenger RNA precursors. Cell 42, 345-353.

Grosveld, G.C., Rosenthal, A., and Flavell, R.A. (1982). Sequence requirements for the transcription of the rabbit β-globin gene *in vivo*: the -80 region. Nucleic Acids Res. 10, 4951-4971.

Hartley, J.L., and Donelson, J.E. (1980). Nucleotide sequence of the yeast plasmid. Nature 286, 860-865.

Hieter, P., Pridmore, D., Hegemann, J.H., Thomas, M., Davis, R.W., and Philippsen, P. (1985a). Functional selection and analysis of yeast centromeric DNA. Cell 42, 913-921.

Hieter, P., Mann, C., Snyder, M., and Davis, R.W. (1985b). Mitotic stability of yeast chromosomes: a colony color assay that measures nondisjunction and chromosome loss. Cell 40, 381-392.

Hinnen, A., Hicks, J.B., and Fink, G.R. (1978). Transformation of yeast. Proc. Natl. Acad. Sci. USA 75, 1929-1933.

Hinnen, A., and Meyhack, B. (1982). Vectors for cloning in yeast. Current Topics in Microbiology and Immunology 96, 101-117.

Hitzeman, R.A., Hagie, F.E., Levine, H.L., Goeddel, D.V., Ammerer, G., and Hall, B.D. (1981). Expression of a human gene for interferon in yeast. Nature 293, 717-722.

Holland, M.J., and Holland, J.P. (1978). Isolation and Identification of Yeast Messenger Ribonucleic Acids Coding for Enolase, Glyceraldehyde-3-phosphate Dehydrogenase and Phosphoglycerate Kinase. Biochemistry 17, 4900-4907.

Hollenberg, C.P. (1979). The expression in *Saccharomyces cervisiae* of bacterial β-lactamase and other antibiotic resistance genes integrated in a 2-μm DNA vector. ICN-UCLA Symp. Mol. Cell. Biol. 15, 325-338.

Hollenberg, C.P. (1982). Cloning with 2-μm vectors and the expression of foreign genes in *Saccharomyces cervisiae*. Current Topics in Microbiology and Immunology 96, 119-144.

Holm, C., Goto, T., Wang, J.C., and Botstein, D. (1985). DNA Topoisomerase II is required at the time of mitosis in yeast. Cell 41, 553-563.

Jiminez, A., and Davies, J. (1980). Expression of a transposable antibiotic resistance element in *Saccharomyces*. Nature 287, 869-871.

Julius, D., Blair, L., Brake, A., Spragne, G., and Thorner, J. (1983). Yeast α factor is processed from a larger precursor polypeptide: The essential role of a membrane-bound dipeptidyl aminopeptidase. Cell 32, 839-852.

Keller, W. (1984). The RNA Lariat: A new ring to the splicing of mRNA precursors. Cell 39, 423-425.

Kingsman, A.J., Clarke, L., Mortimer, R.K., and Carbon, J. (1979). Replication in *Saccharomyces cerevisiae* of plasmid pBR313 carrying DNA from the yeast TRP-1 region. Gene 7, 141-153.

Koshland, D., Kent, J.C., and Hartwell, L.H. (1985). Genetic analysis of the mitotic transmission of minichromosomes. Cell 40, 393-403.

Kozak, M. (1981). Possible role of flanking nucleotides in recognition of the AUG initiator codon by eukaryotic ribosomes. Nucleic Acids Res. 9, 5233-5252.

Kozak, M. (1982). Analysis of ribosome binding sites from the S1 message of Reovirus. J. Mol. Biol. 156, 807-820.

Kozak, M. (1984a). Point mutations close to the AUG initiator codon affect the efficiency of translation of rat preproinsulin *in vivo*. Nature 308, 241-246.

Kozak, M. (1984b). Compilation and analysis of sequences upstream from the translation start site in eukaryotic mRNAs. Nucleic Acids Res. 12, 857-872.

Kramer, R.A., DeChiara, T.M., Schaber, M.D., and Hilliker, S. (1984). Regulated expression of a human interferon gene in yeast: Control by phosphate concentration or temperature. Proc. Natl. Acad. Sci. USA 81, 367-370.

Kurjan, J., and Herskowitz, I. (1982). Structure of a yeast pheromone gene (MFα): A putative α factor precursor contains four tandem copies of mature α factor. Cell 30, 933-943.

Langford, C.J., Klinz, F.J., Donath, C., and Gallwitz, D. (1984). Point mutations identify the conserved, intron-contained TACTAAC Box as an essential splicing Signal Sequence in Yeast. Cell 36, 645-653.

Livingston, D.M. (1977). Inheritance of 2μm DNA plasmid from *Saccaromyces*. Genetics 86, 73-84.

Miller, A.M. (1984). The yeast *MATa1* gene contains two introns. The EMBO J. 3, 1061-1065.

Mortimer, R.K., and Schild, D. (1981). Genetic mapping in *Saccaromyces cerevisiae*. In: "The molecular biology of the yeast Saccaromyces; life cycle and inheritance", pp.11-26. Strathern, J.N., Jones, E.W., and Broach, J.R. (eds.). Cold Spring Harbor Laboratory, ISBN 0270-1847; 11A.

Mortimer, R.K., and Schild, D. (1985). Genetic map of *Saccaromyces cerevisiae*, Edition 9. Microbiol. Rev. 49, 181-212.

Mount, S.M. (1982). A catalogue of splice junction sequences. Nucl. Acids Res. 10, 459-472.

Murray, A.W., and Szostak, J.W. (1983a). Pedigree analysis of plasmid segregation in yeast. Cell 34, 961-970.

Murray, A.W., and Szostak, J.W. (1983b). Construction of artificial chromosomes in yeast. Nature 305, 189-193.

Newman, A.J., Lin, R.J., Cheng, S.C., and Abelson, J. (1985). Molecular consequences of specific intron mutations on yeast mRNA splicing *in vivo* and *in vitro*. Cell 42, 335-344.

Orr-Weaver, T.L., Szostak, J.W., and Rothstein, R.J. (1981). Yeast transformation: a model system for the study of recombination. Proc. Natl. Acad. Sci. USA 78, 6354-6358.

Petes, T.D. (1980a). Molecular genetics of yeast. Ann. Rev. Biochem. 49, 845-876.

Petes, T.D. (1980b). Unequal meiotic recombination within tandem arrays of yeast ribosomal DNA genes. Cell 19, 765-774.

Rothstein, R.J. (1983). One-step gene disruption in yeast. Methods in Enzymology 101, 202-211.

Scherer, S., and Davis, R.W. (1979). Replacement of chromosome segments with altered DNA sequences constructed *in vitro*. Proc. Natl. Acad. Sci. USA 76, 4951-4955.

Shapiro, J.A. (1983). Mobile Genetic Elements. Academic Press; New York and London.

Shortle, D., Novick, P., and Botstein, D. (1984). Construction and genetic characterization of temperature-sensitive mutant alleles of the yeast actin gene. Proc. Natl. Acad. Sci. USA 81, 4889-4893.

Smith, M., Leung, D.W., Gillam, S., Astall, C.R., Montgomery, D.L., and Hall, B.D. (1979). Sequence of the gene for iso-1-cytochrome c in *Saccharomyces cerevisiae*. Cell 16, 753-761.

Stinchcomb, D.T., Struhl, K., and Davis, R.W. (1979). Isolation and characterization of a yeast chromosomal replicator. Nature 282, 39-43.

Stinchcomb, D.T., Thomas, M., Kelly, J., Selker, E., and Davis, R.W. (1980). Eukaryotic DNA segments capable of autonomous replication in yeast. Proc. Natl. Acad. Sci. USA 77, 4559-4563.

Strathern, J.N., Jones, E.W., and Broach, J.R. (1982). The Molecular Biology of the yeast *Saccharomyces*. Cold Spring Harbor Laboratory, Cold Spring Harbor, New York, 11724.

Szostak, J.W., and Blackburn, E.H. (1984). The molecular structure of centromeres and telomeres. Ann. Rev. Biochem. 53, 163-194.

Struhl, K., Cameron, J.R., and Davis, R.W. (1976). Functional genetic expression of eukaryotic DNA in *Escherichia coli*. Proc. Natl. Acad. Sci. USA 73, 1471-1475.

Struhl, K., Stinchcomb, D.T., Scherer, S., and Davis, R.W. (1979). High-frequency transformation of yeast: Autonomous replication of hybrid DNA molecules. Proc. Natl. Acad. Sci. USA 76, 1035-1039.

Thill, G.P., Kramer, R.A., Turner, K.J., and Bostian, K.A. (1983). Comparative analysis of the 5'-end regions of two repressible acid phosphatase genes in *S. cerevisiae*. Mol. Cell. Biol. 3, 570-579.

Tuite, M.F., Dobson, M.J., Roberts, N.A., King, R.M., Burke, D.C., Kingsman, S.M., and Kingsman,

A.J. (1982). Regulated high efficiency expression of human interferon-α in *S. cerevisiae*. The EMBO Journal 1, 603-608.

Wickner, R.B. (1981). Killer Systems in *Saccharomyces cerevisiae*. In: "The molecular biology of the yeast *Saccharomyces*", pp.415-444. Strathern, J.N., Jones, E.W., and Broach, J.R. (eds.). Cold Spring Harbor Laboratory; ISBN 0270-1847; 11A.

Wieringa, B., Meyer, F., Reiser, J., and Weissmann, C. (1983). Unusual splice sites revealed by mutagenic inactivation of an authentic splice site of the rabbit β-globin gene. Nature 301, 38-45.

Zaret, S.K., and Sherman, F. (1982). DNA sequence required for efficient transcription termination in yeast. Cell 28, 563-573.

6 Cloning in Gram-positive Bacteria

Some species of eubacteria live free in the soil and are capable of producing spores. Among these, the *Bacilli*, for example, are either strict aerobic or facultative anaerobic organisms, while *Clostridiae* are obligate anaerobes, although some will grow under microaerophilic conditions. Most of these bacteria are classified as Gram-positive, in accord with their reaction to the Gram stain empirically developed by the Danish physician Christian Gram in 1884. Bacteria are fixed on a microscope slide by heating, and are stained with dilute solutions of a basic dye, *e.g.*, crystal violet. The slides are then treated with an iodine/potassium iodide solution to fix the stain and washed with acetone or ethanol. Gram-positive organisms retain the initial violet stain while Gram-negative organisms are destained by the organic solvent. This is due to differences in the structures of bacterial envelopes. Gram-negative bacteria are surrounded by two membranes, the inner (cytoplasmic) membrane, and the outer membrane (Fig. 6-1) (DiRienzo *et al.*, 1978). Thin sections reveal that both membranes show the structures typical of lipid bilayers. They are separated from each other by a periplasmic space and the peptidoglycan or murein sacculus. This sacculus is approximately 2 nm thick and is an enormous bag-shaped macromolecule consisting of cross-linked polysaccharides and peptides. It is connected to the outer membrane by the lipoprotein. The outer membrane is extremely rich in lipids and proteins (Osborn and Wu, 1980) and covered on the outside by a thick layer of lipopolysaccharides which are responsible for the characteristic antigenic properties of Gram-negative bacteria. Gram-positive bacteria lack the outer membrane found in Gram-negative organisms; instead, Gram-positive bacteria have a rigid layer of murein or peptidoglycan which may be up to 20 nm thick. This outer layer is not only thicker than its equivalent in Gram-negative organisms but also has a much more complex structure. In particular, it is associated with teichoic acid, which may constitute up to 50% of the total mass of the coat structure. With the exception of

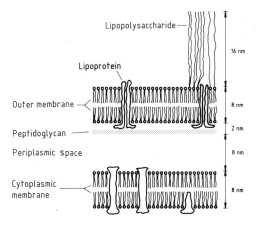

Fig. 6-1. Schematic representation of the structure of cell walls of Gram-negative bacteria.
The different terms are explained in the text. The outer membrane contains lipoprotein molecules embodied into the peptidoglycan moiety. Other proteins, also integrated into the peptidoglycan moiety, have been omitted. The periplasmic space also contains proteins.

nutrients, this murein coat is an impenetrable barrier for a variety of substances, including as a rule, DNA molecules. Only under certain favourable growth conditions can a state of competence be reached which allows DNA molecules to penetrate the murein layer. It can be hydrolysed away by lysozyme treatment, which, under isotonic conditions, yields protoplasts *i.e.*, bacteria without a cell wall, merely surrounded by a cytoplasmic membrane. These protoplasts are viable as long as isotonic conditions are maintained. Protoplasts are more easily penetrated by extracellular molecules than are undamaged bacteria.

Gram-positive and Gram-negative organisms do not only differ in their cell wall structures. A comparison of 16S ribosomal RNA sequences, for example, has shown that there is hardly any phylogenetic relationship between these two classes of bacteria and that, in the course of evolution, Gram-positive micro-organisms must have developed much earlier than Gram-negative bacteria.

The two genera *Micrococcus* and *Staphylococcus*, with its pathogenic variants, as well as the genus *Bacillus* are also Gram-positive. *Staphylo-*

coccus aureus plasmids have played an important role in the development of cloning techniques.

6.1 Cloning in B. subtilis

The prototype of the *Bacilli* is *Bacillus subtilis*. Use of this organism in gene technology is regarded to be important and to have several advantages over use of *E. coli*. *B. subtilis* has already been used in industry for a long time and is fermented on a large scale; in addition, the *Bacilli* synthesise such important proteins as amylases, β-lactamases, and a series of proteases which are excreted into the growth medium (Priest, 1977). These and other proteins can thus be obtained directly from culture fluids and do not have to be isolated from bacterial extracts. This particular feature may offer enormous advantages for the expression of heterologous proteins which are often degraded inside bacteria.

6.1.1 Transformation Techniques

The first successful transformation experiments in *B. subtilis* were carried out by Spizizen (1958) more than 25 years ago. Ehrlich (1977) observed that plasmids of *Staphylococcus aureus* were able to replicate in *B. subtilis*. This expansion of transformation methods to plasmid vectors has been an essential prerequisite for the application of cloning techniques in Gram-positive bacteria. Experiments carried out with *B. subtilis* soon revealed that the ability of plasmid DNAs to transform these bacteria depended very much on their molecular structure: while monomers were inactive, multimeric plasmids transformed well, and the efficiency of transformation increased

Stanislav D. Ehrlich
Paris

with an increasing degree of oligomerisation. This phenomenon can be understood if the fate of transforming DNAs in competent bacteria is considered in detail (Fig. 6-2). De Vos *et al.* (1981a) have shown that monomeric double-stranded DNA is completely converted to single-stranded forms after transformation (Fig. 6-2, Reaction A); in contrast, partially double-stranded DNA molecules are observed after transformation with multimeric forms (Fig. 6-2, Reaction B). Of course, the single-stranded molecules of a monomeric plasmid cannot be replicated; however, multimeric DNA molecules contain overlapping and complementary regions and can form double strands, or at least partially double-stranded molecules which can be replicated if they bear a functional replicon. It is therefore essential that *B. subtilis* be transformed with multimeric

plasmids. Such multimers are easily obtained if plasmids are linearised *in vitro* with a restriction endonuclease and subsequently ligated in the presence of DNA ligase.

Under certain conditions, however, competent cells can also be transformed by monomeric DNA molecules, if these monomers carry an insertion of *B. subtilis* DNA. Successful transformation then depends on the presence of functional recombination systems in the competent host bacteria, in particular the *recE* function (Canosi *et al.*, 1981). This finding suggests that the single-stranded donor DNA interacts with chromosomal DNA by forming partially double-stranded DNA regions (Fig. 6-2, Reaction C). In principle, this situation resembles transformation by multimeric plasmids; the only difference is that the complementary DNA sequences required for the forma-

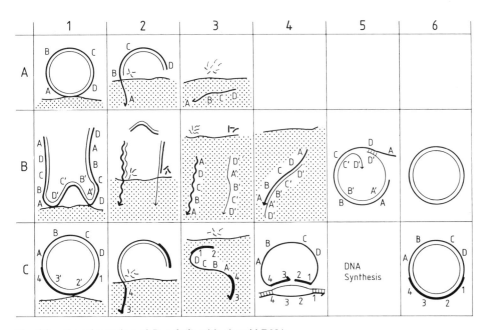

Fig. 6-2. Transformation of *B. subtilis* with plasmid DNA.
This scheme shows three possible fates of superhelical DNA molecules entering competent *B. subtilis* cells. (A) monomeric, heterologous plasmid DNA; (B) oligomeric plasmid DNA; (C) monomeric plasmid DNA with partial homologies to *B. subtilis* chromosomal DNA. Chromosomal DNA in (C) could also be replaced by a second, partially homologous plasmid (*cf.* Gryczan *et al.*, 1980, for example). Regions of partial homology in (C) are represented as thick lines. Numbers and letters represent sequence regions on individual strands. (After Canosi *et al.*, 1978).

tion of double-stranded molecules are provided by homologous chromosomal sequences rather than plasmid DNA itself.

The reaction mechanisms shown in Fig. 6-2 also explain why plasmids introduced into competent cells by transformation are unstable and frequently acquire deletions. The DNA appears to be extremely labile during its conversion into a single-stranded form and the subsequent recombinational events. At best, only those regions whose presence can be selected for will survive these processes.

Practical applications were long hampered not only by the instabilities of plasmid DNAs but also by unsatisfactory transformation frequencies. A milestone in the development of cloning techniques for *B. subtilis*, then, was the observation made by Chang and Cohen (1979) that *B. subtilis* protoplasts obtained by lysozyme treatment can be effectively transformed with plasmid DNAs in

the presence of polyethylene glycol. Yields of transformants often approach 80% and transformation frequencies are often higher than 10^7 transformants per microgram of DNA. It is not only the high transformation frequencies that distinguish this system from the transformation of competent cells: above all, the great advantage is that monomeric DNAs are also biologically active in this system, because they are taken up by protoplasts as circular molecules which then replicate as such. Reactions which are usually responsible for the conversion of double-stranded DNAs into single-stranded molecules and for the recombination of these molecules apparently do not occur with chromosomal DNA in competent cells; accordingly, protoplasts can also be transformed in the absence of functional recombination systems, *i.e.*, suitably mutated strains could also be used (de Vos and Venema, 1981b). Since plasmid instabilities have not been reported so far, protoplast transformation appears to be the method of choice for the transformation of *B. subtilis*.

6.1.2 Plasmids and Vectors

Most strains of *Bacilli* contain extrachromosomal DNA molecules (Kreft and Hughes, 1982); however, many strains obtained from culture collections harbour only cryptic plasmids which are devoid of any selectable markers. The first transformation experiments were carried out with *Staphylococcus aureus* plasmids carrying antibiotic resistance genes (Ehrlich, 1977). Subsequently, tetracycline-resistant plasmids have been isolated from *B. cereus* and other *Bacilli* and transferred successfully into *B. subtilis* (Bernhard *et al.*, 1978). Most of the plasmids currently used for cloning in *B. subtilis* are derived from four *S.*

Thomas A. Trautner
Berlin

aureus plasmids. Their structure is shown in Fig. 6-3. Although these plasmids already carry one selectable antibiotic resistance marker they have been manipulated further so that they can be used as vectors for insertional inactivation and as shuttle vectors for *E. coli* and *B. subtilis*. The additional resistance markers have been derived from other *S. aureus* plasmids, chromosomal DNA of other strains of *Bacilli*, or from the *E. coli* vector pBR322; the *E. coli* replicon required for the development of shuttle vectors has also been obtained from pBR322. Vector pHV11 for example, which expresses tetracycline and chloramphenicol resistances in *B. subtilis*, is a recombinant between the chloramphenicol resistance plasmid pC194 cut by *Hind* III and the largest of three *Hind* III fragments of pT127 (Fig. 6-4)

(Ehrlich, 1978). It can be employed as an insertional vector by making use of its unique *Kpn* I site. Recombinants with insertions in this site retain their chloramphenicol resistance and become tetracycline-sensitive. The major components of vector pPL603 are derived from plasmid pUB110. pPL603 expresses kanamycin resistance (Williams *et al.*, 1981a); in addition, this vector contains a large section of chromosomal DNA approximately 1 000 bp in length which originates from a chloramphenicol-resistant strain of *B. pumilus* (Fig. 6-4). Due to the lack of a suitable promoter the chloramphenicol acetyltransferase (CAT) encoded by this fragment is not expressed. Heterologous *Eco* RI fragments which carry a promoter, and are therefore able to switch on genes, can be identified by cloning in the single

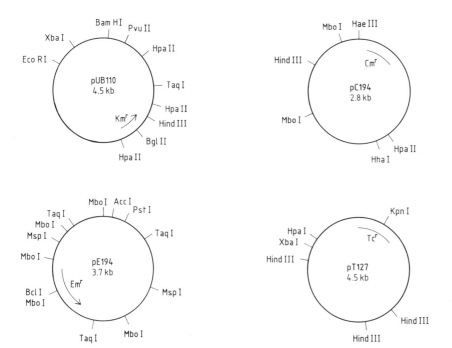

Fig. 6-3.   Structure of four naturally occurring *S. aureus* plasmids.
pUB110 (Keggins *et al.*, 1978) codes for kanamycin/neomycin resistance (Km^r), pC194 for chloramphenicol resistance (Cm^r), pE194 (Horinouchi and Weisblum, 1980) for erythromycin resistance (Em^r), and pT127 for tetracycline resistance (Tc^r) (Ehrlich *et al.*, 1982). Enzymes such as *Bgl* II, *Bcl* I, and *Kpn* I, which cut pUB110, pE194, and pT127 only once within the corresponding resistance gene, can be used for cloning of foreign DNA according to the principle of insertional inactivation.

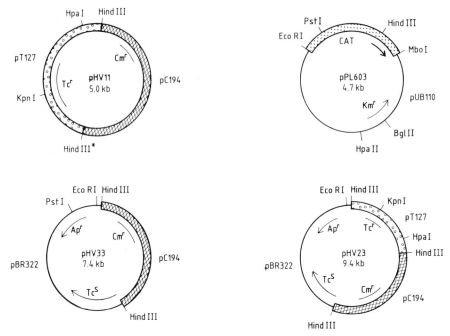

Fig. 6-4. Structure of four chimaeric vectors.
pHV11 consists of pC194 (cross-hatched bar), and parts of pT127 (dotted bar; *cf.* also Fig. 6-3). pPL603 contains parts of pUB110 devoid of the small *Eco* RI-*Bam* HI fragment, and an insertion of 1 000 bp with the chloramphenicol acetyltransferase (*CAT*) gene of *B. pumilus* (dotted bar). This *CAT* gene lacks its own promoter; however, a promoter can be provided by inserting suitable DNA fragments into the single *Eco* RI site as described in the text. pHV33 and pHV23 are shuttle vectors containing parts of pBR322. pHV33 is a recombinant between pBR322 and pC194; pHV23 a chimaeric molecule of pHV11 and pBR322. Resistance regions are indicated by curved lines or arrows within the circles. The star(*) in pHV11 marks the position of the *Hin*d III site which is opened for generating pHV23.

Eco RI site of pPL603 (see also Fig. 6-7) (Williams *et al.*, 1981b).

Vectors pHV33 and pHV23 (Fig. 6-4) are genuine shuttle vectors (Michel *et al.*, 1980). pHV33 contains a functional copy of pC194 linearised by *Hin*d III cloned into the *Hin*d III site of pBR322. In *E. coli*, pHV33 confers chloramphenicol and ampicillin resistance but tetracycline sensitivity while *B. subtilis* cells are only transform-ed to chloramphenicol resistance. These and other examples demonstrate that the transcriptional machinery of *E. coli* is capable of recognising heterologous control sequences. In contrast, *B. subtilis* usually only recognises homo-

logous signals; the β-lactamase gene of pBR322, for example, is not expressed in *B. subtilis*.

Another useful shuttle vector is pHV23 which contains a copy of pHV11 partially digested with *Hin*d III and ligated to pBR322 linearised by *Hin*d III. The tetracycline resistance region is derived from pT127 (Fig. 6-3). pHV23 contains a *Kpn* I site which can be used for cloning and permits selection for tetracycline sensitivity both in *B. subtilis* and *E. coli*. A disadvantage of these vectors is the size limitation for the insertion of foreign DNA molecules, and the instability of DNA inserts, a problem which has not yet been eliminated. Random cloning of 3 MDa DNA fragments (approximately 4.7 kb) from *B. amylo-liquefaciens*, for example, in vectors pHV33 and pHV23 yielded transformants with recombinant plasmids containing only inserts with an average length of approximately 1 MDa (Michel *et al.*, 1980). This bias against large insertions may be due to a very efficient recombination system which allows moderately homologous DNA sequences to interact with each other. The probability of such limited homologies occuring between inserted DNA and, for example, chromosomal sequences decreases as insertions become smaller. Restriction systems also may play a role (Tanaka, 1979), and if so, the observed smaller insertions are probably the result of a selection against the presence of restriction sites. Considerable efforts will still be required to gain an understanding of these phenomena, and to develop vectors which do not acquire deletions. The use of protoplasts is certainly a first step in this direction; at least it solves problems connected with the uptake of DNA molecules.

The plasmids described above have copy numbers between 20 and 50 per cell. It would be highly desirable to raise these numbers to values obtained with relaxed *E. coli* plasmids, numbers which may reach 3 000 copies per cell. A first step in this direction was the isolation of the copy number mutants of the *S. aureus* plasmid pT181 described by Khan *et al.* (1981), which is normally present in 20 copies per cell. A single deletion of 179 bp raises the copy number to 800-1 000. The molecular basis of this phenomenon is not yet known, but the possibility exists that the copy numbers of other plasmids could be raised by similar manipulations.

6.1.3 Expression Vectors

E. coli is rather promiscuous as far as the recognition of transcription and translation signals from other organisms is concerned, and even expresses yeast genes. *B. subtilis*, on the other hand, is much more stringent. The only example of heterologous gene expression in this organism is the expression of antibiotic resistance genes from other Gram-positive species, such as *S. aureus* (Kreft *et al.*, 1978). The reasons for this selectivity must be sought at the levels of both transcription and translation.

6.1.3.1 Transcription Mechanisms in B. subtilis

The correct recognition of promoters by RNA polymerases is mediated by protein factors known as σ factors. In *E. coli* there are at least two known σ factors which confer different promoter specificities to the same RNA polymerase core enzyme. σ factor σ^{70} with its molecular weight of 70 kDa was the first factor to be discovered and recognises a broad spectrum of *E. coli* promoters. σ factor σ^{32} recognises a small set of promoters involved in the heat shock response of *E. coli* (Cowing *et al.*, 1985). The consensus "-10" and "-35" region of these promoters are quite different from those recognised by RNA polymerase holoenzyme containing σ^{70} (Figs. 7-2, 7-3, and Fig. 6-5). *B. subtilis* possesses at least five σ subunits which become effective in various stages of cell growth or in the course of viral infections (Losick and Pero, 1981; Table 6-1). σ^{43}, the major σ factor in *B. subtilis*, which has a molecular weight of 43 kDa, is replaced by other factors under the following circumstances: in cells infected with *B. subtilis* phage SPO1, σ^{43} recognises the promoters of the early class of phage

Table 6–1. Sigma factors in *B. subtilis*.

Growing cells	Sporulating cells	SP01-virus-infected cells
σ^{43}	σ^{29}	σ^{gp28}
σ^{37}		$\sigma^{gp33-34}$
σ^{28}		

Superscript numbers indicate molecular weights (in kDa); (Losick and Pero, 1981).

Holoenzyme		-35	Spacing, bp	-10
E. coli				
$E\sigma^{70}$	(168)	TTGACA	16–18	TATAAT
$E\sigma^{32}$	(6)	TNtCNCcCTTGAA	13–15	CCCCATtTa
$T4E\sigma^{gp55}$	(4)			TATAAATA
B. subtilis				
$E\sigma^{43}$	(9)	TTGACA	17–18	TATAAT
$E\sigma^{29}$	(4)	TTNAAA	14–17	CATATT
$E\sigma^{28}$	(2)	CTAAA	16	CCGATAT
$E\sigma^{37}$	(4)	AGNNTT	13–16	GGNATTNTT
$E\sigma^{32}$	(2)	AAATC	14,15	TANTGNTTNTA
SPO1 $E\sigma^{gp28}$	(5)	TNAGGAGANNANTT	12–13	TTTNTTT
SPO1 $E\sigma^{gp33-34}$	(5)	CGTTAGA	17–19	GATATT

Fig. 6-5. Consensus sequence of *E. coli* and *B. subtilis* promoters. The consensus sequences of the *E. coli* σ^{70} system (Hawley and McClure, 1983) and the *B. subtilis* σ^{43} system (Moran *et al.*, 1982) are identical. Also shown are sequences characteristic of four other σ factors from *B. subtilis* vegetative, sporulating and virus-infected cells (Losick and Pero, 1981; Moran *et al.*, 1981; Gilman *et al.*, 1981) as well as sequences for *E. coli* heat-shock promoters and promoters from T4-infected cells. "-35" and "-10" indicate DNA regions located 35 bp or 10 bp proximal to the transcription initiation site. The total number of sequences analysed for each class of promoters is shown in parenthesis (Cowing *et al.*, 1985).

genes; intermediate and late genes are recognised by phage-encoded factors σ^{gp28} and $\sigma^{gp33-34}$ which replace σ^{43}. In sporulating cells σ^{43} is replaced by the sporulation-induced factor σ^{29}. Even vegetative cells do not only contain σ^{43}, but also factors σ^{37} and σ^{28} (Haldenwand and Losick, 1980; Wiggs *et al.*, 1979). The various σ factors, listed in Table 6-1, recognise different sequences on the DNA (Fig. 6-5). A comparison of nine promoter sequences recognised by σ^{43} has revealed highly conserved hexanucleotide sequences in the "-35" and "-10" regions which are identical with corresponding regions in the *E. coli* consensus sequences (*cf.* Fig. 7-3); furthermore, the distance between these two regions is identical in *B. subtilis* and *E. coli*. Promoter sequences controlled by the other σ factors also possess highly conserved sequences in regions "-35" and "-10"; however, these sequences differ markedly from those recognised by σ^{43} (Fig. 6-5). These differences are also reflected in the substrate specificities of different holoenzymes containing either σ^{43} or σ^{28} (Table 6-2). The enzyme containing σ^{43} recognises *B. subtilis* as well as *E. coli* promoters, the latter with reduced activity. Polymerases

Table 6–2. Substrate specificity of two *B. subtilis* RNA polymerase holoenzymes.

DNA	Activity of holoenzyme	
	σ^{43} (U/mg)	σ^{28} (U/mg)
B. subtilis	2900	1700
E. coli	800	70
Φ29	7800	140
T7	3400	200
pBR322	1100	30
pHV14	2600	70
pMG102	1500	4400
pMG201	1900	4800

The two different RNA-polymerase holoenzymes used in this study contained either sigma factor σ^{43} or sigma factor σ^{28}. Enzymes were incubated with different DNA templates. One unit (U) corresponds to the incorporation of 1 nmole ^{32}P-CMP per hour at 37°C. Results are not corrected for differences in template size but nevertheless permit an approximate comparison. Φ29 stands for the double-stranded DNA genome of *B. subtilis* phage Φ29. Plasmid pHV14 is identical with plasmid pHV33 shown in Fig. 6–4. Plasmid pMG102 and pMG201 are derived from pHV14 and carry *B. subtilis* DNA insertions in their PstI-sites. These insertions were previously shown to interact with the σ^{28}-holoenzyme (Wiggs *et al.*, 1981).

containing σ^{28}, however, are completely inactive on *E. coli* DNA (Gilman *et al.*, 1981). It is, as yet, completely unknown why σ^{43} holoenzymes initiate less effectively on *E. coli* DNA than on *B. subtilis* DNA although their consensus sequences are identical. Regions other than "-35" and "-10" sequences may be essential for promoter recognition and the effect of deviations from consensus sequences may be more pronounced for *E. coli* than *B. subtilis* promoters. At present, it is difficult to answer these questions because too little is known about such sequences.

It appears that on the level of transcription, σ factors are the only factors responsible for the recognition and specificity of promoters. A well-documented case has been described by Ernst *et al.* (1982) who showed that an active hybrid enzyme consisting of *E. coli* RNA polymerase subunits β, β', α, and the σ subunit of *Micrococcus luteus* RNA polymerase possessed the same promoter specificity for ΦX174 RFI DNA as functional *M. luteus* holoenzymes. In this particular case *M. luteus* RNA polymerase used a promoter in the gene E region of ΦX174 which is usually not recognised by *E. coli* polymerases, but which was recognised quite well by hybrid *E. coli* RNA polymerases containing the *M. luteus* σ factor. It is still not known, however, how σ factors can alter the interaction between RNA polymerases and promoters.

6.1.3.2 Translation in B. subtilis

Initiation sites of protein biosynthesis in Gram-negative bacteria are characterised by a nucleotide sequence known as the ribosomal binding site, comprising the AUG start codon and a short sequence, the Shine-Dalgarno (S/D) sequence which is complementary to the 3' end of 16S ribosomal RNA. The distance between AUG codon and S/D sequence, and mRNA secondary structure play a crucial role in determining the efficiency of the initiation reaction. The primary structures of only a few genes from Gram-positive organisms have been analysed so far. Sequences from mRNAs in the vicinity of the start codon have revealed a high degree of homology with the 3' ends of 16S ribosomal RNA. Homologies in 12 sequenced ribosomal binding sites are much more conserved than those in binding sites recognised by *E. coli* ribosomes (Fig. 6-6) (McLaughlin *et al.*, 1981). The free energy of the interaction between S/D sequences and 16S ribosomal RNAs is therefore increased to values between -50 and -88 kJ/mol while the corresponding values in *E. coli* vary widely between -17 and -92 kJ/mol. This may explain why *B. subtilis* ribosomes recognise mRNAs from Gram-positive, but not from Gram-negative organisms in *in vitro* translation systems; on the other hand, *E. coli* ribosomes recognise binding sites of, and initiate on, *B. subtilis* mRNA. *E. coli* ribosomes *in vitro* require certain protein fractions though, which are obtained from a ribosomal wash and show little activity in extracts from Gram-positive bacteria.

As has been demonstrated, for example, for the β-lactamase gene of *S. aureus*, a further difference between the structures of ribosomal binding sites of Gram-positive and Gram-negative organisms is the occasional occurrence, in the former, of an UUG instead of an AUG start codon (Fig. 6-6). Such deviations from the use of the normal AUG start codon are very rare in *E. coli*, where UUG has never been observed although a GUG codon has been identified in 4 of 123 analysed cases. More genes from Gram-positive organisms will have to be analysed before the differences between translation mechanisms in *E. coli* and *B. subtilis* will be fully understood.

6.1.3.3 Transport Phenomena

One useful characteristic of *B. subtilis* is its tendency to excrete synthesised proteins into the growth medium; in contrast, many proteins synthesised and excreted by *E. coli* only reach the

B.amyloliquefaciens α-AMYLASE	5'--- AAGAAAAUGA<u><u>GAGGGA</u></u>GAGGAAAC<u>AUG</u> AUU CAA AAA---
S.aureus β-LACTAMASE	5'--- AACUGUAAUAUC<u><u>GGAGGG</u></u>UUUA<u>UUUUG</u> UUG AAA AAG---
B.licheniformis β-LACTAMASE	5'--- UUCAAAC<u><u>GG AGGG</u></u>AGACGAUUUUG<u>AUG</u> AAU UAU GGU---
Φ29 22, kk PROTEIN	5'--- AACCAAUCAUA<u><u>GGAGG</u></u>AAUUACAC<u>AUG</u> AAU AAC UAU---
S.aureus Emʳ-PEPTIDE	pppAUUUUAUAA<u><u>GGAGG</u></u>AAAAAAU<u>AUG</u> GGC AUU UUU---
S.aureus Emʳ-29k PROTEIN	5'--- AUAACCAAAUUAAAGAGGGUUAUA<u>AUG</u> AAC GAG AAA---
16S RNA B.subtilis	3' UCUUU<u>CCUCC</u>ACUAG---
complementary strand	5' AGAAA<u>GGAGG</u>UGAUC---
16S RNA E.coli	3' AUU<u>CCUCC</u>A---
complementary strand	5' UA<u>AGGAGG</u>U---

Fig. 6-6.   Ribosomal binding sites of some mRNA species of Gram-positive organisms. The start codons are underlined, S/D sequences are indicated by double underlines. The data are from Moran *et al.* (1982). Also shown are the 3′ ends of 16S rRNAs of *B. subtilis* and *E. coli* and their complementary sequences.

periplasmic space due to the particular organisation of its cell wall. Secretion of proteins in eukaryotic and prokaryotic cells is mediated by the presence of an *N*-terminal signal peptide. Two secreted proteins have been analysed so far in Gram-positive organisms. These two proteins, penicillinase (penP) of *Bacillus licheniformis* (Neugebauer *et al.*, 1981) and α-amylase of *B. amyloliquefaciens* (Palva *et al.*, 1981) have signal peptides of 34 and 31 amino acids, respectively, which are removed from the exopeptide precursor proteins by successive proteolytic cleavage. *B. licheniformis* penicillinase is expressed poorly in *E. coli*, but this expression is still good enough to allow the corresponding gene to be used as a selective marker; in contrast, *B. subtilis* expresses and processes this enzyme very efficiently (Gray and Chang, 1981).

Since only a few signal sequences from Gram-positive organisms have been analysed so far the true structural requirements for the secretion of proteins in these organisms are unknown; how-

ever, *B. subtilis* signal peptides appear to be as functional in *E. coli* as are *E. coli* secretory signals in *E. coli*. It has been shown, for example, that the transport into the periplasm of the thermostable *B. coagulans* α-amylase is as effective when fused to the prepeptide of β-lactamase encoded by pBR322 as under the control of its own *Bacillus* signal sequences (Cornelis *et al.*, 1982). It appears that, in contrast to transcription and translation, the problem of protein transport through membranes has been solved in similar ways in these different organisms (see also Section 7.5).

6.1.3.4 Inducible Promoters

The optimal expression of heterologous genes often does not only require the correct regulatory sequences but also the presence of inducible promoters. In view of the strong proteolytic activities found in *B. subtilis*, it may be of considerable importance to express a gene only in certain growth phases. While the *E. coli* K12 system provides a considerable variety of inducible genes (see Chapter 7), the only genes known to be inducible in *B. subtilis* are plasmid-encoded genes for various antibiotic resistances. Expression vectors have been developed on the basis of chloramphenicol and erythromycin resistance genes derived from *S. aureus*. Vector pPL603 (Fig. 6-4), for example, essentially consists of parts of pUB110 and carries a 1000 bp fragment of *B. pumilus* DNA encoding the gene for chloramphenicol acetyltransferase (*CAT*). This gene cannot be expressed, because it lacks a suitable promoter. Plasmid pPL603 contains a unique *Eco*RI site which allows suitable promoter sequences to be inserted immediately upstream of the *CAT* gene. Vector pPL608, for example, contains a 350 bp *Eco*RI fragment derived from *B. subtilis* phage SP02 which, at this position, shows promoter activity and allows the expression of the *CAT* gene. This vector therefore can confer chloramphenicol resistance upon *B. subtilis* (Fig. 6-7). Resistance is inducible; the specific activity of the chloramphenicol acetyltransferase is 10 to 20-fold increased in the presence of 5 μg/ml of chloramphenicol as compared to the basal level in the absence of chloramphenicol. This increase of activity is also observed for other genes interrupting or replacing the *CAT* gene. When DNA fragments are cloned into the *Pst*I or *Hind*III site of pPL608, CAT activity is destroyed by insertional inactivation. Williams *et al.* (1981b) used pPL608 for cloning and expressing an *E. coli* DNA fragment of 2.55 kb containing the *trpC* gene, and observed a sevenfold increase of the level of the enzyme indole glycerolphosphate synthetase in the presence of chloramphenicol.

It is quite remarkable that the inducible expression of the *CAT* gene in pPL608 and related constructs is independent of the nature of the promoter inserted upstream of the *CAT* gene as long as a promoter is present. The observed inducibility in the presence of chloramphenicol therefore may be due more to translational than

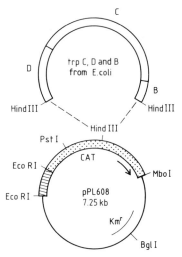

Fig. 6-7. Structure of expression vector pPL608. pPL608 is derived from pPL603 (Fig. 6-4) and contains an additional 350 bp *Eco*RI fragment with promoter sequences of *B. subtilis* phage SPO2 (hatched bar). The *CAT* gene (1000 bp) is represented as a stippled bar. The incorporation of a 2.55 kb *E. coli* DNA fragment, coding for *trpC* and parts of the *trpD* and *trpB* genes, yields vector pPL608-5. This vector does not express chloramphenicol resistance because the *CAT* gene has been interrupted by the insertion; expression of *trpC*, however, is inducible by chloramphenicol. (Williams *et al.*, 1981b).

transcriptional controls. This phenomenon has been investigated in greater detail for the inducible erythromycin resistance (Horinouchi and Weisblum, 1980). Bacteria become resistant to concentrations of more than 100 μM of erythromycin after a 40 to 60 minutes incubation in the presence of 10 to 100 nM of the antibiotic. Resistance is mediated by a protein of 29 kDa; the corresponding gene has been mapped to DNA around the *Bcl*I cleavage site of plasmid pE194 (Fig. 6-3). The DNA sequence in this region contains two open reading frames coding for proteins of 19 and 243 amino acids, respectively (Fig. 6-8). The reading frame for the 19 amino acid-long peptide terminates 58 bp in front of the *N* terminus of the 243 amino acids (29 kDa) protein. The sequence for each protein possesses a promoter, an S/D sequence, and a stop codon.

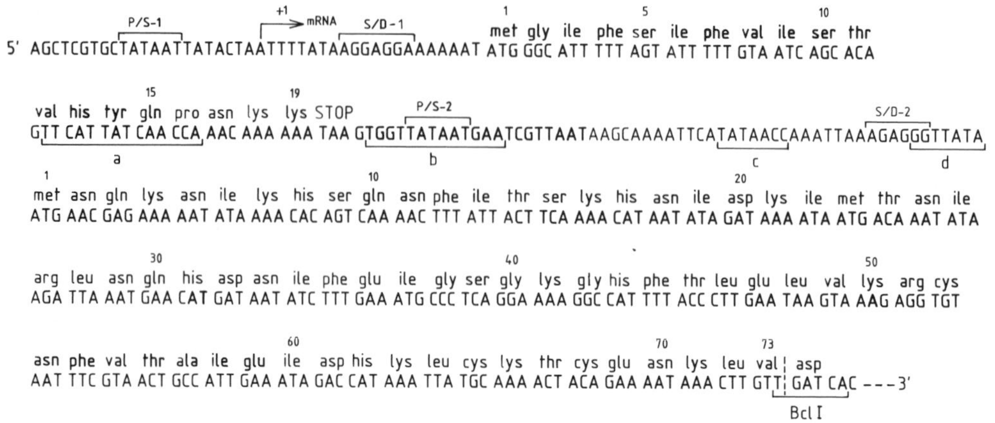

```
        P/S-1                      +1  mRNA    S/D-1              1              5                    10
                                                        met gly ile phe ser ile phe val ile ser thr
5' AGCTCGTGCTATAATTATACTAATTTTATAAGGAGGAAAAAAT ATG GGC ATT TTT AGT ATT TTT GTA ATC AGC ACA

            15               19
     val his tyr gln pro asn lys lys STOP    P/S-2                                              S/D-2
     GTT CAT TAT CAA CCA AAC AAA AAA TAA GTGGTTATAATGAATCGTTAATAAGCAAAATTCATATAACCAAATTAAAGAGGGTTATA
          a                                  b                               c                        d
      1                              10                            20
     met asn gln lys asn ile lys his ser gln asn phe ile thr ser lys his asn ile asp lys ile met thr asn ile
     ATG AAC GAG AAA AAT ATA AAA CAC AGT CAA AAC TTT ATT ACT TCA AAA CAT AAT ATA GAT AAA ATA ATG ACA AAT ATA

               30                              40                                 50
     arg leu asn gln his asp asn ile phe glu ile gly ser gly lys gly his phe thr leu glu leu val lys arg cys
     AGA TTA AAT GAA CAT GAT AAT ATC TTT GAA ATG CCC TCA GGA AAA GGC CAT TTT ACC CTT GAA TAA GTA AAG AGG TGT

                         60                              70          73
     asn phe val thr ala ile glu ile asp his lys leu cys lys thr cys glu asn lys leu val  asp
     AAT TTC GTA ACT GCC ATT GAA ATA GAC CAT AAA TTA TGC AAA ACT ACA GAA AAT AAA CTT GTT GAT CAC ---3'
                                                                                            Bcl I
```

Fig. 6-8. Part of the *Em^r* gene sequence.
This sequence which is part of pE194 (*cf.* Fig. 6-3) codes for a leader peptide (1) of 19 amino acids, and a protein (2) of 29 kDa (243 amino acids). The sequence shown reaches from the promoter region to the *Bcl* I site (T/GATCA) at amino acid 73. P/S and S/D indicate the promoter and S/D regions for proteins 1 and 2. Letters a, b, c, and d mark regions of inverted repetitions which could form hairpin structures. (Horinouchi and Weisblum, 1980).

The region between the two reading frames contains inverted repetitions which might allow DNA strands to form hairpin structures. It has been postulated that the S/D sequence of the 29 kDa protein is blocked by intramolecular base-pairing in the absence of erythromycin and that only the small peptide is synthesised. In the presence of erythromycin, however, the S/D sequence of the 29 kDa protein gene would be free to interact with the ribosomal subunit. This behaviour reminds one of attenuator models, although there are no termination signals, and it remains an enigma how erythromycin should effect the postulated alterations of secondary structures. The erythromycin system has been used for the development of an expression vector (Fig. 6-9) (Hardy *et al.*, 1981) by starting with a chimaeric plasmid containing parts of pUB110 and pE194, and therefore carrying kanamycin, neomycin, and erythromycin resistance genes (Fig. 6-3). The single *Bcl* I site in the 29 kDa erythromycin resistance gene (Fig. 6-8) was used for the insertion of a 4.08 kb DNA fragment, resulting in the production of a fusion protein consisting of 73 amino acids from the 29 kDa protein and 284 amino acids from the coat protein

of foot and mouth disease virus (Fig. 6-9). The fusion protein can be expressed after induction with erythromycin, and then amounts to 1% of the total protein in induced cells. The system has also been used to express the hepatitis B virus core antigen and human leukocyte interferon.

Another inducible expression system exploits the *lac* repressor/operator control elements from *E. coli* (Yansura and Henner, 1984). The system involves a hybrid promoter in which the *E. coli lac* operator DNA region has been placed immediately downstream from the penicillinase gene promoter from *B. licheniformis*. *Lac* repressor is produced from the *E. coli lacI* gene under the control of a promoter and ribosomal binding site which allow its expression in *B. subtilis*. Upon induction with IPTG, expression of the penicillinase increases 100-fold, which indicates that the *lac* control system from *E. coli* indeed can be introduced successfully into *B. subtilis*. This approach has been exploited for the IPTG-inducible expression of human leukocyte interferon.

Finally, there are indications that a system analogous to the strong early promoter of coliphage λ might be within reach for *B. subtilis*

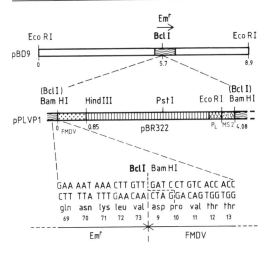

Fig. 6-9. Structure of vector pPLVP1 for the expression of the major antigen of foot-and-mouth disease virus.
The top shows the *Eco* RI-linearised vector pBD9 which is 8.9 kb in length, and contains parts of pUB110 and pE194 (Gryczan and Dubnau, 1978). pPLVP1 (*cf.* also Fig. 7-48) is linearised with *Bam* HI and cloned into the *Bcl* I site of the *Em*[r] gene (wavy lines) of pBD9 (Fig. 6-8). pPLVP1 is 4.08 kb in length and codes for amino acids 9–293 of the major capsid antigen of foot-and-mouth disease virus (FMDV). The resulting fusion protein consists of 73 *N*-terminal amino acids encoded by the erythromycin resistance gene, 284 amino acids of FMDV VP1 antigen (dotted bar), and 13 amino acids encoded by pBR322 sequences (vertical lines). (Hardy *et al.*, 1981).

(Dhaese *et al.*, 1984). The "promoter-probe" vector pPL603 has been used for the isolation, from the temperate *B. subtilis* phage Φ105, of a DNA fragment carrying a promoter which is controlled by a phage-encoded repressor. In order to obtain an inducible expression system the Φ105 repressor gene has been cloned into *Em*[r] plasmid pE194, which has the remarkable property of being naturally temperature-sensitive for replication. The repressor gene can thus be turned off by a shift-up to the non-permissive temperature (45 °C), even though the repressor itself is not temperature-sensitive. As yet, no applications involving the expression of foreign proteins have been described.

6.1.3.5 Excretion Vectors

It has already been mentioned previously that most bacteria belonging to the genus *Bacillus* have the ability to excrete large amounts of various proteins into the culture media. In principle, it should be possible to exploit this property not only for the improved expression of these enzymes but also for the construction of expression vectors which would allow heterologous proteins to be excreted under the control of *Bacillus* excretion signals. Such vectors have, indeed, been derived from *B. amyloliquefaciens* genes encoding α-amylase (Palva *et al.*, 1982) and neutral protease (Honjo *et al.*, 1985), and the α-amylase gene of *B. subtilis* (Ohmura *et al.*, 1984). *B. amyloliquefaciens* exoamylase is preceded by a signal sequence of 31 amino acids which is cleaved off specifically after the Ala residue 31 to yield the mature protein. In plasmid pKTH38, this sequence is carried by a 560 bp DNA fragment insert in pUB110 which extends from an *Mbo* I site approximately 400 bp upstream of the pre-α-amylase coding region to an *Eco* RI linker inserted after codon 30 of the α-amylase sequence (Palva *et al.*, 1983 (Fig. 6-10). Initiating from the *Eco* RI site, the coding sequence was removed by *Bal*31 digestion to yield vectors pKTH53 and pKTH51 which carry the *Hind* III linkers necessary to accept the human interferon α2 gene. Appropriate insertions into pKTH51 and pKTH53 yielded pKTH68 and pKTH93a, respectively. *B. subtilis* cultures carrying these plasmids produced and excreted active interferon polypeptides which were cleaved exactly after Ala 31. The polypeptide from pKTH68 thus carries additional six amino acids, the polypeptide from construction pKTH93 one single additional alanine residue. Accurate processing of these hybrid preproteins is therefore determined solely by the signal peptide which also provides a signal sufficient for efficient excretion; in contrast, the prepropenicillinase-preproinsulin fusions from *E. coli* are also cleaved properly but are not excreted into the medium (Fig. 7-77). Unfortunately, it is not clear

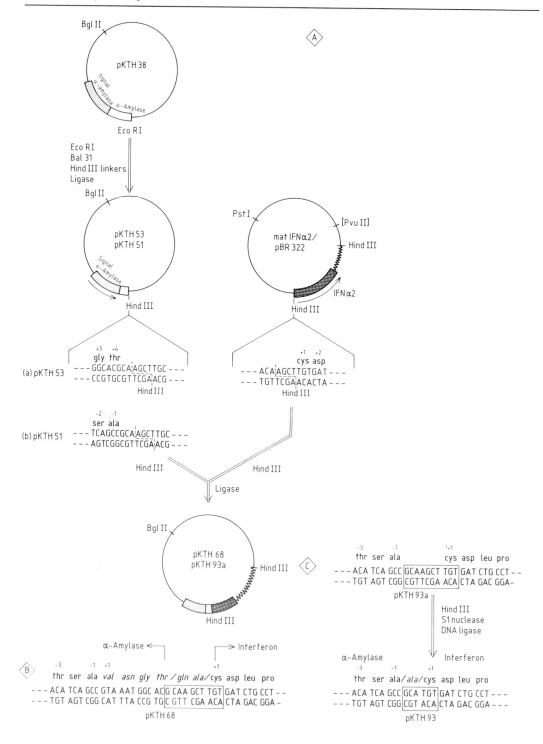

at present whether *B. subtilis* constructions will consistently achieve the high yield of heterologous enzyme production reported for the expression of the cloned α-amylase gene itself. In addition, it should be noted that the stability of the excreted proteins is not unlimited, due to extracellular protease activities (Ohmura *et al.*, 1984). Recent progress, however, indicates that these difficulties can be overcome and that *B. subtilis* will become an attractive host for the expression of heterologous proteins.

◄— **Fig. 6-10.** Interferon-α2 secretion plasmids. (A) Construction of pKTH68 and pKTH93a plasmids. pKTH38 contains the kanamycin resistance gene and a replicon derived from pUB110 (thin line), and the promoter, ribosome binding site (hatched box) and the coding sequence (open bar) of pre-α-amylase from *B. amyloliquefaciens* up to codon 61, where an *Eco* RI linker is inserted. pKTH38 was cleaved with *Eco* RI, the ends nibbled back to varying extents and rejoined *via* *Hind* III linkers to yield pKTH51 and pKTH53. The DNA encoding the mature IFN-α2 sequence was excised from pmatIFN-α2::pBR322, inserted into the *Hind* III site of pKTH51 and pKTH53 to yield pKTH68 and pKTH93a, respectively. Crosshatched bar, IFN coding sequence; open box, α-amylase or linker sequences; hatched bar, α-amylase signal (promoter) sequence; wavy line, IFN 3'-non-coding sequence. (a) and (b), nucleotide sequence at the *Hind* III sites of plasmids pKTH53 and pKTH51, respectively. (B) Structure of pKTH68 at the α-amylase-interferon junction. (C) Construction of pKTH93. pKTH93a was partially cleaved with *Hind* III and the ends were blunted and rejoined to give pKTH93. (Palva *et al.*, 1983).

6.2 Cloning in Streptomycetes

The members of the genus *Streptomyces* are Gram-positive, mycelial- and spore-forming bacteria. *Streptomycetes* are common inhabitants of soil and produce more than 60% of the known antibiotics as well as numerous extracellular enzymes and enzyme inhibitors (Martin and Gil, 1984). The genomes of *Streptomycetes* comprises 10.5×10^3 kb, approximately three times as much as that of *E. coli*.

Spores of *Streptomyces* can be germinated in nutrient solutions. Within two hours this process leads to the emergence of germ tubes, eventually evolving into a complex "substrate mycelium" composed of branching multinuclear hyphae which are characteristic of vegetative growth. Induction of spore formation is thought to be mediated by a variety of low-molecular weight factors, *e.g.* ppGpp. This differentiation process occurs *via* aerial hyphae developing from the substrate mycelium. These hyphae are subsequently converted, by septation, into long chains of spores. During this stage the organism is particularly sensitive to invading micro-organisms

Werner Goebel
Würzburg

which are attracted by the lysing mycelium. The enormous diversity of antibiotics that can be isolated from these bacteria may be the result of the evolution of strong chemical defense mechanisms to protect this vulnerable stage.

Streptomycetes provide interesting model systems for genetic and molecular biological analyses since, in addition to their substantial value as producers of antibiotics, they possess many attractive properties: they are genetically more complex than most single-celled prokaryotes, their genomes have very high GC contents, and they undergo cellular differentiation. In addition, *Streptomycetes* are also capable of extensive DNA amplification.

The development of an endogenous cloning system for *Streptomyces* appears highly justified for several reasons. Many of the products produced in this genus, for example antibiotics, may be lethal to other bacteria. Several commercial applications may simply require an increase in gene dosage to improve the yields of existing products. Finally, the recent characterisation of gene clusters coding for entire biosynthetic pathways for various antibiotics (Chater and Bruton, 1985; Hopwood *et al.*, 1985), and their transfer between different strains of *Streptomycetes* have resulted in the formation of novel antibiotics and thus may be of considerable commercial interest.

6.2.1 Transformation Techniques

The natural exchange of genes between hyphae of *Streptomycetes* occurs mainly through conjugation. This phenomenon is mediated by recombination systems which in turn depend on the activity of certain sex plasmids (see below). The

David A. Hopwood
Norwich

transfer of sex plasmids between neighbouring cells within these hyphae is an extremely efficient process and can proceed over a distance of several millimetres in a single day. This transfer is often accompanied by chromosomal mobilisation, and therefore *Streptomycetes* frequently display a high frequency of chromosomal recombination (Chater and Hopwood, 1983).

High frequency of recombination between chromosomal markers can also be induced by fusing bacterial protoplasts in the presence of polyethylene glycol (PEG). The fused protoplasts which contain the two complete parental genomes can regenerate efficiently into substrate mycelia.

The efficient transfer of plasmid DNA into cells of *Streptomyces* was an enigma until Bibb *et al.* (1978) showed that the uptake of plasmids into protoplasts was greatly enhanced by a treatment with polyethylene glycol. In *S. coelicolor* A3(2) and *S. lividans* for example, for which conditions have been optimised, transformation frequencies of 10^7 clones per microgram of plasmid DNA are readily obtained. Lower frequencies observed in other strains may either be due to the presence of restriction systems or to the ignorance on proper and optimal transformation conditions.

Transformants are generally identified by the selection for appropriate phenotypes, for example antibiotic resistance. Clones harbouring conjugative (sex) plasmids can also be detected by the visualisation of so-called "pocks". The property of "pock" formation, also known as lethal zygosis (ltz), is exhibited if a strain containing a plasmid, *e.g.* SCP2, SLP1, or pIJ101, is replica-plated onto a lawn of the corresponding plasmid-free strain. Under these conditions, clones containing plasmids are surrounded by a narrow zone in which the growth of the plasmid-free strain is retarded (Chater and Hopwood, 1983). Although the origin of this phenomenon and its underlying mechanism are still unknown, this particular feature has been used to identify the presence of many cryptic plasmids devoid of any identifiable phenotypic markers.

6.2.2 Plasmids and Plasmid Vectors in Streptomyces

Plasmid vectors in *Streptomyces* were developed on the basis of three different naturally occurring plasmid molecules, SCP2*, SLP1.2 and pIJ101 (Hopwood *et al.*, 1983). In order to obtain useful cloning vehicles, these plasmids had to be furnished with selectable marker genes, *i.e.*, with determinants for antibiotic resistance. The dominant markers commonly used (Thompson *et al.*, 1982) include resistance to the antibiotics thiostrepton (*tsr*; from *S. azureus* ATCC 1492), viomycin (*vph*; from *S. vinaceus* NCIB 8852), neomycin (*aph*; from *S. fradiae* ATCC 10745), and methylenomycin (*mnv*; from *S. coelicolor* A3(2)). The restriction maps of two of the corresponding genes are shown in Fig. 6-11.

Plasmid SCP2* (Fig. 6-12A; SCP denotes S. coelicolor A3(2) plasmid) is a derivative (Lydiate *et al.*, 1985) of the sex plasmid SCP2 (Schrempf and Goebel, 1977; Biff *et al.*, 1977). Both plasmids have a size of 31.4 kb and are physically indistinguishable, although SCP2* exhibits a much more pronounced lethal zygosis reaction. SCP2* and its derivatives have a low copy number (1-4) and are of particular interest because they can accomodate and stably inherit large DNA inserts of up to 35 kb in length. All known biological functions of SCP2* are located on the two *Pst* I fragments A and B. The two smaller *Pst* I fragments C and D can thus be deleted from SCP2* to form the 24.5 kb plasmid pSCP103 (Fig. 6-12B) which lacks restriction sites 19 to 36 present in SCP2* (Fig. 6-12A).

A variety of cloning vectors were derived from pSCP103 by the insertion of antibiotic resistance genes. Plasmids pIJ919 and pIJ922, for example, both carry an insertion of the *tsr* gene (Fig. 6-11A) within *Bcl* I site 12 of pSCP103. Plasmid pIJ922, in addition, lacks a region between sites 2 and 7 of pSCP103 and thus represents a cloning vector with a single *Bam* HI cloning site (Fig. 6-13).

pIJ922 was used by Malpartida and Hopwood (1984) to clone a 32.5 kb DNA fragment from *S.*

(A) Streptomyces azureus–tsr gene

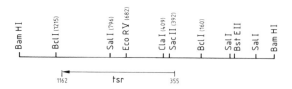

(B) Streptomyces vinaceus – vph gene

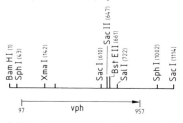

Fig. 6-11.   Cleavage map of the *tsr* and *vph* genes.
The rRNA methylase (*tsr*) gene of *S. azureus* (A) confers resistance to thiostrepton by methylation of 23S rRNA. It resides on a 1.9 kb *Bam* HI fragment and carries a promoter region between positions 160 and 352, followed by an open reading frame (arrow) coding for a 28 864 Da protein. The AT-rich sites *Cla* I and *Eco* RV are convenient sites for insertional inactivation. The viomycin (*vph*) phosphotransferase gene from *S. vinaceus* carries its promoter towards the left-hand end of a 1 930 bp long *Bam* HI fragment. The only open translational reading frame between position 97 and 957 codes for a protein of 30 513 Da. *Xma* I, *Sac* I and *Sal* I sites are candidates for insertional inactivation. Co-ordinates are taken from the sequence analysis of Bibb *et al.* (1985).

coelicolor A3(2) DNA carrying the complete cluster of genes directing the biosynthesis of the antibiotic actinorhodin. The fact that such a plasmid could be introduced into a variety of *Streptomycetes* other than *S. coelicolor* attests to the broad host range of pSCP103-based vectors (Hopwood et a., 1985; Lydiate *et al.*, 1985).

The SLP1 family of autonomously replicating plasmids was obtained after the transfer of chromosomal DNA from *S. coelicolor* A3(2) into *S. lividans* (Bibb *et al.*, 1981) and was detected by the lethal zygosis reaction on *S. lividans* lacking exogenous DNA (see above).

The prototype vector SLP1.2 (14.5 kb) has an intermediate copy number (4-5) and displays a limited host-range. Its derivative, pIJ61, lacks two non-essential segments of SLP1.2 and carries two resistance determinants, the *tsr* gene from *S. azureus* (*cf.* Fig. 6-11) and the *aph* gene (amino-

glycoside phosphotransferase) from *S. fradiae* (Thompson *et al.*, 1982). Insertional incactivation is accomplished by cloning into the *Bam* HI or *Pst* I sites within the *aph* gene, or the *Eco*R V and *Cla* I sites within the *tsr* gene (Fig. 6-14).

High-copy number cloning (40-300 copies) is rendered possible by employing the broad host-range vector pIJ101 and its derivatives (Kieser *et al.*, 1982) (Fig. 6-15). Plasmid pIJ702, for example, contains inserts with the *tsr* gene from *S.*

Fig. 6-12.   Physical maps of plasmids SCP2* (A) and →
pSCP103 (B).
Pst I fragments A and B of SCP2* which make up pSCP103 are shown as a black bar. The location of various plasmid functions are indicated along the periphery of the circular map of SCP2*. The *Bcl* I site 12 in pSCP103 used for the insertions of antibiotic resistance genes (*cf.* Fig.6-13) is marked in boldface (Lydiate *et al.*, 1985).

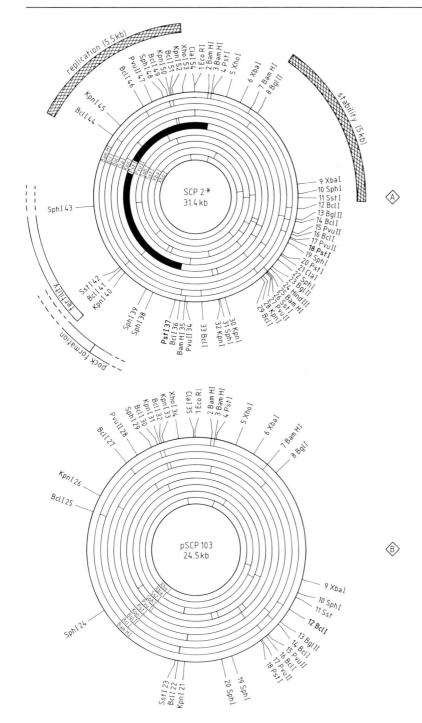

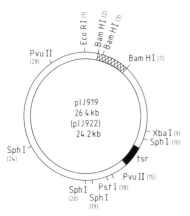

Fig. 6-13. Structures of vectors pIJ919 and pIJ922. Vector pIJ919 is derived from pSCP103 (open bar) by insertion of a 1,900 bp *Bam* HI fragment with the *tsr* gene from *S. azureus* (black bar) into *Bcl* I site 12 (Fig. 6-12B). Vector pIJ922 lacks the two small *Bam* HI fragments (cross-hatched bar) from pIJ919. Numbers in parenthesis are restriction site numbers from pSCP103 (Fig. 6-12B) (Lydiate *et al.*, 1985).

azureus and the *mel* gene coding for tyrosinase, from *S. antibioticus* IMRU 3720. These inserts replace a set of non-essential *Bcl* I fragments from pIJ101 (Katz *et al.*, 1983). Useful cloning sites are those for *Sph* I, *Sst* I, and *Bgl* II within the *mel* gene fragment. Transformants in which the plasmid-borne *mel* gene is inactivated by the insertion of foreign DNA are no longer able to convert tyrosine into melanin, and they therefore form white colonies when grown on agar containing tyrosine. In contrast, pIJ702-containing transformants form black colonies. Recent applications of this vector include the cloning of genes involved in undecylprodigiosin and clavulanic acid biosynthesis (Feitelson and Hopwood, 1983; Bailey *et al.*, 1984).

6.2.3 Phage Vectors in Streptomyces

The temperate phage ΦC31 is the most extensively studied and most widely used *Streptomyces* phage. The construction of cloning vectors derived from ΦC31 follows similar principles described already for the temperate *E. coli* phage λ. (*cf.* Section 4.2).

The linear double-stranded genome of ΦC31 has a size of 41.2 kb; the minimum and maximum sizes for packaging are 36.2 and 42.4 kb, respectively (Harris *et al.*, 1983). The right-hand end of

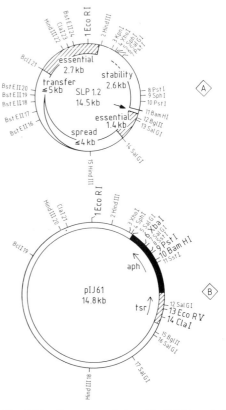

Fig. 6-14. Structures of plasmids SLP1.2 and pIJ61. The non-essential (stability) region between SLP1.2 sites *Sal* I-6 and *Bam* HI-11 is deleted in pIJ61. The insertions in pIJ61 of the *aph*- and *tsr* genes are depicted as black and hatched bars, respectively. The *aph*-portion is 2 190 bp, the *tsr*-portion 1 055 bp and the SLP1.2 portion (open bar) 11 580 bp long. Arrows indicate directions of transcription. The segments in SLP1.2 marked "essential" represent the minimal replicon while "spread" and "transfer" sections are responsible for the lethal zygosis reaction (see text). The exact positions of these two functions have not been delineated (Thompson *et al.*, 1982).

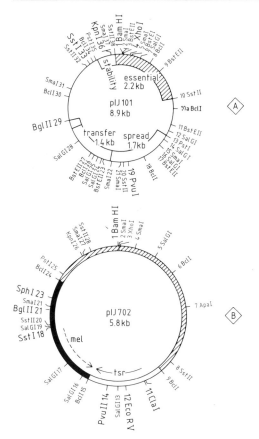

Fig. 6-15. Structures of plasmids pIJ101 and pIJ702. The pIJ101 derived portion in pIJ702 (dashed bar) includes the "essential" and "stability" regions between pIJ101 sites *Bcl*I-34 and *Bcl*I-10a. These sites are numbered *Bcl*I-24 and *Bcl*I-9, respectively, in pIJ702. "Essential" and "stability" regions are required for replication, maintenance, and stable inheritance of plasmid copies by daughter cells. The property of pIJ101 to spread within a plasmid-free mycelial culture to produce lethal zygosis zones (pocks) is coded for by a segment marked "spread" while "transfer" indicates a *cis*-function required for plasmid transfer. Useful sites for marker inactivation in the *mel*-gene (black bar) in pIJ702 and the *tsr*-gene (open bar) are marked in bold type (Kieser et al., 1982; Hopwood *et. al.*, 1985).

the phage genome contains regions which are inessential for plaque formation. In vector ΦC31KC400, for example, and many other de-

rivatives, a piece of DNA to the right of the *Eco* RI site flanking *Eco* RI fragments G and C was removed and replaced by plasmid pBR322 DNA and the *vph* gene from *S. vinaceus* (Fig. 6-16). The molecular ends of this and many other constructions carry complementary single-stranded termini or cohesive ends (*cos*) which permit circularisation of the DNA. The resulting circular molecule can therefore replicate as a plasmid vector in *E. coli*. For cloning of foreign DNA, a 3.9 kb *Pst*I fragment is removed from the circularised vector and replaced by *Pst*I-cleaved donor DNA before transfection of the recombinant molecules into *Streptomyces* protoplasts. The average size of inserts lies between 5.1 and 6.1 kb, as expected (Chater and Hopwood, 1983; Harris *et al.*, 1983).

Since most vectors lack the attachment site, *attP*, which mediates the integration of the phage DNA into the host chromosome, they are unable to form lysogens. However, this problem was solved in two ways. The first procedure employs ΦC31 lysogens as indicator strains. *AttP*-deleted derivatives can then integrate by homologous recombination between DNA sequences on the resident phage genome and the superinfecting vector genome, giving rise to double lysogens. Alternatively, the region of homology required for integration can be provided by host DNA sequences homologous to the recombinant DNA fragment present on the superinfecting vector.

The first strategy permits the use of *c* gene deletions (which provide additional cloning space). In analogy to bacteriophage λ, the product of the ΦC31 *c* gene controls the prophage state of the viral genome. Phages with deletions in this gene cannot lysogenise and are thus readily recognised by their clear plaque morphology. However, on lysogenic indicator strains, the product of the resident *c* gene can act *in trans* to stabilise the incoming *c*-deleted genome in its prophage state. *c* deletions have been mapped to a 3 kb DNA segment between positions 18.72 and 21.74 kb of the ΦC31 genome and thus accomodate an additional two to three kb of donor DNA

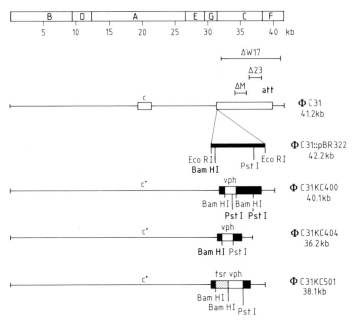

Fig. 6-16. Structure of phage ΦC31 and some replacement vectors.
The linear, double-stranded genome of ΦC31 comprises 41.2 kb. Non-essential regions for plaque formation are indicated as open bars. The seven *Eco* RI fragments are shown above the ΦC31 map. Two spontaneous deletions, ΔM and Δ23, cover almost 4 kb within the non-essential *Eco* RI-C fragment. Double deletion mutants with both the ΔM and the Δ23 deletions were used to construct a ΦC31::pBR322 chimaera which contains an *Eco* RI-linearised pBR322 insertion (black bar) within the *Eco* RI site of ΦC31 flanking wild-type fragments G and C. The *Bam* HI site within the pBR322 insert can be used to insert a 1 900 bp *Bam* HI fragment carrying the viomycin phosphotransferase gene of *S. vinaceus*. The ΦC31::pBR322::vph recombinant used for the construction of ΦC31KC400 lacks additional DNA sequences derived from the pBR322 insert as well as *Eco* RI fragments C and F including the *attP* site (Chater *et al.*, 1982). Vector ΦC31KC404, the parent of ΦC31KC501, lacks the *Pst* I insert from ΦC31KC400. The single *Bam* HI site in ΦC31KV404 was used to insert a 1 900 *Bam* HI fragment with the *tsr* gene to yield KC501. This vector and a familiy of vectors derived from it (Rodicio *et al.*, 1985) may be more useful than the KC400 family since viomycin is not readily available. Vector KC401 (not shown) differs from KC400 by the absence of the *c* gene. The Δc3 deletion accomodates 2 kb more donor DNA but can only be used in the double lysogen procedure in conjunction with a c⁺ ΦC31 lysogen (see text). For mutational cloning, only c⁺ vectors can be employed.

(Harris *et al.*, 1983; Rodicio *et al.*, 1985). While the size of the DNA which can be inserted is limited to approximately 6 kb in the case of ΦC31KC400, 8 kb of foreign DNA can be incorporated into ΦC31KC401 which carries the ΔC3 deletion.

Lysogens carrying *Pst* I-generated inserts can be selected for by their viomycin-resistant phenotype since the *Pst* I site in the *vph* gene insertion is located downstream of the *vph* gene reading frame which thus remains unaffected. *Bam* HI deletions of the vector which could accomodate DNA fragments of up to 9 kb would result in a loss of the *vph* marker (Fig. 6-16; ΦC31KC400).

The other strategy for the integration of a vector into chromosomal DNA relies on DNA sequences cloned within the vector DNA and not on phage DNA. The integration event is thought to lead to a duplication of the homologous sequences which are separated only by the vector

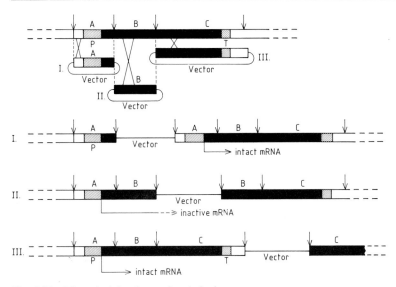

Fig. 6-17. The principle of mutational cloning.
The figure depicts a chromosomal gene with promoter (*P*; dashed bar), coding region (black bar) and terminator (*T*; stippled bar). Three vectors contain different restriction fragment insertions carrying either the promoter region (A), an internal region (B) or the terminator region (C). Upon recombination *via* homologous sequences on vector and chromosomal DNA, three different results can be expected. In cases I and III, when the cloned segment includes promoter or terminator sequences, an intact version of the gene remains upon integration. In case II, when the cloned segment only contains coding sequences, integration disrupts the chromosomal gene in such a way that only a shortened, inactive message is obtained. Similar techniques are being used in mutational analyses of yeast genes (*cf.* Figs. 5-6 and 5-7) (Chater and Bruton, 1983).

genome (Fig. 6-17). If the cloned fragment carries either promoter or terminator elements of a gene, the single-crossover events reconstitute one intact copy of the gene (Fig. 6-17 I and III). The particular gene thus continues to be expressed. If, however, the insert is derived from an internal part of the gene, the gene is destroyed by mutational inactivation. It is no longer expressed and transductants will not only be viomycin-resistant (due to the presence of the vector DNA) but will also express a mutant phenotype. This approach which is known as "mutational cloning" was used to isolate genes involved in antibiotic production (Chater and Bruton, 1983). In a first step, DNA segments from the desired organism producing antibiotic X were cloned into vector ΦC31KC400. Recombinant phages were transduced into the same strain from which the inserts had been derived originally. Transductants were then screened and selected both for viomycin resistance to ensure the presence of recombinant phages and for their inability to synthesise antibiotic X. This strategy can be applied to any antibiotic-producing strain of *Streptomyces* as long as it is sensitive to ΦC31 infections. Approximately 70% of all *Streptomycetes* are sensitive to ΦC31.

The strategy of mutational cloning has also been extended to the shotgun cloning of other chromosomal genes of *Streptomyces*. In these cases, wild-type DNA cloned in ΦC31 vectors was transduced into a mutant strain. Transductants were subsequently screened for the restoration of the wild-type phenotype. Using this strategy it has been possible to isolate a set of genes, the *bld* (bald colony appearance) class of genes, involved in the formation of aerial mycelia (Piret and Chater, 1985). Whether this method will be applicable to other organisms will depend on the frequency of insert-directed prophage insertion and the stability of the resulting inserts.

6.2.4 Gene Expression in Streptomyces

Only very few sequences controlling the transcription and translation of *Streptomyces* genes have been identified. The molecular mechanisms underlying gene expression are therefore largely unknown. They are particularly intriguing because of the average high GC content of 70 to 74 per cent of *Streptomyces* DNA. Two specific problems come to mind immediately. The promoter regions of other prokaryotic organisms, notably *E. coli* and *B. subtilis*, show a relatively high AT content (Fig. 6-5), which differs considerably from the overall AT content of *Streptomyces* DNA. In addition, the paucity of A and T residues, and the concomitant infrequent occurrence of out-of-frame translational stop codons

complicates the identification of protein-encoding DNA sequences in *Streptomyces* DNA.

Bibb *et al.* (1984) have devised an algorithm which permits an examination of the GC content of a DNA sequence as a function of the position of G and C residues within triplet codons. An analysis of codon usage revealed that the distribution of nucleotides within codons was non-random in protein-encoding regions but much less so in other regions of the DNA of organisms showing a high GC content. In particular, it was found that only the GC content of the first position reflected the average GC content of the organism's DNA, while the second position was always of a lower, and the third position of a considerably higher GC content (Table 6-3; Fig. 6-18). The pronounced bias in the distribution of G and C residues across codons does not only permit the unambiguous localisation of a correct reading frame out of a total of six possible reading frames, but immediately also predicts the direction of transcription. A recent example of this

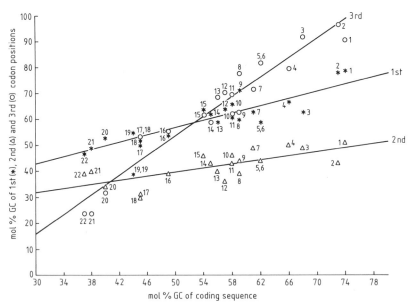

Fig. 6-18. Linear regression analysis of mol% GC at the first, second and third codon positions of various DNA sequences plotted against overall mol% GC of the corresponding coding sequences. Numbers 1-22 refer to the sequences listed in Table 6-3. Correlation and regression coefficients are described in Bibb *et al.*, 1984).

Table 6–3.   G + C base composition at the three codon positions of 22 bacterial genes.

Gene	Mol% G + C of coding sequence				L*	Position of codons for correct	
	Overall	Codon position				Start	Stop
		1	2	3			
1 *Streptomyces vinaceus* – vph	74	79	51	91	1119	97	958
2 *Streptomyces fradiae* – aph	73	78	43	97	1372	399	1203
3 *Streptomyces plicatus* – endoglycosidase H gene	68	63	49	92	1102	70	1009
4 *Pseudomonas aeruginosa* – merA	66	67	50	80	1747	29	1712
5 *Streptomyces azureus* – tsr	62	59	44	82	1521	355	1162
6 *Halobacterium halobium* – bacteriorhodopsin gene	62	59	44	82	1229	361	1147
7 *Escherichia coli* tetracycline resistance gene	61	63	49	72	1321	86	1274
8 *Rhizobium parasponia* – nifH	59	60	39	78	2029	576	1458
9 *Escherichia coli* – neo	59	71	44	63	1300	151	943
10 *Salmonella typhimurium* – trpE	58	66	46	62	1563	1	1561
11 *Anacystis nidulans* – ribulose-1,5-biphosphate carboxylase/oxygenase gene	58	61	43	70	1616	101	1517
12 *Pseudomonas putida* – xylE	57	64	36	71	958	30	948
13 *Rhizobium trifolii* – nifH	56	59	40	69	1696	277	1168
14 *Escherichia coli* – proC	55	62	43	59	968	88	895
15 *Escherichia coli* – trpR	54	64	46	62	558	110	434
16 *Escherichia coli* – dam	49	54	39	55	1134	195	1029
17 *Escherichia coli* – chloramphenicol acetyltransferase gene	45	51	31	53	1142	244	901
18 *Streptococcus faecalis* – aminoglycoside phosphotransferase gene	45	52	30	53	1489	535	1327
19 *Bacillus licheniformis* – penicillinase gene	44	55	39	39	1300	266	1187
20 *Bacillus sphaericus R* – modification methylase gene	40	53	34	32	1613	132	1404
21 *Bacillus cereus* – penicillinase gene	38	49	40	24	1218	264	1185
22 *Bacillus thuringensis* – crystal protein gene	37	47	39	24	1175	177	?

* L indicates the total length (in nucleotides) of each sequence analysed (from Bibb *et al.*, 1984).

type of analysis is the identification of the gene coding for isopenicillin N synthetase from *Cephalosporium acremonium* (Samson *et al.*, 1985). The coding region of this gene has a GC content of 63 % such that the rules of Bibb *et al.* (1984) could be applied in this case.

The analyses of nucleotide sequences involved in transcription initiation in *Streptomyces* have revealed at least two types of sequences. By employing promoter-probe vectors (Bibb and Cohen, 1982) which either depended on the chloramphenicol acetyltransferase or the *ampC* β-lactamase gene of *E. coli*, Jaurin and Cohen (1984, 1985) were able to identify a class of promoters which function both in *Streptomyces* and in *E. coli*. These promoters, which are known as *SEP* promoters (*Streptomyces-E.coli*-type promoters), contain sequence elements (*cf.* Fig. 6-5) which are similar to those defining the strength of promoters in *E. coli*, *i.e.*, "-10", "-35" regions and the intervening space. The other class of promoters does not function in *E. coli* and lacks the sequence determinants of an *E. coli* consensus promoter. In contrast to these determinants these promoters, for example the promoters of the rRNA methylase (*tsr*) gene, the aminoglycoside phosphotransferase (*aph*) gene, and the viomycin phophotransferase (*vph*) gene display and reflect the high GC content of *Streptomyces* (Bibb *et al.*, 1985).

The observed heterogeneity of *Streptomyces* promoters finds its parallel in the fact that there are two forms of RNA polymerase holoenzymes (Westphaling *et al.*, (1985). One of these holoenzymes in *S. coelicolor* with a transcription (σ) factor of 35 kDa recognises the *B. subtilis* vegetative (*veg*) promoter. The sequence of this promoter strongly corresponds with the consensus *E. coli* promoter recognised by σ factor 70 (*cf.* Fig. 6-5). The other form of RNA polymerase with a σ factor of 49 kDa does not recognise the *B. subtilis veg* promoter although it recognises the promoter of a developmentally regulated gene from *B. subtilis* known as the *ctc* gene. It is not recognised by the σ factor σ^{43} of *B. subtilis* but by

a minor form of *B. subtilis* RNA polymerase holoenzyme with σ factor σ^{37} (*cf.* Fig. 6-5). Possibly, *B. subtilis* and certain strains of *Streptomycetes*, although very different in many biochemical aspects, employ similar strategies for the regulation of expression of developmental genes. It remains to be seen, however, whether the same plethora of transcription factors known from *B. subtilis* will also be observed in *Streptomycetes*.

In *B. subtilis* (*cf.* Section 6.1.3.2), efficient translation depends on a high degree of complementarity between the 3′ end of the 16S ribosomal RNA and that of the mRNA 5′ to the translational start codon. Although the sequence of the 3′ end of the 16S ribosomal RNA in *S. lividans* is identical with that of *B. subtilis*, an inspection of thirteen translational start sites reveals that the degree of complementarity is much smaller in the mRNAs of *Streptomycetes*. In fact, the requirements for complementary sequences may even be less stringent than in *E. coli*, considering, for example, the remarkable fact that the transcriptional start site of the *aph* gene from *S. fradiae* (Bibb *et al.*, 1985) coincides with its translational start site. In this case, efficient translation must depend on sequences within the coding region of the *aph* gene (where there is no significant homology to the 3′ end of the 16S rRNA). Detailed conclusions, however, will have to await the identification of additional genes and their control elements as well as functional analyses in homologous and heterologous systems.

References

Bailey, C.R., Butler, M.J., Normansell, I.D., Rowlands, R.T., and Winstanley, D.J. (1984). Cloning of a *Streptomyces clavuligerus* genetic locus involved in clavulanic acid biosynthesis. Biotechnology 2, 808-811.

Bernhard, K., Schrempf, H., and Goebel, W. (1978). Bacteriocin and antibiotic resistance plasmids in

Bacillus cereus and *Bacillus subtilis*. J. Bact. 133, 897-903.

Bibb, M.J., Freeman, R.F., and Hopwood, D.A. (1977). Physical and genetical characterisation of a second sex factor, SCP2, for *Streptomyces coelicolor* A3(2). Mol. Gen. Genet. 154, 155-166.

Bibb, M.J., Ward, J.M., and Hopwood, D.A. (1978). Transformation of plasmid DNA into *Streptomyces* at high frequency. Nature 274, 398-400.

Bibb, M.J., Ward, J.M., Kieser, T., Cohen, S.N., and Hopwood, D.A. (1981). Excision of chromosomal DNA sequences from *Streptomyces coelicolor* forms a novel family of plasmids detectable in *Streptomyces lividans*. Mol. Gen. Genet. 184, 230-240.

Bibb, M.J., and Cohen, S.N. (1982) Gene expression in *Streptomyces*: Construction and application of promoter-probe plasmids in *Streptomyces lividans*. Mol.Gen. Genet. 187, 265-277.

Bibb, M.J., Findlay, P.R., and Johnson, M.W. (1984). The relationship between base composition and codon usage in bacterial genes and its use for the simple and reliable identification of protein-coding sequences. Gene 30, 157-166.

Bibb, M.J., Bibb, M.J., Ward, J.M., and Cohen, S.N. (1985). Nucleotide sequences encoding and promoting expression of three antibiotic resistance genes indigenous to *Streptomyces*. Mol. Gen. Genet. 199, 26-36.

Canosi, U., Morelli, G., and Trautner, T.A. (1978). The relationship between molecular structure and transformation efficiency of some *S. aureus* plasmids isolated from *B. subtilis*. Mol. Gen. Genet. 166, 259-267.

Canosi, U., Iglesias, A., and Trautner, T.A. (1981). Plasmid Transformation in *Bacillus subtilis*: Effects of Insertion of *Bacillus subtilis* DNA into Plasmid pC194. Mol. Gen. Genet. 181, 434-440.

Chang, S., and Cohen, S.N. (1979). High Frequency Transformation of *Bacillus subtilis* Protoplasts by Plasmid DNA. Mol. Gen. Genet. 168, 111-115.

Chater, K.F., Hopwood, D.A., Kieser, T., and Thompson, C.J. (1982). Gene cloning in *Streptomyces*. Curr. Top. Microbiol. Immunol. 96, 69-95.

Chater, K.F., and Bruton, C.J. (1983). Mutational cloning in *Streptomyces* and the isolation of antibiotic production genes. Gene 26, 67-78.

Chater, K.F., and Hopwood, D.A. (1983). *Streptomyces* Genetics. In "Biology of the *Actinomycetes*", Goodfellow, M., et al., eds.; pp. 229-285. Academic Press, London.

Cornelis, P., Digneffe, C., and Willemot, K. (1982). Cloning and expression of a *Bacillus coagulans* amylase Gene in *Escherichia coli*. Mol. Gen. Genet. 186, 507-511.

Cowing, D.W., Bardwell, J.C.A., Craig, E.A., Wool-ford, C., Hendrix, R.W., and Gross, C.A. (1985). Consensus sequence for *Escherichia coli* heat shock gene promoters. Proc. Natl. Acad. Sci. USA 82, 2679-2683.

Dhaese, P., Hussey, C., and Van Montagu, M. (1984). Thermo-inducible gene expression in *Bacillus subtilis* using transcriptional regulatory elements from temperate phage Φ105. Gene 32, 181-194.

de Vos, W.M., and Venema, G. (1981a). Fate of plasmid DNA in transformation of *Bacillus subtilis* protoplasts. Mol. Gen. Genet. 182, 39-43.

de Vos, W.M., Venema, G., Canosi, U., and Trautner, T.A. (1981b). Plasmid Transformation in *Bacillus subtilis*: Fate of plasmid DNA. Mol. Gen. Genet. 181, 424-433.

DiRienzo, J.M., Nakamura, K., and Inouye, M. (1978). The outer membrane proteins of Gram-negative bacteria: Biosynthesis, assembly and functions. Ann. Rev. Biochem. 47, 481-532.

Ehrlich, S.D. (1977). Replication and expression of plasmids from *Staphylococcus aureus* in *Bacillus subtilis*. Proc. Natl. Acad. Sci. USA 74, 1680-1682.

Ehrlich, S.D. (1978). DNA cloning in *Bacillus subtilis*. Proc. Natl. Acad. Sci. USA 75, 1433-1436.

Ehrlich, S.D., Niaudet, B., and Michel, B. (1982). Use of plasmids from *Staphylococcus aureus* for cloning of DNA in *Bacillus subtilis*. Current Topics in Microbiol. and Immunol. 96, 19-29.

Ernst, H., Hartmann, G., and Domdey, H. (1982). Species specificity of promotor recognition by RNA Polymerase and its transfer by the σ Factor. Europ. J. Biochem, 124, 427-433.

Feitelson, J.S., and Hopwood, D.A. (1983). Cloning of a *Streptomyces* gene for an O-methyltransferase involved in antibiotic biosynthesis. Mol. Gen. Genet. 190. 394-398.

Gilman, M.Z., Wiggs, J.L., and Chamberlin M.J. (1981). Nucleotide sequence of two *Bacillus subtilis* promotors used by *Bacillus subtilis* σ^{28} RNA polymerase. Nucleic Acids Res. 9. 5991-6000.

Gray, O., and Chang, S. (1981). Molecular cloning and expression of *Bacillus licheniformis* Lactamase Gene in *Escherichia coli* and *Bacillus subtilis*. J. of Bacteriol. 145, 422-428.

Gryczan, T.J., and Dubnau, D. (1978). Construction and properties of chimeric plasmids in *Bacillus subtilis*. Proc. Natl. Acad. Sci. USA 75, 1428-1432.

Gryczan, T., Contente, S., and Dubnau, D. (1980). Molecular cloning of heterologous chromosomal DNA by recombination between a plasmid vector and a homologous resident plasmid in *Bacillus subtilis*. Mol. Gen. Genet. 177, 459-467.

Haldenwand, W.G., and Losick, R. (1980). A novel RNA polymerase σ factor from *Bacillus subtilis*. Proc. Natl. Acad. Sci. USA 77, 7000-7004.

Hardy, H., Stahl, S., and Küpper, H. (1981). Production in *B. subtilis* of hepatitis B core antigen and of major antigen of foot and mouth disease virus. Nature 293, 481-483.

Harris, J.E., Chater, K.F., Bruton, C.J., and Piret, J.M. (1983). The restriction mapping of *c* gene deletions in *Streptomyces* bacteriophage ΦC31 and their use in cloning vector development. Gene 22, 167-174.

Hawly, D.K., and McClure, W.R. (1983) Compilation and analysis of *Escherichia coli* promoter DNA sequences. Nucleic Acids Res. 11, 2237-2255.

Honjo, M., Manabe, K., Shinada, H., Mita, I., Nakayama, A., and Furutani, Y. (1984). Cloning and expression of the gene for neutral protease of *Bacillus amyloliquefaciens* in *Bacillus subtilis*. J. Biotechnology 1, 265-277.

Hopwood, D.A., Bibb, M.J., Bruton, C.J., Chater, K.F., Feitelson, J.S., and Gil, J.A. (1983). Cloning *Streptomyces* genes for antibiotic production. Trends in Biotechnology 1, 42-48.

Hopwood, D.A., Malpartida, F., Kieser, H.M., Ikeda, H., Duncan, J., Fujii, I., Rudd, B.A.M., Floss, H.G., and Omura, S. (1985). Production of hybrid antibiotics by genetic engineering. Nature 314, 642-644.

Hopwood, D.A., Bibb, M.J., Chater, K.F., Kieser, T., Bruton, C.J., Kieser, H.M., Lydiate, D.J., Smith, C.P., Ward, J.M., and Schrempf, H. (1985). Genetic manipulation of *Streptomyces*. A laboratory manual. The John Innes Foundation, Norwich, 1985. ISBN 0-7084-0336-0.

Horinouchi, S., and Weisblum, B. (1980). Posttranscriptional modification of mRNA conformation: Mechanism that regulates erythromycin-induced resistance. Proc. Natl. Acad. Sci. USA 77, 7079-7083.

Jaurin, B., and Cohen, S.N. (1984). *Streptomyces lividans* RNA polymerase recognizes and uses *Escherichia coli* transcriptional signals. Gene 28, 83-91.

Jaurin, B., and Cohen, S.N. (1985). *Streptomyces* contain *Escherichia coli*-type A+T-rich promoters having novel structural features. Gene 39, 191-201.

Katz, E., Thompson, C.J., and Hopwood, D.A. (1983). Cloning and expression of the tyrosinase gene from *Streptomyces antibioticus* in *Streptomyces lividans*. J. Gen. Microbiol. 129, 2703-2714.

Keggins, K.M., Lorett, P.S., and Duvall, E.J. (1978). Molecular cloning of genetically active fragments of *Bacillus* DNA in *Bacillus subtilis* and properties of the vector plasmid pUB110. Proc. Natl. Acad. Sci. USA 75, 1423-1427.

Khan, S.A., Carleton, S.M., and Novick, R.P. (1981). Replication of plasmid pT181 DNA *in vitro*: Requirement for a plasmid-encoded product. Proc. Natl. Acad. Sci. USA 78, 4902-4906.

Kieser, T., Hopwood, D.A., Wright, H.M., and Thompson, C.J. (1982). pIJ101, a multi-copy broad host-range *Streptomyces* plasmid: Functional analysis and development of DNA cloning vectors. Mol. Gen. Genet. 185, 223-238.

Kreft, J., Bernhard, K., and Goebel, W. (1978). Recombinant plasmids capable of replication in *B. subtilis* and *E. coli*. Mol. Gen. Genet. 162, 59-67.

Kreft, J., and Hughes, C. (1982). Cloning vectors derived from plasmids and phage of *Bacillus*. Current Topics in Microbiol. and Immunol. 96, 1-17.

Losick, R., and Pero, J. (1981). Cascades of σ Factors. Cell 25, 582-584.

Lydiate, D.J., Malpartida, F., and Hopwood, D.A. (1985). The *Streptomyces* plasmid SCP2*: its functional analysis and development into useful cloning vectors. Gene 35, 223-235.

Malpartida, F., and Hopwood, D.A. (1984). Molecular cloning of the whole biosynthetic pathway of a *Streptomyces* antibiotic and its expression in a heterologous host. Nature 309, 462-464.

Martin, J.F., and Gil, J.A. (1984). Cloning and expression of antibiotic production genes. Biotechnology 2, 63-72.

McLaughlin, J.R., Murray, C.L., and Rabinowitz, J.C. (1981). Unique features in the ribosome binding site sequence of the Gram-positive *Staphylococcus aureus* β-Lactamase gene. J. Biol. Chem. 256, 11283-11291.

Michel, B., Palla, E., Niaudet, B., and Ehrlich, S.D. (1980). DNA cloning in *Bacillus subtilis*. III. Efficiency of random segment cloning and insertional inactivation vectors. Gene 12, 147-154.

Moran, C.P., Lang, N., and Losick, R. (1981). Nucleotide sequence of a *Bacillus subtilis* promotor recognized by *Bacillus subtilis* RNA polymerase containing σ[37]. Nucleic Acids Res. 9, 5979-5990.

Moran, C.P., Lang, N., LeGrice, S.F.J., Lee, G., Stephens, M., Sonenshein, A.L., Pero, J., and Losick, R. (1982). Nucleotide sequences that signal the initiation of transcription and translation in *Bacillus subtilis*. Mol. Gen. Genet. 186, 339-346.

Neugebauer, K., Sprengel, R., Schaller, H. (1981). Penicillinase from *Bacillus licheniformis*: nucleotide sequence of the gene and implications for the biosynthesis of a secretory protein in a Gram-positive bacterium. Nucleic Acids Res. 9, 2577-2588.

Ohmura, K., Nakamura, K., Yamazaki, H., Shiroza, T., Yamane, K., Jigami, Y., Tanaka, H., Yoda, K., Yamasaki, M., and Tamura, G. (1984). Length and structural effects of signal peptides derived from *Bacillus subtilis* α-amylase on secretion of *Escherichia coli* β-lactamase in *B. subtilis* cells. Nucleic Acids Res. 12, 5307-5319.

Osborn, M.J., and Wu, H.C.P. (1980). Proteins of the

outer membrane of Gram-negative bacteria. Ann. Rev. Mikrobiol. 34, 369-422.

Palva, I., Pettersson, R.F., Kalkkinen, N., Lehtovaara, P., Sarvas, M., Söderlund, H., Takkinen, K., and Kääviäinnen, L. (1981). Nucleotide sequence of the promotor and NH$_2$-terminal signal peptide region of the α-amylase gene from *Bacillus amyloliquefaciens*. Gene 15, 43-51.

Palva, I., Sarras, M., Lehtovaara, P., Gibakor, M., and Kääviäinnen, L. (1982). Secretion of *Escherichia coli* β-lactamase from *Bacillus subtilis* by the aid of α-amylase signal sequences. Proc. Natl. Acad. Sci. USA 79, 5582-5586.

Palva, I., Lehtovaara, P., Kääriäinnen, L., Gibakor, M., Cantell, K., Schein, C.H., Kashiwagi, k., and Weissmann, C. (1983). Secretion of interferon by *Bacillus subtilis*. Gene 22, 229-235.

Piret, J., and Chater, K.F. (1985). Phage-mediated cloning of *bldA*, a region involved in *Streptomyces coelicolor* morphological development, and its analysis by genetic complementation. J. Bacteriol. 163, 965-972.

Priest, F.G. (1977). Extracellular enzyme synthesis in the genus *Bacillus*. Bacteriol. Rev. 41, 711-753.

Rodicio, M.R., Bruton, C.J., and Chater, K.F. (1985). New derivatives of the *Streptomyces* temperate phage ΦC31 useful for the cloning and functional analysis of *Streptomyces* DNA. Gene 34, 283-292.

Samson, S.M., Belagaje, R., Blankenship, D.T., Chapman, J.L., Perry, D., Skatrud, P.L., VanFrank, R.M., Abraham, E.P., Baldwin, J.E., Queener, S.W., and Ingolia, T.D. (1985). Isolation, sequence determination and expression in *Escherichia coli* of the isopenicillin N synthetase gene from *Cephalosporium acremonium*. Nature 318, 191-194.

Schrempf, H., and Goebel, W. (1977). Characterization of a plasmid from *Streptomyces coelicolor* A3(2) J. Bacteriol. 131, 251-258.

Spizizen, J. (1958). Transformation of biochemically deficient strains of *Bacillus subtilis* by deoxyribonucleate. Proc. Natl. Acad. Sci. USA 44, 1072-1078.

Tanaka, T. (1979). Restriction of plasmid-mediated transformation in *Bacillus subtilis*. Mol. Gen. Genet. 175, 235-237.

Thompson, C.J., Kieser, T., Ward, J.M., and Hopwood, D.A. (1982). Physical analysis of antibiotic resistance genes from *Streptomyces* and their use in vector construction. Gene 20, 51-62.

Westphaling, J., Ranes, M., and Losick, R. (1985). RNA polymerase heterogeneity in *Streptomyces coelicolor*. Nature 313, 22-27.

Wiggs, J.L., Bush, J.W., and Chamberlin, M.J. (1979). Utilization of promotor and terminator sites on bacteriophage T7 DNA by RNA polymerase from a variety of bacterial orders. Cell 16, 97-109.

Wiggs, J.L., Gilman, M.Z., and Chamberlin, M.J. (1981). Heterogeneity of RNA polymerase in *Bacillus subtilis*: Evidence for an additional σ factor in vegetative cells. Proc. Natl. Acad. Sci. USA 78. 2762-2766.

Williams, D.M., Duvall, E.J., and Lovett, P.S. (1981a). Cloning restriction fragments that promote expression of a gene in *Bacillus subtilis*. J. Bacteriol. 146, 1162-1165.

Williams, D.M., Schoner, R.G., Duvall, E.J., Preis, L.H., and Lovett, P.S. (1981b). Expression of *Escherichia coli trp* genes and the mouse dihydrofolate reductase gene cloned in *Bacillus subtilis*. Gene 16, 199-206.

Yansura, D.G., and Henner, D.J. (1984). Use of the *Escherichia coli lac* repressor and operator to control gene expression in *Bacillus subtilis*. Proc. Natl. Acad. Sci. USA 81, 439-443.

7 Expression Vectors in Prokaryotes

An important goal of applied gene technology is the production of proteins from recombinant DNA molecules. This requires a special class of vectors which have been designated expression vectors. Such vectors not only provide the structural prerequisites of cloning, transfer, and multiplication of recombinant DNA, but also allow this DNA to be translated into protein. Several points have to be taken into consideration when such vectors are constructed:

- the optimal choice of regulatory sequences such as promoters, operators, and ribosomal binding sites;
- the selection and distribution of codons;
- the copy numbers of recombinant molecules and stability of the plasmid;
- the nature and stability of translated proteins;
- the localisation of translated proteins;
- the problem of incompatible DNA sequences;
- the choice of host cells.

Not all of these items have been analysed in detail. Some have been, or will be, discussed in other chapters. This chapter will mainly address the nature of regulatory sequences and the problems of stability and transport of translated proteins.

The expression of a cloned eukaryotic gene in bacteria depends upon its correct transcription into functional mRNA and its efficient translation into the desired protein.

While the initiation of transcription is defined by promoter sequences, the initiation of protein biosynthesis depends on the presence of a suitable ribosome binding site. Eukaryotic genes lack the sequences recognised in prokaryotes as regulatory sites since the mechanisms of transcription and translation in eukaryotic cells are different. Promoters and ribosome binding sites must therefore be placed in front of a cloned eukaryotic gene to ensure that bacterial cells will recognise and translate them.

7.1 E. coli Promoters

A promoter is a region of DNA which is recognised by bacterial RNA polymerases. It is the structural prerequisite for the initiation of transcription of any prokaryotic gene, or groups of genes known as operons. The promoter is part of a larger DNA unit, known as the transcription unit, which contains the promoter, the coding regions and a terminator sequence (Fig. 7-1; see also Section 7.1.4). Promoter specificity of an RNA polymerase molecule is mediated by particular protein subunits known as (sigma) σ factors (cf. Section 6.3.1). In E. coli, most promoters are recognised by an RNA polymerase containing a σ factor subunit with a molecular weight of 70 kDa. The topography of the corresponding RNA polymerase binding site is well known, since more than 160 prokaryotic promoters have been sequenced and analysed in detail (Rosenberg and Court, 1979; Hawley and McClure, 1983). All promoters studied so far have in common a DNA

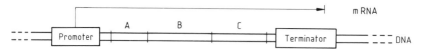

Fig. 7-1.   Structure of a prokaryotic transcription unit.
The synthesis of a polycistronic mRNA encoding polypeptides A, B, and C is controlled by promoter and terminator sequences which interact with DNA-dependent RNA polymerases and regulatory proteins such as repressors or antitermination factors (not to scale).

Fig. 7-2.   Structure of a prokaryotic promoter.
Shown are the highly conserved "-10" region, *i.e.*, the Pribnow-Schaller box, and other conserved regions around positions "-35", and "-43". Less conserved bases are printed in lower case (see text).

sequence which is approximately 80 bp in length and which can be divided into four characteristic regions (Fig. 7-2). The transcriptional start site at position "+1" is extremely specific, but less conserved than the other regions. It lies approximately six to ten nucleotides downstream from

a characteristic sequence known as the Pribnow-Schaller box or "-10" region. This region is an extraordinarily conserved hexameric sequence first described by Schaller *et al.* (1975) and Pribnow (1975). The consensus sequence 5'-TATAAT-3' comprises the most frequently occurring nucleotides. It does not occur in all promoters in this form, but every promoter sequence contains at least the last nucleotide T. The first two positions, TA, are conserved in more than 90% of all promoters, and it is only the internal three nucleotides which show a certain degree of variation (Fig. 7-3). Another region of homology occurs at position "-35" and contains the highly conserved trinucleotide TTG and some other less conserved nucleotides. The distance between the "-10" and "-35" regions is thought to influence the strength of a promoter. In all cases studied so far this distance is between 15 and 18 bp.

Different genes in a bacterial or phage genome are usually transcribed with different frequencies. This is partly due to the different strengths of the promoters, which is directly correlated with the rate of complex formation between RNA polymer-

Heinz Schaller
Heidelberg

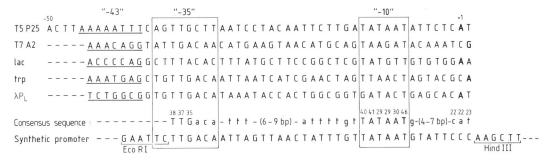

Fig. 7-3. DNA sequences of promoter regions.
A comparison of 168 promoter sequences (Rosenberg and Court, 1979; Hawley and McClure, 1983) revealed the existence of a consensus sequence (boxed) with marked homologies within a region of six base pairs around position "-10". This consensus sequence was first recognised by Pribnow (1975) and Schaller *et al.* (1975) and is known as the Pribnow-Schaller box. Other regions of homology can be found around positions "-35" (boxed) and "-43". The AT-rich region around position "-43" (underlined) has been found in a variety of strong promoters. Numbers above the consensus sequence indicate how often a particular base occurs at this position among 46 promoters selected here. Homologies of less than 70% are indicated by the use of small letters. A single T residue has been found to be conserved in all promoter sequences. Apart from the consensus sequence, sequences of five selected promoters are shown with their corresponding transcription initiation sites (marked by bold-face letters and position "+1"). Two of the promoters shown are from bacteriophages T5 and T7, the others from the *lac*, *trp* and λ systems discussed in the text. The bottom line shows a synthetic "consensus" promoter which exhibits a remarkable heparin resistance (Rossi *et al.*, 1983; see Section 7.1.5).

ase and the promoter sequence. Strong promoters, such as those in the early region of phage T5, for example, do not only contain the prototypic regions mentioned above, but also an additional AT-rich region at approximately position "-43" (Fig. 7-3) (Bujard, 1980).

In spite of these extensive and characteristic homologies, the question of what the structure of an ideal promoter is cannot be answered without hesitation since the strength of a promoter does not only depend upon the interaction with RNA polymerase. Promoter activities are also modulated by interactions with other proteins. In contrast to RNA polymerase which interacts with all promoters, these proteins, the CAP protein of the *lac* promoter, for example, interact with certain promoters, but not with others. In these cases it is difficult to recognise clearcut correlations since the recognition sequences of such proteins are superimposed upon those of RNA polymerase; in addition, the situation is rendered even more complex since the DNA sequences within promoter regions may possess secondary structures of an unknown nature and uncertain role in DNA-protein interactions.

"Good" promoters for the expression of foreign proteins do not only lead to the synthesis of high levels of mRNA. In particular, "good" promoters are also inducible, *i.e.*, their activity can be influenced by the experimenter. It is conceivable that the continuous high level expression of a foreign protein may be detrimental, if not toxic, for bacterial cells. It is therefore important to control promoter activities in such a way that the foreign protein is expressed at certain times only, and not for example, when the micro-organisms in question are actively dividing. There are several naturally occurring promoters which meet these requirements, *i.e.*, they lead to high levels of mRNA and they can be controlled. The *lac* and *trp* promoters of *E. coli* and the P_L promoter of phage λ appear to be most suitable. The biology of these systems has been reviewed in detail by Miller and Reznikoff (1980). In addition, hybrid promoters have been developed which combine sequences from constitutively

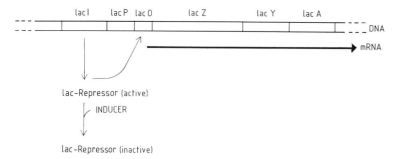

Fig. 7-4. Genetic elements of the *lac* operon of *E. coli*.
Active *lac* repressor encoded by the *lacI* gene interacts with *lac* operator DNA and thus prevents the initiation of transcription. The repressor is inactivated and released from the DNA upon interaction with inducers such that transcription proceeds unimpaired. *lacZ*, *lacY*, and *lacA* are the structural genes of the *lac* operon, coding for β-galactosidase, *lac* permease, and transacetylase.

expressed strong promoters with controlling regions from natural inducible promoters, forming inducible hybrid promoters.

7.1.1 The lac Promoter

The *lac* promoter is part of the *lac* operon, a gene complex comprising three structural genes, *Z*, *Y*, and *A*, coding for β-galactosidase, lactose permease, and thiogalactoside transacetylase, respectively. Co-ordinate expression of these genes is regulated by the promoter, *lacP*, and the interaction between the *lacI* gene product (the repressor) and the *lac* operator, *lacO* (Fig. 7-4). In the absence of lactose, which is an inducer, repressor molecules bind to the operator region and prevent RNA polymerase from binding to the promoter region, thus blocking initiation of transcription. If lactose or another suitable inducer is present the repressor is released from the DNA and *lac* mRNA is synthesised. This system is advantageous for two reasons. Firstly, isopropyl-thiogalactoside (IPTG) is a potent inducer, but not a substrate for β-galactosidase (Fig. 7-5), and secondly, one of the induced enzymes, β-galactosidase, is a hydrolytic enzyme. An induced cell

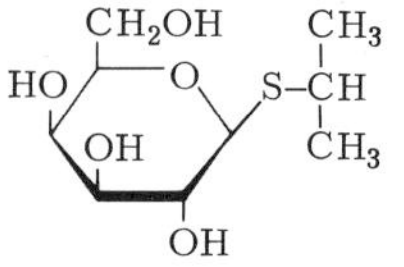

Isopropylthiogalactoside (IPTG)

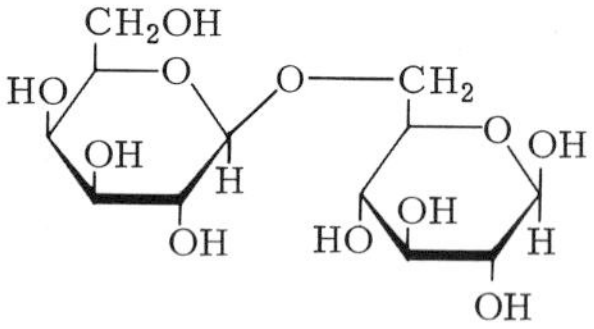

1,6-Allolactose

Fig. 7-5. Structure of inducers of the *lac* operon.
1,6-allolactose is both an inducer of the *lac* operon and a substrate for β-galactosidase. IPTG is a gratuitous inducer which is not metabolised.

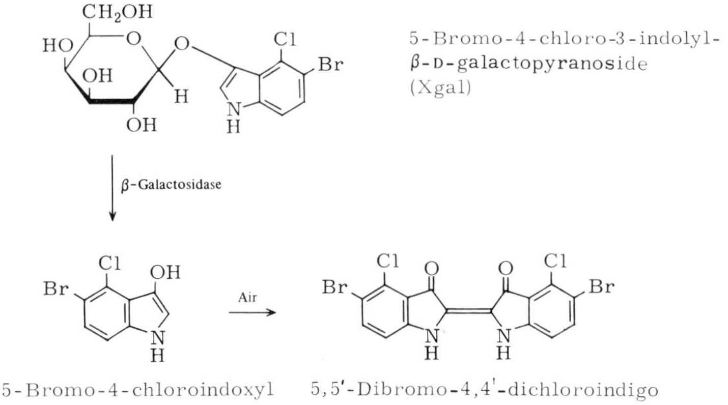

Fig. 7-6. Detection of β-galactosidase activity.

can easily be distinguished from an uninduced (repressed) one using chromogenic non-inducing substrate analogues. The colourless compound 5-bromo-4-chloro-3-indolyl-β-D-galactopyranoside (Xgal) is converted by β-galactosidase into its indoxyl derivative which in turn is oxidised to the blue dye 5,5'-dibromo-4,4'-dichloro-indigo. Bacterial colonies producing β-galactosidase are therefore easily recognised by their blue colour (Fig. 7-6) (Miller, 1972). Another indicator substrate 4-methylumbelliferyl-β-D-galactoside (MUG) which is converted by β-galactosidase to the highly fluorescent 4-methylumbelliferone was shown to be even considerably more sensitive than Xgal (Youngman *et al.*, 1985).

A different way to differentiate between bacterial colonies which ferment lactose (Lac$^+$) and those that do not (Lac$^-$) is the use of McConkey agar, although this method is much less sensitive. Since sugar fermentation leads to the production of acid which in turn causes a drop in pH , Lac$^+$ colonies can be recognised if the growth medium contains lactose and a suitable pH indicator dye. McConkey agar uses neutral red for this purpose and Lac-positive colonies therefore appear dark red, while Lac-negative colonies are colourless.

The *lac* promoter region does not only contain a binding site for RNA polymerase, but also one for the CAP (catabolite activator) protein, which is a positive regulator of *lac* operon transcription. In the presence of high levels of cAMP this protein binds to a site within the promoter region, known as the CAP site, where it facilitates the interaction of RNA polymerase and the promoter (Fig. 7-7).

The *lac* promoter thus is subject to two different regulatory mechanisms, one involving a negative regulation by *lac* repressor which inhibits transcription, and another involving positive regulation by the CAP protein. Regulation by the CAP protein also explains the well-known "glucose effect". *E. coli* cells do not react to inducers of the *lac* operon in the presence of glucose. This phenomenon is also referred to as catabolite repression. It is caused by a marked decrease of intracellular cAMP levels brought about by glucose and its metabolites. This in turn influences the binding of CAP protein to the *lac* promoter, and hence the initiation of transcription. Not even the inactivation of the *lac* repressor by suitable inducers is sufficient to reverse this effect.

Sequence analysis and binding studies have allowed the *lac* promoter region to be mapped in detail (Fig. 7-7). The promoter comprises 122 bp and is located between the 3' end of the *lacI* gene and the 5' end of the *lacZ* gene. The operator, the binding site of the repressor, lies between the promoter (which covers the first 85 bases of this

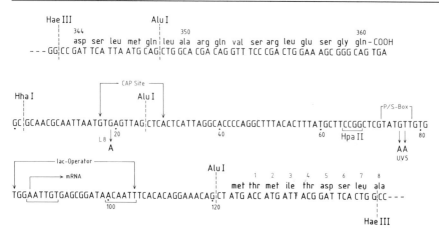

Fig. 7-7. DNA sequence of a 203 bp *Hae* III fragment coding for the *C*-terminal sequences of the *lacI* gene, regulatory regions, and the *N*-terminal region of the *lacZ* gene of the *E. coli lac* operon.
Bases protected by the CAP protein are marked as *CAP* site. *P/S* box marks the Pribnow-Schaller box. Numbers refer to amino acid sequences of the *lacI* gene, to co-ordinates within the promoter region between the stop codon of the *lacI* gene and the start codon of the *lacZ* gene, and to the *N*-terminal amino acids of β-galactosidase. *L8* and *UV5* are mutations discussed in the text. Regions of symmetry within the *lac* operator are underlined.

region) and the start signal of the *lacZ* gene (see also Fig. 7-3). The operator region is 24 bp in length and contains regions with two-fold rotational symmetry (Fig. 7-7). LacI repressor molecules recognise this and only this region among the 3×10^6 base pairs of the *E. coli* chromosome.

The *lac* promoter sequence contains a Pribnow-Schaller box between positions "-7" and "-12" (Fig. 7-7). The sequence which is TATGTT differs from the consensus sequence TATAAT in two nucleotides. One mutant allele of the *lac* promoter, known as *lacUV5*, contains the canonical Pribnow-Schaller box. The exchange of the two bases at position "-8" and "-9" increases the efficiency of transcription from the *UV5* promoter by a factor of 2.5. For historical reasons the *UV5* allele is associated with an additional mutation in the CAP-site (shown in Fig. 7-7 as *L8*) which renders the promoter CAP-independent (Silverstone *et al.*, 1970). Because of this convenient alteration the *lacUV5* promoter is frequently used in expression vectors.

The expression of the *lac* operon is delicately regulated. An *E. coli* cell usually contains approx-

imately ten repressor molecules, which are able to reduce *lac* expression by a factor of 1000 as compared to the fully induced state. This value pertains to one copy of the *lac* operon per cell. If, however, the *lac* operator/promoter region is carried on a high copy number plasmid, the existing repressor molecules are titrated out by the many copies of the operator sequence. *Lac* mRNA synthesis then becomes constitutive. In this case it is possible to use repressor-overproducing, i^q strains of *E. coli*, but even such strains are frequently only partially inducible.

A good source of *lac* control region DNA is the family of pOP plasmids (Fig. 7-8; Fuller, 1982). The pOP203 plasmids contain a 203 bp *Hae* III fragment with *lac* regulatory sequences cloned in pBR322 or pMB9, the pOP95 family a 95 bp *Alu* I fragment covering a region extending from the CAP-site to two base pairs in front of the start codon of the *lacZ* gene (Fig. 7-7). The *Hae* III fragments are obtained from the transducing phage λh80cI857dlacUV5. The λ DNA is sonicated first and *lac* promoter DNA sequences are removed from the mixture of DNA fragments by complexing them with *lac* repressor and binding the complexes to nitrocellulose filters (Fig. 7-8). The DNA is eluted, treated with *Hae* III and joined to filled-in ends of pMB9 DNA linearised with *Eco* RI. Clones containing vectors with *lac* regulatory sequences are easily identified since the many plasmid copies with *lac* DNA titrate the existing repressor, and a chromosomal *lacZ* gene is therefore expressed constitutively (see above). *LacZ*-constitutive cells are recognised by their blue colour on Xgal indicator agar in the absence of inducer (see also Fig. 7-6).

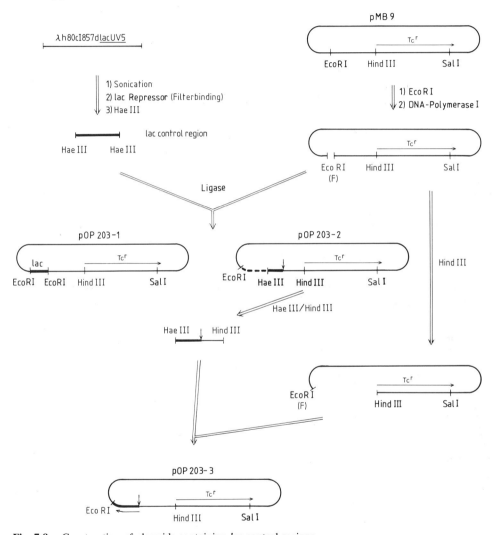

Fig. 7-8. Construction of plasmids containing *lac* control regions.
Cloning of the 203 bp *Hae* III fragment comprising the *lac* control region in vector pMB9 yields three different plasmids. pOP203-1 contains the original *Hae* III fragment flanked by *Eco* RI sites. In pOP203-2, the distal *Eco* RI site (arrow) in the *lacI* region is lost; an unidentified λ DNA fragment is located between the *Hae* III and *Eco* RI sites (dashed line). This fragment is removed during the construction of pOP203-3. The *lacUV5* promoter fragment is represented as a heavy line. *Eco* RI(F) indicates filled-in *Eco* RI sites. (Fuller, 1982).

The fusion of filled-in *Eco* RI sites in pMB9 with flush *Hae* III ends regenerates the *Eco* RI site. Indeed, this strategy yielded one clone, pOP203-1, which, upon *Eco* RI digestion, gave rise to a 207 bp fragment containing the *lac* control region, and linearised pMB9 DNA. Another derivative, pOP203-2, contained the desired *Hae* III fragment, but also some unidentified phage DNA. This could be removed by a combined *Hae* III/*Hind* III digestion. The remaining DNA fragment was joined to pMB9 DNA possessing one filled-in *Eco* RI and one *Hind* III end. The new vector, pOP203-3, contains only one *Eco* RI site. As expected, this site lies in the ninth codon of the *lacZ* gene. The distal *Eco* RI site in the *lacI* region of the fragment has been lost. Thus by fortunate chance an expression vector was obtained, the only *Eco* RI site of which can be used for cloning foreign DNA and expressing it under the control of *lac* regulatory sequences. Expression leads to the synthesis of a fusion peptide which consists of the protein in question and eight *N*-terminal amino acids of β-galactosidase.

The *Hind* III-*Eco* RI fragment of pOP203-3, which is 530 bp in length, has also been cloned in pBR322 (Fig. 7-9). In this new vector, pOP203-13, the *lacUV5* promoter points in the direction of the β-lactamase gene and away from the tetracycline resistance gene.

The 95 bp *Alu* I fragment mentioned above (Fig. 7-7) has been cloned in a similar way by annealing it with an *Eco* RI-*Pvu* II fragment of pBR322. The *Eco* RI end of this fragment was filled-in for this purpose. As shown in Fig. 7-10, the insertion of the *Alu* I fragment regenerates the *Eco* RI and the *Pvu* II sites (Lauer *et al.*, 1981; Johnsrud, 1978).

7.1.2 The trp Promoter

The tryptophan (*trp*) operon consists of five adjacent structural genes coding for the enzymes responsible for tryptophan biosynthesis from chorismate, and of appropriate control sequences

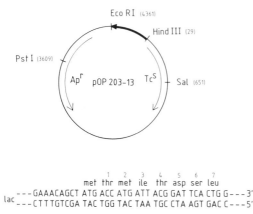

Fig. 7-9. Structure of vector pOP203-13.
The sequence shown in the bottom part of this figure is the region to the left of the *Eco* RI site around codons 8 and 9 of the *lacZ* gene with the wild-type *lac* sequence shown above. The *lac* promoter insertion is depicted as a heavy line; the arrows indicate directions of transcription. Numbers in brackets are pBR322 co-ordinates.

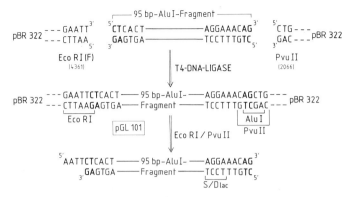

Fig. 7-10. Structure of vector pGL101.
The two ligated fragments are a 95 bp *Alu* I fragment containing the *lac* control region, and an *Eco* RI-*Pvu* II fragment of pBR322 coding for ampicillin resistance. Numbers in brackets are pBR322 co-ordinates. Due to its particular construction, the *Eco* RI and the *Pvu* II sites are regenerated in this vector. Transcription initiating at the *lacUV5* promoter is rightwards. S/D*lac* signifies the Shine-Dalgarno sequence. (Lauer *et al.*, 1981).

(Fig. 7-11). The five genes, *trpE, D, C, B*, and *A* are transcribed as a polycistronic mRNA, expression of which is negatively controlled by a repressor, the product of the *trpR* gene. In contrast to the *lac* system the *trpR* gene is not located in the vicinity of the *trp* operon.

The strength of binding of the *trp* repressor to the operator *trpO* is influenced by intracellular concentrations of tryptophan. At low levels of tryptophan the repressor dissociates from the operator and transcription begins. Some competitive inhibitors of *trp* repressor binding, such as indole-3-propionic acid or β-indolylacrylic acid, mimic low levels of tryptophan and inactivate the

repressor (Squires *et al.*, 1975). Therefore, such compounds serve as inducers of the *trp* system.

The distance between the 5' end of the mRNA and the first codon of the first structural gene, 38 bp in the *lac* system, is 162 nucleotides in the *trp* operon. This region, the leader sequence, contains a structure known as the attenuator which is a DNA sequence capable of terminating transcription approximately 140 bp downstream from its initiation site (Fig. 7-12 and 7-13). The region in question also codes for the *trp* leader peptide which is 14 amino acids in length and contains two tryptophan residues (Fig. 7-14). Deletions in this region enhance mRNA synthesis by a factor of

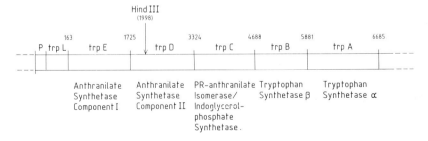

Fig. 7-11. Organisation of the L-tryptophan operon.
Numbering begins at the initiation site of transcription; numbers mark the positions of the individual start codons. The stop codon of the *trpA* gene is at position 6685; termination occurs 290 bp beyond this position. PR stands for the N-5'-phosphoribosyl residue. (Yanofsky *et al.*, 1981).

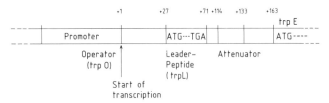

Fig. 7-12.   Regulatory sequences of the tryptophan operon.
The numbers refer to the start site of transcription at position "+1". The leader peptide resides between positions "+27" and "+73"; the attenuator between "+114" and "+133". The reading frame for the *trpE* product starts at position "+163". The operator (*trpO*) is part of the promoter (*trpP*).

8-10 above even the level observed in the absence of tryptophan. In order to understand and explain the regulatory function of this region it has been postulated that the *trp* leader mRNA is capable of forming extensive secondary structures (Fig. 7-13) (Crawford and Stauffer, 1980; Yanofsky, 1981); in addition, the two tryptophan residues within the leader peptide are thought to serve as a sensor of intracellular tryptophan levels. At high tryptophan concentrations the cell contains comparatively high levels of charged tRNA^{trp}. A

ribosome closely following RNA polymerase stops at the stop codon of the leader peptide as expected. The formation of a double-stranded terminator structure within the attenuator region, between positions 114 and 133, then provides a termination signal for RNA polymerase. If, however, tryptophan and hence tRNA^{trp} concentrations are low, the ribosome stops at the first trp codon. This prevents the formation of the characteristic secondary structure of the terminator. RNA polymerase then proceeds with transcription in the direction of the structural genes. This model of regulation of transcriptional termination and attenuation proposed by Yanofsky postulates an interaction of the RNA polymerase (present on the DNA) with an already synthesised region of mRNA which can form secondary structures. The model appears decidedly attractive since it can also be applied to other operons involved in amino acid biosynthesis, such as the histidine operon. Although there are a number of deletions in the region of the leader peptide which alter the secondary structure of the leader mRNA and influence *trp* operon transcription as a whole, it should be noted that this model cannot yet be regarded as strictly proven (Stroynowski and Yanofsky, 1982).

The leader peptide region has a ribosomal binding site (S/D-*trpL*) of its own, which makes it

Charles Yanofsky
Stanford

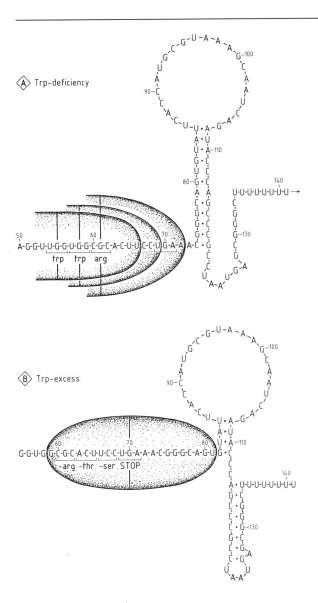

Fig. 7-13. Structures of tryptophan mRNA-ribosome complexes near the attenuator.
When tryptophan is scarce (A), the ribosome pauses at the Trp codons within the leader peptide mRNA, thus allowing the formation of mRNA secondary structures as indicated. In the presence of tryptophan (B) the ribosome moves forward to the stop codon. Base pairing within the attenuator (terminator) region (positions 114-133) leads to premature termination of *trp* operon transcription. (Yanofsky, 1981).

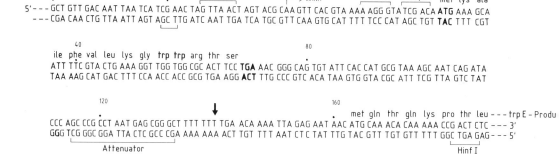

```
      -40                          -20              HpaI                  -1·+1                    S/D trpL      TaqI      met lys ala
       .                           .             ┌─────┐              ┌──→ mRNA                 ┌────┐    ┌────┐
5'---GCT GTT GAC AAT TAA TCA TCG AAC TAG TTA ACT AGT ACG CAA GTT CAC GTA AAA AGG GTA TCG ACA ATG AAA GCA
   ---CGA CAA CTG TTA ATT AGT AGC TTG ATC AAT TGA TCA TGC GTT CAA GTG CAT TTT TCC CAT AGC TGT TAC TTT CGT
                                              └─────┘

        40                                                    80
      ile phe val leu lys gly trp trp arg thr ser                       .
      ATT TTC GTA CTG AAA GGT TGG TGG CGC ACT TCC TGA AAC GGG CAG TGT ATT CAC CAT GCG TAA AGC AAT CAG ATA
      TAA AAG CAT GAC TTT CCA ACC ACC GCG TGA AGG ACT TTG CCC GTC ACA TAA GTG GTA CGC ATT TCG TTA GTC TAT

        120                              ↓                                   160        met gln thr gln lys pro thr leu---trp E-Produ
       .                                                                      .
      CCC AGC CCG CCT AAT GAG CGG GCT TTT TTT TGA ACA AAA TTA GAG AAT AAC ATG CAA ACA CAA AAA CCG ACT CTC---3'
      GGG TCG GGC GGA TTA CTC GCC CGA AAA AAA ACT TGT TTT AAT CTC TAT TTG TAC GTT TGT GTT TTT GGC TGA GAG---5'
       └──────────────────────────────┘                                                                   └─────┘
                  Attenuator                                                                                HinfI
```

Fig. 7-14. Sequence of the *trp* promoter-leader peptide region. Position "+1" marks the start site of transcription. The attenuator/terminator region is underlined. The arrow marks the 3' terminus of the leader transcript. The start and stop codons of the leader peptide (*trpL*) are printed in boldface. S/D*trpL* marks the Shine-Dalgarno sequence of the *trpL* peptide. The *Taq* I, *Hpa* I and *Hinf* I sites discussed in the text are bracketed.

particularly interesting for certain applications (Fig. 7-14). The regulatory sequences of the *trp* operon can be isolated by starting with a 5.4 kb *Hind* III fragment of *E. coli* DNA which contains the *trp* operator/promoter region, the entire *E* gene (1 725 bp), and part of gene *D* (273 bp) (Fig. 7-15) (Hallewell and Emtage, 1980). Cloning into the *Hind* III site of pBR322 yields plasmid ptrpED-3 (Fig. 7-16). This or similar plasmids can

be identified in a mixture of other plasmids by selecting for *trpE* complementation of a TrpE-negative strain of *E. coli* in the absence of tryptophan.

Plasmid ptrpED-3 was not a useful expression vector since it contained another *Hind* III site in the vicinity of the single *Eco* RI site. In order to remove this site and to shorten the molecule, the plasmid was first linearised with *Eco* RI and then successively treated with exonuclease III and S1 nuclease. The latter enzyme removes the 5' protruding ends generated by exonuclease III. Plasmid ptrpED5-1 was obtained after circularisation with T4 ligase and selection for *trpE* complementation and ampicillin resistance (Fig. 7-16). This new plasmid has lost almost all the upstream regions present in the original *Hind* III fragment (Fig. 7-15) and contains only a single *Hind* III site in the *trpD* region which can be used for cloning.

Such a vector synthesises remarkable amounts of protein. Its use as an expression vector is nevertheless limited since cloning allows only one of three possible reading frames. *Hind* III cleaves

Terry Platt
New Haven

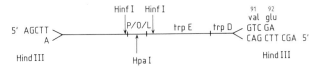

Fig. 7-15. Structure of a 5.4 kb *E. coli* DNA fragment carrying *trp* regulatory sequences (*P/O/L*), the entire *trpE* gene, and the first 273 bp of the *trpD* gene.
Cleavage at the *Hind* III site in the *trpD* gene occurs between the second and the third base of the codon for amino acid 92 (glutamic acid). This position in the coding strand is known as "-1" (*cf.* Fig. 7-17), and also occurs in the name given to the expression vector, namely ptrpED5-1. Arrows indicate the positions of *Hpa* I and *Hinf* I sites discussed in the text.

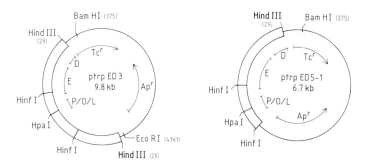

Fig. 7-16. Structures of two plasmids carrying *trp* regulatory regions, gene *E*, and parts of gene *D*.
Shown are the *Bam* HI and *Eco* RI sites of the DNA section derived from pBR322, the *Hpa* I site within the *trp* promoter, and the flanking *Hinf* I sites (*cf.* also Fig. 7-14). Numbers in brackets are pBR322 co-ordinates.

ptrpED-5 DNA at a position designated "-1" between the second and third base of the codon for amino acid 92. The nomenclature is schematically explained in Fig. 7-17. A vector DNA cleaved at one of the three positions "-1", "0", or "+1" and ligated to a DNA fragment with compatible termini will always lead to one in-frame fusion permitting the correct expression of a fusion protein.

A variety of *trp* vectors have been developed for just this purpose. They are derived from a 492 bp *Hinf* I fragment (Tacon *et al.*, 1980). As shown in Fig. 7-14, *Hinf* I (G/ANTC) cleaves at position 180 of the *trp* operon within the coding region of

Spencer Emtage
Slough, GB

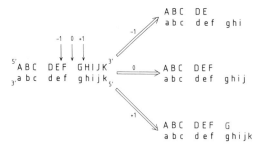

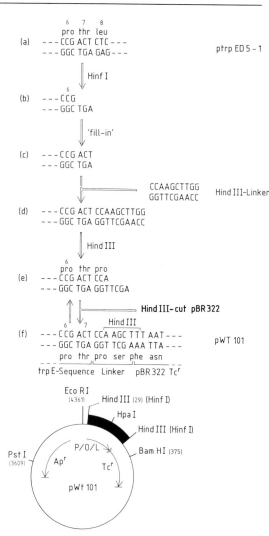

Fig. 7-17. Nomenclature for the construction of expression vectors.
Designations "-1", "0" and "+1" signify cleavages after the second, the third, and the first nucleotide of a codon.

the *trpE* gene. The resulting *Hinf* I fragment carries the entire regulatory region and codons for the first six amino acids of the *trpE* gene product. The *Hinf* I fragment is obtained from plasmid ptrpED5-1, the *Hinf* I ends are filled-in, and *Hind* III linkers are added with ligase (Fig. 7-18a to e). This fragment can now be cloned into pBR322 DNA cleaved with *Hind* III. The resulting plasmid pWT101 contains the *trp* regulatory region oriented in the same reading direction as the tetracycline resistance region (Fig. 7-18f). Plasmid pWT101 can be used as a *Hind* III expression vector if the second *Hind* III site in the immediate vicinity of the *Eco* RI site is eliminated. This is achieved by *Eco* RI cleavage and a careful exoIII/S1 nuclease digestion starting from the *Eco* RI site. The new plasmid pWT111 lacks the *Eco* RI and the *Hind* III site.

As shown in Fig. 7-19, the remaining *Hind* III site in the *trpE* region of pWT111 is opened by *Hind* III cleavage, filled-in, and annealed with *Hind* III linkers, yielding plasmid pWT121. This process is repeated and yields another plasmid, pWT131. Fig. 7-19 also shows that plasmid pWT111 allows cloning in position "0", pWT121 in position "+1", and pWT131 in position "-1" (see also Fig. 7-17). This set of plasmids now permits cloning of any DNA fragments with *Hind* III ends in a way which preserves the

Fig. 7-18. Construction of plasmid pWT101.
Hinf I digestion of ptrpED5-1 yields a 492 bp *Hinf* I fragment containing the regulatory sequences and the codons for the first six amino acids of the *trpE* gene product (black bar). A section from the coding region of *trpE* is shown at the top (line a). Line b shows one end of the indicated *Hinf* I fragment. The *Hind* III sites in pWT101 are marked *Hind* III(*Hinf* I) since the DNA fragment flanked by them represents the original *Hinf* I fragment mentioned above. The sequence around the *Hind* III site (line f) is located adjacent to the *Bam* HI site of pBR322. Numbers in brackets are pBR322 co-ordinates.

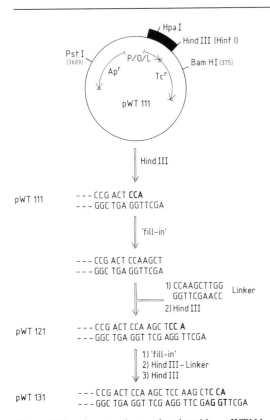

pWT 111   ---CCG ACT CCA
            ---GGC TGA GGTTCGA

            'fill-in'

            ---CCG ACT CCAAGCT
            ---GGC TGA GGTTCGA

            1) CCAAGCTTGG   Linker
               GGTTCGAACC
            2) Hind III

pWT 121   ---CCG ACT CCA AGC TCC A
            ---GGC TGA GGT TCG AGG TTCGA

            1) 'fill-in'
            2) Hind III - Linker
            3) Hind III

pWT 131   ---CCG ACT CCA AGC TCC AAG CTC CA
            ---GGC TGA GGT TCG AGG TTC GAG GTTCGA

Fig. 7-19. Construction of plasmids pWT111, pWT121, and pWT131.
Plasmid pWT111 was obtained from pWT101 (Fig. 7-18) following linearisation with *Eco* RI and treatment with a combination of exoIII and S1 nuclease; pWT111 therefore lacks the *Eco* RI site and the neighbouring *Hind* III site. pWT121 is obtained by cloning a *Hind* III linker into the filled-in *Hind* III site of pWT111. A repetition of these manipulations yields plasmid pWT131. The sequence section only depicts sequences near the *Hind* III site. For the nomenclature as expression vector see legend to Fig. 7-17. Relevant bases are printed in bold-face.

reading frame at the transition from *trpE* to the recombinant region.

An important practical application has been described by Emtage *et al.* (1980) who cloned the haemagglutinin (HA) glycoprotein gene of fowl plague virus. The gene was cloned with *Hind* III linkers and contains additional 19 AT pairs between the linker and a 21 bp untranslated

region. Using plasmid pWT121 as vector, the HA protein obtained would be a fusion protein consisting of seven *N*-terminal amino acids of the *trpE* product (anthranilate synthetase), six amino acids which derive from the particular construction in pWT121, six phenylalanine residues, seven amino acids of the 5' untranslated region, 558 amino acids of the haemagglutinin protein including prepeptide sequences, and five amino acids derived from a *Hind* III linker at the *C* terminus (Fig. 7-20). Such a fusion protein with a total of 589 amino acids was indeed observed after induction in transformed bacteria.

It should be noted that genes cloned using the *trp* cloning system are always partially derepressed, *i.e.*, transcription, albeit limited, is observed even in the presence of high concentrations of tryptophan. As in the *lac* system the reason is the abundance of operator sequences relative to trp repressor molecules, which is caused by the multiplication of such sequences in multicopy plasmids.

This situation may cause problems if pAT153, rather than pBR322 derivatives, is used since the copy number of pAT153 is two to three times higher than that of pBR322. Windass *et al.* (1982), for example, observed that approximately one third of the *trp* promoters in a cell were active in the presence of tryptophan when they studied the expression in pAT153 of the interferon α1 gene controlled by a chemically synthesised *trp* promoter. It would be desirable to increase the number of *trp* repressor molecules in such cells. One way to achieve this would be to place the *trp* repressor gene on plasmids with high copy numbers (see also Gonsalus and Yanofsky, 1980).

7.1.3 The λ P_L Promoter

Early transcription of λ DNA initiates at two positions, promoters P_L and P_R, immediately after infection of bacterial cells with bacteriophage λ (Fig. 7-21; see also Section 4.2). Eventually, this leads to the synthesis of two proteins, N, and Cro. N protein is an anti-termination factor. Its

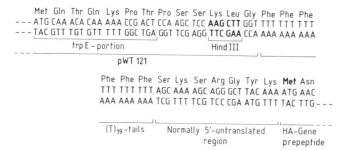

```
       Met Gln Thr Gln Lys Pro Thr Pro Ser Ser Lys Leu Gly  Phe Phe Phe
---ATG CAA ACA CAA AAA CCG ACT CCA AGC TCC AAG CTT GGT TTT TTT TTT
---TAC GTT TGT GTT TTT GGC TGA GGT TCG AGG TTC GAA CCA AAA AAA AAA
    |_____|   |_____|
         trpE - portion                 Hind III
---|_____|
                 pWT 121
```

```
         Phe Phe Phe Ser Lys Ser Arg Gly Tyr Lys Met Asn
         TTT TTT TTT AGC AAA AGC AGG GCT TAC AAA ATG AAC
         AAA AAA AAA TCG TTT TCG TCC CGA ATG TTT TAC TTG ---
```

```
  |_____|  |_____|  |_____
  (T)₁₉-tails    Normally 5'-untranslated      HA-Gene
                       region                 prepeptide
```

Fig 7-20. Construction of an expression vector for the haemagglutinin (HA) protein of fowl plague disease virus using vector pWT121 (Fig. 7-19).
Shown are the *trpE* part of the vector, the *Hind* III site, and the first methionine codon of the HA prepeptide. The remaining 16 amino acids of the prepeptide and the entire HA gene have been omitted. The entire fusion protein consists of 589 amino acids. (Emtage *et al.*, 1980).

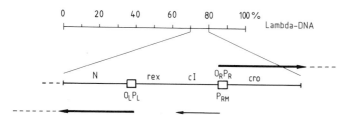

Fig. 7-21. Early transcription of bacteriophage λ.
Shown are the leftward operator/promoter region, O_LP_L, with the transcript of the *N* gene, and the rightward operator/promoter region, O_RP_R, which controls the transcription of two mRNA species. Rightward transcription leads to the synthesis of Cro mRNA; leftward transcription, starting at P_{RM}, to the synthesis of λ cI repressor (*cf.* also Fig. 4.2-5).

interaction with certain sites on λ DNA, designated *nut* sites (N utilisation), allows early transcription to proceed beyond *N* and *cro* into flanking regions. Cro is a repressor protein essential for the lytic infection cycle. It interacts with the right operator/promoter region O_RP_R and blocks the synthesis of λ repressor cI which would be synthesised by transcription starting at the P_{RM} promoter region (see also Fig. 4.2-5). Once formed, *cI* repressor binds to O_LP_L and O_RP_R and thus inhibits the synthesis of mRNA for genes *N* and *cro*. The left operator/promoter region O_LP_L only binds *cI* repressor while the other region binds cI and Cro protein as well. The left promoter P_L is preferred in gene technology because it interacts with only one repressor and

appears functionally simpler. In order to control, *i.e.*, switch on and off promoter P_L, bacterial host strains are used which contain a mutated *cI* gene (*cI857*, for example), which renders the repressor thermolabile. At 32°C the presence of approximately 20-30 repressor molecules produced from a single chromosomal *cI* gene is sufficient to block completely transcription controlled by P_L. At the non-permissive temperature, 41°C, the repressor is inactive and cannot bind to the operator (Fig. 7-22). A temperature shift, therefore, allows transcription to be initiated and this leads to the synthesis of the desired gene products. In contrast to the *lac* and *trp* systems discussed above, induction does not require the addition of special inducers or inhibitors.

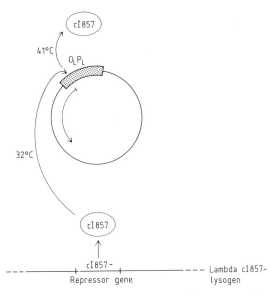

Fig. 7-22. Use of a temperature-sensitive repressor for controlling promoter P_L.

At 32 °C, the thermo-labile repressor, cI857, encoded by a *cI* gene on the chromosome binds to the operator region O_L on a plasmid and prevents transcription from initiating at P_L. By raising the temperature to 41 °C, the repressor is inactivated and released from the operator region (dotted bar) so that transcription can initiate (arrow).

Several strategies have been used to clone promoter P_L which is located around position 73.4 of the λ genome. Shimatake and Rosenberg (1981) and Rosenberg *et al.* (1983) used a *Hind* III-*Bam* HI fragment (positions 36900-34504) of the λ region in question and inserted it between the *Hind* III and *Bam* HI sites within the tetracycline resistance gene of pBR322. The insertion within the new plasmid is 2396 bp in length and contains the entire *N* gene, the promoter/operator region $O_L P_L$, the entire *rexB* gene, and a large part of *rexA* (Fig. 7-23). Vector pKC30 contains a single *Hpa* I site (GTT/AAC) in the middle of the *N* gene, 321 bp away from the start of P_L transcription and is therefore well suited for the insertion of foreign DNA (Fig. 7-24). When the λ *cII* gene is cloned in this site, the corresponding gene product may constitute 4-5% of the total cellular protein after temperature induction. This value is obtained approximately 60-90 minutes after induction. Plasmid pKC30 can be used only to overexpress bacterial genes that contain their own translational signal sequences. The expression of N protein fusion products obtained by translational

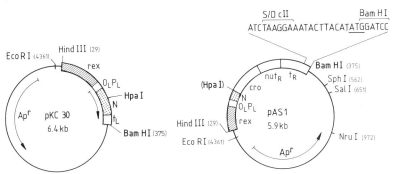

Fig. 7-23. Construction of vectors pKC30 and pAS1.

The 2,396 bp insertion of pKC30 contains parts of the λ genes *rexA* and *rexB* (hatched bars), the left operator promoter region, $O_L P_L$, the entire *N* gene (stippled bars), and the left early terminator t_L. The direction of transcription beginning at $O_L P_L$ is marked by an arrow. The blunt *Hpa* I site within the *N* gene (bold-face) is used for cloning foreign DNA. Plasmid pAS1 carries an insertion of a 1318 bp λ *Hae* III fragment with parts of *cro* and the entire *cII* gene cloned between the *Hpa* I and *Bam* HI sites of pKC30. Shown are the recognition site *nutR*, the right terminator t_R, and the Shine-Dalgarno (S/D) sequence of the *cII* gene. The start codon ATG of the *cII* gene is part of a *Bam* HI site. Numbers in brackets are pBR322 co-ordinates. (Shimatake and Rosenberg, 1981; Rosenberg *et al.*, 1983).

read-through from the *N* portion into the inserted DNA fragment has never been described.

A derivative of pKC30 is pAS1 which was obtained by the insertion into the *Hpa* I site of pKC30 of a 1318 bp *Hae* III fragment containing the *cII* region of λ DNA. The *cII* gene was shortened by cutting back from its single *Hinc* II site (position 38548) until the only remaining codon was the ATG start codon (Rosenberg *et al.*, 1983). A *Bam* HI site (G/GATCC) the first G residue of which is the G of the ATG codon can be used for cloning and expression of eukaryotic genes that do not normally carry regulatory signals for their translation in *E. coli*. An alternative in this case is the conversion of the 5' protruding ends obtained after *Bam* HI digestion to flush ends by nuclease treatment. This flush end, which terminates with an ATG codon, can then be annealed to suitably treated ends of the gene to be expressed.

A widely used derivative of pAS1, plasmid pOTS, contains an additional 189 bp *Dpn* I fragment with the efficient λ terminator cloned

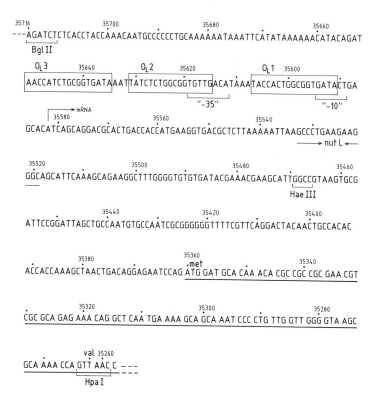

Fig. 7-24. DNA sequence of the r strand (r for rightward transcription) of bacteriophage λ DNA around the O_LP_L operator/promoter region and gene *N*.

This sequence, comprising 453 base pairs, extends from the *Bgl* II site at position 35715 to the *Hpa* I site at position 35262 behind codon 33 of gene *N*. Direction of transcription proceeds from left to right, although it is conventionally drawn from right to left in the λ genome. Shown are the three regions O_L3, O_L2, and O_L1 which interact with λ repressor, and the promoter regions "-10" and "-35" (*cf.* also Fig. 7-3). The *N* gene transcript which initiates at position 35582 contains a 221 bp leader extending towards the ATG codon at position 35360. The *N* gene region which codes for 33 amino acids has been underlined. Amino acids 1 (met) and 33 (val) are marked. The numbering is that of Sanger *et al.* (1982).

into its *Nru* I site; in addition it carries a 59 bp-long multiple cloning site (*Sac* I, *Xho* I, *Xba* I) in its *Sph* I site. pOTS has been used for the expression of the human *c-myc* gene product and of the PDGF-related transforming protein of Simian Sarcoma Virus (Devare *et al.*, 1984; Watt *et al.*, 1985).

A different approach was used by Remaut *et al.* (1981) who cloned a 243 bp DNA fragment containing promoter P_L. The starting material for this experiment was a 4.7 kb *Eco* RI-*Bam* HI λ DNA fragment comprising the region between positions 71.3 and 81.02% of the λ genome (Fig. 7-25). By cloning this fragment between the *Eco* RI and *Bam* HI sites of pBR322, plasmid pPLa2 was obtained. The only *Bgl* II sites of this vector lie within the λ DNA insertion. Since there are four *Bgl* II sites, *Bgl* II digestion and re-

ligation allows the two outermost *Bgl* II sites at positions 73.77 and 80.28% to be joined, thus eliminating a portion of the λ DNA insert between these two positions. The new plasmid, pPLa20, contains just one *Bgl* II site a few bases upstream of the P_L promoter region (see also Fig. 7-24). A *Bgl* II-*Hae* III digestion of pPLa20 yields the desired 243 bp fragment which can be inserted between the normal *Bam* HI site and a filled-in *Eco* RI site of pBR322. The result of this fusion is plasmid pPLa23. It codes for the first 114 nucleotides of the P_L transcripts, but is not yet an ideal expression vector.

If pPLa23 is digested with *Hae* II (PuGCGC/Py), a fragment of approximately 1 900 bp in length is obtained which contains the P_L promoter region and the entire β-lactamase gene of pBR322. This fragment can be annealed

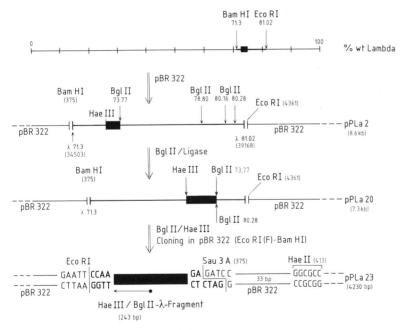

Fig. 7-25. Strategy for cloning the λ P_L promoter.
The top shows a physical map of λ DNA with the *Eco* RI (81.02%; position 39168) and *Bam* HI sites (71.3%; position 34503) in question. This λ region is shortened stepwise by subcloning into plasmids pPLa2, pPLa20, and pPLa23. The black bar represents the *Bgl* II-*Hae* III fragment containing the P_L promoter. Numbers in brackets refer to the positions of restriction sites in the individual pBR322 constructions. In pPLa23, the *Hae* II site of pBR322 used for further cloning (see Fig. 7-26) is also shown.

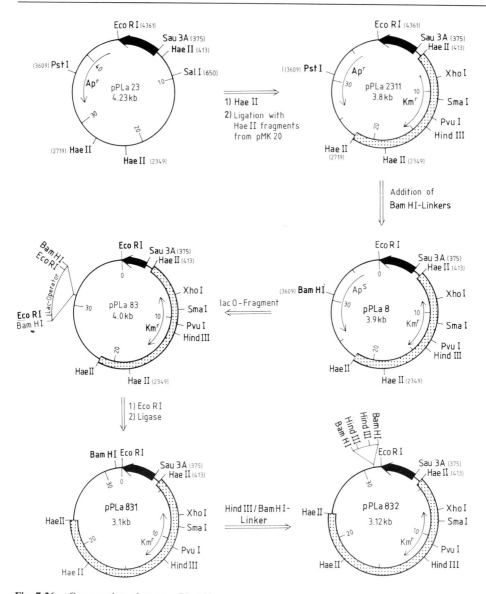

Fig. 7-26. Construction of vector pPLa832.

The construction of pPLa832 from pPLa23 is achieved in five steps discussed in detail in the text. An important feature is the introduction of a kanamycin resistance gene (*Km^r*) from pMK20 (Kahn *et al.*, 1979), which provides a necessary selection marker and allows the ampicillin resistance region to be manipulated. The tetracycline resistance of these plasmids was destroyed by the initial insertion of the λ *Eco* RI-*Bam* HI fragment (Fig. 7-25). The ampicillin resistance region is well suited for manipulations because it lies downstream from the P_L promoter. The most salient feature of the various cloning steps is the introduction of a *Bam* HI site into the vicinity of the *Eco* RI site immediately downstream from the promoter. As exemplified in pPLa832, this site can be exploited for introducing a polylinker which allows additional insertions to be introduced. λ DNA sequences are represented as thick black arrows, pMK20 sequences as stippled bars, and pBR322 DNA as thin lines.

with two other *Hae* II fragments derived from the kanamycin resistance plasmid pMK20 (Kahn *et al.*, 1979). These two *Hae* II fragments code for the kanamycin resistance gene and the origin of DNA replication of pMK20, respectively. By selecting for ampicillin and kanamycin resistance, plasmid pPLa2311 is obtained (Fig. 7-26).

The *Pst* I site of pPLa2311 is cleaved, trimmed with S1 nuclease, and converted to a *Bam* HI site by the addition of a suitable linker. The insertion of the decamer 5'-CCGGATCCGG-3' conserves the reading frame, but it replaces the alanine residue at position 182 of the β-lactamase with the sequence arg-ile-arg. The new *Bam* HI site is used for the insertion of a *lac* operator DNA fragment flanked with *Eco* RI and *Bam* HI sites, which yields plasmid pPLa83. Plasmid pPLa831 is obtained from pPLa83 by *Eco* RI cleavage and subsequent re-ligation. The goal of this sequence of reactions was the introduction of a *Bam* HI site in the immediate vicinity of the *Eco* RI site located downstream from the P_L promoter. This versatile vector can be improved by the insertion of a *Hind* III-*Bam* HI linker, which yields pPLa832 (Fig. 7-26).

In the plasmids described so far, the direction of transcription from promoter P_L is anti-clockwise. In plasmid pPLa2311, for example, this leads to an increased transcription of the β-lactamase gene, and hence to a dramatic increase of β-lactamase activity after temperature induction; however, a peculiarity of the vectors described above is that although total protein synthesis increases after temperature induction as expected, it ceases completely within less than one hour. The cause of this rather strange phenomenon is unknown. β-lactamase should not have anything to do with it since the cessation of protein synthesis is also observed with pPLa832 which lacks both the start signals and the codons for the first 182 amino acids of β-lactamase. A new series of vectors was therefore constructed in which transcription from promoter P_L proceeds in clockwise direction. The starting material for these vectors was a *Bam* HI-*Hind* III fragment of

pPLa832, approximately 1 200 bp in length, which was inserted into pBR322 cleaved with *Bam* HI and *Hind* III (Fig. 7-27). Due to the location of the *Bam* HI and *Hind* III sites in pBR322 this fragment could be inserted only in an orientation which allows clockwise transcription from P_L. Plasmid pPLc2 (c stands for clockwise, a for anti-clockwise) required modification before it could be used in practice. One *Eco* RI site in the vicinity of the *Hind* III site derived from pBR322 was removed by treatment with *Bal*31 after opening the vector with *Hind* III and *Xho* I. The resulting deletion is longer than 800 bp, but the entire P_L promoter region and the ampicillin resistance are retained in plasmid pPLc23. This plasmid was modified by the introducion of a *Hind* III-*Bal* I-*Hind* III-*Bam* HI linker into the *Bam* HI site. Approximately 1 700 bp were removed from pPLc236 by cleavage with *Bal* I/*Pvu* I and subsequent ligation of the resulting flush ends (Fig. 7-28). The λ DNA insert in pPLc28 contained no ribosome binding site (see Section 7.2). An *Eco* RI-*Bam* HI DNA fragment comprising the first 98 amino acids of the MS2 replicase and the corresponding ribosome binding site was inserted into this plasmid to allow the expression of eukaryotic fusion proteins. This fragment, which is 429 bp in length, was derived from plasmid pMS2-7 which contains a full-length copy of MS2 (Devos *et al.*, 1979). The *Bam* HI site of pPLc24 is located in the immediate vicinity of the codon for amino acid 98 and can be used for the insertion of a coding DNA fragment. The resulting plasmid could express a fusion protein containing 98 amino acids of replicase at its N-terminus (Fig. 7-29; see also Fig. 7-48).

In addition to the O_LP_L region, all vectors described so far also contain a recognition site, *nutL*, which is utilised by N protein in its role as anti-terminator (Fig. 7-24). As in phage λ itself, anti-termination is of consequence only if the DNA fragment to be expressed contains termination sequences. It cannot be predicted whether eukaryotic genes contain such terminator sequences, therefore the use of λ lysogenic host

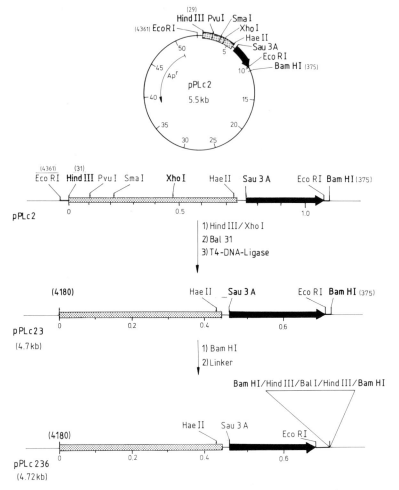

Fig. 7-27. Construction of plasmids showing clockwise P_L transcription.
Plasmid pPLc2 contains a *Bam* HI-*Hind* III fragment of pPLa832 (Fig. 7-26). The two other plasmids are derivatives of pPLc2 which have been shortened and provided with a polylinker; only the insertions containing the $O_L P_L$ region are shown. λ insertions are represented as thick black arrows, the arrow head indicating the direction of transcription. The treatment with *Bal*31 extends up to position 4180 of pBR322 and leaves intact the β-lactamase gene with its start codon at position 4155 and the Shine-Dalgarno sequence. Thin lines are pBR322 sequences, stippled bars represent pMK20 sequences.

bacteria coding for a thermoinducible repressor (cI857) *and* a functional N gene product is advisable. It is possible, of course, to use the vectors described above in λ lysogenic hosts which do not synthesise N protein; however in this case premature termination of transcription may occur (see also Rosenberg *et al.*, 1983).

7.1.4 Hybrid Promoters

The search for ideal promoters for the expression of foreign proteins in *E. coli* also led to the construction of artificial hybrid promoters. An important example is the hybrid *tac* promoter which consists of parts of the *trp* and *lac* promot-

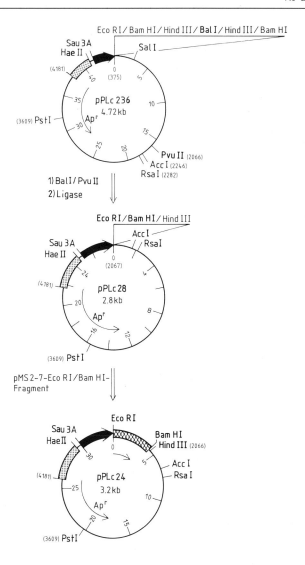

Fig. 7-28. Expression vectors with λ promoter P_L.

pPLc28 is derived from pPLc236 by removal of a *Bal* I-*Pvu* II fragment. The *Eco* RI-*Bam* HI-*Hind* III linker in pPLc28 is used for the insertion of an *Eco* RI-*Bam* HI DNA fragment coding for the replicase of bacteriophage MS2 (cross-hatched bar) (*cf.* also Fig. 7-29). Since this fragment contains a ribosomal binding site for replicase mRNA, insertions into the adjacent *Bam* HI and *Hind* III site of vector pPLc24 can be exploited for the expression of fusion proteins. Black bars represent λ DNA, stippled bars pMK20 sequences, and thin lines pBR322 sequences. Numbers in brackets are pBR322 co-ordinates. The orientation of the MS2 region is indicated by an arrow.

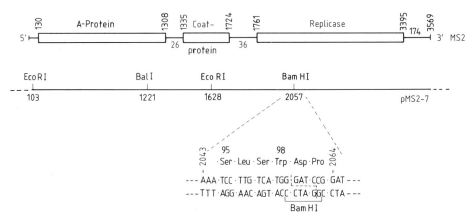

Fig. 7-29. Genetic map of plasmid pMS2-7.
The schematic drawing at the top shows the 3 569 bases of the RNA genome of bacteriophage MS2, which codes for three gene products, namely the A protein, the coat protein, and the replicase. Numbers above the map refer to the distances from the 5' end of the RNA. Plasmid clone pMS2-7 contains a DNA copy of MS2 coding for the entire RNA sequence apart from 14 bases at the 5' end. The DNA was cloned into the *Pst*I site of pBR322 *via* homopolymeric poly(dA)-poly(dT) tails. The *Eco* RI-*Bam* HI fragment shown in Fig. 7-28 is composed of the 3' terminal portion of the coat potein gene, the 5' terminal part of the replicase gene, and the intergenic region between the two genes. The sequence section at the bottom shows the *Bam* HI site at position 2057 and adjacent amino acid codons of the replicase gene. (Fiers *et al.*, 1976; Devos *et al.*, 1979).

ers (De Boer *et al.*, 1983; Amann *et al.*, 1983). At their respective "-21" positions the *trp* promoter (Fig. 7-30) possesses a *Taq* I site (T/CGA) and the *lacUV5* promoter a *Hpa* II site (C/CGG). The protruding ends obtained by *Taq* I and *Hpa* II cleavage are compatible, and the "-35" region of the *trp* promoter could therefore be fused with the "-10" region of the *lac* promoter. The new hybrid promoter is the *tac* promoter mentioned above. While the distance between the "-10" and "-35" region is 17 bp in the *trp* and *lac* promoters, it is 16 bp in the *tac* promoter (Fig. 7-30). The *tac* promoter is, therefore, characterised by canonical "-10" and "-35" regions and a distance between these two sites which is regarded as optimal, since it is also observed in strong ribosomal RNA

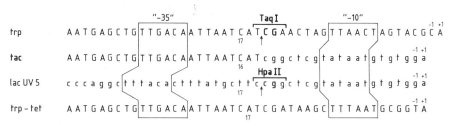

Fig. 7-30. Structure of hybrid promoters.
Capital letters represent *trp*, and lower case letters *lacUV5* promoter sequences. The *Taq* I and *Hpa* II cleavage sites are marked by arrows. The hybrid *tac* promoter consists of a "-35" region derived from the *trp* promoter, and a "-10" region of the *lac* promoter. Transcription begins at position "+1" and proceeds rightward. The hybrid *trp-tet* promoter was constructed by ligating the "-35" region of a *Taq* I-restricted *trp* promoter to pBR322 DNA linearised with *Cla* I. For unknown reasons this chimeric molecule cannot be cloned (see text). Numbers below the sequences indicate the distances between the "-10" and the "-35" regions.

promoters. Experiments designed to express the *cI* gene of λ and the gene for human growth hormone in suitable expression vectors have shown that the *tac* promoter is at least five times stronger than the *lacUV5* promoter. It is therefore as efficient as the λP_L promoter and may, in fact, even surpass it. Two derivatives of the *tac* promoter, the *trc* promoter with a 17 bp separation and the *tic* promoter with a 18 bp spacer, are on average only 90% and 65% as active as the *tac* promoter (Brosius *et al.*, 1985).

Due to its construction the *tac* promoter contains the *lac* operator region. Its strength therefore can be regulated in *E. coli* strains overproducing *lac* repressor. Induction is achieved, for example by the commonly used inducer of the *lac* operon, IPTG.

The starting material for the construction of the *tac* promoter was plasmid ptrpH1 which carries the *trp* promoter region (see also Fig. 7-14). A fragment of 192 bp obtained by cleavage with *Taq* I (T/CGA) was inserted into pBR322 DNA linearised with *Cla* I (AT/CGAT). One of the two possible orientations is shown in plasmid pEA301 (Fig. 7-31). Since the *Taq* I site in the *trp* promoter region is flanked by an adenine residue, the *Cla* I site was reconstituted after cloning. This is very useful since, in contrast to *Taq* I and *Hpa* II sites, a *Cla* I site occurs only once on the plasmid. Vector pEA301 was linearised with *Eco* RI, the *Eco* RI ends were filled-in, and the molecule was subsequently digested with *Cla* I. The resulting large DNA fragment was modified in such a way that it could take up a 55 bp *Hpa* II-*Pvu* II fragment containing the *lac* "-10" region and the *lac* S/D sequence (see Section 7.2) in the desired orientation. The suitable *lac* fragment could be obtained from vector pGL101, for example, by digestion with *Hind* III/*Pvu* II, isolation of the 550 bp fragment and further digestion with *Hpa* II (Fig. 7-10) (Lauer *et al.*, 1981). Vector ptac11 contained the hybrid *tac* promoter with an *Eco* RI site at its distal end. It deserves mention that the *trp* promoter fragment in pEA301 was always obtained only in the orientation shown in Fig.

7-31, and never in the other orientation. In the latter case a hybrid *trp-tet* promoter would have been created which evidently cannot be cloned (see Fig. 7-30). The reasons for this difficulty are unknown; such a construction would alter neither the "-35" sequence nor the distance between the "-10" and "-35" regions.

In order to construct a *tac* promoter which directs transcription proceeding in clockwise direction, a pBR322 derivative with a deletion in the *tet* promoter region was used. This derivative, pKK84-1, lacks base pairs 34-63, but still retains the *Cla* I and the *Hind* III sites (Fig. 7-32). Vector pKK84-1 first was cleaved with *Cla* I and then annealed with the 192 bp fragment containing the "-35" region (see above), which yields pEA300. This was modified to yield ptac12 by cleavage with *Cla* I/*Pvu* II and insertion of the 55 bp "-10" *lac Hpa* II-*Pvu* II fragment as has already been described for the construction of pEA301 (Fig. 7-31). In contrast to ptac11, ptac12 contains a *Pvu* II site which allows the insertion of DNA fragments with flush ends. The ptac family of vectors has recently been extended to include plasmids with different polylinker insertions, thus increasing the versatility of this promoter (Hallewell *et al.*, 1985).

Vectors pKK84-1 and pEA300 also carry a terminator in the direction of transcription, which in these cases is the strong terminator of the ribosomal RNA operon *rrnB*. Terminators are DNA sequences which signal the end of transcription. They are usually characterised by two structural peculiarities, a GC-rich inverted repetition which is capable of forming intramolecular stem-loop structures, and several consecutive U (T) residues. These features can easily be recognised in the terminators of phage fd DNA (Fig. 7-33) and the *trp* attenuator (Fig. 7-13B).

It has been observed that strong promoters usually can be cloned only in conjunction with strong terminators (Gentz *et al.*, 1981). In constructions involving the *tac* promoter, it is therefore advisable to provide means for the insertion of appropriate termination signals, although such

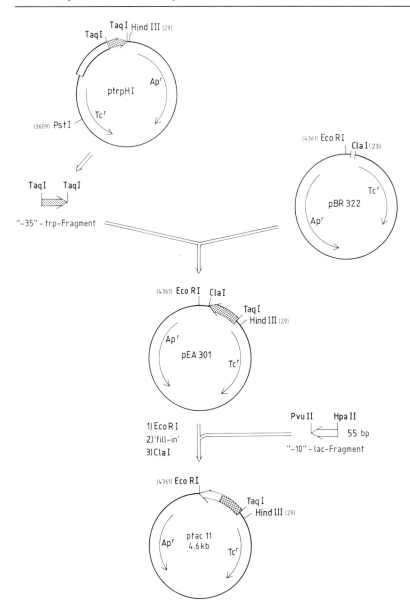

Fig. 7-31. Construction of vector ptac11 with a hybrid *trp-lac* promoter.
Vector pEA301 was constructed by cloning the "-35" *trp* fragment with *Taq* I ends (stippled bar) into pBR322 DNA linearised with *Cla* I. The insertion of a 55 bp "-10" *lac* fragment (open bar), obtained from either pGL101 or pLJ3 (Fig. 7-10), into pEA301 linearised with *Cla* I yields vector ptac11. The only insertions observed are those with the promoter inserted in a direction opposite to that of the tetracycline gene (see text). The *tac* promoter lacks the *Taq* I and *Hpa* II sites (Fig. 7-30) between the "-10" and the "-35" regions. (Amann *et al.*, 1983; De Boer *et al.*, 1983).

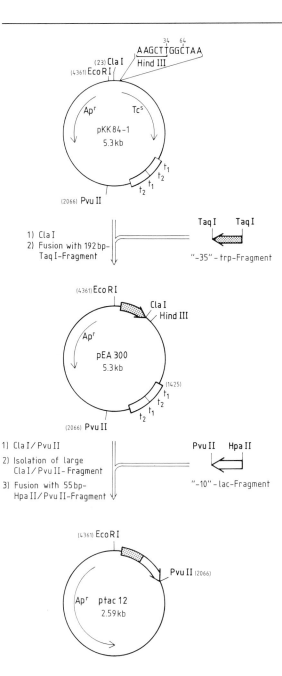

Fig. 7-32. Construction of vector ptac12.

In contrast to pEA301 (Fig. 7-31), the *tac* promoter can be obtained in both orientations in vectors pKK84-1 and pEA300 since they lack the tetracycline promoter; in addition, they carry two copies of a 500 bp DNA fragment with the transcription terminators t_1 and t_2 of the *rrn* operon (Brosius *et al.*, 1981) inserted into the *Ava* I site of the pBR322 portion at position 1425. (Amann *et al.*, 1983).

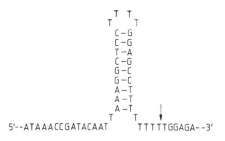

```
        T T
      T     T
      C - G
      C - G
      T - A
      C - G
      G - C
      G - C
      A - T
      A - T
      A - T          ↓
      T      T
5'--ATAAACCGATACAAT     TTTTTGGAGA--3'
```

Fig. 7-33. Structure of the central terminator of bacteriophage fd.

The sequence shown covers the nucleotides between positions 1522 and 1570 of the fd genome as determined by Beck and Zink (1981). The arrow marks the site of termination of transcription which is located between genes *VIII* and *III* on the genetic map of fd.

signals may not be essential for the expression of all proteins.

The *tac* promoter construction links consensus "-35" and "-10" regions from the *trp* and *lac* promoters, respectively, in an optimal distance to yield a highly expressed and inducible promoter. Another approach simply fuses promoter elements of a constitutively expressed promoter with the regulatory elements of the *lac* promoter. An impressive and widely used example is based on the outer membrane lipoprotein (lpp), the most abundant protein in *E. coli* (Nakamura and Inouye, 1982). The *lpp* gene, coding for a prolipoprotein of 78 amino acids, is constitutively expressed from an extremely AT-rich promoter

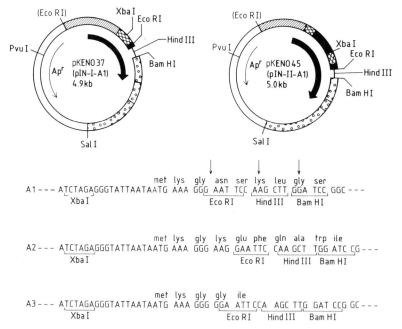

Fig. 7-34. Construction of expression vectors based on the *E. coli lpp* promoter.

Both vectors carry the large *Eco* RI-*Sal* I fragment of pBR322 (3.7 kb; open bar) and an *lpp* gene insertion (1.2 kb) consisting of the *lpp* promoter (positions -210 to -1; hatched bar), the 5' untranslated region (positions +1 to +39; cross-hatched bar) with the first three codons of the *lpp* protein (positions +40 to +45; black bar), an *Eco* RI/*Hind* III/*Bam* HI linker (solid line) and a 950 bp fragment carrying a 3' terminal part of the coding region, the terminator of the *lpp* gene and some adjacent *E. coli* DNA sequences (stippled bar). Sequences A1, A2, and A3 consist of parts of the 5' untranslated region including the *Xba* I site, and an *Eco* RI/*Hind* III/*Bam* HI linker with its cleavage sites arranged in different reading frames. The *lpp* mRNA is represented by a thick arrow. Plasmid pKEN045 contains a 95 bp insertion within the *Xba* I site of pKEN037 with the *lac* promoter-operator sequences.

region. The prototype expression vector pKEN037 (Fig. 7-34) contains the *lpp* promoter, the 5′ untranslated region of the *lpp* mRNA, the coding region for the first three amino acids of the prelipoprotein, and the 3′ portion of the *lpp* mRNA with the transcription termination signals. The single *Eco* RI site can be used to insert foreign DNA, leading to the expression of a fusion protein containing the four *N*-terminal amino acids of the *lpp* protein. The reading frame variants A2 and A3 permit expression in the two other possible reading frames. Constitutive expression of a cloned gene in the pIN-I-A series of vectors can be rendered inducible by insertion of the 95 bp *lac* DNA fragment containing the *lacUV5* promoter/operator region (*cf.* Fig. 7-10) into the 5′ untranslated region of the *lpp* gene. The repressor binding site in the pIN-II series of vectors acts as a transcriptional switch for transcription initiated from the *lpp* promoter (Nakamura *et al.*, 1982). It is completely repressed in recipient cells carrying the *lacI* gene, but strongly induced after addition of IPTG. A similar approach was taken to control gene expression in *B. subtilis* (Yansura and Henner, 1984).

The *E. coli* lipoprotein is made as a precursor which has a signal peptide consisting of 20 *N*-terminal amino acids. This structure guides the prelipoprotein into the bacterial outer membrane, where it is subsequently cleaved off by a signal peptidase. In order to exploit this phenomenon for the expression of fusion proteins, two other series of expression vectors were constructed in which the same multiple cloning sites were inserted in three different reading frames, either immediately after the signal peptide region (pIN-I-B or pIN-II-B types) or inside the structural gene of the mature lipoprotein (pIN-I-C and pIN-II-C types). It thus becomes possible to guide a fusion protein not only into the cytoplasm (pIN-I-A and pIN-II-A types), but also into the periplasmic space or the outer membrane, respectively.

7.1.5 Other Promoters

Occasionally, and for special purposes, the use of promoters other than those described above is considered. Rossi *et al.* (1983), for example, synthesised an artificial promoter based on the most statistically favoured bases in the promoter consensus sequence described by Rosenberg and Court (1979) (see Fig. 7-3). This promoter is unusually strong *in vitro* and directs the production of run-off transcripts even in the presence of heparin. The remarkable property of heparin resistance makes it an alternative to the SP6 promoter system currently used for the large-scale preparation of transcripts as hybridisation probes and substrates for studies of RNA processing (*cf.* Section 11.2.2.2).

Another expression system exploits the very specific bacteriophage T7 promoters. The linear, double-stranded DNA molecule of bacteriophage T7 DNA is 39 936 bp long (Dunn and Studier, 1983) and codes for approximately 50 gene products (Fig. 7-35). Transcription is regulated by three strong promoters for *E. coli* RNA polymerase (see Fig. 7-3), and 17 promoters for the phage-specific T7 RNA polymerase. The presence of this enzyme, a monomeric protein with a molecular weight of 98.8 kDa, together with a T7 promoter is lethal to an *E. coli* cell, since this highly efficient combination serves as a sink for ribonucleoside triphosphates.

A system of two coupled plasmids exploits this T7 RNA polymerase/promoter combination for the high-level expression of foreign proteins in *E. coli* (Fig. 7-36; Tabor and Richardson, 1985). One component, vector pGP1-2, contains the T7 RNA polymerase gene (gene *1* of T7) driven by the bacteriophage λP_L promoter. Its expression is controlled by a temperature-sensitive cI857 λ repressor present on the same plasmid (Fig. 7-22). In addition, this vector contains a kanamycin resistance gene and an origin of replication from plasmid pACYC177, which belongs to a different incompatibility group than plasmid pBR322 (Chang and Cohen, 1978; *cf.* Section 4.1.5). It

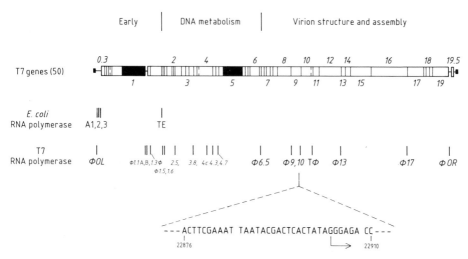

Fig. 7-35. Genetic and physical map of T7 DNA.
The map shows the coding positions of the 50 T7 genes (open boxes). Among those with integral numbers are the T7 RNA polymerase (gene *1*), and the T7 DNA polymerase (gene *5*) (black bars). Also indicated are the positions of *E. coli* RNA polymerase promoters, (*A1, 2, 3*), of the 17 T7 promoters, and of the corresponding terminators *TE* and *TΦ*. The sequence section is from the strong T7 Φ10 promoter used for the construction of plasmid pT7-1. The 23 bp conserved sequence is set off from adjacent sequences by a space; numbers are from the sequence of Dunn and Studier (1983).

therefore, permits the coexistence of pGP1-2 with the ColE1-derived origin of the second component, plasmid pT7-1. This pBR322 derivative contains the strong T7 Φ10 promoter (Fig. 7-36) upstream from a polylinker into which the gene to be cloned can be inserted. At 30 °C the system is repressed by the λ cI857 repressor. Upon temperature induction, however, the synthesis of the T7 RNA polymerase is induced, and this in turn initiates transcription of the gene inserted into pT7-1 and driven by the T7 Φ10 promoter. A problem may arise with proteins toxic for *E. coli*,

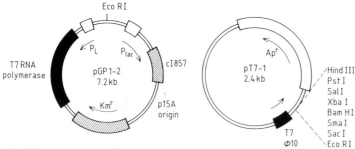

Fig. 7-36. A coupled T7 RNA polymerase/promoter system.
Vector pGP1-2 is composed of the T7 RNA polymerase gene (gene *1*; positions 3106-5840) driven by the λ P_L promoter (from plasmid pKC30; Fig. 7-23), a kanamycin resistance gene, the *p15A* origin of DNA replication from plasmid pACYC177 (*cf.* Fig. 4.1-14; Chang and Cohen, 1978), and the mutated λ repressor gene *cI857*. Plasmid pT7-1 contains the large *Pvu* II-*Eco* RI fragment of pBR322, a 40 bp *Taq* I-*Xba* I fragment from T7 DNA with the Φ10 promoter (positions 22879-22928), and the polylinker region of pUC12 (Fig. 7-58). In pT7-2 (not shown) the orientation of this polylinker region is reversed with respect to the T7 promoter (Tabor and Richardson, 1985).

since the cI857 repressor does not repress the P_L promoter tightly enough to shut off the synthesis of T7 RNA polymerase completely. A further reduction of the uninduced level of T7 RNA polymerase might be obtained in this case by the introduction of a terminator for *E. coli* RNA polymerase into pGP1-2 to prevent the synthesis of transcripts several times the length of the plasmid.

7.2 Ribosome Binding Sites

Initiation of translation, *i.e.*, of protein biosynthesis, requires a ribosome binding site. The efficiency of translation is affected by the primary and secondary structures of mRNAs in the region of the binding site of the 30S ribosome subunit. Shine and Dalgarno (1975) have compared the ribosome binding sites in the start codon regions of various genes. They hypothesised that the specificity of ribosome binding is due to base pairing of a sequence at the 3' end of the 16S ribosomal RNA with a purine-rich sequence immediately preceding the first AUG codon of mRNAs (Fig. 7-37). This hypothesis is now supported by a large body of biochemical and genetic data (Gold *et al.*, 1981). The Shine-Dalgarno sequence is four to nine bases in length and is positioned three to eleven bases upstream from the initiation codon. Shine-Dalgarno (S/D) sequences together with initiation codons AUG or GUG form the ribosome binding site (RBS). Many ribosome binding sites also contain the sequence PuPuPuUUUPuPu which is thought to be recognised by ribosomal protein S1. This sequence occurs predominantly in ribosome binding sites of mRNA molecules coding for proteins which are expressed at high levels, such as ribosomal proteins L11, L12, S12, and the capsid

proteins of phages ΦX174, fd, Qβ, R17, MS2, etc., but is not highly conserved *i.e.*, it usually does not occur as such in ribosome binding sites. The following rules for the primary structure of a ribosome binding site are based on a comparison of the start sequences of 124 genes (Scherer *et al.*, 1980; Stormo *et al.*, 1982):

Rule I:
The typical primary structure is $AGGN_{6-9}ATG$, with N signifying any of the four nucleotides.

Rule II:
G should not occur at position "-3" (A in the ATG codon occupies position "+1"; the position of the next distal nucleotide, *i.e.*, towards the 5' end, occupies position "-1" (Fig. 7-37)).

Rule III:
Less than two G residues should occur between positions "-1" and "-7".

Rule IV:
Positions "+5" and "+10" should be occupied by either A or T.

It should be noted that, as a corollary of rule IV, the sequence in the coding region is also important for efficient translation. Indeed, Taniguchi

Hermann Bujard
Heidelberg

16S rRNA	3' _{OH} A U U C C U C C A C U - - - 5'
complementary strand	5' T A A G G A G G T G A - - - 3'
lac Z	5'- - - C A C A C <u>A G G A</u> A A C A G C T <u>A T G</u> A C C A T G A T T A C G - - - 3'
trp Leader	5'- - - A A A <u>A A G G G</u> T A T C G A C A <u>A T G</u> A A A G C A A A T T T T - - - 3'
L7/L12	5'- - - T T C <u>A G G A</u> A C A A T T T A A A <u>T G</u> T C T A T C A C T A A A - - - 3'
QB - Replicase	5'- - - A C T <u>A A G G A</u> T G A A A T G C <u>A T G</u> T C T A A G A C A G C A - - - 3'

Start codon positions: -15   -10   -5   -1 +1   +5

Fig. 7-37. Primary structure of translation initiation sites.
The top line shows the sequence of the 3' end of *E. coli* 16S rRNA. The DNA sequence below is that of the corresponding complementary strand which is the prototype of a Shine-Dalgarno sequence. Shown also are the structures of initiation sites of four selected genes, namely β-galactosidase (*lacZ*), *trp* leader peptide, ribosomal proteins L7/L12, and bacteriophage Qβ replicase. The sequences are aligned with respect to the position of their ATG codons. Start codons and regions of homology with 16S rRNA are underlined.

and Weissmann (1978) have shown that the efficiency of ribosome binding at the beginning of the Qβ coat protein gene can be increased by a factor of three if the sequence in the region of the start codon is changed from 5'-AUGG-3' to 5'-AUGA-3' (*cf.* also Fig. 12-17). Although the initiation sequence 5'-AUGA-3' has a higher affinity to the anticodon and its 5' neighbouring

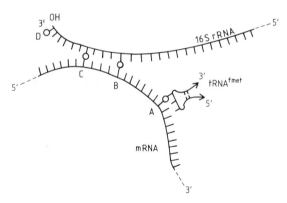

Fig. 7-38. Initiation complex of 16S rRNA, mRNA, and tRNA^{fmet} at the time of initiation of protein biosynthesis.
Position A describes the interaction between the 5' A residue of the start codon AUG with the anticodon. Positions B, C, and base pairs located between these positions represent the interactions between mRNA and 16S rRNA postulated by Shine and Dalgarno. D signifies the 3' end of 16S rRNA. The typical distances between positions A, B, C, and D are summarised in Table 7-1 (Stormo *et al.*, 1982).

nucleotide (3'-UACU-5') of formylmethionyl tRNA, no general rules could be derived from this single observation.

The relative distance between the S/D sequence and the AUG initiation codon seems to be as important as the primary structure of this region. Several general rules for this parameter were also derived from the analysis of the 124 initiation sequences mentioned above. Fig. 7-38 shows a model for the initiation complex consisting of mRNA, tRNA^{fmet} and 16S rRNA, in accordance with the Shine-Dalgarno hypothesis. As can be seen from Table 7-1, which summarises the observed interactions, the standard interaction between the S/D sequence and 16S rRNA requires 5 base pairs. The distance between the

Table 7-1. Configuration of ribosomal binding sites.

Coordinates	Mean distance (nucleotides)	Standard deviation (nucleotides)
A – B	6.9	2.0
A – C	11.9	2.1
A – D	15.1	2.2
B – C	5.0	1.2
B – D	8.1	1.5
C – D	3.1	1.6

The distances refer to coordinates given in Fig. 7–38. They are derived from a statistical analysis of 124 ribosomal binding sites (Stormo *et al.*, 1982).

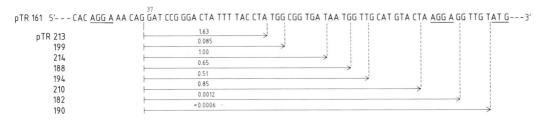

Fig. 7-39. Structures of a set of nine plasmids with a *lac* S/D region fused to different deletions of the λ *cro* gene.
All deletions start shortly behind the *lac* S/D sequence at position 37 of *lac* mRNA, *i.e.*, at the *Alu* I site (*cf.* also Fig. 7-7). Only the start codon of the *cro* gene is shown. S/D sequences and start codons are underlined. Numbers refer to the amount of Cro protein expressed as the percentage of the total cellular protein (Roberts *et al.*, 1979a).

S/D sequence and the start codon is approximately 7 nucleotides and complies with rule I. These "geometrical" parameters have also been intensively studied experimentally. Roberts *et al.* (1979a), for example, have fused DNA sequences comprising the leader and the S/D sequence of the *lacZ* gene with the *cro* gene of phage λ and varied the distance between these two fusion partners (Fig. 7-39). The nine plasmids shown in Fig. 7-39 only differ by their 5'-terminal leader sequences preceding the *cro* gene. The starting plasmid pTR161 codes for 36 nucleotides of the *lac* mRNA leader and 58 nucleotides of λ mRNA. All deletions begin immediately after the S/D sequence at position 37 of the *lac* mRNA. Six of the nine plasmids contain an intact *cro*-S/D sequence and the distance to the *cro* initiation codon is the same. Although all plasmids direct the synthesis of large amounts of Cro protein, expression varies by a factor of up to 1 000. This suggests that the distance between promoter and ATG codon, *i.e.*, the length of the leader sequence, influences the efficiency of expression. In two plasmids, pTR182 and pTR190, *cro* expression is controlled by the *lac* S/D sequence. Extremely little Cro protein is made and presumably the reason is the suboptimal distance between *lac* S/D and ATG-*cro* (10 nucleotides in pTR182 and 5 in pTR190).

Of particular interest are fusions between S/D sequences and eukaryotic genes, since such genes lack S/D sequences. SV40 small t-antigen (Fig.

8-12) has been particularly well studied and was fused in a series of experiments with *lac* and λ promoter sequences (Roberts *et al.*, 1979b; Thummel *et al.*, 1981; Derom *et al.*, 1981). Small t-antigen is 174 amino acids in length and presumably plays an important role for the transformation of non-permissive cells with SV40 (Volckaert *et al.*, 1978). The primary structures of various ribosome binding sites involving fusions with t-antigen are shown in Table 7-2. Plasmids pTR436 and HP1 only differ in one dinucleotide. Although the distance between the *S/D* sequence and the ATG codon is identical in these two plasmids, their t-antigen expression differs by a factor of approximately twenty. Even larger variations were observed with plasmids in which t-antigen expression is controlled by λ promoters. Expression of t-antigen from plasmids pPLcSVt5-374 and -72, for example, differs by a factor of 100. In these two cases the primary structures could not easily be correlated with expression. Secondary structures of mRNAs therefore have been postulated to play an important role in the initiation of protein biosynthesis. It has been known for some time that the small RNA phages possess ribosome binding sites at the beginning of individual genes which meet all requirements of the Shine-Dalgarno hypothesis and yet are inactive *in vivo*. Fiers and associates have succeeded in formulating models for the secondary structures of initiator regions for pro-

Table 7–2.   Production of SV40 t-antigen from different plasmid constructions.

Production of t-antigen (% of total protein)	Distance S/D-ATG (nucleotides)	Sequences of translation initiation region	Plasmid
0.068	9	AGGAAACAGAAAGATGGAT	pTR436
1.0	9	AGGAAACAGCCAGATGGAT	HP1
0.01	5	GTCGAGGAATTCCATGGAT	pPLcSVt5–372
0.1	5	ATTGGAATTCCATGGAT	pPLcSVt5–37
2.5	8	TTGGAATTATTCCATGGAT	pPLcSVt5–379
1.0	9	TTGGAATTAATTCCATGGAT	pPLcSVt5–374
0.01	9	AGGAATTCCAAAGATGGAT	pPLcSVt5–72

This table displays a structural and functional comparison of different recombinant plasmids which carry the SV40 t-antigen gene under the control of either *lac*- or λ derived sequences. The S/D-sequences in pTR436 (Fig. 7–59) and HP1 (Fig. 7–61) are derived from *lac*Z; in the other cases they orginate from regions around the pBR322 EcoRI-site. These sequence stretches display partial homology to the S/D consensus sequence (Gheysen *et al.*, 1982).

tein synthesis. They have shown that efficient translation requires an easily accessible AUG start codon residing in a single-stranded rather than a double-stranded region (Iserentent and Fiers, 1980; Gheysen *et al.*, 1982). The secondary structures of the mRNA molecules listed in Table 7-2 confirm the hypothesis that expression is always comparatively low if the AUG codon is part of a double-stranded region (Fig. 7-40). The substitution of AA for CC in plasmid HP1, for example, exposes the start codon by bringing it into an easily accessible single-stranded region, and hence expression is higher in HP1 than in pTR436 (Fig. 7-40A and B). The differences between pPLcSVt5-374 and -72 can be explained in a similar way (Fig. 7-40C and D). More recent work has established that translation also is reduced if parts of the S/D sequence are involved in mRNA secondary structure (Hallewell *et al.*, 1985; Buell *et al.*, 1985). The latter interactions, in general, derive from complementarity between the S/D region and the initial codons of the gene to be expressed. It is obvious now that nucleotide sequences at the beginning of the coding regions of a foreign gene have to be taken into consideration for the design of suitable leader sequences for optimal gene expression (*cf.* also Rule IV above).

The secondary structures of mRNA molecules can be predicted by following the rules described by Tinoco *et al.* (1973) and Boser *et al.* (1974) which allow the thermodynamic stability of RNA secondary structures to be estimated. According to these rules, double-stranded stems of hairpin structures enter the calculations as stabilising factors while bulges or interruptions within these stem structures are destabilising. The amounts of free energy, ΔG (25 °C), contributed by these single structural elements towards the secondary structure of an RNA molecule is shown in Table 7-3. The example given in Fig. 7-41 demonstrates how the contributions of individual elements add up to provide a general impression of the stability of a secondary structure. Although such calculations appear to be relatively straightforward, the researcher is not saved the trouble of formulating a secondary structure of the regions in question in the first place. This requires a good deal of experience and/or a suitable computer program (Mount, 1984; Martinez, 1984).

Stable secondary structures in the region of the initiator AUG which reduce the efficiency of translation initiation can be eliminated by the introduction of appropriate mutations. Often this can be accomplished by *in vitro* mutagenesis randomising the nucleotides preceding the start

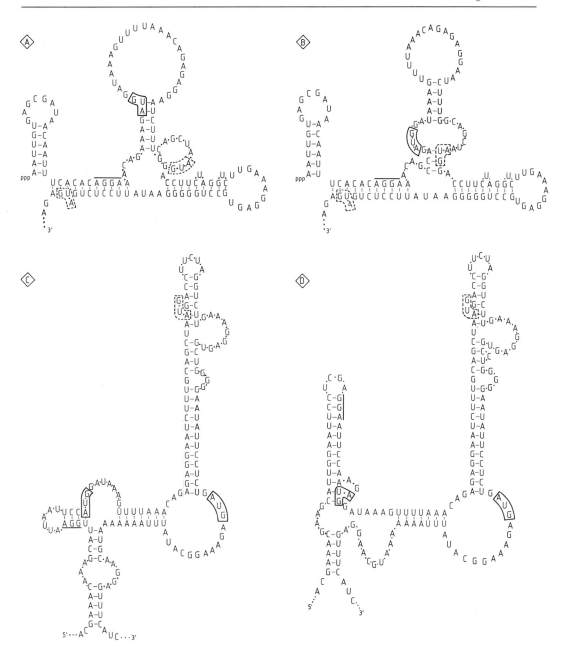

Fig. 7-40. Secondary structures of mRNA sequences near the start codon of small t-antigen in different *lac* or λ plasmids.
Shown are sequences of plasmids pTR436 (A) (Roberts *et al.*, 1979b), HP1 (B) (Thummel *et al.*, 1981), pPLcSVt5-374 (C), and pPLcSVt5-72 (D). (*cf.* also Table 7-2) (Gheysen *et al.*, 1982). S/D sequences are under- or overlined.

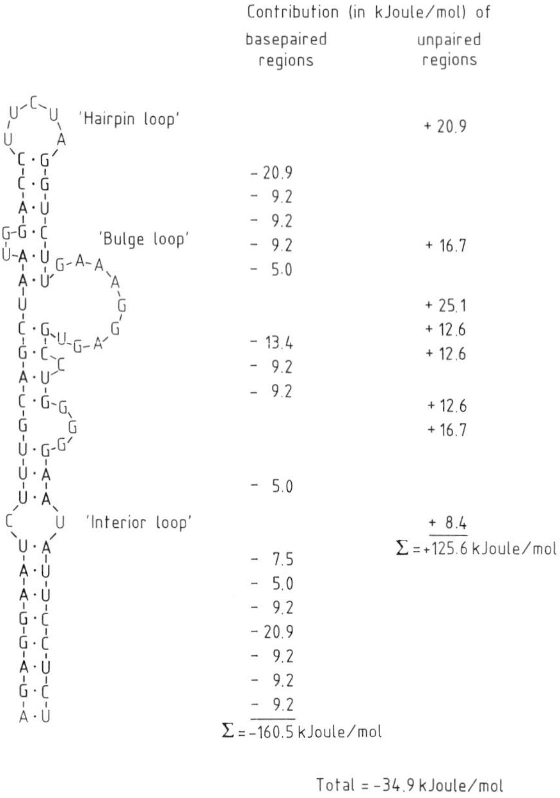

Contribution (in kJoule/mol) of

basepaired regions	unpaired regions

'Hairpin loop'   + 20.9

− 20.9
− 9.2
− 9.2
'Bulge loop'   − 9.2   + 16.7
− 5.0

+ 25.1
+ 12.6
− 13.4   + 12.6
− 9.2
− 9.2

+ 12.6
+ 16.7

− 5.0

'Interior loop'   + 8.4

Σ = +125.6 kJoule/mol

− 7.5
− 5.0
− 9.2
− 20.9
− 9.2
− 9.2
− 9.2

Σ = −160.5 kJoule/mol

Total = −34.9 kJoule/mol

Fig. 7-41. Contribution of various structural elements such as base pairs, hairpin loops, bulge loops, and interior loops to the total free energy of RNA secondary structure.
The structure shown is a postulated secondary structure for the 5′ end of a mRNA encoded by pPLcSVt5-72. The sum obtained from all individual contributions is −34.9 kJoule/mol. The entire structure is shown in Fig. 7-40D.

codon and/or the silent third positions of the initial codons within the coding region (Tessier *et al.*, 1984; Buell *et al.*, 1985; Hallewell *et al.*, 1985).

Another approach, described by Schoner *et al.* (1984), employs a two-cistron system to optimise expression of bovine growth hormone (bGH). In plasmid constructions using both the *E. coli* lipoprotein and tryptophan promoters, expression was initially observed to be rather poor. Increased levels of the protein could subsequently be obtained by the insertion of eight codons between the AUG initiation codon and the genuine second codon of bGH (Fig. 7-42, Part A). Since this manipulation alters the amino-terminal region of the protein, a small number of base changes were introduced into the region of the extra codons to create a termination (TAA) and an initiation codon (ATG) as well as a Shine-Dalgarno sequence for the second (bGH) cistron. Two protein products were expected to be synthesised from this construction (Fig. 7-42, Part B), *i.e.*, a seven-amino-acid polypeptide and methionine-bGH as the product of the second cistron.

Table 7–3. Contribution of secondary structure elements to the free energy of mRNA molecules.

Structural elements				Contributions of single elements to the total free energy of on intact molecule (kJ/mol; ±10%)
Basepaired regions				
–A–A– –U̇–U̇–				– 5.0
–A–U– –U̇–Ȧ–	–U–A– –Ȧ–U̇–			– 7.5
–A–C– –U̇–Ġ–	–C–A– –Ġ–U̇–	–A–G– –U̇–Ċ–	–G–A– –Ċ–U̇–	– 9.2
–C–G– –Ġ–Ċ–				–13.4
–G–C– –Ċ–Ġ–	–G–G– –Ċ–Ċ–			–20.9
–G–U– –U̇–Ġ–				– 1.3
–G–X– –U̇–Ẏ–	–X–G– –U̇–Ẋ–			+ 0

Unpaired regions	
Number of bases unpaired	**Interior loops**
2 – 6	+ 8.4
7 – 20	+12.6
m (>20)	+ 4.19 + 8.4 log m
	Bulge loops
1	+12.6
2 – 3	+16.7
4 – 7	+20.9
8 – 20	+25.1
m (>20)	+16.7 + 8.4 log m

	Hairpin loops	
	closed by $G \cdot C$	closed by $A \cdot U$
3	+33.5	>33.5
4 – 5	+20.9	+29.3
6 – 7	+16.7	+25.1
8 – 9	+20.9	+29.3
10–30	+25.1	+33.5
m (>30)	+14.6 + 8.4 log m	+23 + 8.4 log m

The rules are derived from thermodynamic studies on the stability of oligoribonucleotides of defined structure (Tinoco et al., 1973). The free energies (at 25 °C) for the basepaired regions refer to the free energy of adding a base pair to a preexisting helix; the magnitude thus depends on the sequence of two base pairs. Basepairing is indicated by dots. X and Y stand for different bases; m = number of unpaired bases.

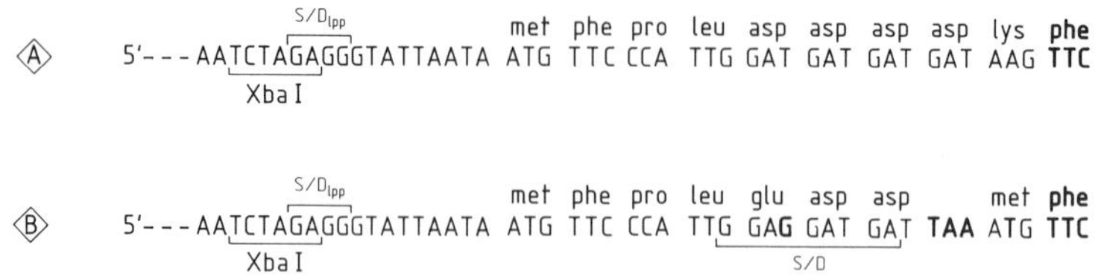

```
              S/D_lpp                  met phe pro leu asp asp asp asp lys phe
(A)   5'---AATCTAGAGGGTATTAATA ATG TTC CCA TTG GAT GAT GAT GAT AAG TTC
          XbaI
```

```
              S/D_lpp                  met phe pro leu glu asp asp     met phe
(B)   5'---AATCTAGAGGGTATTAATA ATG TTC CCA TTG GAG GAT GAT TAA ATG TTC
          XbaI                                            S/D
```

Fig. 7-42.   A two-cistron construction for the expression of Methionyl-bovine growth hormone (bGH). (A) shows an expression vector with an S/D sequence from the *E. coli lpp* gene, and the *N*-terminal sequence of a fusion protein of bGH with eigth additional *N*-terminal amino acids. The phenylalanine (Phe) residues towards the right in both (A) and (B) represent the first residue of mature bGH. Construct (B) displays a set of mutations (bold-face) which introduce a second cistron coding for Met-bGH (Schoner *et al.*, 1984).

With a yield of approximately 19% of the total cell protein the two-cistron construction (Fig. 7-42, Part B) was only slightly less active than the parent construct, which yielded a fusion protein accounting for 30% of the total cell protein. This two-cistron structure may be functionally analogous to that of certain late genes of bacteriophage λ and that of genes of the single-stranded bacteriophages, for example MS2. In the latter case, the expression of the replicase gene depends on the translation of the preceding coat protein cistron (Kastelein *et al.*, 1983). In bacteriophage λ, a variety of late genes, which are expressed from a polycistronic mRNA, overlap in the nucleotide sequence 5'-ATGA-3', which contains the termination codon of the upstream gene and the initiation codon of the downstream gene (Kröger and Hobom, 1982). Ribosomes initiating at the first cistron of such an mRNA disrupt potentially inhibitory secondary structures of the second cistron mRNA while moving along the mRNA; furthermore, they may not even fall off from the message if termination and initiation signals are overlapping or positioned in close proximity, and thereby increase the efficiency of translation.

7.3 Codon Selection

The genetic code comprises 64 codons: 61 of these specify a total of twenty amino acids; while three codons, UAA (ochre), UAG (amber) and UGA (opal) are used as termination signals (Fig. 7-43). Only tryptophan and methionine are coded for by single corresponding triplets. Each of the other eighteen amino acids are specified by at least two

Walter Fiers
Gent

Ala	Arg	Asp	Asn	Cys	Glu	Gln	Gly	His	Ile
GCA	AGA	GAC	AAC	UGC	GAA	CAA	GGA	CAC	AUA
GCC	AGG	GAU	AAU	UGU	GAG	CAG	GGC	CAU	AUC
GCG	CGA						GGG		AUU
GCU	CGC						GGU		
	CGG								
	CGU								

Leu	Lys	Met	Phe	Pro	Ser	Thr	Trp	Tyr	Val	STOP
UUA	AAA	AUG	UUC	CCA	AGC	ACA	UGG	UAC	GUA	UAA
UUG	AAG		UUU	CCC	AGU	ACC		UAU	GUC	UAG
CUA				CCG	UCA	ACG			GUG	UGA
CUC				CCU	UCC	ACU			GUU	
CUG					UCG					
CUU					UCU					

Fig. 7-43. Genetic code.
Amino acids appear in alphabetical order with corresponding codons listed below. All codons begin with the 5′ terminal nucleotide.

codons. Indeed, arginine, leucine, and serine have six codons each. Codons that specify the same amino acid are called synonymous.

The first sequence data, obtained from the RNA bacteriophages, had already suggested that synonymous codons are not used with equal frequencies in a given genome (Min Jon *et al.*, 1971). This uneven distribution was later observed to occur even within different genes of a single organism, and to correlate with the extent of gene expression. Several explanations and rules for optimal codon usage were subsequently formulated by comparing the frequencies of codons in frequently expressed genes (present only once on the E. coli chromosome but expressed in up to 5×10^5 copies of mRNA) with those of codons used in weakly expressed genes (Grosjean and Fiers, 1982). Two situations are generally discussed in which codon choice is manifested, i.e., in the interaction of the unique anticodon of a single charged tRNA with synonymous codons, and in the existence of tRNA populations with different anticodons charged with the same amino acid. Let us consider first the case relating to the use of synonymous, i.e., different codons for the same amino acid read by a single charged tRNA. In this case codon selection is influenced by the strength of the interaction between codon and anticodon, which is variable since the third base in

a codon "wobbles". Such interactions should neither be too strong nor too weak and should not vary abruptly from codon to codon in order to guarantee optimal expression of an mRNA molecule. If, for example, the nucleotides A and/or U occupy the first two positions in a codon and hence base pairing is comparatively weak, the interaction between codon and anticodon could be improved if the third base were C. In contrast, a stronger interaction between anticodon and codon due to the presence of G and/or C residues in the first two positions could be modified by the destabilising effects of U as the third base. As far as codons for the amino acids phenylalanine (113 occurrences of UUC as opposed to 39 occurrences of UUU), isoleucine (262 AUC versus 67 AUU), tyrosine (98 UAC versus 34 UAU), and asparagine (159 AAC versus 13 AAU) are concerned, strongly expressed genes appear, indeed, to have a clearcut preference for NNC rather than NNU codons; on the other hand, NNU appears more often than NNC in the codons for proline (21 CCU versus 2 CCC), alanine (173 GCU versus 48 GCC), and glycine (226 GGU versus 174 GGC). These values have been obtained from an analysis of the *E. coli* genes for RNA polymerase, several ribosomal proteins, elongation factors for protein synthesis, and membrane proteins. All these proteins occur

Table 7–4. Comparison of codon usage in strongly and weakly expressed genes of *E. coli*.

		U			C			A			G		
		strong	weak		strong	weak		strong	weak		strong	weak	
U	Phe	39	151	Ser	93	36	Tyr	34	96	Cys	13	34	U
	Phe	113	102	Ser	87	49	Tyr	98	65	Cys	23	39	C
	Leu	12	71	Ser	6	37	ochre			opal			A
	Leu	16	64	Ser	12	62	amber			Trp	25	66	G
C	Leu	26	73	Pro	21	29	His	19	95	Arg	223	99	U
	Leu	33	69	Pro	2	46	His	75	59	Arg	101	133	C
	Leu	→ 3	22	Pro	26	45	Gln	38	90	Arg	→ 3	27	A
	Leu	345	294	Pro	162	101	Gln	169	166	Arg	→ 1	42	G
A	Ile	67	156	Thr	103	46	Asn	13	101	Ser	10	56	U
	Ile	262	118	Thr	137	119	Asn	159	98	Ser	49	61	C
	Ile	→ 2	27	Thr	15	32	Lys	259	163	Arg	→ 3	28	A
	Met	140	130	Thr	28	76	Lys	106	44	Arg	→ 1	17	G
G	Val	192	108	Ala	173	87	Asp	116	183	Gly	226	124	U
	Val	41	66	Ala	48	178	Asp	204	106	Gly	174	140	C
	Val	119	48	Ala	119	107	Glu	333	210	Gly	→ 4	42	A
	Val	83	123	Ala	129	149	Glu	106	98	Gly	→ 14	66	G

The strongly expressed genes represent 24 mRNA species and a total of 5 253 codons. Among these are the genes *rpo*A-D for RNA polymerase, the genes for 12 ribosomal proteins, several outer membrane proteins, and protein synthesis initiation and elongation factors. Weakly expressed genes are represented by 18 mRNAs with 5 231 codons. Among these are several repressor genes, genes for *lac*-permease and the λ-Int protein as well as those of the transposase and the β-lactamase of transposon Tn3. Codons which are read by a single tRNA only and the choice of which is dependent on the nature and strength of codon/anticodon interactions are boxed. Arrows point to codons which are used only rarely and thus may possess modulating properties (Grosjean and Fiers, 1982).

in at least 1 000 and usually in more than 10^5 copies per cell (*cf.* Grosjean and Fiers, 1982) (Table 7-4).

The observed preferences for the usage of different codons also are mirrored in the usage of anticodon sequences. The codons for asparagine, AAC, and AAU, for example, are read by a tRNA with a QUU anticodon, where Q is a G derivative and, therefore, interacts much better with the C wobble base of AAC than with U in AAU. In the case of the preferred codon for glycine (GGU), however, the anticodon sequence is GCC. Since the first two bases already form two stable G/C pairs, a mismatch (G/U) is preferred at the position of the wobble base. It is interesting to note that the observed codon preferences in weakly expressed genes (such as repressor genes) are just the reverse, although the effects are not as distinctive as is the case for strongly expressed genes. It is this observation which supports the hypothesis that the strength of anticodon-codon interaction must be extremely finely regulated in order to achieve the proper efficiency of translation. These rules are much less clear for viral RNAs, due to the interfering and overshadowing influences of other processes such as replication and packaging which require particular sequences.

The second manifestation of codon usage preference occurs in the selection of synonymous codons by different tRNA species charged with the same amino acid. Such a population of tRNAs is known as a family of "iso-accepting" tRNAs. As would be expected for strongly expressed genes, there is a clear correlation between the relative amounts of different iso-accepting tRNA species and the use of corresponding codons (Ikemura, 1981). In other words, the higher the concentration of a particular iso-accepting tRNA, the more often the corresponding codon appears in the sequence of the strongly expressed gene; on the other hand, this means that translation may be modulated and controlled by rare codons, for which the corresponding tRNAs occur in trace amounts only. Such codons may be AUA, coding for isoleucine, CUA (leucine), CGG and CGA (arginine), and GGA (glycine). These codons are marked by an arrow in Table 7-4, and, indeed, they are hardly used at all. It is not known whether organisms really use this mechanism to control gene expression.

It should be noted here that the choice of a codon for which there is a limited supply of a corresponding charged tRNA would inevitably cause an imbalance in the tRNA population of an organism. This in turn would not only slow down translation but would also make the system more prone to errors. One may imagine, for example, a competition between correct and false tRNAs at the ribosome A site prepared for the entry of an aminoacylated tRNA. If, due to its low concentration, the correct tRNA were too slow to interact, the false tRNA would associate with the ribosome and, hence, a false amino acid would be incorporated into the growing polypeptide chain. This process may even be associated with alterations of the reading frame if the structure of the false tRNA prevents the proper entry of the next tRNA, and this has, indeed, been observed for several suppressor tRNAs. It is the basis of the phenomenon known as frameshift suppression. Suboptimal translation conditions of this kind have been artificially induced *in vivo* by starving bacterial cells for certain amino acids or *in vitro* by the addition of certain tRNAs to cell-free systems (Roth, 1981; Weiss und Gallant, 1983).

The significance of an appropriate codon choice for the expression of foreign genes in heterologous organisms has never been convincingly documented; nevertheless, especially since other unknown parameters may affect heterologous gene expression, the rules mentioned in this section should be followed as closely as possible in order to approach natural conditions. For chemically synthesised genes, for example, codons should be selected in accordance with the frequencies with which such codons occur in the desired host organism (*cf.* Section 11.2.2.1).

7.4 Construction of Expression Vectors

Several strategies using regulatory sequences discussed in the preceding sections have been pursued to optimise the expression of genes. In principle, these strategies are aimed at the construction of vectors allowing the synthesis either of fusion proteins comprising vector and insertion sequences (Fig. 7-44A) or of pure proteins exclusively encoded by the insertion (Fig. 7-44B). The first construction is referred to as a translational fusion, the second as a transcriptional fusion. The following selected examples will clarify this distinction.

7.4.1 Synthesis of Fusion Proteins

In order to obtain a hybrid protein, the foreign DNA must be inserted into an expressable vector gene in such a way that the reading frame in this

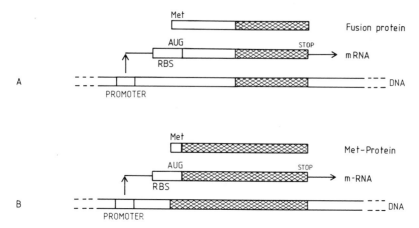

A

B

Fig. 7-44. Construction of expression vectors.
Two approaches are shown, namely the formation of fusion proteins (A), and the formation of native proteins (B) from recombinant DNA. RBS signifies a ribosomal binding site. Met-protein indicates that proteins obtained from recombinant DNA by approach (B) always carry an *N*-terminal methionine residue. Bacterial sequences are represented as open, eukaryotic sequences as hatched bars.

gene is conserved. The synthesis of hybrid mRNA is initiated by the prokaryotic promoter and its translation is controlled by the corresponding ribosome binding site. The first practical application of fusion proteins allowed the expression of rat insulin, rat growth hormone, and human growth hormone, and demonstrated for the first time that bacteria are, indeed, capable of expressing eukaryotic coding sequences.

7.4.1.1 Expression of Rat Insulin

The starting point in this case was the insertion of a rat insulin cDNA into the *Pst* I site of pBR322 by homopolymeric poly(dG)-poly(dC) tailing (Villa-Komaroff *et al.*, 1978). The variable lengths of these tails guaranteed that at least one in three clones contained the right reading frame; however, since the cDNA could be inserted in two

Howard M. Goodman
Boston

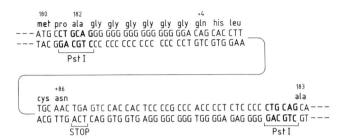

Fig. 7-45. Insertion of the rat insulin DNA sequence into the *Pst* I site of the β-lactamase gene of pBR322. The two *Pst* I sites, and amino acids 182 and 183 of β-lactamase, which are separated by the insertion, are printed in bold-face. The insulin insertion begins with amino acid Gln (position +4) in the B chain, and ends with asparagine 86 of the proinsulin. The order of the insulin peptides is pre-B-C-A. (Villa-Komaroff *et al.*, 1978).

different orientations, only one sixth of the clones containing the desired insulin insertion would also make insulin. In spite of these obvious limitations, cloning by homopolymeric tails was the method of choice because the exact sequence of the cDNA was not known and the desired constructions therefore, could not be planned in advance (*cf.* also Section 3.2). The structure of one the rat insulin clones is shown in Fig. 7-45. Starting with position 182 (ala) the sequence of the β-lactamase gene then proceeds with polyglycine and eventually reaches the insulin sequence at amino acid "+4" (gln) of proinsulin. The desired fusion protein was detected by immunological techniques (see Section 11.2.3.2).

7.4.1.2 Expression of Rat Growth Hormone and the Structural Protein VP1 of Foot and Mouth Disease Virus

A much more direct strategy was pursued for the construction of vectors coding for rat growth hormone (Seeburg *et al.*, 1978). The rat growth hormone cDNA possesses a single *Pst* I site at position "-24" of the prepeptide region, which allowed it to be annealed with the *Pst* I site of the β-lactamase gene of pBR322 in such a way that the reading frame was conserved (Fig. 7-46); in addition, the strategy employed for the construc-

tion of this expression vector also allowed a direct selection for clones containing the desired recombinant.

The starting material was a cDNA cloned into the *Hind* III site of pBR322. The insert was excised and recloned into the *Hind*III site of plasmid pMB9 which lacks the β-lactamase gene. The resulting plasmid, pMB9-RGH, expressed only low levels of tetracycline resistance since the *Hind*III site is located within the tetracycline promoter region (Fig. 7-47; *cf.* also Fig. 4.1-10). Tetracycline resistance was restored and the coding sequence of the rat growth hormone gene brought under the control of the β-lactamase promoter by replacing a small *Pst*I-*Bam*HI fragment of pMB9-RGH with a corresponding fragment of pBR322 (*cf.* also Fig. 7-46). This expression vector could be distinguished from pMB9-RGH by its increased tetracycline resistance (20 µg/ml instead of 5 µg/ml). The hybrid gene in this expression vector coded for a chimaeric protein

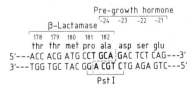

Fig. 7-46. DNA sequence in the vicinity of the *Pst* I site of a hybrid vector containing a fusion of the β-lactamase gene with the gene for rat growth hormone.

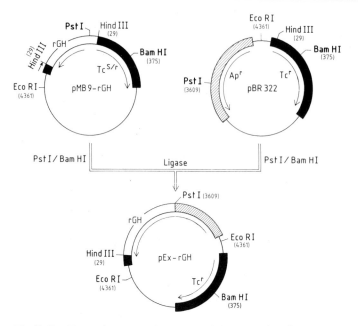

Fig. 7-47.   Expression vector for rat growth hormone (rGH).
The expression plasmid pEx-rGH is constructed by replacing the small *Pst*I-*Bam*HI fragment of plasmid pMB9-rGH by the smaller *Pst*I-*Bam*HI DNA fragment of pBR322. Numbers in brackets refer to co-ordinates in plasmid pBR322. Arrows indicate the direction of transcription or translation. In pMB9-rGH, the promoter of the tetracycline resistance region is interrupted by the rGH insertion; expression of tetracycline resistance is therefore markedly reduced. Transformants are resistant to 5 μg tetracycline/ml, while full expression in pBR322 or the expression vector pEx-RGH allows selections with 20 μg/ml. Tc^r regions are represented as black, Ap^r regions as hatched, and rGH regions as open bars. (Seeburg *et al.*, 1978).

395 amino acids in length. It comprises 181 *N*-terminal amino acids derived from the β-lactamase gene and 214 *C*-terminal amino acids of the pre-growth hormone and was, indeed, detected as a protein with a molecular weight of 46 000 in a mini-cell test system (see Section 11.2.3.3); however, the amount of hybrid protein was only one-fifth the amount of β-lactamase produced by pBR322.

Another example is the synthesis of a fusion protein containing a part of protein VPI of Foot and Mouth Disease Virus (Fig. 7-48; Küpper *et al.*, 1981). In this case, cloning started with the insertion, into vector pPLc24 cut with *Bam*HI and *Hind*III, of an 849 bp *Bam*HI-*Hind*III fragment coding for amino acids 9 to 292 of the desired protein (Fig. 7-28). The fusion protein obtained

was 395 amino acids in length and consisted of 98 N-terminal amino acids of MS2 replicase, 284 amino acids of the desired viral protein, and thirteen plasmid-derived amino acids added because of read-through into neighbouring vector sequences.

7.4.1.3 Expression of Human Growth Hormone

Suitable restriction sites are rarely positioned such that they are located at the beginning of a structural gene and also allow this gene to be inserted into the vector gene in the correct reading frame. Quite frequently it is necessary to design special constructions in which linker molecules play an important role. The following

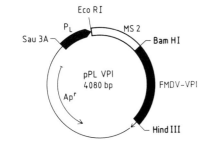

The expression plasmid pPLVP1 is derived from plasmid pPLc24 which contains the *N*-terminal part of MS2 replicase under the control of λ promoter P_L (Fig. 7-28). The structure of the FMDV cDNA and the position of the VP1 structural protein with flanking *Bam* HI and *Hind* III sites is indicated in the centre. Shown at the bottom are the sequences around the *Bam* HI and *Hind* III site in expression vector pPLVP1. The stop codon is at position 2105 of the pBR322 sequence.

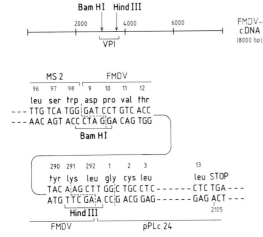

example of a human growth hormone gene (hGH) expressed under the control of the *trp* promoter may illustrate the point (Martial *et al.*, 1979). The starting material in this case was vector ptrpED5-1 (Fig. 7-16) with a *Hind* III site within the codon for amino acid 92 of the *trpD* gene. If this *Hind* III site had been joined with a DNA fragment flanked by a *Hind* III site in the 5′ untranslated region of the cDNA for human growth hormone, the correct reading frame would have been lost (Fig. 7-49). It was therefore necessary to manipulate the *Hind* III site in ptrpED5-1 to shift the reading frame by one base in the recombinant molecule. This was accomplished by filling-in the 5′ protruding ends with Klenow fragment of *E. coli* DNA polymerase I and adding a synthetic DNA decamer which contained a *Hind* III site. As shown in Fig. 7-50, cleavage of the new plasmid ptrpED50 with *Hind* III and subsequent ligation with *Hind* III-cut hGH cDNA conserved the correct reading frame

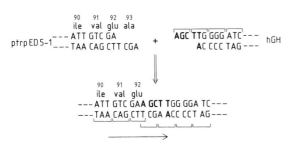

Fig. 7-49. Fusion of a *Hind* III site in plasmid ptrpED5-1 with a *Hind* III-flanked cDNA fragment coding for human growth hormone (hGH).
Brackets indicate the reading frames, the arrow denotes the direction of translation. The hGH section shown is derived from the 5′ untranslated region.

in the recombinant molecule. After induction of transformed bacteria with 3-β-indolylacrylic acid, the expected fusion protein with a molecular weight of 34 kDa (322 amino acids) was obtained with a yield corresponding to 3% of the total cellular protein. This value, however, is only one sixth of that expected from induction of the non-recombinant plasmids ptrpED50 or ptrpED5-1.

7.4.1.4 Expression of Somatostatin

Somatostatin is a peptide hormone consisting of fourteen amino acids, which controls the secretion of a number of other hormones, such as insulin and glucagon. The starting material for cloning and expression of a somatostatin fusion protein was plasmid pBR322 and a synthetic gene with the coding sequence of somatostatin (Itakura *et al.*, 1977). The synthetic gene was pieced together by annealing eight different oligonucleotides (A to H) and contained terminal protrud-

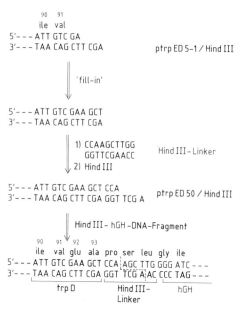

Fig. 7-50. Construction of vector ptrpED50 and cloning of the gene for human growth hormone (hGH).
In order to move the reading frame at the *Hind* III site of ptrpED5-1 by one base, the 5' ends were filled-in and fused with a decameric *Hind* III linker. The fusion peptide contains the first 93 amino acids of the *trpD* protein, three amino acids encoded by the *Hind* III linker, nine amino acids of the 5' untranslated region, and 217 amino acids of pre-hGH (total of 322 amino acids). (Martial *et al.*, 1978).

ing *Eco* RI and *Bam* HI ends, respectively. As shown in Fig. 7-51, an additional methionine codon was introduced directly in front of the *N*-terminal alanine codon; the carboxy-terminal cysteine codon is followed by two stop codons.

Since the somatostatin gene was to be expressed under the control of the *lac* system, the first step was to generate a suitable vector containing the *lac* regulatory region. This was accomplished

Axel Ullrich
San Francisco

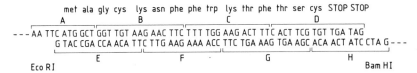

```
          met ala gly cys  lys asn phe phe trp  lys thr phe thr ser cys  STOP STOP
          ┌──────────────┐┌──────────────────┐┌──────────────────────┐┌──────────
               A       B              C                      D
---AA TTC ATG GCT GGT TGT AAG AAC TTC TTT TGG AAG ACT TTC ACT TCG TGT TGA TAG
      G TAC CGA CCA ACA TTC TTG AAG AAA ACC TTC TGA AAG TGA AGC ACA ACT ATC CTA G---
      └────────────┘└──────────────┘└──────────────┘└─────────────────┘
            E              F              G                    H
   Eco RI                                                        Bam HI
```

Fig. 7-51.   A chemically synthesised somatostatin gene.
Shown are the eight oligonucleotides, A-H, used as building blocks, the *Eco* RI and *Bam* HI termini, the amino acid codons, the start codon, and the two stop codons.

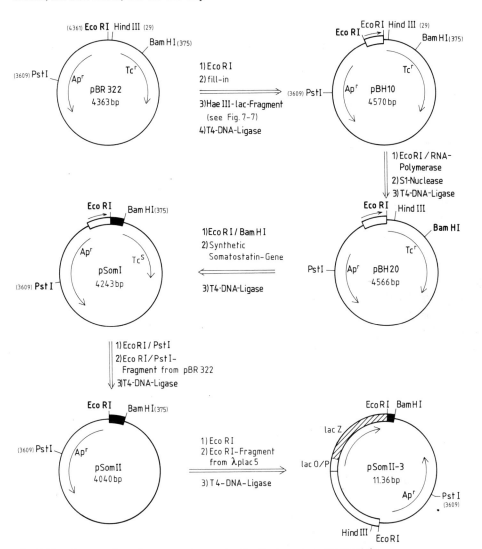

Fig. 7-52.   Construction of an expression vector for the hormone somatostatin.
Open bars represent the *lac* control region; hatched bars the *lacZ* gene and black bars the somatostatin gene. Arrows indicate the direction of transcription.

by partially digesting λ*plac5* DNA with *Hae* III. The resulting mixture of DNA fragments contained a 203 bp *Hae* III fragment coding for the entire *lac* control region and the first seven amino acids of β-galactosidase (*cf.* Fig. 7-7). The mixture of DNA fragments was then ligated with pBR322 DNA which had been linearised with *Eco* RI and subsequently filled-in to convert its protruding 5' ends to flush ends. Ligation of filled-in *Eco* RI ends with blunt *Hae* III ends generated new *Eco* RI ends in the recombinant molecules at the sites of fusion (see also Section 2.1.2.1).

Transformants obtained from this DNA mixture containing the desired *Hae* III fragment were identified as blue colonies on agar plates containing Xgal (Fig. 7-6). This screen did not distinguish between the two possible orientations of the *Hae* III fragment; however, since there was an asymmetrically positioned *Hha* I site directly following the stop codon of the *lacI* gene on the *Hae* III fragment (Fig. 7-7), it was easy to determine the orientation of the inserted *Hae* III fragment. The desired orientation was found in vector pBH10 (Fig. 7-52), in which *lac* transcription proceeds toward the tetracycline resistance region.

An unusual procedure was used to selectively remove the distal *Eco* RI site in pBH10. *E. coli* RNA polymerase binds to promoter regions in the absence of nucleoside triphosphates. In pBH10 binding occurs at the *lac* promoter and

also at the tetracycline promoter region, 20 bp away from the proximal *Eco* RI site (Fig. 7-53). A region spanning approximately 20 bp around a Pribnow-Schaller box is protected by RNA polymerase, blocking further enzymatic attack. Therefore the proximal, but not the distal *Eco* RI site in pBH10 was protected, and only the distal recognition site was cleaved by *Eco* RI in the presence of RNA polymerase. Subsequent digestion with nuclease S1 was used to convert this *Eco* RI terminus into a blunt end. Ligase treatment yielded plasmid pBH20 with only one *Eco* RI site (Fig. 7-52).

Plasmid pBH20 was prepared for the incorporation of the synthetic somatostatin gene by first digesting it with *Eco* RI and *Bam* HI and separating the resulting two fragments by gel electrophoresis. The larger of the two fragments was treated with phosphatase and annealed with the synthetic somatostatin gene. Transformants were selected for ampicillin resistance and screened for tetracycline sensitivity. Fig. 7-54A shows the DNA sequence from the region of the insertion in clone pSomI. This clone should yield a fusion protein of 24 amino acids, the expression of which is controlled by the ribosome binding site of the *lacZ* gene. Suitably induced bacteria were then treated with cyanogen bromide in order to cleave somatostatin from the entire mixture of proteins. Since cyanogen bromide cleaves peptides specifically at the carboxyl group of methionine residues

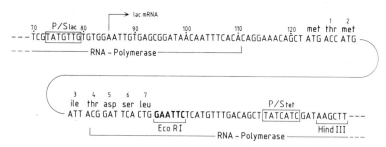

Fig. 7-53. DNA sequences between the *lac* and *tet* promoter regions in plasmid pBH10. The Pribnow-Schaller boxes are framed, the proximal *Eco* RI site of the *tet* promoter and the *Hind* III site are bracketed. Binding sites for RNA polymerase, which extend approximately 35 bp to the left and right of the Pribnow-Schaller boxes, are indicated by brackets. It is apparent that the proximal *Eco* RI site lies in a region protected by RNA polymerase. For the numbering in the *lac* region see legend to Fig. 7-7.

```
(A)
        1   2   3   4   5   6   7
        met thr met ile thr asp ser leu glu phe Met Ala Gly Cys Lys Asn Phe Phe Trp Lys Thr Phe Thr Ser Cys STOP STOP
---CAGGAAACAGCT ATG ACC ATG ATT ACG GAT TCA CTG GAA TTC ATG GCT GGT TGT AAG AAC TTC TTT TGG AAG ACT TTC ACT TCG TGT TGA TAGGATCC---   pSOM 1
   S/D lac                                      Eco RI                                                                 Bam HI

(B)
        1                            1005
        met thr ---1003 amino acids--- glu phe Met Ala Gly Cys Lys Asn Phe Phe Trp Lys Thr Phe Thr Ser Cys STOP STOP
---CAGGAAACAGCT ATG ACC ---3009 bp------- GAA TTC ATG GCT GGT TGT AAG AAC TTC TTT TGG AAG ACT TTC ACT TCG TGT TGA TAGGATCC---   pSOM II-3
   S/D lac                                Eco RI                                                                 Bam HI
```

Fig. 7-54. Nucleotide sequence of Lac-somatostatin fusion genes.
(A) shows a fusion with only seven *N*-terminal amino acids of β-galactosidase, (B) a fusion with 1003 amino acids of β-galactosidase. Both fusions contain two additional amino acids derived from the *Eco* RI site (Itakura *et al.*, 1977).

this procedure should have yielded functional somatostatin; nevertheless, all attempts to detect the hormone in various bacterial extracts were unsuccessful. These negative results probably resulted from proteolytic cleavage of the hormone (*cf.* Section 7.5).

A new plasmid, pSomII, was constructed in the hope that the presence of a large peptide would prevent this proteolytic attack (Fig. 7-52). In this plasmid the smaller *Eco* RI-*Pst* I fragment with the *lac* region of pSomI was replaced by a corresponding *Eco* RI-*Pst* I fragment of pBR322. Transformants were selected for ampicillin resistance and screened on Xgal plates for the absence of *lac* operator DNA. A *lac* region containing the entire control region and the codons for 1 003 out of 1 021 amino acids of β-galactosidase, rather than the first seven amino acids, was then chosen to replace the missing *lac* region. A suitable *Eco* RI fragment of 7.45 kb was obtained from λ*plac5* DNA. As shown in Fig. 7-54B, the correct reading frame was retained in this construction and a fusion protein of 1 020 amino acids with the amino acid sequence of somatostatin at the *C*-terminal end was obtained. When total cellular proteins of suitably induced bacteria were treated with cyanogen bromide, somatostatin activity was indeed detectable. The yield in uninduced cells was estimated to be on the order of 0.001-0.03% of the total protein. This low yield reflects the low basal level of transcription from a fully repressed *lac* promoter. Induction with IPTG led to a three- to sevenfold increase in somatostatin yields. This induction experiment confirmed the DNA sequence data indicating that in pSomII-3 the synthesis of somatostatin was regulated by the *lac* control sequences. However, the induction ratio was approximately tenfold lower than had been

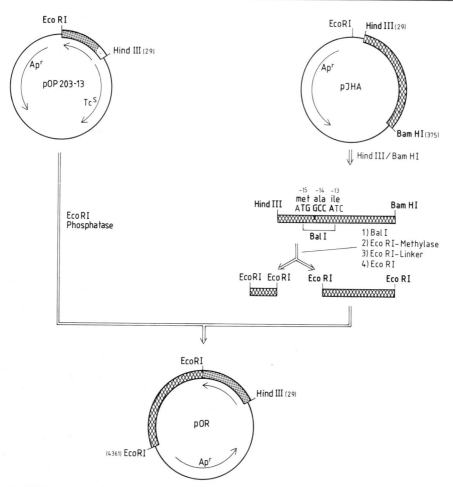

Fig. 7-55. Construction of a plasmid for the expression of the haemagglutinin (HA) gene of human influenza virus type A/Japan/305/57 (subtype H2).
The modifications of the HA DNA fragment which is obtained from pJHA by *Hind* III-*Bam* HI digestion are described in detail in the text. *Lac* control regions are indicated by a stippled bar, HA DNA sequences by a cross-hatched bar (Heiland and Gething, 1981).

expected. Similar observations subsequently have been made with other expression plasmids based on *lac* control elements. There are several possible explanations for this phenomenon, including the selective cleavage of the foreign protein by bacterial proteases, insufficient solubility of the fusion protein during cyanogen bromide cleavage and the instability of the recombinant plasmid.

7.4.1.5 Construction of Expression Plasmids for Influenza Virus Specific Sequences

This case deals with the expression of a DNA copy of an RNA fragment coding for the haemagglutinin (HA) protein of human influenza virus strain A/Japan/305/57 (subtype H2) (Heiland and

Gething, 1981). The vector used was plasmid pOP203-13 (Fig. 7-9) which contains, between the *Eco* RI and *Hind* III sites of pBR322, the same 203 bp of the *lac* control region as pHB10 (Fig. 7-52). The direction of *lac* transcription is anticlockwise, *i.e.*, in the direction of the β-lactamase gene (Fuller, 1982), which means that any DNA inserted into the single *Eco* RI site of this plasmid will be controlled by the *lac* promoter.

The haemagglutinin gene to be expressed was inserted between the *Hind* III and *Bam* HI sites of pBR322 in plasmid pJHA (Fig. 7-55). It codes for the entire 560 amino acids of the haemagglutinin protein and eleven nucleotides of the 5' untranslated region. This sequence must be modified before it can be inserted into the *Eco* RI site of pOP203-13. A *Bal* I site comprising the ATG start codon of the HA gene is important. The DNA fragment obtained by *Hind* III and *Bam* HI digestion is further cleaved by *Bal* I treatment to yield two sub-fragments. The mixture of fragments is first treated with *Eco* RI methylase in order to methylate internal *Eco* RI sites and to render them resistant to *Eco* RI digestion. *Eco* RI linkers are then added by ligation. Following *Eco* RI digestion the sub-fragments are separated from each other and the larger fragment is inserted into the *Eco* RI site of the expression vector pOP203-13 (Fig. 7-55).

Cloning of the large *Bal* I fragment was expected to yield a fusion protein with the structure shown in Fig. 7-56, containing seven *N*-terminal amino acids derived from β-galactosidase, three amino acids coded for by the linker, and 560 amino acids of the haemagglutinin gene. Two *N*-terminal amino acids of the leader sequence of the HA gene were removed by this cloning procedure. Again, the two possible orientations for the inserted gene could be easily distinguished by suitable digestions. Three of the clones obtained expressed antigenic determinants of haemagglutinin, as shown by solid phase radioimmunoassay. The nucleotide sequences of all three clones confirmed that they preserved the correct reading frame, but also showed that they did not have the expected sequence of the hypothetical vector pOR (Fig. 7-56) at the site of fusion. For unknown reasons fifteen amino acids of the signal peptide and the first ten to fifteen amino acids of the mature protein were missing. Perhaps the eukaryotic hydrophobic signal sequences were not tolerated by the *E. coli* host organism in this particular case. By way of contrast, other hydrophobic signal sequences, such as that of human preproinsulin, have been found to be quite stable in *E. coli* (Chan *et al.*, 1981).

7.4.1.6 General Technique for the Construction of Expression Vectors for Fusion Proteins

In the examples discussed so far, a restriction site within the region of the ATG start codon was always positioned in such a way that the insertion into the *lacZ* gene (or another suitable gene) either conserved the correct reading frame or could be easily arranged to fit into frame. In the case of the somatostatin gene this was accomplished by suitably planning the chemical synthesis, while it was mere coincidence in the case of the HA gene. In most cases, however, a convenient restriction site will not be available; the following procedure is therefore recommended for cloning and expression: the DNA to be expressed, for example a cDNA, is cut out and isolated from a parent plasmid. The example shown in Fig. 7-57 uses a *Pst* I digestion. The next step places a suitable restriction site in the vicinity of the ATG start codon. The DNA is first treated with a combination of the enzymes *Exo* III and S1 or with *Bal* 31 (*cf.* also Fig. 2.1-9). Digestion conditions depend on the distance between the original cleavage site (*Pst* I in our example) and the ATG start codon, and must be determined separately for every individual case. In our example, *Eco* RI linkers are then added to the fragment, so that it can be inserted into the *Eco* RI site of a suitable vector. Before the fragment containing the linkers is cloned, it is digested with *Eco* RI and another restriction enzyme which cleaves at site X within

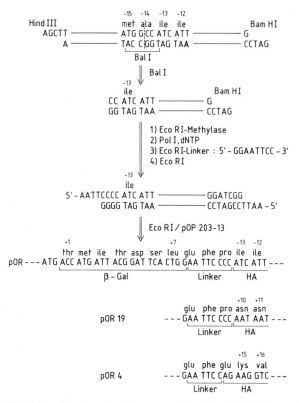

Fig. 7-56. Construction of an expression plasmid by linker technology.
A *Hind* III-*Bam* HI fragment of plasmid pJHA (Fig. 7-55) is cleaved into two fragments by digestion with *Bal* I. Only the flanking sequences of the larger of the two fragments are shown. Following an *Eco* RI methylase treatment, the protruding *Bam* HI site is filled-in in order to allow the subsequent addition of *Eco* RI linkers. A further *Eco* RI digestion only attacks the *Eco* RI sites within the linker, but not the internal methylated *Eco* RI site. This DNA fragment is cloned into the *Eco* RI site of plasmid pOP203-13 (Fig. 7-9). Shown is the expected structure of the fusion protein (pOR) consisting of seven amino acids of β-galactosidase, three amino acids encoded by the linker, and two amino acids out of a total of 560 from the haemagglutinin. The actual experiment did not yield clones with the expected structure; instead, plasmids were obtained, which begin with sequences of the mature HA protein and which do not contain the hydrophobic leader sequence. Part of the structures of two of these plasmids, pOR19 and pOR4, are also shown.

the gene to be cloned. This yields a defined right-hand molecular end which can be used at a later stage to reconstruct the entire gene. Although the left end of the fragment is defined by an *Eco* RI site, the distance between this site and the ATG start codon varies in different molecules. Ligation with a suitable vector will therefore yield a wide range of different clones with varying distances between start codon and *Eco* RI site. In addition, the insertions may not be

in the correct reading frame. Those clones in the mixture which contain the correct reading frame, can be identified by exploiting the phenomenon of α-complementation (*cf.* Section 2.4.2.3). As in the case of M13 cloning (Section 2.4.2), a number of plasmids, known as pUC plasmids, have been developed for this purpose (Fig. 7-58). These plasmids contain the *lac* regulatory region and a part of the *lacZ* gene which codes for the 59 *N*-terminal amino acids of β-galactosidase (Vieira and Messing, 1982). The corresponding host strain (JM83) carries the deletion *M15* of the *lac* operon, which removes amino acids 11-41 of β-galactosidase, but retains the entire *C*-terminal part of the enzyme. Each incomplete *lacZ* gene will direct the synthesis of an inactive polypeptide. Together, these polypeptides will be capable of complementing each other by forming aggregates. The resulting enzymatic activity can be detected on Xgal indicator plates as described above (Fig. 7-6).

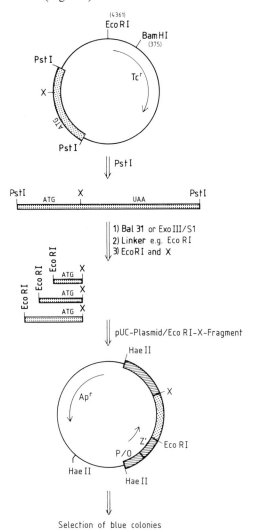

Selection of blue colonies

Plasmids of the type pUC7, 8, 9, 12, 13, 18 and 19 contain polylinkers within the region of the *lacZ* gene corresponding to the 59 *N*-terminal amino acids. These polylinkers allow cloning of DNA fragments with a variety of different ends. It is interesting to note that the inserted polylinkers (and other insertions) do not interfere with α-complementation, as long as they preserve the correct reading frame.

When the cDNA fragments with different left ends described above (Fig. 7-57) are inserted into a polylinker of a pUC plasmid, only those clones with insertions in the correct reading frame for β-galactosidase will yield blue colonies. Since the maximum length of an insertion which still allows α-complementation is not known, it is advisable to clone only relatively small DNA fragments. The intensity of the blue colour of different clones is usually quite variable. Clones showing the most intense colouration are those giving the highest levels of expression, and can be used as recipients of the missing part of the gene, in order to obtain the entire fusion protein.

Fig. 7-57. General approach for the construction of expression vectors directing the synthesis of fusion proteins.
The starting material can be a cDNA clone whose insertion has been removed from the plasmid (by *Pst* I digestion in this example). By treatment with exonuclease *Bal*31 or a combined ExoIII/S1 nuclease digestion a restriction site is positioned close to the start codon. The example of *Eco* RI linkers shown here may necessitate the use of *Eco* RI methylase if the DNA fragment contains an internal *Eco* RI site. The next step is an *Eco* RI digestion followed by digestion with another restriction enzyme (X) which should preferably cut asymmetrically. The mixture of DNA fragments obtained is then cloned into a pUC vector (*cf.* also Fig. 7-58). Of course, the cleavage site of endonuclease X must be present within the polylinker of the pUC plasmid. Since a wide spectrum of pUC plasmids is available, it should not be difficult to find a suitable vector containing the desired cleavage site. Once a suitable clone is identified, site X can be used for the insertion of the missing portion of the gene in question, which can be obtained from the original cDNA clone. The *lacO/P* and *lacZ'* regions are represented by hatched bars; the insertion by stippled bars.

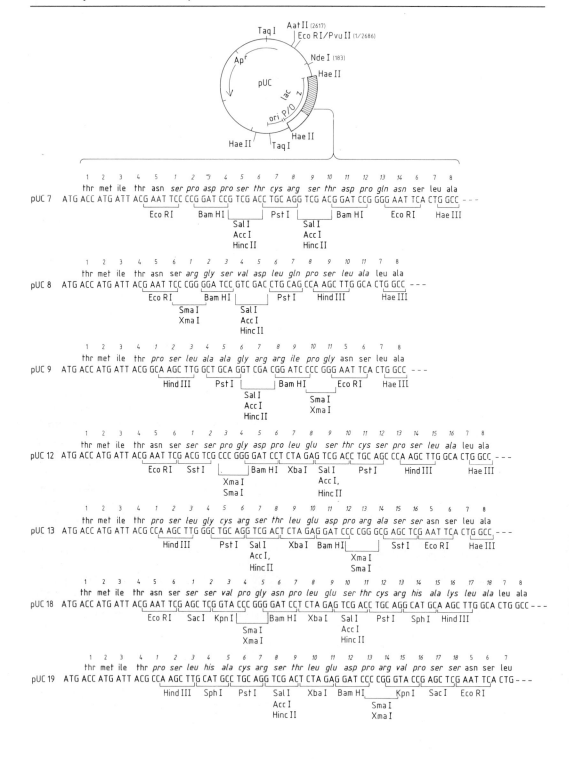

In summary, the successful synthesis of hybrid proteins with prokaryotic and eukaryotic components has been described in detail for several systems. This approach has considerable advantages, particularly the fact that fusion proteins with the large, 1 000 amino acids-long *N* terminus from β-galactosidase often are insoluble within the bacterial cell. Such fusion proteins thus are protected from proteolytic degradation (see below) and are easily purified; however, it should be kept in mind that this strategy also has its limitations. There is no doubt that it permits the detection of antigenic determinants in the fusion protein. Actually, it allowed the initial demonstration of the possibility of expressing eukaryotic DNA sequences in prokaryotes; however, if the eukaryotic proteins in question are to be obtained in a pure form, the original protein must be separable from the bacterial component of the chimaeric fusion products. Cyanogen bromide cleavage, which was used in the case of somatostatin, is restricted to proteins, such as some proinsulins, which do not contain internal methionine residues. In other cases enzymatic cleavage must be employed. Since the codons for suitable specific amino acids (for example arg and lys residues for tryptic cleavage) usually are not found in desired positions on the vector, this

opens important vistas for linker technology. The strategy is similar to that described above (Fig. 7-57). Instead of *Eco* RI linkers, more complex linkers, coding for protease-sensitive amino acids, would have to be employed.

Even in these cases difficulties may be experienced if internal protease-sensitive amino acids occur somewhere along the polypeptide chain of the fusion protein. Although it will depend largely on the individual case, it can be safely said that the synthesis of fusion proteins will be particularly useful for the production of small proteins and peptides. A very good example is the production of endogenous opiate peptides (*cf.* Ohsuye *et al.*, 1983).

7.4.2 Synthesis of Unique Proteins in Bacteria

In contrast to procedures described above, this process, known as transcriptional fusion, aims directly at the production of unique, non-fusion polypeptides. In this case protein synthesis should not initiate from the first methionine residue of a prokaryotic leader peptide, such as *lacZ* or *trpE*, but from the first methionine of the desired polypeptide itself (Fig. 7-44B). Biologically active constructions, therefore, usually contain an inducible prokaryotic promoter and a hybrid ribosomal binding site consisting of a bacterial Shine-Dalgarno sequence and a correctly spaced ATG codon which does not have to be of bacterial origin. In the case of a eukaryotic protein this ATG may correspond to the initiation codon of the eukaryotic gene itself. The individual elements of such constructions will be described for the *lac*, *trp* and λ systems as examples.

7.4.2.1 The lac System

M. Ptashne and co-workers have developed the concept of a portable *lac* promoter which can be placed at a suitable distance in front of a desired

←— **Fig. 7-58.** Structure of pUC plasmids.
pUC plasmids (*cf.* Fig. 2.4-22) are derived from the 2297 bp *Pvu* II-*Eco* RI fragment of pBR322, which contains the origin of DNA replication (*ori*) and the coding region of β-lactamase (Ap^r). *Pst* I, *Hind* III and *Acc* I sites were removed by mutagenesis. pUC plasmids carry a 433 bp *Hae* II fragment with *lac* control elements (*lac* promoter (P) and operator (O); open bars) inserted into the *Hae* II site in the immediate vicinity of the replication origin at position 2352; in addition, they contain the coding region for a functional β-galactosidase α-peptide (*lacZ*) (hatched bar). Short polylinker regions within this regions provide multiple recognition sites for various restriction endonucleases. Amino acids encoded by polylinker insertions are printed in italics. Numbers in parenthesis are pUC18 co-ordinates (Appendix B-4; Vieira and Messing, 1982; Yanisch-Perron *et al.*, 1985).

structural gene to allow the synthesis of the corresponding gene product as a pure unfused protein. The *lac* operon contains a suitable *Alu* I site between the ribosomal binding site and the start codon of the *lacZ* gene. A 95 bp long *Alu* I fragment containing almost the entire *lac* promoter region, the initiation site for mRNA synthesis, the S/D sequence, and five additional base pairs therefore can be isolated from chromosomal DNA (Fig. 7-7). This fragment also can be obtained from plasmid pGL101, in which this fragment is flanked by an *Eco* RI and a *Pvu* II site

(Fig. 7-10) (Lauer *et al.*, 1981). Since *Pvu* II (CAG/CTG) recognises a hexanucleotide sequence comprising the *Alu* I recognition sequence (AG/CT) and produces blunt ends at the same position as *Alu* I, *Eco* RI/*Pvu* II cleavage of pGL101 yields a DNA fragment with the desired blunt *Alu* I ends immediately downstream from the *lac* S/D sequence. This fragment contains an S/D sequence, but lacks an ATG codon, and must be placed at the proper distance upstream from a structural gene. For this purpose, the gene in question should preferably contain a unique

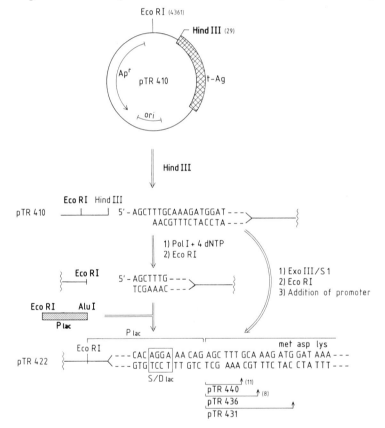

Fig. 7-59. Expression plasmid for SV40 t-antigen.
Vector pTR410 consists of parts of pBR322 comprising the coding region for β-lactamase (Ap^r) and the origin of DNA replication; in addition, it contains SV40 sequences with the entire coding region for the small t-antigen (cross-hatched bar). Manipulations near a *Hind* III site upstream of the start codon for t-antigen are described in detail in the text. P$_{lac}$ signifies the portable *lac* promoter fragment (strippled bar; see also text). Small numbers at the arrow heads of deletions pTR440 and pTR436 indicate the distances between the S/D*lac* regions and the start codon.

recognition sequence immediately 5′ of its ATG codon, which is, of course, rare. In the case of small t-antigen of SV40, however, the *Hind* III site in the SV40 *Hind* III-B fragment (which is 1 169 bp in length), is only twelve bp upstream of the ATG start codon of t-antigen. When this DNA fragment with filled-in *Hind* III ends was annealed with a portable promoter, the construction designated pTR422, shown in Fig. 7-59, was obtained (Roberts *et al.*, 1979b).

In order to obtain clones in which the distance between the ATG codon and the *lac* S/D sequence is shortened, the DNA first was digested with *Hind* III and then subjected to a partial digestion with exonuclease III. Blunt ends were generated by S1 nuclease treatment and the molecule was circularised after *Eco* RI cleavage and addition of the portable promoter. This procedure yielded

inserts of variable length due to the unspecificity of the exonuclease III reaction. The clones were then screened for expression of the desired protein. In this example, clone pTR436, in which the distance between the S/D sequence and the ATG start codon was 8 bp, was particularly active (Fig. 7-59). Plasmid pTR440 is only weakly active and the starting plasmid pTR422 and a deletion, pTR431, were completely inactive.

In a similar case the starting materials were the same as those described above, namely the portable *lacUV5* promoter, and the SV40 *Hind* III-B fragment with the coding region for the small t-antigen (Thummel *et al.*, 1981). The strategy employed, however, differed from that described above in that a *Hind* III linker was introduced between the S/D sequence and the ATG start codon (Fig. 7-60). This *Hind* III site in

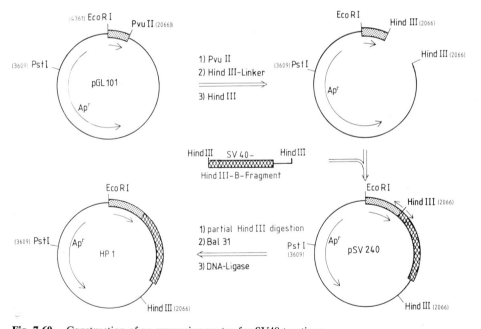

Fig. 7-60. Construction of an expression vector for SV40 t-antigen.
The *Pvu* II site of pGL101 (Fig. 7-10) is converted into a *Hind* III site by using *Hind* III linkers, which allows the subsequent insertion of the SV40 DNA fragment. In contrast to the construction shown in Fig. 7-59, the insertion, *i.e.*, the t-antigen DNA fragment (cross-hatched bar), as well as the S/D sequence within the portable *lac* promoter (stippled bar) are shortened by exonuclease treatment (indicated by arrows extending from the *Hind* III site in pSV240). Numbers in brackets are pBR322 co-ordinates. Directions of transcription are indicated by arrows within the plasmid circle.

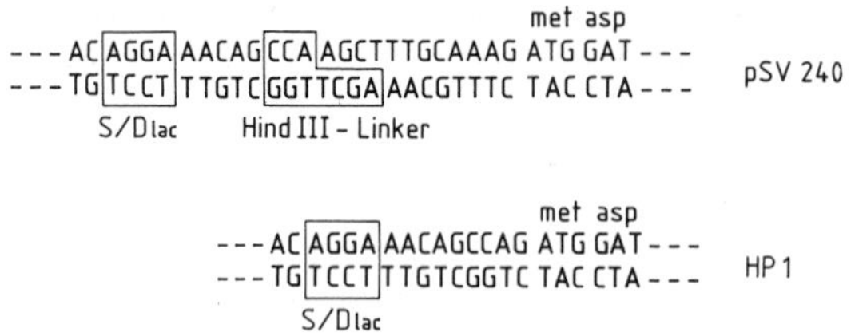

```
                       met asp
- - - AC AGGA AACAG CCA AGCTTTGCAAAG ATG GAT - - -
- - - TG TCCT TTGTC GGTTCGA AACGTTTC TAC CTA - - -       pSV 240
      S/Dlac    Hind III – Linker
```

```
                       met asp
      - - - AC AGGA AACAGCCAG ATG GAT - - -
      - - - TG TCCT TTGTCGGTC TAC CTA - - -       HP 1
            S/Dlac
```

Fig. 7-61. Structure of hybrid ribosomal binding sites comprising the S/D sequences of the *lac* operon and the start codon of SV40 t-antigen.
S/D sequences and a part of the *Hind* III linker in pSV240 are boxed. In contrast to pSV240, 11 base pairs of the 5' untranslated region of the SV40 t-antigen are missing in HP1; the distance between S/D sequence and ATG codon is therefore reduced to nine base pairs. The expression of t-antigen in HP1 is 40-fold higher than in pSV240 (*cf.* also Table 7.2 and Fig. 7-60).

plasmid pSV240 was used to shorten the distance between the S/D sequence and the ATG codon by *Bal*31 treatment (Fig. 7-61). The properties of different clones, in particular their activity with respect to production of t-antigen, and the secondary structures of the ribosome binding sites are summarised in Table 7-2 and discussed in Section 7-2.

In principle, this procedure can also be used for any other gene to be expressed from the *lac* promoter (Fig. 7-62) (Guarente *et al.*, 1980a). A cDNA copy of the desired gene is first cloned in pBR322. A cleavage site is then introduced in the vicinity of the 5' end by using a synthetic linker with a recognition site which does not occur in the cDNA itself. In the example shown in Fig. 7-62, a *Bam* HI linker was used. Other recommended and suitable cleavage sites are situated clockwise from the single *Eco* RI site in pBR322, *i.e.*, *Cla* I, *Hind* III, *Eco* RI, *Bam* HI (*cf.* example in Fig. 7-62) or *Sal* I. The resulting plasmid is opened at the desired site by partial digestion and the DNA is treated with exonuclease III before the promoter fragment is annealed. This cloning approach is simple because there is an efficient screen for those plasmids which contain the *lac* promoter region. The operator sequence on the insertion titrates the repressor molecules within the bacterial cells, the chromosomal β-galactosidase is expressed constitutively, and blue colonies appear on Xgal medium.

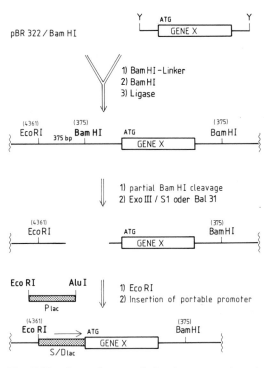

Fig. 7-62. General approach for the construction of expression vectors with foreign genes expressed under the control of a portable *lac* promoter.
Shown are the reactions to be carried out for bringing a hypothetical gene (X) under the control of a portable promoter. See text for details. Numbers in parentheses are pBR322 co-ordinates (Guarente *et al.*, 1980a).

One disadvantage of this procedure is that constructions which maximally express the inserted gene are not immediately identified. It is necessary to carry out functional or immunological tests for the expression of the desired protein with each clone. Since this may be comparatively laborious, a method was developed to allow identification of not only insertions of the *lac* operator/promoter region, but also optimal positioning of the hybrid ribosome binding site (Guarente *et al.*, 1980b). This technique exploits a particular property of the enzyme β-galactosidase. Large carboxy-terminal peptide fragments are enzymatically active irrespective of the nature of the N-terminal end. This fact was first observed with fusions of *lac* repressor and β-galactosidase (*lacI/lacZ*) (Müller-Hill and Kania, 1974). In order to generate such fused genes, a bacterial strain was constructed which contained a *lacI* gene promoter mutation (i^q) and an ochre mutation in *lacZ* (*U118*). The i^q mutation causes an overproduction of repressor protein, and the ochre mutation terminates transcription in *lacZ* at the position of the seventeenth amino acid so that

a truncated, enzymatically inactive β-galactosidase of sixteen amino acids is synthesised (*cf.* Fig. 7-63). A selection for revertants expressing β-galactosidase activity led to gene fusions which lacked the termination signal for *lacI*, the *lac* promoter region, and variable parts of the N-terminal region of *lacZ*. This indicated that the N-terminal part of β-galactosidase could be replaced by parts of the *lac* repressor without influencing the enzymatic activity of β-galactosidase. This important observation, which is also the basis of α-complementation, has later been applied to fusions of *lacZ* with other prokaryotic genes.

It is important to realise that this principle can also be applied to N-terminal regions of eukaryotic genes (Guarente *et al.*, 1980b). Plasmid pLG contains a large C-terminal part of the *lacZ* gene (*lacZ'*), but this region is not expressed, since the corresponding promoter regions are absent. As shown in Fig. 7-64, the 5' terminal part of gene, gene X, can be fused with the *lacZ* region on the plasmid in such a way that a functional reading frame is generated. In the case of a eukaryotic

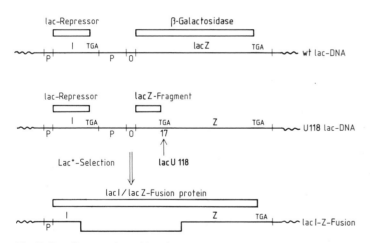

Fig. 7-63. Construction of *lacI-lacZ* fusions.
The top line shows a part of the *E. coli* chromosome with the arrangement of genes for *lac* repressor and wild-type β-galactosidase. Mutation *U118lac* is characterised by a stop codon at position 17 of β-galactosidase (*lacZ*). The *lacI-lacZ* fusion bypasses the stop codon, TGA, of the *lac* repressor gene (*lacI*), and also the nonsense codon in *lacZU118*. Proteins encoded by the DNA regions in question are shown as open bars above the maps (Müller-Hill and Kania, 1974).

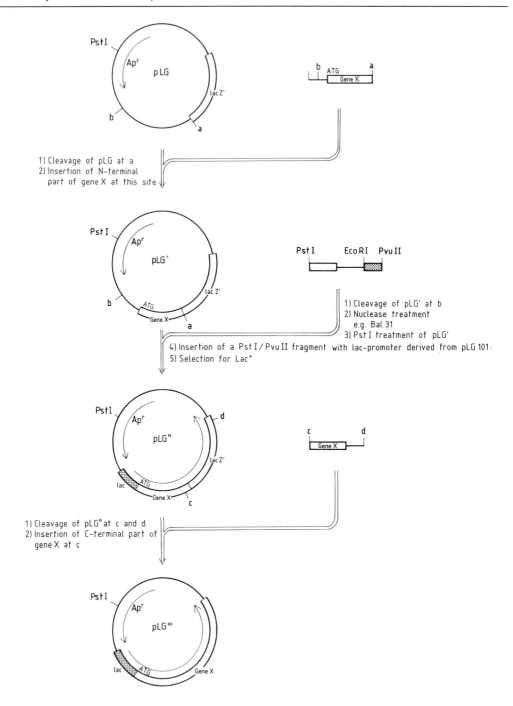

β-Galactosidase	---T A A C A A T T T C A C A C	A G G A	A A C A G C T	A T G ---
SV 40 t-Ag (pTR 436)	---T A A C A A T T T C A C A C	A G G A	A A C A G	A A A G A T G ---
β-Globin (pLG 302-2)	---T A A C A A T T T C A C A C	A G G A	A A C A G	A C A G A A T G ---
pre FIF (pLG 104)	---T A A C A A T T T C A C A C	A G G A	A A C A G	A C A T G ---
FIF (pLG 117)	---T A A C A A T T T C A C A C	A G G A	A A C A G	C C A T G ---

S/Dlac

Fig. 7-65. Ribosomal binding sites in different expression vectors. All structures contain the same S/D sequence also found in the *lac* system (boxed). The point of transition to sequences specific for various eukaryotic genes is indicated by a vertical line (Guarente *et al.*, 1980a).

◄ --- Fig. 7-64. General approach for constructing optimally expressing clones in the *lac* system.
The 5' terminal part of a gene X to be expressed is introduced into a restriction site "a" of a plasmid containing the 3' terminal portion of the *lacZ* gene. Plasmid pLG' is then opened at "b", modified by nuclease treatment and ligated with a *lac* promoter fragment (obtained, for example, from pGL101 Fig. 7-10) which is flanked by *Pst*I and *Pvu*II sites. Maximally expressing clones are identified by selection for Lac⁺ in a growth medium containing Xgal. The *lacZ'* region in these clones is then replaced by the 3' terminal portions of X by using site "c", which restores gene X (Guarente *et al.*, 1980).

gene this fusion in plasmid pLG' is neither transcribed nor translated, since no promoter signals are available. A combination of restriction enzyme digestions and nuclease treatments similar to those described above (Fig. 7-62) can be used to insert a portable promoter in front of the first ATG of gene X. If the resulting ribosome binding site in plasmid pLG'' has the correct structure, it will direct the synthesis of an enzymatically active gene X/β-galactosidase fusion protein. The enzymatic activity of this protein, which contains an *N*-terminal portion of gene X protein, is easily identified on Xgal plates after transformation of Lac-negative bacteria. Finally, the *lacZ* part of the hybrid gene in the plasmid is replaced by the 3' terminal DNA fragment of gene X in order to obtain the entire gene X in a correct configuration (pLG'''). A comparison of β-galactosidase activities of the fusion proteins (X-*lacZ'*) obtained from pLG'' with the amount of functional X protein in pLG''' shows that these values correlate quite well. A clone expressing the β-galactosidase fusion protein generally will produce the intact eukaryotic gene product as well. Good producers can therefore be identified solely on the basis of β-galactosidase activity expressed from gene X/β-galactosidase fusions. At least four eukaryotic genes have been expressed using this technique: small t-antigen of SV40, rabbit β-globin, and human fibroblast interferon (FIF; with and without its pre-sequence). The struc-

Benno Müller-Hill,
Cologne, FRG

Table 7–5.   Expression yields of IGF-1/*lacZ* fusion proteins from different mutated plasmid constructions.

Plasmid		2 Pro	3 Glu	4 Thr	5 Leu	6 Cys	β-gal activity units/cell (JM83)	SMC ng/10^7 cells (HB101)
original sequence		x	x	x	x x	x		
pUCmuSMCA$_{ori}$		CCA	GAA	ACC	CTG	TGC	0.4	1.4
blue colonies								
pUCmuSMCA	1	CCC	GAA	ACT	CTG	TGT	3.1	33
	2	CCT	GAA	ACT	TTG	TGC	2.6	45
	3	CCA	GAG	ACG	TTG	TGC	0.9	35
	4	CCA	GAG	ACG	TTG	TGT	0.9	43
	5	CCT	GAA	ACT	TTG	TGT	2.9	33
	6	CCT	GAG	ACG	TTG	TGT	1.2	58
	7	CCG	GAA	ACG	TTA	TGT	1.9	50
	8	CCG	GAA	ACA	TTG	TGT	1.2	65
	9	CCA	GAA	ACG	TTG	TGT	1.1	32
	10	CCT	GAG	ACT	CTA	TGT	2.3	22
white colonies								
pUCmuSMCA	11	CCC	GAA	ACC	CTC	TGT	<0.1	0.10
	12	CCT	GAA	ACC	CTC	TGT	<0.1	0.11
	13	CCG	GAA	ACC	CTC	TGT	<0.1	0.10
	14	CCA	GAA	ACC	CTC	TGT	<0.1	0.09

The table relates to plasmid constructions described in Figs. 7–66 and 7–67. SMC stands for somatomedin-C, a 70 amino acid protein found in human serum, also known as insulin-like growth factor I (IGF-I). Numbers 1 to 14 refer to fourteen plasmid colonies, ten of which displayed a blue and four of which displayed a white phenotype following plating on JM83 cells in the presence of Xgal plus ampicillin. x indicates the positions of mutations introduced in codons 2 to 6 of the IGF-I gene (Buell *et al.*, 1985).

tures of the respective ribosome binding sites are shown in Fig. 7-65. At equilibrium the corresponding bacteria synthesise between 5 000 and 15 000 molecules of the desired protein per cell. In individual cases the yields may be lower since the various proteins may differ in their stability within the host bacteria (Guarente *et al.*, 1980a) (see Section 7.5).

This concept can also be applied to the pUC family of vectors. In order to optimise expression of the IGF-I gene in *E. coli*, Buell *et al.* (1985) inserted a 165 bp fragment containing a ribosome binding site and the coding region of the first 33 amino acids of IGF-I into the 65 bp polylinker region of pUC8 (Fig. 7-67). Expression in this construction was under the control of the *lac* promoter, while translation could initiate either at the *lacZ* gene or at the IGF-I gene start codon. However, since translation from the *lacZ* AUG would quickly encounter a stop codon (Fig. 7-67),

Fig. 7-66.   Construction of IGF-I/*lacZ* fusion vectors ⟶ for improved expression of the IGF-I protein. Vector pPLmuSMCori contains a synthetic IGF-I gene (parts "A" and "B", hatched and stippled bars) preceded by a 66 bp fragment derived from bacteriophage mu (see Fig. 7-67) which provides the S/D sequence (black bar). The construction is driven by the λ P_L promoter, but results in only low level expression of the desired protein. To improve expression, the N-terminal part "A" of the IGF-I gene is cloned into pUC8 to yield pUCmuSMCAori. An *Ava*II-*Hae*II fragment, (G/G(AorT)CC) (PuGCGC/Py), coding for amino acids 2-8 of IGF-I, is replaced by a synthetic mixture of DNA fragments containing all possible base substitutions which retain the amino acid sequence. The *Eco*RI-*Bam*HI fragments from plasmids isolated from blue colonies were isolated and reconstructed into pPLmuSMC1-14, as indicated. The open bar sector within part "A" of the IGF-I gene in plasmids pUCmuSMCA1-14 and pPLmuSMC1-14 represents the synthetic fragment. N = one of the four possible bases, P = purines, and Y = pyrimidines (Buell *et al.*, 1985).

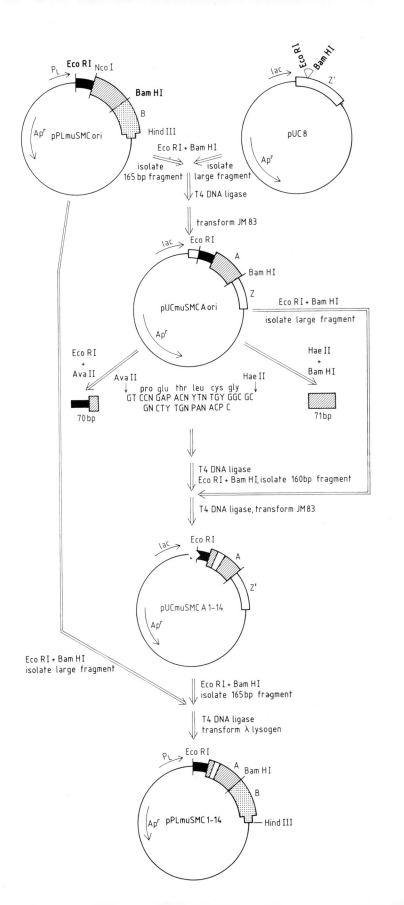

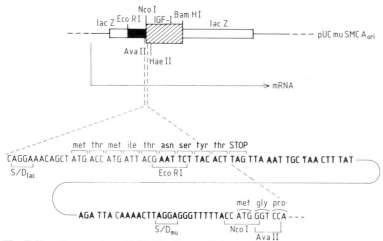

Fig. 7-67.   Construction of IGF-I/*lacZ* gene fusions.
The figure shows a section of plasmid pUC8 with an insertion between its *Eco* RI-*Bam* HI polylinker sites within the *lacZ* α-peptide (open bar). The insertion comprises a 66 bp *Eco* RI-*Nco* I fragment containing the S/D sequence from the *nerI* gene of bacteriophage mu (Gray *et al.*, 1984) (black bar), and a 98 bp long *Nco* I-*Bam* HI fragment with the first half (part "A" in Fig. 7-66) of the coding region of the human IGF-I gene (hatched bar). Transcription in this construction starts at the *lacZ* promoter present in the pUC8 portion of the vector, and covers the *lacZ* portion and the insert. Translation from this message can initiate both at the *lacZ* ribosomal binding site (S/D$_{lac}$) and the ribosomal binding site (S/D$_{mu}$) of the insert derived from bacteriophage mu. Ribosomes initiating at the *lacZ* ribosomal binding site will soon encounter a stop codon, while the second translation product continues through the IGF-I coding region into the *lacZ* portion of pUC8. This construct yields only white plaques on *E. coli* strain JM83. Mutations introduced into the coding region within an *Ava* II-*Hae* II fragment (*cf.* Fig. 7-66) result in a blue phenotype, indicating expression of the IGF-I/*lacZ* fusion peptide (Buell *et al.*, 1985).

the only protein formed was derived from a fusion between the IGF-I portion and the distal *lacZ* gene region (Fig. 7-67). Transfection of a construction containing genuine IGF-I sequences into JM83 yielded only white colonies, indicating little or no β-galactosidase activity. In order to increase expression, a large number of mutants were generated by synthesising a mixture of oligonucleotides that included all the 256 possible sequences encoding amino acids 2-6 of IGF-I (Fig. 7-66). After re-insertion into the proper position in the fusion and transformation of *E. coli* strain JM83, approximately 500 out of 5 000 colonies were pale blue. The best of these, after reconstruction of the whole IGF-I gene, produced more than 20 times more IGF-I than did the wild-type construction (Table 7-5). These mutations affected a secondary stem structure around a ribosome binding site region and the increased

β-galactosidase activity in the mutants thus confirms some of the conclusions mentioned in Section 7-2.

7.4.2.2 The trp System

As in the *lac* system, the regulatory sequences of the *trp* operon can be used to create hybrid ribosome binding sites. In the *trp* system, the site of transcription initiation and the start codon of the TrpE protein are separated by 162 bp known as the leader sequence (Fig. 7-14). This region codes for a peptide which is fourteen amino acids in length and plays a decisive role in the control of the *trp* operon. A *Taq* I site is situated between the corresponding S/D sequence and the ATG codon, allowing both parts of the ribosome binding site to be separated from each other. The ATG of the leader peptide can therefore be

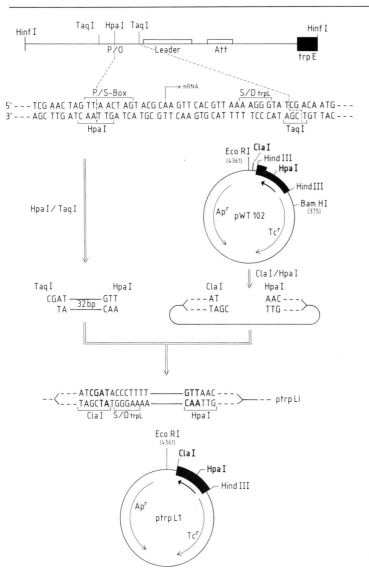

Fig. 7-68. Construction of an expression vector with *trp* regulatory sequences.

The top shows a portion of a *Hinf*I DNA fragment with *trp* regulatory sequences (*cf.* also Fig. 7-18). The region between a *Hpa*I and a *Taq*I site contains the Pribnow-Schaller box (P/S) and the S/D sequence of the *trp* leader peptide (S/D$_{trpL}$). This 32 bp *Hpa*I-*Taq*I fragment is inserted into vector pWT102 opened with *Hpa*I and *Cla*I. pWT102 and pWT101 (Fig. 7-18) are identical, but contain the *trp* insertion in opposite directions (arrows). Quite unexpectedly, pWT102 shows a considerable tetracycline resistance, which is presumably due to the presence of a cryptic promoter in the *trp* Hinf I fragment. By a fortunate coincidence, the *Taq*I site of the insertion contains an AT pair, and this regenerates the *Cla*I site. Vector ptrpL1 lacks the *trp* coding regions, and the *Cla*I site can therefore be used directly for cloning of foreign DNA (Edman *et al.*, 1981).

replaced by the ATG of a eukaryotic gene.

A suitable plasmid which contains the entire *trp* promoter region, the S/D sequence for the leader peptide, but no other parts of the *trp* operon was constructed from plasmid pWT102 (Fig. 7-18) (Edman *et al.*, 1981). Digestion with a combination of *Hpa* I (GTT/AAC) and *Taq* I (T/CGA) yielded a 32 bp fragment which contained the desired S/D sequence and the initiation site for transcription (Fig. 7-68). In a parallel experiment all *trp*-specific regions upstream of the *Hpa* I site were removed from the same plasmid, pWT102, by digestion with *Hpa* I and *Cla* I and replaced by the short *Hpa* I-*Taq* I fragment. This construction presented no problems since the tetrameric *Taq* I recognition sequence (T/CGA) is part of the hexameric recognition sequence of *Cla* I (AT/CGAT), and since both enzymes cleave in the same pattern with protruding 5'-CG termini. Since there was an A/T pair in the immediate vicinity of the original *Taq* I site, the *Cla* I site was regenerated. The resulting plasmid, ptrpL1, was opened at its unique *Cla* I site to allow insertion of a foreign gene in the immediate vicinity of the S/D sequence. The *Cla* I site is particularly well suited for this purpose since it can accomodate various DNA fragments with protruding 5'-CG ends, such as those obtained by *Hpa* II (C/CGG), *Taq* I (T/CGA) and *Acc* I (GT/CGAC) cleavage. This strategy was tested and employed for cloning and expressing hepatitis B virus core antigen (HBcAg). The gene for this protein, which is 183 amino acids in length, was obtained from a suitable plasmid by *Hha* I cleavage (Fig. 7-69). The start codon for HBcAg on a *Hha* I fragment of 1 005 bp was 15 bp away from one of the molecular ends of this fragment. The distance between the ATG and the S/D sequence would have been too long, and therefore the usual modifications were carried out as described above. The DNA was first treated with exonuclease III to remove approximately ten base pairs and then made blunt-ended with S1 nuclease before *Bam* HI linkers were added. The commercially available decameric linkers do not only contain a

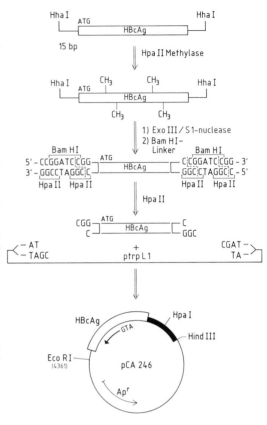

--- AAA AAG TTG CAT GGT GCT GGT C⎮CG ATA CCC TTT TTA ---
--- TTT TTC AAC GTA CCA CGA CCA GGC⎮TAT GGG AAA AAT ---
    phe leu gln met   S/D trp
                         ←

Fig. 7-69. Application of expression vector ptrpL1. A *Hha* I DNA fragment containing the coding sequence for HBcAg was modified by a combined exonuclease III S1 nuclease treatment. Following addition of *Bam* HI linkers, the construct was digested with *Hpa* II in preparation for cloning into the *Cla* I site of ptrpL1. The sequence around the ribosomal binding site at the bottom shows that the distance between S/D sequence and ATG codon is 16 bp. The direction of transcription of the HBcAg sequences is indicated by an arrow.

Bam HI site but also two *Hpa* II sites. A subsequent *Hpa* II digestion completely removes the *Bam* HI recognition site and creates 5' overhanging ends which are compatible with *Cla* I ends. Since the tetrameric *Hpa* II recognition site occurs

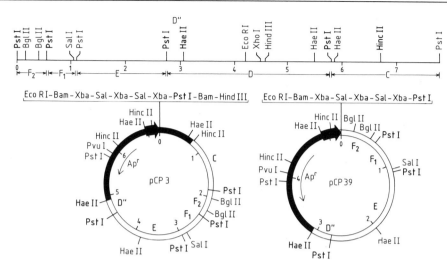

Fig. 7-70. Physical maps of pKN402 and its derivatives.
Plasmid pKN402 (shown on top in a linear presentation) is a miniderivative of a temperature-sensitive runaway replication mutant of plasmid R1drd19. Plasmids pCP3 and pCP39 are derivatives of pKN402, which contain the *Pst* I-F2 fragment required for thermoinducible runaway replication as well as a selectable marker missing in pKN402. Capital letters indicate the *Pst* I fragments of pKN402. Plasmid pCP39 lacks a 1790 bp *Pst* I fragment, present in pCP3, which represents part of the pKN402 *Pst* I-C fragment. The ampicillin resistance gene (black bar) and the λ *P_L* promoter (black bars with arrow) are derived from the pPLa series of plasmids described in Figs. 7-26 and 7-27 (Remaut *et al.*, 1983).

quite frequently, the DNA fragment must be protected by treatment with *Hpa* II methylase prior to *Hpa* II digestion. In principle, this strategy can be used for any other gene. A disadvantage is that it does not directly allow selection of or quick screening for maximally expressing clones. In our example (Fig. 7-69) screening of a large number of transformants yielded the expression vector pCA246, which produces up to 10% of the newly synthesised protein as HBcAg after induction with 3-β-indolylacrylic acid.

7.4.2.3 The λ P_L System

The strong λ P_L promoter has been particularly useful for the high-level expression of proteins detrimental to an *E. coli* cell. In an elegant and most efficient application it is employed in a two plasmid system (Remaut *et al.*, 1983) which also exploits the temperature-sensitive runaway repli-

cation phenomenon alluded to earlier (Section 4.1.1). One plasmid component is derived from plasmid pKN402, a 7.8 kb mini-derivative of a runaway replication mutant of plasmid R1drd19 (Fig. 7-70). This plasmid contains both the temperature-sensitive replicon and the λ P_L promoter; the latter lies upstream from a polylinker, into which the desired gene can be inserted. Expression of the P_L promoter from such a construct can be regulated by the *cI* gene product encoded by a single chromosomal gene copy or, even better, by a *cI* gene on a compatible multicopy plasmid. Such a plasmid, pcI857, is described in Section 4.1.5. It confers kanamycin resistance, carries the *cI857* allele of the λ repressor, and is compatible with the replicon of pKN402 and its derivatives. *E. coli* cells transformed with both the runaway replication P_L vector and the pcI857 vector contain approximately 30-50 copies of each vector at 28°C. At this temperature the active cI857 repressor acts *in trans* to prevent any transcription from the P_L promoter on the other plasmid. A shift to higher temperature (42 °C) leads to two events, a ten- to twentyfold amplification of the runaway replication vector copy number, and a simultaneous derepression of the P_L promoter due to inactivation of the cI857 repressor at 42 °C. This two-plasmid expression system was tested with the T4-derived DNA ligase gene, the expression of which could be induced to levels up to 25% of the total cellular protein. It is effective in many *E. coli* strains and has also proved successful for the expression of the human IGF-I protein (Buell *et al.*, 1985).

7.4.2.4 Synthetic Ribosome Binding Sites

The hybrid ribosome binding sites discussed in Sections 7.4.2.1 and 7.4.2.2 are not necessarily optimal for ribosome binding, and hence for efficient translation (*cf.* also Section 7.2). These binding sites contain naturally occurring S/D sequences which frequently show a relatively low degree of homology with the sequence of the 3' end of 16S ribosomal RNA. In the *lac* system it is only four and in the *trp* leader peptide S/D sequence only three bases which show this homology at all. It was postulated (Jay *et al.*, 1981) that ribosome binding, and hence initiation of protein biosynthesis, would be much more efficient if these regions of homology could be extended. A DNA oligomer containing an S/D sequence of nine base pairs and an additional sequence,

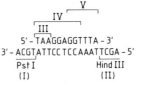

Fig. 7-71. Structure of a synthetic linker with *Pst* I (I) and *Hind* III (II) ends, coding for a stop codon (III), an S/D sequence (IV), and the GGTTTA sequence. (Jay *et al.*, 1981).

5'-GGTTTAA-3', which is important for binding ribosomal proteins (Fig. 7-71; *cf.* also Fig. 7-37; also Jay *et al.*, 1982) therefore was synthesised chemically. The entire synthetic ribosome binding site consists of two oligonucleotides of twelve and twenty bases, respectively. The left-hand 3' protruding end contains a sequence which allows ligation with a *Pst* I site (I), the right-hand 5' protruding end a *Hind* III site (II). A TAA stop codon (III) within this linker molecule is in phase with β-lactamase (see below); in the inner part of this linker lie the S/D sequence of nine bases (IV) and the sequence GGTTTAA (V). Since the linker is asymmetrical it is more universally applicable than conventional symmetrical linkers.

As shown in the example in Fig. 7-72, this linker is positioned at a correct distance in front of the start codon of a gene to be expressed, and inserted together with this gene X into the *Pst* I site within the β-lactamase gene of pBR322 (*cf.* also Fig. 4.1-11). In a bacterial cell, transcription initiates at the promoter of the β-lactamase gene to yield a hybrid mRNA containing the β-lactamase component and sequences of gene X

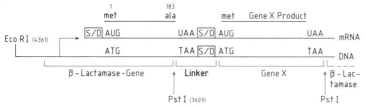

Fig. 7-72. Use of synthetic ribosomal binding sites for the construction of expression vectors. The linker carries a stop codon and a consensus S/D sequence. Although only one hybrid mRNA is transcribed, two proteins are synthesised, one of which is a fragment of β-lactamase with amino acids 1-183; the other is the gene X product with an *N*-terminal methionine residue. Numbers in parenthesis are pBR322 co-ordinates.

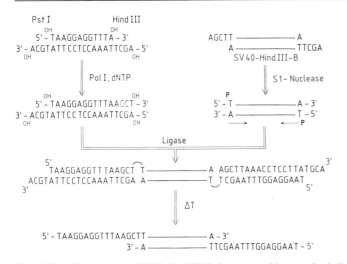

Fig. 7-73. Ligation of an SV40 *Hind*III-B fragment with a synthetic linker.
Parentheses in the sequence in line 3 indicate that phosphodiester bonds are formed between adjacent TT residues, but not between the complementary AA bases. ΔT indicates a heat step which serves to remove the 20 bases of the oligonucleotide which is only stabilised by hydrogen bonds. The arrows below the SV40 sequences point out the fact that the SV40 DNA fragment could be shortened by a suitable exonuclease treatment before the addition of the linker, which would bring the ATG start codon into a more favourable distance to the S/D sequence on the linker.

with its new ribosome binding site. This new mRNA is polycistronic, and translation therefore initiates at two sites, the natural ribosome binding site for β-lactamase, and the synthetic ribosome binding site for gene X. The synthetic ribosome binding site contains a stop codon which allows the synthesis of only an *N*-terminal 183 amino acid fragment of β-lactamase.

This strategy was tested for the SV40 t-antigen (Fig. 7-73). The corresponding gene was located on a *Hind*III fragment, 1 169 bp in length. One end of this fragment was only twelve base pairs away from the 5′-terminal ATG of this gene. The 5′ protruding *Hind*III end was converted to blunt ends by S1 nuclease treatment, reducing to eight base pairs the distance to the first ATG (*cf.* also Fig. 7-59). The synthetic linker was also made blunt-ended at its protruding *Hind*III end by a DNA polymerase I reaction. Ligation of the blunt SV40 ends with the blunt ends of the linker created an SV40 fragment flanked by the linker molecule. Under these conditions phosphodiester

bonds were formed only between the 5′ phosphate ends of SV40 DNA and the 3′ hydroxyl end on the sixteen base pairs long strand of the linker. Such bonds were not formed between the 3′ hydroxyl end of SV40 DNA and the 5′ hydroxyl end of the linker represented by the twenty base pairs long oligonucleotide (Fig. 7-73).

In principle, this recombinant DNA fragment could be cloned into the *Pst*I site of pBR322; however, low yields were expected, since self-ligation of the vector or of the recombinant DNA molecules could occur. This problem was circumvented by providing the molecular ends of both molecules with non-complementary ends like those created by homopolymeric tailing. Under these circumstances vector and recombinant DNA fragments are capable of base pairing only with each other, not with themselves. Since only the 16-mer and not the 20-mer was ligated to SV40 DNA, the complementary 20-mer therefore could be removed by carefully melting the DNA (Fig. 7-73). The resulting SV40 DNA fragment then

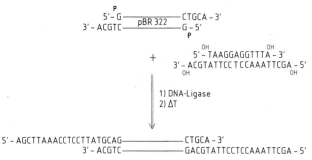

Fig. 7-74. Ligation of *Pst* I-linearised pBR322 DNA with a synthetic linker. (*cf.* also Fig. 7-71). The linker only carries hydroxyl ends; stable phosphodiester bonds can therefore be formed only with the 5'-G terminus of the pBR322 molecules, but not with their 3' ends; the oligonucleotide which is 12 bases long can therefore be removed by a heat step (indicated by ΔT). (Jay *et al.*, 1981).

possessed two non-complementary 5' protruding ends.

Plasmid pBR322 was prepared for cloning in a similar way (Fig. 7-74) by adding the linker shown in Fig. 7-71 to vector DNA linearised with *Pst* I. In this case, the pBR322 molecules are flanked by the linker, and covalent bonds were only formed between the 5' phosphate end of the plasmid and

the strand of the linker which is twenty bases in length. A vector with two protruding ends twenty bases in length could be obtained by melting the DNA. These ends were identical and could not form base pairs with themselves, but only with the ends of the SV40 DNA fragments (Fig. 7-75). When the transformation experiment was carried out, 93% of all transformants were ampicillin-

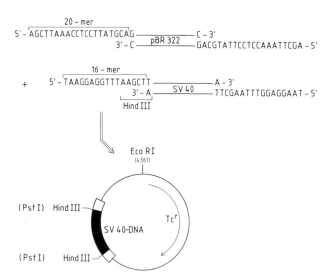

Fig. 7-75. Construction of an expression vector by ligation of DNA fragments shown in Figs. 7-73 and 7-74. SV40 sequences are represented as a black bar. These sequences are flanked by *Hind* III sites. *Pst* I in parentheses indicates that the *Hind* III sites are now located at the original *Pst* I site in the β-lactamase gene of pBR322. Linker regions are represented as open bars. (Jay *et al.*, 1981).

sensitive. Practically all transformants contained SV40 insertions, and both orientations were obtained.

The SV40 component can be cut out from these plasmids and can be replaced by other DNA. In this case the only other *Hind* III site of pBR322, which is located in the promoter region of the tetracycline resistance gene, can be protected against digestion by using *E. coli* RNA polymerase (*cf.* also Fig. 7-53), or another cleavage site can be used. If a suitable site is not available, the first step could be to correctly position the start codon by digestion with *Exo*III/S1 or *Bal*31 as described above although this step probably would sacrifice some convenience. The fusion of a filled-in *Hind* III site with an S1-treated *Hind* III site regenerates this site, and this can hardly be expected when a non-specific nuclease treatment is used.

Cloning of the SV40 t-antigen DNA fragment directly *via Hind* III ends as described above did not result in t-antigen expression. One clone, however, which contained a deletion of four base pairs between the *Hind* III site and the t-antigen start codon, yielded up to 0.4% of the total protein as t-antigen. This yield is lower than might be obtained with other, similar constructions, since t-antigen by itself is rather unstable and since the β-lactamase gene does not possess an efficient promoter. It is to be expected that combinations of efficient inducible promoters and efficient synthetic ribosome binding sites should lead to much higher yields.

7.5 Maturation, Transport, and Stability of Eukaryotic Proteins in Bacteria

Two alternative approaches to the expression of eukaryotic proteins in bacteria have been discussed in the preceding sections. One procedure yields a hybrid protein consisting of bacterial and eukaryotic protein sequences, while the other leads to the synthesis of the genuine eukaryotic protein. In the first case, the undesired bacterial residues must be specifically removed by enzymatic or chemical cleavage at an amino acid especially introduced for this purpose. This is only possible if this amino acid (a methionine residue, for example, in the case of a cyanogen bromide treatment) does not occur elsewhere within the eukaryotic part of the hybrid protein.

If a direct synthesis of the eukaryotic protein is envisaged, this protein necessarily starts with *N*-formyl methionine at the *N* terminus. The corresponding AUG start codon is a component of the hybrid ribosome binding site required for correct translation. In most cases the formyl group is not found in the isolated protein and is therefore correctly removed in the bacterial cells. Many eukaryotic proteins, in particular those proteins which are either incorporated into, or transported and secreted through, the plasma membrane, contain a signal peptide at their *N*-terminal ends (Fig. 7-76). Post-translational cleavage usually removes these signal sequences from the respective precursor proteins. It is only in exceptional cases (human fibroblast interferon; Fig. 7-76, for example) that the sequence of the mature and active protein obtained after enzymatic cleavage of the signal peptide also starts with a methionine residue. There are two alternatives when such proteins are to be expressed in *E. coli*. Either a DNA sequence is constructed, which contains the bacterial ribosome binding site with the original ATG of the precursor peptide, or an additional ATG codon is introduced immediately before the codon of the first amino acid of the mature gene product. In the first case the protein will be synthesised with a pre-sequence. The latter construction will yield a protein which, in addition to the normal sequence, will contain an *N*-terminal methionine residue. The case of insulin (Fig. 7-76) may serve as an example. Bacterial cells normally do not remove the additional methionine residue, but there will be many cases in which this additional

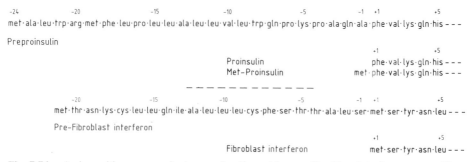

Fig. 7-76. Amino acid sequence of rat preproinsulin and human fibroblast interferon prepeptides.
In order to achieve expression of the proinsulin, an additional methionine residue had to be incorporated. In contrast, the amino acid at position "+1" of the mature fibroblast interferon is methionine, the codon of which is part of a ribosomal binding site.

amino acid will not influence the biological activity of the product.

This approach appears to be quite unsatisfactory for the production of proteins of medical interest, which may have to be administered for long periods of time. Several observations made in the past few years suggest that the mechanisms of secretion of proteins in bacteria and eukaryotic cells are similar in that they both seem to require the presence of hydrophobic signal peptides (Davis and Tai, 1980; Michaelis and Beckwith, 1982).

Even eukaryotic signal sequences may be used to direct a correct processing of proteins and to effect the transport of secretory proteins into the periplasm of bacteria. In the case of rat insulin it has been possible to obtain clones which synthesise parts of the bacterial prepenicillinase and the insulin precursor (Talmadge et al., 1980a,b). The amino acid sequence of these hybrid proteins is shown in Fig. 7-77.

Clone I contains the entire penicillinase presequence but no eukaryotic signal peptide. The other three clones contain various deletions of the bacterial component and a complete eukaryotic signal sequence. In clone I which contains the functional prokaryotic pre-sequence and lacks the corresponding eukaryotic pre-sequence, the hybrid protein is cleaved correctly downstream of the prokaryotic signal sequence. In the other

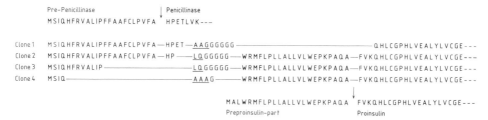

Fig. 7-77. Amino acid sequences of hybrid proteins containing parts of prepenicillinase and preproinsulin signal sequences.
The original penicillinase and proinsulin sequences are shown on top and bottom, respectively. Each hybrid sequence begins with the fMet residue of the prepenicillinase and ends with amino acid 21 of proinsulin. Apart from presequences, some of the hybrid proteins also contain parts of mature penicillinase. Ligation with the preproinsulin moiety was achieved by short peptide bridges consisting of a few amino acids, which are encoded by Pst I linkers (underlined) or poly(dG)-poly(dC) tails (glycine residues). A=Ala; R=Arg; N=Asn; D=Asp; C=Cys; Q=Gln; E=Glu; G=Gly; H=His; I=Ile; L=Leu; K=Lys; M=Met; F=Phe; P=Pro; S=Ser; T=Thr; W=Trp; Y=Tyr; V=Val (Talmadge et al., 1980b).

	-34	-33	-32	-31	-30	-29	-28	-27	-26	-25	-24	-23	-22	-21	-20	-19	-18	-17	-16	-15	-14	-13	-12	-11	-10	-9	-8	-7	-6	-5	-4	-3	-2	-1	+1
Preproinsulin (Rat)(S)												met	ala	leu	trp	arg	met	phe	leu	pro	leu	leu	ala	leu	leu	val	trp	glu	pro	lys	pro	ala	glu	ala	phe
Pre-Immunoglobulin-K-light chain (Mouse)(S)														met	asp	met	arg	ala	pro	ala	gln	ile	phe	gly	phe	leu	leu	leu	phe	pro	gly	thr	arg	cys	asp
Pre-Lysozyme (Chicken)(S)																	met	arg	ser	leu	leu	ile	leu	val	leu	cys	phe	leu	pro	leu	ala	ala	leu	gly	lys
Pre-Growth hormone (Rat)(S)								met	ala	ala	asp	ser	gln	thr	pro	trp	leu	leu	thr	phe	ser	leu	leu	cys	leu	leu	trp	pro	gln	glu	ala	gly	ala	leu	
Pre-Glycoprotein (Vesicular Stomatitis Virus)(M)																met	lys	cys	leu	leu	tyr	leu	ala	phe	leu	phe	ile	his	val	asn	cys	lys			
Pre-β-Lactamase (pBR 322)(S)												met	ser	ile	gln	his	phe	arg	val	ala	leu	ile	pro	phe	phe	ala	ala	phe	cys	leu	pro	val	phe	ala	his
Pre-Maltose-Binding protein (E.coli)(S)									met	lys	ile	lys	thr	gly	ala	arg	ile	leu	ala	leu	ser	ala	leu	thr	thr	met	met	phe	ser	ala	ser	ala	leu	ala	lys
Pre-OmpA-Protein (E.coli)(M)														met	lys	lys	thr	ala	ile	ala	ile	ala	val	ala	leu	ala	gly	phe	ala	thr	val	ala	gln	ala	ala
Pre-Coat protein (Phage fd)(M)											met	lys	lys	ser	leu	val	leu	lys	ala	ser	val	ala	val	ala	thr	leu	val	pro	met	leu	ser	phe	ala	ala	ala
Pre-β-Lactamase (Bacillus licheniformis) (S)		met	leu	lys	lys	phe	ser	thr	leu	lys	lys	leu	ala	ala	ala	val	leu	leu	phe	ser	cys	val	ala	leu	ala	gly	cys	ala	asn	asn	gln	thr	asn	ala	ser

Fig. 7-78. Amino acid sequences of *N*-terminal signal sequences of various precursors for membrane proteins (M), and a variety of secretory proteins (S). The vertical line indicates the transition from hydrophilic to hydrophobic regions within the signal sequences, the arrow the start of the mature proteins. (Davis and Tai, 1980).

three clones (clones 2-4) cleavage occurs at the eukaryotic sequence so that functional proinsulin is synthesised. It appears that it is of no consequence whether the site of cleavage is positioned 52 (clone 2), 40 (clone 3), or 29 (clone 4) amino acids away from the *N* terminus. In all four cases the yields are above 90%, *i.e.*, more than 90% of the antigenic activity is transported into the periplasm and processed correctly. Similar results have been obtained with human insulin (Chan *et al.*, 1981). A number of *N*-terminal sequences of secreted or membrane proteins of eukaryotes and prokaryotes are listed in Fig. 7-78. They differ in length and can be divided into two regions, namely a relatively short *N*-terminal hydrophilic region, and a neighbouring hydrophobic part. There do not appear to be any obvious differences between prokaryotic and eukaryotic sequences although the *N*-terminal hydrophilic part seems to be larger in Gram-positive bacteria.

The observed overrepresentation of certain amino acids, such as lysine and arginine in the hydrophilic part, and leucine and proline in the hydrophobic part of the signal sequence cannot be taken as evidence for sequence homology. Just as the secondary structure of RNA around the S/D sequence may be important for the function of the ribosome binding site, the secondary structure of the signal peptide, as well as sequences in the mature part of the protein may determine its secretory properties. Alkaline phosphatase of *E. coli* is a good example. Chimaeric peptides which contain the signal sequence and up to 80% of the sequence of the mature functional protein are processed correctly; however, these proteins are not transported into the periplasmic space (Ohsuye *et al.* 1983). For practical purposes it may be advisable to test several signal sequences in order to optimise secretion and transport of secreted proteins in bacteria.

Apart from cleavage of signal sequences from secreted proteins, bacterial proteases have other physiological functions (Mount, 1980) among which are (i) cleavage of abnormal proteins, (ii) cleavage of normal proteins under conditions of stress, (iii) utilisation of exogenous proteins as a source of amino acids, (iv) inactivation of regulatory proteins and colicins, and (v) morphogenesis of bacteriophage coats. One may surmise that the first two functions are particularly important for applications of gene technology.

Bacterial proteins are usually very stable and have a very slow turnover *in vivo*. This situation is dramatically changed when nitrogen and/or carbon sources become scarce. One reason for this may be the need to recruit new amino acids from the intracellular pool of proteins under conditions of impaired *de novo* synthesis. Enhanced breakdown is also observed under normal growth conditions for proteins which are synthesised as fragments due to the presence of nonsense mutations and for proteins which contain amino acid analogues. It is conceivable that this mechanism is also responsible for the breakdown of short peptides expressed from recombinant DNA molecules. A good example is somatostatin which is quite unstable if it is part of a short fusion protein. It could only be isolated when the chimaeric protein contained a *C*-terminal β-galactosidase portion of more than 1 000 amino acids in length (Fig. 7-54; *cf.* Section 7.4.1.4). In the latter case, the fusion protein is presumably protected from degradation because it aggregates; in fact it forms granules which are easily visible in phase contrast. This phenomenon which is quite common when the concentration of an overproduced protein exceeds its solubility (Fig. 7-79), does not only confer stability upon a protein, but also simplifies its purification (Simons *et al.*, 1984; Schoner *et al.*, 1985).

The origin of proteases in *E. coli* and their mechanisms of action are only partially known. The fractionation of proteolytically active cell-free extracts of *E. coli* has so far revealed eight different activities (Swamy and Goldberg, 1981). One of these proteins, a protease known as La, can be stimulated up to fortyfold by the addition of ATP (Voellmy and Goldberg, 1981) and has been known for some time to be involved in the first endoproteolytic steps involved in the break-

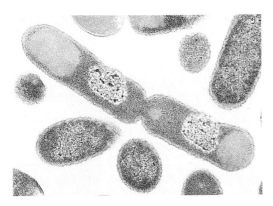

Fig. 7-79. Thin sections of *E. coli* bacteria showing granules of β-galactosidase/proinsulin fusion peptides (fixed in formaldehyde/glutaraldehyde according to Karnofsky; embedded in Epon; stained with uranylacetate/lead hydroxide; 38 000:1; Courtesy of Dr. W. Wetekam, Hoechst AG).

down of larger intracellular proteins. The enzyme is encoded by the *lon* gene of *E. coli*, and in *lon* mutants which contain half the normal level of protease La, the degradation of abnormal proteins is reduced by a factor of two to four. Expression of *lon* is stimulated by the presence of abnormal proteins, originating for example from cloned eukaryotic genes (Goff and Goldberg, 1985). It is regulated, at least in part, by the htpR gene product (Goff *et al.*, 1984) which is involved in the "heat-shock" response of *E. coli*. Upon shift to high temperature, *E. coli* cells, like eukaryotic cells (*cf.* Section 8.3.3), increase the synthesis of a limited number of proteins (approximately seventeen), one of which is protease La. This appears to be highly advantageous for *E. coli* since it prevents the accumulation of denatured or otherwise aberrant polypeptides at the high temperature. Cells carrying mutations in the *htpR* gene do not show the expected increase in transcription of heat shock genes, and thus of protease La at the high temperature, and even display reduced *lon* transcription at low temperature (Goff *et al.*, 1984). Double mutants of the *lon⁻ htpR⁻* genotype therefore should be of considerable value for cloning and expression of

foreign proteins in *E. coli*. In fact, this has already been demonstrated for the production of IGF-I in *E. coli* (Buell *et al.*, 1985).

7.6 Conclusions

The factors influencing the successful construction of expression vectors in *E. coli* are quite numerous, and interact in complex ways. There are now many examples of successful and efficient expression of eukaryotic proteins in *E. coli*. Other proteins, however, present unexpected difficulties. Some of the possible reasons have been discussed in the preceding sections, others are still unknown. It should be noted, however, that the situation is unsatisfactory, not because there are occasional difficulties, but because most of these difficulties are not anticipated. Much experimental effort is still needed to provide a solid basis for this field of application. At present, the best strategy is *flexibility:* the use of various host strains, promoters, and possibly the transition from prokaryotic to eukaryotic vector systems such as yeast vectors.

The reader may be astonished that only *E. coli* vectors have been discussed in this chapter about prokaryotic expression vectors. Indeed, progress in the construction of suitable vectors for other Gram-negative organisms is slow, and the results obtained so far are relatively meagre (*cf.* also Section 4.1.4). The situation for Gram-positive bacteria has been discussed in Chapter 6. There are only few data about expression vectors in these bacteria, apart from special applications, *e.g.*, the production of antibiotics. One of the reasons may be the instability of recombinant DNA molecules and the strong proteolytic activities observed in these systems.

In spite of all difficulties experienced with the expression of heterologous proteins in *E. coli*, it should be remembered that enormous scientific progress and economic success has already been achieved.

References

Amann, E., Brosius, J., and Ptashne, M. (1983). Vectors bearing a hybrid *trp-lac* promotor useful for regulated expression of cloned genes in *E. coli*. Gene 25, 167-178.

Backman, K., Ptashne, M., Gilbert, M. (1976). Construction of plasmids carrying the *cI* gene of bacteriophage λ. Proc. Natl. Acad. Sci. 73, 4174-4178.

Beck, E., and Zink, B. (1981). Nucleotide sequence and genome organization of filamentous bacteriophage f1 and fd. Gene 16, 35-58.

Boser, P.N., Dengler, B., Tinoco, Jr. I., and Uhlenbeck, O.C. (1974). Stability of ribonucleic acid double-stranded helices. J. Mol. Biol. 86, 843-853.

Brosius, J., Dull, T.J., Sleeter, D.D., and Noller, H.F. (1981). Genome organization and primary structure of a ribosomal RNA Operon from *Escherichia coli*. J. Mol. Biol. 148, 107-127.

Brosius, J., Erfle, M., and Storella, J. (1985). Spacing of the -10 and -35 regions in the tac promoter. J. Biol. Chem. 260, 3539-3541.

Buell, G., Schulz, M.F., Selzer, G., Chollet, A., Morra, N.R., Semon, D., Escanez, S., and Kawashima, E. (1985). Optimizing the expression of a synthetic gene encoding somatomedin-C (IGF-I). Nucleic Acids Res. 13, 1923-1938.

Bujard, H. (1980). The interaction of *E. coli* RNA polymerase with promoters. Trends in Biochem. Sciences 5, 274-278.

Chan, S.J., Weiss, J., Konrad, M., White, T., Bahl, C., Yu, S.D., Marks, D., and Steiner, D.F. (1981). Biosynthesis and periplasmic segregation of human proinsulin in *E. coli*. Proc. Natl. Acad. Sci. USA 78, 5401-5405.

Chang, A.C.Y., and Cohen, S.N. (1978). Construction and characterization of amplifiable multicopy DNA cloning vehicles derived from the P15A cryptic miniplasmid. J. Bacteriol. 134, 1141-1156.

Charnay, P., Perricaudet, M., Galibert, F., and Tiollais, P. (1980). Bacteriophage λ and plasmid vectors allowing fusion of cloned genes in each of the three translation phases. Nucleic Acids. Res. 5, 4479-4494.

Crawford, I.P., and Stauffer, G.V.A. (1980). Regulation of tryptophan biosynthesis. Ann. Rev. Biochem. 49, 163-195.

Davis, B.D., and Tai, P.C. (1980). The mechanism of protein secretion across membranes. Nature 283, 433-438.

De Boer, H.A., Comstock, L.J., and Vasser, M. (1983). The *tac* promotor: A functional hybrid derived from the *trp* and *lac* promotors. Proc. Natl. Acad. Sci. USA 80, 21-25.

Derom, C., Gheysen, D., and Fiers, W. (1982). High level synthesis in *E. coli* of the SV40 small t-antigen under controll of the bacteriophage λ P_L promotor. Gene 17, 45-54.

Devare, S.G., Shatzman, A.R., Robbins, K.C., Rosenberg, M., and Aaronson, S.A. (1984). Expression of the PDGF-related transforming protein of Simian Sarcoma Virus in *E. coli*. Cell 36, 43-49.

Devos, R., Emmelo, J., Contreras, R., and Fiers, W. (1979). Construction and characterization of a plasmid containing a nearly full-size copy of bacteriophage MS2 RNA. J. Mol. Biol. 128, 595-619.

Dunn, J.J., and Studier, F.W. (1983). Complete nucleotide sequence of bacteriophage T7 DNA and the location of T7 genetic elements. J. Mol. Biol. 166, 477-535.

Edman, J.C., Hallewell, R.A., Valenzuela, P., Goodman, H.M., Rutter, W.J. (1981). Synthesis of hepatitis B surface and core antigens in *E. coli*. Nature 291, 503-506.

Emtage, J.S., Tacon, C.A., Catlin, G.h., Jenkins, B., Porter, A.G., and Carey, N.H. (1980). Influenza antigenic determinants are expressed from haemagglutinin genes cloned in *Escherichia coli*. Nature 283, 171-174.

Fiers, W., Contreras, R., Duerinck, F., Haegeman, G., Iserentant, D., Merregaert, J., Min Jon, W., Molmans, F., Raeymackers, A., Van den Berghe, A., Volckaert, G., and Ysebaert, M. (1976). Complete nucleotide sequence of bacteriophage MS2 RNA primary and secondary structure of the replicase gene. Nature 260, 500-507.

Fuller, F. (1982). A family of cloning vectors containing the *lacUV5* promotor. Gene 19, 43-54.

Gentz, R., Langner, A., Chang, A.C.Y., Cohen, S.N., and Bujard, H. (1981). Cloning and analysis of strong promotors is made possible by the downstream placement of a RNA termination signal. Proc. Natl. Acad. Sci. USA 78, 4936-4940.

Gheysen, D., Iserentent, D., Derom, C., and Fiers, W. (1982). Systematic alteration of the nucleotide sequence preceding the translational initiation codon and the effects on bacterial expression of the cloned SV40 small t-antigen gene. Gene 17, 55-63.

Goff, S.A., Casson, L.P., and Goldberg, A.L. (1984). Heat shock regulatory gene *hptR* influences rates of protein degradation and expression of the *lon* gene in *E. coli*. Proc. Natl. Acad. Sci. USA 81, 6647-6651.

Goff, S.A., and Goldberg, A.L. (1985). Production of abnormal proteins in *E. coli* stimulates transcription of *lon* and other genes. Cell 41, 587-595.

Gold, L., Pribnow, D., Schneider, T., Shinedling, S., Singer, B.S., and Stormo, G. (1981). Translation initiation in prokaryotes. Ann. Rev. Microbiol. 35, 365-403.

Gonsalus, R.P., and Yanofsky, C. (1980). Nucleotide sequence and expression of *Escherichia coli trpR*, the structural gene for the *trp* aporepressor. Proc. Natl. Acad. Sci. USA 77, 7117-7121.

Gray, G., Selzer, G., Buell, G., Shaw, P., Escanez, S., Hofer, S., Voegeli, P., Thompson, C.J. (1984). Synthesis of bovine growth hormone by *Streptomyces lividans*. Gene 32, 21-30.

Grosjean, H., and Fiers, W. (1982). Preferential codon usuage in prokaryotic genes: the optimal codon-anticodon interaction energy and the selective codon usage in efficiently expressed genes. Gene 18, 199-209.

Guarente, L., Roberts, T.M., Ptashne, M. (1980a). A technique for expressing eukaryotic genes in bacteria. Science 209, 1428-1430.

Guarente, L., Lauer, G., Roberts, T.M., Ptashne M. (1980b). Improved methods for maximizing expression of a cloned gene: A bacterium that synthesises rabbit β-globin. Cell 20, 543-553.

Hallewell, R.A., and Emtage, S. (1980). Plasmid vectors containing the tryptophan operon promoter suitable for efficient regulated expression of foreign genes. Gene 9, 27-47.

Hallewell, R.A., Masiarz, F.R., Najarian, R.C., Puma, J.P., Quiraga, M.R., Randolph, A., Sanchez-Pescador, R., Scandella, C.J., Smith, B., Steimer, K.S., and Mullenbach, G.T. (1985). Human Cu/Zn superoxide dismutase cDNA: isolation of clones synthesizing high levels of active or inactive enzyme from an expression library. Nucleic Acids Res. 13, 2017-2034.

Hawley, D.K., and McClure, W.R. (1983). Compilation and analysis of *Escherichia coli* promoter DNA sequences. Nucleic Acids Res. 11, 2237-2255.

Heiland, I., and Gething, M.J. (1981). Cloned copy of the haemagglutinin gene codes for human influenza antigenic determinants in *E. coli*. Nature 292, 851-852.

Horwitz, J.P., Chua, J., Curby, R.J., Thomson, A.J., Da Rooge, M.A., Fischer, B.E., Mauricio, J., Klundt, I. (1964). Substrates for cytochemical demonstration of enzyme activity. I. Some substituted 3-indolyl-β-D-glycopyranosides. J. Med. Chem. 7, 547-548.

Ikemura, T. (1981). Correlation between the abundance of *Escherichia coli* transfer RNAs and the occurrence of the respective codons in its protein genes: A proposal for a synonymous codon choice that is optimal for the *E. coli* translational system. J. Mol. Biol. 151, 389-409.

Ineichen, K., Shepherd, J.C.W., and Bickle, T.A. (1981). The DNA sequence of the phage λ genome between P_L and the gene *bet*. Nucleic Acids Res. 9, 4639-4653.

Iserentent, D., and Fiers, W. (1980). Secondary structure of mRNA and efficiency of translation initiation. Gene 9, 1-12.

Itakura, K., Hirose, T., Crea, R., Riggs, A.D., Heyneker, H.L., Bolivar, F., and Boyer, H. (1977). Expression in *Escherichia coli* of a chemically synthesized gene for the hormone Somatostatin. Science 198, 1056-1063.

Jay, G., Khoury, G., Seth, A.K., and Jay, E. (1981). Construction of a general vector for efficient expression of mammalian proteins in bacteria: Use of a synthetic ribosome binding site. Proc. Natl. Acad. Sci. USA 78, 5543-5548.

Jay, E., Seth, A.H., Rommens, J., Sood, A., and Jay, G. (1982). Gene expression: Chemical synthesis of *Escherichia coli* ribosome binding sites and their use in bacteria. Nucleic Acids Res. 10, 6319-6329.

Johnsrud, L. (1978). Contacts between *Escherichia coli* RNA Polymerase and a *lac* operon promotor. Proc. Natl. Acad. Sci. USA 75, 5314-5318.

Kahn, M., Koller, R., Thomas, C., Figurski, D., Meyer, R., Remaut, E., and Helinski, D.R. (1979). Plasmid cloning vehicles derived from plasmids ColE1, F, R6K and RK2. Methods in Enzymol. 68, 268-280.

Kastelein, R.A., Berkhout, B., and van Duin, J. (1983). Opening the closed ribosome binding site of the lysis cistron of bacteriophage MS2. Nature 305, 741-743.

Kröger, M., and Hobom, G. (1982). A chain of interlinked genes in the ninR region of bacteriophage λ. Gene 20, 25-38.

Küpper, H., Keller, W., Kurz, C., Forss, S., Schaller, H., Franze, R., Strohmaier, K. Marquardt, O., Zaslarsky, V.G., and Hofschneider, P.H. (1981). Cloning of cDNA of major antigen of foot and mouth disease virus and expression in *Escherichia coli*. Nature 289, 555-559.

Lauer, G., Pastrana, R., Sherley, J., and Ptashne, M. (1981). Construction of overproducers of the bacteriophage 434 repressor and cro proteins. J. Mol. Appl. Genet. 1, 139-147.

Martial, J.A., Hallewell, R.A., Baxter, J.D., and Goodman, H.M. (1979). Human growth hormone: Complementary DNA. Cloning and expression in bacteria. Science 205, 602-607.

Martinez, H.M. (1984). An RNA folding rule. Nucleic Acids Res. 12, 323-334.

Messing, I., and Vieira, I. (1982). A new pair of M13 vectors for selecting either DNA strand of double digest restriction fragments. Gene 19, 269-276.

Michaelis, S., and Beckwith, J. (1982). Mechanism of incorporation of cell envelope proteins in *Escherichia coli*. Ann. Rev. Microbiol. 36, 435-465.

Miller, I.H. (1972). Experiments in Molecular Genetics; ISBN 0-87969-106-9. Cold Spring Harbor Laboratory, Cold Spring Harbor, N.Y., 11724.

Miller, I.H., and Reznikoff, W.S. (1980). The Operon. Cold Spring Harbor Laboratory, ISBN 0-87969-133-6; Cold Spring Harbor, N.Y., 11724.

Min Jon, W., Haegeman, G., and Fiers, W. (1971). Studies on the bacteriophage MS2: nucleotide fragments from the coat protein cistron. FEBS Letters 13, 105-109.

Mount, D.W. (1980). The genetics of protein degradation in bacteria. Ann. Rev. Genetics 14, 279-319.

Mount, D.W. (1984). Modeling RNA structure. Biotechnology 2, 791-795.

Müller-Hill, B., and Kania, I. (1974). *Lac* repressor can be fused to β-galactosidase. Nature 249, 561-563.

Nakamura, K., and Inouye, M. (1982). Construction of versatile expression cloning vehicles using the lipoprotein gene of *Escherichia coli*. The EMBO J. 1, 771-775.

Nakamura, K., Masui, Y., and Inouye, M. (1982). Use of a *lac* promoter-operator fragment as a transcriptional control switch for expression of the constitutive *lpp* gene in *Escherichia coli*. J. Mol. Appl. Gen. 1, 289-299.

Ohsuye, K., Nomura, M., Tanaka, S., Kubota, I., Nakazote, H., Shinagawa, H., Nakata, A., and Noguchi, T. (1983). Expression of chemically synthesized β-neo-endorphin gene fused to *E. coli* alkaline phophatase. Nucleic Acids Res. 11, 1283-1294.

Pribnow, D. (1975). Bacteriophage T7 early promoters: Nucleotide sequences of two RNA polymerase binding sites. J. Mol. Biol. 99, 419-443.

Remaut, E., Staussens, P., and Fiers, W. (1981). Plasmid vectors for high-efficiency expression controlled by the P_L promoter of coliphage λ. Gene 15, 81-93.

Remaut, E., Tsao, H., and Fiers, W. (1983). Improved plasmid vectors with a thermoinducible expression and temperature-regulated runaway replication. Gene 22, 103-113.

Roberts, T.M., Kacich, R., and Ptashne, M. (1979a). A general method for maximizing the expression of a cloned gene. Proc. Natl. Acad. Sci. USA 76, 760-764.

Roberts, T.M., Bikel, I., Yocum, R., Livingstone, D. and Ptashne, M. (1979b). Synthesis of simian virus 40 t-antigen in *E. coli*. Proc. Natl. Acad. Sci. USA 76, 5596-5600.

Rosenberg, M., and Court, D. (1979). Regulatory sequences involved in the promotion and termination of RNA transcription. Ann. Rev. Genet. 13, 319-353.

Rosenberg, M., Ho, Y.S., and Shatzman, A. (1983). The use of pKC30 and its derivatives for controlled expression of genes. Methods in Enzymology 101, 123-138.

Rossi, J.J., Soberon, X., Marumoto, Y., McHation, J., and Itakura, K. (1983). Biological expression of an *Escherichia coli* consensus sequence promoter and some mutant derivatives. Proc. Natl. Acad. Sci. USA 80, 3203-3207.

Roth, R.R. (1981). Frameshift suppression. Cell 24, 601-602.

Sanger, F., Coulson, A.R., Hong, G.F., Hill, D.F., and Petersen, G.B. (1982). Nucleotide sequence of bacteriophage λ DNA. J. Mol. Biol. 162, 729-773.

Schaller, H., Gray, C., and Herrmann, K. (1975). Nucleotide sequence of an RNA polymerase binding site from the DNA of bacteriophage fd. Proc. Natl. Acad. Sci. USA 72, 737-741.

Scherer, G.F.E., Walkinshaw, M.D., Arnott, S., and Morre, D.J. (1980). The ribosome binding sites recognized by *E.coli* ribosomes have regions with signal character in both the leader and protein coding segments. Nucleic Acids. Res. 8, 3895-3907.

Schoner, B.E., Hsiung, H.M., Belagaje, R.M., Mayne, N.G. and Schoner, R.G. (1984). Role of mRNA translational efficiency in bovine growth hormone expression in *E. coli*. Proc. Natl. Acad. Sci. USA 81, 5403-5407.

Schoner, R.G., Elks, L.F., and Schoner, B.E. (1985). Isolation and purification of protein granules from *E. coli* cells overproducing bovine growth hormone. Biotechnology 3, 151-154.

Seeburg, P.H., Shine, J., Martial, J.A., Ivarie, R.D., Morris, J.A., Ullrich, A., Baxter, J.D., and Goodman, H.M. (1978). Synthesis of growth hormone by bacteria. Nature 276, 795-798.

Shimatake, H., and Rosenberg, M. (1981). Purified λ regulatory protein cII positively activates promoters for lysogenic development. Nature 292, 128-132.

Shine, J., and Dalgarno, L. (1975). Determinants of cistron specificity in bacterial ribosomes. Nature 254, 34-38.

Silverstone, A.E., Arditti, R.R., and Magasanik, B. (1970). Catabolite-insensitive revertants of *lac* promotor mutants. Proc. Natl. Acad. Sci. USA 66, 773-779.

Simons, G., Remaut, E., Allet, B., Devos, R., and Fiers, W. (1984). High level expression of human interferon-γ in *E. coli* under control of the P_L promoter of bacteriophage λ. Gene 28, 55-64.

Squires, C.L., Lee, F.D., Yanofsky, C. (1975). Interaction of the *trp* repressor and RNA polymerase with the *trp* Operon. J. Mol. Biol. 92, 93-111.

Stroynowski, I., and Yanofsky, C. (1982). Transcript secondary structure regulate transcription termination at the attenuator of *S. marcescens* tryptophan operon. Nature 298, 34-38.

Stormo, G.D., Schneider, T.D., and Gold, L.M. (1982). Characterization of translational initiation sites in *E. coli*. Nucleic Acids Res. 10, 2971-2995.

Swamy, K.H.S., Goldberg, A.L. (1981). *E. coli* contains eight soluble proteolytic activities, one being ATP-dependent. Nature 292, 652-654.

Tabor, S., and Richardson, C.C. (1985). A bacteriophage T7 RNA polymerase/promoter system for controlled exclusive expression of specific genes. Proc. Natl. Acad. Sci. USA 82, 1074-1078.

Tacon, W., Carey, N., and Emtage, S. (1980). The construction and characterization of plasmid vectors suitable for the expression of all DNA phases under the control of the *E. coli* tryptophan promotor. Molec, Gen. Genet. 177, 427-438.

Talmadge, K., Stahl, S., Gilbert, W. (1980a). Eukaryotic signal sequence transports insulin antigen in *E. coli*. Proc. Natl. Acad. Sci. USA 77, 3369-3373.

Talmadge, K., Kaufman, J., Gilbert, W. (1980b). Bacteria mature preproinsulin to proinsulin. Proc. Natl. Acad. Sci. USA 77, 3988-3992.

Taniguchi, T., and Weissmann, C. (1978). Site-directed mutagenesis in the initiatior region of the bacteriophage Qβ coat protein cistron and its effect on ribosome binding. J. Mol. Biol. 118, 533-565.

Tessier, L.H., Sondermeyer, P., Faure, Th., Dreyer, D., Benavente, A., Villeral, D., Courtney, M., and Lecocq, J.D. (1984). The influence of mRNA primary and secondary structure on human IFN-γ gene expression in *E. coli*. Nucleic Acids Res. 12, 7663-7676.

Thummel, C.S., Burgess, T.L., and Tjian, R. (1981). Properties of simian virus 40 small t-antigen overproduced in bacteria. J. Virol. 37, 683-597.

Tinoko, Jr.I., Borer, P.N., Dengler, B., Levine, M.D., Uhlenbeck, O.C., Crothers, D.M., and Gralla, J. (1973). Improved estimation of secondary structure in ribonucleic acid. Nature New Biology 346, 40-41.

Vieira, I., and Messing I. (1982). The pUC plasmids, an M13mp7-derived system for insertion mutagenesis and sequencing with synthetic universal primers. Gene 19, 259-268.

Villa-Komaroff, L., Efstratiadis, A., Broome, S., Lomedico, P., Tizard, R., Naber, S.P., Chick, W.L.,

and Gilbert, W. (1978). A bacterial clone synthesising proinsulin. Proc. Natl. Acad. Sci. USA 75, 3727-3731.

Voellmy, R.W., and Goldberg, A.L. (1981). ATP-stimulated endoprotease is associated with the cell membrane of *E. coli*. Nature 290, 419-421.

Volckaert, G., Van de Voorde, A., and Fiers, W. (1978). Nucleotide sequences of the simian virus 40 small t-gene. Proc. Natl. Acad. Sci. USA 75, 2160-2164.

Watt, R.A., Shatzman, A.R., and Rosenberg, M. (1985). Expression and characterization of the human *c-myc* DNA-binding protein. Mol. Cell. Biol. 5, 448-456.

Weiss, R., and Gallant, J. (1983). Mechanism of ribosome frameshifting during translation of the genetic code. Nature 302, 389-393.

Windass, J.D., Newton, C.R., De Maeyer-Guignard, I., Moore, V.E., Markham, A.F., and Edge, M.D. (1982). The construction of a synthetic *Escherichia coli trp* promotor and its use in the expression of a synthetic interferon gene. Nucleic Acids Res. 10, 6639-6657.

Yanisch-Perron, C., Vieira, J., and Messing, J. (1985). Improved M13 phage cloning vectors and host strains: nucleotide sequences of the M13mp18 and pUC19 vectors. Gene 33, 103-119.

Yanofsky, C. (1981). Attenuation in the control of expression of bacterial operons. Nature 289, 751-758.

Yanofsky, C., Platt, T., Crawford, I.P., Nichols, B.P., Christie, G.E., Horowitz, H., Van Cleemput, M., and Wu, A.M. (1981). The complete nucleotide sequence of the tryptophan operon of *E. coli*. Nucleic Acids Res. 9, 6647-6668.

Yansura, D.G., and Henner, D.J. (1984). Use of the *Escherichia coli lac* repressor and operator to control gene expression in *Bacillus subtilis*. Proc. Natl. Acad. Sci. USA 81, 439-443.

Youngman, P., Zuber, P., Perkins, J.B., Sandman, K., Igo, M., and Losick, R. (1985). New ways to study developmental genes in spore-forming bacteria. Science 228, 285-291.

8 Eukaryotic Vectors

Recombinant DNA technology has found means to transfer DNA molecules not only into bacterial, but also into eukaryotic cells. In the latter case the transferred genes may be of bacterial origin, and in conjunction with suitable vectors, may be used to select for transformation of higher cells. The re-introduction of eukaryotic genes into their original host cells is of particular interest, since it is only in such cells that the specific mechanisms of gene expression can be studied in detail. In principle, this technology would also allow the introduction of mutated genes into higher cells, *i.e.*, the development of what has been termed surrogate genetics. The reverse, *i.e.*, the introduction of functional genes into cells with genetic defects, appears also feasible. The latter will open new vistas for an important potential application of these techniques, namely, gene therapy (see also Chapter 13).

Several methods are available for the introduction of DNA molecules into higher cells. They include the use of suitable eukaryotic vectors and viruses (Elder *et al.*, 1981), DNA-mediated gene transfer (Scangos and Ruddle, 1981), fusions of artificial and naturally occurring cell membranes, and direct microinjection of genes (Capecchi, 1980). However, these techniques can be employed only if suitable selection systems are available, allowing identification of transformed cells which have taken up recombinant DNA molecules. The practical application of such selection systems is still hampered by some fundamental limitations since the genetic analysis of higher cells is still in its infancy: only a few selectable markers derived from metabolic pathways of amino acids and nucleic acids are available. Dominant markers, effective in normal cells and capable of selectively altering cell phenotypes would be particularly important; however, the use of antibiotics, for example, which play a significant role in bacterial genetics, is restricted to a small number of highly toxic compounds which can be employed in eukaryotic cells. Nevertheless, some selection markers are available and the biochemical bases of their application will bedescribed in the following sections.

8.1 Selectable Marker for Eukaryotic Cells

8.1.1 Thymidine Kinase (TK)

Two pathways, *de novo* biosynthesis and the salvage pathway, provide higher cells with purine and pyrimidine nucleotides. Thymidylate synthetase (TS) which catalyses the conversion of deoxyuridine monophosphate (dUMP) to TMP plays a key role in the biosynthesis of thymidine monophosphate (TMP) (Fig. 8-1). This enzyme requires tetrahydrofolic acid and is inhibited by folic acid analogues such as methotrexate or aminopterin (Fig. 8-5). Higher cells can also

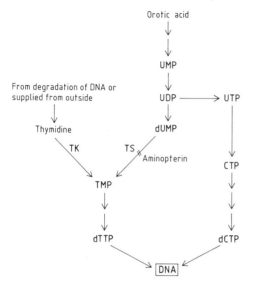

Fig. 8-1. Simplified scheme of pyrimidine metabolism.

The enzymes thymidylate synthetase and thymidine kinase are abbreviated TS and TK, respectively. Biosynthesis starts with orotic acid. Thymidine used in the salvage pathway is either derived from degraded DNA, or can be supplied by adding it to the culture media.

Table 8–1. Selection systems for the enzymes thymidine kinase (Tk; part A) and hypoxanthine guanine phosphoribosyltransferase (HGPRT; part B).

A	HAT	BrdU	
Tk$^+$	+++	– – –	
Tk$^-$	– – –	+++	

B	HAT	6-Thioguanine 8-Azaguanine	XAT
HGPRT$^+$	+++	– – –	– – –
HGPRT$^-$	– – –	+++	– – –
HGPRT$^+$ (Ecogpt$^+$)	+++ .	– – –	+++

(– – –) indicates that cells do not grow and do not survive under these conditions. (+++) indicates that cells either grow or only survice as in the case of Tk$^-$ cells in the presence of bromodeoxyuridine (BrdU). HAT stands for hypoxanthine-aminopterin-thymidine medium, X for xanthine.

synthesise TMP directly from thymidine which they may obtain, for example, from the growth medium. The phosphorylation reaction is catalysed by the enzyme thymidine kinase (TK) (Fig. 8-1). Thymidine kinase positive (Tk$^+$) and negative (Tk$^-$) phenotypes are easily distinguishable if growth media contain hypoxanthine, aminopterin, and thymidine. Such media, known as HAT media, allow only Tk$^+$ cells to grow. It has been mentioned already that thymidine kinase is an enzyme of the salvage pathway. Cells containing this enzyme (Tk$^+$ cells) are capable of converting exogenously added thymidine into TMP and therefore can grow if *de novo* biosynthesis is blocked by aminopterin. Thymidine kinase negative (Tk$^-$) cells cannot grow in HAT medium since both pathways of TMP biosynthesis are blocked, *de novo* biosynthesis by the presence of aminopterin and the salvage pathway by the absence of

thymidine kinase (Table 8-1A). Since aminopterin also blocks purine biosynthesis, the presence of hypoxanthine in HAT medium is required as a source for purine nucleotides (*cf.* Section 8.1.2).

The use of the thymidine kinase system not only allows a selection for Tk$^+$ transformants (Littlefield, 1964); if the thymidine analogue bromodeoxyuridine (BrdU) is used, the Tk$^-$ phenotype can also be selected for. Tk$^+$ cells incorporate exogenous bromodeoxyuridine into their DNA and can be selectively destroyed by irradiation with light of a suitable wave length. Tk$^-$ cells will survive this treatment since bromodeoxyuri-

Fig. 8-2. Primary structure of the HSVI thymidine →
kinase gene and its flanking regions.
Shown are characteristic cleavage sites, the initiation site for transcription (position 409), start and stop codons (double underlining), and control signals such as the TATA box, and the polyadenylation site (single underlining). There are no cleavage sites for *Bcl* I, *Cla* I, *Hind* III, *Hpa* I, *Pvu* I, *Sal* I, *Xba* I, and *Xho* I. Minor sequence alterations are observed in the DNAs of different HSVI strains (McKnight, 1980; Wagner *et al.*, 1981).

```
 10         20         30         40         50         60
GGGTCCTAGG CTTCCATGGG ACCGTATACG TGGACAGGCT CTGGAGCATC GCACGACTGC

 70         80         90        100        110        120
GTGATATTAC CGGAGACCTT CTGCGGGACG AGCCGGGTCA CGCGGCTGAC GGAGGCGTCCG

130        140        150        160        170        180
TTGGGCGACA AACACCAGGA CGGGGCACAG GTACACTATC TTGTCACCCG GAGGCGGAGG

190        200        210        220        230        240
GACTGCAGGA GCTTCAGGGA GTGGCGAGGC TGCTTCATCC CCGTGGCCCG TTGCTGCGGT
    PstI                      PvuII

250        260        270        280        290        300
TTGCTGGCGG TGTCCCCGGA AGAAATATAT TTGCATGTCT TTAGTTCTAT GATGACACAA

310        320        330        340        350        360
ACCCCGCCCA GCGTCTTGTC ATTGGCGAAT TCGAACACGC AGATGCAGTC GGGGCGGGGC
                              EcoRI

370        380        390        400        410        420
GGTCCCAGGT CCACTTCGGA TATTAAGGTG ACGGGTGTGG CCTCGAACAC CGAGCGACCC
                                                          → mRNA

430        440        450        460        470        480
TGCAGGCGGAC CGCTTAACAG CGTCAACAGC GTGCCGCAGA TCTTGGTGGC GTGAAACTCC
    PstI                              BglII

490        500        510        520        530        540
CGCACCTCTT TGGCAAGCGC CTTGTAGAAG CGCGTATGGC TTCGTACCCC TGCCATCAAC
                  HhaI

550        560        570        580        590        600
ACGCGGTCTGC GTTCGACCAG GCTGCCGGTT CTCGGCGGCCA TAGCAACCGA CGTACGGCGT

610        620        630        640        650        660
TGCGCCCTCG CCGGCAGCAA GAAGCCACGG AAGTCCGCCT GGAGCGAGAAA ATGCCCACGC

670        680        690        700        710        720
TACTGCGGGT TTATATAGAC GGTCCTCACG GGATGGGGAA AACCACCACC ACGCAAACTGC

730        740        750        760        770        780
TGGTGGCCCT GGGTTCGCGC GACGATATCG TCTACGTACC CGAGCCGATG ACTTACTGGC

790        800        810        820        830        840
AGGTGCTGGG GGCTTCCGAG ACAAATCGGA ACATCTACAC CACACAACAC CGCCTCGACC

850        860        870        880        890        900
AGGGTGAGAT ATCGGCCGGG GACGCGGCGG TGGTAATGAC AAGCGCCCAG ATAACAATGG

910        920        930        940        950        960
GCATGCCTTA TGCCGTGACC GACGCCGTTC TGGCTCCTCA TGTGTGGGGG GAGGCTGGGGA
```

```
970        980        990       1000       1010       1020
GTTCACATGC CCCGCCCCG GCCCTCACCC TCATCTTCGA CCGGCCATCC ATCGCCGCCC

1030       1040       1050       1060       1070       1080
TCCTGTGTA CCCGGCCGCG CGGTACCTTA TGGGCAGCAT GACCCCCAG GCCGTGCTGG
                           KpnI

1090       1100       1110       1120       1130       1140
CGGTCGTGGC CCTCATCCCG CCGACCTTGC CCGGCACACC AACATCGTGC TTGGGGCCCT

1150       1160       1170       1180       1190       1200
TCCGGAGGAC AGACACATCG ACCGGCCTGGC CAAAACGCAG CGCCCCGGCG AGCGGCTGGA

1210       1220       1230       1240       1250       1260
CCTGGCTATG CTGGCTGCGA TTCGGCGCGT TTACGGGCTA CTTGCCAATA CGGTGCGGTA
                    HinfI

1270       1280       1290       1300       1310       1320
TCTGCAGTGC GGCGGGTCGT GGCGGGGAGGA CTGGGGACAG CTTTCGGGGA CGGCCGTGCC
    PstI

1330       1340       1350       1360       1370       1380
GCCCCAGGGT GCCGAGCCCC AGAGCAACGC GGGCCCAGGA CCCCATATCG GGGACACGTT

1390       1400       1410       1420       1430       1440
ATTTACCCTG TTTCGGGCCC CCGAGTTGCT GGCCCCCAAC GGCGACCTGT ATAACGTGTT

1450       1460       1470       1480       1490       1500
TGCCTGGGGC TTGGACGTCT TGGCCAAACG CCTCCGTCCC ATGCAGTCT TTATCCTGGA

1510       1520       1530       1540       1550       1560
TTACGACCAA TCGCCCGCCG GCTGCCGGGA CGCCCTGCTG CAACTTACCT CCGGGATGGT
                                             HaeIII

1570       1580       1590       1600       1610       1620
CCAGACCCAC GTCACCACCC CCGGGCTCCAT CCGGCTCCAT TGGGACCTGG CGCGCACGTT

1630       1640       1650       1660       1670       1680
TGCCCGGGAG ATGGGGGAGG CTAACTGAAA GACCATACCG GACAATACCG GAAGGAACCC
    SmaI

1690       1700       1710       1720       1730       1740
GGCTATGAC GGCAATAAAA AGACAGAATA AAACGCACGG GTGTTGGGTC GTTTGTTCAT

1750       1760       1770       1780       1790       1800
AAACGCGGGG TTCGGTCCCA GGGCTGGCAC TCTGTCGATA CCCCACCGAG ACCCCATTGG
```

dine will not be phosphorylated, due to the lack of thymidine kinase, and will not be incorporated into DNA. Stable Tk⁻ mutants, however, are only obtained if the cells are kept for several passages in growth medium containing BrdU.

The thymidine kinase encoded by herpes simplex virus type I (HSVI) is of particular interest because the incorporation of the herpes virus gene will complement for the defect in Tk⁻ cells; in addition, the viral gene product differs from the cellular thymidine kinase in a number of biochemical and serological parameters so that the viral and cellular kinases can be easily distinguished from each other. The herpes virus thymidine kinase gene was first identified by Wigler *et al.* (1977). It lies on a 3.4 kb *Bam* HI fragment of the viral DNA between 26 and 31% of the physical map, and has been completely sequenced (McKnight, 1980; Wagner *et al.*, 1981). The gene does not contain introns, and the mRNA which is approximately 1 300 nucleotides in length codes for 376 amino acids (Fig. 8-2). Plasmid ptkM2 which is 9.57 kb in length, is of practical importance since it contains the HSVI *tk* region cloned in pBR322 (Fig. 8-3) (Wilkie *et al.*, 1979).

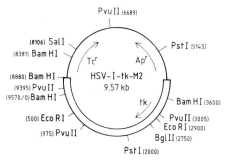

Fig. 8-3.   Structure of vector HSVI-tk-M2.
This vector contains a 5.2 kb *Kpn* I fragment of HSVI DNA (open bar) cloned into the *Eco* RI site of pBR322 (single line) *via* poly(dA)-poly(dT) tails. Co-ordinates are relative to the *Bam* HI site at position 0/9570 and run counterclockwise. The *tk* gene is located between positions 2300 and 3400. The only *Sal* I site of the vector is at position 8106 within the pBR322 portion (Wilkie *et al.*, 1979).

The unique *Sal* I site of this plasmid at position 8 106 is frequently exploited for the generation of recombinants containing the DNA to be employed in transformation assays. Stable Tk⁺ cells can be isolated after transformation of Tk⁻ mouse cells with ptkM2 DNA and salmon sperm or calf thymus DNA as carrier. Transformed cells contain a copy of the HSV thymidine kinase gene integrated at different positions on the chromosomal DNA (Pellicer *et al.*, 1978). Wigler *et al.* (1979b) have shown that not only the *tk* gene but also any other DNA, which does not even have to be covalently linked to *tk* DNA, can be introduced into cells and become integrated into its chromosomes. This approach which has been termed "co-transformation" quite frequently leads to the expression of co-transformed DNA. Mantei *et al.* (1979) have shown, for example, that rabbit β-globin mRNA is processed correctly in co-transformed cells. Breathnach *et al.* (1980) have described the expression of chicken ovalbumin mRNA in co-transformed mouse cells; in addition, Mantei and Weissmann (1982) have observed that co-transformed human α1 interferon genes were not only expressed in mouse L cells, but that this expression was also inducible. Infection of transformed cells with suitable viruses resulted in an approximately twenty-fold increase in α1 interferon-specific mRNA.

Yields of Tk⁺ transformants obtained in a co-transformation experiment are usually on the order of 1 000 transformants per microgram of input DNA; the number of stably integrated genes is small and varies between 0.25 to 30 per haploid genome. Of course, the technique can be employed only if suitable Tk⁻ cells are available.

8.1.2 Hypoxanthine Guanine Phosphoribosyl Transferase (HGPRT)

The enzyme HGPRT catalyses the conversion of the purine bases hypoxanthine and guanine into

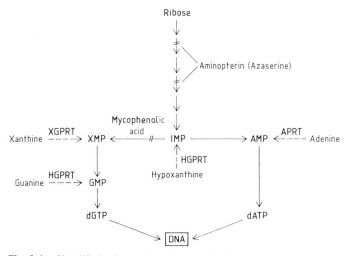

Fig. 8-4. Simplified scheme of purine metabolism.
HGPRT = hypoxanthine guanine phosphoribosyl transferase; XGPRT = xanthine guanine phosphoribosyl transferase; APRT = adenine phosphoribosyl transferase. Solid arrows indicate *de novo* biosynthesis, broken arrows the salvage pathway.

the corresponding nucleotides IMP and GMP (Fig. 8-4). Since HGPRT is also an enzyme of the salvage pathway for purine nucleotides, HAT medium can be used as a positive selection for the HGPRT-positive phenotype. Although aminopterin will block *de novo* synthesis of purines in two places, purines still can be obtained from hypoxanthine through a bypass reaction catalysed by HGPRT; on the other hand, HGPRT-negative cells, for which *de novo* synthesis of purines is blocked by aminopterin, and which cannot utilise hypoxanthine, are unable to grow in HAT medium (Table 8-1B).

HGPRT-negative cells can be selected by treatment with 8-azaguanine or 6-thioguanine (see Fig. 8-5), which are toxic analogues of guanine and substrates for HGPRT; only HGPRT-negative cells will survive this treatment. The human disease known as Lesch-Nyhan syndrome is a genetic defect of HGPRT. Cells from homozygous patients afflicted by the disease lack HGPRT and are unable to grow in HAT medium.

Fibroblasts from such patients, derived from skin biopsies and immortalised by SV40 transformation (Croce *et al.*, 1973), are frequently used by cell biologists and gene technologists.

The enzyme XGPRT or Ecogpt is the *E. coli* analogue of HGPRT (Nüesch and Schimperli, 1984). The bacterial enzyme also catalyses the condensation of phosphoribosyl pyrophosphate with hypoxanthine or guanine, yielding inosinic acid (IMP) and guanylic acid (GMP), respectively; in addition, it converts xanthine to XMP, a reaction not catalysed by the mammalian enzyme. The *E. coli* enzyme therefore can be used to considerable advantage as a selectable marker (Mulligan and Berg, 1981a). It does not only complement the enzymatic defect in HGPRT-negative mammalian cells, but also confers on them a new property, namely, the ability to utilise xanthine in the biosynthesis of GMP, irrespective of the HGPRT phenotype of the cells (Fig. 8-4). Normal cells and cells expressing the *Ecogpt* gene product therefore can easily be distinguished from

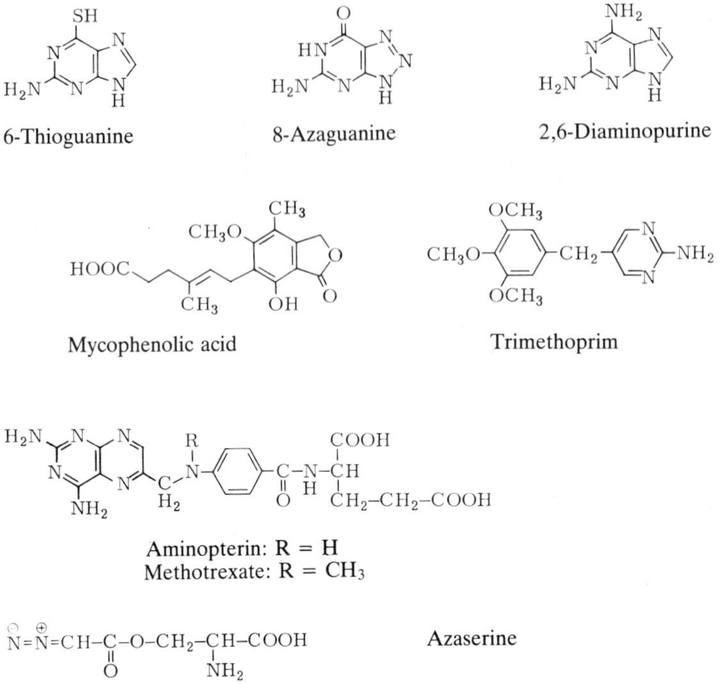

6-Thioguanine   **8-Azaguanine**   **2,6-Diaminopurine**

Mycophenolic acid   **Trimethoprim**

Aminopterin: R = H
Methotrexate: R = CH$_3$

Azaserine

Fig. 8-5. Inhibitors of nucleotide biosynthesis.

each other by this new dominant phenotype: cells transformed by *Ecogpt* do not only grow in HAT medium but also in a growth medium in which hypoxanthine is replaced by xanthine (Table 8-1B). This selection system can be improved by using mycophenolic acid, an inhibitor of IMP dehydrogenase (Fig. 8-5). Normal, *i.e.*, Ecogpt-negative cells, can survive in the presence of aminopterin and mycophenolic acid only if they are supplied with adenine and guanine; they do not survive if they are forced to make do with adenine and xanthine. Cells expressing Ecogpt, however, will grow because they can utilise xanthine. The addition of mycophenolic acid prevents cells from using hypoxanthine which is constantly being resynthesised in the cells. This selection system, therefore, completely depends on xanthine utilisation (Mulligan and Berg, 1981a).

8.1.3 Adeninephosphoribosyl Transferase (APRT)

The enzyme APRT, another enzyme of the purine salvage pathway, catalyses the conversion of adenine to AMP (Fig. 8-4). APRT-positive cells therefore can be selected in a medium containing, for example, the glutamine analogue azaserine which prevents *de novo* synthesis of purines (Wigler *et al.*, 1979a). The use of azaserine (Fig. 8-5) is preferred to that of aminopterin since it inhibits *de novo* synthesis of purine, but not of thymidine. APRT-negative cells cannot grow in a medium containing azaserine and adenine and can be selected by treatment with 2,6-diaminopurine. This compound is extremely toxic for normal cells, but APRT-negative cells survive because they do not incorporate it.

8.1.4 Dihydrofolate Reductase (DHFR)

Dihydrofolate reductase catalyses the reduction of dihydrofolic acid to tetrahydrofolic acid (Fig. 8-6), which in turn is required for the *de novo* synthesis of purines, TMP, and glycine from serine. Cells lacking the gene therefore can grow only in the presence of purines, glycine, and thymidine. The mammalian enzyme is extremely sensitive to folate antagonists, such as aminopterin and methotrexate (amethopterin) (Fig. 8-5). These drugs are inhibitory at concentrations of 10^{-9} M and easily overcome normal intracellular concentrations of dihydrofolic acid.

There are several mechanisms by which cells may acquire methotrexate resistance: they may be unable to take up methotrexate, they may contain many copies of the *DHFR* gene and synthesise large amounts of the enzyme, or they may be mutated in a way that reduces the affinity of the enzyme for methotrexate. Cells adapted to methotrexate by a stepwise increase of methotrexate concentrations usually show an increase of DHFR levels, which is due to the amplification of the *DHFR* gene from 2 to 50 or even 1 000 copies per cell (Alt *et al.*, 1978). Increased resistance in this case is due to a genuine gene dosage effect, and is not the result of structural mutations; nevertheless, the phenomenon of gene amplification can be exploited and there are a variety of interesting applications (*cf.* also Section 8.8).

A population of methotrexate-resistant cells will sometimes contain cells with DHFR exhibiting a reduced affinity to methotrexate. An enzyme with a binding affinity to methotrexate 270-fold lower than that of the wild-type enzyme has been isolated from cultured mouse cells after a suitable selection (Simonsen and Levinson, 1983). The analysis of the corresponding cDNA clone revealed that this difference had been caused by a single amino acid change from arginine to leucine at position 22 of the protein which is 186 amino acids in length. The expression vector pFR400 which expresses *DHFR* cDNA under the control of the SV40 early promoter efficiently transforms DHFR-negative cells and also wild-type cells. Transformants can be selected for at methotrexate concentrations which inactivate the cellular enzyme. This mutant DHFR therefore can be used as another dominant selectable marker. Furthermore, a number of other genes can be co-transformed with this marker. Co-transformation yields stable transformants which also express the co-transformed gene (Gitschier *et al.*, 1984).

Another dominant selectable marker is trimethoprim resistance. The affinity of mammalian dihydrofolate reductase towards trimethoprim is approximately three orders of magnitude lower than that of bacterial enzymes (Fig. 8-5). Bacteria expressing the mammalian enzyme are resistant to concentrations of trimethoprim which would

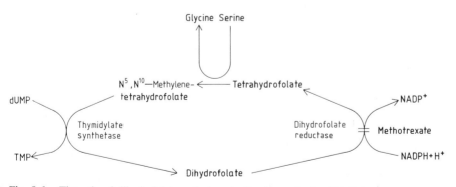

Fig. 8-6. The role of dihydrofolate reductase in the biosynthesis of TMP from dUMP.

normally inhibit cell proliferation completely (Chang *et al.*, 1978). This property has been used for cloning a mouse dihydrofolate reductase gene in *E. coli*.

DHFR-negative cells can be selected using ^{3}H-deoxyuridine of high specific activity as a selective agent (Urlaub and Chasin, 1980; Urlaub *et al.*, 1981). The tritium-labelled compound is first converted to TMP and subsequently incorporated into DNA. DNA molecules are very sensitive to the radioactive decay of incorporated tritium atoms. Since the thymidylate synthetase reaction which catalyses the conversion of dUMP to TMP requires methylene-tetrahydrofolic acid, only DHFR-negative cells will be resistant to the toxic effects of ^{3}H-deoxyuridine.

DHFR-negative cells are auxotrophic, requiring exogenous glycine, hypoxanthine and thymidine, and can multiply only if these three compounds are supplied with the growth medium; in addition, such cells are methotrexate-resistant. Therefore they can be selected in a medium containing glycine, hypoxanthine, thymidine, and methotrexate. DHFR-positive cells can be distinguished from DHFR-negative cells because the former do not require glycine, hypoxanthine, and thymidine.

8.1.5 Bacterial Antibiotic Resistance Genes

It has been shown that bacterial antibiotic resistance genes may also be functional in mammalian cells. Higher concentrations of antibiotics normally used in the treatment of bacterial infections are usually not toxic for eukaryotic cells; however, not only yeast, but also mouse, simian, and human cell lines are sensitive to compounds such as the aminoglycoside antibiotic G418. This 2'-deoxystreptamine antibiotic, a derivative of gentamycin, is inactivated through phosphorylation by aminoglycoside 3'-phosphotransferase activities (APH(3')I and APH(3')II) which are encoded by the well-known kanamycin or neomycin resistance genes localised on transposons such as Tn5 or Tn601 (Tn903) (Davies and Smith, 1978). These genes can be transferred to higher cells, and their expression results in an antibiotic resistance which can be easily selected for as a stable dominant marker (Colbère-Garapin, 1981). Many mammalian cell lines, such as mouse 3T3, L, CV1, TC7, and HeLa cells, are sensitive to 100 µg/ml of G418. Since cytotoxic effects at this concentration are only observed after several days, the use of 400 and even 800 µg/ml is recommended. Even at these concentra-

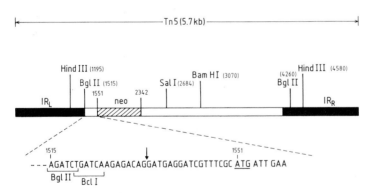

Fig. 8-7. Physical map of transposon Tn5.
Transposon Tn5 is 5.7 kb in length and carries the *APH(3')II* gene (hatched bar). It is flanked by two inserted terminal repetitions, IR$_L$ and IR$_R$, each of which is 1 523 bp in length (black bars). The lower line shows a sequence section with parts of the 5' untranslated region, and the first 3 of a total of 264 codons of the *APH(3')II* gene (*neo*). The arrow marks the transition to the leftward IR$_L$ region. Numbers are Tn5 co-ordinates (Beck *et al.*, 1982).

tions sensitive cells will go through one or two cycles of cell division before cell death ensues (Southern and Berg, 1982). The frequency of spontaneous resistance to G418 in mammalian cells is lower than 10^{-7}. G418 resistance is also expressed in plant and *Drosophila* cells and can therefore also be used as a dominant marker in these systems.

Sequence analysis of the Tn5-derived *APH(3')II* and the Tn903-derived *APH(3')I* kanamycin resistance genes has shown that there are no homologies between these two genes (Velten *et al.*, 1984; Beck *et al.*, 1982). The *APH(3')II* gene is contained in a 1.1 kb *Bgl* II-*Bam* HI fragment from Tn5, which has been cloned into the *Bam* HI site of pBR322 to yield pKC7 (Rao and Rogers, 1979). The *Bgl* II site is located in the 5′ untranslated region, 35 bp upstream of the initiation codon (Fig. 8-7). It is conveniently used to express the *APH(3')II* gene under the control of various eukaryotic promoters, *e.g.*, the herpes thymidine kinase promoter (Colbère-Garapin *et al.*, 1981) or the Ti plasmid-derived nopaline synthetase promotor (Herrera-Estrella *et al.*, 1983; *cf.* Chapter 10). The *APH(3')II* gene codes for a 29 kDa protein, the active form of which is a dimer of 58 kDa. A very sensitive and quantitative assay has been described by Reiss *et al.* (1984).

Another useful marker is chloramphenicol resistance. The coding sequence of the resistance gene, carried on transposon Tn9, is 1 102 bp in length and is flanked by two IS1 elements of 768 bp (Shaw, 1975). Resistance is mediated by chloramphenicol acetyltransferase (CAT) which inactivates chloramphenicol by converting it into mono- and bi-acetylated derivatives (Shaw, 1967) (Fig. 8-8). These derivatives can be detected easily, for example, by thin layer chromatography (Fig. 8-9). This enzyme is of great advantage because it does not occur in mammalian cells; it is easily detectable and is expressed in eukaryotic cells (Gorman *et al.*, 1982). The gene can be obtained from a derivative of pBR322 carrying transposon Tn9 by cleavage with suitable enzymes (Fig. 8-10). It is then introduced into vector pSV2, yielding plasmid pSV2-cat^s (Fig. 8-11), which expresses the *CAT* gene from the early promoter of SV40 (see Section 8.3.2). Since enzyme activities are easily identified, a CAT assay is ideal for the quantitation of eukaryotic promoter functions (Laimins *et al.*, 1982). In principle it should also be possible to develop a positive selection system based on chloramphenicol acetyltransferase because this enzyme inhibits mitochondrial protein biosynthesis in higher cells. Especially in the presence of high concentrations of glucose (6.5 mM), chloramphenicol treatment results in cell death after only a few cell generations; cells harbouring suitable CAT vectors would be able to detoxify intracellular chloramphenicol by acetylation and thus prevent the antibiotic from binding to mitochondrial ribosomes (Ziegler and Davidson, 1979; Gorman *et al.*, 1982).

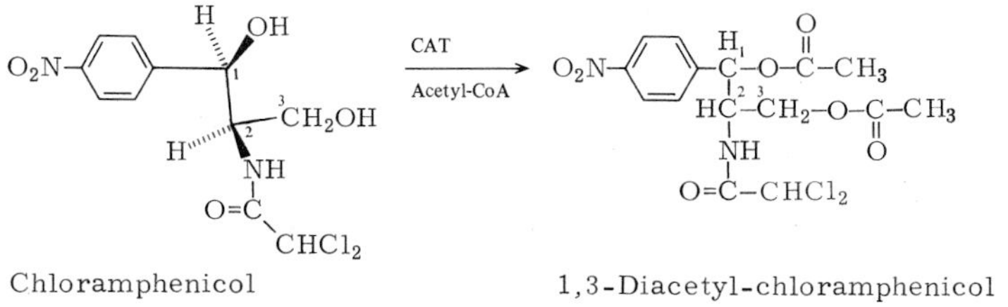

Fig. 8-8. Reactions of chloramphenicol acetyl transferase (CAT).
The major product of the reaction of chloramphenicol with acetyl-CoA is 3-acetyl-chloramphenicol. The diacetyl derivative is formed with a rate approximately two orders of magnitude slower than the monoacetyl derivatives (Shaw, 1967).

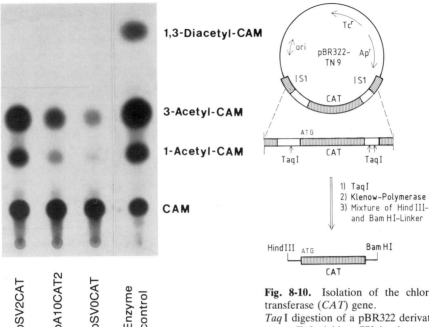

1,3-Diacetyl-CAM

3-Acetyl-CAM

1-Acetyl-CAM

CAM

pSV2CAT pA10CAT2 pSV0CAT Enzyme control

Fig. 8-9. CAT assay.
The CAT assay measures the activity of the bacterial enzyme chloramphenicol acetyltransferase (CAT) following transfer of the corresponding gene into eukaryotic cells. In appropriate plasmid constructions the expression of the *CAT* gene is dependent upon the presence of a eukaryotic promoter preceding this gene. Plasmid pSV2CAT carries the entire SV40 early promoter region in front of the CAT coding region (*cf.* Fig. 8-11); plasmid pSV0CAT was generated by removing the small *Hind* III-*Acc* I fragment from pSV2CAT, and by adding *Hind* III linkers to the junction. It thus lacks the entire SV40 early promoter. Plasmid pA10CAT2 lacks the 72 bp repeat elements but retains the other elements of the SV40 early promoter extending from positions 5171 to 130 (*cf.* Fig. 8-13; Laimins *et al.*, 1982). Following transfection, cells are incubated for 2 days at 37°C, lysed, and incubated with ^{14}C-chloramphenicol (CAM) and acetyl-CoA. Reaction products (1-acetyl-CAM, 3-acetyl-CAM and 1,3-diacetyl-CAM) are analysed by ascending thin layer chromatography (chloroform/methanol = 95:5) and subsequent autoradiography (Courtesy of Dr. Martin Lipp, Munich).

Fig. 8-10. Isolation of the chloramphenicol acetyl transferase (*CAT*) gene.
Taq I digestion of a pBR322 derivative carrying transposon Tn9 yields a 773 bp fragment with protruding ends which can be converted to blunt ends by treatment with the Klenow fragment of *E. coli* DNA polymerase I. The modified fragment is ligated with a mixture of equimolar amounts of *Hind* III and *Bam* HI linkers such that 25% of the resulting DNA molecules contain a *Hind* III site at their 5' end and a *Bam* HI site at their 3' end (Gorman *et al.*, 1982). Coding regions for the *CAT* gene and flanking IS sequences are shown as hatched bars, non-coding regions as open bars.

8.2 Simian Virus 40

The viral genome of simian virus 40 (SV40) is 5 243 bp in length and has been sequenced completely. It offers extraordinary advantages for the development of vectors. SV40 contains a small circular DNA molecule (Fiers *et al.*, 1978; Reddy *et al.*, 1978) which is easily obtainable in large quantities; the various viral functions have been accurately located with respect to a detailed physical map of the DNA. Since the genuine aim of vector development is the expression of insert-

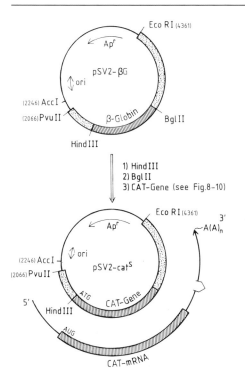

Fig. 8-11. Construction of a CAT-SV40 plasmid vector.

The β-globin sequences of vector pSV2-βG (Fig. 8-22) are removed by *Hind* III/*Bgl* II digestion, and replaced by the coding sequence for the *CAT* gene (Fig. 8-10). The resulting plasmid is called pSV2-cat^s (s for sensitive) since it does not confer chloramphenicol resistance to transformed bacteria. This is due to the lack of a prokaryotic promoter in front of the *CAT* gene insertion which therefore cannot be expressed in *E. coli*. In eukaryotic cells, however, the SV40 promoter between the *Pvu* II and the *Hind* III site is active and promotes transcription of the *CAT* gene. The interruption indicated in the mRNA shows the 5′ and 3′ splice sites of the intron from the SV40 small t-antigen, which are present in vector pSV2 at the sites indicated (Gorman *et al.*, 1982). SV40 sequences are shown as stippled bars, CAT sequences as hatched bars, and pBR322 sequences as thin lines. Numbers are pBR322 co-ordinates.

ed genes, the mechanisms of gene expression in higher cells in general, and the functional organisation of SV40 DNA in particular, have to be taken into account. Some basic principles of SV40 replication and gene expression will therefore be discussed in the following sections before the development of SV40 vectors will be addressed.

8.2.1 The Structure of SV40 DNA

SV40 DNA is isolated from virus particles in the form of a double-stranded superhelical DNA which has been named Form I DNA. Form I DNA is converted into Form II DNA by opening a single phosphodiester bond in the DNA molecule. Form II DNA sediments much slower than Form I DNA under alkaline conditions (Table 8-2). The introduction of a double strand break, for example, by digestion of the DNA with a restriction endonuclease, yields linear DNA molecules known as Form III DNA. All three forms of SV40 DNA can be distinguished from each other by their sedimentation behaviour (Table 8-2) and also by their mobilities in agarose gels. Form I DNA is superhelical because of a particular tension in the entire DNA molecule; this tension results from the fact that SV40 DNA in infected cells is not naked but complexed with histone proteins. This form of the DNA is often referred to as the SV40 minichromosome and consists of 24 nucleosomes showing the typical pearl-bead structure of chromatin. When nucleosome proteins are removed during DNA purification this DNA superstructure is destroyed and the naked DNA reacts by forming superhelical twists. The number of twists is identical to the number of nucleosomes, *i.e.* 24. The distribution of nucleosomes on SV40 DNA is not entirely random. Electron microscopic observations (Saragosti *et al.*, 1980) and experiments with DNase I (Elgin, 1981; Varshavsky *et al.*, 1979) have revealed that a region of approximately 400 bp around the origins of DNA replication and of late transcription is entirely free of nucleosomes. Since it is generally assumed that the DNase I sensitivity of DNA sequences in chromatin is a sign of transcriptional activity (Weisbrod, 1982), a particular biological role has been ascribed to this region of the SV40 DNA, namely a specific interaction with SV40 T-antigen.

Table 8–2.   Properties of SV40 DNA.

DNA	Sedimentation rate	
	neutral pH	alkaline pH
form I	20S (ccc DNA)	53S
form II	16S (relaxed circle)	18S circular single strand
		16S linear single strand
form III	14S (linear DNA)	16S circular single strand

from Vinograd *et al.* (1965); ccc DNA stands for closed circular DNA.

T-antigen is a protein with a molecular weight of 90 kDa which is absolutely required for viral DNA replication. However, its exact role in this process still remains to be elucidated. The replication of SV40 DNA is bidirectional and begins at a specific origin of replication at position 0/5243 near a unique *Bgl*I cleavage site. Replication does not require particular termination signals and ends where the two replication forks meet. An essential contribution to the elucidation of the mechanism of SV40 DNA replication has been made by Hirt (1967) who devised a procedure for the rapid and quantitative isolation of viral DNA and its replicative forms from infected cells by extraction with sodium dodecylsulfate (SDS). This procedure is known as Hirt extraction.

The primary structure of SV40 DNA can be found in Appendix B.2. It will be referred to quite frequently in the following sections. Unfortunately, two different sequencing approaches have resulted in two different numbering systems; the system known as the "BBB" nomenclature (Buchmann *et al.*, 1981) will be used in this book. Its reference point is a central G/C base pair of a palindromic sequence of 27 bp around the *Bgl*I recognition site. This base pair has been designated 0/5243 (Fig. 8-13). Numbering proceeds clockwise in the same direction in which the late RNA is read. The reference point for the "W" numbering system (after S. Weissman) of Reddy *et al.* (1978) is shifted by 82 bp in a counterclockwise direction with respect to the "BBB" system. Both

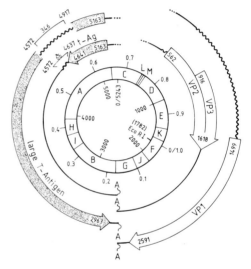

Fig. 8-12.   Genetic map of SV40 DNA.
The reference point for this map is the central nucleotide in the 27 bp palindromic sequence in the vicinity of the single *Bgl*I site (GCCNNNN/NGGC) at position 5243. Late transcription proceeds clockwise, early transcription counterclockwise. Shown are the mRNAs (thin lines) and their products (thick arrows). Numbers at the ends of arrows and at arrow heads, indicate the positions of start and stop codons, respectively. The positions of the two introns in the region coding for the two tumour antigens are indicated by numbers according to the "BBB" nomenclature (see text). Wavy lines indicate intron sequences. The two inner circles show the positions of the 13 *Hinc*II-*Hind*III fragments A to M. Older representations frequently make use of the single *Eco*RI site at position 1782 as a zero reference point. In this case the genome is divided into fractions ranging from 0 to 1. The corresponding positions are marked at the outer periphery (Fiers *et al.*, 1978).

systems can be aligned by applying the following formulas:

"BBB" = "W" + 65 (if "W" < 95);
"BBB" = "W" + 82 (if 95 ≤ "W" ≤ 5 161);
"BBB" = "W" − 5 160 (for "W" < 5 161).

The SV40 map is also traditionally divided into map units (m.u.). The reference point in this case is the *Eco* RI site at position 1782 = 0 or 1 m.u.. Co-ordinates and map units can be calculated by applying the formula

m.u. = ("BBB" + 346) : 5 243

For historical reasons, graphical representations of map units usually place the *Eco* RI site at position "12 o'clock".

8.2.2 The Lytic Infection Cycle of SV40

SV40 infection with virus particles or transfection with naked circular DNA induces a lytic infection cycle in permissive monkey cells. The first step in this cycle is the synthesis and expression of the early mRNA. This is followed by viral DNA replication which, in turn, triggers the synthesis of late mRNA. After the synthesis of late proteins has been completed, the viral DNA is packaged and intact virus particles are formed and released; infected cells do not survive. This cycle of early and late phases is also reflected in the organisation of the viral genome which can be divided into two parts of almost equal lengths (Fig. 8-12), the left part representing the early, and the right part the late regions.

8.2.2.1 Early mRNA Synthesis

The early region codes for two partially overlapping genes which direct the synthesis of T-antigen (large T) with a molecular weight of 90 kDa, and t-antigen (small t) with a molecular weight of 18 kDa. T-antigen has a variety of biological functions, in particular the initiation of DNA replication and the control of its own transcription. The

latter is achieved by binding of T-antigen to the origin of DNA replication, located within a region of approximately 400 bp, which also comprises control regions for early and late transcription (Fig. 8-13). Small t-antigen is required for the interaction of the virus with non-permissive cells. Both early proteins are encoded by the same primary transcript, but each protein possesses its own mRNA, the 5' and 3' ends of which map close to each other. The differences in the sizes of these proteins are explained by the fact that the two mRNA molecules arise by two different splicing events. They alter the reading frames on the mRNA molecules in such a way that a stop codon occurs after 525 of a total of 2 600 bases in the 2.6 kb mRNA of t-antigen (see also Fig. 8-12). The two introns with their common 5', but different 3' splice sites differ in length: 66 bp for small t-antigen, and 346 bases for large T-antigen (Fig. 8-12). In the case of large T-antigen this generates a long open reading frame for the synthesis of a 90 kDa protein. 82 amino acids of the *N*-terminal regions of large T-antigen and small t-antigen are identical, but due to the splicing events large T-antigen lacks the C-terminal part of small t-antigen.

The synthesis of early proteins not only requires correct splicing, but also depends on the organisation of the DNA within the transcriptional start region. While the Goldberg-Hogness or TATA box, which lies 25 bp downstream from the transcriptional start, can be removed without major consequences, other upstream sequences within a region containing two 72 bp repeats are regarded as essential for transcription (Fig. 8-13). These repetitions must not be removed from recombinant DNA constructions (Gruss *et al.*, 1981c). It is quite remarkable that the influence of these *cis*-acting elements, known as enhancer sequences, on the expression of early genes is independent of their position and orientation (see also Section 8.4).

The expression of early mRNA is subject to autoregulation (Fig. 8-13C) (Hansen *et al.*, 1981). T-antigen binds as a tetramer to two regions

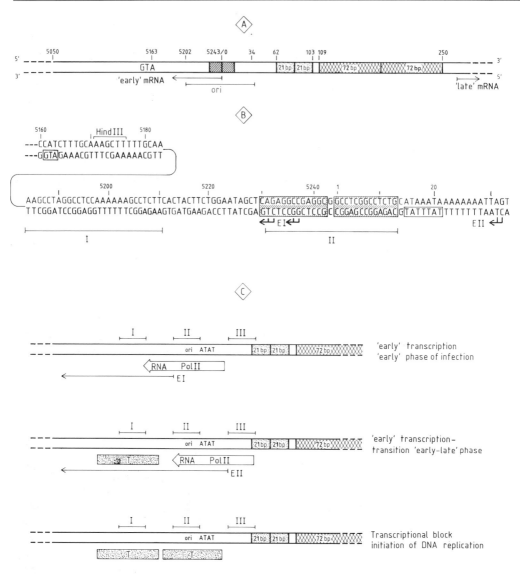

Fig. 8-13. Autoregulation of T-antigen synthesis.
(A) Shown is a 350 bp fragment of SV40 DNA near the origin of DNA replication with the two 72 bp and the 21 bp repeats, and the 27 bp palindromic sequence which defines the origin of DNA replication (position 0/5243). The start codon for the two early proteins, large T- and small t-antigen, is at position 5163.
(B) DNA sequence near the T-antigen binding sites I and II. *EI* marks the position of the 5' ends of transcripts in the early phase of the lytic cycle. *EII* is the initiation site for transcription of early mRNA in a later phase of infection when T-antigen binds to position I. The start codons for the tumour antigens and the TATA box for transcripts starting at *EI* are boxed. The 27 bp palindromic sequence is boxed and hatched.
(C) Schematic representation of regulatory mechanisms which allow T-antigen to control its own transcription. Shown are binding sites I, II, and III of T-antigen, binding sites for RNA polymerase II, including the Goldberg-Hogness box (TATA box), and start sites *EI* and *EII* for transcription (Hansen *et al.*, 1981).

(regions I and II), which are 20 to 30 bp in length and comprise both the origin of DNA replication and the initiation sites for early transcription (Tjian, 1978). Early after infection when no or little T-antigen is available transcription initiates in the *EI* region (positions 5231 and 5237). Newly synthesised T-antigen binds first to region I for which the affinity is highest. Bound T-antigen blocks RNA polymerase II entry and causes the transcriptional start point to be moved upstream to region *EII* (positions 30 and 32). When the concentration of T-antigen increases, *i.e.*, in a later stage of infection, region II also becomes saturated with bound T-antigen molecules. Since the transcriptional start point cannot be moved further to the right, transcription eventually is repressed and DNA replication is initiated at region II near the *Bgl* I site. This model does not differ very much from those developed for the interaction of repressor molecules and DNA in certain prokaryotic systems such as the *lac* operon or phage λ.

8.2.2.2 Late mRNA Synthesis

Late mRNA is synthesised in SV40-infected cells from the time of onset of DNA replication. Two species of mRNA are made in this phase, namely, a smaller 16S mRNA coding for the major capsid protein VP1, and a larger 19S mRNA coding for the structural proteins VP2 and VP3 (Fig. 8-12). Although both mRNAs come from the same transcriptional unit between positions 18 and 2674, complex splicing patterns provide different reading frames for the synthesis of the structural proteins ((Fig. 8-14). The leader sequence of 16S mRNA is linked directly at position 526 with the main exon which extends from positions 1463 to 2674. The intron region of 19S mRNA is much smaller and includes the region between positions

373 and 558. The 19S RNA continues from the latter position to position 2674. The late region contains at least two functional 5′ splice sites at positions 373 (19S) and 526 (16S and 19S) and two acceptor sites at positions 558 (19S) and 1463 (16S). There is also a sub-class of 19S mRNA which is never spliced; moreover, the 5′ ends of the late mRNA species, unlike the 3′ polyA site common to all late messages, are extremely heterogeneous. Both 16S and 19S mRNA species are synthesised in large amounts in the late phase of the lytic infection cycle. The 16S mRNA amounts to approximately 10%, and 19S mRNA to 1% of the total cytoplasmic mRNA in an infected cell. In permissive cells the lytic cycle ends with the release of infectious viruses and cell death.

In considering the use of SV40 as a vector, it is important to note that the length of exogenous DNA which can be packaged into viral particles is rather limited; it ranges between 70% and 101% of the length of SV40 DNA, *i.e.*, between 3.7 and 5.3 kb. Since many eukaryotic genes are much longer, genes can be cloned in SV40 only if one dispenses with packaging of DNA into viral

David S. Hogness
Stanford

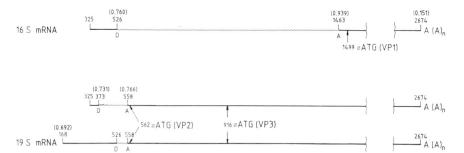

Fig. 8-14. Structure of late SV40 mRNA.
There are two classes of late mRNA, namely 16S and 19S mRNA. Only the most abundant species of these two mRNAs are shown. D and A indicate the 5' and 3' donor-acceptor splice sites. All mRNA species use the same polyadenylation site at position 2674. Intron sequences are represented by wavy lines.

particles and employs episomal vectors for the amplification and expression of cloned genes (see also Section 8.3.2).

While the infection of monkey cells with SV40 results in a lytic infection, the lytic cycle is interrupted before the onset of DNA replication when non-permissive cells, such as mouse or rat cells, are infected. Early mRNA synthesis in these cells proceeds normally, but the late functions cannot be initiated in the absence of DNA replication. The majority of non-permissive cells will survive an infection and some cells will be converted into transformed cells containing viral DNA integrated into their chromosomes. Integration sites are nonspecific. Since the integration event itself is sometimes accompanied by rearrangements of the viral genome and abutting cellular DNA sequences, the use of SV40 as an integrating vector in non-permissive systems is connected with certain problems.

8.3 SV40 Vectors

There are two fundamental systems which allow the use of SV40 DNA as a vector system. In a lytic infection cycle in permissive cells, hybrid DNA molecules can be amplified and packaged into virus particles, provided they are not too long. In non-permissive cells such hybrid mole-

Phillip A. Sharp
Cambridge, Massachusetts

cules can also be amplified under conditions which do not lead to the production of virus particles. In such non-permissive systems the hybrid DNA is replicated either as an episome or in an integrated state. The system of choice will, of course, depend on the individual experimental problem.

8.3.1 Vectors for the Lytic Cycle

There are major differences between permissive and non-permissive systems. Since a lytic infection always leads to cell death, only events that take place within approximately 48 hours can be studied in permissive systems; in addition, experiments are restricted to permissive monkey cells. Finally, the recombinant genome must be smaller than 5.3 kb so that it can be packaged into viral particles. Nevertheless, the lytic system may offer inestimable advantages: firstly, viral DNA can be introduced into susceptible cells very efficiently *via* virus particles; secondly, the DNA is amplified up to 10^5-fold in the nuclei of infected cells, and thirdly, late SV40 genes are efficiently expressed in infected cells.

The only *cis*-acting element required for the replication of viral DNA is the origin of DNA replication. This origin is defined as a region around the *Bgl* I site (position 0/5243) with a length of approximately 75 bp (Fig. 8-12). All other functions required for DNA replication and viral morphogenesis are supplied by *trans*-acting elements. In principle, then, any recombinant DNA carrying the SV40 origin of DNA replication should be replicated and packaged in monkey cells as long as the other early and late viral functions are provided by a suitable helper virus. Such a system would allow the incorporation of up to 5 kb of foreign DNA into a virus particle; however, since it is necessary to employ suitable

selection systems, vectors are used in which only one of the two functional regions, *i.e.*, either the early or the late region, has been replaced by foreign DNA. Vectors with a substitution of the late region require helper viruses providing a functional late region, while a substitution of the early region must be complemented by helper viruses with a functional early region.

8.3.1.1 Deletions of the Late Region

The late region can be removed conveniently from the SV40 genome due to the fortunate position of several unique restriction enzyme cleavage sites. A double digestion of SV40 DNA with either *Hpa* II/*Hae* II or *Hpa* II/*Bam* HI generates deletions, the shortest of which is only 485 bp in length, while the longest removes the entire late region (Fig. 8-15). The large sub-genomic fragment obtained after a simultaneous digestion with *Hpa* II/*Bam* HI is vector SVGT1 which replicates in permissive cells, but does not generate mature virus particles. Infectious virus is obtained only if a complementing helper virus

Robert Tjian
Berkeley

provides the necessary late regions. The use of a suitable early temperature-sensitive (ts) mutant of SV40 is recommended for such experiments because a mixed infection with ts-helper virus and vector carried out at 41 °C then allows selection of those cells which have been simultaneously infected with virus and the recombinant: the vector will support the helper virus by providing functional early gene products which the latter cannot

synthesise due to the ts mutation; the helper virus will complement the vector by providing late gene functions. This system has been employed for cloning the *E. coli* thymidine kinase gene and the gene coding for tRNA[tyr] from *Saccharomyces cerevisiae* in vector SVGT1 (Goff and Berg, 1979). In these two cases the foreign DNA was inserted using homopolymeric poly(dA)-poly(dT) tails (see Chapter 3).

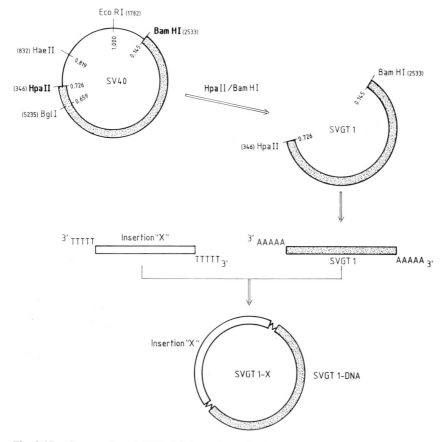

Fig. 8-15.   Construction of SV40 deletion mutants in the late region.
The prototype vector SVGT1 is generated by *Hpa* II/*Bam* HI digestion of SV40 DNA. The *Hpa* II site (346) lies in the vicinity of the late leader; the *Bam* HI site (2533) only 54 bp before the stop codon for VP1 at position 2591. By employing the technique of homopolymeric tailing, insertions (X) can be incorporated into SVGT1. This technique was used for cloning the gene for the *E. coli* enzyme thymidine kinase (*Ecotk*), and a *S. cerevisiae* gene coding for a tRNA[tyr] (Goff and Berg, 1979). Since the vector does not contain any prokaryotic sequences, its use is restricted to higher cells.

8.3.1.2 Deletions of the Early Region.

The bipartite functional organisation of the SV40 genome also allows the development of vectors in which the entire early region can be replaced by foreign DNA. Early functions are then provided by a helper virus with a functional early and a defective late region. Suitable late mutants are of the *tsB* type, which synthesise a defective VP1 protein at the non-permissive temperature; however, it is also possible to complement the deficiencies of vectors with deletions in their early region with *trans*-acting elements provided by the infected cells. Special cells, known as COS cells (for <u>C</u>V1 <u>O</u>rigin <u>S</u>V40), have been devised for this purpose by Gluzman (1981). COS cells are permissive CV1 monkey cells which have been transformed with an SV40 mutant possessing a defective origin of replication. The mutation G-1 used in this experiment deletes 6 bp between positions 5236 and 5243 near the *Bgl*I site (Gluzman *et al.*, 1980). This mutation does not influence the expression of T-antigen, and COS cells therefore synthesise biologically active T-antigen encoded by the integrated G-1 genome. Due to the mutation in the origin of DNA replication, however, T-antigen cannot interact with binding site II in this region (Fig. 8-13) and viral DNA replication cannot be initiated. Thus the late functions are not expressed and the cells survive. COS cells are permissive for SV40. An infection with SV40 mutants carrying deletions in the early region leads to a normal lytic cycle since the T-antigen from the integrated viral genome interacts *in trans* with early deletions as long as they possess a functional origin of DNA replication. Indeed, this system allows the development of vectors in which the early region has been replaced by a heterologous gene. Frequently, however, yields are low. It appears that only a few cells in a population of COS cells synthesise enough T-antigen to allow virus replication at a level corresponding to that obtained in a lytic infection (Gething and Sambrook, 1981). Nevertheless, COS cells have been an important tool, for example, for the analysis of the early SV40 promoter (see Fromm and Berg, 1982). Other COS cell lines have been developed which express a temperature-sensitive T-antigen (Rio *et al.*, 1985). These cells allow DNA replication to be switched on and off by a simple temperature shift, and hence also allow the copy number of suitable plasmids to be controlled. It was not an easy task to generate these cells since the conventional approach of integrating an SV40 mutant defective in the origin of DNA replication failed (see above). The attempt was sucessful, however, when SV40 mutant *tsA1609* was employed and the early SV40 promoter was replaced by the *LTR* region of Rous sarcoma virus. The *LTR* region (see Section 8.7) is much more efficient in monkey cells, and the SV40 mutant provided an extremely temperature-sensitive T-antigen which was still stable enough at the permissive temperature and was not too rapidly proteolysed. Both features increased the concentration of active T-antigen at the permissive temperature and therefore allowed the establishment of permanently transformed cell lines.

As a general word of caution it should be added that any direct attempts to isolate genes from mixtures of DNA using SV40 vectors of the kind described above are hardly worth the effort since the selection techniques available for higher cells are too limited. These experiments would be too difficult, too time-consuming and too expensive. If vectors of this kind are used, it is almost imperative that the initial cloning and selection experiments be carried out in bacteria and that higher cells be transformed only after all modifications have been carried out successfully. Of course, these vectors then require additional DNA sequences which allow their replication in micro-organisms (see Section 8.3.2).

8.3.1.3 Expression Vectors

SV40 vectors, such as SVGT1-X (Fig. 8-15) described in Section 8.3.1.1, replicate very efficiently in monkey cells in the presence of a suitable helper virus; however, they do not synthesise stable cytoplasmic mRNA. The reason is that these vectors do contain all the signals for the initiation and termination of transcription, but not for processing of RNA. New data acquired after these vectors had been constructed suggested that the synthesis of a functional mRNA does not only require the presence of promoter and polyadenylation signals (Fig. 5-10) at the 5' and 3' ends of the transcriptional unit. In fact, correct transcription also depends on the presence of splice signals, *i.e.*, DNA sequences in the regions of the 5' and 3' splice sites of intervening sequences or introns. This topic is discussed in detail in Section 5.6.

If the region of the late 16S mRNA is used for the expression of a foreign gene the 5' splice site at position 526 and the 3' splice site at position 1463 must be conserved (Fig. 8-14). Insertions therefore must occur only in a region of approximately 1 000 bp within the coding sequence of the VP1 gene (positions 1499-2591) (Fig. 8-12). In 19S mRNA, it is the 3' splice site at position 558 and the two donor sites at positions 526 and 373 that must be retained, and in this case approximately 2 000 bp of foreign DNA can be inserted (Fig. 8-12).

It is not surprising that vector SVGT1, for example, does not synthesise stable mRNA from these regions since it carries a deletion between positions 346 and 2533, and hence lacks the splice sites discussed above. Other vectors, designated SVGT5 and SVGT7, were developed which express either 16S and 19S mRNA, or 19S mRNA exclusively (Mulligan *et al.*, 1979). The *Hind* III site at position 1493 and the *Bam* HI site at position 2533 (Fig. 8-16) were used in SVGT5 for this purpose. The *Hind* III site (0.945 m.u.) lies 30 bp from the 3' splice site (1463) and 6 bp from the initiation codon (1499) of the VP1 gene, the *Bam* HI site 58 nucleotides away from the UGA stop codon of the VP1 gene and 141 bp away from its polyadenylation site (2674) (Fig. 8-17). The construction of SVGT5 was difficult since SV40 DNA contains six *Hind* III sites. The starting material for the construction of SVGT5 was a collection of full-length linearised SV40 DNA molecules obtained by partial *Hind* III digestion of SV40 DNA and isolated by gel electrophoresis (Fig. 8-16). These molecules were then digested with *Bam* HI and/or *Eco* RI. *Eco* RI cleavage eliminated a potential contaminating fragment of 4 280 bp, which could have been generated by cleavage at the *Hind* III and *Bam* HI sites at m.u. 0.325 and m.u. 0.145, respectively. The resulting vector, SVGT5, is 4 203 bp in length and can receive inserts of foreign DNA with suitably modified ends. The *Bam* HI site is very useful since its ends are compatible with the termini generated by cleavage with *Bgl* II and *Sau* 3A. The cDNAs of rabbit β-globin (Fig. 8-17), mouse dihydrofolate reductase (Subramani *et al.*, 1981) and the bacterial gene coding for xanthine guanine-phosphoribosyl transferase (Ecogpt) (Mulligan and Berg, 1981a) have been cloned using SVGT5 as a vector. In all three cases infection of permissive CV1 cells with these vectors and suitable helper viruses led to the expected synthesis of two species of mRNA corresponding to 16S

Paul Berg
Stanford

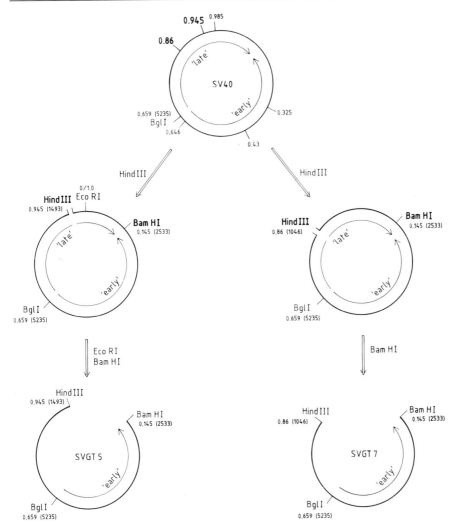

Fig. 8-16. Construction of vectors SVGT5 and SVGT7.
Shown are the positions of six *Hind* III sites on SV40 DNA (positions in m.u.). Linear molecules which are opened at either 0.945 or 0.86 m.u. are obtained by partial cleavage with *Hind* III. Further digestions with *Bam* HI and/or *Eco* RI yield the desired vectors (Mulligan *et al.*, 1979).

and 19S viral mRNAs. A typical cloning strategy is illustrated in Fig. 8-18 showing the construction of SVGT5-dhfr which carries a cDNA insert of mouse dihydrofolate reductase.

The second vector, SVGT7 (Mulligan *et al.*, 1979), contains the same *Bam* HI site as pSVGT5, but a different *Hind* III site at position 1046 (m.u.

0.86) instead of 0.945 (Fig. 8-16). This vector lacks the 3' acceptor site for 16S mRNA at position 1463 and, indeed, does not express this mRNA. In infected permissive monkey cells SVGT7 directs the synthesis of two species of 19S mRNA which are equivalent to genuine SV40 19S mRNAs. As shown in Fig. 8-14 and 8-18, the two

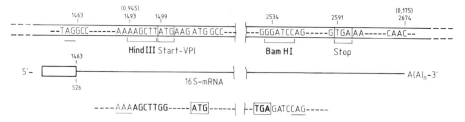

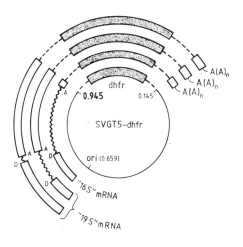

Fig. 8-17. Detailed structure of the late SV40 16S mRNA and the corresponding DNA.
Shown is a sequence region with the start and stop codon of the VP1 structural protein, the *Hind* III and *Bam* HI sites (positions 1493 and 2533, respectively), and the polyadenylation site at position 2674. The AG dinucleotide which is characteristic of the structure of a 3′ splice site is underlined. In the mRNA shown nucleotide 1463 is linked to nucleotide 526 of the leader. The exact position of the 5′ end of the mRNA cannot be shown because it is variable (*cf.* also Fig. 8-14). The sequence at the bottom shows a section of vector SVGT5-RaβG. Between its *Hind* III and *Bam* HI sites this vector contains a 477 bp rabbit β-globin cDNA fragment with the entire coding region of the β-globin gene including start and stop codon (boxed). Three proximal and three distal nucleotides which belong to flanking SV40 sequences are underlined.

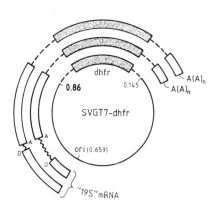

splice signals for these mRNAs have been conserved in this vector. It is interesting to note, with regard to splicing mechanisms, that additional mRNA molecules smaller than the equivalent of the 19S mRNAs are observed with SVGT7 vectors containing the β-globin or the *Ecogpt* gene; they are not observed with the cDNA clone of mouse dihydrofolate reductase. These smaller mRNAs are generated by abnormal splicing events at the 3′ splice sites within the 5′ untranslated regions of the β-globin gene or the coding region of the *Ecogpt* gene (Mulligan and Berg, 1981b).

It is also possible to use SV40 vectors for the expression of chromosomal eukaryotic DNA sequences containing their own promoter and splice signals. In some of these constructions

Fig. 8-18. Structure of mRNA species controlled by the late SV40 promoter expected in vectors SVGT5-dhfr and SVGT7-dhfr.
SV40 sequences in the mRNA are shown as open bars, non-coding sections of the insertion as broken lines, and coding dhfr sections as stippled bars. dhfr denotes the insertion of a cDNA of the gene for mouse dihydrofolate reductase. Apart from the 16S mRNA, two mRNAs characteristic of 19S mRNA species are formed in SVGT5-dhfr. These mRNAs possess different 5′ donor signals (D) but identical 3′ acceptor sites (A). In SVGT7-dhfr, only mRNA species equivalent to 19S mRNA are formed (Subramani *et al.*, 1981).

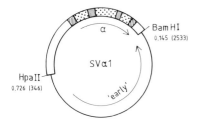

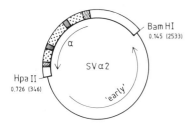

Fig. 8-19. Structure of two SV40 recombinants containing the mouse α-globin gene.
In both vectors the late SV40 region between 0.726 and 0.145 m.u. was deleted; both vectors therefore lack all donor and acceptor splice sites for late SV40 mRNA (*cf.* also Fig. 8-14). Instead, these vectors contain a 2.1 kb DNA fragment corresponding to the entire mouse α-globin gene, and additional 1 000 bp of 5′, and 250 bp of 3′ flanking sequences. In SVα1, transcription from the α-globin insertion points in the same direction as late SV40 transcription. SVα2 contains the insertion in opposite direction, *i.e.*, in the direction of early transcription. The missing SV40 splice sites were replaced by corresponding α-globin DNA sequences. Coding regions are represented as finely stippled bars, the two introns as coarsely stippled bars, and untranslated regions as hatched bars. Non-transcribed flanking sequences are shown as open bars (Hamer *et al.*, 1980).

stable mRNA is formed not only under the control of the SV40 promoter, but also from promoters located within the inserted foreign DNA. Hamer *et al.* (1980) constructed SV40 vectors carrying the mouse α-globin gene (with its own promoter), inserted in either of the two possible orientations with respect to the late SV40 promoter (Fig. 8-19). In SVα1, the SV40 late promoter has the same orientation as the α-globin promoter such that, in principle, the globin gene could be read from either promoter. In SVα2, the orientation of the globin gene is reversed and its coding strand is colinear with the early SV40 promoter. The late SV40 promoter is therefore unable to direct the synthesis of a functional transcript. Nevertheless, both vectors produce equal amounts of α-globin-specific mRNA after infection of permissive monkey cells. This finding suggests that not only a heterologous promoter, but also the mouse mRNA processing signals can be recognised in monkey kidney cells, despite differences in species and cell type. It has not yet been established, however, whether this conclusion can be generalised (Elder *et al.*, 1981).

Richard C. Mulligan
Cambridge, Massachusetts

In vectors SVGT5, SVGT7, SVα1, and SVα2 the late region of SV40 DNA has been replaced by foreign DNA. Vectors in which foreign DNA is inserted into the early region are rarely used. There are two reasons for this: firstly, the early transcriptional unit is expressed at a lower level than is the late one in lytic infections of permissive cells, and secondly, the construction of such vectors is much more difficult since it requires the correct positioning of splice signals. In contrast to the late 16S and 19S mRNAs, early mRNAs contain splice signals in coding rather than non-coding regions. Therefore these signals are lost when such vectors are constructed and the coding region of a foreign DNA is substituted for T-antigen coding sequences. No difficulties are encountered if the inserted DNA itself carries an intron. If the foreign DNA is a cDNA, for

example, care must be taken to provide a suitable intron with 5′ and 3′ splice sites. The problem is generally solved by inserting the intron of the small SV40 t-antigen downstream from the coding region, but upstream from the polyadenylation site. A suitable DNA fragment containing the desired intron (position 4571-4638) is the *Mbo* I fragment between co-ordinates 4100-4710 (see also Fig. 8-12). Gething and Sambrook (1981) constructed such a vector carrying the influenza virus haemagglutinin (HA) gene between the *Hind* III site at position 5171 and the *Bam* HI site at position 2533 (Fig. 8-20). The intron-free HA DNA is 1773 nucleotides in length and codes for a precursor protein of 547 amino acids. It does not contain a promoter of its own and therefore must be expressed under the control of heterologous promoters. The *Hind* III site of the vector mentioned above lies downstream from the 5′ untranslated region of the early SV40 mRNA. This plasmid generates a hybrid-mRNA which begins at the normal initiation site of early SV40 mRNA, and contains 61 bases of SV40-specific RNA before the junction with the HA message. Transcription is terminated at the polyadenylation site of early mRNA at position 2587.

Findings relevant to the design of expression vectors are summarised in the following structural prerequisites for such vector constructions:

– A functional promoter, with an initiation site (cap site) for mRNA and the necessary 5′ untranslated regions is required. In the case of the early SV40 promoter this comprises a region approximately spanning the *Hpa* II site at position 346 and the *Hind* III site at position 5171 (Fromm and Berg, 1982).

– A restriction cleavage site which occurs only once on the vector DNA and which is located upstream from the first AUG start codon is required. This site can be used for the insertion of foreign DNA which carries its own start codon.

– 5′ and 3′ splice sites either must be provided by the gene to be expressed, or, as is the case with cDNA clones, must be supplied by the vector itself. In some early region SV40 vectors such

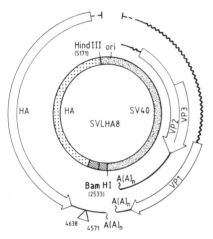

Fig. 8-20. Structure of the early vector SVLHA8. A haemagglutinin gene (HA) from influenza virus is inserted between the *Hind* III site at position 5171 (immediately before the start codon for T-antigen at position 5163) and the *Bam* HI site at position 2533. The late SV40 region remains intact and leads to the formation of the expected late mRNA species. Two SV40-specific regions can be seen at the 3′ end of the HA insertion (hatched bars); they contain a DNA fragment with the intron of small t-antigen, and an SV40 section with the polyadenylation site for early mRNA fused to it. Co-ordinates of the introns correspond to their natural co-ordinates in SV40 DNA (Gething and Sambrook, 1981).

sites are supplied by the intron of SV40 small t-antigen which can be cloned as an *Mbo* I fragment (position 4100-4710). Depending on the specific construction, this or another intron is ligated to the 5' or 3' proximal end of the gene to be expressed in the vector (Okayama and Berg, 1983).

- A polyadenylation site must be provided by the vector or the DNA to be expressed. The polyadenylation site of the late SV40 mRNA lies at position 2674. The sequence AAUAAA which is found in the vicinity of the 3' ends of all eukaryotic polyadenylated mRNA species is only a few bases upstream of this position, and is essential for the generation of functional mRNAs (Fitzgerald and Shenk, 1981; Cole and Santangelo, 1983).

Even if all these requirements are met, one should expect that the degree of expression of different genes in the vectors described above will be extremely variable. Data reported in the literature thus vary between 10^6 copies of rat preproinsulin and 6×10^8 copies of influenza virus haemagglutinin or β-globin per cell. In contrast to prokaryotic expression vectors, little is known about other parameters apart from those mentioned above which influence the extent of gene expression in eukaryotic cells.

8.3.2 Plasmid Vectors

Papovavirus plasmid vectors are defined as cloning vehicles which can be used not only in permissive cells (*i.e.* monkey or mouse cells, in the cases of SV40 and polyoma virus vectors, respectively), but also in other mammalian cells. In permissive cells these vectors are propagated lytically if they possess a functional origin of replication and if they are provided with functional T-antigen. They are even packaged into virus particles when the DNA is shorter than 5.3 kb and capsid proteins are available. Such vectors do not, of course, replicate in non-permissive cells such as human cells; however, a transient and possibly strong expression of genes cloned in such vectors can be observed shortly after transfection and before the onset of cell division. As cell division proceeds transfected vector DNA molecules are diluted out since they cannot replicate, and the initially observed gene expression is concomitantly reduced. Under these conditions cells containing recombinant DNA usually can be identified only if a selectable marker, such as a TK gene, also resides on the vector carrying the gene in question. Little is known about the fate of vector DNA after transfection into non-permissive cells. According to several reports, however, transformed cells may harbour such vectors in a stably integrated form (see Buetti and Diggelman, 1981, for example).

These vectors cannot be employed for cloning since they do not replicate as plasmids. They must be furnished with DNA sequences derived from pBR322, for example, which allows them to be used as shuttle vectors for *E. coli*. The actual cloning experiments are then carried out in *E. coli* and expression is studied in mammalian cells.

It may seem at first that the development of plasmid vectors for non-permissive cells would be less desirable than the construction of vectors for the efficient lytic system; however, due to newly developed transfection techniques, the initial transient expression observed after transfection with vectors for non-permissive cells is usually good enough to allow, for example, a detailed analysis of promoters.

In the SV40 system, vectors expressing recombinant DNA under the control of the early SV40 promoter region are distinguished from those controlled by the late promoter function. Plasmids pSV1GT5 and pSV1GT7 (Fig. 8-21) contain the fragments SVGT5, and SVGT7, respectively (Fig. 8-16). The expression of inserted DNAs in these cloning vehicles is controlled by the late SV40 promoter. These vectors have been ligated to a pBR322 derivative in which an SV40 *Hind* III-*Bam* HI fragment with co-ordinates 0.145-0.325 was substituted for a *Hind* III-*Bam* HI fragment between positions 29 and 375 of

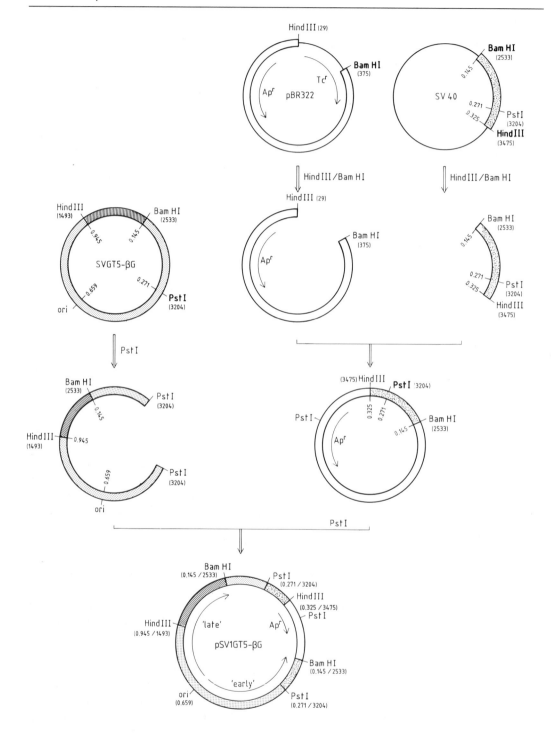

pBR322 (Fig. 8-21). The pBR322 derivative contains two *Pst* I sites, one in the β-lactamase gene of pBR322, and the other in the inserted SV40 *Hind* III-*Bam* HI fragment. It was opened at the *Pst* I site within the SV40 portion by partial digestion with *Pst* I and then inserted into *Pst* I-linearised SVGT5 or SVGT7 molecules. Molecules which retained the *Pst* I site in the β-lactamase gene were isolated by transforming *E. coli* strain HB101 and selecting for ampicillin resistance. The introduction of another SV40 fragment into the pSV1GT vectors *via* the pBR322 derivative reconstituted the entire early SV40 region of the vector. Such vectors therefore replicate in permissive monkey cells but are too large (9 kb) to be packaged *in vivo*. They can be introduced into eukaryotic cells by common DNA transfection techniques. These plasmid vectors have been shown to express, for example, the *E. coli* gene for xanthine guanine phosphoribosyl transferase (pSV1GT5-gpt) (Mulligan and Berg, 1981b) and mouse dihydrofolate reductase cDNA (pSV1GT5-dhfr) (Subramani *et al.*, 1981). The successful transformation of non-permissive cells, such as DHFR-negative CHO cells with pSV1GT5-dhfr, which results in a DHFR-positive phenotype, is surprising. One would have assumed that the late promoter region would be inactive in these cells since the viral DNA cannot be replicated. Transformation frequencies are, indeed, approximately one order of magnitude lower than those observed with early region

---

Fig. 8-21. Construction of SV40-plasmid vectors. Plasmid vectors of the pSV1GT5 type are derived from the lytic vector SVGT5; they contain an insertion of pBR322 DNA, approximately 4 kb in length, which codes for β-lactamase and contains the origin of DNA replication of pBR322 DNA. These vectors can therefore be used as shuttle vectors and can be propagated both in monkey cells and *E. coli*. All vector constructions reconstitute the entire early SV40 region. A DNA fragment with the rabbit β-globin gene (*cf.* also Fig. 8-17) was inserted into the region of the late transcription unit (hatched bar). pBR322 sequences are shown as open bars, SV40 sequences as stippled or dotted bars, and β-globin sequences as hatched bars.

vectors (see below); it may very well be that the expression of late SV40 genes is usually lethal for these cells, but that the expression of the DHFR gene is tolerated.

A second class of vectors contains SV40 fragments which bring inserted foreign genes under the control of the early promoter region (Mulligan and Berg, 1981a). In addition to the early promoter region, such vectors also contain the intron of small t-antigen, which allows transcripts to be processed correctly. The prototype vector pSV2-βG (Fig. 8-22) contains a 2.3 kb fragment of pBR322 with *Eco* RI and *Pvu* II ends which carries the origin of DNA replication and the β-lactamase gene. Linked to the pBR322 DNA is an SV40 fragment which has *Pvu* II and *Hind* III ends and contains the early promoter region and the SV40 origin of DNA replication (co-ordinates 270-5171). Immediately adjacent to the SV40 sequences is a *Hind* III-*Bgl* II rabbit β-globin cDNA fragment of 477 bp (see also Fig. 8-17). The *Hind* III site is 37 bases upstream of the β-globin translation initiation codon, while the *Bgl* II site follows immediately after the termination codon. Next to the *Bgl* II site is a DNA fragment with *Mbo* I sites at either end which carries the small intron (66 bp) of SV40 small t-antigen. The *Mbo* I fragment can be inserted in either of two possible orientations; however, only one orientation will regenerate a *Bgl* II site (A/GATCT) since only the *Mbo* I site at position 0.558 possesses the correct sequence (GATC*T*), while the other site (position 0.44) has the sequence GATCC. Finally, the vector includes an *Eco* RI-*Bcl* I SV40 fragment comprising the region between positions 1782 and 2770 (0-0.19), which carries the polyadenylation site (2587) for early SV40 mRNA. A *Hind* III/*Bgl* II digestion of this construct forms the linearised vector pSV2. The *Hind* III-*Bgl* II sites can be used, for example, for the insertion of a *Hind* III-*Bam* HI Ecogpt fragment of 1 000 bp in length to yield plasmid pSV2-gpt. Although *Bam* HI and *Bgl* II sites are compatible the *Bgl* II site is lost (see Fig. 8-23) in the course of the insertion since the flanking

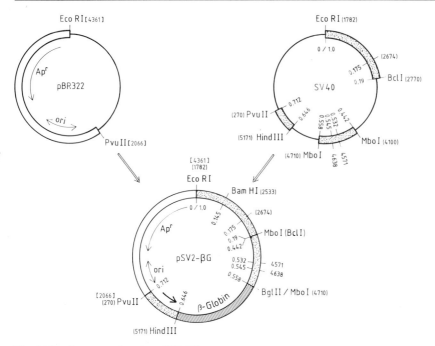

Fig. 8-22. Structure of vector pSV2-βG.
This vector is composed of a pBR322 segment (open bar), three SV40 DNA fragments (stippled bars), and the β-globin cDNA (hatched bar). Numbers on the inside of the circles refer to the positions (in m.u.) of the SV40 fragments occupied in SV40 DNA. Numbers in parentheses are SV40 co-ordinates, numbers in brackets pBR322 co-ordinates. The thick arrow indicates the direction of transcription from the early SV40 promoter. The position of the small t-antigen intron is indicated by its co-ordinates, 4638/4571, in the *Bgl* II-*Mbo* I fragment (Mulligan and Berg, 1981a).

sequences are not symmetrical. The neomycin resistance gene of transposon Tn5 was introduced into pSV2 in a similar way to yield vector pSV2-neo (Southern and Berg, 1982).

Another useful eukaryotic expression vector is pKCR described by O'Hare *et al.* (1981) (Fig. 8-24). It contains the pBR322 large *Eco* RI-*Bam* HI DNA fragment (3 987 bp) which carries the β-lactamase gene and the origin of DNA replication. Proceeding clockwise from the pBR322-derived *Eco* RI site, there follows a 418 bp *Hpa* II-*Hind* III DNA fragment containing the early SV40 promoter (co-ordinates 346 to 5171). The *Eco* RI site was preserved by annealing a filled-in *Eco* RI site (G/AATTC) with a filled-in *Hpa* II site (C/CGG). The terminal *Hind* III site was converted into a *Bam* HI site by filling-in and then ligating with *Bam* HI linkers. Next to the *Bam* HI site follows a 1 055 bp fragment of the rabbit β-globin gene. Its ends are defined by a *Bam* HI site located in the second exon and a *Pvu* I site located in the 3' untranslated region of the β-globin gene (Hardison *et al.*, 1979). This β-globin section is followed by a 133 bp SV40 DNA fragment which carries the polyadenylation site of the early SV40 mRNA (position 2587). Foreign DNA can be inserted into the single *Bam* HI site of this vector, which lies downstream from the early SV40 promoter. The transcripts do not contain the β-globin intron. They terminate at the β-globin polyadenylation site, and not at the early SV40 site. The vector was initially used for the expression of the gene for a prokaryotic methotrexate-resistant dihydrofolate reductase.

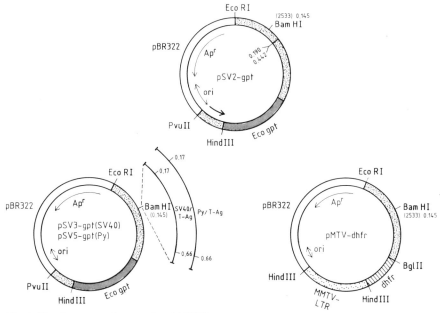

Fig. 8-23. Structure of some plasmid SV40 vectors.
Vector pSV2-gpt is obtained from pSV2-βG (Fig. 8-22) by digesting with *Hind*III and *Bgl*II and replacing the β-globin insert by an Ecogpt fragment of approximately 1 000 bp flanked by *Hind*III and *Bam*HI ends; the *Bgl*II site is lost. Vectors pSV3-gpt and pSV5-gpt are derived from pSV2-gpt by inserting DNA fragments carrying the entire early region of SV40 (pSV3), or polyoma virus (pSV5). In pMTV-dhfr, the SV40 promoter region of a pSV2-dhfr clone was replaced by an *LTR* region from MMTV. This region of approximately 1 000 bp displays promoter activity and also makes the expression of the adjacent gene, the gene for dihydrofolate reductase in this example, sensitive to glucocorticoids (see text).

More recently, pKCR has also been successfully used for the expression of eukaryotic genes such as human interleukin-2 (Taniguchi *et al.*, 1983).

Plasmid vectors pSV2 and pKCR lack the coding region for T-antigen and are unable to replicate even in permissive monkey cells. However, these and other plasmids can be maintained as episomes in eukaryotic cells in three different ways:

(1) Plasmid vectors are provided with a marker gene which either complements a genetic defect of the host cells or which can be selected for in normal cells. Human Lesch-Nyhan cells, for example, which lack the hypoxanthine guanine phosphoribosyl transferase gene (HGPRT) gain a gpt-positive phenotype and are able to grow in HAT medium after transformation with pSV2-gpt, since the bacterial *Ecogpt* gene is expressed in these cells (Mulligan and Berg, 1980). The presence of the bacterial xanthine guanine phosphoribosyl transferase (XGPRT) also allows a positive selection because it enables transformed cells to utilise adenine and xanthine in a growth medium containing aminopterin and mycophenolic acid (see Section 8.1.2).

The promoter controlling the expression of a selectable marker like gpt does not necessarily have to be the early SV40 promoter. Promoter functions can also be provided by the inserted gene itself or by another heterologous promoter region. The latter option allows DNA sequences to be tested for their promoter activities. In order

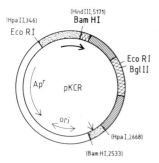

Fig. 8-24. Structure of the eukaryotic expression vector pKCR.

pBR322 sequences are represented as open bars, SV40 sequences as stippled bars. The section of the rabbit β-globin gene corresponds to approximately 18 bp of the second exon, the intron between the second and third exon, the entire third exon, and a 3′ flanking region. Exons are represented as cross-hatched bars, introns as hatched bars. Restriction sites in parentheses are not present in the vector but indicate their position on the SV40 genome. The thick arrow marks the direction of transcription from the early SV40 promoter. The polyadenylation site is marked by an arrow head pointing towards the stippled bar representing SV40 sequences (O'Hare *et al.*, 1981).

to ascertain whether *LTR* regions of mouse mammary tumour virus (MMTV), which are longer than 1 000 bp and flank MMTV DNA, possess promoter activity, such *LTR* sequences were substituted for the SV40 promoter region in pSV2 (Lee *et al.*, 1981). A key question in this experiment was whether the presumed promoter

activities could be controlled by glucocorticoids. The starting point for this experiment was vector pSV2-dhfr, which carries a mouse dihydrofolate reductase cDNA insertion. The early SV40 promoter was removed by *Pvu* II/*Hind* III digestion and the blunt *Pvu* II site was converted into a *Hind* III site by using *Hind* III linkers to permit the insertion of the *LTR* region with *Hind* III ends (Fig. 8-23). Transfection of dhfr-negative CHO cells yielded transformants which expressed the *DHFR* gene and were able to grow in media without glycine. The addition of the glucocorticoid dexamethasone stimulated expression two- to ten-fold, proving conclusively that the MMTV *LTR* regions do indeed possess the required regulatory signals.

(2) It may be of importance in some situations not only to select for episomal DNA, but also to replicate such DNA molecules. Of course, this requires the presence of a functional SV40 origin of DNA replication and functional T-antigen. The latter is supplied by the vector itself or by the host genome. Plasmids such as pSV2-gpt, pSV2-dhfr, pSV2-neo etc. (Fig. 8-23) are able to replicate in *E. coli*, but not in monkey cells. Replication in monkey cells is allowed by inserting the entire early SV40 region into the single *Bam* HI site of these vectors. The resulting new vectors (pSV3-gpt(SV40), pSV3-dhfr(SV40), pSV3-neo(SV40), etc.) now contain two origins of DNA replication (Fig. 8-23). The presence of two origins has also

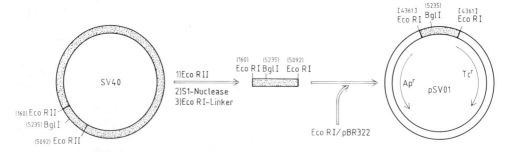

Fig. 8-25. Construction of vector pSV01.

pSV01 contains one of 17 SV40 DNA *Eco* RII (*Bst* NI) fragments to which *Eco* RI linkers were added. This fragment was cloned into the *Eco* RI site of pBR322. Numbers in parenthesis are SV40, numbers in brackets pBR322 co-ordinates. (Myers and Tjian, 1980).

been observed in naturally occurring variants of SV40. Vector pSV2 can also be made to grow in mouse cells by using its *Bam* HI site to introduce the entire early region of polyoma virus. The corresponding vectors have been designated pSV5-gpt(Py) (dhfr, neo, etc.). In these cases the SV40 origin remains silent, and replication occurs *via* the origin of polyoma virus (Mulligan and Berg, 1981a).

(3) An alternative method employs host cells supplying the T-antigen. This is rendered possible by the development of cell lines which have been transformed by SV40 mutants defective in the origin of DNA replication (see also Section 8.3.1.2). These COS cell lines allow the replication of SV40 vectors lacking the entire early region and even replication of plasmids carrying only the SV40 origin. Vector pSVO1, for example, replicates efficiently in COS cells. It contains the 311 bp *Eco* RII-G fragment of SV40 DNA (position 5092-160). This sub-fragment carries the SV40 origin with the T-antigen binding sites, and is inserted into the *Eco* RI site of pBR322 (Myers and Tjian, 1980) (Fig. 8-25).

region of SV40 DNA for a cDNA copy of this region (Fig. 8-26). The starting point in this experiment was to anneal 16S mRNA (which, by definition, lacks all intron sequences) with a single-stranded DNA fragment of SV40 DNA from which the region between positions 0.726 and 1.0 had been removed by *Eco* RI/*Hpa* II cleavage. This deletion removes the DNA sequence corresponding to the intron between map position 0.76 and 0.935. The 3' hydroxyl end of the resulting RNA-DNA hybrid served as a primer for reverse transcription. The DNA obtained after alkali treatment (to remove the RNA) represented the entire SV40 genome except for the intron of VP1 between positions 0.76 and 0.935 (co-ordinates 526 and 1463). Single-stranded DNA from *Hin*d II/III fragment C comprising the region between co-ordinates 0.665 and 0.74 was used to bridge the gap between the molecular ends of the DNA fragment, and the circularised molecules were then used to infect monkey cells in conjunction with helper virus carrying an early *ts* lesion. Infected cells did not contain stable cytoplasmic 16S mRNA, nor did they synthesise

8.3.3 Application of SV40 Vectors

SV40 vectors have been employed for various purposes. They play a very important role in analysing the mechanisms of transcriptional and post-transcriptional control of mammalian cells. In fact, it was only after the importance of splicing had been recognised that these vectors could be used successfully for the expression of genes. Expression vectors, therefore, must contain 5' and 3' splice sites. pSV2-type vectors, for example, contain the intron of SV40 small t-antigen for this purpose. Gruss *et al.* (1979) have studied the role of introns by substituting the entire late

Peter Gruss
Göttingen, FRG

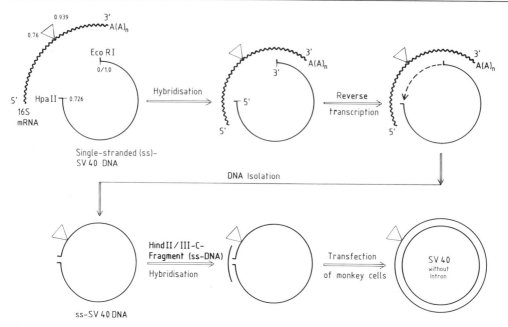

Fig. 8-26.   Construction of intron-free SV40 mutants.
Solid- and dashed lines indicate DNA, wavy lines RNA sequences. The triangle marks the intron deletion in the 16S mRNA (Gruss *et al.*, 1979).

the corresponding VP1 gene product. The genetic defect in this mutant could be overcome by introducing another intron, *i.e.*, the first intron of mouse β-globin gene (Gruss and Khoury, 1980). It is therefore possible to substitute one intron for another and thus to restore the biological function, namely the synthesis of a stable mRNA.

It appears that a heterologous intron can also be inserted at positions other than the normal intron site, but only in a certain orientation. The β-globin intron mentioned above had been introduced near the 5' end of the 16S mRNA leader sequence; however, in order to be functional the orientation of the intron with respect to the direction of transcription of the VP1 gene had to be the same as that in the β-globin gene. The experiments described above indicated that the presence of an intron and suitable signals for

George Khoury
Bethesda

RNA processing are essential for the generation of stable cytoplasmatic mRNA. Subsequent experiments with SV40 vectors showed that the situation is much more complex. A number of genes, the histone genes, the adenovirus protein IX gene, α- and β-interferon genes (not the gamma-interferon gene which contains three introns, for example, are expressed efficiently although they do not contain any introns (Taya *et al.*, 1982; Gray and Goeddel, 1982). In addition, there are some eukaryotic vectors which allow the protein encoded by the inserted gene to be translated from an unspliced mRNA. In a series of experiments similar to those described above, Gruss *et al.* (1981b) constructed an SV40 vector with an intron-free cDNA of the rat preproinsulin gene under the control of the late SV40 promoter. This vector expressed stable mRNA. Incidentally, it is interesting to note that the monkey cells were able to process the prepeptide, but not the propeptide, which enabled them to secrete proinsulin. Using a strategy like that employed for the

construction of vector SVGT1 (Fig. 8-15), Gething and Sambrook (1981) constructed a late SV40 vector which expressed the influenza virus haemagglutinin gene. The starting point was an SV40 derivative cut with *Hpa* II (346) and *Bam* HI (2533), which thus lacked all splice signals for late transcription (Fig. 8-14). The 5' end of the inserted haemagglutinin gene, therefore, lay in a leader region of the late mRNA species which usually would constitute the 5' untranslated region of the mRNA. This vector, designated SVEHA3, therefore was expected to allow the synthesis of an mRNA with 5' untranslated SV40 sequences and the entire coding region for the haemagglutinin gene (Fig. 8-27). Indeed, the protein was synthesised in large amounts, with yields on the order of 10^8 molecules per cell (80 µg per 5×10^6 cells). These yields exceeded those observed in infected cells, which was surprising since the transcription unit does not contain an intron; in addition, the protein was glycosylated and deposited at the cell surface of monkey cells in a form indistinguishable from genuine haemagglutinin protein.

These and other experiments demonstrate that processing of mRNA may be absolutely dependent on splicing in certain combinations of promoters and coding regions, while it may be equally independent in other combinations. Much work lies ahead to clarify the issue and to define exactly what an intron is.

The experiments concerned with the expression of the haemagglutinin gene of influenza virus described above suggest some interesting applications in virology. Such SV40-based expression systems do not only allow the production of large amounts of biologically active and glycosylated viral components. They also allow the analysis of the role of such proteins in the production of other viral proteins. It will now be possible to investigate which sequences are responsible for the deposition of proteins into cell membranes and which additional factors, whether of viral or cellular origin, are responsible, for example, for the excretion of haemagglutinin into the ambient

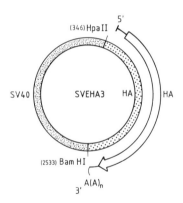

Fig. 8-27. Structure of the lytic SV40 vector SVEHA3.
The haemagglutinin (*HA*) gene of influenza virus is located between the *Hpa* II site (346) and the *Bam* HI site (2533) of SV40 DNA, replacing the late region of SV40 DNA. The early mRNA species are ommitted, because they are identical with those of the wild-type (*cf.* also Fig. 8-12). mRNA synthesised is equivalent to late SV40 16S mRNA (open bar). The polyadenylation site is unknown, because two such sites are available, one within the *HA* gene, and one for polyadenylation of late SV40 mRNA at position 2674.

medium as is the case when influenza viruses are released from infected cells.

Cloning in eukaryotic cells has also been used to address some biologically relevant questions about hepatitis B virus (HBV). This is all the more important since it has not been possible to cultivate HBV in tissue culture. Perhaps it will be possible now for the first time to identify hitherto unknown gene products of HBV by cloning and expressing HBV DNA fragments in monkey cells. A promising start was made by cloning the gene for HBV surface antigen (HBsAg), which occurs in the serum of hepatitis patients in the form of 22 nm aggregates (Moriarty *et al.*, 1981). When this gene was brought under the control of the late SV40 promoter, it was expressed, albeit weakly (2.5 μg per 2×10^7 cells), and the product aggregated in the form of 22 nm particles. A more efficient expression was observed in amplifiable vectors (*cf.* Section 8.8) with HBsAg under the control of the SV40 early promoter (Michel *et al.* 1984). This particular construction included a 55 amino acids long precursor of HBsAg and led to the formation of HBsAg 22 nm particles which elicited a strong immune response in mice. Incidentally, progress in the molecular biology of hepatitis B virus with an emphasis on vaccine production has recently been summarised by Tiollais *et al.* (1985).

The analysis of oncogenes is another promising field for the application of SV40 vectors. Gruss *et al.* (1981a) were able to express, for example, the transforming protein p21 of Harvey murine sarcoma virus (Ha-MuSV) in SV40 vectors. Similar work with other oncogene products has been reported.

It has been pointed out that it is very important that eukaryotic expression vectors should contain controllable, *i.e.* inducible, promoters. It is conceivable that the unrestricted expression of certain genes would be toxic for the host organisms. From the experimenter's point of view it would be more advantageous to be able to switch on such genes at will. In prokaryotic vectors this is achieved by exploiting naturally occurring control mechanisms, namely repressors. In eukaryotic cells gene expression could be controlled in at least six different ways, namely by:

– the autoregulatory properties of SV40 T-antigen;
– metallothionein genes;
– heat shock genes;
– control by glucocorticoid hormones;
– methylation of DNA;
– anti-sense RNA.

It has already been mentioned that SV40 T-antigen regulates its own transcription. It is actively transcribed early after infection and later prevents transcription by binding to its own promoter region (Fig. 8-13). This repressor function of T-antigen could be exploited for the control of expression in expression vectors using suitable temperature-sensitive mutants of T-antigen. Like *cI*857 repressor mutants of phage λ, vectors with a temperature-sensitive SV40 T-antigen mutant are actively transcribed at the elevated non-permissive temperature, but not at normal permissive temperatures (Rio *et al.*, 1985; see Section 8.3.1.3). Since the autoregulation of T-antigen involves the early promoter, such a system would be restricted to vectors carrying the early SV40 region. Such vectors have one great disadvantage: the yields of early mRNA are approximately 20-fold lower than those of late mRNA. It may be possible though, to circumvent these problems by constructing hybrid promoters.

Metallothioneins are cysteine-rich proteins with low molecular weights (on the order of 6 000 Da), which bind cadmium, zinc and other heavy metals. A variety of cells respond to treatment with heavy metals by synthesising these proteins. Their induction is accompanied by elevated levels of the corresponding mRNAs (Enger *et al.*, 1979). At cadmium concentrations of 4 μM, Hamer and Walling (1982) observed a stimulation of metallothionein synthesis in cultured monkey cells by factors of up to 50, with a corresponding rise in metallothionein mRNA levels. Induction ratios certainly could be increased considerably if it were possible to use growth media absolutely free of heavy metal contaminants. Several metallothionein (MT) genes have been isolated from the genomes of man, mouse, sheep and *Drosophila*. Mutational analysis of *MT* gene promoter regions have delineated 15 bp long heavy metal regulatory regions which are homologous among *MT* genes of different origin and which, in general, come in multiple copies. Synthetic copies of these metal regulatory elements (MRE) which comprise the consensus sequence

5'-CTNTGCPuCPyCGGCCC-3'

are able to confer heavy-metal regulation on heterologous promoters, *e.g.* the HSV *tk* promoter (Stuart *et al.*, 1985). Such sequences may thus be exploited to control the expression of genes other than metallothionein genes. In plants, the response to heavy metals does not yield metallothioneins but rather a family of low molecular weight peptides known as phytochelatins (Grill *et al.* 1985). Phytochelatins have the general structure $(\gamma\text{-glu-cys})_{3\text{-}6}$-gly and thus do not represent primary gene products.

A sudden increase in temperature causes dramatic changes in the protein composition of eukaryotic cells. The best studied system is *Drosophila melanogaster*, in which a temperature shift from 25 °C to 37 °C provokes an abrupt termination of normal protein biosynthesis and leads to the synthesis of a characteristic set of new proteins, known as heat shock (HS) proteins. The molecular causes of this phenomenon must be sought both at the level of translation and, in particular, of transcription (Ashburner and Bonner, 1979). The involvement of transcription is indicated by the rapid disappearance of old, and the rapid appearance of new puffs on polytene chromosomes (Ritossa, 1962). Several *Drosophila* heat shock genes have already been cloned as cDNAs, and certain similarities have been observed not only in the 5' untranslated regions, but also in the coding regions of these genes (Holmgren *et al.*, 1981; Pelham, 1985). The most abundant heat shock protein is usually hsp70,

which has a molecular weight of 70 kDa. Although its function is not known in detail, hsp70 is thought to be involved in the repair of heat-damaged ribonuclear protein particles (RNPs). The mechanism of transcriptional activation appears to be highly conserved since a cloned *Drosophila hsp70* gene introduced into mouse or monkey cells can be transcribed only after a heat shock. Deletion analysis of several *Drosophila* heat shock protein genes has revealed a minimal consensus sequence,

5'-CNNGAANNTTCNNG-3',

located between 70 and 250 bp upstream of the TATA box. Using a novel exonuclease protection assay, Wu (1985) has been able to fractionate a protein factor that binds to this heat-shock gene control element.

Similar phenomena have also been observed in plants (Key *et al.*, 1981; Schöffl and Key, 1982; Schöffl *et al.*, 1984). A striking application of the heat-shock system was described by Wurm *et al.* (1986) who described the inducible overexpression of the mouse c-myc protein in a mammalian cell line. In this construction, the *c-myc* gene coding region was fused to a *Drosophila hsp70* promoter which possesses an extremely low basal level of expression. The chimaeric gene was introduced into DHFR-deficient CHO cells using a *DHFR* gene as a selectable marker. The c-myc gene could then be amplified in a silent form (*cf.* Section 8.8) resulting in stable cell lines containing approximately 3 000 copies of the recombinant *c-myc* gene. Upon induction at 43 °C and a recovery period of two hours at 37 °C, the induced cells were shown to overexpress the c-myc protein in significant amounts (3 mg protein per 10^9 cells per induction experiment; cells died within eight hours after induction). The low basal level of expression as well as the high degree of inducibility (at least 100-fold) suggest that this system may be applicable to the expression of other mammalian proteins, cytotoxic or not.

In higher organisms steroids elicit several responses which usually are attributed to alterations in transcriptional processes. The current theory is that steroids penetrate the cellular membrane and react with cytoplasmic receptor proteins, inducing conformational changes which allow these complexes to enter the nucleus and to bind to specific DNA regions (Tata, 1982). In vertebrates these receptors can be isolated and identified by sucrose gradient centrifugation of complexes of receptor proteins and radio-labelled steroids because the steroids have a very high affinity for such receptors ($K_d = 10^{-10} - 10^{-9}$ M). The affinity of the ecdysteroid receptor of *Drosophila* for ecdysone is much lower ($K_d = 10^{-7} - 10^{-6}$ M; Bonner, 1982). This receptor can therefore be identified only by using ecdysone analogues which bind receptor more tightly, or by employing a technique known as photoaffinity labelling (Schaltmann and Pongs, 1982). Mouse mammary tumour virus (MMTV) has proved particularly useful for the study of interactions of receptor-steroid complexes with DNA (Payvar *et al.*, 1981). The experiments of Lee *et al.* (1981) demonstrated that the *LTR* region of this virus, which is approximately 1 000 bp in length, could be used to replace the early SV40 promoter in

SV40 expression vectors and that *LTR* promoter activities can be stimulated two- to tenfold by glucocorticoids (see Fig. 8-23). The use of DNA sequences which bind glucocorticoid receptor proteins more tightly should allow considerably higher levels of stimulation.

Another parameter which has been associated with the control of gene expression in eukaryotic and prokaryotic cells is DNA methylation (Razin and Riggs, 1980; Doerfler, 1983; Felsenfeld and McGhee, 1982).

The only methylated base identified so far in vertebrate DNAs is 5-methylcytosine (Fig. 8-28).

5 - Methylcytosine   5 - Azacytosine

Fig. 8-28.   Structure of cytosine derivatives.

In mammalian cells, more than 90% of this methylated nucleotide occur in the dinucleotide sequence 5'-CpG-3'. The extent of methylation can be analysed by comparing DNA cleavage patterns obtained by digestion with a pair of isoschizomeric enzymes, such as *Hpa* II and *Msp* I, which recognise sites (*e.g.* 5'-CCGG-3') containing the 5'-CG-3' sequence. This analysis is based on the observation that certain restriction endonucleases are unable to cleave methylated recognition sequences. *Hpa* II, for example, cleaves only unmethylated CCGG sequences, while *Msp* I cleaves both unmethylated and methylated recognition sites (Bird and Southern, 1978; see also Section 2.1). The comparison of DNA methylation patterns using these two enzymes has revealed an inverse relationship between the extent of methylation in the vicinity of a

Richard A. Flavell
Cambridge, Massachusetts

promoter and the rate of transcription of the corresponding gene: only weak transcription occurs from methylated DNA, while unmethylated DNA is strongly transcribed.

It is also possible to alter methylation patterns at will, in order to study the influence of such changes on gene expression. Vardimon *et al.* (1982), for example, cloned the gene for the adenovirus type 2 DNA binding protein in pBR322 and injected this DNA into *Xenopus* oocytes after specific *in vitro* methylation with *Hpa* II methylase. They observed that unmethylated DNA directed the synthesis of adenovirus-specific mRNA while methylated DNA was inactive.

Methylation patterns of DNA can also be altered *in vivo* using 5-azacytidine which not only inhibits DNA methylation, but also demethylates DNA (Fig. 8-28; Jones, 1985). Employing this technique, Jones and Taylor (1980) were able to show that demethylation of DNA in mouse embryo cells *in vivo* is accompanied by an activation of gene expression.

As yet, the analysis of methylation patterns is restricted to those sites which occur within the recognition sequences of several endonucleases. The actual methylation patterns and the biochemical basis of methylation-induced alterations in gene expression are mainly unknown. The new technique of genome sequencing, however, will make these problems more amenable to investigation (Ephrussi *et al.*, 1985; *cf.* Fig. 2.4-10). If these mechanisms were known in greater detail they could be exploited most certainly also for the construction of eukaryotic expression vectors.

Anti-sense transcription may prove an alternative method for influencing gene expression in eukaryotic cells. This method is based on the observation that mRNA is not translated *in vitro* if it exists in a double-stranded form, *e.g.*, if it is complexed with its complementary sequence,

i.e. anti-sense RNA. Indeed, there are many indications that a specific control on the levels of both transcription and translation can be exerted by an anti-sense mechanism in prokaryotes as well as in eukaryotes (Weintraub *et al.*, 1985). Anti-sense RNA, for instance, has been implicated in RNA splicing (Lerner *et al.*, 1980) and in the control of gene expression in bacteria, in particular of the IS10 transposase gene (Simons and Kleckner, 1983) and of the outer membrane protein gene *ompF* of *E. coli* (Coleman *et al.*, 1984). Izant and Weintraub (1984) co-injected Tk⁻ mouse cells with plasmids containing the HSVI thymidine kinase gene in the proper and in a reverse orientation, respectively. It was demonstrated that introduction of the flipped gene reduced, but did not eliminate, *TK* gene expression in such a system. A similar effect was observed upon expression of an anti-sense thymidine kinase RNA as part of a chimaeric DHFR-TK anti-sense transcript in mouse L-cells overproducing DHFR (Kim and Wold, 1985). It remains to be seen, however, whether this anti-message

Walter Doerfler
Cologne, FRG

approach will be a useful method for the analysis of gene expression, for instance in embryogenesis.

8.4 Stimulating and Inhibiting DNA Sequences

SV40 vectors containing additional pBR322 sequences replicate poorly even in the presence of T-antigen encoded either by the vector itself or by the host chromosome. A vector consisting of pBR322 and SV40 DNAs annealed *via* their *Bam* HI ends, for example, replicates 100 times less efficiently in monkey CV1 or COS cells than genuine SV40 DNA (Lusky and Botchan, 1981). If plasmid DNA is isolated from COS cells transformed with such a shuttle vector, plasmids which replicate as efficiently as SV40 DNA are obtained at a low frequency. Such molecules possess specific deletions in their pBR322 component, indicating that certain prokaryotic sequences interfere with the replication of extrachromosomal DNA in monkey cells. The molecular basis of this interference by so-called poison sequences is far from clear; however, all the deletions known to relieve the replication defect cover the *nic-bom* site of pBR322 DNA. This site is cleaved before plasmid DNA is mobilised during bacterial conjugation (see also Section 4.1.3). Plasmid pBR322 does not code for mobilisation proteins, but it possesses a specific site between its origin of DNA replication and the *Pvu* II site (positions 2207-2263) at which these relaxation proteins can bind (bom = basis of mobility). Vectors pBR327 and pAT153 lack this site and are preferred in the development of shuttle vectors for *E. coli* and monkey cells (Twigg and Sherratt, 1980; Soberon *et al.*, 1980). As shown in Fig. 8-29, pBR327 lacks an *Eco* RII-B fragment between positions 1442 and 2502 of pBR322, while pAT153 lacks two *Hae* II fragments between positions 1644 and 2349. These deletions remove the *nic-bom* site. Furthermore, in *E. coli* these deletions increase

approximately threefold the plasmid copy number relative to the parent molecule, pBR322.

Of course there are activating DNA sequences as well as inhibitory ones. Transcription of early SV40 genes is less dependent on the presence of the Goldberg-Hogness box than on the existence of several DNA repetitions which may lie as far as 200 bp upstream from the transcriptional start signal (Fig. 8-13) (Benoist and Chambon, 1981; Fromm and Berg, 1982). These reiterations include two 21 bp repeats and two 72 bp repeats (Fig. 8-30). The region of the two 21 bp repeats can also be seen as a sixfold repetition of the sequence 5'-PyPyCCGCCC-3'. It was shown by Gidoni *et al.* (1984) to bind transcription factor Sp1.

While the removal of one of the two 72 bp repetitions appears to be permissible, the elimina-

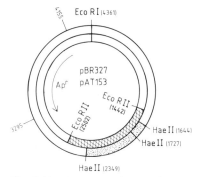

Fig. 8-29. Structure of pBR322 derivatives lacking sequences which inhibit replication in mammalian cells.

The *nic-bom* sequence of pBR322 DNA between positions 2207 and 2263 represents the so-called poison sequence. Vector pBR327 (inner circle) lacks the *Eco* RII-B fragment of pBR322 DNA (cross-hatched bar in the inner circle) and some flanking base pairs. pBR327 is 3 273 bp in length and is therefore 1 090 bp smaller than pBR322 (*cf.* also Fig. 4.1-8). Vector pAT153 (outer circle) lacks two *Hae* II fragments from pBR322 DNA between positions 1644 and 2349, comprising a total of 705 bp. The ampicillin and tetracycline resistance markers are conserved in both deletion derivatives. Shown are the start and stop codons of the β-lactamase gene at positions 4155 and 3297, respectively. The reading frame of the tetracycline resistance gene (not shown) is located between positions 86 (start) and 1273 (stop) (*cf.* Section 4.1.3 and Fig. 4.1-5).

```
      109           120                140              160              180
       |             |        |         |      |         |         |      |
5'---GTTGCTGACTAATTGAGATGCATGCTTTGCATACTTCTGCCTGCTGGGGAGCCTGGGGACTTTCCACACCTG---3'
```

Fig. 8-30. Primary structure of the 72 bp repetition of SV40 DNA (co-ordinates 109 to 180). Only the sequence of one strand, and of one of the two identical 72 bp repetitions, is shown. The second 72 bp repetition is located between positions 181-251. Early transcription initiates approximately 100 bp away, around position 0/5243, and proceeds leftward (*cf.* Fig. 8-13).

tion of both repetitions is lethal. It is quite remarkable that the presence of these enhancers or stimulating sequences does not only influence transcription of early SV40 proteins, but also of heterologous genes (reviewed by Serfling *et al.*, 1985). After transfection of HeLa cells, a rabbit β-globin gene cloned in an SV40 vector is transcribed from its own promoter region 200-fold more efficiently if the plasmid carries the 72 bp repetitions of SV40 DNA (Banerji *et al.*, 1981). The stimulatory effect of enhancers on transcription of an adjacent gene is independent of their respective orientation and of distance. Enhancers may even act if they are located downstream from the transcribed region. However, they are *cis*- and not *trans*-acting and thus have to be present on the genome which is to be transcribed. The functional sequence on the 72 bp repetitions responsible for enhancement has been mapped by deletion mapping and *in vitro* mutagenesis. It is the sequence 5'-GGTGTGGAAAG-3', situated at the 5' end of the 72 bp repeats (Weiher *et al.*, 1982).

Functionally similar sequences have also been observed in other viruses, polyoma virus (deVilliers and Schaffner, 1981), bovine papilloma virus (Lusky *et al.*, 1983), BK virus, and the large terminal repetitions (*LTR*) of Moloney sarcoma virus (MSV), for example. The *LTR* region of MSV contains two 72 bp repetitions which can be substituted for the corresponding repetitions of SV40 DNA although there are no sequence homologies between these two different repetitive sequences (Levinson *et al.*, 1982). Human

papova virus BKV contains three 68 bp repetitions which also show no homology with the 72 bp repetitions of SV40 or MSV, but are capable of activating heterologous genes to the same extent (Seif *et al.*, 1979). The strongest enhancer identified so far was isolated from the DNA genome of human cytomegalovirus (Boshart *et al.*, 1985). It is located upstream of the transcription initiation site of the major immediate-early gene of this virus and was isolated by a procedure known as "enhancer-trap" assay (Weber *et al.*, 1984). Linearised SV40 DNA lacking its own enhancer is co-transfected with short and randomly fragmented heterologous DNA into permissive cell cultures. Incorporation of foreign DNA with enhancer functions will then give rise to lytically growing SV40-like recombinants. The cytomegalovirus enhancer is several-fold more active than

Pierre Chambon
Strasbourg

the SV40 enhancer and may substitute for it in the future as a component of expression vectors.

Polyoma virus DNA does not contain any recognisable repetitions in the early promoter region. It was possible, however, to isolate a DNA fragment from the early promoter region which is 244 bp in length and shows enhancing properties similar to those observed with the 72 bp repeats of SV40 (de Villiers and Schaffner, 1981). The mechanism of action of this fragment, which is also *cis*-acting, is unknown. With the exception of enhancer sequences of bovine papilloma virus, which are located at the distal end of the early transcription unit, viral enhancers are usually located approximately 100-500 bp upstream of a transcriptional start signal.

The uninitiated observer will have difficulties with the interpretation of data related to transcriptional enhancement since three different test systems are used to detect such activating DNA sequences. The first tests whether a particular DNA sequence can replace the 72 bp repetition of SV40 DNA. Since the SV40 repetitions influence the expression of T-antigen, and this in turn influences viral DNA replication, an activating activity can be recognised indirectly by studying the DNA replication of the vector in question. It may even be possible to correlate this activity with the yield of viral particles produced in an infection. This test was used, for example, by Levinson *et al.* (1982) to search for the presence of enhancer sequences in Moloney sarcoma virus DNA. The second test measures the activation of transcription after transfection of non-permissive cells with SV40 expression vectors containing, for example, a β-globin gene. Since such vectors exist in high copy numbers for only a limited period of time, any observed gene expression is, of course, only transient. With transcription rates enhanced by factors of up to and exceeding 200, however, enhancing activities are easily recognised (see Banerji *et al.*, 1981, for example). The third test measures the increase in frequency of stable transformants obtained after transfection of cells with vectors containing selectable genes such as the thymidine kinase gene (Capecchi, 1980) and fragments of the DNA in question. The number of such stable transformants is influenced by the activating properties of the 72 bp repetition of SV40. Although these three test systems differ from each other, each identified sequence which shows enhancer activity in any one of these assays is also active in the other two assays.

Several genomic sequences have been identified in the meantime which appear to be related to the SV40 origin of DNA replication (Conrad and Botchan, 1982). Their enhancer activity has been inferred so far from sequence homologies, but has not been demonstrated in any of the three test systems mentioned above. Since almost no sequence homology is observed among viral enhancer elements, the search for homologous genomic sequences may be one promising approach obtaining such genomic enhancer sequences, but it will hardly be sufficient. One will have to take into consideration that, in contrast to viral enhancers, cellular enhancer sequences may be,

Walter Schaffner
Zürich

and probably are, cell-type-specific. This conclusion is suggested especially by observations pertaining to the genes coding for immunoglobulin light and heavy chains (Banerji et al., 1983; Queen and Baltimore, 1983; Picard and Schaffner, 1984).

One can only speculate that the mechanism of activation is perhaps caused by an association of these sequences with the nuclear matrix, an alteration of chromatin structure, and/or the binding activities of RNA polymerase II. There are indications that enhancers represent regions of interaction with site-specific DNA binding proteins since different enhancers compete with each other for limiting components in the cell (Schöler and Gruss, 1984). Eventually, it will be possible to solve these questions only by developing suitable in vitro transcription systems which respond to the presence of enhancer elements. Initial progress has been reported in this field (Sassone-Corsi et al., 1985).

Naked DNA can be introduced into cells by using substances such as DEAE dextran (McCutchan and Pagano, 1968) or calcium phosphate precipitates (Graham and van der Eb, 1973; Pellicer et al., 1980), which raise the usual low transformation frequencies to 1-5%. Calcium phosphate precipitates are obtained by adding a $DNA/CaCl_2$ solution to an isotonic buffered phosphate solution. The precipitates which form after approximately 30 minutes are used to transfect suitable tissue culture cells. In comparison to viral infections, yields obtained with linear DNA molecules, such as adenovirus, or herpes virus DNA in particular, are disappointing. This is usually explained by assuming that only a small fraction of cells are competent for DNA uptake (Wigler et al., 1979a). The proportion of these cells should rise with prolonged incubation, and indeed the percentage of SV40-infected cells increases to up to 50% when the DEAE dextran technique is used with incubation times of up to sixteen hours (Sompayrac and Danna, 1981).

8.5 Methods for Transfection of DNA

The vectors described above can be employed successfully only if efficient techniques for the introduction of vector DNA molecules into suitable cells are available. The most efficient method is, of course, viral infection. Unfortunately, there are no in vitro packaging systems for SV40 DNA or DNAs of other animal viruses. However, recombinant DNA molecules are packaged in vivo, as long as they conform with requirements imposed by size limitations. In nature, in vivo packaging is also used for defective viral genomes.

Frank Graham
Hamilton, Ontario

Similar improvements are achieved by additions of DMSO and glycerol (Stow and Wilkie, 1976; Frost and Williams, 1978; Copeland and Cooper, 1979). Chu and Sharp (1981) observed that transfections are more efficent when cells are grown in suspension rather than monolayer cultures. A shock treatment with 60-70% polyethylene glycol was recommended by Chen *et al.* (1982). These authors also pointed out that the nature and the physiological state of cells used for transfection may be an important variable (see also Corsaro and Pearson, 1981; Klobutcher and Ruddle, 1981). The hamster cell line BHKtk-ts13 seems to be particularly well-suited for transfection experiments, since transfection yields may be on the order of 70%. Subsequently, Wigler *et al.*

(1979b) demonstrated that the calcium phosphate technique also can be used to transfect non-selectable genes. If, for example, a selectable marker such as thymidine kinase and a non-selectable gene, which may be physically unlinked, are co-transfected, *i.e.*, transfected simultaneously, up to 90% of the thymidine kinase-positive transformants may contain the non-selectable gene. The molar ratio of non-selectable and selectable marker DNA used for transfection is usually on the order of 1 000:1 (see also Section 8.1.1).

DNA and RNA molecules also can be incorporated into phospholipid vesicles, which usually contain phosphatidylserine and cholesterol. These vesicles, known as liposomes, are generated by suspending lipids and nucleic acids in ether and subjecting the mixture to ultrasonic treatment, which converts the micelles into liposomes containing DNA molecules (Fraley and Papahadgopoulos, 1982). The DNA is introduced into cells by fusing the liposomes with the plasma membrane. Liposomes have great advantages since they protect the DNA against attack by nucleases. In addition, they are comparatively non-toxic, and allow surprisingly high transformation yields. Schaefer-Ridder *et al.* (1982) studied the transformation of thymidine kinase-negative mouse L cells by a plasmid encoding the thymidine kinase gene. The comparison of yields obtained by four different transfection techniques, shown in Table 8-3, demonstrates that the efficiency of transfection by liposomes compares favourably to other methods of gene transfer.

Schaffner (1980) has devised a method which allows a direct and efficient transfer of DNA between bacterial and mammalian cells. The advantage of this technique is that SV40 DNA and other vector DNAs do not have to be isolated and purified from bacteria. Bacteria containing the recombinant plasmid are first converted into

A. J. van der Eb
Leiden, The Netherlands

Table 8–3. Transformation frequencies with different transfection methods.

Method	% Tk⁺-cells	Stable colonies per 10^6 cells
Liposomes	10	200
Calcium phosphate	3	500
DEAE-Dextran	0.1–1	0
Microinjection	50 –100	200–1 000

These experiments were performed with plasmid pA60 which carries the HSV-*tk* gene as a 2 kb *Pvu*II insert within the single *Pvu*II site of pBR322. Liposomes were prepared from 250 nmoles phospholipid and 500 ng pA60 DNA prior to transfection of 10^6 Tk⁻-L-cells. For the other methods, 50 ng pA60 DNA were employed per 10^6 cells (from Schaefer-Ridder *et al.*, 1982).

protoplasts by treatment with lysozyme. These protoplasts are subsequently fused with eukaryotic cells by polyethylene glycol treatment. A careful observation of optimal fusion conditions, and the use of suitable bacterial strains stimulated with chloramphenicol, is said to result in the transfer of SV40 and polyoma vectors which is as efficient as that observed in viral infections. This technique would therefore be approximately ten to twenty times more efficient than the calcium phosphate method (Rassoulzadegan *et al.*, 1982). It has been pointed out, however, that the application of this technique may also result in the transfer of bacterial chromosomal DNA sequences. One should be aware of this complication and not underestimate the potential toxicity of contaminating prokaryotic DNA sequences.

A physical procedure for the introduction of nucleic acids into mammalian cells, known as electroporation, utilises short electric pulses of a certain field strength, which alter the permeability of membranes in such a way that DNA molecules can enter the cell (Neumann *et al.*, 1982). When plasmids carrying the herpes virus thymidine kinase gene were transfected into tk-negative mouse cells, transformation frequencies were on the order of 500 transformants per 10^6 cells for 5 micrograms of DNA. This method recently permitted the introduction of DNA into mouse lymphoid cell lines which previously had been refractory to transfection by more classical methods (Falkner and Zachau, 1984).

A very important technique employs glass capillaries to microinject nucleic acid molecules directly into the nuclei of tissue culture cells (Capecchi, 1980; Graessmann and Graessmann, 1976; Shen *et al.*, 1982). This technique has several advantages. Transformation frequencies reach almost 100%, the use of carrier DNA becomes unnecessary, transformants are stable, and the amount of injected DNA can be exactly controlled. Obviously the number of cells that can be injected is limited. This limitation notwithstanding, the technique is widely used in conjunction with other sensitive detection systems such as immunofluorescence, radioimmunoassays, and autoradiography. In conclusion, one may say that a wide spectrum of efficient techniques for the introduction of DNA into tissue culture cells is available. Genetic manipulations are therefore not fundamentally hampered by this step.

8.6 Papilloma Viruses as Vectors

Papilloma viruses are members of the same papova virus group that also comprises polyoma virus and SV40. They have a broad host range and induce warts in a variety of mammals. Bovine papilloma viruses types I and II (BPV-1, BPV-2) however, also transform rodent cells (mice, hamsters, rats) in tissue culture and induce tumours in hamsters. To date, there is no tissue culture system which allows the propagation of these viruses, thus viral DNA must be isolated from naturally occurring warts and tumours. Transformed cells usually contain approximately 50-200 copies of the BPV genome which persists exclusively in an unintegrated episomal form. Transformed cells grow more efficiently than do normal cells and can be easily selected for by growing them in agar. These features are an excellent basis for the development of a vector system.

A physical map of BPV-1 DNA, which is 8 kb in length, is now available (Fig. 8-31). It was

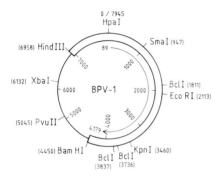

Fig. 8-31. Physical map of the 7945 bp long DNA genome of BPV-1.
The section between the *Hind* III and *Bam* HI site (open bar) represents the transforming region. The arrow extending from position 89 to 4179 indicates the postulated transcription unit expressed in transformed cells (Chen *et al.*, 1982).

shown that a subgenomic *Bam* HI-*Hind* III fragment comprising 69% of the total genome is still tumorigenic (Lowy *et al.*, 1980). There are as yet unpublished data which seem to demonstrate that a much smaller *Bam* HI-*Bcl* I fragment, which is only 600 bp in length, still retains tumorigenic capacities. Sarver *et al.* (1981) incorporated a rat preproinsulin gene and its promoter and splice sites into the 69% subgenomic BPV-1 fragment. They showed that insulin is indeed expressed by hybrid DNA molecules persisting in the transformed rat cells in an unintegrated form.

It has been difficult to transform cells with hybrid BPV vectors containing parts of pBR322. While these vectors replicated efficiently in *E. coli*, they transformed sensitive rat cells with frequencies two orders of magnitude lower than did normal BPV-1 DNA. These problems have not been resolved satisfactorily and there are still controversies and conflicting results. DiMaio *et al.* (1982) showed that hybrid DNA molecules composed of the transforming BPV fragment, a fragment of pBR322 which lacks the region inhibiting SV40/pBR322 hybrids, and a 7.6 kb fragment of the human β-globin gene transformed efficiently. Transformed mouse cells were stable and expressed the corresponding globin gene product from its own promoter. Binetruy *et al.* (1982) postulated that low efficiencies were due to inefficient transfection techniques. They devised a method of polyethylene glycol-induced fusion of bacterial protoplasts with sensitive mammalian cells (see above) and reported remarkable transformation frequencies with BPV-pBR322 hybrids. At present, it is not totally clear whether the data reported by these two groups are compatible, nor whether the difficulties occasionally experienced with BPV vectors really are due only to unsatisfactory transformation techniques or to certain unknown sequence requirements which were met only by chance. DiMaio *et al.* (1982) identified an enhancer sequence in the human β-globin gene which may be of great advantage for further cloning experiments with BPV-1 DNA.

The sequence analysis of BPV-1 DNA reveals that all five open reading frames are found on one strand of the DNA molecule (Chen *et al.*, 1982). This analysis also predicts a single transcriptional unit within the transforming region, initiating at position 82 and terminating with a polyadenylation site at position 4179. In order to explain the different lengths of the five mRNA species, it was postulated that these mRNAs possess a common leader sequence which is only 6 bp in length and is spliced to the main bodies of the mRNA molecules.

The region between the *Bam* HI and *Hind* III site of BPV-1 is not expressed in transformed cells. It contains a long open reading frame between positions 4171 and 7095 which is interrupted by a stop codon at position 5593. It is known that this region codes for a 53 kDa capsid protein. The region between positions 7146 and 49 does not contain open reading frames, and it has been assumed that this region contains the origin of DNA replication. In addition to the human β-globin gene mentioned above, the BPV vector system has been successfully used to express the rat preproinsulin gene (Sarver *et al.*, 1981, 1985), the *E. coli gpt* gene (Law *et al.*, 1982) and the human interferon β1 gene (Zinn *et al.*, 1982; Mitrani-Rosenbaum *et al.*, 1983). The latter achievement is of particular significance since it introduces the highly inducible β-interferon promoter. This promoter is totally inactive in mouse cells transformed by a hybrid plasmid containing the transforming region of BPV-1 and a 1.6 kb fragment of the human IF-β gene. Upon induction with either Newcastle disease virus or poly(rI)-(rC), IF-β transcripts which initiate from the authentic cap site of the IF-β gene are produced with high efficiency. This inducible promoter, the regulatory site of which has been assigned to a region stretching from "-40" to "-105" upstream from the cap site of the IF-β gene (Fujita *et al.*, 1985) thus retains its function in an extrachromosomal plasmid system. It may be of considerable interest for expressing cytotoxic gene products.

8.7 Retroviruses as Vectors

Retroviruses are single-stranded avian or mammalian RNA viruses which replicate *via* DNA intermediates. After entering the cell, the single-stranded RNA is copied by reverse transcriptase into double-stranded linear and circular DNA and integrated into the cellular genome at specific sites (Fig. 8-32).

The provirus which is the integrated form of the viral DNA serves as a primer for the synthesis of viral RNA. This RNA is either packaged into virus particles or used as mRNA for the translation of virus-specific proteins (Varmus, 1982). The topic has been reviewed in a book by Weiss *et al.* (1982).

The prototype virus is Rous sarcoma virus (RSV). Its genome is approximately 9 300 bases in length and carries four genes flanked by non-coding terminal regions known as long terminal repeats (*LTR*s) (Fig. 8-33). Two of these genes, *gag* and *env*, code for viral structural proteins, the third gene, *pol*, codes for reverse transcriptase, an RNA-dependent DNA polymerase. The fourth gene, *src* (derived from sarcoma), codes for a 60 kDa phosphoprotein with protein kinase activity. This gene is responsible for the transformation of fibroblasts. *Src* belongs to a group of genes called oncogenes which are usually found in retroviruses defective in replication. In order to replicate, these defective viruses require the presence of a helper virus supplying the necessary functions for replication and packaging from functional *gag*, *pol*, and *env* genes. Helper viruses are generally non-transforming. The defective, highly oncogenic viruses are usually called sarcoma viruses, the replicating helpers leukaemia viruses. Leukaemia viruses are also oncogenic but are non-acute; tumours occur only after very long latency periods. Mice, for example, harbour mouse leukaemia virus (MLV) which contains *gag*, *pol*, and *env* genes but no transforming *onc* gene (Fig. 8-33). The transforming Abelson murine leukaemia virus (Ab-MLV) contains only 1.2 kb of the 5'- and 0.72 kb of the 3'-terminal ends of mouse leukaemia virus (Fig. 8-33). The central region of MLV, which is 6.5 kb in length and contains the replication functions, has been substituted for a 3.6 kb fragment with an *onc* gene designated *abl*. Ab-MLV therefore can replicate only in conjunction with MLV providing the polymerase and the two structural proteins *gag* and *env* as helper functions. Rous sarcoma virus must be regarded as an exception since it

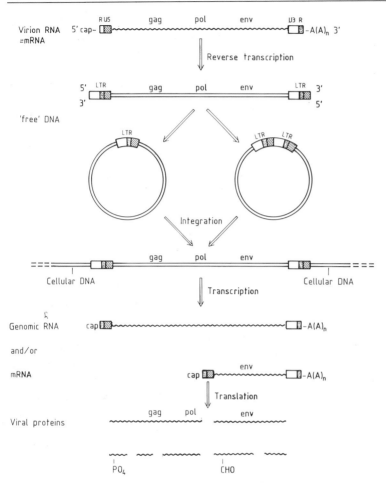

Fig. 8-32.   Life cycle of retroviruses.
Shown at the top is one of the two single-stranded RNA genomes within a virus particle with its characteristic structural elements, namely the 5' cap nucleotide, sequence *R* (20-80 nucleotides in length) which is repeated at either end, region *U5* (80-100 nucleotides) which is characteristic of the 5' terminus, the coding regions for the genes *gag, pol,* and *env*, and sequence *U3* (0.2-1 kb) which is characteristic of the 3' end. Double-stranded linear and circular forms which can be inserted into the host DNA are generated by reverse transcription. These forms are flanked by long terminal repeats (*LTR* regions) which are composed of *U3, R,* and *U5* sequences. The integrated retrovirus, known as the provirus, directs the synthesis of genomic RNA and mRNA species. The *env* mRNA is spliced. Mature proteins which are phosphorylated and/or glycosylated are derived from polyproteins by proteolytic cleavage.

carries the oncogene in addition to genes for replication and packaging functions, and thus does not require a helper virus for its propagation.

As shown in Table 8-4, there are at present approximately 25 different known *onc* genes derived from a variety of acutely transforming viruses of birds and mammals (Bishop, 1983). Viral *onc* genes *(v-onc)* have cellular counterparts *(c-onc)*, called proto-oncogenes. Ab-MLV, for example, codes for a 120 kDa transforming protein (p120) which requires a DNA coding

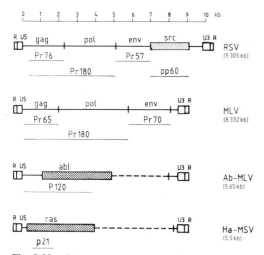

Fig. 8-33. Structure of some retroviruses.
RSV = Rous sarcoma virus; MLV = mouse leukaemia virus; Ab-MLV = Abelson mouse leukaemia virus; Ha-MSV = Harvey mouse sarcoma virus. Horizontal lines represent the translation products with molecular weights in kDa. Hatched bars indicate *onc* gene insertions. "p", "P", "pp", and "Pr" stand for protein or polypeptide. The terminal sequences, *RU5* and *U3R*, with their signal sequences for replication and transcription are represented as open squares. Broken lines represent deletions; solid lines at the 5′ and 3′ ends of defective viruses indicate short homologies with the helper virus MLV. Defective transforming viruses either express the *onc* gene (Ha-MSV), or, like Ab-MLV, a fusion protein consisting of *onc* and *gag* gene sequences.

capacity of approximately 2.4 kb. This genetic information is also found in the genome of mice but it is spread over a region covering approximately 20 kb (Goff *et al.*, 1980). Proto-oncogenes thus contain introns which presumably were eliminated when they were transduced by retroviruses. Retroviruses may have acquired such genes by copying cellular mRNA sequences which are by definition free of introns, using the enzyme reverse transcriptase. It is thought that linkage of these cDNA copies to retroviral promoters, as well as mutations of these cellular sequences within retroviral genomes resulted in activation of the malignant properties of the proto-oncogenes. Proto-oncogenes as such have been highly conserved in evolution and thus are thought to play an important role in normal cellular growth and development (Shilo and Weinberg, 1981).

Retrovirus vectors should resemble defective transforming viruses. Instead of, or in addition to, *onc* genes, these viruses would carry foreign DNA sequences. Like v-*onc* genes this foreign DNA should be flanked by terminal *LTR* sequences derived from leukaemia virus, which are essential *cis* elements for replication. Helper viruses could be functional, replication-competent leukaemia viruses. Such retrovirus vector systems consisting of vector and helper virus would offer a number of advantages compared to vectors derived from

Table 8–4. Name and origin of some viral *onc* genes.

Name	Virus	Species	Protein product
src	Rous sarcoma virus (RSV)	chicken	pp60^{src}
myb	avian myeblastosis virus (AMV)	chicken	P35^{myb}
myc	myelocytomatosis virus (MC29)	chicken	P110^{myc}
erb (A+B)	avian erythroblastosis virus (AEV)	chicken	P75$^{gag-erbA}$
ros	UR-2 virus	chicken	Hp65$^{gag-ros}$
fps	Fujinami sarcoma virus (FuSV)	chicken	P140$^{gag-fps}$
yes	Y73 sarcoma virus	chicken	P140$^{gag-yes}$
rel	reticuloendotheliosis virus (REV-T)	turkey	?
abl	Abelson mouse leukeamia virus (Ab-MLV)	mouse	P120$^{gag-abl}$
mos	Moloney mouse sarcoma virus (Mo-MSV)	mouse	?
ras	rat sarcoma virus (RaSV)	rat	P29$^{gag-ras}$
	Kirsten/Harvey sarcoma virus (Ki-MSV)	rat	P21^{ras}
fes	Snyder-Theilen feline sarcoma virus (SM-FeSV)	cat	P85$^{gag-fes}$
fms	McDonough feline sarcoma virus (SM-FeSV)	cat	P180$^{gag-fms}$
sis	Simian sarcoma virus (SSV)	monkey	P28^{sis}

DNA tumour viruses. Retroviruses have an extremely broad host range while SV40, for example, replicates only in monkey cells. In addition, retroviruses do not kill infected cells although the cells produce a large amount of infectious virus. Infected cells integrate retroviral DNA sequences into their genomes, thus genetic information introduced by retrovirus vectors would be stably inherited in the progeny of infected cells. It is important to note that it is usually only one copy of the retroviral DNA which becomes integrated into the host genome. The helper virus which is required for the replication of the defective vector induces tumours only after long latency periods. The effects of transduced genes therefore can be studied long before the onset of the disease.

It would be more convenient, of course, to employ and obtain helper-free vector stocks. Such stocks can indeed be prepared in special cell lines which harbour an endogenous retrovirus genome

the mRNA of which is not encapsidated and which provides only the necessary replication and packaging functions for the vector. The basis for the development of such cell lines is the observation that packaging of retrovirus (RNA) genomes requires not only coat proteins, but also the presence of a certain sequence on the RNA itself. In the Moloney murine leukaemia virus (Mo-MuLV) proviral genome these *cis* elements are located on a 351 bp DNA fragment, desigated ψ (psi), between a *Bal*I site, 6 bp downstream from the *env* mRNA donor splice site, and a *Pst*I site approximately 50 bp upstream of the start codon for the *gag* precursor (p65^{gag}) (Fig. 8–34). A plasmid lacking this ψ fragment, called pMOV-ψ⁻, thus retains the 5' splice site for the subgenomic *env* mRNA, and, in spite of the deletion, is able to express all three genes of the virus correctly (Fig. 8-35). It was therefore possible to transform NIH/3T3 cells with plasmid

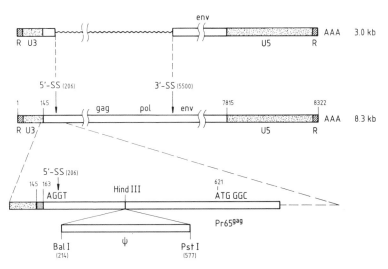

Fig. 8-34.   Principle of the construction of vector pMOVψ⁻.
Shown (in the middle) is the structure of the 8.3 kb Mo-MLV mRNA with the arrangement of genes *gag, pol*, and *env*, and the two splice sites at positions 206 and 5500, which are responsible for the synthesis of the subgenomic *env* mRNA (3.0 kb; shown at the top). The expanded region at the bottom depicts the exact position of the ψ⁻ deletion between a *Bal*I site (position 214) and a *Pst*I site (position 577) immediately upstream from the start codon of the 65 kDa precursor of the *gag* gene product (Pr^{65gag}). Since this construction retains the 5' splice site (5'-ss) of the env mRNA at position 206 the synthesis of this subgenomic mRNA species is unaffected. Numbers refer to the sequence published by Shinnick *et al.* (1981). The termini of the mRNA molecules are characterised by specific sequence sections named *R, U3*, and *U5*. The wavy line indicates the sequence deleted upon creation of the *env* mRNA.

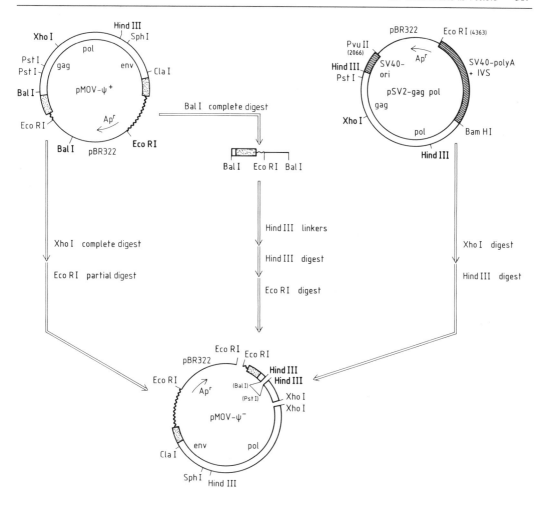

Fig. 8-35. Construction of vector pMOVψ⁻.

This vector is a composite of three segments. Two segments are derived from pMOVψ⁺ which contains the intact Mo-MLV provirus flanked by some cellular sequences (wavy lines) cloned into the *Eco* RI site of pBR322 (Chumakov *et al.*, 1982). The third segment is derived from vector pSV2-gag-pol. This plasmid is composed of the large *Hind* III-*Bam* HI fragment from pSV2 (*cf.* Fig. 8-22) carrying the SV40 polyadenylation site, the *Eco* RI-*Pvu* II fragment of pBR322 DNA, and the SV40 origin of replication plus a *gag-pol* fragment from Mo-MLV in which the *Pst* I site at position 577 was converted into a *Hind* III site. A small *Xho* I-*Hind* III fragment from pSV2 gag-pol can thus be isolated and used in the three-part construction which restores the *gag*-encoding region in pMOVψ⁻, but deletes the ψ sequence (marked by *Bal* I and *Pst* I in parenthesis). Flanking cellular sequences are represented as wavy lines, pBR322 sections as thin lines, the coding regions of Mo-MLV as open bars, the flanking *LTR* regions as stippled bars and SV40 sections as hatched bars. In pMAVψ⁻ (see text) the *Sph* I-*Cla* I fragment from pMOVψ⁻ with the *env* gene is replaced by a corresponding fragment from an amphotropic retrovirus genome (Cone and Mulligan, 1984). Numbers in parentheses represent pBR322 co-ordinates. For details of the construction see Mann *et al.* (1983).

pMOV-ψ⁻, and to obtain a cell line called ψ-2 which does not produce virus but provides *in trans* all functions which are required for the propagation of defective retroviruses (Mann *et al.*, 1983). Transfection of this cell line with recombinant defective retroviruses leads to a high titer of packaged retroviruses in the supernatant which are entirely free of helper viruses. These recombinant retroviruses are capable of transmitting the defective recombinant viral genome to suitable target cells in a way which is characteristic of retroviruses, *i.e.*, by integration into the cellular DNA. For the target cell the recombinant virus is simply a new transcription unit which expresses a foreign gene under the control of the viral *LTR* region. Virus particles could only be generated in the target cell if this cell contained coat proteins derived, for example, from a cryptic endogenous retrovirus. A cell line similar to ψ-2 has also been described for the reticuloendotheliosis virus of chicken (Watanabe and Temin, 1983).

The vectors developed so far contain well-known dominant markers such as *gpt* or *neo*; in addition, these vectors are shuttle vectors which can also be replicated in *E. coli*. Vector pSVX-neo, for example (Fig. 8-36), is flanked by *LTR*

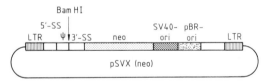

Fig. 8-36. Structure of vector pSVX(neo).
This vector contains a section of approximately 1 000 bp from the left end of Mo-MLV, comprising the *LTR* region (hatched bars), the 5′ splice site for the *env* mRNA (open bars) and the ψ region. Adjacent to it are the 3′ splice site of the M-MuLV genome (nucleotides 5409-5768), the *neo* gene derived from transposon Tn5 (stippled bars; *cf.* Fig. 8-7), the SV40 origin of replication (nucleotides 160-5171) (dashed bar), the pBR322 origin (nucleotides 2521-3102), and the right terminus of Mo-MLV. The presence of the Tn5-derived *APH(3′)II* gene leads to kanamycin resistance in bacterial cells, and G418 resistance in mammalian cells. Similar vectors which contain the *E. coli gpt* gene (Hellermann *et al.*, 1984), or the mouse dihydrofolate reductase gene (Williams *et al.*, 1984) have also been constructed.

regions and carries the first 1 000 bp of Mo-MLV with the 5′ splice site, the region of the 3′ splice site of the subgenomic mRNA, and the neomycin resistance gene derived from transposon Tn5. The pBR322 origin permits replication in *E. coli*, while the SV40 origin allows replication as a plasmid in COS cells. Additional DNA sequences can be inserted into the *Bam* HI site of this vector. If genomic DNA is inserted, it is converted to cDNA since the vector goes through an RNA phase during its replication in ψ-2 cells. This provides an easy method, designated "gene collapsing", for the conversion of genomic DNA into cDNA.

The retrovirus vector system described above, which is based on the ψ-2 packaging cell line, can be used only in mouse or closely related rodent cells. Fortunately, the phenomenon of pseudotype formation in retroviruses offers the possibility of extending its applications to other mammalian cells. Retroviruses can be classified as ecotropic, xenotropic, and amphotropic viruses. Ecotropic viruses, such as Moloney mouse leukaemia virus, replicate only in their specific host cells, *i.e.*, in mouse cells. Xenotropic viruses replicate in all cell types but not in those cells from which they were isolated. Amphotropic viruses have a broad host range specificity and infect both their own host as well as cells of other organisms. If a mouse cell which harbours a pSVX genome (for example, pSVX-neo) is infected with an amphotropic virus, the supernatant obtained from these cells contains not only the amphotropic helper virus but also a pSVX-neo genome packaged in the coat of the amphotropic virus. These pseudotypes can be used for infecting monkey cells. If COS cells are used, vector pSVX-neo will be replicated as a plasmid with approximately 2-200 copies per cell, since pSVX-neo contains an SV40 origin. An analogous vector which does not contain an SV40 origin will be integrated into the host genome with one copy per cell. Again in this case, helper-free virus stocks would be highly desirable. Cone and Mulligan (1984) have thus developed a ψ-2-like packaging cell line, termed ψ-AM. Since it is well established that host range differences of retroviruses are a consequence of differences in receptor specificity of the viral envelope protein gp70. The sequences coding for the ecotropic envelope protein in pMOV-ψ⁻ were therefore replaced by an amphotropic *gp70* sequence derived from the amphotropic 4070A viral genome. Transfection of NIH/3T3 fibroblasts with this recombinant clone, called pMAV-ψ⁻, yielded a series of ψ-AM lines which produced high viral titers after transfection with recombinant retroviral vectors. These viruses, in turn, permitted the infection of a broad range of mammalian cell types, including a variety of human cell lines. It is important to note that most of the transfected ψ-AM lines remained helper-free, even after months in culture. Although they contained high levels of two viral transcripts derived from the transfected recombinant vector and the endogenous leukemia virus lacking the ψ sequences, these two transcripts did not seem to generate fully wild-type amphotropic virus by recombination. Instead, they only encapsidated the transcript from the recombinant vector. Safe methods are thus now available for the introduction of foreign genes *via* murine retrovirus vectors of the pSVX-type.

The SV40 origin in the pSVX-type vectors also serves another purpose. It has been known for a long time that the fusion of monkey cells with mouse cells harbouring an integrated copy of the SV40 genome will activate the SV40 origin. This process is accompanied by the amplification of the DNA in this region and ultimately leads to a release of this DNA. Flanking DNA sequences are usually also affected. This system therefore allows identification of the cellular integration sites of the provirus. In the long run, this technique will allow the investigation of position effects, *i.e.*, a study of the role of integration sites of a gene on its expression.

The vectors described above can be used not only in tissue culture cells but also in hematopoietic stem cells of bone marrow (Williams *et al.*, 1984) and in postimplantation mouse embryos

(Stuhlmann *et al.*, 1984). It can therefore be expected that these systems will be employed in the initial approaches to gene therapy (Anderson, 1984).

An early application of retrovirus vector systems to obtain cDNA from genomic DNA was described by Shimotohno and Temin (1982). They constructed a defective spleen necrosis virus (SNV) which contained a thymidine kinase gene

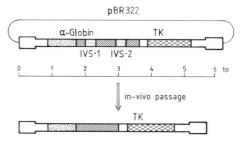

Fig. 8-37. Structure of retrovirus vector pSNV-TK-α-globin.
Apart from pBR322 sequences (Shimotohno and Temin, 1981), this vector contains a copy of the HSV *tk* gene, and the mouse α-globin gene, flanked by the terminal sequences of SNV (Spleen Necrosis Virus). The mouse α-globin gene lacks 3′ terminal non-coding sequences, including the polyadenylation site, because such deletions increase virus yields obtained with retrovirus vectors. pBR322 sequences are represented as a thin line, SNV sequences as an open bar, the *tk* gene as a cross-hatched bar. The 5′ untranslated region of the mouse α-globin gene is represented by dotted bars, the coding regions are hatched. Coarsely stippled regions are introns (IVS-1 and IVS-2). Unintegrated viral DNA isolated several days after infection lacks the two introns.

and mouse α-globin sequences (Fig. 8-37). The mouse α-globin gene contains two introns of 122 and 134 bp in its coding sequence, both of which were present in the DNA used for this cloning experiment. This vector was introduced into tk-negative cells together with reticuloendotheliosis virus as a helper. When the unintegrated linear DNA was analysed several days after infection, the virus contained only α-globin DNA without introns. These and other data demonstrate the effectiveness of retroviral vectors in the conversion of genomic DNA into cDNA. In addition, these findings also support the hypothesis that intron-free genes, known as pseudogenes, which are sometimes found in the vicinity of their normal cellular counterparts, may be generated by reverse transcription of mRNA (see Vassin *et al.*, 1980, for example). This flow of information from genomic DNA *via* spliced mRNA to intron-free genomic DNA (Fig. 8-37) could also explain the generation of *v-onc* genes which, among other things, differ from *c-onc* genes by the absence of introns.

8.8 Gene Expression by Gene Amplification

The term gene amplification describes the phenomenon of increased gene copy numbers which is observed for certain genes in tissue culture cells under suitable selection conditions (Stark and Wahl, 1984). One example is the *CAD* gene (Wahl *et al.*, 1984) which codes for a multifunctional protein catalysing the *de novo* synthesis of uridine. The three enzymes involved are carbamylphosphate synthetase, aspartate transcarbamoylase, and dihydro-orotase. Gene amplification occurs upon selection for resistance to *N*-phosphonoacetyl-L-aspartate (PALA) a potent inhibitor of this enzyme. Another example of importance in genetic engineering is the amplification of the dihydrofolate reductase gene as the result of a selection for methotrexate resistance (Schimke, 1984). Selection usually employs stepwise increases in methotrexate concentrations for several months from 0.01 µM to 0.02, 0.1, 1.0 etc. to 50-100 µM. Depending on the cell type, cell lines with up to several 1 000 copies of the *DHFR* gene may be obtained. In freshly selected cells this phenotype is unstable, *i.e.*, the phenotype is lost after several cell cycles in the absence of selective pressure. In many cases the unstable configuration of *DHFR* genes becomes stable

under continuous selective pressure and is only lost after many months (several hundred cell divisions) in the absence of methotrexate. Stably amplified cells contain the amplified genes within their chromosomes, where they can be identified by suitable techniques as so-called homogeneously stained regions (HSRs). Unstable *DHFR* genes are usually located on small extrachromosomal elements which frequently occur in pairs and which are known as double-minute chromosomes (DMC). It is still unknown why certain cells seem to favour the establishment of an unstable state (mouse cells) while other cells give rise to stable phenotypes (hamster cells). Since the behaviour of cells cannot be predicted, the establishment of new cell lines carrying amplified genes must be confirmed by karyotype analysis, blotting techniques and/or an instability of the resistant phenotype in the absence of selective pressure.

Gene amplification can be exploited for gene technology, since not only the *DHFR* gene but any other gene on the same vector will be stably integrated and amplified under suitable conditions, leading to increased expression of dihydrofolate reductase and also of the co-transformed gene. Kaufmann and Sharp (1982a,b), for example, constructed a vector which carries a *DHFR* cDNA under the control of a suitable promoter, suitable splice and polyadenylation signals, and the entire SV40 genome. This vector transforms DHFR-negative hamster cells to a DHFR-positive phenotype. After selecting for methotrexate resistance by a stepwise increase of the drug concentration, the resulting cell variants contained up to 1 000 copies of the *DHFR* gene and of the SV40 DNA linked to it. These cells express the small SV40 t-antigen in amounts which correspond to up to 10% of the total cellular protein. The vector used for this experiment contains the *DHFR* gene under the control of the late adenovirus promoter. This promoter is efficient in the late phase of infection in cells lytically infected with adenoviruses, but should generally be regarded as a weak promoter. The weakly transcribed *DHFR* gene must therefore be highly amplified in order to obtain a methotrexate-resistant pheno-type. Nevertheless, the much more efficient SV40 early promoter has also been used success-fully in several instances (Lau *et al.*, 1984).

It can be expected that increased gene expres-sion mediated by gene amplification will gain significance, especially for eukaryotic genes which are not available as cDNAs and for large proteins which cannot be expressed in *E. coli* even if the corresponding cDNAs are used. Neverthe-less, word of caution appears to be appropriate. A review of the literature demonstrates that trans-formants may be extremely unstable and that the desired genes may not be expressed even if the constructions are correct. There are several plau-sible explanations for these observations. The expressed gene product should not interfere with gene amplification as such, and of course the gene product should not be toxic for the cell. As in *E. coli*, the use of controllable promoters, such as the promoters of the metallothionein, β-interferon or heat-shock genes, is recommended (*cf.* also Sec-tion 8.4).

8.9 Gene Transfer into Drosophila

Until very recently manipulation of *Drosophila melanogaster* by specific and directed genetic techniques was not possible. This has changed dramatically since Rubin and Spradling (1982) developed a transformation system for *D. mela-nogaster* which allows discrete fragments of DNA to be integrated into the chromosomes of germ cells. The transformation system employs a specif-ic class of mobile genetic elements, known as *P* elements, in *D. melanogaster*. Normally these elements are stable, but they can be induced to

transpose at high frequencies under conditions of a genetic syndrome called P-M hybrid dysgenesis. This term describes the occurrence of unusual and frequent mutations in the progeny of crosses between certain strains of *D. melanogaster*, in particular between female *M* strains (*M* = mater-nal) and male *P* strains (*P* = paternal). *P* stocks contain intact *P* elements, 3 kb long stretches of DNA flanked by 31 bp inverted repeats. *M* stocks lack functional copies of this element, although they may carry defective ones. When *P* males and *M* females are crossed, the *P* elements enter the *M* cytoplasm and are mobilised into new chromo-somal sites, generating insertion mutations and rearrangements. The genomes of *P* strains also carry smaller *P* elements which arise by internal

Gerald M. Rubin
Berkeley

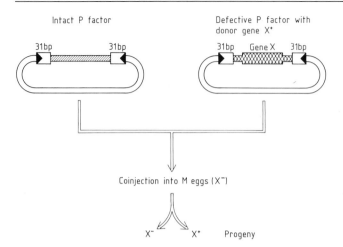

Fig. 8-38. Principle of *P*-mediated germ line gene transfer in *D. melanogaster*.
A mixture of an intact *P* factor vector and of a defective *P* factor vector carrying gene *X* flanked by the characteristic 31 bp repeats is injected into M mutant (X⁻) eggs. The complete *P* factor complements the defective one by providing the necessary transposase activity (Gehring, 1984).

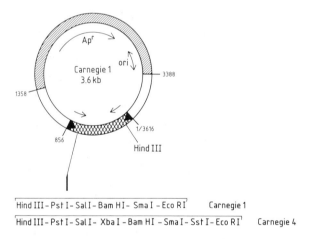

Fig. 8-39. Structure of the Carnegie series of vectors.
Carnegie 1 is 3 616 bp long and consists of a sequence derived from pUC8 (hatched bar) and a *Sal* I fragment with the non-autonomous *P* element 6.1 (cross-hatched bar) flanked by DNA sequences from the white locus of *D. melanogaster* (open bars). The polylinker within the *P* element portion with the 31 bp flanking repeats (arrows) can be used for the insertion of additional DNA. Carnegie 20, for example, carries an additional 7.2 kb *Hind* III fragment with the *rosy* gene cloned into the *Hind* III site of the polylinker. After gene transfer into rosy⁻ flies, transformants can easily be detected by their eye colour (Rubin and Spradling, 1983).

deletions of the intact *P* element, and which can be mobilised *in trans* by intact *P* factors provided they have retained intact termini. This observation was exploited by Rubin and Spradling (1982) in the development of a gene transfer system (Fig. 8-38), which permits introduction of any DNA present within a *P* element into *D. melanogaster* or other strains of *Drosophila* (Brennan *et al.*, 1984). The Carnegie series of *P* vectors contains a non-autonomous *P* element cloned into pUC plasmids, allowing the vector to be shuttled between *E. coli* and *D. melanogaster* (Rubin and Spradling, 1983). Additional DNA can be cloned into a multiple polylinker cloning site within the *P* element region (Fig. 8-39). DNA transfer into *D. melanogaster* is achieved through microinjection of embryonic cells with a mixture of vector DNA and a plasmid carrying a fully autonomous *P* factor. The *P* factor is 2 907 bp in length, and is thought to code for a transposase which can act *in trans* to mediate the integration of the non-autonomous *P* vector into the chromosomal DNA. Transformed flies are obtained from injected embryo cells at a frequency varying between 5 and 20%. The detection of transformed flies requires a genetic marker which either produces an easily discernable phenotype, or which can be efficiently selected for. Rubin and Spradling (1983) initially used the *rosy* gene which can serve as a visible marker if injected flies carry a *rosy⁻* mutation. The *rosy* gene codes for xanthine dehydrogenase, the expression of which, even in minute amounts, alters the fly's eye colour. More recent experiments have alternatively employed the bacterial neomycin resistance gene (*APH3′ (II)*) which can render *Drosophila* larvae resistant to the antibiotic G418 (Steller and Pirrotta, 1985). This dominant selectable marker obviates the use of a mutated recipient organism and may eventually allow this system to be employed in organisms other than *Drosophila*, e.g. nematodes or even plants. The corresponding cosmid vectors, which will allow efficient recovery of the integrated DNA, are currently being developed (Steller and Pirotta, 1985).

8.10 Vaccinia Virus Vectors

Vaccinia virus is a large double-stranded DNA virus which replicates within the cytoplasm of infected cells. The 187 kb genome possesses a 10 kb inverted terminal repetition and codes for 75 early polypeptides which are produced within the first minutes of infection, and 40 late polypeptides which are synthesised after the onset of DNA replication. Isolated naked DNA is not infectious, since vaccinia virus has evolved its own transcriptional regulatory sequences which are recognised by a virus-encoded RNA polymerase packaged into the infectious virus particle. Insertion of foreign DNA into the vaccinia virus genome can therefore be accomplished only *in vivo* by exploiting the principle of general homologous recombination. A general protocol devised by Mackett *et al.* (1984) employs a plasmid which carries the foreign gene to be introduced into the vaccinia genome flanked by an active vaccinia promoter on its 5′ end, and by DNA from a non-essential vaccinia gene on its 3′ end. This plasmid is transfected into cells infected with wild-type vaccinia virus so that homologous recombination can take place between the vaccinia DNA sequences flanking the foreign gene on the plasmid and the corresponding sequences on the wild type virus genome. The foreign gene thus is integrated into the vaccinia virus genome, thereby inactivating the vaccinia gene which possesses the flanking sequences present in the plasmid construction. If the infectivity of the recombinant virus is to be maintained, this gene must be non-essential, and if possible selectable, to facilitate selection for the recombinant virus. The most suitable marker for vaccinia virus is thymidine kinase. Insertion of foreign DNA into the thymdine kinase coding sequences of vaccinia virus results in the inactivation of the *TK* gene and in the concomitant formation of a TK⁻ recombinant vaccinia virus. This can be selected from infected cell lysates by plaque assay on TK⁻ host cell in the presence of 5-bromodeoxyuridine (see Section 8.1.1 and Table 8-1).

A useful plasmid vector for the insertion of foreign DNA is plasmid pMM5 which contains a pUC9-derived polylinker sequence flanked by the vaccinia *TK* gene promoter and a 1 000 bp *Eco* RI fragment containing the 3' end of the *TK* gene together with adjacent sequences. There appears to be no practical size limitations for the insertion of foreign DNA into the vaccinia virus genome since it has been possible to produce infectious vaccinia virus containing 25 000 bp of bacteriophage λ DNA (Smith and Moss, 1983).

The strategy of employing *TK* insertion vectors was used to insert the genes encoding hepatitis B virus surface antigen (Smith *et al.*, 1983a), influenza virus haemagglutinin (Smith *et al.*, 1983b), herpes simplex virus type 1 glycoprotein D (Paoletti *et al.*, 1984), Rabies virus glycoprotein (Kieny *et al.*, 1984), and the *Plasmodium knowleri* sporozoite antigen (Smith *et al.*, 1984). In each case, inoculation of rabbits with the respective live recombinant vaccinia virus induced high titers of neutralising antibodies.

It has been proposed that these recombinant viruses may be used not only as live vaccines for human use, but also for veterinary applications. Since vaccinia virus can accomodate at least 25 000 bp of foreign DNA it may even be possible to prepare a polyvalent vaccine against a number of different pathogenic agents within a single recombinant virus.

The use of vaccinia virus vaccination as a method of preventing smallpox dates back to 1796 when the British physician Edward Jenner observed that vaccinia was sufficiently related to variola and could prevent smallpox disease (Jenner 1798; see also Baxby, 1981). Application of this live vaccine has been such a tremendous success that in 1980 the World Health Organisation declared the worldwide eradication of smallpox. Nevertheless, there have been a few serious side effects following vaccinia vaccination. Among these are the adverse side effects observed in immunosuppressed subjects, individuals with certain skin conditions, and the serious, albeit rare, complication of post-vaccinial encephalitis. The development of attenuated strains of vaccinia virus has continuously reduced the incidence of adverse reactions. Nevertheless, it appears that vaccinia-based vaccines will not be of much use in highly developed countries. In less developed countries, however, the risks of contracting a variety of parasitic and viral diseases, *e.g.* malaria, rabies, hepatitis etc., combined with the low cost of the vaccine may greatly outweigh the low incidence of side effects of a vaccinia vaccination.

References

Alt, F.W., Kellems, R.E., Bertino, J.R., and Schimke, R.T. (1978). Selective multiplication of dihydrofolate reductase genes in methotrexate-resistant variants of cultured murine cells, J. Biol. Chem. 253, 1357-1370.

Anderson, W.F. (1984). Prospects for human gene therapy. Science 226, 401-409.

Ashburner, M., and Bonner, J.J. (1979). The induction of gene activity in *Drosophila* by heat shock. Cell 17, 241-254.

Baxby, D. (1981). Jenner's smallpox vaccine: the riddle of vaccinia virus and its origin. London, Heinemann Educational Books Ltd.

Banerji, I., Russoni, S., and Schaffner, W. (1981). Expression of a β-globin gene is enhanced by remote SV40 DNA sequences. Cell 27, 299-308.

Banerji, I., Olson, L., and Schaffner, W. (1983). A lymphocyte-specific cellular enhancer is located downstream of the joining region in immunoglobulin heavy chain genes. Cell 33, 729-704.

Beck, E., Ludwig, G., Auerswald, E.A., Reiss, B., and Schaller, H. (1982). Nucleotide sequence and exact localization of the neomycin phosphotransferase gene from transposon Tn5. Gene 19, 327-336.

Benoist, C., and Chambon, P. (1981). *In vivo* sequence requirements of the SV40 promotor region. Nature 290, 304-310.

Binetruy, B., Meneguzzi, G., Breathnach, R., and Cuzin, F. (1982). Recombinant DNA molecules comprising bovine papilloma virus type 1 DNA linked to plasmid DNA are maintained in a plasmidial state both in rodent fibroblasts and in bacterial cells. The EMBO Journal 1, 621-628.

Bird, A.P., and Southern, E.M. (1978). Use of restriction enzymes to study eukaryotic DNA methylation: I. The methylation pattern in ribosomal DNA from *Xenopus laevis*. J. Mol. Biol. 118, 27-47.

Bishop, J.M. (1983). Cellular oncogenes and retroviruses. Ann. Rev. Biochem. 52, 301-354.

Bonner, I.J. (1982). An Assessment of the ecdysteroid receptor of *Drosophila*. Cell 30, 7-8.

Boshart, M., Waber, F., Jahn, G., Dorsch-Häsler, K., Fleckenstein, B., and Schaffner, W. (1985). A very strong enhancer is located upstream of an immediate early gene of human cytomegalovirus. Cell 41, 521-530.

Breathnach, R., Mantei, N., and Chambon, P. (1980). Correct splicing of a chicken ovalbumin transcript in mouse L-cells. Proc. Natl. Acad. Sci. USA 77, 740-744.

Brennan, M.D., Rowan, R.G., and Dickinson, W.J. (1984). Introduction of a functional P-Element into the germ-line of *Drosophila hawaiiensis*. Cell 38, 147-151.

Buetti, E., and Diggelmann, H. (1981). Cloned mouse mammary tumor virus DNA is biologically active in transfected mouse cells and its expression is stimulated by glucocorticoid hormones. Cell 23, 335-345.

Buchman, A.R., Burnett, L., and Berg, P. (1981). The SV40 nucleotide sequence. In "DNA Tumor Viruses. Molecular Biology of Tumor Viruses", 2nd edition, revised, J. Tooze, (ed.), Cold Spring Harbor, N.Y., Cold Spring Harbor Laboratory, pp. 799-841.

Capecchi, M.R. (1980). High efficiency transformation by direct microinjection of DNA into cultured mammalian cells. Cell 22, 479-488.

Cepko, C.L., Roberts, B.E., Mulligan, R.C. (1984). Construction and applications of a highly transmissable murine retrovirus shuttle vector. Cell 37, 1053-1062.

Chang, A.C.Y., Nunberg, I.H., Kaufman, R.J., Erlich, H.A., Schimke, R.T., and Cohen, S.N. (1978). Phenotypic expression in *E. coli* of a DNA sequence coding for mouse dihydrofolate reductase. Nature 275, 617-624.

Chen, E.Y., Howley, P.M., Levinson, A.D., and Seeburg, P.H. (1982). The primary structure and genetic organization of the bovine papilloma virus type 1 genome. Nature 299, 529-534.

Chu, G., and Sharp, P.A. (1981). SV40 DNA transfection of cells in suspension: analysis of the efficiency of transcription and translation of T-Antigen. Gene 13, 197-202.

Colbère-Garapin, F., Horodniceanu, F., Khourilsky, P., Garapin, A.C. (1981). A new dominant hybrid selective marker for higher eukaryotic cells. J. Mol. Biol. 150, 1-14.

Cole, C.N., and Santangelo, G.M. (1983). Analysis in cos-1 cells of processing and polyadenylation signals

by using derivatives of the Herpes Simplex Virus Type 1 Thymidine Kinase Gene. Mol. Cell. Biol. 3, 267-279.

Coleman, J., Green, P.J., and Inouye, M. (1984). The use of RNAs complementary to specific mRNAs to regulate the expression of individual bacterial genes. Cell 37, 429-436.

Cone, R.D., and Mulligan, R.C. (1984). High-efficiency gene transfer into mammalian cells: Generation of helper-free recombinant retrovirus with broad mammalian host range. Proc. Natl. Acad. Sci. USA 81, 6349-6353.

Conrad, S.E., and Botchan, M.R. (1982). Isolation and characterization of human DNA fragments with nucleotide sequence homologies with the SV40 regulatory region. Mol. Cell. Biol. 2, 949-965.

Copeland, N.G., and Cooper, G.M. (1979). Transfection by exogenous and endogenous murine retrovirus DNAs. Cell 15, 347-356.

Corsaro, C.M., and Pearson, M.L. (1981). Enhancing the efficiency of DNA-mediated gene transfer in mammalian cells. Somatic Cell Genet, 7, 603-616.

Croce, C.M., Bakey, B., Nyhan, W.L., and Koprowski, H. (1973). Re-expression of the rat hypoxanthine phosphoribosyl transferase gene in rat-human hybrids. Proc. Natl. Acad. Sci. USA 70, 2590-2594.

Davies, J., and Smith, D.I. (1978). Plasmid determined resistance to antimicrobial agents. Ann. Rev. Microbiol. 32, 469-518.

DiMaio, D., Treisman, R., and Maniatis, T. (1982). Bovine papilloma virus vector that propagates as a plasmid in both mouse and bacterial cells. Proc. Natl. Acad. Sci. USA 79, 4030-4034.

Doerfler, W. (1983). DNA methylation and gene activity. Ann. Rev. Biochem. 52, 93-124.

Elder, I.T., Spritz, R.A., and Weissman, S.M. (1981). Simian Virus 40 as eukaryotic cloning vehicle. Ann. Rev. Genet. 15, 295-340.

Elgin, S.C.R. (1981). DNAaseI-hypersensitive sites of chromatin. Cell 27, 413-415.

Enger, M.D., Rall, L.B., and Hidebrand, C.E. (1979). Thionein gene expression in Cd variants of the CHO cell: correlation of thionein synthesis rates with translatable mRNA during induction, deinduction and superinduction. Nucleic Acids Res. 7, 271-288.

Ephrussi, A., Church, G.M., Tonegawa, S., and Gilbert, W. (1985). B lineage-specific interaction of an immunoglobulin enhancer with cellular factors *in vivo*. Science 227, 134-140.

Falkner, F.G., and Zachau, H.G. (1984). Correct transcription of an immunoglobulin κ gene requires an upstream fragment containing conserved sequence elements. Nature 310, 71-75.

Felsenfeld, G., and McGhee, J. (1982). Methylation and gene control. Nature 269, 602-603.

Fiers, W., Contreras, R., Haegeman, G., Rogiers, R., van der Voorde, A., van Heuverswyn, H., van Herreweghe, J., Volckaert, G., and Ysebaert, M. (1978). The complete nucleotide sequence of SV40 DNA. Nature 273, 113-120.

Fitzgerald, M., and Shenk, T. (1981). The sequence 5'-AAUAAA-3' forms part of the recognition site of polyadenylation of late SV40 mRNAs. Cell 24, 251-260.

Fraley, R., and Papahadgopoulos, D. (1982). Liposomes: The development of a new carrier system for introducing nucleic acids into plant and animal cells. Current Topics in Microbiol. and Immunol. 96, 171-191.

Fromm, M., and Berg, P. (1982). Deletion mapping of DNA regions required for SV40 early region promotor function in vivo. J. Mol. Appl. Genet. 1, 457-481.

Frost, E., and Williams, J. (1978). Mapping temperature-sensitive and host-range mutations of adenovirus type 5 by marker rescue. Virology 911, 39-50.

Fujita, T., Ohno, S., Yasumitsu, H., and Taniguchi, T. (1985). Delimitation and properties of DNA sequences required for the regulated expression of human Interferon-β gene. Cell 41, 489-496.

Gething, M.J., and Sambrook, J. (1981). Cell-surface expression of influenza haemagglutinin from a cloned DNA copy of the RNA gene. Nature 293, 620-625.

Gidoni, D., Dynan, W.S., and Tjian, R. (1984). Multiple specific contacts between a mammalian transcription factor and its cognate promoter. Nature 312, 409-413.

Gitschier, J., Wood, W.I., Goralka, T.M., Wich, K.L., Chen, E.Y., Eaton, D.H., Vehar, G.A., Capon, D.J., and Lavin, R.M. (1984). Characterization of the human factor VIII gene. Nature 312, 326-330.

Gluzman, Y., Sambrook, J., and Frisque, R.J. (1980). Expression of early genes of origin-defective mutants of SV40. Proc. Natl. Acad. Sci. USA 77, 3898-3902.

Gluzman, Y. (1981). SV40-transformed simian cells support the replication of early SV40 mutants. Cell 23, 175-182.

Goff, S.P., and Berg, P. (1979). Construction, propagation and expression of SV40 recombinant genomes containing the E. coli gene for thymidine kinase and a Saccharomyces cerevisiae gene for tyrosine transfer RNA. J. Mol. Biol. 133, 359-383.

Goff, S.P., Gilboa, E., Witte, O.N., and Baltimore, D. (1980). Structure of the Abelson Murine Leukemia Virus Genome and the homologous cellular gene: Studies with cloned viral DNA. Cell 22, 777-785.

Gorman, C.M., Moffat, L.F., and Howard, B.H. (1982). Recombinant genomes which express chloramphenicol acetyltransferase in mammalian cells. Mol. Cell. Biol. 2, 1044-1051.

Graessmann, M., and Graessmann, A. (1976). Early simian virus 40-specific RNA contains information for tumor antigen formation and chromatin replication. Proc. Natl. Acad. Sci. USA 73, 366-370.

Graham, F.L., and van der Eb, A.J. (1973). A new technique for the assay of infectivity of human adenovirus 5 DNA. Virology 52, 456-457.

Gray, P.W., and Goeddel, D.V. (1982). Structure of the human immune interferon gene. Nature 298, 859-863.

Grill, E., Winnacker, E.L., and Zenk, M.H. (1985). Phytochelatins: The principal heavy-metal complexing peptides of higher plants. Science 230, 674-676.

Gruss, P., Lai, C.-J., Dhar, K., and Khoury, G. (1979). Splicing as a requirement for biogenesis of functional 16S mRNA of simian virus 40. Proc. Natl. Acad. Sci. USA 76, 4317-4321.

Gruss, P., and Khoury, G. (1980). Rescue of a splicing defective mutant by insertion of an heterologous intron. Nature 286, 634-637.

Gruss, P., Ellis, R.W., Shih, T.Y., Konig, M., Scolnick, E.M., and Khoury, G. (1981a). SV40 recombinant molecules express the gene encoding p21 transforming protein of Harvey murine sarcoma virus. Nature 293, 486-488.

Gruss, P., Efstratiadis, A., Karathanasis, S., Konig, M., and Khoury, G. (1981b). Synthesis of stable unspliced mRNA from an intronless simian virus 40 rat preproinsulin gene recombinant. Proc. Natl. Acad. Sci. USA 78, 6091-6095.

Gruss, P., Dhar, R., Khoury, G. (1981c). Simian Virus 40 tandem repeated sequences as an element of the early promotor. Proc. Natl. Acad. Sci. USA 78, 943-947.

Hamer, D.H., Kaehler, M., and Leder, P. (1980). A mouse globin gene promotor is functional in SV40. Cell 21, 697-698.

Hamer, D.H., and Walling, M.J. (1982). Regulation in vivo of a cloned mammalian gene: Cadmium induces the transcription of a mouse metallothionein gene in SV40 vectors. J. Mol. Appl. Genet. 1, 273-288.

Hansen, U., Tenen, D.G., Livingston, D.M., and Sharp. P.A. (1981). T Antigen Repression of SV40 early transcription from two promotors. Cell 27, 603-612.

Hardison, R.C., Butler, III. E.T., Lacy, E., Maniatis, T., Rosenthal, N., and Efstratiadis, A. (1979). The structure and transcription of four linked rabbit β-globin genes. Cell 18, 1385-1297.

Hellerman, I.G., Cone, R.C., Potts, I.T., Rich, A., Mulligan, R.C., and Kronenberg, H.M. (1984). Secretion of human parathyroid hormone from rat pituitary cells infected with a recombinant retrovirus encoding preproparathyroid hormone. Proc. Natl. Acad. Sci. USA 81, 5340-5344.

Hirt, B. (1967). Selective extraction of polyoma DNA from infected mouse cell cultures. J. Mol. Biol. 26, 365-369.

Holmgren, R., Corces, V., Morimoto, R., Blackman, R., Meselson, M. (1981). Sequence homologies in the 5'-regions of four *Drosophila* heat shock genes. Proc. Natl. Acad. Sci. USA 78, 3775-3778.

Izant, J.G., and Weintraub, H. (1984). Inhibition of thymidine kinase gene expression by anti-sense RNA: a molecular approach to genetic analysis. Cell 36, 1007-1015.

Jenner, E. (1798). An inquiry into the causes and effects of the variolae vaccinaé, a disease discovered in some western counties of England, particularly Clouce-stershire, and known by the name of cow pox. (Reprint: Cassell, London, 1896).

Jones, P.A., Taylor, S.M. (1980). Cellular differentia-tion, cytidine analogs and DNA methylation. Cell 20, 85-93.

Jones, P.A. (1985). Altering gene expression with 5-Azacytidine. Cell 40, 485-486.

Kaufmann, R.J., and Sharp, P.A. (1982a). Amplifica-tion and expression of sequences cotransfected with a modular dihydrofolate reductase complementary DNA gene. J. Mol. Biol. 159, 601-621.

Kaufmann, R.J., and Sharp, P.A. (1982b). Construc-tion of a modular dihydrofolate reductase cDNA gene: Analysis of signals utilized for efficient expres-sion. Mol. Cell. Biol. 2, 1304-1319.

Key, I.L., Lin, C.Y., and Chen, Y.M. (1981). Heat shock proteins of higher plants. Proc. Natl. Acad. Sci. USA 78, 3526-3530.

Kieny, M.P., Lathe, R., Drillien, R., Spehner, D., Skory, S., Schmitt, D., Wiktor, T., Koprowski, H., and Lecocq, J.P. (1984). Expression of rabies virus glycoprotein from a recombinant vaccinia virus. Nature 312, 163-166.

Kim, S.K., and Wold, B.J. (1985). Stable reduction of thymidine kinase activity in cells expressing high levels of anti-sense RNA. Cell 42, 129-138.

Klobutcher, L.A., and Tuddle, F.H.(1981). Chromo-some mediated gene transfer. Ann. Rev. Biochem. 50, 533-554.

Laimins, L.A., Khoury, G., Gorman, C., Howard, B., and Gruss, P. (1982). Host specific activation of transcription by tandem repeats from SV40 and Moloney Murine Sarcoma Virus. Proc. Natl. Acad. Sci. USA 79, 5453-5457.

Lau, Y.F., Lin, C.C., and Kan, Y.W. (1984). Amplifi-cation and expression of human α-globin genes in Chinese hamster ovary cells. Mol. Cell. Biol. 4, 1469-1475.

Law, M.F., Howard, B., Sarrer, N., and Howley, P.M. (1982). Expression of selective traits in mouse cells transformed with a BPV DNA derived hybrid mole-cule containing E. coli gpt. In "Eukaryotic viral vectors", ed. Y. Gluzman, Cold Spring Harbor Laboratory, Cold Spring Harbor, New York, p.79-86.

Lee, F., Mulligan, R., Berg, P., and Ringold, G. (1981). Glucocorticoids regulate expression of dihydrofolate reductase cDNA in mouse mammary tumor virus chimaeric plasmids. Nature 214, 228-232.

Lerner, M.R., Boyle, J.A., Mount, S.M., Wolin, S.L., and Steitz, J.A. (1980). Are snRNPs involved in splicing? Nature 283, 220-224.

Levinson, B., Khoury, G., Van de Woude, G., and Gruss, P. (1982). Activation of SV40 genomes by 72 bp tandem repeats of Moloney Sarkoma Virus. Nature 295, 568-572.

Littlefield, J.W. (1964). Selection of hybrids from mating of fibroblasts *in vitro* and their presumed recombinants. Science 145, 709-710.

Lowy, D.R., Drovetzky, I., Shober, R., Law, M.F., Engel., L., and Howley, P.M. (1980). *In vitro* tumorigenic transformation by a defined sub-genomic fragment of bovine papilloma virus DNA. Nature 287, 72-74.

Lusky, M., and Botchan, M. (1981). Inhibition of SV40 replication in simian cells by specific pBR322 DNA sequences. Nature 293, 79-81.

Lusky, M., Weiher, H., and Botchan, M. (1983). Bovine papilloma virus contains an activator of gene expression at the distal end of the early transcription unit. Mol. Cell. Biol. 3, 1108-1122.

Mackett, M., Smith, G.L., and Moss, B. (1984). General method for production and selection of infectious vaccinia virus recombinants expressing foreign genes. J. Mol. Virol. 49, 857-864.

Mann, R., Mulligan, R.C., and Baltimore, D. (1983). Construction of a retrovirus packaging mutant and its use to produce helper-free defective retrovirus. Cell 33, 153-159.

Mantei, N., Boll, W., and Weissmann, C. (1979). Rabbit β-globin mRNA production in mouse L-cells transformed with cloned rabbit β-globin chromoso-mal DNA. Nature 281, 40-46.

Mantei, N., and Weissmann, C. (1982). Controlled transcription of a human α-interferon gene introduc-ed into mouse L-cells. Nature 297, 128-132.

McCutchan, J.H., and Pagano, J.S. (1968). Enhance-ment of the infectivity of simian virus 40 deoxyribo-nucleic acid with diethylamino-ethyl-dextran. J. Natl. Cancer Inst. 41, 351-357.

McKnight, S.L. (1980). The nucleotide sequence and transcription map of the herpes simplex virus thymi-dine kinase gene. Nucleic Acids Res. 8, 5949-5964.

Michel, M.L., Pontisso, P., Sobczak, E., Malpiece, Y.,

Streek, R.E., and Tiollais, P. (1984). Synthesis in animal cells of hepatitis B surface antigen particles carrying a receptor for polymerised human serum albumin. Proc. Natl. Acad. Sci. USA 81, 7708-7712.

Mitrani-Rosenbaum, S., Maroteaux, L., Mory, Y., Revel, M., and Howley, P.M. (1983). Inducible expression of the human interferon β1 gene linked to a bovine papilloma virus DNA vector and maintained extrachromosomally in mouse cells. Mol. Cell. Biol. 3, 233-240.

Moriarty, A.M., Hoyer, B.H., Shih, I.W., Gerin, J.L., and Hamer, D.H. (1981). Expression of the hepatitis B virus surface antigen gene in cell culture by using simian virus 40 vector. Proc. Natl. Acad. Sci. USA 78, 2606-2610.

Mulligan, R.C., Howard, B.H., and Berg, P. (1979). Synthesis of rabbit β-globin in cultured monkey kidney cells following transfection with a SV40 β-globin recombinant genome. Nature 277, 108-114.

Mulligan, R.C., and Berg, P. (1980). Expression of a bacterial gene in mammalian cells. Science 209, 1422-1427.

Mulligan, R.C., and Berg, P. (1981a). Selection for animal cells that express the *Escherichia coli* gene coding for xanthine-guanine phosphoribosyl transferase. Proc. Natl. Acad. Sci.USA 78, 2072-2076.

Mulligan, R., and Berg, P. (1981b). Factors governing the expression of a bacterial gene in mammalian cells. Mol. Cell. Biol. 1, 449-459.

Myers, R.M., and Tijan, R. (1980). Construction and analysis of SV40 origins defective in tumor antigen binding and DNA replication. Proc. Natl. Acad. Sci. USA 77, 6491-6495.

Neumann, E., Schaefer-Ridder, M., Wang, Y., and Hofschneider, P.H. (1982). Gene transfer into mouse L-cells by electroporation in high electic fields. The EMBO Journal 1, 841-845.

Nüesch, J., and Schümperli, D. (1984). Structural and functional organization of the *gpt* gene region of *Escherichia coli*. Gene 32, 243-249.

O'Hare, K., Benoist, C., and Breathnach, R. (1981). Transformation of mouse fibroblasts to methotrexate resistance by recombinant plasmid expressing a prokaryotic dihydrofolate reductase. Proc. Natl. Acad. Sci. USA 78, 1527-1531.

Okayama, H., and Berg, P. (1983). A cDNA cloning vector that permits expression of cDNA inserts in mammalian cells. Mol. Cell. Biol. 3, 280-289.

Paoletti, E., Lipinskas, B.R., Samsonoff, C., Mercer, S., and Panicali, D. (1984). Construction of live vaccines using genetically engineered poxviruses: Biological activity of vaccinia virus recombinants expressing the hepatitis B virus surface antigen and the herpes simplex virus glycoprotein D. Proc. Natl. Acad. Sci. USA 81, 193-197.

Payvar, F., Wrange, Ö., Carlstedt-Duke, J., Okret, S., Gustafsson, J.A., and Yamamoto, K.R. (1981). Purified glucocorticoid receptors bind selectively *in vitro* to a cloned DNA fragment whose transcription is regulated by glucocorticoids *in vivo*. Proc. Natl. Acad. Sci. USA 78, 6628-6632.

Pelham, H. (1985). Activation of heat-shock genes in eukaryotes. Trends in Genetics 1, 31-35.

Pellicer, A., Wigler, M., Axel, R., and Silverstein, S. (1978). The transfer and stable integration of the HSV thymidine kinase gene into mouse cells. Cell 14, 133-141.

Pellicer, A., Robins, D., Wold, B., Sweet, R., Jackson, I., Lowry, I., Roberts, J.M., Sim, G.K., Silverstein, S., and Axel., R. (1980). Altering genotype and phenotype by DNA-mediated gene transfer. Science 209, 1414-1422.

Picard, D., and Schaffner, W. (1984). A lymphocyte-specific enhancer in the mouse immunoglobulin kappa gene. Nature 307, 80-82.

Queen, C., and Baltimore, D. (1983). Immunoglobulin gene transcription is activated by downstream sequence elements. Cell 33, 741-748.

Rao, R.N., and Rogers, S.G. (1979). Plasmid pKC7. A vector containing ten restriction sites suitable for cloning DNA segments. Gene 7, 79-82.

Razin, A., and Riggs, A.D. (1980). DNA methylation and gene function. Science 210, 604-610.

Rassoulzadegan, M., Binetruy, B., and Cuzin, F. (1982). High frequency of gene transfer after fusion between bacteria and eukaryotic cells. Nature 295, 257-259.

Reddy, V.B., Thimmappaya, B., Dhar, R., Subramanian, K.N., Zain, B.S., Pan, J., Ghosh, P.K., Celma, M.L., and Weissman, S.M. (1978). The genome of simian virus 40. Science 200, 494-502.

Reiss, B., Sprengel, R., Will, H., and Schaller, H. (1984). A new sensitive method for qualitative and quantitative analysis of neomycin phosphotransferase in crude cell extracts. Gene 30, 217-233.

Rio, D.C., Clark, S.G., and Tjian, R. (1985). A mammalian host-vector system that regulates expression and amplification of transfected genes by temperature induction. Science 227, 23-28.

Ritossa, F. (1962). A new puffing pattern induced by heat shock and DNP in *Drosophila*. Experientia 18, 571-573.

Rubin, G.M., and Spradling, A.C. (1982). Genetic transformation of *Drosophila* with transposable element vectors. Science 218, 346-353.

Rubin, G.M., and Spradling, A.C. (1983). Vectors for P-element-mediated gene transfer in *Drosophila*. Nucleic Acids Res. 11, 6341-6351.

Saragosti, S., Moyne, G., and Yaniv, M. (1980). Absence of nucleosomes in a fraction of SV40 chromatin between the origin of replication and the region coding for the late leader RNA. Cell 20, 65-73.

Sarver, N., Gruss, P., Law, M.F., Khoury, G., and Howley, P.M. (1981). Bovine Papilloma Virus Deoxyribonucleic Acid: a novel eukaryotic cloning vector. Mol. Cell. Biol. 1, 486-496.

Sarver, N., Mischel, R., Byrne, J.C., Khoury, G., and Howley, P.M. (1985). Enhancer-dependent expression of the rat preproinsulin gene in bovine papilloma type I vectors. Mol. Cell. Biol. 5, 3507-3516.

Sassone Corsi, P., Wildeman, A., and Chambon, P. (1985). A *trans*-acting factor is responsible for the simian virus 40 enhancer activity *in vitro*. Nature 313, 458-463.

Scangos, G., and Ruddle, F.H. (1981). Mechanisms and applications of DNA mediated gene transfer in mammalian cells; a review. Gene 14, 1-10.

Schaefer-Ridder, M., Wang, Y., and Hofschneider, P.H. (1982). Liposomes as Gene Carriers: Efficient Transformation of Mouse L-cells by Thymidine Kinase Gene. Science 215, 166-168.

Schaffner, W. (1980). Direct transfer of cloned genes from bacteria to mammalian cells. Proc. Natl. Acad. Sci. USA 77, 2163-2167.

Schaltmann, K., and Pongs, O. (1982). Identification and characterization of the ecdysterone receptor in *Drosophila melanogaster* by photoaffinity labeling. Proc. Natl. Acad. Sci. USA 77, 6-10.

Schimke, R.T. (1984). Gene amplification in cultured animal cells. Cell 37, 705-713.

Schöffl, F., and Key, J.L. (1982). An analysis of mRNAs for a group of heat shock proteins of soybean using cloned cDNAs. J. Mol. Appl. Genet. 1, 301-314.

Schöffl, F., Raschke, E., and Nagaxo, R.T. (1984) The DNA sequence analysis of soybean heat-shock genes and identification of possible regulatory promoter elements. The EMBO J. 3, 2491-2497.

Schöler, H.R., and Gruss, P. (1984). Specific interaction between enhancer-containing molecules and cellular components. Cell 36, 403-411.

Seif, I., Khoury, G., and Dhar, R. (1979). The genome of human papovavirus BKV. Cell 187, 963-977.

Serfling, E., Jasin, M., and Schaffner, W. (1985). Enhancers and eukaryotic gene transcription. Trends in Genetics 1, 224-230.

Shaw, W. (1967). The enzymatic acetylation of chloramphenicol by extracts of R-factor-resistant *Escherichia coli*. J. Biol. Chem. 242, 687-693.

Shaw, W. (1975). Chloramphenicol acetyltransferase from resistant bacteria. Methods in Enzym. 53, 737-754.

Shen, Y.M., Hirshhorn, R.R., Mercer, W.E., Surmacz E., Tsutsui, Y., Soprano, K.J., and Baserga, R. (1982). Gene transfer: DNA microinjection compared with DNA transfection with a very high efficency. Mol. Cell. Biol. 2, 1145-1154.

Shilo, B., and Weinberg, R.A. (1981). DNA sequences homologous to vertebrate oncogenes are conserved in *Drosophila melanogaster*. Proc. Natl. Acad. Sci. USA 78, 6789-6792.

Shimotohno, K., and Temin, H.M. (1981). Formation of infectious progeny virus after insertion of herpes simplex thymidine kinase gene into DNA of an avian retrovirus. Cell 26, 67-77.

Shimotohno, K., and Temin, H.M. (1982). Loss of intervening sequences in genomic mouse α-globin DNA inserted in an infectious retrovirus vector. Nature 299, 265-268.

Shinnick, T.M., Lerner, R.A., and Sutcliffe, I.G. (1981). Nucleotide sequence of Moloney murine leukaemia virus. Nature 293, 543-548.

Simons, R.W., and Kleckner, N. (1983). Translational control of IS10 transposition. Cell 34, 683-691.

Simonsen, C.C., and Levinson, A.D. (1983). Isolation and expression of an altered mouse dihydrofolate reductase cDNA. Proc. Natl. Acad. Sci. USA 80, 2495-2499.

Smith, G.L., and Moss, B. (1983). Infectious pox virus vectors have capacity for at least 25000 bp of foreign DNA. Gene 25, 21-28.

Smith, G.L., Godson, G.N., Nussenzweig, V., Nussenzweig, R.S., Barnwell, J., and Moss, B. (1984). *Plasmodium knowlesi* sporozoite antigenes: Expression of infectious recombinant vaccinia virus. Science 224, 397-399.

Smith, G.L., Mackett, M., and Moss, B. (1983a). Infectious vaccinia virus recombinants that express hepatitis B virus surface antigen. Nature 302, 490-495.

Smith, G.L., Murphy, B.R., and Moss, B. (1983b). Construction and characterization of an infectious vaccinia virus recombinant that expresses the influenza hemagglutinin gene and induces resistance to influenza virus infectious in hamsters. Proc. Natl. Acad. Sci. USA 80, 7155-7159.

Soberon, X., Covarrubias, L., and Bolivar, F. (1980). Construction and characterization of new cloning vehicles. IV. Deletion derivatives of pBR322 and pBR325. Gene 9, 287-305.

Sompayrac, L.M., and Danna, J.K.J. (1981). Efficient infection of monkey cells with DNA of simian virus 40. Proc. Natl. Acad. Sci. USA 78, 7575-7578.

Southern, P.J., and Berg, P. (1982). Transformation of mammalian cells to antibiotic resistance with a bacterial gene under control of the SV40 Early Region Promotor. J. Mol. Appl. Genet. 1, 327-341.

Stark, G.R., and Wahl, G.M. (1984). Gene amplification. Ann. Rev. Biochem. 53, 447-491.

Steller, H., and Pirrotta, V. (1985). A transposable P-vector that confers selectable G418 resistance to *Drosophila* larvae. The EMBO J. 4, 167-171.

Stow, N.D., and Wilkie, N.M. (1976). An improved technique for obtaining enhanced infectivity with Herpes Simplex Virus Type 1 DNA. J. Gen. Virol. 33, 447-458.

Stuart, G.W., Searle, P.F., and Palmiter, R.D. (1985). Identification of multiple metal regulatory elements in mouse metallothionein-II promoter by assaying synthetic sequences. Nature 317. 828-831.

Stuhlmann, H., Cone, R., Mulligan, R.C., and Jaenisch, R. (1984). Introduction of a selectable gene into different animal tissue by a retrovirus recombinant vector. Proc. Natl. Acad. Sci. USA 81, 7151-7155.

Subramani, S., Mulligan, R., and Berg, P. (1981). Expression of the mouse dihydrofolate reductase complementary deoxyribonucleic acid in simian virus 40 vectors. Mol. Cell. Biol. 1, 854-864.

Tabin, C.J., Hoffmann, J.W., Goff, S.P., and Weinberg, R.A. (1982). Adaptation of a retrovirus as a eukaryotic vector transmitting the Herpes Simplex Virus Thymidine Kinase Gene. Mol. Cell. Biol. 2, 426-436.

Takeda, S., Naito, T., Hama, K., Noma, T., and Honjo, S. (1985). Construction of chimaeric processed immunoglobulin genes containing some variable and human constant region sequences. Nature 314, 452-454.

Taniguchi, T., Matsui, H., Fujita, T., Takaoka, C., Kashima, N., Yoshimoto, R., and Hamuro, J. (1983). Structure and expression of a cloned cDNA for human interleukin-2. Nature 302, 305-310.

Tata, J.R. (1982). Do steroid receptors recognize DNA sequences? Nature 298, 707-708.

Taya, Y., Devors, R., Tavernier, J., Cheronte, H., Engler, G., and Fiers, W. (1982). Cloning and structure of the human immune interferon chromosomal gene. The EMBO Journal 1, 953-958.

Tiollais, P., Pourcel, C., and Dejean, A. (1985). The hepatitis B virus. Nature 317, 489-495.

Tjian, R. (1978). The binding site on SV40 for T-antigen related protein. Cell 13, 165-179.

Twigg, A.J., and Sherratt, D. (1980). Transcomplementable copy-number mutants of plasmid ColE1. Nature 283, 216-218.

Urlaub, G., and Chasin, L.A. (1980). Isolation of Chinese hamster cell mutants deficient in dihydrofolate reductase activity. Proc. Natl. Acad. Sci. USA 77, 4216-4330.

Urlaub, G., Landzberg, M., and Chasin, L.A. (1981). Selective killing of methotrexate-resistant cells carrying amplified dihydrofolate reductase genes. Cancer Res. 41, 1594-1601.

Vardimon, L., Kressmann, A., Cedar, H., Maechler, M., and Doerfler, W. (1982). Expression of a cloned adenovirus gene is inhibited by *in vitro* methylation. Proc. Natl. Acad. Sci. USA 79, 1073-1077.

Varmus, H.E. (1982). Form and function of retroviral proviruses. Science 216, 812-820.

Varshavsky, A.J., Sundin, O.H., and Bohn, M. (1979). A stretch of "late" SV40 viral DNA about 400 bp long which includes the origin of replication is specifically exposed in SV40 minichromosomes. Cell 16, 453-466.

Vassin, E.F., Goldberg, G.I., Tucker, P.W., and Smithies, O. (1980). A mouse α-globin related pseudogene lacking intervening sequences. Nature 286, 222-226.

Velten, J., Velten, L, Hain, R., and Schell, J. (1984). Isolation of a dual plant promoter fragment from the Ti plasmid of *Agrobacterium tumefaciens*. The EMBO J. 3, 2723-2730.

deVilliers, J., and Schaffner, W. (1981). A small segment of polyoma virus DNA enhances the expression of a cloned β-globin gene over a distance of 1400 base pairs. Nucleic Acids Res. 9, 6251-6264.

Vinograd, J., Lebowitz, I., Radloff, R., Watson, R., and Laipis, P. (1965). The twisted circular form of polyoma viral DNA. Proc. Natl. Acad. Sci. USA 53, 1104-1111.

Weber, F., de Villiers, J., and Schaffner, W. (1984). An SV40 "enhancer trap" incorporates exogeneous enhancers or generates enhancers from its own sequences. Cell 36, 983-992.

Wagner, M.J., Sharp, I.A., Summers, W.C. (1981). Nucleotide sequence of the thymidine kinase gene of herpes simplex virus Type 1. Proc. Natl. Acad. Sci. USA 78, 1441-1445.

Wahl, G.M., de Saint Vincent, R., and DeRose, M.L. (1984). Effect of chromosomal position on amplification of transfected genes in animal cells. Nature 307, 516-520.

Watanabe, S., and Temin, H.M. (1983). Construction of a helper cell line for avian reticuloendothelious virus cloning vectors. Mol. Cell. Biol. 3, 2241-2249.

Weiher, H., König, M., and Gruss, P. (1982). Multiple point mutations affecting the simian virus 40 enhancer. Science 219, 626-631.

Weintraub, H., Izant, J.G., and Harland, R.M. (1985). Anti-sense RNA as a tool for genetic analysis. Trends in Genetics 1, 22-25.

Weisbrod, S. (1982). Active chromatin. Nature 297, 289-295.

Weiss, R., Teich, N., Varmus, H., and Coffin, J. (1982). RNA Tumor viruses. Cold Spring Harbor Laboratory, Cold Spring Harbor, New York, 11724, USA.

Wigler, M., Silverstein, S., Lee, L.S., Pellicer, A., Cheng, Y.C., and Axel, R. (1977). Transfer of purified herpes virus thymidine kinase gene into cultured mouse cells. Cell 11, 223-232.

Wigler, M., Pellicer, A., Silverstein, S., Axel, R., Urlaub, G., and Chasin, L. (1979a). DNA-mediated transfer of the adenine phosphoribosyl transferase locus into mammalian cells. Proc. Natl. Acad. Sci. USA 76, 1373-1376.

Wigler, M., Sweet, R., Sim, G.K., Wold, B., Pellicer, A., Lacy, E., Maniatis, T., Silverstein, S., and Axel, R. (1979b). Transformation of mammalian cells with genes from prokaryotes and eukaryotes. Cell 16, 777-785.

Wigler, M., Perucho, M., Kurtz, D., Dana, S., Pellicer, A., Axel, R., and Silverstein, S. (1980). Transformation of mammalian cells with an amplifiable dominant-acting gene. Proc. Natl. Acad. Sci. USA 77, 3567-3570.

Wilkie, N.M., Clements, J.B., Boll, W., Mantei, N., Lonsdale, D., and Weissmann, C. (1979). Hybrid plasmids containing an active thymidine kinase gene of herpes simplex virus I. Nucleic Acids Res. 7, 859-877.

Williams, B.G., and Blattner, F.R. (1979). Construction and characterization of the hybrid bacteriophage λ charon vectors for DNA cloning. J. Virol. 29, 555-575.

Williams, D.A., Lemischka, I.R., Nathan, D.G., and Mulligan, R.C. (1984). Introduction of new genetic material into pluripotent haematopoietic stem cells of the mouse. Nature 310, 476-480.

Wu, C.(1985). An exonuclease protection assay reveals heat-shock element and TATA box DNA-binding proteins in crude nuclear extracts. Nature 317, 84-87.

Wurm, F.M., Gwinn, K.A., and Kingston, R.E. (1986). Inducible overexpression of the mouse c-myc protein in mammalian cells. Proc. Natl. Acad. Sci. USA 83, 5414-5418.

Ziegler, M., and Davidson, R. (1979). The effect of hexose on chloramphenicol sensitivity and resistance in Chinese hamster cells. J. Cell Physiol. 98, 627-636.

Zinn, K., Mellon, P., Ptashne, M., and Maniatis, T. (1982). Regulated expression of an extrachromosomal human β-interferon gene in mouse cells. Proc. Natl. Acad. Sci. USA 79, 4897-4901.

9 Genomic Libraries

9.1 Isolation of Genomic DNA Fragments

A genomic library is a collection of plasmid clones, or phage lysates, which contain recombinant DNA molecules. Together the individual DNA inserts in this collection of molecules represent the entire genetic information of an organism. The probability P of finding a specific gene in such a collection depends on the length of the DNA inserts and the complexity of the entire genome. If insert lengths remain constant, the number of clones required to find a particular gene increases with increasing complexity of the genome: the larger and more complex a genome, the more clones of a specific size are needed for a complete representation of the entire genetic information. The number N of such clones can be determined by applying the formula

$$N = \frac{\ln (1-P)}{\ln (1-f)}$$

where P is the probability of obtaining a particular sequence, and f is the ratio of the length of the insert to the entire genome (Clarke and Carbon, 1976). Table 9-1 gives the values of N for an assumed insert length of 20 kb and probabilities of 90% and 99% for obtaining a particular clone. These values must be regarded as lower estimates, because they were obtained by assuming that each hybrid molecule transforms equally well and that each colony represents an independent transformation event.

In principle, any gene for which there is a suitable detection system, e.g., a hybridisation

Louise Clarke
John Carbon

Santa Barbara,
California

Table 9–1. Colony size of a genomic library.

| Organism | Size | | Size of | Number N of colonies | |
	kb	M_r	insertion	$P = 0.90$	$P = 0.99$
E. coli	3.75×10^3	2.4×10^6	20 kb	430	1 157
S. cerevisiae	1.5×10^4	1×10^7	20 kb	1 726	3 462
Drosophila	1.65×10^5	1.1×10^8	20 kb	19 000	38 000
Man	3×10^6	2×10^9	20 kb	345 409	690 819

The number N of colonies of a genomic library is determined from the equation in the text. It depends on the desired probability P with which a given gene is to be found in the library and on the ratio f of the size of the insert to the total genome size of the organism.

probe, can be isolated from a genomic library provided that the library fulfills the criteria outlined above, and is indeed completely representative. If the genomic library contains DNA fragments obtained by partial digestion, such fragments will overlap and the library will therefore allow an examination of the structure not only of single genes, but also of neighbouring sequences. Such an analysis is often referred to as "chromosome walking" (*cf.* Section 11.2.2.3).

Two questions usually arise in practice, namely, the procurement of appropriately sized random DNA fragments and suitable vectors. It is not advisable, for example, to start cloning with double-stranded genomic DNA which has been digested to completion with restriction endonucleases, because the resulting DNA fragments are too heterogeneous in size; some of these DNA fragments could be too large for cloning; if they were too small, they might escape detection. Suitable starting materials would be DNA fragments generated by mechanical shearing, or by limited digestion with restriction enzymes. In the latter case, enzymes with hexameric recognition sequences should not be employed, because the distance between two cleavage sites may be too long to allow cloning of the resulting DNA fragments, for example, in λ vectors. This problem usually can be avoided by using enzymes with tetrameric recognition sequences. On the average such enzymes cleave DNA sixteen times more frequently than enzymes with hexameric recogni-

tion sites. The probability of generating different fragments which are, for example, approximately 20 kb in length, increases with the number of potential cleavage sites on a given DNA. A collection of DNA fragments obtained from a partial digestion with an enzyme utilising a tetrameric cleavage site will therefore be much closer to the statistically ideal situation than a similar collection generated by hexameric enzymes such as *Eco* RI. Single or mixed digestions with the enzymes *Alu* I (AG/CT), *Hae* III (GG/CC), or *Sau* 3A (/GATC) have been described in the literature. Constant amounts of DNA were either digested with various amounts of enzymes for a fixed incubation period, or incubated for various times with constant amounts of enzymes. Digestions were monitored by gel electrophoresis and suitably digested samples were pooled. DNA fragments of the desired size class were then isolated from such pools by agarose gel electrophoresis or separation in sucrose gradients. In spite of all the care usually employed in the production of mixtures of DNAs of random lengths, certain DNA fragments should be expected to be over- or under-represented, or even missing, in genomic libraries. A commonly used human library, for example, which was prepared from DNA partially digested with *Hae* III and *Alu* I, does not contain clones of the γ-globin region, nor does it contain a region between the γ- and δ-globin genes (Maniatis *et al.*, 1978; Lawn *et al.*, 1978); yet, these DNA regions are detected in a library of human DNA partially digested with *Mbo* I (Van der Ploeg *et al.*, 1980). The causes for these discrepancies are unknown and difficult to explain. It is well known, however, that endonuclease cleavage sites are often not recognised equally well. Certain DNA fragments may therefore never appear in a limited digest. For several reasons certain DNA sequences also may be lost during the necessary amplification of a genomic library: for example, a fragment might code for a toxic product, or might be replicated slowly, or might have been altered by recombinational events.

9.2 Vectors

In principle, all vectors employed in genetic engineering can be used to construct genomic libraries. In practice, however, λ vectors and cosmids in particular have proven most effective for several reasons (see also Section 4.2). If plasmids are used, the efficiency of transformation decreases with increasing length of the inserts; inserts of >15 kb can hardly ever be cloned. The situation is just the reverse in phage and cosmid systems. Foreign DNAs of up to 23-25 kb can be inserted in suitably adapted λ genomes, in particular replacement vectors (Williams and Blattner, 1979; Tiemeier *et al.*, 1976); in addition, such DNA molecules can be incorporated into infectious phage particles *in vitro*, or used directly, since the DNA itself is highly infectious and may yield up to $10^5 - 10^6$ pfu per microgram of recombinant phage DNA. Finally, phage DNAs can be stored and screened more easily than plasmid DNAs. Compared to λ vectors of the replacement type, cosmids offer an advantage because they can accommodate even larger DNA fragments. Cosmid MUA-3, for example, (Meyerowitz *et al.*, 1980) was constructed specifically for preparing genomic libraries. MUA-3 consists essentially of plasmid pBR322 with a 403 bp *cos* fragment inserted at the *Pst* I site. The new vector

(Fig. 9-1) is only 4.8 kb in size, and hence insertions of 40 to 48 kb can be incorporated and packaged with high efficiency. The yield of non-recombinant clones is exceptionally low because cosmid dimers are too small for packaging.

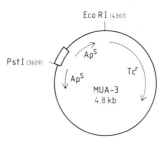

Fig. 9-1. Structure of cosmid MUA-3.
This cosmid contains a 403 bp *Hinc* II fragment with the *cos* region of bacteriophage λ (open bar) inserted into the *Pst* I site of pBR322 via homopolymeric tailing. Numbering refers to the position of cleavage sites in pBR322 DNA.

Another advantage of cosmids is that they need to be packaged only once during formation of the transducing particles. In contrast, recombinant λ DNA molecules have to be packaged in each amplification step. In the long run, those recombinants which are at the borderline of being either too small or too large for packaging will be under-represented in a λ library (Feiss *et al.*, 1977).

9.3 Ligation Techniques

Basically, the methods employed for the ligation of gene fragments and vector molecules very much depend on the size and complexity of the genome to be cloned. Libraries can be obtained using restriction enzyme digests and subsequent ligation steps, by homopolymeric tailing, or *via* linker technology. In each case, the most important factor will be the number of plaques or colonies obtained from one microgram of recombinant DNA. Approximately 10^4 colonies per microgram of DNA can be obtained if cosmid vectors are used and the foreign DNA is ligated with the help of linkers or homopolymeric tails. This may be sufficient for a *Drosophila* library; however, this number of clones will certainly be inadequate for a human library with a complexity approximately thirtyfold greater. Somewhat higher yields, on the order of up to 5×10^5 colonies per microgram of DNA, can be obtained for cosmid libraries by simply cutting and rejoining the DNA molecules. Optimal yields may be obtained by applying the linker technology to λ DNA vectors in conjunction with a suitable *in vitro* packaging system. Such experiments are almost exclusively carried out with replacement vectors, such as λgtWES, Charon phages, or the EMBL series of vectors, which usually contain two restriction sites allowing the removal of the

central non-essential section of the λ genome. The remaining arms contain the essential genes and comprise approximately 25-30 kb. Since the upper limit for packagable DNA is approximately 52 kb, replacement vectors allow the insertion of 20-25 kb of foreign DNA.

The insertion of DNA into the Charon 4A phage vector may illustrate the technique (Fig. 9-2; Maniatis *et al.*, 1978). DNA molecules of approximately 20 kb are isolated by sucrose gradient centrifugation of cellular DNA having been fragmented, for example by sonication. The ends of such molecules, which are later to be provided with suitable linkers, are first trimmed with S1 nuclease. Since S1 treatment often leads to unsatisfactory results, the DNAs frequently are subjected to a limited digestion with endonucleases such as *Hae* III (GG/CC), or *Alu* I (AG/CT), which cut frequently and create blunt ends. To allow addition of *Eco* RI linkers and digestion of the fragments with *Eco* RI endonuclease without further fragmentation of the DNA, pre-existing *Eco* RI sites are first methylated by *Eco* RI methylase.

The Charon 4A vector is prepared for the cloning experiment by circularisation at the *cos* site and *Eco* RI treatment, which yields a large 31 kb fragment containing the *cos* site and the two arms, and two smaller internal fragments of 7 and 8 kb, respectively. The large fragment can be purified by gradient centrifugation. Vector DNA and eukaryotic DNA are now ligated and packaged *in vitro*. The resulting phage population is purified by caesium chloride density gradient centrifugation. This technique generally yields $10^5 - 10^6$ clones per microgram of DNA.

It is of considerable practical importance that such phage populations can be amplified and re-used several times. It has been mentioned

Tom Maniatis
Cambridge, Massachusetts

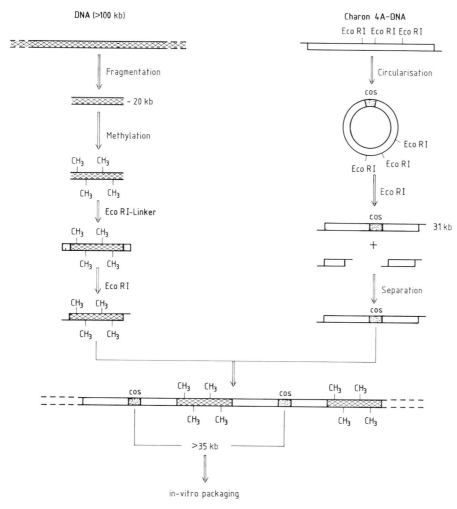

Fig. 9-2. Scheme depicting the construction of a genomic library in vector Charon 4A; see text for details (Maniatis *et al.*, 1978).

already that one may run the risk of losing certain recombinant phages during amplification, and that phages with particular insertions may be over-represented. There are no universal designs which avoid these problems, and it is important therefore to be aware of these difficulties. If need be, these problems may be circumvented by employing other restriction enzymes and other ligation strategies.

9.4 Selection Techniques

9.4.1 Selection for Recombinant DNA Molecules

There are three different techniques which allow recombinant DNA molecules to be selected during the construction of a gene library. In many instances, in particular if replacement vectors or small cosmids are used, the size selection of DNA during *in vitro* packaging plays an important role. It has been mentioned that approximately 35 kb of DNA are required for efficient packaging. After linearisation and re-ligation, the distance between two *cos* sites in the vector itself should therefore be less than 35 kb. This would ensure that vectors without insertions could not be packaged. The central replaceable fragment of some vectors contains one, or several, genes which can easily be screened for, either by their presence or their absence (see also Fig. 4.2-8A and B). Some Charon phages, for example, carry a substitution from the *lac* region of *E. coli* with a functional β-galactosidase gene. Such phages give dark blue plaques on Lac⁻ indicator bacteria when grown on plates containing Xgal, a colourless substrate for β-galactosidase; however, in recombinant phages, the substitution of the central fragment with the *lac* region abolishes β-galactosidase production. Recombinants therefore appear colourless.

The replacement vector λ1059 employs a marker of quite a different nature (Karn *et al.*, 1980). This vector is a so-called phasmid, *i.e.*, a phage which can also be replicated like a plasmid. In this case, the vector carries a replaceable central *Bam* HI fragment with the two λ genes *red* and *gam* which render the vector Spi⁺. Spi⁺ phages are unable to grow on P2-lysogenic bacteria, because they are sensitive to P2 interference. The phenotype of recombinant phages in which the central fragment has been replaced by foreign DNA is Spi⁻. Recombinant phages can be easily detected, because they are the only ones that can be propagated on P2 lysogens. For practical purposes it is advisable even in this system, to purify the vector arms from the internal DNA fragments in the same way that has already been described for vector Charon 4A. Cloning of cDNA libraries in immunity vectors, such as λgt10, is particularly easy, because only recombinants will form plaques if a selection on *hfl* hosts is used to force the parental vector phages quantita-

tively into the lysogenic state (*cf.* also Section 4.2.2.4).

9.4.2 Determination of the Complexity of DNA Insertions

The genome size of human DNA is approximately 3×10^9 bp (Table 9-1). If a library contains insertions with an average length of 20 kb, only one out of approximately 700 000 plaques or clones should contain an insertion with a DNA sequence which occurs only once in the haploid genome. This fraction of uniquely represented sequences, and hence the quality of the library, can be checked by determining the complexity of the library. This can be achieved by saturation hybridisation of unlabelled DNA obtained from the library (or from the organism from which the library was constructed) with the organism's own tritium-labelled DNA. When such a reassociation kinetic analysis is performed, the two hybridisation curves obtained with the library and authentic DNA obtained, for example, from embryonic cells, should be identical and reach the same $C_0 t$ value. This would indicate that the genomic library had been constructed without major losses of certain DNA fractions. A reduction in the complexity of the library, *i.e.* the loss of certain DNA sequences, would be indicated by a fraction of reassociated DNA obtained at the end of the hybridisation reaction with library DNA which is smaller than that obtained in the control experiment with genomic DNA (see also Section 3.2).

9.4.3 Screening by Nucleic Acid Hybridisation

Gene libraries can be screened for the presence of the desired gene if a suitable nucleic acid probe for this gene is available (see also Section 11.2.2.1). The technique employed is colony hybridisation which can be employed equally well for bacterial colonies and phage plaques (Grunstein and Hogness, 1975). The transformed bacteria to be screened are plated onto nitrocellulose filter discs laid on agar. A reference set of these colonies on the master plate is retained. The pattern of colonies is transferred to other nitrocellulose filters by replica plating. DNA prints of the colonies on the filters are prepared by treatment with alkali which lyses the bacteria and denatures the DNA. The DNA is fixed firmly by baking the filters at 80 °C. Such filters can be hybridised with radioactive probe DNAs, for example, ^{32}P-labelled cDNA, or chemically synthesised oligonucleotides. Hybridisation results can be monitored by autoradiography. A DNA print which gives a positive autoradiographic signal, *i.e.*, a black spot on the X-ray film, contains sequences which are complementary to the probe. The corresponding colony can be picked from the master plate for further cultivation and analysis. The same principle applies to phage plaques (Benton and Davis, 1977), but in comparison with the analysis of bacterial colonies, *in situ* hybridisation of phage plaques allows screening of more recombinants. Up to 40 000 plaques can be analysed on a Petri dish with a diameter of 15 cm. An additional asset is a much more favourable signal-to-noise ratio. The problem of nonspecific hybridisation is lessened by carefully selecting and maintaining hybridisation conditions; in addition, hybridisation signals can be improved considerably by employing a modification introduced by Woo (1979), who transferred naked phage DNA, which is always found in phage plaques, onto nitrocellulose filters and amplified these minute amounts of DNA by soaking the filters with bacteria. After passing through a complete cycle of lytic infection, the initial phage DNA will have been considerably amplified. The DNA on such filters is denatured with alkali and treated as described above.

Nonspecific hybridisation may not only be the result of too little recombinant DNA on the filters, but also of reduced specificity of the hybridisation probes. Especially if chemically synthesised hybridisation probes are used, it should be remembered that there is a minimal length for such probes, below which there is no guarantee that the sequence in question occurs only once or, at most, a few times in the genome.

The sizes of suitable probes are determined by the complexity of the gene library to be analysed. If the size of the haploid human genome is taken as approximately 3×10^9 bp, a suitable hybridisation probe should be at least sixteen nucleotides in length ($4^{16} = 4\,295\,255\,296$). Genomes with lower complexities do not require probes this long; for example, an *E. coli* library can be probed with a probe of eleven nucleotides in length ($4^{11} = 4\,194\,304$). It has been possible to isolate the gene for the enzyme iso-1-cytochrome c of *Saccharomyces cerevisiae* with a probe of only 13 nucleotides (Montgomery *et al.*, 1978); however, in this case the difference between the temperature at which specific hybridisation occurs and the temperature at which the two strands of the hybrid between probe and genomic DNA begin to dissociate is very small. It is therefore advisable to use longer probes whenever possible.

9.4.4 Direct Detection of the Desired Phenotype

This method is based on the selection for foreign DNA sequences which complement certain mutations in *E. coli*. Of course, there are several prerequisites: suitable *E. coli* mutants must be available, and the cloned DNA sequences in the library should contain functional genes which can be expressed in *E. coli*. Unfortunately, this technique cannot be used for many eukaryotic genes because they contain intervening sequences which cannot be spliced in *E. coli*. Complementation analysis also fails, of course, if the genes in question, for example, immunoglobulin genes, do not exist in *E. coli*. Nevertheless, this screening technique is invaluable for the detection of genes from lower eukaryotes such as yeasts (see also Chapter 5). Indeed, the first example of cloning and expression of a eukaryotic gene in *E. coli* was that of the *HIS3* gene of *S. cerevisiae*. The *HIS3* gene product, imidazole-glycerolphosphate (IGP)-dehydratase, is involved in the biosynthesis of histidine in *S. cerevisiae* and can complement *E. coli* mutations in the corresponding bacterial gene. The genes for the loci *LEU2*, *TRP1*, *URA3*, and *ARG4* also were isolated by direct selection in *E. coli* (see also Section 5.3 and Table 5-1). The desired genes were isolated by transforming a suitable auxotrophic *E. coli* strain with a mixture of recombinant plasmids and by plating the transformants on suitable selective agars, for example lacking histidine. Genomic libraries constructed in λ phages can be treated similarly by infecting an appropriate *E. coli* mutant with a mixture of recombinant phages and plating on selective agar. In this case, plaques will only be formed if the recombinant phage carries the complementing DNA sequence; there will be no bacterial growth, and hence no plaques, if the bacteria have not been infected with the "right" phage carrying the desired insertion. Cloning by complementation in *E. coli* is used infrequently nowadays for the analysis of yeast genes, because the development of techniques for the direct transformation of yeast cells (see Chapter 5) allows cloning by complementation in yeast itself.

9.4.5 Selection by Genetic Recombination

Generalised recombination is the exchange of genetic information between homologous chromosomes (Dressler and Potter, 1982) and is instigated by DNA strand break(s) in homologous, paired chromosomes (Holliday, 1964). As shown in Fig. 9-3, the process of recombination proceeds by the separation of the double-stranded DNA of either chromosome and the formation of

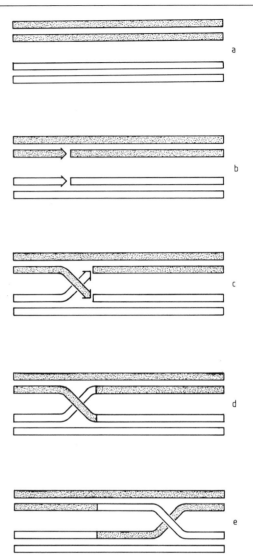

Fig. 9-3. Holliday model of homologous genetic recombination.
(a) shows two aligned double strands. Breaks occur in the positive or negative strand (b). The free ends are dissociated from their complementary strands and associate with the complementary strand of the neighbouring double strand (c). Two phosphodiester bonds are closed by a repair process (d), which stabilises the linkage between the two chromosomes. The cross-over point can migrate along the two chromosomes (e) (Holliday, 1964).

a new double helix comprising one DNA strand from each chromosome.

The frequency of such recombinational events differs in various biological systems and depends on the length of the homologous regions. In phage λ, for example, which possesses a linear genome with a length of approximately 50 kb, recombination between genetic markers at either end of the molecule occurs with frequencies of 10% to 20%. A mixed infection with two mutated phages, each carrying one of the hypothetical markers A and B, with A at one end of the phage genome and B at the other, would then yield between 10% and 20% wild-type phages (Fig. 9-4). In this case,

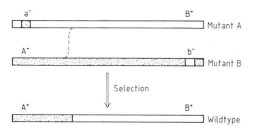

Fig. 9-4. Recombinational events between two mutated λ phages.
Wild-type phages are generated by recombination between the two mutants with a frequency of 10-20%. In principle, any phosphodiester bond along the entire region between the two hypothetical genes A and B can be used for the initiation of recombination. Only a single recombinant is shown.

recombination can take place almost anywhere along the 50 kb of the λ DNA. The relationship between the size of the homologous region and the frequency of recombination is usually linear. If the λ system is taken as an example, approximately 0.1% to 0.2% recombinants (i.e., a frequency of $1-2 \times 10^{-3}$) could still be expected even if the region of homology were only 500 bp in length.

In principle, the technique of homologous recombination can also be applied to the screening of λ phage gene libraries; however, there are three prerequisites: firstly, a suitable probe for

the gene to be identified must be available; secondly, this probe must be present in all host cells in which such a library is amplified; thirdly, the recombination events must be detectable, for example, by a selection system for recombinant phages. In principle, suitable bacteria are first transformed with a plasmid containing the cDNA of the desired gene. Transformants are subsequently infected with a phage genomic library. The desired recombination event between phage and plasmid DNA can occur only in those cells which have been infected with a phage containing the gene being sought. Newly recombined phages must be selected for in the progeny of this infection, for example, by the presence of an additional marker from the plasmid DNA which has become integrated into the phage DNA in the course of the recombinational event.

A useful marker which has proven particularly effective in practice, is the suppressor allele *supF* (Seed, 1983; Maniatis et al., 1982). Suppressor alleles code for abnormal tRNA molecules which are capable of reading stop codons as amino acid codons in the course of protein biosynthesis (Fig. 9-5). Among the E. coli tRNA genes are five different loci, at which mutations can give suppressor phenotypes. Three of these alleles, *supD*, *E*, and *F* suppress the amber codon UAG, while the other two, *supC* and *supG*, read either UAG, or the ochre codon UAA (Garen, 1968). The *supF* gene codes for a tyrosine-tRNA and thus in the mutant the stop signal UAG is read as a

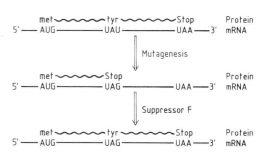

Fig. 9-5. Mechanism of action of suppressor tRNAs. The *supF* tRNA reads the stop codon generated by a mutation as a tyrosine codon.

tyrosine codon. The amber codon, UAG, which would normally cause chain termination is suppressed in the presence of the suppressor tRNA because the mutant tRNA recognises UAG as a tyrosine codon and incorporates tyrosin at the position of the nonsense mutation (Fig. 9-5). λ phages which carry amber mutations in their late genes cannot be propagated in strains of *E. coli* which do not contain suppressor genes (sup^0); however, the phages will grow normally in *supF* strains since the amber mutation is suppressed. It is of no consequence whether the suppressor gene in such systems is located on the bacterial chromosome itself or whether it resides on plasmid or phage DNA. If the suppressor gene is located on the phage DNA, a phage carrying amber mutations in a structural gene will also grow on suppressor-free host bacteria, which is the basis of the following very efficient selection system for phage genomic libraries.

The system developed by Seed (1983) depends on ΠVX, a small 902 bp multicopy plasmid which is composed of a ColE1-derived origin of DNA replication (*cf.* Fig. 4.1-3), a polylinker with several hexameric restriction enzyme cleavage sites for the insertion of foreign DNA, and the suppressor gene *supF* (Fig. 9-6). An improved version of vector ΠVX is ΠAN7 (Huang *et al.*, 1985; *cf.* Appendix for its sequence). This new vector not only contains a different polylinker, but also gives much higher yields in plasmid DNA preparations. This may be due to presence in ΠAN7 of the entire promoter region (both the "-10" and "-35" regions) of the ColE1 replication primer RNA. In ΠVX, only part of this promoter (the "-10" region) is present.

A DNA sequence to be identified in a gene library, a certain cDNA for example, is inserted into the polylinker of these miniplasmids, and

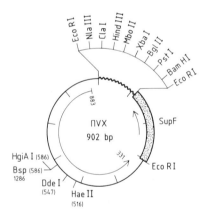
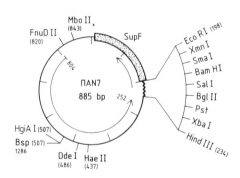

Fig. 9-6.   Structure of miniplasmids ΠVX and ΠAN7.
ΠVX consists of three DNA regions of different origin, namely, a 580 bp *Fnu* DII DNA fragment from plasmid pKB413 (Backman *et al.*, 1978) with the ColE1 replicon (*cf.* Fig. 4.1-3), a 203 bp synthetic tRNA nonsense suppressor gene (stippled bar; Ryan *et al.*, 1979; Brown *et al.*, 1979), and a 109 bp polylinker providing suitable restriction sites as indicated. With the exception of the *Eco* RI site all cleavage sites are represented only once and can therefore be used, for example, for cloning cDNA.
ΠAN7 contains a different polylinker which is located upstream of the suppressor region instead of being positioned between the suppressor gene and the origin of DNA replication. This new location considerably increases the copy numbers of this plasmid. In contrast to ΠVX, the polylinker in ΠAN7 contains a unique *Eco* RI site. (After Huang *et al.*, 1985). Numbers refer to ΠVX or ΠAN7 co-ordinates. The *Fnu* DII and *Mbo* I sites in ΠAN7 are at positions 3102 and 3125, respectively, of the pBR322 map (*cf.* Fig. 4.1-3). Long arrows indicate the direction of transcription of the RNAII primer RNA.

recombinant miniplasmids are then used to transform suitable strains of *E. coli* (Fig. 9-7). *SupF*-containing transformants can be selected in bacterial strains carrying the Tra⁻ derivative, p3, of plasmid RP1. p3 carries amber mutations in its tetracycline and ampicillin resistance genes and a functional kanamycin resistance gene. The phenotype of suppressor-free bacteria harbouring this plasmid is therefore $Kn^r/Tc^S/Ap^S$, but becomes $Kn^r/Tc^r/Ap^r$ upon transformation with the miniplasmid carrying the suppressor (Fig. 9-7). Bacteria containing the miniplasmid are infected with λ library phages which carry amber mutations in their late genes *A* and *B*, *e.g.*, Charon 4A. Phage DNA is replicated and, if it contains the desired gene will recombine with plasmid DNA (Fig. 9-8). In this way the phage DNA will acquire a copy of the miniplasmid, including the suppressor gene. The resulting lysates are plated on suppressor-free (sup^0) hosts to guarantee that only those phages which carry a suppressor gene, and therefore suppress the nonsense mutations in their late genes, will be able to multiply. The frequency with which recombination events occur in such a system can be estimated. If the region of homology is 500 bp in length, recombination frequencies should be on the order of 10^{-3}; if the gene to be isolated occurs in approximately one out of 10^5 phages in the library, at least 10^8 phage clones will have to be analysed. The desired recombinants should be detectable because practical experience has shown that approximately 5 × 10^9 phages can be plated on a Petri dish with a diameter of 10 cm. It is obvious that it is very important to start with λ phages which definitely carry amber mutations and have not reverted back to wild-type.

The system described above was initially applied and worked satisfactorily, for Charon 4A genomic libraries. It sometimes fails if EMBL3A-derived libraries are used. There are speculations (Huang *et al.*, 1985) that the cause of these failures may be a recombination function encoded by the *rep* gene which is localised in the *nin5* region (Fig. 4.2-8A and B). Charon 4A and

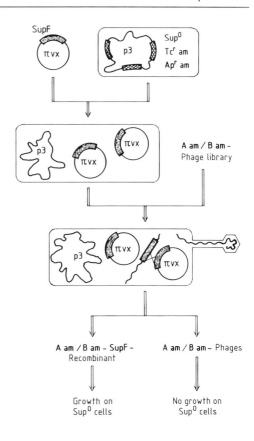

Fig. 9-7. Scheme of a selection technique based on homologous recombination.
Plasmid ΠVX carries the suppressor *supF* and the DNA probe (stippled bar) the counterpart of which is to be identified in a library. Plasmid p3 is a derivative of the 57 kb RP1 plasmid pLM2, which lacks all transfer functions. p3 contains a functional kanamycin resistance gene (Kn), and ampicillin (Ap) and tetracycline (Tc) genes carrying amber mutations. In the absence of amber suppressors, the presence of p3 can be selected for by kanamycin. If the cells also acquire a minivector coding for *supF*, transformed cells become, in addition, tetracycline-, and ampicillin-resistant. Two types of phage are generated after infection with a λ library. Firstly, homologous recombination yields a population of *Aam-Bam* phage which carry their own suppressor and are therefore capable of growing in sup^0 cells, and secondly, a population of unrecombined *Aam-Bam* phages which cannot grow on sup^0 cells and therefore do not yield plaques (Seed, 1983; Maniatis *et al.*, 1982).

a)
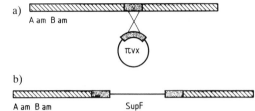
A am B am

πVX

b)

A am B am   SupF

Fig. 9-8. Structure of πVX-phage recombinants. Homologous recombination leads to the duplication (a) of the gene in question (stippled bar), and to the insertion of the suppressor gene (b). The *supF* tRNA suppresses the amber mutations in the left λ arm and therefore allows growth on *sup⁰* cells.

EMBL3A both carry the *nin5* deletion; however, Charon 4A contains an additional region in its right arm, derived from the lambdoid phage Φ80, which may have retained this or a similar recombination function (B. Seed, personal communication). Until this point has been clarified, it seems advisable to use Charon 4A rather than EMBL3A genomic libraries. Another complication of EMBL3A is the low but detectable recombination frequency between the polylinkers of EMBL3A and πAN7. This problem may be circumvented, however, by disruption of the polylinker insertion in πAN7 by suitable insertion of recombinant DNA, or by the elimination of phage vector sequences from the library by a rigorous and repeated selection for the loss of the central fragment (carrying the *red* and *gam* genes) by propagation on P2 lysogens.

A modification of this technique essentially based on the same principles was used by Goldfarb *et al.* (1982) to isolate the transforming gene of a human bladder carcinoma.

The method of cloning and selecting by recombination has also been extended to cosmid libraries (Poustka et al., 1984).

References

Backman, K., Betlach, M., Boyer, H.W., and Yanofsky, S. (1978). Genetic and physical studies on the replication of ColE1-type Plasmids. Cold Spring Harbor Symp. Quant. Biol. 43, 69-76.

Brown, E.L., Belagaje, R., Ryan, M.J., and Khorana, H.G. (1979). Chemical synthesis and cloning of a tyrosine tRNA gene. Methods Enzymol. 68, 109-151.

Benton, W.D., and Davis, R.W. (1977). Screening λgt recombinant clones by hybridization to single plaques *in situ*. Science 196, 180-182.

Clarke, L., and Carbon, J. (1976). A colony bank containing synthetic ColE1 Hybrid Plasmids representative of the entire *E. coli* genome. Cell 9, 91-99.

DiMaio, D., Corbin, V., Sibley, E., and Maniatis, T. (1984). High-level expression of a cloned HLA heavy chain gene introduced into mouse cells on a bovine papillomavirus vector. Mol. Cell. Biol. 4, 340-350.

Dressler, D., and Potter, H. (1982). Molecular Mechanisms in Genetic Recombination. Ann. Rev. Biochem. 51, 727-761.

Feiss, M., Fisher, R.A., Crayton, M.A., and Egner, C. (1977). Packaging of bacteriophage chromosome: Effect of chromosome length. Virology 77, 281-293.

Garen, A. (1968). Sense and nonsense in the genetic code. Science 160, 149-159.

Grunstein, M., and Hogness, D. (1975). Colony hybridization: A method for the isolation of cloned DNAs that contain a specific gene. Proc. Natl. Acad. Sci. USA 72, 3961-3965.

Goldfarb, M., Shimizu, K., Perucho, M., and Wigler, M. (1982). Isolation and preliminary characterization of a human transforming gene from T24 ladder carcinoma cells. Nature 296, 404-409.

Holliday, R. (1964). A mechanism for gene conversion in fungi. Genet. Res. 5, 282-303.

Huang, H., Little, P., and Seed, B. (1985). Suppressor tRNA cloning vehicles and Plasmid-Phage Recombination. In "Vectors: A survey of molecular cloning vectors and their applications", Rodriguez, R., ed., Butterworth Press, Stoneham, MA, USA.

Karn, J., Brenner, S., Barnett, L., and Cesareni, G. (1980). Novel bacteriophage λ cloning vector. Proc. Natl. Acad. Sci. USA 77, 5172-5176.

Lawn, R.M., Fritsch, E.F., Parker, R.C., Blake, G., Maniatis, T. (1978). The isolation and characterization of linked δ- and β-globin genes from a cloned library of human DNA. Cell 15, 1157-1174.

Leavitt, J., Gunning, P., Porrera, P., Ng, S.-Y., Lin, C.-S., and Kedes, L. (1984). Molecular cloning and characterization of mutant and wild-type human β-actin genes. Mol. Cell. Biol. 4, 1961-1965.

Maniatis, T., Hardison, R.C., Lacy, E., Lauer, J., O'Connell, C., Quon, D., Sim, G.K., and Efstradiatis, A. (1978). The isolation of structural genes from libraries of eucaryotic DNA. Cell 15, 687-701.

Maniatis, T., Fritsch, E.F., and Sambrook, J. (1982). In

"Molecular Cloning", Cold Spring Harbor Laboratory, Box 100, Cold Spring Harbor, N.Y. 11724: ISBN 087969-136-0; pp.335-361.

Meyerowitz, E.M., Guild, G.M., Prestidge, L.S., and Hogness, D.S. (1980). A new high capacity cosmid vector and its use. Gene 11, 271-282.

Montgomery, D., Hall, B., Gillam, S., and Smith, M. (1978). Identification and isolation of the yeast cytochrome c gene. Cell 14, 673-680.

Poustka, A., Rackwitz, H.R., Frischauf, A.M., Hohn, B., and Lehrach, H. (1984). Selective isolation of cosmid clones by homologous recombination in Escherichia coli. Proc. Natl. Acad. Sci. USA 81, 4129-4133.

Ryan, M.J., Brown, E.L., Sekiya, T., Kupper, H., and Khorana, G. (1979). Total synthesis of a tyrosine suppressor tRNA gene. XVIII. Biological activity and transcription in vitro of the cloned gene. J. Biol. Chem. 254, 5817-5826.

Seed, B. (1983). Purification of genomic sequences from bacteriophage libraries by recombination and selection in vivo. Nucleic Acids Res. 11, 2427-2445.

Tiemeier, D., Enquist, L., and Leder, P. (1976). Improved derivative of a phage λ EK2 vector for cloning recombinant DNA. Nature 263, 526-529.

Van der Ploeg, L.H.T., Groffen, J., and Flavell, R.A. (1980). A novel type of secondary modification of two CCGG residues in the human γ-, δ-, β-globin gene locus. Nucleic Acids Res. 8, 4563-4574.

Williams, B.G., and Blattner, F.R. (1979). Construction and characterization of the hybrid bacteriophage λ Charon vectors for DNA cloning. J. Virol. 29, 555-575.

Woo, S.L.C. (1979). A sensitive and rapid method for recombinant phage screening. Methods in Enzymology 68, 389-395.

10 Cloning in Plants

In past decades the classical methods of plant breeding have brought enormous successes in the improvement of yields per acre (Table 10-1). It is not without reason that the term "green revolution" was coined in view of the development, for example, of new varieties of wheat by Norman Borlaug and his colleagues in Mexico. In spite of these successes it will be difficult in the future to obtain further fundamental improvements by applying classical techniques. Currently available techniques of plant breeding are generally hampered by the fact that species barriers cannot be crossed. It is, therefore, a considerable advantage that two new techniques are at hand today which complement classical methods by allowing such problems to be circumvented, and novel combi-

nations of genes to be created at will. One of the new techniques is somatic cell fusion, the other recombinant DNA technology (see also Cocking *et al.*, 1981; Kleinhofs and Behki, 1977). Somatic cell fusion allows whole groups of genes, and even chromosomes, of different species which may even be sexually incompatible to be brought together. Recombinant DNA technology provides means of introducing specific genes into plant cells. Both techniques normally require the use of protoplasts, *i.e.*, plant cells surrounded only by a plasma membrane, from which the cell walls have been removed by enzymatic treatment. It is this lack of a cell wall which makes protoplasts amenable to genetic manipulation in the first place (Flores *et al.*, 1981). Many protoplasts

Table 10–1. Average yields of major crops in 1930 and 1975.

Plant	Yield per acre 1930	1975	unit	% increase
Wheat	14.2	30.6	bushel	115
Rye	12.4	22.0	bushel	77
Rice	46.5	101.0	bushel	117
Corn	20.5	86.2	bushel	320
Oats	32.0	48.1	bushel	50
Barley	23.8	44.0	bushel	85
Cotton	157.1	453.0	pounds	185
Soybeans	13.4	28.4	bushel	112
Potatoes	61.0	251.0	Cwt	311
Tomatoes	7.1	30.0	tons	322
Peanuts	650	2 565	pounds	295

(1 acre = 0.404 hectare; 1 bushel = 35 l; 1 Cwt = 112 pounds = 51 kg)

From "Impacts of Applied Genetics", Congress of the United States. Office of Technology Assessment. April 1981, pp. 137–164.

can be kept and propagated in tissue culture, serving also as producers of a variety of important plant components, such as alkaloids and certain amino acid derivatives (Vasil, 1984). As far as gene manipulation is concerned, it is of considerable importance that whole plants, in particular dicotyledons such as vegetables, fruits, nuts and flowers, can be regenerated from individual protoplasts. Unfortunately, the numerous and economically important monocotyledonous cultivated plants such as wheat, rice, oats and barley, still present some difficulties. It can be expected, however, that these problems will be overcome sooner or later.

10.1 Fusion of Somatic Cells

Cell fusion unites two intact protoplasts and leads to a single hybrid cell. The methods employed are similar to those commonly used for the fusion of animal cells, for example, polyethyleneglycol treatment (Flores et al., 1981). When cells of two sexually compatible species are fused, the resulting hybrid cells are stable and contain both sets of parental chromosomes. This composition of chromosomes is also retained in whole hybrid plants regenerated from the fused cells. Cells from two sexually incompatible species can also be fused with success. The resulting hybrids are truly unique since they could never have been obtained by sexual reproduction. Unfortunately, such hybrids occasionally may, over many cell divisions, completely lose one of the parental genomes so that these hybrid cells may be very unstable for many generations. Sometimes only a few genes or chromosomes belonging to one of the fusion partners are retained. It is very difficult to control these parameters, and the initial enthusiasm with which this technique was accepted has sobered considerably. These problems notwithstanding, there are some very interesting applications of commercial interest, in particular with tobacco plants (Melchers et al., 1978).

Protoplast fusion unites not only the nuclei, but of course also the cytoplasms which contain genetically independent organelles, i.e., plastids such as mitochondria and chloroplasts, the genomes of which direct the synthesis of important enzymes (Kung, 1977; Bogorad, 1979; Hoober, 1984). Chloroplasts, for example, carry genetic information for the larger of two sub-units of ribulose-1,5-biphosphate carboxylase which plays a central role in photosynthesis; the smaller sub-unit is encoded in the cell nucleus (Broglie et al., 1983).

Plant breeding research has to consider the co-operation and interaction between nuclear and plastid functions. It is obvious that the interaction between plastids themselves or plastids and cell nuclei in fused cells can influence the phenotype of the resulting plants. When a fusion of protoplasts leads to the inactivation of one of the cellular genomes, fused cells possess a nucleus and a heterologous plastid complement. Such cells are known as cybrids, and may indeed possess new characteristics. The sterility of male plants of Nicotiana tabacum, which is caused by genes carried in the mitochondrial DNA, has been successfully transferred to Nicotiana silvestris by fusion of appropriate protoplasts (Zelcher et al., 1978). The use of such cytoplasmic functions in plant breeding will increase in the future since the characterisation of plastids on a molecular level is advancing rapidly (see also Leaver et al., 1983).

10.2 DNA Transfer

In principle, the prerequisites for the genetic manipulation of plant cells are the same as those already known for other systems. It is, therefore, essential to develop methods for cloning plant genes, to identify selection techniques, and to construct suitable vectors.

The isolation of plant genes has only just begun. In principle, the isolation of mRNA and cDNA clones does not differ from other biological systems. Early examples for the application of these techniques are the isolation of cDNA and genomic clones of the Zein storage protein family of maize (Hu *et al.*, 1982) and the isolation of genomic clones of sucrose synthetase from the same organism (Geiser *et al.*, 1982). The gene for phaseolin, the main storage protein of *Phaseolus vulgaris L.*, possesses three introns in its coding region (Sun *et al.*, 1981). It appears, therefore, that plant genes, like animal genes comprise both introns and exons.

Selection techniques currently available are extremely limited. The only stable auxotrophic marker known today is the gene for nitrate reductase *(nin)* (Müller and Grafe, 1978). Nin-negative mutants could be isolated since they are resistant to chlorate and are unable to grow on nitrate as the sole source of nitrogen (Cocking *et al.*, 1981). Due to recent developments, positive selection systems are now available, or will be at hand in the near future. These include selectable markers, such as resistance to antibiotics, for example, to the aminoglycoside antibiotic G418, or to heavy metals and herbicides. To accomplish the transfer of DNA into plant cells, current practice employs three main systems, namely, the so-called Ti plasmid, certain plant viruses, and the direct transfer of DNA into plant cells.

10.2.1 Ti plasmids of Agrobacteria

10.2.1.1 The Principle of Genetic Colonisation

The principle of genetic colonisation centres around a neoplastic disease known as crown gall which is ubiquitous in dicotyledonous plants. These tumours are caused by an infection of wounded plants with *Agrobacterium tumefaciens*, a species of Gram-negative soil bacteria. Investigations carried out in various laboratories have established that the induction of tumours is caused by certain plasmids, Ti (tumour-inducing) plasmids, found in oncogenic strains of *Agrobacterium tumefaciens* (for early reviews, see Schell *et al.*, 1979; Nester and Kosuge, 1981). Plasmid-free strains are no longer oncogenic, and non-oncogenic strains into which the plasmid is introduced become oncogenic. Three major groups of oncogenic Ti plasmids, *i.e.*, octopine, nopaline, and agropine plasmids are known today. They are classified partly by DNA homology, but mainly by their capacity to synthesise opines, which are specific derivatives of basic amino acids. As shown in Fig. 10-1, octopine and nopaline are the prototypes of a whole family of such derivatives; the octopine family is derived from pyruvate, the nopaline family from α-ketoglutarate.

Jeff Schell
Cologne, FRG

Agrobacteria use these compounds as a source of carbon and/or nitrogen. The gene products required for the catabolism of opines are encoded by the corresponding Ti plasmids. The three specific characteristics of Ti plasmids, *i.e.*, tumour induction, biosynthesis, and catabolism of opines have led Schell *et al.* (1979) to propose a novel concept of parasitism: genetic colonisation (see Chilton, 1983, for example). Agrobacteria which require opines for growth, but are unable to synthesise them, infect susceptible plants, transfer Ti plasmids into plant cells, and thus induce a tumour. This process is coupled with the synthesis of the urgently required opines. Such tumours, therefore, constitute an ecological niche for these bacteria, and it is solely the bacteria, and not the plant, which profit from the synthesis of opines. Agrobacteria thereby gain a growth advantage over other soil bacteria.

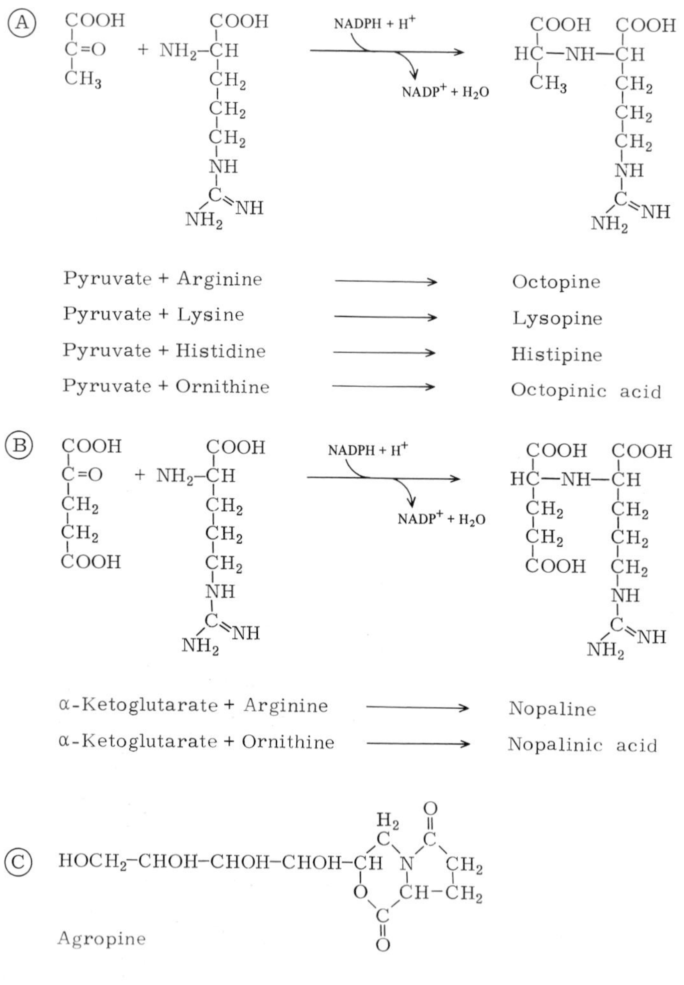

(A)

$$\begin{array}{l}\text{COOH}\\ |\\ \text{C=O}\\ |\\ \text{CH}_3\end{array} + \begin{array}{l}\text{COOH}\\ |\\ \text{NH}_2\text{-CH}\\ |\\ \text{CH}_2\\ |\\ \text{CH}_2\\ |\\ \text{CH}_2\\ |\\ \text{NH}\\ |\\ \text{C}\diagdown^{\text{NH}}\\ \text{NH}_2\end{array} \xrightarrow[\text{NADP}^+ + \text{H}_2\text{O}]{\text{NADPH} + \text{H}^+} \begin{array}{l}\text{COOH}\quad\text{COOH}\\ |\qquad\quad|\\ \text{HC}-\text{NH}-\text{CH}\\ |\qquad\quad|\\ \text{CH}_3\quad\;\text{CH}_2\\ \qquad\quad|\\ \qquad\quad\text{CH}_2\\ \qquad\quad|\\ \qquad\quad\text{CH}_2\\ \qquad\quad|\\ \qquad\quad\text{NH}\\ \qquad\quad|\\ \qquad\quad\text{C}\diagdown^{\text{NH}}\\ \qquad\;\;\text{NH}_2\end{array}$$

Pyruvate + Arginine   ⟶   Octopine

Pyruvate + Lysine   ⟶   Lysopine

Pyruvate + Histidine   ⟶   Histipine

Pyruvate + Ornithine   ⟶   Octopinic acid

(B)

$$\begin{array}{l}\text{COOH}\\ |\\ \text{C=O}\\ |\\ \text{CH}_2\\ |\\ \text{CH}_2\\ |\\ \text{COOH}\end{array} + \begin{array}{l}\text{COOH}\\ |\\ \text{NH}_2\text{-CH}\\ |\\ \text{CH}_2\\ |\\ \text{CH}_2\\ |\\ \text{CH}_2\\ |\\ \text{NH}\\ |\\ \text{C}\diagdown^{\text{NH}}\\ \text{NH}_2\end{array} \xrightarrow[\text{NADP}^+ + \text{H}_2\text{O}]{\text{NADPH} + \text{H}^+} \begin{array}{l}\text{COOH}\quad\text{COOH}\\ |\qquad\quad|\\ \text{HC}-\text{NH}-\text{CH}\\ |\qquad\quad|\\ \text{CH}_2\quad\;\text{CH}_2\\ |\qquad\quad|\\ \text{CH}_2\quad\;\text{CH}_2\\ |\qquad\quad|\\ \text{COOH}\quad\text{CH}_2\\ \qquad\quad|\\ \qquad\quad\text{NH}\\ \qquad\quad|\\ \qquad\quad\text{C}\diagdown^{\text{NH}}\\ \qquad\;\;\text{NH}_2\end{array}$$

α-Ketoglutarate + Arginine   ⟶   Nopaline

α-Ketoglutarate + Ornithine   ⟶   Nopalinic acid

(C) $\text{HOCH}_2\text{-CHOH-CHOH-CHOH-CH}\diagup\overset{\displaystyle \text{H}_2}{\text{C}}\diagdown\text{N}\diagup\overset{\displaystyle \overset{\text{O}}{\|}\; \text{C}}{}\diagdown\text{CH}_2$

Agropine

Fig. 10-1. Structure of opines.
(A) shows the family of octopines, (B) of nopalines, and (C) of agropine.

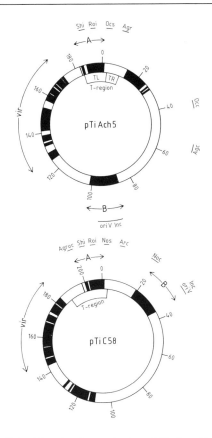

10.2.1.2 Structure and Function of T-DNA

Originally, the concept of genetic colonisation was purely hypothetical. It required, however, that a piece of DNA, known as tumour or T-DNA, be transferred from the Ti plasmid to the nuclei of plant cells during colonisation, and it has been shown that this is indeed the case (see also Bevan and Chilton, 1982). The sizes of the various Ti plasmids range between 150 and 200 kb (Van Montagu and Schell, 1982). Approximately 10% of the plasmid DNA, corresponding to approximately 22 kb, can be detected as T-DNA in plant cells of the nopaline tumour series (Fig. 10-2). Octopine tumours contain approximately 12 kb. In some octopine plasmids the T-DNA is divided into two adjacent DNA segments, one of 13 kb (T_L = left T-DNA), and one of 7 kb (T_R = right T-DNA), which are transferred to the plant genome either independently or as a contiguous stretch of DNA.

Fig. 10-2. Functional organisation of Ti-plasmids. The octopine plasmid pTiAch5 and the nopaline plasmid pTiC58 are 190 and 210 kb, respectively, in length. Black bars represent regions of homology between the two plasmids, which are observed mainly in the *vir*, the *T* and the *ori* regions. The virulence (*vir*) region is required for T DNA transfer, but is not transferred to the plant cell. The genes responsible for opine synthesis are indicated as *ocs* (octopine synthetase), *agroc* (agrocinopine synthetase), and *nos* (nopaline synthetase). The abbrevations *shi* and *roi* indicate the positions of genes involved in the inhibition of shoot and root formation (Schell *et al.*, 1984).

Marc Van Montagu
Gent, Belgium

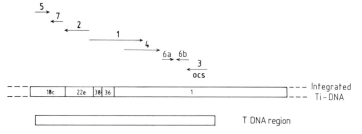

Fig. 10-3.   Transcriptional map of T DNA.

Shown are the positions of the 8 transcripts of a wild-type octopine tumour line. The directions of transcription are indicated by arrows above the map. Transcript 3, at the extreme right-hand end, codes for octopine synthetase (*ocs*). The products of the other transcripts suppress the specific organomorphogenesis in roots and/or shoots (*cf.* also text). The T DNA region is approximately 13 kb in length. The section of the colinear *Hind* III gene map of Ti DNA shows fragments 18c, 22e, 38, 36, and 1 (Willmitzer *et al.*, 1982).

T–DNA contains a region, common to both octopine and nopaline plasmids, which carries the genes responsible for the induction and maintenance of crown-gall tumours. A series of experiments were carried out to investigate the possible biological functions of T-DNA. Willmitzer *et al.* (1982) have analysed the T-DNA-specific mRNA in cell lines of octopine and nopaline tumours and discovered minute amounts of eight mRNA species in the transformed cells, which together represent hardly 0.001% of the total poly(A)-containing mRNA population. Transcripts 1, 2, 4, 5, and 6a and 6b are conserved in octopine and nopaline tumours, while transcripts 3 and 7 do not hybridise with nopaline T-DNA. These transcripts must, therefore, be specific for octopine tumours. The location of all transcripts has been determined by Northern analysis (Fig. 10-3; Willmitzer *et al.*, 1983) and correlates well with the open reading frames identified by sequence analysis of the T_L-DNA of the octopine plasmid pTiAch5 (Gielen *et al.*, 1984).

The function of these transcripts was elucidated by applying a specially developed mutagenesis technique. Because of their size, Ti plasmids are very difficult to manipulate *in vitro*. The various T-DNA regions were therefore subcloned in pBR322 or its derivatives. These clones were then used to introduce deletions or insertions in many

R. A. Schilperoort
Leiden, The Netherlands

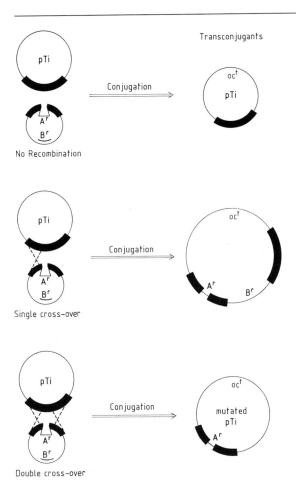

Transconjugants

Conjugation

No Recombination

Single cross-over

Double cross-over

Fig. 10-4. Introduction of a mutated T region into a Ti-plasmid.
The T region of a Ti-plasmid (pTi) is to be replaced by a mutated T region. This mutated T region is located on a plasmid, known as the intermediate vector, into which it was inserted as a transposon insertion, mediated *e.g.*, by transposon Tn7. The intermediate vector contains two resistance markers, one of which was provided by the inserted transposon (A^r), the other by the vector itself (B^r). A single cross-over event leads to the insertion of the entire intermediate vector into the Ti-plasmid; a double cross-over to the right and left of the transposon insertion only inserts the mutated T region (see text; Leemans *et al.*, 1981).

regions of the T-DNA (Leemans *et al.*, 1981, 1982). These "intermediate" vectors were then introduced into Agrobacteria together with intact Ti plasmids. As shown in Fig. 10-4, *in vivo* recombination then lead to substitutions of mutated sequences for homologous T region fragments in the Ti plasmid. Use of suitable vectors for these experiments allowed selection of plasmids, the formation of which required a double cross-over event (see also Section 12.2, Fig. 12-10). In the examples described above, the intermediate vector carried the mutated T region, and, in addition, two antibiotic resistance genes A^r and B^r. One gene, A^r, for example, resided within the T region (Fig. 10-4) and was introduced by nonspecific transposon mutagenesis (DeGreve *et al.*, 1981), while the other gene, B^r, was located outside. In principle, this intermediate vector could interact with the Ti plasmid in a donor cell in three different ways which could be distinguished from each other after conjugation with an appropriate recipient cell. In the absence of recombination, only the original Ti plasmid would be transferred to the recipient; thus, transconjugants could synthesise opine (oc^+) only, and would carry no resistance marker. A single cross-over in one of the homologous regions would lead to recombinants in which the entire intermediate vector had integrated into the Ti plasmid. Such recombinants carry the mutated T region of the intermediate vector and the intact T region of the Ti plasmid and are easily recognised since they contain both resistance markers. Two recombinational events, one on each side of the insertion, would yield a recombinant Ti plasmid carrying only the mutated T region. These conjugants can be recognised as B-sensitive clones in a population of A-resistant transconjugants.

The various mutated Ti plasmids were introduced into plant cells and their phenotype could be correlated with the loss of individual T-DNA transcripts (see also Garfinkel et al., 1981). The data obtained from such experiments can be summarised as follows (see also Fig. 10-3):

(1) T-DNA transfer and tumour growth are encoded by different independent functions. T-DNA is also transferred and integrated into the plant genome if the regions responsible for tumour formation are deleted or inactivated (see also item 4).

(2) Transcript 3 codes for octopine synthetase.

(3) Mutations in transcript 7 do not influence the formation of tumours or the synthesis of octopine. Its biological function remains to be clarified.

(4) Transcripts 1, 2, 4, 5, and 6a/6b which are observed in both octopine and nopaline tumours, influence the morphogenesis of plants by suppressing the organogenesis of shoots and roots (see below). Transcripts 1, 2, and 5 are involved in the

negative control of shoot organogenesis. Transcript 4 specifically inhibits the formation of roots. Transcripts 6a and 6b become effective at a later state of differentiation and repress the development of already existing shoots into normal plants. This block is responsible for the generation of teratomas, i.e., tumour masses which contain differentiated tissues.

Taken together, these data implicate that the transcripts listed in item 4 are involved in the regulation of plant growth. It was postulated that in transformed plant cells these transcripts code for enzymes involved in the biosynthesis of auxins, cytokinins, and other phytohormones, causing phenotypic transformation into crown-gall tumour cells; in contrast to normal healthy plant cells, transformed cells harbouring T-DNA and growing in tissue culture indeed do not require such factors. Subsequent biochemical analyses have confirmed these conclusions. Transcript 2 in particular codes for an enzyme which hydrolyses indole-3-acetamide to indole-3-acetic acid, an auxin (Schröder et al., 1984). The product of gene 1 is thought to represent a mono-oxygenase converting tryptophan into indole-3-acetamide, the substrate for the gene 2 product (Kemper et al., 1985). Gene 4 is involved in the biosynthesis of cytokinin, and codes for an isopentenyl-transferase activity (Buchmann et al., 1985). The products of these three genes are responsible for the hormone independence of crown-gall tumours and appear to be sufficient for the establishment of the tumour phenotype. The functions of the other transcripts remain unknown. There is no doubt, however, that a detailed study of crown gall transformation will make considerable contributions to our knowledge of plant growth control (Bevan, 1982).

10.2.1.3 Ti plasmids as Vectors

The functional studies described above already indicate that T-DNA is not just an arbitrary piece of DNA, but a very defined region on the Ti plasmid. This conclusion is supported by addition-

```
--- G C T G G |T G G C A G G A T A T A T T G| T G |G T G T A A A C| A A A T T --- Nopaline L
--- G T G T T |T G A C A G G A T A T A T T G| G C |G G G T A A A C| C T A A G --- Nopaline R
--- A G C G G |C G G C A G G A T A T A T T C| A A |T T G T A A A T| G G C T T --- Octopine A
--- C T G A C |T G G C A G G A T A T A T A C| C G |T T G T A A T T| T G A G C --- Octopine B
--- A A A G G |T G G C A G G A T A T A T C G| A G |G T G T A A A A| T A T C A --- Octopine C
--- A C T G A |T G G C A G G A T A T A T G C| G G |T T G T A A T T| C A T T T --- Octopine D
```

Fig. 10-5. Terminal sequences flanking the T DNAs of octopine and nopaline Ti plasmids. Region with the highest degree of homology are boxed; the 2 bp regions between these boxes are not conserved. Nopaline L (left) and R (right) are the repeats flanking the T DNAs of plasmids pTiC58 and pTi37. Octopine A/B and C/D represent the left hand and right hand terminal repeats flanking the T_L or T_R regions in plasmids pTiAch5 and pTiA5955 (Schell *et al.*, 1984).

al data obtained from the sequence analysis of junction fragments between T-DNA and plant DNA from various tumour cells (Zambryski *et al.*, 1982). These analyses show that the region of homology between sequences in the tumour (plant) DNA and sequences in the Ti plasmid DNA always ends within a 25 bp direct repeat sequence flanking the T region (Fig. 10-5). A stretch of 12 bp within these 25 bp repeats is completely conserved both in octopine and nopaline tumours. In contrast to the defined boundaries of integrated T-DNA, the integration sites within the plant genome appear to be absolutely random. These results suggest that only the defined and preserved boundaries of T-DNA are important for the integration event. In principle then any DNA should become integrated into the plant genome if it were inserted between these boundaries. Of course, such an insertion should not inactivate those functions on the Ti plasmid which are required *in trans* for integration and maintenance of the transformed state, and which are known as *vir* functions (Klee *et al.*, 1983). These functions, coded for by the 40 kb virulence *(vir)* region of Ti plasmids (*cf.* Fig. 10-2), are beginning to be analysed in detail. They represent at least six different complementation groups which are organised in a single operon. Activation of *vir* expression which precedes and initiates T-DNA transfer, is mediated by acetyl-dimethoxyphenol produced by actively growing plant cells (Stachel *et al.*, 1985). It is also known that the *vir* region is not found in the transformed cell and

that the T-DNA does not have to be physically linked to the *vir* region as long as the latter is maintained in the bacterial cell.

Hernalsteens *et al.* (1980) originally investigated the question of a possible transfer of foreign DNA with a mutant T-DNA clone carrying a Tn7 insertion in the nopaline synthetase gene (transcript 3). Although this long insertion (9.2 kb) inactivated the nopaline synthetase gene the corresponding T-DNA was transferred into the plant and integrated in an intact form together with the transposon. Tn7 carries the gene for a methotrexate-resistant dihydrofolate reductase; plant cells harbouring the insertion actually expressed methotrexate resistance and, unlike normal plants, were able to grow in the presence of 2 µg/ml of methotrexate. More recent experiments have shown other regions on the T-DNA which also can tolerate insertions of foreign DNA. The principal problems associated with the transfer of DNA and its stable integration into plant genomes are therefore solved.

Since plant cells containing these insertions still developed into tumours, it was of major importance to investigate whether it would be possible to separate the tumour-inducing from the T-DNA transfer functions, and to regenerate normal fertile plants from transformed plant cells. In order to answer these questions a series of transposon insertion mutants in the T-DNA of an octopine plasmid were isolated: one of these mutants, which carried an insertion that inactivated transcript 2, was much less oncogenic than the

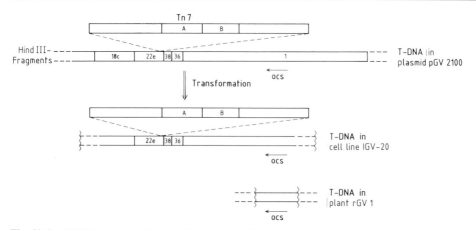

Fig. 10-6. T DNA content of plant cell cultures and regenerated plants.
Starting material is the tumour line IGV-20, which was obtained by transformation with plasmid pGV2100. This plasmid contains a Tn7 insertion in the *Hind*III fragment 22e within transcript 2 (*cf.* also Fig. 10-3). Due to the transposon insertion, which destroys the transcript, the tumour derived from IGV-20 yields shoots, one of which gave rise to a plant (rGV1). This plant only contained the flanking T DNA sequences and the region coding for octopine synthetase. The plant contains T DNA, synthesises octopine, but does not produce crown-galls. Wavy lines indicate the insertion of the plasmid DNA into the chromosomal DNA (broken line) of the plant (De Greve *et al.*, 1982).

wild-type plasmid (DeGreve *et al.*, 1982). The insertion in transcript 2 abolished the inhibition of shoot organogenesis (Fig. 10-6). When the resulting shoots were analysed for their capacity to regenerate whole plants, one shoot was found to give rise to a normal fertile plant which also expressed octopine synthetase. The analysis of T-DNA in these plant cells revealed that the T-DNA had consecutively lost regions from the left end and then regions from the middle of the T-DNA, and that whole plants only contained those parts of T-DNA representing the octopine synthetase (*ocs*) gene (Fig. 10-6). It is therefore possible to produce normal and healthy plant cells which have lost all properties of tumour cells, but still harbour and express certain parts of T-DNA. Several experiments of this kind have led to the conclusion that loss of the neoplastic, phytohormone-encoding regions on the T-DNA is a prerequisite for the regeneration of normal plants from transformed tissues. This conclusion is also based on the observation that healthy revertants of crown gall tumours have retained the right

flanking sequences, but lost the middle and left parts of T-DNA which code for the tumour genes (Yang and Simpson, 1981). The tumour-inducing capacity and the property of T-DNA transfer can therefore be separated from each other, which is an essential prerequisite for the generation of healthy plants from transformed cells. Since, in fact, none of the genes carried on T-DNA are required for T-DNA transfer or integration, any DNA to be transferred into the plant genome would have to be flanked only by the T-DNA border sequences; moreover, in the nopaline system, the right end of nopaline T-DNA alone is sufficient to direct the integration of foreign DNA into plant cells (An *et al.*, 1985).

A widely used vector which is based on the previously discussed findings is pGV3850 (Zambryski *et al.*, 1982). It contains the T-DNA border regions including the nopaline synthetase coding region from the right border, and all other Ti plasmid sequences except those T-DNA regions which are responsible for tumour formation. Transformants can therefore be identified by their

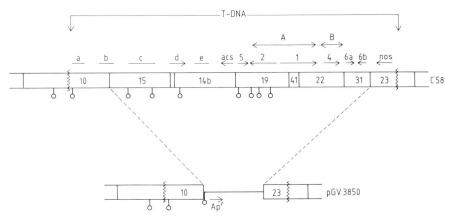

Fig. 10-7. Structures of the T DNA regions of plasmid pGV3850 and Ti-plasmid C58.
Transcripts of the region common to nopaline and octapine T DNAs are labelled 1, 2, 4, 5, 6a and 6b (see also Fig. 10-2). Transcripts originating from the non-conserved regions are named a to e. They have not been assigned any function. The two transcripts designated *acs* and *nos* code for agrocinopine and nopaline synthesis. Numbers in the maps refer to *Hind* III restriction sites, and the open circles to *Eco* RI sites. The notched lines represent T DNA borders, the thin line in pGV3850 indicates pBR322 sequences (Willmitzer *et al.*, 1983).

ability to synthesise nopaline (Fig. 10-7). The oncogenic T-DNA sequences have been replaced by pBR322 DNA, which allows any other gene cloned in a pBR322-derived plasmid to be inserted within T-DNA border sequences by homologous recombination (see below).

Strains of *Agrobacteria* containing pGV3850 are able to insert the DNA contained between the T-DNA terminal sequences into the chromosomal DNA of plant cells. These sequences include the nopaline synthetase (*nos*) gene which is located on *Hind* III fragment 23 of Ti plasmid C58. Although there is no direct selection for pGV3580-transformed cells, such cells can be obtained easily by screening for nopaline production.

The introduction of dominant selectable marker genes into plants has been accomplished through the construction of suitable expression vectors (Zambryski *et al.*, 1984) which express various bacterial resistance genes under the nopaline synthetase (*nos*) promoter, e.g., genes for neomycin (*neo*), chloramphenicol (*CAT*), or methotrexate resistance. The *nos* promoter is particularly useful since it is functio-

nal in cells derived from all dicotyledonous plants tested so far, and in all tissues of regenerated *nos*-containing plants. The *nos* gene is encoded entirely within the *Hind* III fragment 23 of Ti plasmid C58 and codes for a protein of 413 amino acids. Its transcription signals are typical of a eukaryotic gene although it is of prokaryotic origin. Some low level of transcription of the *nos* region has been observed in Agrobacteria, but since opines have never been detected in bacteria it is assumed that the RNA is defective, and that this prokaryotic gene is active only in transformed plants.

The expression vector pLGV2381 was constructed from a 345 bp *Sau* 3A fragment containing the *nos* promoter region from position "-264" up to codon 16 of the *nos* reading frame (Depicker *et al.*, 1982). As shown in Fig. 10-8, pLGV2381 contains a unique *Bam* HI site conveniently placed behind the cap site of *nos* mRNA. It has been used, for example, to introduce the 1.5 kb *Bgl* II-*Bam* HI fragment with the Tn5-derived *APH(3')II* gene conferring neomycin resistance (*npt*) (cf. Fig. 8-7). Other markers included the chloramphenicol acetyltransferase gene from

CATAAATCCCCTCGGTATCCAATTAGAGTCTCATATTCACTCTCAATCCAAATAATCTGCA ATG GCA

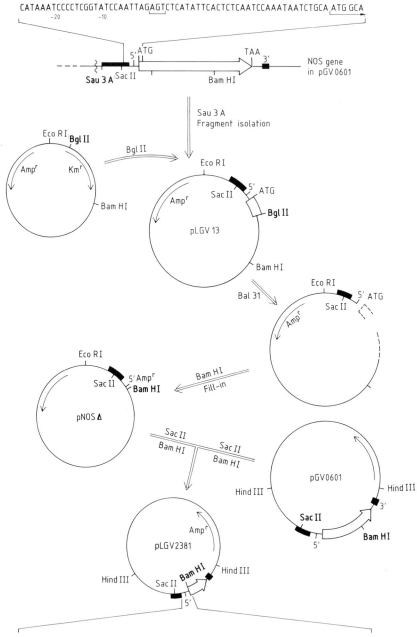

CATAAATCCCTCGGTATCCAATTAGAGTCTCATATTCACTCTCAATCCAAATAATCTGGATCCAACACT

pBR325, or the plasmid R67-derived dihydrofolate reductase gene conferring methotrexate resistance.

Chimaeric genes in plasmid pLGV2381 can be introduced into *Agrobacteria* by a method involving direct mobilisation from *E. coli*. In the first step, two helper plasmids producing transfer and mobilisation functions are transferred by conjugation into *E. coli* containing the pLGV2381-derived chimaeric plasmid. One of these helper plasmids, R64drd11, provides the transfer functions, the other plasmid, pGJ28, the *mob* functions necessary for the transmission of the ColE1-type plasmid with the chimaeric gene marker (Van Haute *et al.*, 1983). The resulting strain containing all three plasmids is conjugated with *Agrobacteria* containing either a wild-type Ti plasmid, pTiC58, or Ti plasmid pGV3850 as a recipient. With pTiC58 as recipient, exconjugants carrying the *Ap^r* marker result from co-integration events involving cross-overs through homologous T region DNA sequences in the acceptor Ti plasmid and vector pLGV2831; with pGV3850 as recipient (Fig. 10-9), exconjugants arise from a single cross-over event between pBR322 sequences in the two plasmids. The resulting duplication of pBR322 sequences does not lead to instability of the insertions. Recombinants have been used to infect plant (tobacco) protoplasts and to

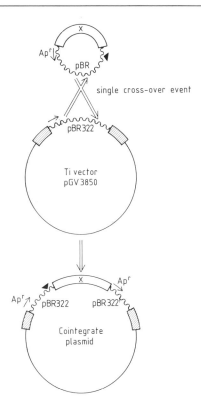

Fig. 10-9. Use of the Ti-plasmid vector pGV3850 as an acceptor of foreign DNA.
As described in the text, a pBR322-type plasmid (*e.g.* pLGV2381) containing the foreign gene (*X*) is mobilised into *Agrobacterium* containing the Ti-plasmid vector pGV3850 (*cf.* Fig. 10-7). A single recombination event creates a cointegrate plasmid structure containing gene *X*, which can subsequently be transferred to plants. Wavy lines are pBR322 sequences. Hatched bars represent the remaining T DNA border sequences in pGV3850 (Zambryski *et al.*, 1983).

◄— **Fig. 10-8.** Construction of the expression vector pLGV2381.
The top of the figure shows a section of the *nos* gene present in clone pGV0601. A 350 bp *Sau*3A fragment from this plasmid is cloned into the *Bgl*II linearised vector pKC7 to yield pLGV13. This plasmid is shortened from its *Bgl*II site with *Bal*31 in order to remove the *nos*-derived ATG codon. Plasmid pNOSΔ is a DNA molecule containing a regenerated *Bam*HI site derived from ligation of a filled-in *Bam*HI end and the terminus of a *Bal*31 deletion. Plasmid pLGV2381 is eventually produced from ligation of the small *Sac*II-*Bam*HI fragment from pNOSΔ and the large *Sac*II-*Bam*HI fragment from pGV0601. The section of the promoter sequence shows the 5' cap site and the *Bam*HI cloning site. Black bars indicate the *nos* promoter region, open bars the *nos* coding region (Herrera-Estrella *et al.*, 1983).

isolate, for instance, kanamycin-resistant cells, which were regenerated into complete plants containing the inserted DNA and expressing the expected resistance phenotype, *e.g.*, neomycin phosphotransferase activity. It is obvious that other genes can be linked to the selectable marker genes and then be transferred into plants. The choice of these genes will depend on the desired applications.

A new generation of vectors avoids all manipulations requiring intermediate vectors (An *et al.*, 1985). These advances are based on the observation that the virulence (*vir*) region, although required for T-DNA transfer, does not have to be physically linked to the T-DNA. It is thus possible to develop two-plasmid systems in which one component, the Ti plasmid, supplies the *vir* functions while the other component carries selectable markers and/or foreign genes. In the system described by An *et al.* (1985) the second vector contains a chimaeric *nos::npt* fusion as described above, a ColE1 replicon, and a wide host range replicon for replication in both *E. coli* and *Agrobacterium,* the T-DNA border sequences, and the *cos* site of bacteriophage λ. These vectors are maintained in *Agrobacterium* and are transferred to plant cells in the presence of a wild-type Ti plasmid which supplies the necessary *vir* functions in *trans*. Due to their small size, these cosmids allow DNA fragments of up to 35 kb to be packaged *in vitro*. This will permit the establishment of plant gene libraries and the direct isolation of plant genes with selectable phenotypes. With a plant like *Arabidopsis thaliana*, which has a genome size of only 7×10^7 bp (as compared to the 8×10^9 bp of the wheat genome, for example), only 10^4 clones will be required for a complete library. It remains to be seen whether and how this cosmid system can also be exploited for the transfer and expression of foreign genes within plants.

Arabidopsis thaliana, a member of the *cruciferae* (mustards), has other properties which make it an ideal object for molecular geneticists (Meyerowitz and Pruitt, 1985). Apart from its small genome size it is amenable to cell culture, regeneration and transformation mediated by *Agrobacterium tumefaciens*. It has an extremely short generation time (only 4-5 weeks) and produces extraordinarily large amounts of tiny seeds, of which tens of thousands can be planted in a single petri dish.

A widely discussed industrial application is the engineering of herbicide tolerance, by introducing genes conferring either increased resistance or permitting specific degradation of a given herbicide (Nester 1984; Sandermann, 1984). This subject has received considerable attention since, in contrast to other potential applications such as nitrogen fixation, it may require the transfer of only a single gene. Figure 10-10 shows the chemical structures of four important herbicides which interact with key enzymes in important metabolic pathways. The broad spectrum herbicide glyphosate, for instance, inhibits an enzyme in the biosynthesis of aromatic amino acids, 5-enolpyruvyl-shikimate-3-phosphate-synthetase (Amrhein *et al.*, 1983). A resistant enzyme with a greatly decreased affinity for glyphosate has been obtained from *Salmonella*, the mutated gene being available on a cosmid vector (Comai *et al.*, 1983). Chlorosulfuron, a very potent and selective herbicide, inhibits the enzyme acetolactate synthetase, the first enzyme in valine and leucine biosynthesis (Ray, 1982). The target of L-phosphinotricin (L-PPT) was shown to be the enzyme glutamine synthase (GS) which is involved in the regulation of nitrogen metabolism in plants. In alfalfa cells, resistance to L-PPT is due to an amplification of the *GS* gene and to a concomitant overproduction of a normal GS enzyme (Donn et a., 1984). Finally, the widely used herbicide atrazine interacts with a 32 kDa plastoquinone-binding membrane protein of photosystem II. In this case, resistance was shown to be associated with a structural mutation (Hirschberg and McIntosh, 1983). The transfer of such a resistance gene into a plant cell poses a particular challenge since it involves the genetic manipulation of chloroplasts. At present, direct engineering of the chloroplast genome itself appears to be impossible; however, most chloroplast proteins are coded for in the nuclear genome of the plant cell and are translocated as precursors from the cytoplasm through the plastid envelope membrane into the interior of the chloroplast. This transport requires an amino-terminal transit peptide which is subsequently removed by proteolysis to yield mature proteins. One solution to the problem of directing

Herbicide	Chemical strukture	Sensitive enzyme
atrazine	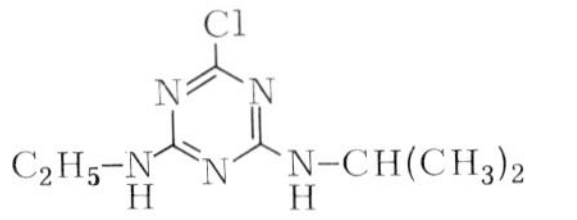	32 K protein from photosystem II
L-phosphinotricin	$CH_3-\overset{O}{\overset{\|}{P}}-CH_2-CH_2-\underset{NH_2}{CH}-COOH$ $\underset{OH}{\|}$	glutamine synthetase
chlorsulfuron	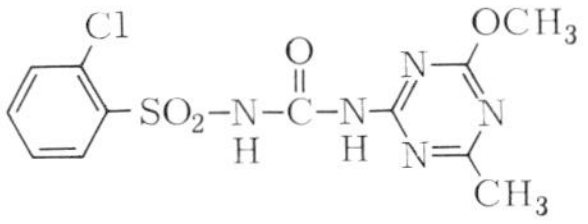	acetolactate synthetase
glyphosate	$HO-\overset{O}{\overset{\|}{P}}-CH_2-\underset{H}{N}-CH_2-COOH$ $\underset{OH}{\|}$	5-enolpyruvyl-shikimate-3-phosphate synthetase

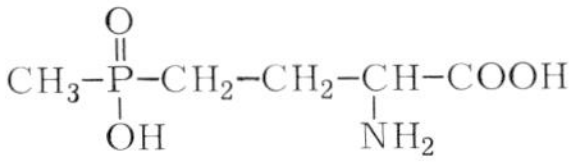

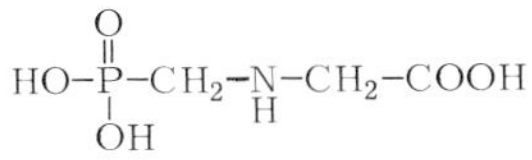

Fig. 10-10. Structure and target enzymes for selected herbicides.
Engineering of tolerance against herbicides may not only be of economical significance but may also be of interest as a means to obtain selectable markers since, for instance, selection for G418 or kanamycin resistance in plants is impractical.

a foreign protein into a chloroplast thus could be a fusion of a transit peptide to the gene product to be expressed within the chloroplast. Following introduction of the fusion gene into the plant genome *via* Ti plasmid-mediated transformation, the foreign polypeptide would subsequently be translocated into chloroplasts as a fusion protein. Indeed, coupling of the transit peptide gene from the small subunit of ribulose-1,5-biphosphate carboxylase to a bacterial neomycin phospho-transferase II gene, results in the expression of the corresponding fusion protein within chloroplasts, conferring kanamycin resistance on transformed tobacco plants (Schreier *et al.*, 1985; Van den Broeck *et al.*, 1985). Current technology thus not only permits manipulation of the nuclear plant genome, but also of cell organelles.

10.2.2 DNA Viruses as Vectors

In addition to the *Agrobacterium* system, the use of plant DNA viruses has also been considered as a method for the transformation of plants. In principle, there are only two types of viruses which could be used for this purpose, namely, the double-stranded Caulimo viruses, the prototype of which is cauliflower mosaic virus (Hohn *et al.*, 1982), and the single-stranded gemini viruses, such as bean golden mosaic virus.

10.2.2.1 Caulimo Viruses

Infections with cauliflower mosaic virus are restricted to the plant families *Cruciferae* and

Solanaceae. The double-stranded circular DNA of this virus is approximately 8 kb in length. The genomes of at least three serotypes, D/H, Cabb-S, and CM1841 have already been sequenced (Franck *et al.*, 1980; Gardner *et al.*, 1981; Balazs *et al.*, 1982). A unique feature of their DNA is the presence of three single-stranded breaks, one occurring in the so-called α strand, the other two in the complementary β strand. Regions containing such discontinuities form short triple helices (Fig. 10-11). Replication occurs *via* a large 35S transcript as the key replicative intermediate and thus would require reverse transcription in order to produce a double-stranded DNA genome (Pfeiffer and Hohn, 1983; Varmus, 1983). The necessary RNA-dependent DNA polymerase activity (the product of gene *V*) has been identified in CaMV-infected *Brassica* (rape seed) leaves (Volovitch *et al.*, 1984). Apart from these prerequisites and similarities the CaMV life cycle is certainly more than just a permutated version of the retrovirus life cycle. This is particularly evident in its transcriptional mechanism: while a set of mRNAs are produced by splicing from a single primary transcript in retroviruses, several unspliced transcripts originate from different promoters in CaMV.

The (-) strands of Caulimo viruses contain six open reading frames, but only three of them have been assigned to virus-specific proteins. The gene *IV* product is a lysine-rich 42 kDa polypeptide which has been identified as the major coat protein. The gene *V* product is reverse transcriptase. Region *VI* presumably encodes a 66 kDa protein constituting the major structural protein of inclusion bodies found in infected plant cells. With respect to the development of these viruses as vector systems, the transcripts will have to be analysed further in order to find possible nonessential regions and also those regions which are responsible for virulence. Data are beginning to accumulate which suggest that the gene *II* region is non-essential. It has been observed that naturally occurring deletions in this region cannot be distinguished from wild type viruses as far as

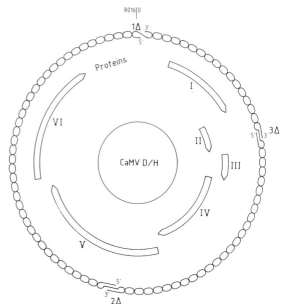

Fig. 10-11. Genetic map of cauliflower mosaic virus. The genome of serotype D/H is 8 016 bp in length. 1Δ, 2Δ, and 3Δ mark the position of the three strand breaks. Roman numerals near the open arrows signify various open reading frames. The direction of transcription points towards the arrow heads.

virulence and growth properties are concerned. Gronenborn *et al.* (1981) have succeeded in inserting a 250 bp DNA fragment containing the *lac* operator into an *Xho*I site of the gene *II* region without altering the phenotype of the virus; moreover, Brisson *et al.* (1984) have recently replaced the entire open reading frame II (*orf II*) by a gene coding for a methotrexate-resistant dihydrofolate reductase derived from the RG7 R factor of *E. coli*. The new transducing virus with the 240 bp insertion rendered infected plants resistant to methotrexate. It is not known, however, whether there are other sites on the viral genome which would allow larger DNA fragments to be inserted without affecting the packaging capacity of the virus. The packaging problem could be circumvented by introducing the CaMV genome not as virus, but as naked

DNA, through transfection into plant cells. In fact, cloned viral DNA which lacks the naturally occurring discontinuities is extremely infectious: one microgram of DNA is sufficient for the infection of ten plants. Using this approach, Gronenborn (1984) has replaced *orf VI*, the coding region of the viral inclusion body protein, by the Tn5-derived neomycin phosphotransferase (*APH(3)II*) gene) (*cf.* Fig. 8-7). A number of kanamycin-resistant cells and calli were observed after transfection of tobacco protoplasts. It is thus possible not only to use sections of the CaMV genome other than *orf II* for a replacement with foreign DNA, but also to extend the host range of CaMV from cruciferous plants to other species, such as tobacco plants. In a somewhat different approach, Odell *et al.* (1985) have isolated the 35S promoter region from CaMV DNA as a 1149 bp *Bgl* II fragment and fused it to a human growth hormone gene. The chimaeric gene was introduced into tobacco cells *via* a Ti plasmid carrying the nopaline synthetase promoter upstream from the neomycin phosphotransferase II coding region. Plants regenerated from kanamycin-resistant calli expressed the human growth hormone gene in leaves, stems, roots and petals. This demonstrates again that promoter regions from the CaMV genome are active even when introduced into non-cruciferous plants.

10.2.2.2 Gemini Viruses

Our knowledge of gemini viruses is even more limited than that of Caulimo viruses. The subject was reviewed by Goodman (1981). Gemini virus particles, which have a molecular weight of 3.8×10^6 Da, are the only known eukaryotic viruses which possess a single-stranded circular DNA genome. This genome is approximately 2 000 nucleotides long (molecular weight = $7\text{-}8 \times 10^5$ Da). Each virus particle contains two DNA molecules which are identical in length, but differ in their sequence (Haber *et al.*, 1981). Both components of the tomato golden mosaic virus

have been cloned (Bisaro *et al.*, 1982). Some sequences have also been reported for the two components of cassava latent virus (CLV) (Stanley and Gray, 1983). The two components of CLV DNA, which were isolated from infected plant material as double-stranded molecules of 2 279 and 2 725 bp, respectively, share a sequence of only 200 base pairs. The positions of the open reading frames suggest that transcription probably proceeds in both directions, *i.e.*, from the viral and the complementary strand, on each component. Although Ikegami *et al.* (1981) have observed replicative forms, there is a long way to go before the first vectors will be available. One particular feature of gemini viruses is their extremely broad host range. They are capable of infecting both monocotyledons and dicotyledons plants and in the long run these viruses may provide very useful plant vectors.

10.2.3 Direct Transfer of DNA into Plant Cells

Agrobacterium tumefaciens is currently the most efficient system for transferring foreign genetic material into plant cells (Schell *et al.*, 1984). Plant cells and/or whole plants transformed by engineered Ti plasmids can generally be obtained by either of the following two methods (*cf.* also Zambryski *et al.*, 1984):

(1) Inoculation *in vivo* of whole plants with *Agrobacteria* followed by subsequent culture *in vitro* of the wounded surface on media allowing the regeneration of shoots and roots. An extension of this method uses co-infection in the presence of other strains of *Agrobacteria* containing mutant Ti plasmids which induce shoot-producing crown gall tumours. These shoots are normally composed of untransformed cells; in mixed infections with pGV3850-containing *Agrobacteria,* for example, some shoots contain the plasmid pGV3850, and some the shorter mutant Ti plasmid. Transformed cells are easily distin-

guished by the marker on the pGV3850 vector, *e.g.*, nopaline;

(2) Co-cultivation of single leaf cell protoplasts with *Agrobacterium*. This method allows the isolation of transformed tissues derived from single cells, but is restricted to plants which can be regenerated from protoplasts.

The natural gene vector system of *Agrobacterium tumefaciens* is restricted to the host range of *Agrobacterium*, which includes dicotyledonous plants, while most monocotyledonous plants, *e.g.*, the most important crops, are not considered susceptible. There are some recent exceptions, for instance, the detection of opines in narcissus plants transformed with *Agrobacterium tumefaciens* (Hooyskaas-Van Slogteren *et al.*, 1985), or the Ti plasmid-mediated transformation of *Asparagus officinalis*, a member of the *Liliaceae* family (Hernalsteens *et al.*, 1984). The important *Gramineae*, however, remain immune.

In principle, the natural and efficient, but restricted, system of *Agrobacterium* could be

circumvented by the artifcial introduction of genetic material into plant cells using the methods employed for mammalian cells (*cf*. Section 8.5). Early experiments by Krens *et al.* (1982) have demonstrated the feasability of a direct transformation of isolated Ti plasmid DNA into plant protoplasts. A limited success has been described by Paszkowski *et al.* (1984) who used a vector derived entirely from Ti plasmid DNA sequences in which the protein coding sequence of the *APH(3′)II* gene was placed under the control of the cauliflower mosaic virus (CaMV) gene *VI* promoter and termination signals. It produced an APH(3′)II fusion protein containing an additional 23 amino acids from the CaMV gene *VI* product. This hybrid gene could be transferred into tobacco protoplasts and conferred kanamycin resistance not only upon protoplasts and cells, but also on haploid plantlets (see also Potrykus *et al.*, 1985a). Hain *et al.* (1985) have transformed tobacco protoplasts with a pLGV2381-derived vector, likewise containing the *APH(3′)II* gene, with transformation frequencies on the order of 10^{-5} or 10^{-4} (10 or 100 kanamycin-resistant colonies per 10^6 living protoplasts): in contrast, frequencies of up to 10% were obtained by co-cultivation of protoplasts with *Agrobacterium tumefaciens* containing the same vector.

Using similar methods, even graminaceous monocotyledons have recently been transformed by direct gene transfer into protoplasts, albeit at low efficiencies. Potrykus *et al.* (1985b) employed suspension cultures from *Colium multiflorum* (Italian rye grass) while Lörz *et al.* (1985) used protoplasts from *Triticum monococcum*. Transformation frequencies were in the range of one in 4×10^3 in the former, and of one in 5×10^5 in the latter case. A real success in this field will depend on an optimisation of methods for the preparation of dividing protoplasts from cereal plants (Ozias-Akins and Lörz, 1984).

Ingo Potrykus
Basel, Switzerland

A break-through in the direct transfer of genes into plant protoplasts was recently described by Shillito *et al.* (1985). Using electroporation (*cf.* Section 8.5), optimal PEG concentrations, and a short heat-shock treatment, protoplasts from tobacco could be transformed with efficiencies in the range of 2%. Such high efficiencies are a prerequisite for the screening of total genome libraries within plant cells and for the isolation of single-copy genes by direct complementation of mutant cell lines. It remains to be seen whether this efficient method can be adapted to proto-plasts from graminaceous plants.

The technical prerequisites for gene transfer into plants are now available, either through the Ti system or through protoplast transformation with naked DNA. Plant molecular biology thus has finally come of age and will rapidly close the gap on the developments in animal cells.

References

Amrhein, N. Johänning, D., Schab, J., and Schulz, A. (1983). Biochemical basis for glyphosate tolerance in a bacterium and a plant tissue culture. FEBS Letters 157, 191-196.

An, G., Watson, B.D., Stachel, S.D., Gordon, M.P., and Nester, E.W. (1985). New cloning vehicles for transformation of higher plants. The EMBO J. 4, 277-284.

Balazs, E., Guilley, H., Jonard, G., and Richards, K. (1982). Nucleotide sequence of DNA from an altered virulence isolate D/H of the califlower mosaic virus. Gene 19, 239-249.

Bevan, M.W., and Chilton, M.D. (1982). T-DNA of the *Agrobacterium* TI and RI plasmids. Ann. Rev. Genet. 16, 357-384.

Bevan, M. (1982). Crown gall tumors and plant growth regulators – any connection? Nature 299, 299.

Bisaro, D.M., Hamilton, W.D.O., Coutts, R.H.A., and Buck, K.W. (1982). Molecular cloning and characterization of the two DNA components of tomato golden mosaic virus. Nucleic Acids Res. 10, 4913-4922.

Bogorad, L. (1979). The chloroplast, its genome and possibilities for genetically manipulating plants. In: Genetic Engineering (J.K. Setlow, A. Hollaender, eds.), Plenum Press, N.Y., Vol. I., pp.181-203.

Brisson, N., Paszkowski, J., Penswick, J.R., Gronen-born, B., Potrykus, I., and Hohn, T. (1984). Expression of a bacterial gene in plants by using a viral vector. Nature 310, 511-514.

Broglie, R., Coruzzi, G., Lampps, G., Keith, B., and Chua, N.H. (1983). Structural analysis of nuclear genes coding for the precursor to the small subunit of wheat ribulose-1,5-bisphosphate carboxylase. Bio-technology 1, 55-61.

Buchmann, I., Marner, F.J., Schröder, G., Waffen-schmidt, S., and Schröder, J. (1985). Tumour genes in plants: T-DNA encoded cytokinin biosynthesis. The EMBO J. 4, 853-859.

Chilton, M.D. (1983). A vector for introducing new genes into plants. Scientific American 249, No. 6, pp.36-45.

Cocking, E.C., Darey, M.R., Pental, D., and Power, J.B. (1981). Aspects of plant genetic manipulation. Nature 293, 265-270.

Comai, L., Sen, L.C., and Stalker, D.M. (1983). An altered *araA* gene product confers resistance to the herbicide glyphosate. Science 221, 370-371.

Depicker, J., Stachel, S., Dhaese, P., Zambryski, P., and Goodman, H.M. (1982). Nopaline Synthase: Transcript mapping and DNA sequence. J. Mol. Appl. Genet. 1, 561-573.

Donn, G., Tischer, E., Smith, A.J.A., and Goodman, H.M. (1984). Herbicide-resistant alfalfa cells: an example of gene amplification in plants. J. Mol. Appl. Genet. 2, 621-635.

Flores, H.E., Sawhney, R.K., and Galston, A.W. (1981). Protoplasts as vehicles for plant propagation and improvement. Adv. Cell Culture 1, 241-279.

Franck, A., Guilley, H., Jonard, G., Richards, K., and Hirth, L. (1980). Nucleotide sequence of cauliflower mosaic virus. Cell 21, 285-294.

Gardner, R.C., Howarth, A.J., Hahn, P., Brown-Keedi, M., Shephard, R.J., and Messing, J. (1981). The complete nucleotide sequence of an infectious clone of cauliflower mosaic virus by M13mp7 shot gun sequencing. Nucleic Acids Res. 9, 2871-2887.

Garfinkel, D.J., Simpson, R.B., Ream, L.W., White, F.F., Gordon, M.P., and Nester, E.W. (1981). Genetic Analysis of Crown-Gall: Fine structure map of T-DNA by site-directed mutagenesis. Cell 27, 143-153.

Geiser, M., Weck, E., Döring, H.P., Werr, W., Courage-Tebbe, U., Tillmann, E., and Starlinger, P. (1982). Genomic clones of a wild-type allele and a transposable element-induced mutant allele of the sucrose synthase gene of *Zea mays L*. The EMBO Journal 1, 1455-1460.

Gielen, J., De Beuckeleer, M., Seurinck, J., Deboeck,

F., De Greve, H., Lemmers, M., Van Montagu, M., and Schell, J. (1984). The complete nucleotide sequence of the TL-DNA of the *Agrobacterium tumefaciens* plasmid pTiAch5. The EMBO J. 3, 835-846.

DeGreve, H., Decraemer, H., Seurinck, J., Van Montagu, M., and Schell, J. (1981). The functional organization of the octopine *Agrobacterium tumefaciens* plasmid pTiB653. Plasmid 6, 235-248.

DeGreve, H.D., Leemans, J., Hernalsteens, J.P. Toong, L.T., Benckeleer, M.D., Willmitzer, L., Otten, L., V. Montagu, M., and Schell, J. (1982). Regeneration of normal and fertile plants that express octopine synthetase, from tobacco crown galls after deletion of tumour-controlling functions. Nature 300, 752-755.

Goodman, R.M. (1981). Geminiviruses. J. Gen. Virol. 54, 9-21.

Gronenborn, B., Gardner, R.C., Schaefer, S., and Shepherd, R.J. (1981). Propagation of foreign DNA in plants using cauliflower mosaic virus as vector. Natue 294, 772-776.

Gronenborn, B. (1984). Cauliflower Mosaic Virus. A plant gene vector. In "The impact of gene transfer techniques in eukaryotic cell biology", J.J. Schell and P. Starlinger (eds.), Springer Verlag, Berlin – New York – Tokyo.

Hain, R., Stabel, D., Czernilofsky, A.P., Steinbiss, K.H., Herrera-Estrella, L., and Schell, J. (1985). Uptake, integration, expression and genetic transmission of a selectable chimaeric gene by plant protoplasts. Mol. Gen. Genet. 199, 161-168.

Haber, S., Ikegami, M., Bajet, N.B., and Goodman, R.M. (1981). Evidence for a divided genome in bean golden mosaic virus, a geminivirus. Nature 289, 324-326.

Hernalsteens, J.P., van Vliet, F., De Benckeleer, M., Depicker, A., Engler, G., Lemmers, M., Holsters, M., van Montagu, M., and Schell, J. (1980). The *Agrobacterium tumefaciens* Ti plasmid as a host vector system for introducing foreign DNA in plant cells. Nature 287, 654-656.

Hernalsteens, J.P., Thia-Toong, L., Schell, J., and Van Montagu, M. (1984). An *Agrobacterium*-transformed cell culture from the monocot *Asparagus officinalis*. The EMBO J. 3, 3039-3041.

Herrera-Estrella, L., Depicker, A., Van Montagu, M., and Schell, J. (1983). Expression of chimaeric genes transferred into plant cells using a Ti-plasmid-derived vector. Nature, 303, 209-213.

Hirschberg, J., and McIntosh, L. (1983). Molecular basis of herbicide resistance in *Amaranthus hybridus*. Science 222, 1346-1349.

Hohn, T., Richards, K., and Lebeurier, G. (1982). Cauliflower Mosaic Virus on its way to becoming a useful plant vector. Curr. Topics Microbiol. Immunol. 96, 194-236.

Hoober, J.K. (1984). Chloroplasts. Plenum Press; New York; pp. 280.

Hooykaas- Van Slogteren, G.M.S., Hooykaas, P.J.J., and Schilperoort, R.A. (1984). Expression of Ti plasmid genes in monocotyledonous plants infected with *Agrobacterium tumefaciens*. Nature 311, 763-764.

Hu, N.T., Peifer, M.A., Heidecker, G., Messing, J., and Rubenstein, I. (1982). Primary structure of a genomic zein sequence of maize. The EMBO Journal 1, 1337-1342.

Ikegami, M., Haber, S., and Goodman, R.M. (1981). Isolation and characterization of virus-specific double-stranded DNA from tissues infected by bean golden mosaic virus. Proc. Natl. Acad. Sci. USA 78, 4102-4106.

Kemper, E., Waffenschmidt, S., Weiler, E.W., Rausch, T., and Schröder, J. (1985). T-DNA-encoded auxin formation in crown-gall cells. Planta 163, 257-262.

Klee, H.J., White, F.F., Iyer, V.N., Gordon, M.P., and Nester, E.W. (1983). Mutational analysis of the virulence region of an *Agrobacterium tumefaciens* Ti plasmid. J. Bacteriol. 153, 878-883.

Kleinhofs, A., and Behki, R. (1977). Prospects for plant genome modification by nonconventional methods. Ann. Rev. Genet. 11, 79-101.

Krens, F.H., Molendijk, L., Wullems, G.J., and Schilperoort, R.A. (1982). *In vitro* transformation of plant protoplasts with Ti-plasmid DNA. Nature 296, 72-74.

Kung, S. (1977). Expression of chloroplast genomes in higher plants. Ann. Rev. Plant Physiol. 28, 401-437.

Leaver, C.J. (1983). Mitochondrial genes and male sterility in plants. In: DNA makes RNA makes Protein (T. Hunt, S. Prentis, J. Tooze, eds.), Elsevier Biomedical Press, Amsterdam, N.Y., Oxford, pp.99-107.

Leemans, J., Shaw, Ch., Deblaere, R., De Greve, H., Hernalsteens, J.P., Maes, M., Van Montagu, M., and Schell, J. (1981). Site-specific mutagenesis of *Agrobacterium* Ti plasmids and transfer of genes to plants. J. Mol. Appl. Genet. 1, 149-164.

Leemans, J., Deblaere, R., Willmitzer, L., De Greve, H., Hernalsteens, J.P., Van Montagu, M., and Schell, J. (1982). Genetic identification of functions of TL-DNA transcripts in octopine crown galls. The EMBO Journal 1, 147-152.

Lörz, H., Baker, B., and Schell, J. (1985). Gene transfer to cereals mediated by protoplast transformation. Mol. Gen. Genet. 199, 178-182.

Melchers, G., Sacristan, M.D., and Holder, A.A.

(1978). Somatic hybrid plants of potato and tomato regenerated from fused protoplasts. Carlsberg Res. Commun. 43, 203-218.

Meyerowitz, E.M., and Pruitt, R.E. (1985). *Arabidopsis thaliana* and plant molecular genetics. Science 229, 1214-1218.

Müller, A.J., and Grafe, R. (1978). Isolation and characterization of cell lines of *Nicotiana tabacum* lacking nitrate reductase. Molec. Gen. Genet. 161, 67-76.

Nester, E.W., and Kosuge, T. (1981). Plasmids specifying plant hyperplasias. Ann. Rev. Microbiol. 35, 531-565.

Nester, W.J. (1984). Engineering herbicide tolerance: When is it worthwhile; Biotechnology 2, 939-944.

Odell, J.T., Nagy, F., and Chua, N.H. (1985). Identification of DNA sequences required for activity of the cauliflower mosaic virus 35S promoter. Nature 313, 810-812.

Ozias-Akins, P., and Lörz, H. (1984). Progress and limitations in the culture of cereal protoplasts. Trends in Biotechnology 2, 119-123.

Paszkowski, J., Shillito, R.D., Saul, M., Mandak, V., Hohn, T., Hohn, B., and Potrykus, I. (1984). Direct gene transfer to plants. The EMBO J. 3, 2717-2722.

Pfeiffer, D., and Hohn, T. (1983). Involvement of reverse transcription in the replication of cauliflower mosaic virus: A detailed model and test of some aspects. Cell 33, 781-789.

Potrykus, I., Paszkowski, J., Saul, M.W., Petruska, J., and Shillito, R.D. (1985a). Molecular and general genetics of a hybrid foreign gene introduced into tobacco by direct gene transfer. Mol. Gen. Genet. 199, 169-177.

Potrykus, I., Saul, M.W., Petruska, J., Paszkowski, J., and Shillito, R.D. (1985b). Direct gene transfer to cells of a graminaceous monocot. Mol. Gen. Genet. 199, 183-188.

Ray, T.B. (1982). The mode of action of chlorsulfuron: a new herbicide for cereals. Pestic. Biochem. Physiol. 17, 10-17.

Sandermann, H.Jr. (1984). Herbicide resistance through gene transfer? Biochemical and biological aspects. In "The impact of gene transfer techniques in eukaryotic cell biology"; J. Schell and P.Starlinger, eds; Springer Verlag, Berlin-Heidelberg; pp- 167-179.

Schell, J., Van Montagu, M., De Beuckeleer, M., De Block, M., Depicker, A., De Wilde, M., Engler, G., Genetello, G., Hernalsteens, J.P., Holsters, M., Seurinck, J., Silva, B., AVan Vliet, F., and Villarroel, R. (1979). Interactions and DNA transfer between *Agrobacterium tumefaciens*, the Ti-plasmid and the plant host. Proc. Roy. Soc. Lond. B. 204, 251-266.

Schell, J., Herrera-Estrella, L., Zambryski, P., De Block, M., Joos, H., Willmitzer, L., Eches, P., Rosahl, S., and Van Montagu, M. (1984). Genetic engineering of plants. In: "The impact of gene transfer techniques in eukaryotic cell biology", J.S. Schell and P. Starlinger (eds.), Springer Verlag, Berlin – Heidelberg – New York – Tokyo.

Schreier, P.H., Seftor, E.A., Schell, J., and Bohnert, H.J. (1985). The use of nuclear-encoded sequences to direct the light-regulated synthesis and transport of a foreign protein into plant chloroplasts. The EMBO J.

Schröder, G., Waffenschmidt, S., Weiler, E.W., and Schröder, J. (1984). The T-region of Ti plasmids code for an enzyme synthesizing indole-3-acetic acid. Eur. J. Biochem. 138, 387-391.

Shillito, R.D., Saul, M.W., Paszkowski, J., Müller, M., and Potrykus, I. (1985). High efficiency direct gene transfer to plants. Biotechnology 3, 1099-1106.

Sprague, G.F., Alexander, D.E., and Dudley, J.W. (1980). Plant Breeding and Genetic Engineering: A Perspective. Bioscience 30, 17-21.

Stachel, S.E., Messens, E., Van Montagu, M., and Zambryski, D. (1985). Identification of the signal molecules produced by wounded plant cells that activate T-DNA transfer in *Agrobacterium tumefaciens*. Nature 318, 624-629.

Stanley, J., and Gray, M.R. (1983). Nucleotide sequence of cassava latent virus DNA. Nature 301, 260-262.

Sun, S.M., Slightom, J.L., and Hall, T.C. (1981). Intervening sequences in a plant gene. Nature 279, 37-41.

Van den Broeck, G., Timko, M.P., Kausch, A.P., Cashmore, A.R., Van Montagu, M., and Herrera-Estrella, L. (1985). Targeting of a foreign protein to chloroplasts by fusion to the transit peptide of ribulose-1,5-bisphosphate carboxylase. Nature 313, 358-363.

Van Haute, E., Joos, H., Maes, M., Warren, G., Van Montagu, M., and Schell, J. (1983). Intergenic transfer and exchange recombination of restriction fragments cloned in pBR322: a novel strategy for the reversed genetics of the Ti plasmids of *Agrobacterium tumefaciens*. The EMBO J. 2, 411-417.

Van Montagu, M., and Schell, J. (1982). The Ti Plasmids of *Agrobacterium*. Curr. Topics Microbiol. Immunol. 96, 237-254.

Varmus, H.E. (1983). RNA viruses: Reverse transcription in higher plants. Nature 304, 116-117.

Vasil, I.K. (1984). Cell culture and somatic cell genetics of plants. Vol. 1. Laboratory procedures and their applications. Academic Press, New York; pp. 825.

Volovitch, M., Modjtahedi, N., Yot, P., and Brun, G. (1984). RNA-dependent DNA polymerase activity in cauliflower mosaic virus-infected plant leaves. The EMBO J. 3, 309-314.

Willmitzer, L., Simons, G., and Schell, J. (1982). The TL-DNA in octopine crowngall tumors codes for seven well-defined polyadenylated transcripts. The EMBO Journal 1, 139-146.

Willmitzer, L., Dhaese, P., Schreier, P.H., Schmalenbach, W., Van Montagu, M., and Schell, J. (1983). Size, location and polarity of T-DNA-encoded transcripts in nopaline crown gall tumors, common transcripts in octopine and nopaline tumors. Cell 32, 1045-1056.

Yang, F., and Simpson, R.B. (1981). Revertant seedlings from crowngall tumors retain a portion of the bacterial Ti plasmid DNA sequences. Proc. Natl. Acad. Sci. USA 78, 4151-4155.

Zambryski, P., Holsters, M., Kruger, K., Depicker, A., Schell, J., Van Montagu, M., and Goodman, H.M. (1980). Tumor DNA Structure in plant cells transformed by *Agrobacterium tumefaciens*. Science 209, 1385-1391.

Zambryski, P., Depicker, A., Kruger, K., and Goodman, H.M. (1982). Tumor induction by *Agrobacterium tumefaciens*: Analysis of the boundaries of T-DNA. J. Mol. Appl. Genet. 1, 361-370.

Zambryski, P., Herrera-Estrella, L., De Block, M. Van Montagu, M., and Schell, J. (1984). The use of the Ti plasmid of *Agrobacterium* to study the transfer and expression of foreign DNA in plant cells: New vectors and methods. In: "Genetic Engineering, Principles and Methods", Vol. VI; Plenum Press, A. Hollaender and J. Setlow (eds.), pp.253-278.

Zambryski, P., Joos, H., Genetello, C., Leemans, J., Van Montagu, M., and Schell, J. (1983). Ti plasmid vector for the introduction of DNA into plant cells without alteration of their normal regeneration capacity. The EMBO J. 2, 2143-2150.

Zelcher, A., Aviv, D., and Galun, E. (1978). Interspecific transfer of cytoplasmic male sterility by fusion between protoplasts of normal *N. silvestris* and X-ray irradiated protoplasts of male sterile *N. tabacum*. Z. Pflanzenphysiol. 90, 397-407.

11 Identification of Recombinant DNA

The third and final step of a cloning experiment is the identification of recombinant phages, clones of bacteria, or cells. There are several methods which allow recombinant DNAs to be detected in a background of non-recombinant clones, and a specific recombinant to be identified in a population of various other recombinant clones. Direct techniques make use of the fact that clones with recombinant DNA possess a new selectable phenotype *in vivo*, which distinguishes them from other clones which do not contain any recombinant DNA or undesired "false" DNA inserts. Indirect techniques allow the desired recombinant clones to be identified by either detecting the desired DNA sequence or its gene products.

11.1 Direct Methods

11.1.1 Selection by Complementation or Nonsense Suppression

A prerequisite for selection by complementation is the existence of auxotrophic or conditionally lethal mutations in the host organisms. A variety of mutants in the metabolism of carbon sources, amino acids, and nucleic acids are known for *E. coli*, and also for other bacteria and yeasts. Such mutants lack particular gene products, usually an enzyme, and under certain conditions will depend on a supply of intermediates, the normal biosynthesis of which is directly or indirectly affected by the mutation. These intermediates must be added to the growth medium unless a gene which restores the organism's ability to make the required substance is introduced into the cell. The compensation of a genetic defect in the host organism is known as complementation. Nonsense suppression is a special case of complementation operating only in the instance of nonsense mutations. In this case the complementing gene does not provide a complete and enzymatically active protein, but a special tRNA, a suppressor tRNA, which is capable of reading a stop codon in the affected gene as an amino acid codon. The principles of complementation by suppression have been explained in Chapter 9, Fig. 9-5, for tyrosine suppressor tRNA.

Selection for genes by complementation plays a prominent role in cloning *E. coli* genes in *E. coli*. In practice, such cloning attempts begin with the establishment of an *E. coli* library, *i.e.*, a collection of transformants carrying plasmid inserts representing the entire genome of *E. coli* (Clarke and Carbon, 1976; see also Chapter 9). A representative library of *E. coli* clones carrying plasmid inserts of approximately 6 kb would consist of approximately 1500 different clones. Plasmid DNA containing these insertions can be isolated from a mixture of bacterial clones in the form of superhelical DNA. When the cells containing these plasmid are pooled one must beware of the fact that individual clones may differ in their growth rates, and thus might be lost during the amplification of the library. This problem can be overcome by growing up individual colonies which subsequently are pooled, or by using comparatively large inocula for growing the mixture of clones as a whole. The various clones in this mixture will then have to undergo only a few cell divisions before chloramphenicol is added to amplify the plasmids.

Superhelical DNA isolated from these plasmid mixtures is used to transform auxotrophic mutants of *E. coli* which are then plated on selective media. Only those colonies which have taken up a plasmid with an insertion producing the missing gene product will grow. Of course, complementation is successful only if the DNA insertions are sufficiently long that transcription leads to the formation of a complete gene product and if the desired gene is functional in *E. coli*. For obvious reasons, this procedure will work well if *E. coli* genes are cloned in *E. coli* (Clarke and Carbon, 1976); however, this complementation technique has also been used successfully for cloning genes of Gram-positive organisms in *E. coli*. Even yeast genes have been identified in *E. coli* by complementation analysis. This approach has not been successful for cloning *E. coli* genes in Gram-positive organisms (*cf.* Chapter 6). Complementation analyses fail upon cloning genes for which there are no suitable auxotrophic mutations, or genes that do not exist in *E. coli*, such as eukaryotic immunoglobulin genes. It should be noted that gene expression in *E. coli* differs markedly from that of eukaryotes, for example, in the existence of introns in eukaryotic DNA. Complementation of an *E. coli* mutation by a eukaryotic gene thus may fail even if the mutation in question is complementable by an *E. coli* gene. However, a few fruitful attempts have been described in the literature. Chang *et al.* (1978), for example, succeeded in identifying a cDNA of a mouse dihydrofolate reductase in *E. coli* by screening colonies for trimethoprim resistance (see also Chapter 8).

11.1.2 Marker Inactivation Techniques

Unlike techniques described in Section 11.1.1, the techniques which will be described now cannot be used for a direct selection of the desired gene, although they allow the desired recombinant phenotype to be recognised. Many of these methods have been mentioned in sections discussing the construction of various vectors. Suitable cloning vectors usually carry two or even more antibiotic resistance genes. The insertion of a foreign DNA into one of these marker genes inactivates this gene and leads to the expression of a new, selectable phenotype which can be easily distinguished from the original phenotype. Plasmid pBR322 may serve as an example.

By cloning into the *Pst* I site of the β-lactamase gene the original phenotype Ap^r/Tc^r is changed to Ap^S/Tc^r. If colonies are plated on agar containing tetracycline (Tc) and replicated onto agar containing ampicillin (Ap) and tetracycline, the cloning vector and recombinant plasmids derived from it are easily distinguished from each other. Bacteria harbouring recombinants grow on tetracycline plates, but not on plates containing tetracycline and ampicillin. An alternative identification procedure in this example would be cycloserine enrichment. Bacteria are first grown on tetracycline plates to select for all those clones harbouring a plasmid, recombinant or not. If ampicillin is added, recombinant clones will stop growing because they are ampicillin-sensitive. Upon addition of cycloserine, which lyses growing but not resting bacteria, all transformants with plasmids that do not contain insertions will be destroyed, thus providing the desired population of recombinant clones (Rodriguez *et al.*, 1976). Another technique makes use of the histochemical detection of β-lactamase to identify tetracycline-resistant transformants which carry an insertion in the *Pst* I site of plasmid pBR322 (Boyko and Ganschow, 1982). β-lactamase converts penicillin to penicilloic acid. The latter compound is capable of binding iodine, which can be detected if agar test plates contain soluble starch. Under these conditions the addition of iodine/potassium iodide (Lugol's solution) will lead to the formation of a typical dark blue iodine-starch complex: ampicillin-resistant colonies will quickly develop a clear halo since the penicilloic acid will compete for iodine with the starch solution; ampicillin-sensitive colonies will not influence the iodine-starch stain.

Apart from antibiotic resistances a variety of other markers can also be exploited for the

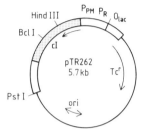

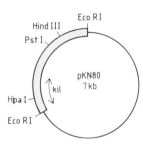

Fig. 11-1. Structure of two vectors which can be used for positive selection.
Plasmid pTR262 contains λ DNA and pBR322 sequences (solid line). The tetracycline resistance gene is under the control of the λ promoter, P_R, which is regulated by the λ *cI* repressor (stippled bar). The vector itself therefore does not confer tetracycline resistance. Insertions into the *Bcl* I (T/GATCA) or *Hind* III (A/AGCTT) sites of the *cI* gene inactivate the repressor, leading to the expression of the tetracycline resistance gene. The *cI* gene is transcribed counterclockwise from its own promoter P_{RM}. O_{lac} signifies a 29 bp insertion with the *lac* operator sequence (Roberts *et al.*, 1980).
Vector pKN80 contains the *Eco* RI-C fragment (stippled bar) of bacteriophage Mu, inserted into the *Eco* RI site of plasmid pRSF2124 (So *et al.*, 1976). The lethal function *kil*, encoded by this fragment, can be inactivated by insertions into the *Hpa* I (GTT/AAC) site.

identification of recombinant clones. λ vectors may serve as an example. Some of these λ vectors, Charon 4A, for example, contain the *E. coli lac* genes which are responsible for the metabolism of lactose, or the *bio* genes responsible for the biosynthesis of biotin. When the *lacZ* gene is inactivated by an insertion, β-galactosidase cannot be synthesised. In the presence of the chromogenic substrate Xgal, vector plaques will be blue, while recombinant plaques will remain white. The histochemical identification of β-galactosidase (see also Section 7.1.1, Fig. 7-6) is also used for vectors derived from single-stranded phages, such as M13mp vectors, and some other plasmids such as pUC, pUR222, and pUR250.

The presence of the *bio* marker allows plaques to be formed on *E. coli* mutants defective in biotin biosynthesis plated on a selective medium, *i.e.*, in the absence of biotin. Under these conditions recombinant phages will not form plaques.

There are also several positive selection systems which are based on marker inactivation and allow recombinant clones, but not parental clones, to grow. Plasmid pTR262, for example, (Roberts *et al.*, 1980) contains tetracycline resistance genes controlled by the right operator-promoter region, and the cI repressor of bacteriophage λ (Fig. 11-1). As long as active repressor is synthesised, the tetracycline gene is inactive and the transformants are sensitive to tetracycline; however, if the *cI* repressor gene is inactivated by an insertion into the *Bcl* II or *Hind* III sites, the tetracycline resistance gene is fully expressed. Transformants containing recombinant DNA are therefore tetracycline-resistant.

Another plasmid, pKN80, contains a 2.2 kb *Eco* RI-C fragment of bacteriophage Mu (Schumann, 1979) coding for the so-called kil protein, the expression of which is lethal for bacteria (Fig. 11-1). The expression of this protein is controlled (*i.e.*, repressed) by the repressor coming from a prophage in a Mu lysogen. Plasmid pKN80 therefore replicates only in bacteria which are lysogenic for Mu and produce repressor; non-lysogens are killed by the protein. As soon as this lethal function on the plasmid is inactivated by the insertion of foreign DNA into its *Hpa*I site, transformants can also be obtained on bacterial strains which do not harbour a Mu prophage.

Vector pHE3 (Hennecke *et al.*, 1982) (Fig. 11-2) contains the structural gene for the α-

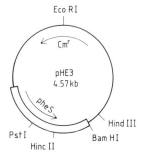

Fig. 11-2. Structure of selection vector pHE3.
Vector pHE3 is composed of plasmid pACYC184 (solid line) (Chang and Cohen, 1978), including a replicon function, the gene for chloramphenicol resistance (Cm^r), and a 1 200 bp long DNA sequence coding for the *E. coli pheS* gene (open bar). Cloning and marker inactivation can be achieved by using the *Pst*I or the *Hinc*II site (Hennecke *et al.*, 1982).

subunit of the *E. coli* phenylalanyl-tRNA synthetase (*pheS*). When this enzyme is present, *E. coli* cells are sensitive to the amino acid p-fluorophenylalanine, which charges and inactivates tRNA^{Phe}. Since the sensitive phenotype is dominant it is also expressed in diploid strains possessing a normal (sensitive) and a mutated structural gene. If the vector pHE3 carrying a functional phenylalanyl-tRNA synthetase gene is introduced into a p-fluorophenylalanine-resistant *E. coli* strain, these bacteria are rendered sensitive to p-fluorophenylalanine. When the plasmid-encoded gene is inactivated by an insertion, however, the resistant chromosomal phenotype becomes effective. Therefore only clones with a DNA insertion will grow in the presence of p-fluorophenylalanine.

Yet another selection system is based upon the instability of palindromic DNA sequences in *E.*

coli. Although the reasons for this phenomenon are unclear, it has been known for a long time that cloning of palindromic DNA sequences (*e.g.* ABCD-D'C'B'A') is either impossible or very difficult (Lilley, 1981). The problem has been investigated intensively with transposon Tn5 (Collins *et al.* 1982) which is 5 700 bp in length and contains a kanamycin resistance gene. The central or unique region containing this gene is flanked by two inverted repetitions of 1 532 bp (Auerswald *et al.*, 1981). Plasmid pGJ53 (Fig. 11-3), which is derived from ColE1, confers chloramphenicol resistance and contains transposon Tn5 (Hagan and Warren, 1982). *Hind* III cleaves this plasmid within the inverted repeats. It should therefore be possible to construct a 2 394 bp long palindrome, composed of two 1 197 bp halves directly facing each other; however, it is impossible to isolate such structures and they must be regarded as lethal. As shown in Fig. 11-3, this observation is the basis for a very elegant selection technique, since it is only possible to obtain transformants with the isolated 6.5 kb DNA fragment of pGJ53 in the presence of heterologous *Hind* III fragments. Although a small number of transformants (0.3% of the transformation frequency of the parental nondeleted plasmid) are observed even in the absence of heterologous DNA, these do not contain the simple *Hind* III deletion, but rather insertions of additional chromosomal DNA sequences. This low background is of course welcome, but rather surprising. Transposons typically exhibit a high frequency of deletions, insertions and internal rearrangements, and this recombinational activity probably reduces the generation of non-viable palindromes (Collins *et al.*, 1982).

Positive selection techniques have also been described for λ vectors. In replacement vectors these techniques are based on the presence or absence of the central DNA fragment, which must carry a suitable biological marker. The Spi$^+$ phenotype, which is expressed if the λ *gam* gene is removed, has been mentioned earlier (Section 4.2).

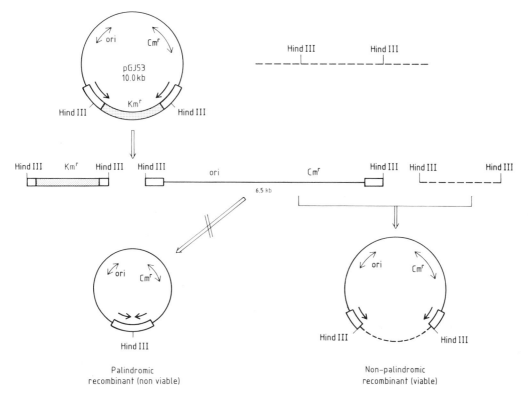

Fig. 11-3. Scheme depicting a positive selection technique for recombinant DNA.
Vector pGJ53 consists of plasmid pAT273 with an insertion of transposon Tn5 coding for kanamycin resistance, Km^r (stippled bar). HindIII digestion of pGJ53 yields two fragments. The larger fragment which is 6.5 kb in length contains the ColE1 region. It cannot give rise to viable vector molecules, because circularisation forms a 2 394 bp long palindromic sequence (open bar). Transformants are generated, however, in the presence of heterologous DNA fragments (broken lines). The inverted repetitions are shown as open bars; their orientations are indicated by arrows (Hagan and Warren, 1982).

The *A3* gene of phage T5 encodes a product which prevents the infection of hosts carrying the ColIb plasmid (Davison *et al.*, 1979). When this T5 gene is present in the central fragment of the λ vector, but absent from the recombinant DNA, its use in conjunction with suitable host bacteria provides a very strong selection for recombinant DNA. Another selection system which is based on the simultaneous incorporation of a suppressor gene and a recombinant DNA insertion was described in Chapter 9. λ vectors with *amber* mutations in late genes can form plaques on suppressor-free cells only if they take up a suppressor gene and express the suppressor tRNA.

In this context one should recall that *in vitro* packaging of λ DNA also can be used as a selection system, because the size requirements for packaging act as a selection for inserts containing phage DNA.

11.2 Indirect Methods

The following sections will describe examples of the detection of specific DNA sequences and their products in transformed cells.

11.2.1 Restriction Enzyme Cleavage Patterns

The characterisation of restriction enzyme cleavage patterns often provides first insights into the nature of isolated recombinant DNA molecules. The position of cleavage sites often can be predicted from the known amino acid sequence encoded by the DNA in question, or from analogies with related genes. Restriction analyses are often carried out with DNA obtained through rapid lysis techniques for the isolation of plasmids from small cultures of single bacterial clones (approximately 5 ml) (Birnboim and Doly, 1979; Holmes and Quigley, 1981; Rüther, 1982). After restriction enzyme digestion, the DNA is electrophoresed on agarose or acrylamide gels. In cases in which only the total size of the plasmid must be ascertained, it is even possible to lyse single plasmid-containing bacterial colonies (2-3 mm in diametre) directly in the slots of a gel (Barnes, 1977), and to subject the DNA to electrophoresis immediately after lysis. DNAs can be stained with ethidium bromide (0.5 µg/ml) or silver, and the sizes of plasmids and their insertions are determined by comparison with the migration of DNA fragments of known sizes. The use of the silver stain is restricted to polyacrylamide gels, but it allows nanogram amounts of DNA to be detected easily (Beidler *et al.*, 1982; Kolodny, 1984; for staining of proteins see Nielsen and Brown, 1984).

11.2.2 Hybridisation Techniques

The specific detection of recombinant DNA molecules by DNA-DNA or RNA-DNA hybridisation plays a prominent role in gene technology. The principles of these methods which comprise screening procedures for bacterial colonies or phage plaques, and blotting techniques, will be described below.

11.2.2.1 Colony and Plaque Hybridisation

The technique known as colony hybridisation of bacterial clones was introduced by Grunstein and Hogness (1975), and has been adapted to high colony densities by Hanahan and Meselson (1980). Colonies obtained from a transformation experiment are transferred from agar plates to nitrocellulose filters by simple replica plating (Fig. 11-4). Since growth media penetrate nitrocellulose filters, colonies can be grown directly on filters placed on top of the agar surface. New copies can then be made on other nitrocellulose filters by further replica plating. Bacterial DNAs are denatured on these filters by incubation in alkali. After several washes with neutral buffers the DNA is fixed to the filters by heating. Such filters are then incubated with a specific radioactively labelled DNA probe. If there are sequence homologies between the DNA probe and DNA on the filters the DNA probe will bind to the DNA on the filters by hybridisation, which is visualised autoradiographically on X-ray films. The positions of black spots on the X-ray films can be correlated with particular clones on the master plate. Incidentally, it is very important to make identical marks on the master plate and the filters. Otherwise one will be in the rather embarassing situation of the wise men of Gotham who sank a bell in the lake and marked its position by making a notch in the railing of their boat. In a 10 cm Petri dish the technique described above allows approximately 1 000 colonies to be tested; modifications introduced by Hanahan and Meselson (1980) even allow up to 10^5 colonies to be screened on a 10 cm Petri dish if the bacteria are plated directly on nitrocellulose filters. In the latter case the colonies are much smaller and the plasmids are amplified by chloramphenicol treatment after copies of the original nitrocellulose filters have been made.

Phage plaques can be tested similarly by a modification of the technique introduced by Benton and Davis (1977) who observed that single plaques contain enough phage particles

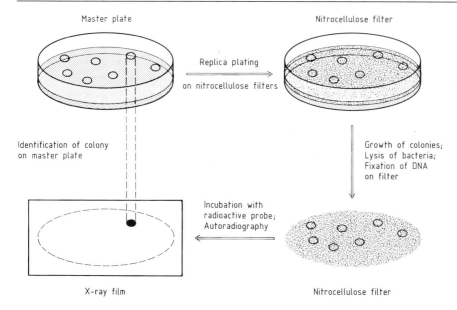

Fig. 11-4.   Schematic representation of the principle of colony hybridisation for the identification of recombinant bacterial clones (Grunstein and Wallis, 1979).

(approximately 10^7) and enough unpackaged DNA to be transferred to a nitrocellulose filter by direct contact between the filter and agar plate. This DNA can be detected by *in situ* hybridisation. Since approximately 20000 plaques can be tested on one filter, a human library in Charon 4A, for example, which would comprise approximately 500000 plaques could be stored and analysed on 20 to 30 filters.

The sensitivity of this method can be increased considerably if the plaques are plated on filters soaked in a suspension of bacteria instead of blank filters. After the transfer, these filters are incubated on the surface of an agar plate to obtain a bacterial lawn which is subsequently infected by the transferred phages. This procedure amplifies

Ronald W. Davis
Stanford

the phage DNA and significantly reduces hybridisation and exposure times, which in turn reduces the background on the films. The sensitivity of this method is several orders of magnitudes higher than that observed under the original conditions (Woo, 1979).

DNAs or RNAs employed as hybridisation probes, such as cDNA clones labelled by nick translation (Rigby *et al.*, 1977; see also Section 11.2.2.2), specific RNA species, or synthetic oligonucleotides are usually labelled with ^{32}P *in vitro*, but also *in vivo*. The specificity of oligonucleotides depends on their lengths and on hybridisation conditions; the length "n" of the probe determines the frequency (P) with which a specific sequence will occur within the DNA to be analysed. By applying the equation $P = 1/4^n$ it is easy to calculate that an oligomer of 12 bases in length should be sufficiently long for the analysis of an *E. coli* library while appropriate probes for a human library would be 17- to 18-mers.

Hybridisation temperatures must be chosen very carefully. The lower the temperature, the higher the chance that the oligonucleotide will not only hybridise to the desired sequence, but also to related sequences; on the other hand, the temperature cannot be raised indefinitely since the efficiency of hybridisation decreases with rising temperature and becomes zero at the melting temperature of the DNA. In practice, the hybridisation temperature will usually lie approximately 20 °C below the melting temperature of the DNA. As a rule of thumb a double-stranded DNA which is 12 base pairs long and consists of 50% GC pairs has a melting temperature of approximately 40 °C in 2 × SSC so that the hybridisation temperature should not exceed 20 °C. Any additional GC pair will raise the melting temperature by 4 °C, an additional AT pair by approximately 2 °C. Since these values depend very much on other parameters of hybridisation, for example the ionic strength of the hybridisation solution, it is advisable to streamline hybridisation conditions by carefully optimising the reaction conditions (see Szostak *et al.*,

1979, for example). This is even more important when simple all-or-nothing results cannot be expected and hybridisations are to discriminate between insertions which differ only in one or a few bases. The example of an *amber* mutation in gene *3* of ΦX174 DNA hybridised with oligonucleotides of a length of 11, 14, and 17 bases demonstrates that a single mismatch will reduce the melting temperature of double-stranded DNAs by up to 12 °C (Fig. 11-5; Wallace *et al.*, 1979). The longer the DNA molecules, the smaller the influence of a single mismatch on the melting temperature. As a rule of thumb, an increase of mismatches in double strands longer than 1 000 bp by one per cent will reduce the melting temperature by approximately 1 °C (Bonner *et al.*, 1973).

The use of oligonucleotides plays a special role in the identification of cDNA clones. In principle, a nucleic acid sequence can be derived from a known amino acid sequence by applying the rules of the genetic code. An oligonucleotide can then be synthesised and used as a hybridisation probe. It is necessary, of course, to take into consideration the fact that the genetic code is degenerate. An instructive example is the identification of cDNA sequences for the human β₂-microglobulin (Suggs *et al.*, 1981) where amino acids 95-99 (Table 11-1) were selected from the known amino acid sequence. This choice was quite ingenious since the sequence contains the single-codon amino acids tryptophan and methionine (see also Fig. 7-43); nevertheless, these five amino acids still predict 24 possible sequences at the level of the mRNA. The problem was solved by synthesising two pools of mixed pentadecanucleotide probes. Probe β₂mI was a mixture of 8, β₂mII a mixture of 16 individual sequences. The cDNA was cloned into the *Pst* I site of pBR322 and yielded 535 tetracycline-resistant, ampicillin-sensitive clones. One of these clones was identified by hybridisation with probe β₂mII. A comparison with the correct sequence of the cDNA revealed that probe β₂mII indeed included the desired sequence. Probe β₂mI did not hybridise with any

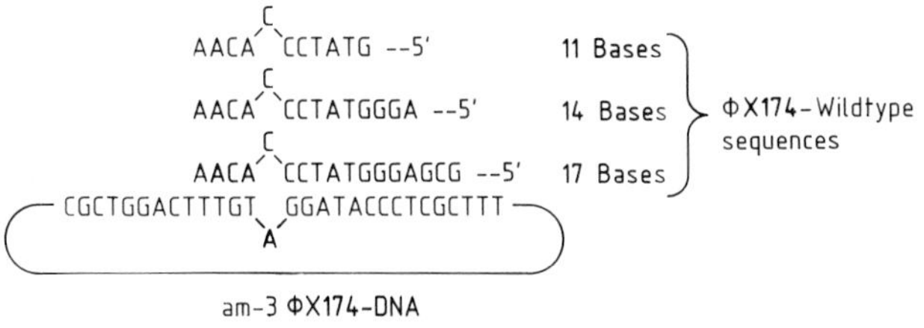

```
              C
      AACA    CCTATG --5'        11 Bases
              C
      AACA    CCTATGGGA --5'     14 Bases       ΦX174-Wildtype
              C                                 sequences
      AACA    CCTATGGGAGCG --5'  17 Bases
  CGCTGGACTTTGT  GGATACCCTCGCTTT
                A

        am-3 ΦX174-DNA
```

Number of bases in probe	Nature of probe	Number of possible bp	% GC in double strand	T_d (°C)
11	wt	11	46	33.2
11	am-3	10	–	–
14	wt	14	50	40.6
14	am-3	13	43	31.1
17	wt	17	59	55.1
17	am-3	16	53	43.5

Fig. 11-5. Influence of mismatched base pairs on DNA melting temperatures. Oligonucleotides of 11, 14, and 17 bases were hybridised with single-stranded wild-type (wt) DNA or *am3* ΦX174 DNA at 12 °C in 6×SSC. T_d is the temperature at which 50% of the double strands are dissociated. Hybridisation of the three oligonucleotides is more inefficient with *am3* DNA than with wild-type DNA. The 11 b oligonucleotide hybridises so weakly that T_d cannot be determined (Wallace *et al.*, 1979).

of the clones. This is all the more remarkable since it was found that one of the sequences in the β₂mI mixture and the actual cDNA sequence differed by only one base (G instead of T at position 7).

Screening with mixed probes is often problematic because the melting temperature T_d depends both on the length of the probes and their GC content. While the lengths of oligonucleotides in a given probe remains constant in this type of application, their GC contents may be very different. Probe β₂mI (Table 11-1) consists of a pool of oligonucleotides ranging in GC content from 40 to 60% (from six GC to nine GC residues out of a total of 15); a temperature that will allow hybridisation of the probe having the lowest GC content may thus also allow hybridisation of regions as short as 11 bp in the probes with the highest GC content, which may lead to a large number of false positives when a complex library

Table 11–1. Oligonucleotides for the identication of β₂-microglobulin specific cDNA clones.

Amino acid sequence	95 Trp	96 Asp	97 Arg	98 Asp	99 Met
	5'–UGG	GA$_C^U$	AG$_G^A$ CGN	GA$_C^U$	AUG–3'
Probe β₂mI	3'–ACC	CT$_G^A$	TC$_C^T$	CT$_G^A$	TAC–5'
Probe β₂mII	3'–ACC	CT$_G^A$	GCN	CT$_G^A$	TAC–5'
Actual sequence	3'–ACC	CTA	GCT	CTG	TAC–5'

From Suggs *et al.*, 1981.

is screened. It is therefore desirable to have hybridisation conditions which are independent of GC content. This can indeed be accomplished by using tetramethylammonium chloride (Wood *et al.*, 1985) which binds selectively to AT sequences, and thereby raises the melting temperature of a given DNA double strand. At a concentration of 3 M tetramethylammonium chloride, the melting temperature of AT base pairs equals that of GC base pairs. As shown in Table 11-2, the T_d values for fragments between 11 and 46 bp rise almost in a linear fashion from 44 °C to 82 °C despite their different GC contents. At a length of above 200 bp, a limiting T_d of 93 °C is reached. For probes ranging in size from 12 to 25 bp, as are commonly used in library screening, the stringency of hybridisation can thus be based solely on the length of the probe. An example of an application is provided by Ullrich *et al.* (1984b) who used this approach to isolate a full-length cDNA clone encoding the γ-subunit of mouse

nerve growth factor. In this experiment the probe consisted of a pool of 32 oligonucleotides, each 14 nucleotides in length and a GC content varying between 58 and 87%.

In addition to the improved methods for the mixed probe approach, the use of unique long probes which are no longer synthesised as mixtures has become increasingly important (Ullrich *et al.*, 1984a). The synthesis of such probes is planned by taking into consideration several criteria suggesting the most probable codons. The two most important parameters for a successful prediction of codons are the distribution of codon frequencies and the preferential occurrence of dinucleotides (Lathe, 1985). Sequence data accumulated over the past years reveal that synonymous codons for individual amino acids are not used with equal frequencies. Table 11-3 is a compilation of such codon frequencies obtained from the sequences of approximately 50 human genes. Ikemura (1985) and Lathe (1985) have discussed these frequencies and the applicability of this distribution to other systems.

The second parameter facilitating a successful prediction of codons is the distribution of dinucleotides. It has been discussed previously (*cf.* Section 2.1.2.3 and Nussinov, 1981) that this distribution is not random and that the dinucleotide 5'-CpG-3' in particular is underrepresented in eukaryotic DNAs. In the present context the discussion can be restricted to intercodon CG dinucleotides. Intracodon CG dinucleotides are taken care of in Table 11-3 and the previous discussion on codon utilisation. Since the observed intercodon CG dinucleotide frequency is, in fact, less than half of the expected frequency, Lathe (1985) proposed that whenever juxtaposition of optimal codons generates a CG pair, the codon NNC should be replaced by the next-favoured codon in the probe sequence. With the exception of threonine and glycine, this codon would be NNT. Based on these criteria, optimum codons are summarised in Table 11-4. Fig. 11-6 demonstrates the actual consequences for the prediction of codons if these observations are

Table 11–2.   Probe G+C content, length, and T_d in 3M tetramethylammonium chloride (Me₄NCl)

Length (bp)	G+C content %	T_d in 3.0 M Me₄NCl °C
11	45	44.5
13	69	47
15	40	49.5
16	44	53.5
18	44	57.5
27	63	70.5
31	58	75
36	53	77
46	54	82.5
91	66	88
105	31	88
207	55	94
1374	55	94

Probes with lengths of 11 to 16 bp are synthetic oligonucleotides described by Wood *et al.*, 1985; those from 18 to 1374 bp are *Sau*3AI fragments of pBR322 DNA. T_d is defined as the temperature at which 50 % of a double stranded DNA fragment have melted.

taken into consideration. The sequence shown is a stretch of 13 amino acids from tumour necrosis factor together with the corresponding cDNA sequence (line B) (Pennica *et al.*, 1984). Line C shows the DNA sequence which would have been obtained by selecting the most frequently used codons from Table 11-3. Without exception, the four errors in this sequence occur at CpG junctions between two codons. Two of the four "false" bases disappear if less frequently used

Table 11–3. Codon frequencies for coding sequences from human genes.

Amino acid	Codon	Codons found	/1000	Fraction	Amino acid	Codon	Codons found	/1000	Fraction
Gly	GGG	320	13.23	0.22	Trp	TGG	338	13.98	1.00
Gly	GGA	363	15.01	0.25	End	TGA	57	2.36	0.59
Gly	GGT	238	9.84	0.16	Cys	TGT	265	10.96	0.44
Gly	GGC	550	22.75	0.37	Cys	TGC	344	14.23	0.56
Glu	GAG	959	39.66	0.60	End	TAG	8	0.33	0.08
Glu	GAA	641	26.51	0.40	End	TAA	31	1.28	0.32
Asp	GAT	499	20.64	0.41	Tyr	TAT	308	12.74	0.41
Asp	GAC	718	29.70	0.59	Tyr	TAC	436	18.03	0.59
Val	GTG	766	31.68	0.49	Leu	TTG	289	11.95	0.11
Val	GTA	156	6.45	0.10	Leu	TTA	142	5.87	0.06
Val	GTT	249	10.30	0.16	Phe	TTT	422	17.45	0.42
Val	GTC	385	15.92	0.25	Phe	TTC	587	24.28	0.58
Ala	GCG	153	6.33	0.09	Ser	TCG	116	4.80	0.06
Ala	GCA	322	13.32	0.19	Ser	TCA	232	9.60	0.12
Ala	GCT	486	20.10	0.29	Ser	TCT	355	14.68	0.18
Ala	GCC	709	29.32	0.42	Ser	TCC	486	20.10	0.25
Arg	AGG	322	13.32	0.27	Arg	CGG	177	7.32	0.15
Arg	AGA	278	11.50	0.23	Arg	CGA	111	4.59	0.09
Ser	AGT	244	10.09	0.12	Arg	CGT	62	2.56	0.05
Ser	AGC	536	22.17	0.27	Arg	CGC	237	9.80	0.20
Lys	AAG	887	36.68	0.62	Gln	CAG	843	34.86	0.74
Lys	AAA	538	22.25	0.38	Gln	CAA	293	12.12	0.26
Asn	AAT	369	15.26	0.41	His	CAT	256	10.59	0.44
Asn	AAC	534	22.09	0.59	His	CAC	330	13.65	0.56
Met	ATG	613	25.35	1.00	Leu	CTG	1154	47.73	0.45
Ile	ATA	130	5.38	0.14	Leu	CTA	150	6.20	0.06
Ile	ATT	288	11.91	0.30	Leu	CTT	251	10.38	0.10
Ile	ATC	537	22.21	0.56	Leu	CTC	556	23.00	0.22
Thr	ACG	127	5.25	0.09	Pro	CCG	150	6.20	0.12
Thr	ACA	340	14.06	0.25	Pro	CCA	285	11.79	0.24
Thr	ACT	321	13.28	0.24	Pro	CCT	366	15.14	0.30
Thr	ACC	553	22.87	0.41	Pro	CCC	411	17.00	0.34

Compiled by Leslie A. Taylor from sequence data of all human genes (exons) available in Genebank as of March 1985. A more recent update is found in Maruyama et al. (1986).

A)
104   116
  glu thr pro glu gly ala glu ala lys pro trp tyr glu

B)   5'-GAG ACC CCA GAG GGG GCT GAG GCC AAG CCC TGG TAT GAG-3'

C)   5'-GAG ACC CC**C** GAG GG**C** **G**C**C** **G**AG GCC AAG CCC TGG TA**C** **G**AG-3'

D)   5'-GAG ACC CC**T** GAG GG**A** GCT GAG GCC AAG CCC TGG TAT GAG-3'

Fig. 11-6. Use of "long" probes for the screening of libraries.
In this example, the probe was designed from a tryptic fragment of human tumour necrosis factor comprising the codons of amino acids 104 to 116 (Pennica *et al.*, 1984). Line (B) shows the correct gene sequence; line (C) represents a sequence derived at by choosing the most frequent codons (*cf.* Tables 11-3 and 11-4). Bold-face letters indicate CpG dinucleotides, and underlined bases are incorrect choices. Line (D) shows a sequence in which the CpG dinucleotide pairs were removed by choosing the respective next most frequent codons. In this probe, 37 out of 39 bases are correct.

codons without a C in position 3 are used. With only two base changes within 39 nucleotides this oligonucleotide should be an excellent hybridisation probe. It is self-evident that the chances of selecting a correct codon are significantly increased if the peptide sequences in question contain tryptophan and/or methionine codons. In some cases, *i.e.* multi-gene families, it may also be helpful to select codons by taking into consideration a small series of related proteins.

For the characterisation of cDNA libraries it is generally accepted today that hybrid screening with synthetic oligonucleotides is superior to specific priming of cDNA synthesis, especially if the mRNA constitutes a minor portion (less than 1%) of the total mRNA population.

If no pure and specific hybridisation probes are available differential hybridisation techniques may be helpful. In order to determine, for example, which cellular genes are induced after a viral infection, a cDNA library from infected cells is first hybridised with unlabelled mRNA from uninfected cells and then with ^{32}P-labelled mRNA of infected cells. Screening with the two different mRNA populations can also be carried out simultaneously by using two identical sets of filters, instead of one, which have been prepared from the same master plates representing cDNA libraries from either infected or uninfected cells. One set of filters is hybridised to ^{32}P-labelled

David Goeddel
San Francisco

Table 11–4. Summary of certainty values and optimal codon choice for probe sequences deduced from human amino acid sequence data[g]

Amino acid	Certainty factor[a]	Optimal codons[b] when subsequent codon starts with	
		A or C or T	G
Methionine	1.00	A T G	nc
Tryptophan	1.00	T G G	nc
Tyrosine	0.93	T A C	T A T
Cysteine	0.93	T G C	T G T
Glutamine	0.91	C A G	nc
Phenylalanine	0.90	T T C	T T T
Aspartic acid	0.90	G A C	G A T
Asparagine	0.90	A A C	A A T
Histidine	0.89	C A C[c]	C A T
Glutamic acid	0.87	G A G	nc
Lysine	0.85	A A G	nc
Alanine	0.84	G C C	G C T
Isoleucine	0.84	A T C	A T T
Threonine	0.84	A C C	A C A[f]
Valine	0.83	G T G[d]	nc
Proline	0.81	C C C[e]	C C T
Glycine	0.81	G G C	nc[f]
Leucine	0.69	C T G	nc
Arginine	0.65	C G G	nc
Serine	0.60	T C C	T C T

a The certainty factor is the overall % homology predicted for the optimal codon and the actual sequence for each amino acid, multiplied by 0.01. Values have been averaged in the weighting 1:2 for the values generated where the subsequent codon starts with G and values where the subsequent codon starts with another nucleotide, respectively, to allow (approximately) for the relative abundance of such following codons.
b The optimal codon is the most frequent codon for all cases except Arg and Ser, where other triplets generate a higher overall homology.
c CAT when followed by C.
d GTC when followed by T.
e CCA when followed by T.
f These cases do not follow the "replace C by T" rule applied when the subsequent codon starts with G.
g from Lathe (1985)
   nc, no change.

mRNA from infected cells, the other (identical) set to ^{32}P-labelled mRNA from uninfected cells. In this case many plaques will light up on both filters but only a few plaques will hybridise exclusively with one (or the other) mRNA. Genes which are expressed only in uninfected cells or in infected cells can then be identified at first sight by comparing the autoradiographs obtained from both sets of filters. This procedure was used for the identification of transformation-specific genes in cells transformed by SV40 (Schutzbank *et al.*, 1982) and the isolation of DNA sequences which can be induced by galactose in *Saccharomyces cerevisiae* (St. John and Davis, 1979).

A special strategy was employed for the isolation of T cell-specific receptor genes (Hedrick *et al.*, 1984). Such receptors are expressed as heterodimeric disulfide-linked glycoproteins on the surface of T cells. The cloning strategy (*cf.* also Fig. 11-7) is based on the assumptions (a) that T

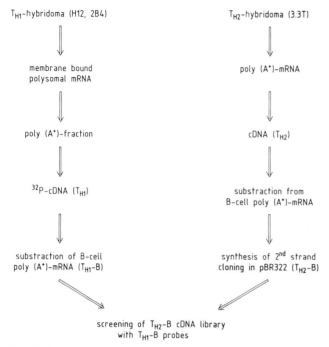

Fig. 11-7. Strategy for cloning T cell specific cDNAs.
Membrane-bound polysomal RNA from one T hybridoma (T_{H1}) was copied into single-stranded ^{32}P-labelled cDNA and repeatedly hybridised with an excess of B cell mRNA. The remaining single-stranded cDNA, T_{H1} minus B (T_{H1}-B) cDNA, was then used as a probe for screening a cDNA library from another T cell hybridoma (T_{H2}) which had likewise been enriched by subtraction with mRNA derived from B cells. Preparative separation of single-stranded DNA from double-stranded RNA/DNA hybrid molecules is performed by hydroxyapatite chromatography (*cf.* Section 2.2.2) (Hedrick *et al.*, 1984).

cell receptor genes are only expressed in T cells, but not in B cells, and (b) that receptor proteins are synthesised by membrane-bound polysomes. Single-stranded cDNA (T_{H1}) prepared from membrane-bound polysomal RNA therefore was hybridised repeatedly with B cell-derived mRNA. Most (61%) of the unhybridised DNA, which itself constituted approximately 2.6% of the input cDNA, was found to be T cell-specific. Since only 15% of the total mRNAs are membrane-bound, the unhybridised fraction comprised only 0.24% of the total cytoplasmic mRNA from T-cells. This value corresponds to an enrichment factor of approximately 400. The enriched cDNA was then screened with a cDNA library from a second T-cell hybridoma (T_{H2}) which had been enriched

approximately 20-fold in T cell-specific sequences by hybridisation with B cell-derived mRNA. Approximately 30 clones out of 5000 were found to be T cell-specific. Of course, this success is mainly due to the fact that T cells and B cells are closely related; in other less related systems application of this technique would hardly make sense.

11.2.2.2 Blotting Techniques

Blotting techniques allow recombinant DNA molecules to be identified and, above all, specific DNA sequences in recombinant DNAs to be mapped and even to be detected in complex mixtures of eukaryotic DNAs. For this purpose DNA fragments are separated in agarose gels according to size, denatured by alkali treatment, and transferred to nitrocellulose filters on which they can be immobilised by heating (Fig. 11-8).

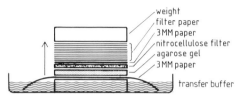

Fig. 11-8. Transfer of DNA or RNA to nitrocellulose filters.
The arrow indicates the flow of buffer solution soaked through the various layers of filter papers by which the transfer of nucleic acids from the gel to the nitrocellulose filter lying on top of it is affected. The speed with which DNA molecules are transferred depends on salt concentration as well as on the size of the DNA. Fragments of up to 1 000 bp require 1-2 hrs in 20×SSC; fragments larger than 12 kb at least 15 hrs (Southern, 1975).

The transfer of DNA to the filters retains the pattern of DNA fragments originally seen in the gel. It is mediated by soaking buffer solution through the gel or by electrophoretic transfer. After the nitrocellulose filters have been incubated with ^{32}P-labelled DNA or RNA probes, hybridising DNA fragments are visualised by autoradiography. The technique was originally developed by Southern (1975) and is generally known as "Southern blotting". It is also possible to transfer RNA or proteins from gels to nitrocellulose filters or chemically activated paper filters, and these techniques are known as "Northern" (Alwine *et al.*, 1977) and "Western" blotting (Symington *et al.*, 1981), respectively. Although RNA and proteins can be transferred and bound to nitrocellulose filters, the use of chemically activated filter papers is preferred since they can be used for several hybridisations (see also Section 11.2.3). Specific bands on Northern filters are detected by hybridisation with ^{32}P-labelled RNA or DNA probes. Proteins can be identified by specifically binding ^{125}I-labelled antibodies.

DNA and RNA probes can be labelled using the enzyme polynucleotide kinase for end-labelling, which involves the transfer of the γ-phosphate residue of γ-^{32}P-labelled ATP to the 5′ hydroxyl end of a DNA or RNA molecule (see Section 2.4.1, Figs. 2.4-3, 2.4-4). For double-stranded DNA, the technique known as nick translation (Rigby *et al.*, 1977) is even more efficient. It exploits the fact that *E. coli* DNA polymerase I is capable of sequentially adding nucleotides to the 3′ terminus of a nick, *i.e.*, an open phosphodiester bond, while it eliminates nucleotides from the 5′ phosphate terminus of this nick. The use of α-^{32}P-labelled deoxynucleoside triphosphates as substrates in the polymerase I reaction yields specific activities on the order of 10^8 cpm per microgram. An interesting alternative uses the biotinylated analogue of dUTP as a substrate (Langer *et al.*, 1981). Biotin has a very high affinity for avidin, a 68 kDa glycoprotein isolated from egg white ($K_{dis} = 10^{-15}$ M; 25 °C). Avidin can be coupled to suitable indicator molecules such as antibodies, enzymes and fluorescent dyes. These complexes allow biotin, and hence nucleic acids, to be detected in minute amounts (Fig. 11-9). Since biotin-labelling is extremely sensitive it is likely to replace ^{32}P-labelling in many applications.

RNA probes labelled to a high specific radioactivity can also be obtained by making use of the so-called SP6 system. SP6 is a *Salmonella typhimurium* bacteriophage which codes for a virus-specific DNA-dependent RNA polymerase (Butler and Chamberlin, 1982). This enzyme consists of a single polypeptide chain with a molecular weight of 96 kDa and is extremely specific for the viral promoters; other bacterial RNA polymerases, such as *E. coli* RNA polymerase, are much less specific and may even initiate transcripts from eukaryotic DNAs. Due to the high specificity and efficiency of SP6 promoters it is possible to produce large amounts of RNA *in vitro* from any DNA sequence cloned in a suitable plasmid vector containing SP6 promoters, as long as the cloned DNA sequences are located downstream

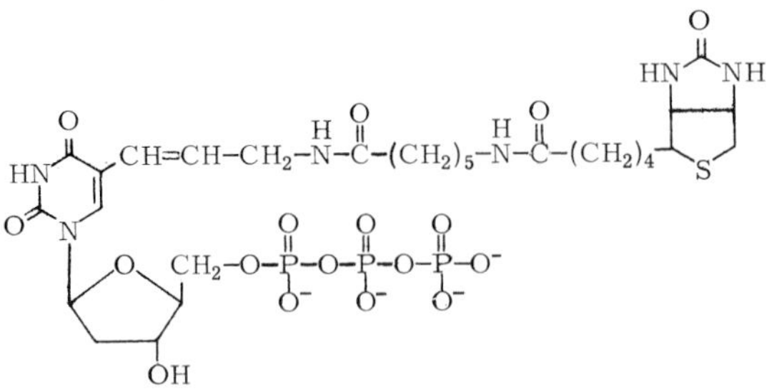

Fig. 11-9. Structure of 5-(*N*-(*N*-biotinyl-ε-aminocaproyl)-3-aminoallyl)deoxyuridine triphosphate.
This compound can replace dTTP in the nick translation reaction of DNA catalysed by DNA polymerase I; in addition, it is a substrate for terminal transferase, T4 DNA polymerase, DNA polymerase α, and herpes simplex type I polymerase, but not for reverse transcriptase. Biotinylated DNA can be detected by a reaction with biotin binding proteins such as avidin (see text), streptavidin, or anti-biotin antibodies coupled with fluorescing dyes or enzymes producing colour reactions (Singer and Ward, 1980).

from these promoters. If the vector DNA downstream from the gene to be transcribed is linearised, one obtains run-off transcripts of defined lengths. Since transcription proceeds conservatively, the synthesised single-stranded RNA corresponds to mRNA or anti-mRNA depending on the orientation of the DNA being transcribed. Under optimal conditions and in the presence of 0.4 mM each of the four ribonucleoside triphosphates, approximately 20 microgram of RNA are synthesised from one microgram of vector DNA within one hour at 40 °C. Since the K_m values for rGTP and rUTP, in particular, are comparatively low, functional and full-length transcripts from inserts of up to 6 kb in length can be obtained at concentrations of only 12 μM. α-^{32}P-labelled undiluted rUTP with a specific activity of 3 000 Ci/mM, for example, thus still yields full-length RNA probes with specific activities of up to 6 × 10^8 cpm/microgram (Krieg and Melton, 1984).

The first vectors described in the literature contained comparatively large SP6 promoter fragments cloned in pBR322 and had only a single restriction enzyme recognition site available for cloning foreign DNA downstream from the promoter (Green *et al.*, 1983). Since then, expression vectors which contain polylinkers located only 5-8 bp away from the SP6 initiation site have been developed. Vectors pSP64 and pSP65, for example, were derived from plasmid pUC12 by substi-

tuting an SP6 promoter fragment for the *lac* α peptide region (Melton *et al.*, 1984) (Fig. 11-10). These constructions provide both orientations of a polylinker with eleven cleavage sites. Presumably, other pUC plasmids will also be adapted as SP6 expression vectors in the future.

Fig. 11-10. Structure of SP6 vectors pSP64 and⟶ pSP65.
These two vectors, derived from pUC12, pUC13, and SP6-specific gene sequences, differ only in the orientation of their polylinkers. In contrast to earlier vectors, the initiation site for transcription lies only a few bases away from the first cleavage site of the polylinker. In order to construct pSP64, pUC12 was linearised with *Nde* I, blunt-ended with Klenow polymerase and then digested with *Hind* III. The large fragment of this digest lacks most of the *lacZ* region, but retains the polylinker (Fig. 7-58). It was joined with a *Sal* I-*Hind* III SP6 promoter fragment (the *Sal* I site of which had been blunt-ended with Klenow polymerase) such that the *Sal* I site of the insert was located adjacent to the *Nde* I site (bold-face). This insert consists of two pieces: (i) a 278 bp segment (hatched bar) derived from pBR322 and flanked by a *Sal* I site at position 651 and the *Bam* HI site at position 375 (the latter site was retained as a *Sau* 3A site) and (ii) a 251 bp segment of SP6 DNA (stippled bar). The *Sau* 3A site as well as a *Nae* I site from pBR322 are included in the sequence section below which, in addition, depicts the SP6-derived sequence up to the polylinker region. Due to the deletion of the *lacZ* region, these vectors a deficient in α-complementation. Numbers in brackets are pUC18 (Appendix B-4), those in parentheses are pBR322 co-ordinates (Melton *et al.*, 1984).

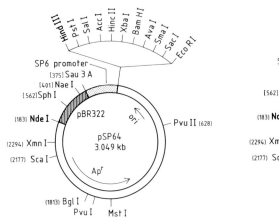

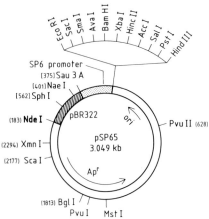

pSP64 Promoter/polylinker region sequence

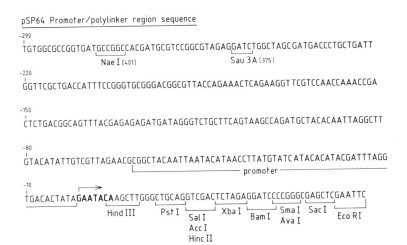

pSP65 Promoter/polylinker region sequence

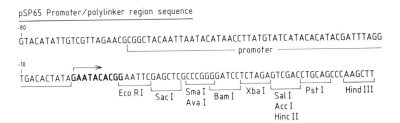

SP6 vectors are particularly useful for the synthesis of highly radioactive RNA probes which can be used, for example, for genomic sequencing or, in general, for Southern and Northern blot hybridisations. Cox *et al.* (1984) used highly labelled RNA probes for localising known mRNAs by *in situ* hybridisation of tissue sections. Subsequently, the insertion of eukaryotic genes, such as the β-globin gene, into SP6 expression vectors facilitated successful *in vitro* synthesis of primary transcripts (*cf.* Section 5.6.1). Such transcripts have allowed the development of cell-free splicing systems, since primary transcripts are required as substrate for these systems, but can be obtained, in minute amounts at best from cells *in vivo*. Constructions similar to SP6 vectors have also been developed for bacteriophage T7 promoters (Tabor and Richardson, 1985; *cf.* Section 7.1.5).

Mapping of RNA, eukaryotic RNA in particular, poses additional questions. The presence or absence of a gene from which a given mRNA is transcribed can be established for example through Northern blot analysis with different DNA probes. Often, however, more complex questions arise, such as whether and where a given RNA is spliced, where it is initiated, and where it terminates. Berk and Sharp (1977) developed a technique (see also Favoloro *et al.*, 1980) for mapping introns. A ^{32}P-labelled double-stranded DNA molecule is denatured and then hybridised with a particular mRNA probe (Fig. 11-11) under conditions which favour the formation of stable RNA-DNA hybrids (80% formamide, 0.4 M NaCl; Casey and Davidson, 1977). If the DNA contains an intron, DNA-mRNA hybrids will possess a single-stranded DNA loop. Such hybrids can be treated with S1 nuclease or exonuclease III. The former cleaves single-stranded DNA and forms 5′ phosphoryl mono- or oligonucleotides, whereas double-stranded DNA and DNA-RNA hybrids remain intact. S1 nuclease treatment would therefore remove the loop and the protruding ends. As shown in Fig. 11-11, the RNA component of these hybrids can be

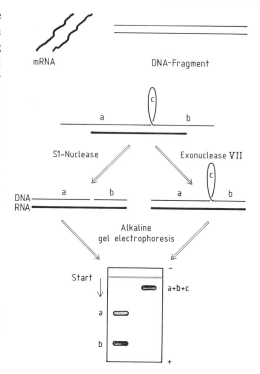

Fig. 11-11.   Berk and Sharp technique for the identification of spliced mRNAs (Berk and Sharp, 1977; *cf.* text for details).

separated from its DNA complement by electrophoresis in denaturing gels. The number of bands obtained depends on the number of introns: if the DNA does not contain introns, there will be no loops, and hence one will observe only a single band; in the case of one intron, digestion of the single DNA loop will produce two DNA fragments (a and b in Fig. 11-11). Intron sizes can be determined by digesting an aliquot of the DNA-RNA hybrid with *E. coli* exonuclease VII, which cleaves only the protruding single-stranded ends of the DNA to 5′ monophosphates, but leaves loops intact. Electrophoresis under denaturing conditions will yield a single band corresponding to the entire region a-b-c. The length of c, *i.e.*, the length of the intron, can be determined by

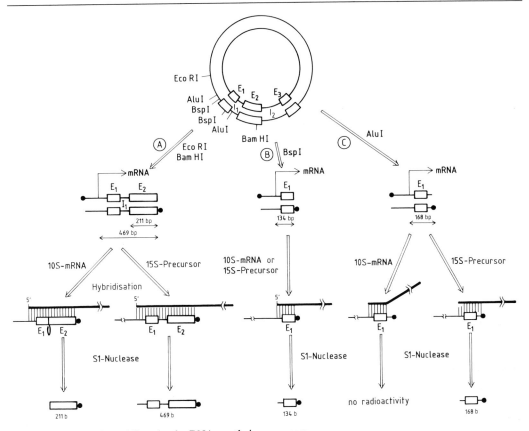

Fig. 11-12.   Mapping of 5' ends of mRNAs or their precursors.
Exons are represented by open bars; introns by solid lines. End-labels are shown as black circles (see text for details;
Weaver and Weissmann, (1979).

substracting the combined lengths of a and b, which are obtained by S1 nuclease treatment.

Weaver and Weissmann (1979) introduced a modification of this procedure, using a DNA probe which is ^{32}P-labelled either at the 3' or at the 5' terminus of one strand (Fig. 11-12). The analysis of the mouse β-globin gene may serve as an example. This gene contains three coding regions separated by two introns; in contrast to mature 10S mRNA, 15S precursor transcripts contain both introns. In order to determine the exact location of the 5' ends of these transcripts, the two transcripts were isolated and hybridised with various DNA probes. Probe A is a DNA fragment of 1 800 bp labelled at the 5' *Bam* HI

end. Denaturation of the probe, followed by hybridisation with unlabelled 10S mRNA (under conditions which favour DNA/RNA hybridisation) and subsequent S1 nuclease treatment yields a single fragment of 211 bases, which covers the region between the *Bam* HI site and the 3' acceptor site of the proximal intron (I_1). Hybridisation with the 15S precursor yields a much larger S1-resistent hybrid molecule of 469 nucleotides, since the precursor also contains the intron regions, and hence does not form an S1-sensitive loop. The 469 nucleotides therefore correspond to the distance between the *Bam* HI site and the 5' end of the 15S precursor. The position of the 5' end of the 10S mRNA can be determined only with an intron-free hybridisation probe (Fig. 11-12; reaction B). A suitable probe in this case is the 210 bp *Bsp* I fragment, probe B, one end of which is located within exon I (E_1). Since both 15S precursor and mature 10S mRNA protect a 134 base-long DNA fragment of this probe, this determines the 5′ ends of each RNA species and shows that both 5′ ends map at the same position.

The same procedure also allows the 5′ ends of mRNA precursors to be mapped precisely if they are present in mixtures with mature mRNA. In comparison to mRNA itself, such precursors may be quite rare, and cannot be recognised easily in an excess of mRNA. The problem can be solved, however, by using a precursor-specific hybridisation probe (Fig. 11-7; reaction C). Such a probe is a 5′ end-labelled *Alu* I fragment, one terminus of which is located within intron I_1. This 5′ label cannot be protected by hybridisation to mRNA, therefore S1-treated mRNA/DNA hybrids will not contain label; however, the mRNA precursor, containing intron sequences, protects the labelled 5′ ends of the DNA fragment so that a base-paired RNA/DNA hybrid fragment of 168 bp remains after S1 digestion.

Labelled S1 resistant RNA-DNA hybrids are preferentially analysed on denaturing polyacrylamide gels of the type used for DNA sequencing (see also Section 2.4.1). If Maxam-Gilbert cleavage products of a DNA probe with the same sequence are loaded simultaneously in a separate slot, the exact location of the cap site can be read immediately from the autoradiograph. This procedure offers an additional advantage in that the direction of transcription can be recognised easily, since only one of the two complementary DNA strands will hybridise with the RNA probes.

11.2.2.3 Chromosome Walking

Genes are identified in libraries by using specific nucleic acid probes. Frequently, however, genes are much longer than a cDNA clone, and thus cannot be obtained as a single insert in one phage or even one cosmid clone. It may also be of interest to identify sequences in the vicinity of a certain gene in order to map whole chromosomal regions, for example, the complex locus comprising the genes for the histocompatibility antigens. If these regions are larger than the usual inserts of a genomic library (*i.e.*, larger than 20-45 kb), a procedure called chromosome walking is the method of choice for physical and genetic characterisation. Screening a library with a known and desired probe and identifying clones which overlap with this probe contitutes a single step in this walk. The overlapping inserts are mapped and those extending farthest in either direction are used as new points of reference, *i.e.*, they are used as probes for screening the library to identify other clones extending further to the left or right. Such steps are repeated until a set of clones has been obtained which cover the entire gene or the desired region of the chromosome. As shown in Fig. 11-13, probe 1 identifies clones λ1 and λ2. Probe 2, derived from λ2, was used to identify clone λ3. A probe derived from the 3′ end of λ3 would yield λ4 and so on. Of course, the same strategy can also be employed for walking towards the 5′ end. In cases of chromosomal rearrangements in the region of interest, it is also possible to jump, rather than to walk, in steps corresponding to the length of a λ or cosmid insertion. Part B of Fig. 11-13 shows a region similar to that in part A; however, the central portion has been inverted. In a library containing this rearrangement, probe 4 will, therefore, identify a new clone, λ5, which covers region A and an additional new chromosomal region which lies in the vicinity of region A after its rearrangment. Probe 5, from the 3′ end of the λ5 insert, then can be used as described above for stepwise chromosome walking. In the example scheme outlined in Fig. 11-13 the jump is clearly discernable; in practice, however, such a jump may end in a distinct chromosomal region or even on a new chromosome, if a translocation has taken place.

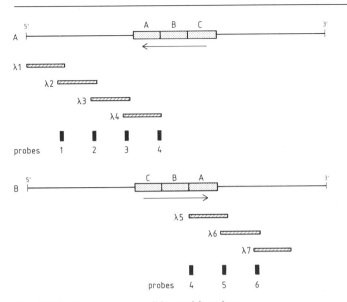

Fig. 11-13.   Chromosome walking and jumping.
(A) and (B) show two sections of a chromosome with genes *A, B, C,* and their flanking regions. Probes are short DNA fragments obtained from the 3′ or 5′ ends of λ clones, which allow the identification of neighbouring sequences. See text for details.

By using a combination of steps and jumps, Bender *et al.* (1983) covered a region of 315 kb from chromosome 3 of *Drosophila melanogaster.* Gitschier *et al.* (1984) starting with a centrally located probe, required 3-4 walking steps in both the 3′ and 5′ directions in order to cover the gene for human clotting factor VIII. A region of approximately 10^6 bp from the major histocompatibility complex of mice was mapped by Steinmetz *et al.* (1982), using cosmid clones for chromosome walking.

Since specific chromosomal rearrangements may not always be available, another method of chromosomal "jumping" has been developed. This strategy involves the isolation of "junction fragments" representing the termini of a large DNA segment (Collins and Weissman, 1984; Poustka and Lehrach, 1986; *cf.* Science (1985) 228, 108). Chromosomal DNA is digested into pieces of 100 kb length and larger. These pieces are then circularised in the presence of a linearised pUC- or ΠAN7-type cloning vector. Clones, however, are not isolated immediately, but only

after the circular molecules have been digested with an enzyme which does not cleave the vector. The latter is thus left with an insert containing both ends, but missing the central portion of a large chromosomal segment. These junction fragment inserts are small enough to be cloned. In a junction fragment library, a probe for one end of a large DNA segment then will identify clones carrying both ends, even though the second end may lie more than 100 kb away. At present the problems associated with preparing and handling high molecular weight DNA make it difficult to create such junction fragment libraries. Once these problems are solved though, this strategy will permit the rapid walks along chromosomes required for the identification of defective genes in human genetic diseases (*cf.* Section 2.1.2.6).

11.2.3 Detection of Specific Proteins

Specific DNA sequences in a mixture of recombinant DNA molecules also can be identified on the basis of their protein products. In principle, the problem can be approached in two different ways. Firstly, recombinant DNA can be used for the selection and isolation, by hybridisation, of specific mRNAs, the products of which can be analysed in an *in vitro* protein synthesis system. Alternatively, the desired protein product can be detected by immunological techniques in cellular extracts of cells replicating and expressing the recombinant DNA.

11.2.3.1 In vitro Translation Techniques

A variety of *in vitro* techniques which allow eukaryotic mRNAs to be translated into the corresponding proteins are known to molecular biologists (see also Section 2.3). The mRNA must be pure and uncomplexed since DNA-RNA hybrids cannot be translated in such *in vitro* systems. This observation can be used as the basis of a test system for cDNA clones. A population of mRNA from which the cDNA clones to be tested were derived is incubated with a mixture of recombinant plasmid DNAs. If these plasmid DNAs contain sequences which are complementary to an mRNA species in the mRNA mixture this mRNA will be converted into DNA-RNA hybrids. In contrast to uncomplexed mRNA molecules such hybrids cannot direct the translation of proteins in an *in vitro* translation system. This technique, which is known as "hybrid arrested translation", provides a negative screening method, since a cDNA clone is detected because the desired protein cannot be obtained *in vitro* (Fig. 11-14). Experimental efforts are simplified if the first hybridisation reactions are carried out with mixtures of plasmid DNA. DNA mixtures which inhibit translation then can be subdivided until the desired clones are identified. In practice this approach usually requires large amounts of mRNA, and it is therefore necessary to use mRNA populations which have already been enriched for the desired mRNA.

Another technique which is more sensitive than hybrid arrested translation is based on the same principles, but exploits the presence, rather than the absence, of an active mRNA species for the identification of clones (Fig. 11-15). This positive screening procedure is known as "hybrid selected translation" (Ricciardi *et al.*, 1979). Purified

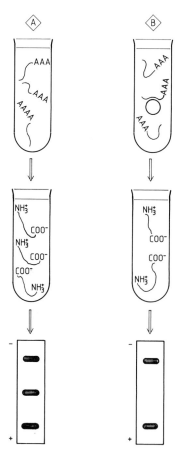

Fig. 11-14. Detection of cDNA clones by inhibition of *in vitro* translation (hybrid-arrested translation). (A) shows three mRNA molecules whose *in vitro* translation products are identified by gel electrophoresis. In (B), plasmid DNA with sequences complementary to one of these mRNAs is added. Under certain reaction conditions RNA/DNA hybrids are formed which are inactive in the *in vitro* translation system. Subsequent electrophoresis of the translation products yields two bands only.

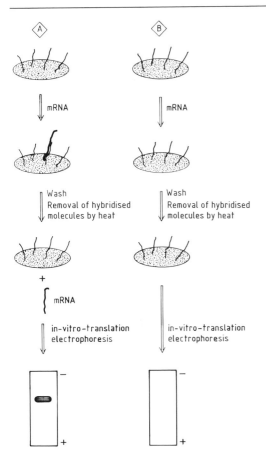

Fig. 11-15.   Identification of cDNA clones by hybrid selection.
Mixtures of plasmid DNAs are bound to nitrocellulose filters and incubated with a suitable mRNA preparation. If DNA complementary to mRNA is bound to the filter, as shown in (A), the mRNA will hybridise. As shown in (B), mRNA does not hybridise if complementary DNA is not present on the filter. Hybridised mRNA (A) can be removed from the filter by heating and can then be used for *in vitro* translations.

plasmid DNA derived from a mixture of cDNA clones is first denatured and then bound to nitrocellulose filters. These filters are incubated with an mRNA population which contains the desired mRNA. Such RNA molecules hybridise only to filters with complementary cDNA clones. After hybridisation the unhybridised mRNA is

washed out; hybridised mRNA is eluted from the filters by heating, and can be translated *in vitro*. The desired products are precipitated by appropriate antibodies and are identified autoradiographically after gel electrophoresis.

There are several modifications of the hybrid selected translation technique. Parnes *et al.* (1981) were able to identify several cDNA clones of mouse β_2-microglobulin using this technique. In this case, the mRNA of the β_2-microglobulin amounted to only 0.03% of the available total mRNA population. Nagata *et al.* (1980) also used filter-bound DNA for the selection of mRNA and the identification of cDNA clones of leukocyte interferon. However, the *in vitro* translated product was finally identified by its biological activity rather than by immunoprecipitation; the eluted mRNA was injected into *Xenopus laevis* oocytes and cellular extracts were analysed in a virus inhibition test. Interferons inhibit the multiplication of certain viruses to such an extent that even the minute amounts of biologically active material obtained from an *in vitro* translation experiment are sufficient to identify positive clones. When these experiments were carried out, nothing was known about the stucture of interferons, and although interferon production was known to be stimulated by certain inducers, extremely low mRNA concentrations had to be expected. The successful isolation and identification of leukocyte interferon cDNA is an outstanding example of cloning a gene when nothing is known about the structure of its product, and when cloning depends entirely upon a biological test system.

DNA probes which are used for the selection of mRNA can be immobilised on nitrocellulose or activated paper filters. Cellulose filters offer a considerable advantage in that they can be used several times since the DNA is bound covalently and irreversibly; in contrast, DNA binds to nitrocellulose by hydrophobic forces which may be weakened under hybridisation conditions. Two techniques are available for the derivatisation of cellulose papers. As shown in Fig. 11-16,

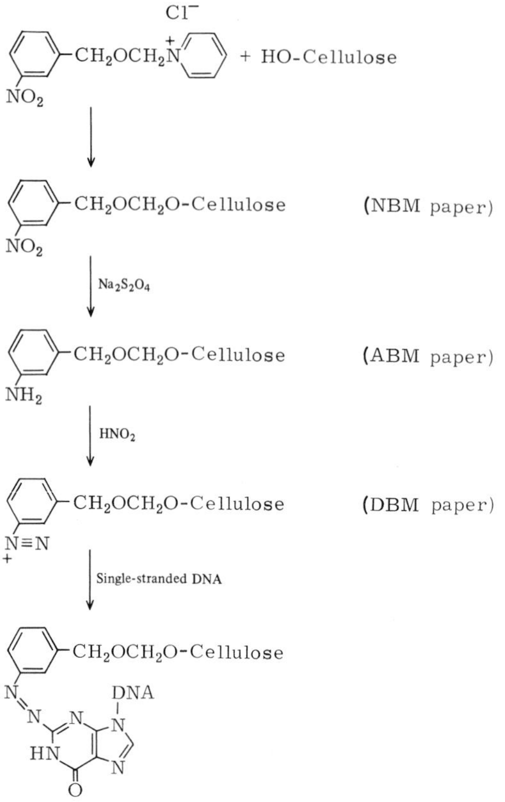

$$\text{NO}_2\text{-}\langle\bigcirc\rangle\text{-CH}_2\text{OCH}_2\overset{+}{\text{N}}\langle\bigcirc\rangle \quad \text{Cl}^- \quad + \text{ HO-Cellulose}$$

↓

CH₂OCH₂O-Cellulose   (NBM paper)

$$\Big\downarrow \text{Na}_2\text{S}_2\text{O}_4$$

CH₂OCH₂O-Cellulose   (ABM paper)

$$\Big\downarrow \text{HNO}_2$$

CH₂OCH₂O-Cellulose   (DBM paper)

$$\Big\downarrow \text{Single-stranded DNA}$$

CH₂OCH₂O-Cellulose

Fig. 11-16. Derivatisation of cellulose by diazobenzyloxymethyl groups (Alwine *et al.*, 1977).

cellulose reacts with m-nitrobenzyloxymethyl pyridinium chloride (NBM) to yield m-nitrobenzyloxymethyl cellulose (NBM paper) (Alwine *et al.*, 1977). Reduction with dithionite yields the m-amino derivative (ABM paper) which in turn can be converted to the diazobenzyloxymethyl paper (DBM paper) in the presence of nitrous acid. The diazonium salt is capable of reacting with single-stranded DNA by covalently binding to guanosine residues. A more recent development is the derivatisation with 1,4-butandiol-diglycidyl ether (Fig. 11-17) which reacts with cellulose to form an oxirancellulose. Coupling with o-aminothiophenol yields APT cellulose, the

diazotised form of which reacts with single-stranded DNAs as described above (Seed, 1982).

11.2.3.2 Immunological Methods

The direct detection of an expected protein product offers the advantage that the desired recombinant DNA molecules need not possess a particular selectable phenotype. However, this approach must be regarded as a more ambitious alternative to the various test systems for recombinant DNA. It does not only require the presence of a recombinant DNA molecule, but also asks that this DNA meet all the requirements concerning structural organisation and orientation which are a prerequisite for efficient expression. Furthermore, random cloning in an expression vector will yield only one in six clones with the correct orientation and reading frame. Finally, the protein can be detected only if a biological test, or at least a specific antibody, is available.

There are two different immunological techniques where the antigens in bacterial extract are either bound to one or to two antibody molecules (Fig. 11-18). If two antibodies are used (Ehrlich *et al.*, 1978; Broome and Gilbert, 1978; Clarke *et al.*, 1979) the first antibody which binds the desired polypeptides in the bacterial extract, *i.e.*, the antigens, is bound to a solid support such as chemically activated papers (Fig. 11-18A). These complexes are then incubated with a second antibody (recognising the same antigen) which either is ^{125}I-labelled itself, or is complexed with ^{125}I-labelled protein A from *Staphylococcus aureus*. Protein A binds to the F_c regions of many, but not all antibodies. This technique has been termed appropriately "sandwich technique" since the antigen lies between two antibody molecules. The reaction of an antigen with two antibodies is unproblematic because an antigen usually possesses several antigenic determinants which can be recognised by the commonly used polyclonal antisera.

$$CH_2^{\frown}CHCH_2O(CH_2)_4OCH_2CH^{\frown}CH_2 \;+\; HO\text{-Cellulose}$$

$$CH_2^{\frown}CHCH_2O(CH_2)_4OCH_2\overset{OH}{\underset{|}{C}}HCH_2O\text{-Cellulose}$$

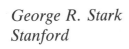

$$\overset{OH}{\underset{|}{}}\qquad\overset{OH}{\underset{|}{}}$$
$$\text{-SCH}_2CHCH_2O(CH_2)_4OCH_2CHCH_2O\text{-Cellulose}$$
$$\underset{NH_2}{}\qquad\qquad\qquad\qquad\text{(APT-Paper)}$$

| HNO₂

$$\overset{OH}{\underset{|}{}}\qquad\overset{OH}{\underset{|}{}}$$
$$\text{-SCH}_2CHCH_2O(CH_2)_4OCH_2CHCH_2O\text{-Cellulose}$$
$$\underset{\overset{N\equiv N}{+}}{}\qquad\qquad\qquad\qquad\text{(DPT-Paper)}$$

Fig. 11-17. Derivatisation of cellulose by O-aminophenylthioethers (Seed, 1982).

The other technique (Fig. 11-18B) is known as a direct immune assay and requires only one antibody (Kemp and Cowman, 1981a, b). In this case, bacterial colonies are lysed *in situ* and the extract containing the antigen is bound covalently to a carrier, usually a paper activated with cyanogen bromide. The antigen is detected as described above, using specific antibodies and autoradiography; again radioactively labelled antibodies or labelled Protein A molecules can be used. In order to avoid nonspecific reactions of the antiserum, it may be helful to purify the antiserum by pre-incubation with extracts made from plasmid-free bacteria, *i.e.* with cells which make bacterial proteins, but not the antigen in question.

George R. Stark
Stanford

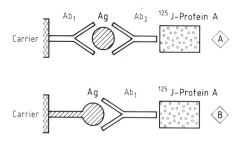

Fig. 11-18. Strategies for *in-situ* detection of antigenic polypeptides.
(A) shows the sandwich technique which employs two antibodies, Ab_1 and Ab_2 directed against the same antigen (Ag). The technique shown in (B) employs only one antibody (Ab_1). Antigens or antibodies can be bound to plastic-coated supports or directly to chemically activated papers (Clarke *et al.*, 1979).

Both techniques have been used successfully for the detection of recombinant bacterial clones expressing, for example, rat insulin and mouse IgM polypeptides. Expression vectors often produce fusion peptides sharing components of a vector protein and the desired protein product. This may prove problematic, since antisera typically are raised against purified proteins, and thus are less active against fusion proteins. Therefore, it may be more difficult to identify positive clones.

It is quite remarkable that the number of positive clones observed is often higher than that predicted on the basis of the possible orientations and reading frames of the insert. One explanation that has been suggested is that internal initiation allows protein synthesis to begin at the junction of vector and insertion. Chang *et al.* (1978), for example, found that the enzyme dihydrofolate reductase is also expressed from constructions in which the cDNA insert is not in phase with the β-lactamase gene of pBR322. The haemagglutinin gene of influenza virus is expressed even when its orientation in the expression vector is reversed (Emtage *et al.*, 1980). In this case it must be postulated that there is not only an internal initiation of protein synthesis, but also a new promoter driving transcription. Such unexpected

initiation processes increase not only the yield of positive clones, but also the sensitivity of the immunological test system, since original rather than fusion proteins are formed. It has been mentioned that antisera may recognise these original proteins much better than fusion proteins. Expression vectors which will allow such unpredictable variables to be controlled probably will be available in the future.

11.2.3.3 Protein Synthesis in Mini-cells

Plasmid- or phage-encoded gene products also can be identified in the mini-cells produced by an atypical cell division near one of the poles of a bacterial cell. Mini-cells are small and almost spherical in shape and do not contain any bacterial chromosomal DNA. Due to their small size, mini-cells can be easily separated from normal bacterial cells, for example, by sucrose gradient centrifugation (Frazer and Curtiss III, 1975). Mini-cells obtained from *E. coli* and *B. subtilis* have proven particularly well-suited to the analysis of the biological properties of high copy number plasmids, since the DNA of such plasmids segregates easily into mini-cells. Plasmids in mini-cells show a variety of macromolecular activities which do not differ from those observed in normal cells, *e.g.*, protein biosynthesis and transcription. Since the chromosomal DNA is absent mini-cells offer the advantage that all RNA and protein biosynthesis in such cells is directed exclusively by the plasmid. It is, therefore, possible to identify and characterise the gene products of recombinant DNA molecules by analysing the newly synthesised proteins on suitable gels. Mini-cells obtained from *E. coli* and *B. subtilis* can also be infected by phages (Reeve, 1979). Although phage DNA is not replicated, it does direct the synthesis of RNA and proteins. Extracts made from mini-cells pulsed with radioactive amino acids after phage infection contain label in phage-encoded proteins only; the confusing background of labelled host proteins is absent from these preparations. The rate of synthesis of virus-encoded polypeptides is lower than that obtained in intact infected bacterial cells, but since viruses are not replicated in mini-cells, these cells are not lysed and protein synthesis may proceed for several hours. In addition, mini-cells offer the advantage of possessing normal bacterial cell walls and cellular membranes, and it is therefore possible to study the cellular distribution of phage-encoded polypeptides. One of the most frequently used mini-cell strains of *E. coli* is strain DS410 originally developed by Sherratt. A widely used mini-cell system for *B. subtilis* is strain CH403 which does not sporulate; it would be difficult to distinguish between spores and genuine mini-cells.

11.2.3.4 Maxi-Cell Technique

Genes encoded by recombinant plasmids can also be identified by employing so-called maxi-cells, which are *recA uvrA* double mutants of *E. coli* (Sancar *et al.*, 1981). The *uvrA* gene, in conjunction with genes *uvrB* and *uvrC*, codes for an endonuclease activity (Lindahl, 1982). This multienzyme complex eliminates pyrimidine dimers (and other UV-induced damage) from DNA molecules, thus initiating the repair process. If cells lacking this enzyme are irradiated with UV light, degradation and inactivation of chromosomal DNA occurs and cell death ensues. DNA molecules from a multicopy plasmid are, of course, also damaged, but provide a smaller target, and accumulate fewer lesions per molecule than does the bacterial chromosome. Under suitable conditions, only the plasmid DNA will survive, and plasmid-encoded proteins will be labelled exclusively in the presence of ^{35}S-methionine. These can be detected, *e.g.*, by SDS gel electrophoresis. One example of the application of the maxi-cell method is the detection of the products of adenovirus genes *EIa* and *EIb* cloned in suitable expression vectors (Ko and Harter, 1984).

11.2.3.5 Exon Cloning

Protein molecules usually possess several antigenic determinants. It is, therefore, not necessary to identify intact gene products in recombinant bacterial colonies; in fact, it is possible to detect small coding regions by identifying immunologically the protein fragments for which they code. One important application of this principle is cloning of exon regions of eukaryotic genes. The average size of such regions is 100-200 bp, yet they may code for several antigenic determinants. This technique could also be important for screening cDNA libraries since cDNA clones frequently do not represent the full length gene. In some cases antibodies for the desired protein may be available, but there may be no sequence data allowing the synthesis of suitable oligonucleotides.

Rüther *et al.* (1982) and Koenen *et al.* (1982) have developed a technique which allows short regions of coding DNA to be identified by inserting such sequences into the *N*-terminal region of β-galactosidase, which is not required for enzymatic activity (see also Section 7.4.2.1, Fig. 7-63). A suitable starting vector is pUK230 (Fig. 11-19) which carries a polylinker insertion destroying the reading frame of the *lacZ* gene between amino acids 5 and 6. The phenotype of bacterial colonies harbouring this vector is Lac⁻, and these cells form white colonies on agar plates containing IPTG and Xgal. If a coding DNA fragment without a stop codon is inserted into the polylinker, the reading frame is restored. Transformants are therefore Lac⁺ and form blue colonies on IPTG/Xgal plates. The histochemical detection of β-galactosidase *via* the substrate Xgal is also used for an immunological test (Fig. 11-20) which allows specific exons to be identified. PVC sheets are first coated with antibodies against the expected antigen and are then allowed to bind proteins from a bacterial extract. Only the specific antigen-β-galactosidase fusion proteins will bind to PVC sheets. These sheets receive a top layer of agar containing Xgal. A lysate of recombinant bacteria expressing the desired exon is identified

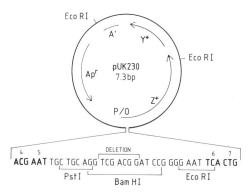

Fig. 11-19. Structure of vector pUK230.
This plasmid is approximately 7.3 kb in length and contains almost the entire *lac* operon, namely, the promoter/operator region *(P/O)*, the *lacZ* gene (Z⁺), the *lacY* gene (Y⁺), and a section of the *lacA* gene (A′). A polylinker with cleavage sites for *Pst*I, *Bam*HI, and *Eco*RI is inserted between the codons for amino acids 5 and 6 of β-galactosidase. The *Bam*HI site is generated by deleting 7 bp (indicated by a bracket) from a suitable precursor plasmid. The deletion destroys the *lacZ* reading frame. *lacZ⁻* bacteria transformed by pUK230 are therefore Lac⁻ and can be identified as white colonies on indicator agars containing the chromogenic dye Xgal. Insertions of exons which restore the reading frame of the *lacZ* gene lead to blue colonies (Koenen *et al.*, 1982).

by the blue colour of the sheet to which it had bound. Rüther *et al.* (1982) were able to identify one specific clone of chicken lysozyme gene fragment among 10⁵ Lac⁺ colonies. This selectivity should also allow eukaryotic gene libraries to

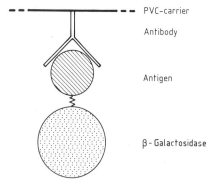

Fig. 11-20. Binding of antigen-β-galactosidase fusion proteins to a fixed antibody (Koenen *et al.*, 1982).

be screened for antigenic determinants, despite the excess of non-coding DNA in such libraries. Insertions in the polylinker region cannot be used for the identification of cDNA clones, since these always carry a stop codon at the end of the coding portion. Cloning vectors have been constructed which permit the insertion of cDNAs into the 3′ end of the *lacZ* gene (Rüther and Müller-Hill, 1983); in this case the screen is for Lac⁻ bacteria.

Exon cloning can also be carried out with phage expression vectors (Young and Davis, 1983a and b; Huynh *et al.*, 1985). In one of these vectors, λgt11 (*lac5 nin5 cI857 S100*) (*cf.* Fig. 4.2-8B), foreign DNA – *e.g.* cDNA; *cf.* Lapeyre and Amalric (1985) for a powerful cDNA synthesis procedure – can be cloned into the *Eco*RI site immediately upstream from the stop codon for the *lacZ* gene. The insertion yields an inactive β-galactosidase fusion protein. The original procedure was designed to permit screening of the expressed antigen in recombinant lysogens. After packaging, plaques are adsorbed to *E. coli hflA* mutant strains under conditions leading to high frequency lysogenisation. Lysogens can be heat-induced at 45 °C by inactivating the temperature-sensitive cI857 repressor. The induced colonies are lysed with chloroform vapour and screened with antibody. Unfortunately, some λgt11 recombinants fail to lysogenise even in *hflA* cells. It is thus preferable and strongly advisable to screen λgt11 recombinant plaques rather than lysogens (bacterial colonies).

After *in vitro* packaging, the recombinant phages are plated on a sensitive *E. coli* strain which preferably should be lon protease-deficient (*e.g.* Y1090). The absence of the lon protease (or protease La) increases the stability of foreign or abnormal proteins in *E. coli* (*cf.* Section 7.5). Since the available lon protease-deficient *E. coli* strains are not defective for host-controlled restriction and modification the initial adsorption and amplification of the packaging mix is performed on a hsdR⁻ hsdM⁺ strain (Y1088). This latter strain also overproduces *lac* repressor which represses the production of potentially toxic polypeptides. Infected cells are incubated at 37 °C in the presence of IPTG in order to induce *lacZ*-directed expression. Recombinants can be identified at this stage as white plaqes in the presence of Xgal and, in general, represent between 10 – 15% of the phage population. For antibody screening, the plates are overlaid with nitrocellulose filters to adsorb the antigen. Filters are then treated consecutively with purified IgG preparations from mono- or polyclonal sera and with ^{125}I-labelled protein A. It is possible to screen approximately 25 000 phage plaques per 8.5 cm plate. Recent examples include the cloning of cDNAs for rat haem oxygenase (Shibahara *et al.*, 1985), firefly luciferase (De Wet *et al.*, 1985), and human estrogen receptor (Walter *et al.*, 1985).

References

Alwine, J.C., Kemp, D.J., and Stark, G.R. (1977). Method for detection of specific RNAs in agarose gels by transfer to diazobenzyloxy-methyl paper and hybridization with DNA probes. Proc. Natl. Acad. Sci. USA 74, 5350-5354.

Auerswald, E.A., Ludwig, G.R., and Schaller, H. (1981). Structural analysis of Tn5. Cold Spring Harbor Symp. Quant. Biol. 45, 107-113.

Barnes, W.M. (1977). Plasmid detection and sizing in single colony lysates. Science 195, 393-394.

Beidler, J.L., Hilliard, P.R., and Rill, R.L. (1982). Ultrasensitive staining of nucleic acids with silver. Anal. Biochem. 126, 374-380.

Bender, W., Spieser, P., and Hogness, D.S. (1983). Chromosomal walking and jumping to isolate DNA from the *Ace* and *rosy* loci and the Bithorax Complex in *Drosophila melanogaster*. J. Mol. Biol. 168, 17-33.

Benton, W.D., and Davis, R.W. (1977). Screening λgt recombinant clones by hybridization *in situ*. Science 195, 180-182.

Berk, A.J., and Sharp, P.A. (1977). Sizing and mapping of early adenovirus mRNAs by gel electrophoresis of S1 endonuclease digested hybrids. Cell 12, 721-732.

Birnboim, H.C., and Doly, J. (1979). A rapid alkaline extraction procedure for screening recombinant plasmid DNA. Nucleic Acids Res. 7, 1513-1523.

Bonner, T.I., Brenner, D.J., Neufeld, B.R., and Britten, R.J. (1973). Reduction in the rate of DNA reassociation by sequence divergence. J. Mol. Biol. 81, 123-135.

Boyko, W.L., and Ganschow, R.E. (1982). Rapid identification of *Escherichia coli* transformed by pBR322 carrying inserts at the *Pst*I site. Anal. Biochem. 122, 85-88.

Broome, S., and Gilbert, W. (1978). Immunological screening method to detect specific translation products. Proc. Natl. Acad. Sci. USA 75, 2746-2749.

Butler, E.T., and Chamberlin, M.J. (1982). Bacteriophage SP6-specific RNA polymerase. I. Isolation and characterization of the enzyme. J. Biol. Chem. 257, 5772-5778.

Casey, J., and Davidson, N. (1977). Rates of formation and thermal stabilities of RNA:DNA and DNA:DNA duplices at high concentrations of formamide. Nucleic Acids Res. 4, 1539-1552.

Chang, A.C.Y., and Cohen, S.N. (1978). Construction and characterization of amplifiable multicopy DNA cloning vehicles derived from the P15A cryptic miniplasmid. J. Bacteriol. 134, 1141-1156.

Chang, A.C.Y., Nunberg, J.H., Kaufman, R.J., Erlich, H.A., Schimke, R.T., and Cohen, S.N. (1978). Phenotypic expression in *E. coli* of a DNA sequence coding for mouse dihydrofolate reductase. Nature 275, 617-624.

Clarke, L., and Carbon, J. (1976). A colony bank containing synthetic hybrid plasmids representative of the entire *E. coli* genome. Cell 9, 91-99.

Clarke, L., Hitzeman, R., and Carbon, J. (1979). Selection of specific clones from colony banks by screening with radioactive antibody. Methods in Enzymol. 68, 436-442.

Collins, J., Volckaert, G., and Nevers, P. (1982). Precise and nearly-precise excision of the symmetrical inserted repeats of Tn5; common features of *recA*-independent deletion events in *Escherichia coli*. Gene 19, 139-146.

Collins, F.S., and Weissman, S. (1984). Directional cloning of DNA fragments at a large distance from an initial probe: A circularization method. Proc. Natl. Acad. Sci. USA 81, 6812-6816.

Cox, V.H., Deleon, D.V., Angerer, L.M., and Angerer, R.C. (1984). Detection of mRNAs in sea urchin embryos by *in situ* hybridization using asymmetric RNA probes. Dev. Biol. 101, 485-502.

Davison, J., Brunel, F., and Merchez, M. (1979). A new host vector system allowing selection for foreign DNA inserts in bacteriophage λgtWES. Gene 8, 69-80.

De Wet, J.R., Wood, K.V., Helinski, D.R., and Deluca, M. (1985). Cloning of firefly luciferase cDNA and the expression of active luciferase in *Escherichia coli*. Proc. Natl. Acad. Sci. USA 82, 7870-7873.

Ehrlich, M., Cohen, S., and McDevitto (1978). A sensitive radioimmunoassay for detecting products from cloned DNA fragments. Cell 13, 681-689.

Emtage, J.S., Tacon, W.C.A., Catlin, G.H., Jenkins, B., Porter, A.G., and Carey, N.H. (1980). Influenza antigenic determinants are expresed from haemagglutinin genes cloned in *Escherichia coli*. Nature 283, 171-174.

Favaloro, J., Treisman, R., and Kamen, R. (1980). Transcription maps of polyoma virus-specific RNA: Analysis by two-dimensional nuclease S1 gel mapping. Methods in Enzymol. 65, 718-749.

Frazer, A.C., and Curtiss, III.R. (1975). Production, properties and utility of bacterial minicells. Current Topics Microbiol. Immunol. 69, 1-84.

Gitschier, J., Wood, W.I., Goralka, T.M., Wion, K.L., Chen, E.Y., Eaton, D.H., Vehar, G.A., Capon, D.J., and Lawn, R.M. (1984). Characterization of the human factor VIII gene. Nature 312, 326-330.

Green, M.R., Maniatis, T., and Melton, D.A. (1983). Human β-globin pre-mRNA synthesized *in vitro* is accurately spliced in *Xenopus* oocyte nuclei. Cell 32, 681-694.

Grunstein, M., and Hogness, D.S. (1975). Colony hybridisation: a method for the isolation of cloned DNAs that contain a specific gene. Proc. Natl. Acad. Sci. USA 72, 3961-3965.

Grunstein, M., and Wallis, J. (1979). Colony hybridization. Methods in Enzymol. 68, 379-389.

Hagan, C.E., and Warren, G.J. (1982). Lethality of palindromic DNA and its use in selection of recombinant plasmids. Gene 19, 147-151.

Hanahan, D., and Meselson, M. (1980). Plasmid screening at high colony density. Gene 19, 147-151.

Hedrick, S.M., Cohen, D.I., Nielsen, E.A., and Davis, M.M. (1984). Isolation of cDNA clones encoding T cell-specific membrane-associated proteins. Nature 308, 149-153.

Hennecke, H., Günther, I., and Binder, F. (1982). A novel cloning vehicle for the direct selection of recombinant DNA in *E. coli*. Gene 19, 231-234.

Holmes, D.S., and Quigley, M. (1981). A rapid boiling method for the preparation of bacterial plasmids. Anal. Biochem. 114, 193-197.

Huynh, T.V., Young, R.A., and Davis, R.W. (1985). Construction and screening cDNA libraries in λgt10 and λgt11. In "DNA Cloning Techniques: A practical approach"; D. Glover, ed.; IRL Press, Oxford.

Ikemura, T. (1985). Codon usage and tRNA content in unicellular and multicellular organisms. Mol. Biol. Evol. 2, 13-34.

Kemp, D.J., and Cowman, A.F. (1981a). Detection of

expressed polypeptides by direct immunoassay of colonies. Methods in Enzymol. 79, 622-630.

Kemp, D.J., and Cowman, A.F. (1981b). Direct immunoassay for detecting *Escherichia coli* colonies that contain polypeptides encoded by cloned DNA segments. Proc. Natl. Acad. Sci. USA 78, 4520-4524.

Ko, J.L., and Harter, M.L. (1984). Plasmid-directed synthesis of genuine adenovirus 2 early-region 1A and 1B proteins in *E. coli*. Mol. Cell. Biol. 4, 1427-1439.

Koenen, M., Rüther, U., and Müller-Hill, B. (1982). Immunoenzymatic detection of expressed gene fragments cloned in the *lacZ* gene of *E. coli*. The EMBO Journal 1, 509-512.

Kolodny, G.M. (1984). An improved method for increasing the resolution and sensitivity of silver staining of nucleic acid bands in polyacrylamide gels. Anal. Biochem. 138, 66-67.

Kramer, W., Drutsa, V., Jansen, H.W., Kramer, B., Pflugfelder, M., and Fritz, H.J. (1984). The gapped duplex DNA approach to oligonucleotide-directed mutation construction. Nucleic Acids Res. 12, 9441-9456.

Krieg, P.A., and Melton, D.A. (1984). Functional messenger RNAs are produced by SP6 *in vitro* transcription of cloned cDNAs. Nucleic Acids Res. 12, 7057-7070.

Langer, P.R., Waldrop, A.A., and Ward, D.C. (1981). Enzymatic synthesis of biotin-labelled polynucleotides. Novel nucleic acid affinity probes. Proc. Natl. Acad. Sci. USA 78, 6633-6637.

Lapeyre, B., and Amalric, F. (1985). A powerful method for the preparation of cDNA libraries: isolation of cDNA encoding a 100 kDa nucleolar protein. Gene 37, 215-220.

Lathe, R. (1985). Synthetic oligonucleotide probes deduced from amino acid sequence data. Theoretical and practical considerations. J. Mol. Biol. 183, 1-12.

Lilley, D.M.J. (1981). *In vivo* consequences of plasmid topology. Nature 292, 370-381.

Lindahl, T. (1982). DNA repair enzymes. Ann. Rev. Biochem. 51, 61-87.

Maniatis, T., Fritsch, E.F., and Sambrook, J. (1982). In: "Molecular Cloning", Cold Spring Harbor Laboratory, Cold Spring Harbor, New York, 11724; pp.342-343.

Maruyama, T., Gojobori, Aota, S., and Ikemura, T. (1986). Codon usage tabulated from the GenBank genetic sequence data. Nucleic Acids Research 14, r151–r197.

Melton, D.A., Krieg, P.A., Rebagliati, M.R., Maniatis, T., Zinn, K., and Green, M.R. (1984). Efficient *in vitro* synthesis of biologically active RNA and

RNA hybridization probes from plasmids containing a bacteriophage SP6 promotor. Nucleic Acids Res. 12, 7035-7056.

Mole, S.E., and Lane, D.P. (1985). Use of simian virus 40 large T-β-galactosidase fusion proteins in an immunochemical analysis of simian virus 40 large T antigen. J. Virol. 54, 703-707.

Nagata, S., Taira, H., Hall, A., Johnsrud, L., Streuli, M., Escödi, J., Boll, W., Cantell, K., and Weissmann, C. (1980). Synthesis in *Escherichia coli* of a polypeptide with human leukocyte interferon activity. Nature 284, 316-320.

Nielsen, B.L., and Brown, L.R. (1984). The basis for colored silver-protein complex formation in stained polyacrylamide gels. Anal. Biochem. 141, 311-315.

Nussinov, R. (1981). Eukaryotic dinucleotide preference rules and their implications for degenerate codon usage. J. Mol. Biol. 149, 125-131.

Parnes, J.R., Velan, B., Felsenfeld, A., Ramanathan, L., Ferrini, U., Appella, E., and Seidman, J.G. (1981). Mouse β-microglobulin cDNA clones: A screening procedure for cDNA clones corresponding to rare mRNAs. Proc. Natl. Acad. Sci. USA 78, 2253-2257.

Pennica, D., Nedwin, G.E., Hayflick, J.S., Seeburg, P.H., Derynck, R., Palladino, M.A., Kohr, W.J., Aggarwal, B.B., and Goeddel, D.V. (1984). Human tumour necrosis factor: precursor structure, expression and homology to lymphotoxin. Nature 312, 724-729.

Poustka, A., and Lehrach, H. (1986). Jumping libraries and linking libraries: the next generation of molecular tools in mammalian genetics. Trends in Genetics 2, 174–179.

Reeve, J. (1979). Use of minicells for bacteriophage-directed polypeptide synthesis. Methods in Enzymol. 68, 493-503.

Ricciardi, R.P., Miller, J.S., and Roberts, B.E. (1979). Purification and mapping of specific mRNAs by hybridization selection and cell free translation. Proc. Natl. Acad. Sci. USA 76, 4927-4931.

Rigby, P.W.J., Dieckmann, M., Rhodes, C., and Berg, P. (1977). Labeeling deoxyribonucleic acid to high specific activity *in vitro* by nick translation with DNA polymerase I. J. Mol. Biol. 113, 237-251.

Roberts, T.M., Swanberg, S.L., Poteete, A., Riedel, G., and Backman, K. (1980)., A plasmid cloning vehicle allowing a positive selection for inserted fragments. Gene 12, 123-127.

Rodriguez, R.L., Bolivar, F., Goodman, H.M., Boyer, H.W., and Betlach, M. (1976). In: "Molecular mechanisms in the control of gene expression", ICN-UCLA Symposia Mol. Cell Biol. (D. P. Nierlich, W.J. Rutter, and C.F. Fox, eds.), Academic Press, New York; Vol. 7, 471-477.

Rüther, U. (1982). pUR250 allows rapid chemical sequencing of both strands of its inserts. Nucleic Acids Res. 10, 5765-5772.

Rüther, U., and Müller-Hill, B. (1983). Easy identification of cDNA clones. The EMBO Journal 2, 1791-1794.

Rüther, U., Koenen, M., Sippel, A.E., and Müller-Hill, B. (1982). Exon-cloning: Immunoenzymatic identification of exons of the chicken lysozyme gene. Proc. Natl. Acad. Sci. USA 79, 6852-6855.

Sancar, A., Wharton, R.P., Seltzer, S., Kacinski, B.M., Clarke, N.D., and Rupp, W.D. (1981). Identification of the *uvrA* gene product. J. Mol. Biol. 148, 45-62.

Schumann, W. (1979). Construction of an *Hpa* I and *Hind* II plasmid vector allowing direct selection of transformants harbouring recombinant plasmids. Molec. Gen. Genet. 174, 221-224.

Schutzbank, T., Robinson, R., Oven, M., and Levine, A.J. (1982). SV40 large Tumor Antigen can regulate some cellular transcripts in a positive fashion. Cell 30, 481-490.

Seed, B. (1982). Diazotizable acrylamine cellulose papers for the coupling and hybridization of nucleic acids. Nucleic Acids Res. 10, 1799-1810.

Shibahara, S., Müller, R., and Yoshida, T. (1985). Cloning and expression of cDNA for rat heme oxygenase. Proc. Natl. Acad. Sci. USA 82, 7865-7869.

Singer, R.H., and Ward, D.C. (1980). Actin gene expression visualized in chicken muscle tissue culture by using *in situ* hybridization with a biotinylated nucleotide analog. Proc. Natl. Acad. Sci. USA 79, 7331-7335.

So, M., Gill, R., and Falkow, S. (1976). The generation of a ColE1-At^r cloning vehicle which allows detection of inserted DNA. Mol. Gen. Genet. 142, 239-249.

Southern, E. (1975). Detection of specific sequences among DNA fragments separated by gel electrophoresis. J. Mol. Biol. 98, 503-517.

Steinmetz, M., Minard, K., Horvath, S., McNicholas, J., Frelinger J., Wake, C., Long, E., Mach, B., and Hood, L. (1982). A molecular map of the immune response region from the major histocompatibility complex of the mouse. Nature 300, 35-42.

St.John, T., and Davis, R. (1979). Isolation of galactose-inducible DNA sequences from *Saccharomyces cerevisiae* by differential plaque filter hybridization. Cell 16, 443-452.

Suggs, S.V., Wallace, R.B., Hirose, T., Kawashima, E.H., and Itakura, K. (1981). Use of synthetic oligonucleotides as hybridization probes: Isolation of cloned cDNA sequences for human β$_2$-microglobulin. Proc. Natl. Acad. Sci. USA 78, 6613-6617.

Symington, J., Green, U., and Brackmann, K. (1981). Immunoautoradiographic detection of proteins after electrophoretic transfer from gels to diazo-paper: Analysis of adenovirus encoded proteins. Proc. Natl. Acad. Sci. USA 78, 177-181.

Szostak, J.W., Stiles, J.I., Tye, B.K., Chin, P., Scherman, F., and Wu, R. (1979). Hybridization with synthetic oligonucleotides. Methods in Enzymol. 68, 419-428.

Tabor, S., and Richardson, C.C. (1985). A bacteriophage T7 RNA polymerase/promotor system for controlled exclusive expression of specific genes. Proc. Natl. Acad. Sci. USA 82, 1074-1078.

Ullrich, A., Berman, C.H., Dull, T.J., Gray, A., and Lee, J.M. (1984a). Isolation of the human insulin-like growth factor I gene using a single synthetic DNA probe. The EMBO J. 3, 361-364.

Ullrich, A., Gray, A., Wood, W.I., Hayflick, J., and Seeburg, P.H. (1984b). Isolation of a cDNA clone coding for the γ-subunit of mouse nerve growth factor using a high-stringency selection procedure. DNA 3, 387-392.

Wallace, R.B., Shaffer, J., Murphy, R.F., Bonner, J. Hirose, T., and Itakura, K. (1979). Hybridization of synthetic oligodeoxyribonucleotides to ΦX174 DNA: the effect of single base pair mismatch. Nucleic Acids Res. 6, 3543-3557.

Walter, P., Green, S., Greene, G., Krust, A., Bornert, J.M., Jeltsch, J.M., Staub, A., Jensen, E., Scrace, G., Waterfield, M., and Chambon, P. (1985). Cloning of the human estrogen receptor cDNA. Proc. Natl. Acad. Sci. USA 82, 7899-7893.

Weaver, R.F., and Weissmann, C. (1979). Mapping of RNA by a modification of the Berk-Sharp procedure: the 5'-termini of 15S β-Globin mRNA precursor and mature 10S β-globin mRNA have identical map coordinates. Nucleic Acids Res. 7, 1175-1193.

Woo, S.L.C. (1979). A sensitive and rapid method for recombinant phage screening. Methods in Enzymol. 68, 389-395.

Wood, W.I., Gitschier, J., Lastey, L.A., and Lawn, R.M. (1985). Base composition-independent hybridization in tetramethyla ammonium chloride: A method for oligonucleotide screening of highly complex gene libraries. Proc. Natl. Acad. Sci. USA 82, 1585-1588.

Young, R.A., and Davis, R.W. (1983a). Efficient isolation of genes by using antibody probes. Proc. Natl. Acad. Sci. USA 80, 1194-1198.

Young, R.A., and Davis, R.W. (1983b). Yeast RNA polymerase II genes: isolation with antibody probes. Science 222, 778-782.

12 Directed Mutagenesis

The study of the function of genes, the essence of scientific endeavours in the field of genetics, requires the availability of suitable mutants. Such mutants have classically been obtained by random mutagenesis and subsequent selection for the desired phenotype. The relative positions of these mutants on a chromosome can then be mapped by employing conventional techniques of complementation and recombination analysis. New methods for the isolation and cloning of genes and, above all, the chemical synthesis of DNA, allow genes to be altered specifically by introducing deletions or insertions, by mutagenising defined parts of a chromosome, or by exchanging individual bases in the DNA. The biological activity of these mutants then can be tested *in vitro* and *in vivo*. Strictly speaking, these new techniques reverse the conventional sequence of events (Fig. 12-1). It is in this context that the term "reverse genetics" (Müller *et al.*, 1978) was coined, although this kind of genetic analysis is much more in agreement with the true sequence of events and mirrors the flow of information from DNA to phenotype (Fig. 12-1).

Molecular geneticists distinguish between three classes of mutations: deletions, insertions, and substitutions. While deletions decrease, and insertions increase, the length of a piece of DNA, substitutions do not influence the length of a DNA molecule, because they constitute replacements of DNA sequences. Single base pair mutations can be classified as either transitions or transversions. A transition mutation exchanges

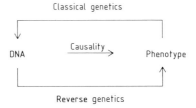

Fig. 12-1. Reverse genetics.

one purine base for another, or one pyrimidine base for another, *e.g.* a CG base pair will change to a TA pair. A transversion replaces a purine base with a pyrimidine for example a CG pair will change to an AT or GC pair. Classical genetics knows several methods and reagents that can be used to induce specific substitutions. Aminopurine, for example, specifically induces transitions from AT to GC (Caras *et al.*, 1982), while nitrosoguanidine leads to transitions from GC to AT (Miller, 1972). The advantage offered by the new methods is two-fold. They not only allow the introduction of specific substitutions, but also their introduction at specific sites on a chromosome. Such sites may be single base pairs, but also larger or smaller stretches of DNA. Several methods for site-specific mutagenesis will be discussed in the following sections. Their choice will, of course, depend on the biological problem to be solved. Further information can be found in an excellent review by Shortle *et al.* (1981) (see also Harris, 1982).

12.1 Deletions

The easiest way to delete a piece of DNA in a circular chromosome, a bacterial plasmid for example, is to digest the DNA with a restriction endonuclease which cuts twice and produces two DNA fragments (Fig. 12-2). These two fragments can be separated from each other. The fragment containing the *ori* sequence is recircularised and cloned by making use of its two cohesive molecular ends. If the linear DNA fragment is introduced directly into suitable host cells without ligation, and circularisation occurs *in vivo*, deletions will occasionally be obtained which cover the recognition site of the endonuclease, because the ends of the linear DNA molecule may be digested away before circularisation. This can be avoided by treatment with DNA ligase prior to transformation.

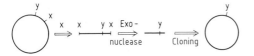

Fig. 12-2. Deletion of a restriction fragment.

Fig. 12-3. Generation of deletions by exonuclease treatment.

It is also possible to generate deletions by exonuclease treatment. A circular DNA molecule is first digested with an endonuclease which cleaves only once. The linearised molecule is then shortened by limited digestion with exonuclease. Shortened linearised DNA molecules (which may have been recircularised *in-vitro*) are then transformed into suitable bacteria, yielding transformants carrying DNA molecules which lack the original recognition site (Fig. 12-3). Carbon *et al.*

(1975) have used this technique to generate deletions in the vicinity of the *Hpa* II and *Eco* RI sites of SV40 DNA (Fig. 12-4). The exonuclease used in these experiments was the phage-encoded exonuclease of phage λ, which specifically removes nucleotides from the 5′ end of a linear DNA (Fig. 2.1-10). The modified DNA probably circularises *in vivo* by base pairing between short homologous regions in the protruding single-stranded ends. The deletions obtained in these experiments covered between 20 and 50 base pairs.

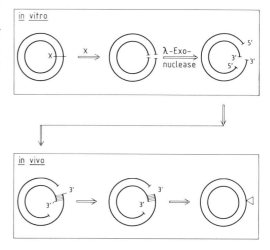

Fig. 12-4. Generation of deletion mutants by combined treatment with endo- and exonucleases. "X" signifies a restriction enzyme with only one cleavage site on the molecule in question. Bacteria are transformed with DNA previously subjected to λ 5′ exonuclease digestion. The sequence of events occurring *in vivo* is purely hypothetical. The triangle marks the position of the deletion (Carbon *et al.*, 1975).

Shenk (1977) has developed a procedure which allows such deletions to be enlarged. The first step is the formation of a heteroduplex between the linear forms of wild-type SV40 DNA and a deletion mutant. A single-stranded loop is formed at the position of the deletion. The heteroduplex DNA is then circularised and treated with S1 nuclease from *Aspergillus oryzae*. This enzyme is single strand-specific and can digest the single-stranded DNA regions in the deletion loop; however, the enzyme usually also removes up to 30 base pairs of the double strand, thus creating new deletions by extending the original deletion (Fig. 12-5). Double-stranded heteroduplexes between intact and deleted DNA can also be used for mapping deletions with respect to recognition sites of endonucleases. Linear heteroduplex molecules are first digested with S1 nuclease. The lengths of the resulting fragments are then compared with those of fragments generated by cleavage of the parental DNA (Shenk *et al.* 1975). Besides the 5' to 3'-specific λ exonuclease there are several other exonucleases which can also be used for the generation of deletions. *E. coli*

exonuclease III (*Exo*III) for example, removes 5' mononucleotides from double-stranded DNA in the 3' to 5' direction. λ exonuclease used in the experiment shown in Fig. 12-4 can be replaced by an *Exo*III/S1 treatment. The linear double strand is first digested at its 3' ends by treatment with *Exo*III. Protruding 5' ends are subsequently removed by S1 nuclease treatment. Since nuclease *Bal*31 possesses 3' and 5' exonuclease activities, the S1 nuclease step can be omitted in experiments using *Bal*31. In either case, the *in vitro* circularisation must proceed *via* blunt end ligation. It is therefore advisable to fill-in remaining protruding 5' ends using DNA polymerase I, before the DNA molecules are ligated with T4 DNA ligase (see Chapter 3).

The use of nuclease *Bal*31, or the combination of *Exo*III and S1 nuclease can be undesirable because both ends of a DNA molecule are digested simultaneously by these enzymes. Quite frequently it is only one end of a DNA molecule which is to be manipulated, and there are two possible solutions to this problem. If the desired sequence is part of a larger recombinant DNA molecule, *i.e.*, if it is linked to vector DNA, digestion at the ends of the linearised molecule will not only remove nucleotides from the desired sequence, but also from the vector. This is of no consequence as long as the other end of the manipulated DNA sequence is so far removed from a recognition site that it will not be digested by the nuclease. The remaining part of the manipulated DNA sequence can then be cut out from the vector and can be recloned *via* synthetic linkers (Fig. 12-6). An early example of this method by Sakonju *et al.* (1980) describes a set of deletions in the 5S ribosomal RNA gene of *Xenopus laevis*. The method has the disadvantage that it cannot distinguish between biological effects exerted through the lack of particular DNA sequences alone or, in addition, through the presence of the newly introduced adjacent DNA sequences, *e.g.* from vector DNA. A variation has thus been described by McKnight and Kingsbury (1982), which avoids and circum-

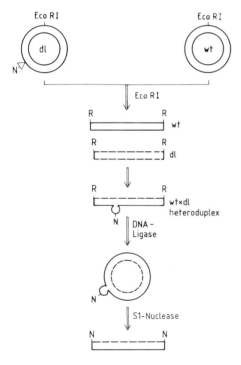

Fig. 12-5. Extension of deletions in circular DNA by S1 nuclease treatment of heteroduplexes formed from wild-type and mutant DNA.
dl denotes a particular deletion mutant; R stands for *Eco* RI ends, and N signifies the position of the deletion (Shenk, 1977).

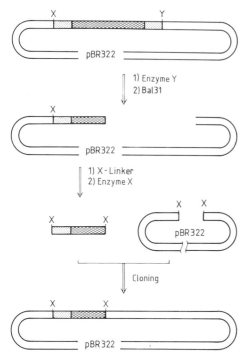

Fig. 12-6. Strategy for generating deletions with an intact end.

X and Y are restriction enzyme cleavage sites on the recombinant DNA molecule. In reality, the DNA fragment with X-specific termini generated in the second reaction step represents a whole population of molecules with variable lengths in the cross-hatched region. These molecules are re-cloned into a vector linearised by digestion with enzyme X. Bal31 digestion thus leads to molecules which are only shortened on the right-hand side.

vents these difficulties. A random set of 5' and 3' deletions mutants is created by sequential treatment of linearised DNA with exonuclease III and S1 nuclease, as described above. In the published example, these were obtained around the single *Eco* RI site within the promoter of the *HSV-tk* gene (*cf*. Fig. 8-2) and were recloned *via* a synthetic *Bam* HI linker ten nucleotides in length. All deletions are identified by DNA sequencing. Matching sets of 5' and 3' deletions the termini of which are separated by ten nuclotides (the length

of the linker) are recombined at their synthetic restriction site such that recombination does not result in net increases or decreases in the total length of the sequence. The procedure thus creates a clustered set of point mutations (represented by the linker) distributed along a desired DNA region. Accordingly, they are referred to as "linker-scanning" mutations.

Another technique makes use of nucleoside α-thiotriphosphates (dNTP-αS; Fig. 12-7). These nucleoside derivatives (Eckstein, 1985) can still be used as substrates for DNA polymerase I, but the phosphothioate bond will protect the 3' ends of a DNA fragment against digestion with exonuclease III. It is therefore possible to introduce an α-thionucleotide at only one of the two 3' ends of a DNA fragment and to trim the other end of the molecule by a combined treatment with exonuclease III and S1 nuclease. Alternatively, exonuclease III treatment alone would yield an intact single strand. Fig. 12-8 shows how this technique is used to obtain a shortened pBR322 derivative (Putney *et al*. 1981). The ends of a pBR322 molecule linearised with *Eco* RI are first filled-in in the presence of dATP-αS and dTTP. Digestion with restriction endonuclease *Hind* III, which cleaves in the vicinity of the other end of the molecule, yields a DNA molecule with α-thionucleotides at only one of the two 3' ends. The other end can then be modified by a combined digestion with exonuclease III and S1 nuclease.

$$HO-\overset{\overset{\displaystyle O}{\|}}{\underset{\underset{\displaystyle O^-}{|}}{P}}-O-\overset{\overset{\displaystyle O}{\|}}{\underset{\underset{\displaystyle O^-}{|}}{P}}-O-\overset{\overset{\displaystyle S}{\|}}{\underset{\underset{\displaystyle O^-}{|}}{P}}-O \qquad Base$$

$$\text{OH} \quad \text{H}$$

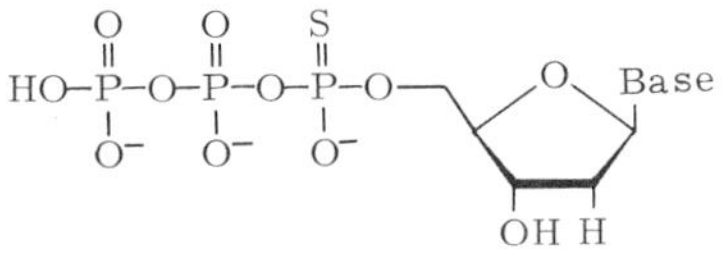

Fig. 12-7. Structure of a 2'-deoxynucleoside-5'-O-(α-thio) triphosphate.

Only the Sp isomer of the two diastereomers is biologically active. The presence of the phosphorothioate diester bonds inhibits the 3' to 5' proof-reading activity of DNA polymerase I (Kunkel *et al.*, 1981); yet, dNTP-αS compounds are normal substrates for the polymerase activity of *E. coli* DNA polymerase I (Vosberg and Eckstein, 1977).

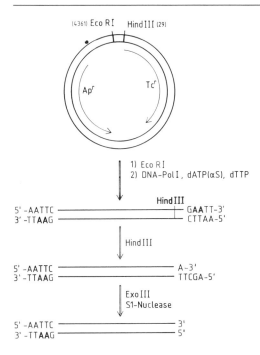

5'-AATTC ———————————— GAATT-3'
3'-TTAAG ———————————— CTTAA-5'

Hind III
↓

5'-AATTC ———————————— A-3'
3'-TTAAG ———————————— TTCGA-5'

Exo III
S1-Nuclease
↓

5'-AATTC ———————————— 3'
3'-TTAAG ———————————— 5'

Fig. 12-8. Use of dATP-(αS) for generating pBR322 molecules shortened at one end only (see text for details).

The methods described so far are limited to introducing deletions only at positions relatively near restriction enzyme cleavage sites. Green and Tibbetts (1980) have described a technique which is independent of any specific cleavage sites. This technique makes use of so-called D-loops between circular double-stranded DNA and a single-stranded DNA fragment covering the region to be mutagenised (Fig. 12-9). Such DNA molecules are substrates for S1 nuclease. Under carefully controlled conditions S1 nuclease will cleave the molecule at either one of the two ends of the loop. Upon circularisation in the presence of T4 DNA ligase, small deletions of approximately 10 base pairs will be generated in the desired region.

Under comparable conditions, i.e., comparable DNA concentrations, the rate of D-loop formation from supercoiled DNA and homolo-gous single-stranded DNA fragments is only 1% of the rate of reassociation between the complementary strands of the same DNA fragment (Beattie et al. 1977). The formation of D-loops in mixtures of superhelical DNA and denatured restriction enzyme fragments in appreciable amounts will thus be negligible. There are, however, three different strategies which help to alleviate these problems. The first employs isolated single strands, e.g. synthetic oligonucleotides, instead of double-stranded DNA fragments. The second strategy exploits the fact that both competing reactions, i.e., reassociation of complemen-

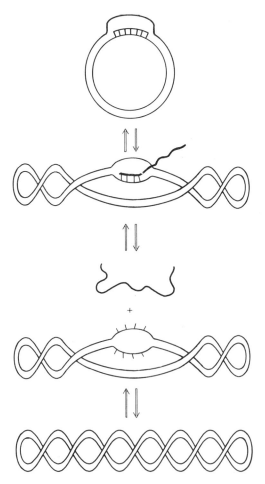

Fig. 12-9.   Generation of D-loops (see text).

tary single strands and formation of D-loops are second-order reactions. If the concentration of superhelical DNA is high and that of denatured DNA fragments low, D-loop formation should be favoured, because the reaction will then follow pseudo-first-order kinetics. Finally, use can also be made of the RecA protein of *E. coli*. In the presence of Mg^{2+} and ATP this enzyme catalyses D-loop formation in reaction mixtures containing single-stranded DNA fragments and superhelical DNA (Radding, 1981).

12.2 Insertions

In principle, insertions can be generated by reversing the reaction described in Fig. 12-2. A large variety of chemically synthesised adaptor and linker molecules are available today, and they can be introduced into a specific recognition site of an endonuclease, thus altering this part of the DNA. Heffron *et al.* (1978) have described a technique for introducing such linker molecules into any site on a circular DNA molecule. Their technique, which has been described in Section 3.1.2.2, Figs. 3-14 and 3-15, yields a whole set of insertion mutations.

Another important method of mutagenesis by insertion is transposon mutagenesis. In principle, any transposon could be used, but transposon mutagenesis is usually carried out with transposons Tn7 and Tn5 (Kleckner, 1981). Transposon Tn5 is 5.7 kb long and codes for kanamycin and/or neomycin resistance. Its structure has been described in detail by Auerswald *et al.* (1981) and Beck *et al.* (1982) (*cf.* Fig. 8-7). Tn5 has been introduced into modified λ phages which are known as λ *kan* hopper phages (Berg *et al.*, 1975). The transposon can jump from these phage genomes to almost any other position, for example to a location on a plasmid DNA, and can inactivate genes. McKinnon *et al.* (1982) have used this technique to mutagenise the transforming genes of adenovirus type 5, which had been cloned in a

plasmid. The various Tn5 insertions were tested for their transforming activities, thus allowing the transforming regions on the adenovirus genome to be identified. An extension of this procedure applicable only in prokaryotes has been described by Ruvkun and Ausubel (1981). According to the latter technique, a DNA molecule cloned in a vector is first mutagenised by Tn5 insertion and then reintroduced into chromosomal DNA to obtain chromosomal mutants. Fig. 12-10 shows the 30 kb plasmid pRK290-nif::Tn5 which carries a 3.9 kb insertion of the *Rhizobium meliloti nif* gene DNA region, into which Tn5 has inserted. In this case the vector pRK290 is used, because no DNA transformation system for *Rhizobium meliloti* is available. Foreign DNA thus can be introduced into these cells only by conjugation. The conjugative plasmid pRK290 possesses a broad host specifity and codes for a tetracycline resistance marker (Ditta *et al.*, 1980) (see also Fig. 4.1-13). pRK290-nif::Tn5 hybrids can be introduced into *R. meliloti* strains by conjugation and replicate stably. Homologous recombination between the *R. meliloti* DNA on the plasmid and chromosomal DNA is observed in a small number of transconjugants. This process replaces the intact chromosomal gene, leaving a Tn5-inactivated gene in its place. In this way the *Rhizobium* genome acquires transposon Tn5, with its neomycin resistance (Nm^r) gene. Recombinants are identified by first eliminating plasmid pRK290-nif by conjugation with plasmid pR751, which is incompatible with the *nif* plasmid, since both belong to incompatibility group P. The new plasmid, *i.e.*, pR751, carries a trimethoprim (*Tp*) resistance marker so that the desired recombinants can be selected by their combined trimethoprim and neomycin resistance. Since pRK290 is a broad host range vector, this strategy should be applicable to a wide spectrum of Gram-negative bacteria.

Boeke (1981) has developed a technique which allows the introduction of insertions with a length of one or two amino acid codons into circular DNA. The technique is based on the fact that

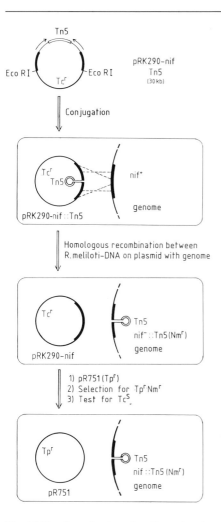

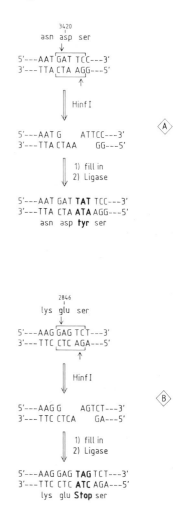

digestion with the enzyme *Hinf*I (G/ANTC) produces a protruding 5′ end which is three bases long. If the 5′ ends are filled-in with DNA polymerase I in the presence of deoxyribonucleoside triphosphates, and the molecule is then circularised, this yields an insertion of three bases at the original cleavage site. If the nucleotide N in

Fig. 12-10. Introduction of Tn5 insertions into chromosomal DNA.
The starting material for the construction is the 8 kb vector pRMR2 which is composed of plasmid pACYC184 and a 3.9 kb *nif* DNA insertion from *R. meliloti*. After mutagenesis with Tn5, the *nif* gene region is transferred to vector pRK290 (20 kb; Fig. 4.1-13; Ditta *et al.*, 1980). The new vector pRK290-nif::Tn5 can be introduced into *R. meliloti* cells by conjugation and transfers its mutated *nif* region into the chromosomal DNA as described in the text. *nif* insertions are shown as black bars, transposon sequences with flanking repetitions as open bars flanked by solid lines. The arrows indicate the orientation of the repetitions. Thick solid lines are pRK290 sequences (Ruvkun and Ausubel, 1981).

Fig. 12-11. Principle of the insertion of three base pairs by *Hinf*I digestions.
Example (A) shows the *Hinf*I site at position 3420 of fd DNA (Beck and Zink, 1981); example (B) a *Hinf*I site at position 2846 at the end of gene *III* of fd. Tyrosine and stop codons are printed in boldface (Boeke, 1981).

the recognition sequence of *Hin*f I was either T or C, and if the sequence GAN was in the right reading frame, the result of mutagenesis will be the insertion of either TAT or TAC, both of which code for tyrosine (Fig. 12-11A). If N is G or A, an amber or ochre stop codon will be obtained (Fig. 12-11B). Since *Hin*f I recognises 25 sites on fd DNA, mutagenesis of DNA fragments obtained from a partial *Hin*f I digestion will yield a series of mutants. Some of these mutants will be nonsense mutants. Such mutants are very valuable, because they are viable in suppressor-positive strains only, and therefore can be easily identified.

12.3 Localised Point Mutations

12.3.1 Bisulfite Mutagenesis

Although it appears to be highly desirable to induce and analyse single point mutations in a gene or a genetic control element, it is rather difficult to predict which base substitutions will be biologically relevant. It is therefore advisable to use a system which allows examination of the biological behaviour of many mutations in a defined chromosomal region before individual point mutations are studied in detail. Shortle and Nathans (1978) have introduced the technique of bisulfite mutagenesis for this purpose. Cytosine residues in single-stranded, but not double-stranded, DNA are easily deaminated to uracil when treated with sodium bisulfite. When the complementary strand is synthesised, these uracil residues will direct the incorporation of adenosine, and hence transitions from GC to AT will be induced. Several methods are available to generate single-stranded regions in double-stranded DNA molecules which will be susceptible to bisulfite treatment.

Shortle and Nathans (1978) took up an observation initially made by Parker *et al.* (1977) who found that restriction endonucleases will only destroy one phosphodiester bond in their cleavage site in the presence of ethidium bromide. Different restriction endonucleases will require totally different conditions (Österlund *et al.*, 1982). Up to 90% of a circular Form I DNA can be converted into Form II DNA. The strand break can then be enlarged by exonucleolytic treatment (Fig. 12-12). DNA polymerase I from *Micrococcus luteus* is particularly well-suited for this purpose since the enzyme only displays a 5' to 3' exonuclease activity under special conditions, *i.e.*, in the presence of only one deoxyribonucleoside triphosphate (dTTP for example). After treatment of the DNA with bisulfite, the gap is closed by the same DNA polymerase, now using its polymerase activity. A subsequent digestion with the initially used restriction endonuclease selects for molecules which are resistant to digestion. The example shown in Fig. 12-12 is the manipulation in the neighbourhood of the single *Hpa* II site of SV40 DNA, which eventually allows the selection of *Hpa* II-resistant SV40 DNA molecules. A sequence analysis of some of the mutants (Fig. 12-13) demonstrated that the technique indeed yields the expected transitions from GC to AT. This procedure, too, has the disadvantage that it is restricted to mutagenesis of regions located around restriction enzyme cleavage sites. This problem can be circumvented using D-loops in the desired DNA region (Shortle *et al.*, 1980). It has been mentioned (Section 12.1; Fig. 12-9) that the RecA protein of *E. coli* catalyses the formation of D-loops in mixtures of superhelical plasmid DNA and a complementary single strand. The single strand displaces a DNA region of the same polarity in the double strand. Carefully controlled S1 nuclease treatment induces a single strand break in the displaced single-stranded region, and the unstable relaxed intermediate displaces the single-stranded short DNA fragment within fractions of a second, yielding a relaxed circular molecule with a single-strand break in the desired region. This region is then mutagenised by bisulfite treatment according to

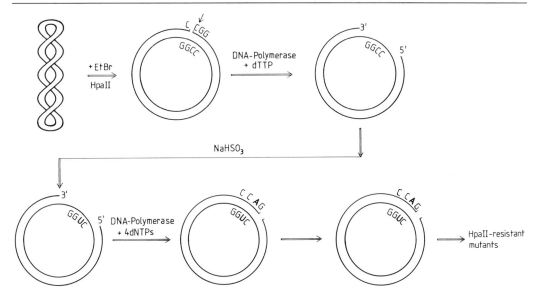

Fig. 12-12. Bisulfite mutagenesis at the *Hpa* II site of SV40 DNA (Shortle and Nathans, 1978). The CG to TA transitions lead to *Hpa* II-resistant mutants.

the scheme depicted in Fig. 12-12. D-loop molecules are stabilised relative to Form I DNA, because D-loop molecules possess fewer negative superhelical turns; one superhelical turn is lost for every 10 bp that are displaced. This stability of D-loops allows them to be identified in agarose gels. In the absence of ethidium bromide they migrate slightly slower than Form I DNA, but faster than Form II DNA. Since the RecA protein of *E. coli* is commercially available, the described technique is practicable. Special care must be taken if S1 nuclease is used, since the reaction conditions must be chosen to prevent further digestion of the desired single-stranded region.

Another approach to the generation of single-stranded regions employs single-stranded phages such as M13 and fd. Since suitable vectors are available, especially in the M13 system (see also Section 2.4), any DNA can be cloned in M13 vectors, and can be obtained as a single-stranded molecule (Fig. 12-14). Of course, this DNA can be mutagenised with bisulfite. A complementary (–) strand can then be synthesised from the

mutagenised strand, using specific primers. The desired mutagenised region is then cut out from the double-stranded RF form and can be recloned in a non-mutagenised vector (Weiher and Schaller, 1982). A simplification of this procedure described by Everett and Chambon (1982) obviates recloning (Fig. 12-15). The first step in this protocol is the isolation of a recombinant single-stranded DNA and the corresponding double-stranded RF form. A DNA fragment containing the desired region, *i.e.*, the site to be mutagenised, is then cut out from the RF DNA. If this linearised DNA fragment is denatured and reassociated in the presence of the original single-stranded DNA, a heteroduplex with a single-stranded region at the desired site is obtained. This site can then be mutagenised with bisulfite as described above. Another technique, which makes use of already existing deletions, is probably even simpler (Fig. 12-16) (Kalderon *et al.* 1982). Wild-type DNA and DNA containing a deletion are first cloned separately in suitable plasmid vectors and linearised with two different

```
                HpaII
         340      ↓       360
5'---TGGTGCTGCGCCGGCTGTCACGCCA---3'
3'---ACCACGACGCGGCCGACAGTGCGGT---5'
            HhaI
```

```
              AluI
    ---TGCGCCAGCTGT---          I
    ---ACGCGGTCGACA---
      HhaI  PvuII
```

```
    ---TGCGTCGGCTGT---          II
    ---ACGCAGCCGACA---
```

```
    ---TGCGCTGGCTGT---          III
    ---ACGCGACCGACA---
        HhaI
```

Fig. 12-13. Sequences of three mutants near the *Hpa* II site of SV40 DNA.
CG to TA transitions are printed in boldface. These transitions were verified by sequencing and can also be identified by the appearance of new, or the disappearance of old, cleavage sites. As expected, the *Hpa* II site is lost in all mutants. Class I molecules show new *Alu* I or *Pvu* II sites; class II molecules do not contain any cleavage sites in the region in question; class III molecules only contain the *Hha* I site (Shortle and Nathans, 1978). Numbers are SV40 co-ordinates according to the "BBB" nomenclature (*cf.* also Chapter 8).

N-glycosylase (Warner and Duncan, 1978). Indeed, controls show that approximately half of the clones are of the wild-type size, and the other half are of the size of the deletion mutant. More than 80% of the full-size clones show the expected transitions from GC to AT. The described tech-

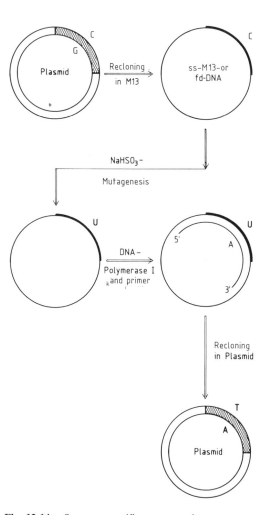

Fig. 12-14. Segment-specific mutagenesis.
The DNA fragment to be mutagenised is cloned in M13 or fd vectors, and bisulfite mutagenesis is carried out on single-stranded M13 or fd DNA. Insertions are indicated as cross-hatched bars or thick, solid lines (Weiher and Schaller, 1982).

enzymes, X and Y. Denaturation and reassociation of a mixture of the linearised plasmids will then yield heteroduplex and homoduplex molecules; the former will be circular because of the staggered position of the cleavage sites, while homoduplexes will be linear. These circular heteroduplexes will contain a single-stranded region, the size of which will be exactly the size of the original deletion. This region can be mutagenised with bisulfite. The mutagenised DNA is then transfected into bacterial Ung⁻ mutants, cells with a defect in the uracil repair enzyme uracil-

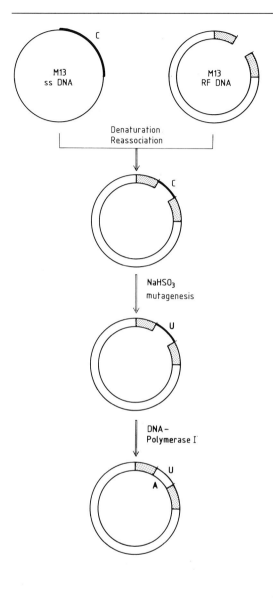

nique is fascinating, because it completely avoids the repair and double strand synthesis required by all other techniques after mutagenesis; however, this is only possible if Ung⁻ strains are used, because the mutagenised single-stranded loop containing the unusual U residues in the DNA otherwise would be cut out after transfection and prior to replication (see below).

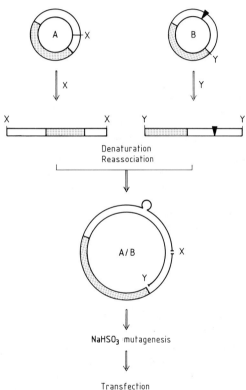

Fig. 12-15. Segment-specific mutagenesis.
Reassociation of a single-stranded M13 clone with denatured RF DNA carrying a suitable deletion yields heteroduplexes with single-stranded regions covering the deletion. These single-stranded sequences can be mutagenised with bisulfite (Everett and Chambon, 1982).

Fig. 12-16. Principle of mutagenesis in deletion loops.
The black triangle in plasmid B indicates the position of a deletion. Denaturation and reassociation of plasmids A and B, linearised with enzymes X and Y, respectively, yield heteroduplexes with a deletion loop (Kalderon et al., 1982).

12.3.2 Incorporation of Nucleotide Analogues

Directed mutagenesis by incorporation of nucleotide analogues into RNA and DNA was introduced by C. Weissmann. N^4-hydroxycytidine triphosphate (N^4-hydroxy CTP) is used for the *in vitro* synthesis of RNA, and N^4-deoxycytidine triphosphate (N^4-hydroxy dCTP) for DNA synthesis. If the synthesis of RNA is catalysed by Qβ replicase for example, the regular C residues are replaced by N^4-hydroxy C residues in the presence of N^4-hydroxy CTP (Fig. 12-17). During the synthesis of the (+) strand, incorporated analogue residues allow and direct the incorporation of either G or A residues. A certain fraction of all molecules (approximately 30%) will then contain transitions of the type GC to AT.

During the synthesis of DNA catalysed by DNA polymerase I in the presence of N^4-hydroxy dCTP, TMP residues will be replaced by N^4-hydroxy dCTP. In turn, they will direct the incorporation of complementary A or G residues

so that eventually AT residues will be replaced by TA or CG (Fig. 12-18).

In order to regulate the incorporation of these analogues, DNA or RNA synthesis must be synchronised. When the Qβ RNA system was used, synchronisation was achieved by exploiting the fact that the purified Qβ RNA (+) strand is capable of binding ribosomes specifically in the regions of the start codon for the coat protein (Taniguchi and Weissmann, 1978). Under suitable conditions the synthesis of the (–) strand, which proceeds from right (5′) to left (3′), is interrupted by ribosomes bound to the (+) strand (Fig. 12-17). Growing RNA chains accumulate at this particular site and can be re-started synchronously by removing the ribosome. In the presence of N^4-hydroxy CTP, the two G residues at positions "+3" and "+4" of the coat protein are complemented by N^4-hydroxy cytidine. (+) strand synthesis then yields wild-type RNA and the expected three classes of mutants, in which either or both of the two G residues are replaced by A (Fig. 12-17).

In vitro DNA synthesis can be started specifically at open phosphodiester bonds within endonuclease cleavage sites (Fig. 12-18; Müller *et al.* 1978) which have been digested in the presence of ethidium bromide so that only one of the two phosphodiester bonds has been cleaved. DNA synthesis will therefore only proceed along one of the two complementary strands, in 5′ to 3′ direction, yielding two different sets of molecules in which dTMP residues are replaced by N^4-hydroxy dCTP in either the (+) or the (–) strand. *In vivo*, incorporated N^4-hydroxy dCTP residues will direct the incorporation of G and A residues, and will produce the expected transitions from TA to CG. Since all mutations in our example occur in the region of an *Eco* RI cleavage site, the transitions can be easily recognised or selected for

Charles Weissmann
Zürich, Switzerland

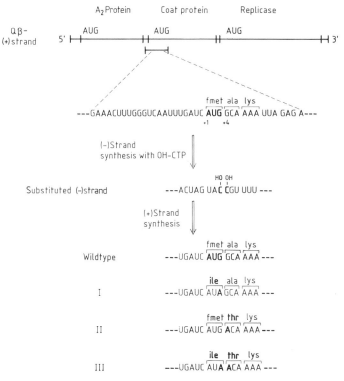

Fig. 12-17.   Specific mutagenesis in bacteriophage Qβ RNA.
Shown is a section of the sequence for the ribosomal binding site near the start codon of the coat protein gene. Synchronised synthesis of (–) strands led to the incorporation of two N⁴-hydroxyl-CTP (OH-CTP) residues at positions "+3" and "+4", which in turn resulted in the three different classes of mutations during the synthesis of the (+) strand (Taniguchi and Weissmann, 1978).

because the mutations lead to DNA molecules which are *Eco* RI-resistant.

The second procedure enforces the incorporation of false nucleotides into growing DNA chains (Shortle *et al.*, 1982) by using special reaction conditions which block correction by DNA polymerase I. Such correction activities can be circumvented in two ways. The first is to use DNA polymerase I and only three of the four deoxyribonucleoside triphosphates to repair gapped DNA; one nucleotide required to repair a gap in the target region is omitted. In the presence of T4 DNA ligase and ATP, only non-complementary

nucleotides can be incorporated at the desired position (Fig. 12-19). The remaining gaps will be closed immediately by the incorporation of the other nucleotides and by ligation. Misincorporated nucleotides are ostensibly held "frozen" in a state in which they are resistant to the 3' to 5' exonucleolytic correcting activities of the enzyme. The use of α-thiophosphate analogues leads to similar results. These compounds are substrates for DNA polymerases, but not for the 3' to 5' exonuclease activity. Once incorporated, a nucleotide analogue cannot be excised by the enzyme, but its 3' terminus can still be used as a primer for

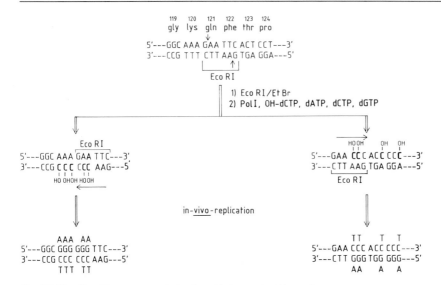

Fig. 12-18. Specific mutagenesis in plasmids by nucleotide analogues.
Shown is a section of a rabbit β-globin cDNA clone near the *Eco* RI site at codons 121 and 122. *Eco* RI treatment in the presence of ethidium bromide leads to a strand break in either one of the complementary strands. The incorporation of N4-hydroxy-dCTP (OH-dCTP), catalysed by *E. coli* DNA polymerase I, therefore can proceed right- or leftwards, as indicated by the arrows. In principle, *in vivo* replication can lead to the mutations shown at the bottom; in practice, not all mutations are observed, and not all mutations are generated at the same time (Müller *et al.*, 1978).

further DNA synthesis so that the gap can be closed.

Upon transformation of DNA mutagenised at a single site, wild-type and mutant DNAs should segregate so that half of the clones should be of the mutant type. In fact, the first method (Fig. 12-19) yielded 5-15% mutant clones, while the second procedure employing thiophosphates yielded up to 40% mutant clones. These values are very close to those predicted theoretically.

12.3.3 Generalised Chemical Mutagenesis

In contrast to bisulfite mutagenesis which exclusively produces GC to AT transitions, the technique described below generates a wide spectrum of transversions and transitions within a given DNA region. The most important feature of this technique (Myers *et al.*, 1985a) is its application in conjunction with a special gel system which allows mutated DNA and unmutated DNA to be separated from each other. The DNA fragment to be mutated should be approximately 50-200 bp in length and is inserted into two special vectors, pGC1 and pGC2, in the vicinity of a GC-rich sequence. pGC1/2 are single-stranded plasmid vectors which carry an M13 origin of replication (Fig. 12-21; *cf.* also Section 4.1.5). Plasmid DNAs therefore can be obtained in a single-stranded form after superinfection with an M13 helper virus. As shown in Fig. 12-20 the isolated DNA is mutagenised by chemical methods (*cf.* also Table 12-1) and is then converted to a double strand by using a specific "GC" primer and AMV reverse transcriptase. The chemical treatment alters the ability of bases to form normal base pairs, so that incorrect nucleotides are incorporated opposite the altered bases when the second strand is synthesised. Since there are no real mismatches,

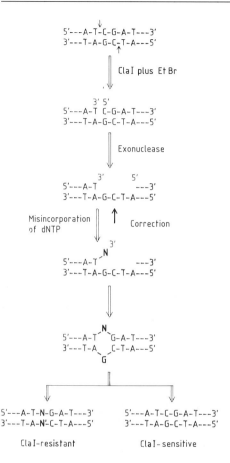

Table 12–1. Reagents in chemical mutagenesis

Chemical	Target	Change
Nitrous acid	C	C→T
	A	A→G
Formic acid	A	all changes
	G	all changes
Hydrazine	C	all changes
	T	all changes
KMnO$_4$	T	all changes

from Myers *et al.* (1985b)

The system described by Fischer and Lerman (1983) exploits the melting behaviour of different DNAs to facilitate the separation of mutant DNA molecules from each other, or of mutant DNAs from wild-type molecules. If the temperature of a solution containing double-stranded DNA is grad-

Fig. 12-19. Induction of point mutations by enzymatic incorporation of false nucleotides and simultaneous inhibition of repair synthesis (Shortle *et al.*, 1982).

the mismatch repair system will not be active when these molecules are propagated in bacteria. After the plasmid DNA has been isolated from pooled colonies, the mutagenised portion is excised, together with the neighbouring GC-rich regions, and separated by gel electrophoresis. Mutagenised DNA then can be recloned into a non-mutagenised vector.

Richard M. Myers
San Francisco

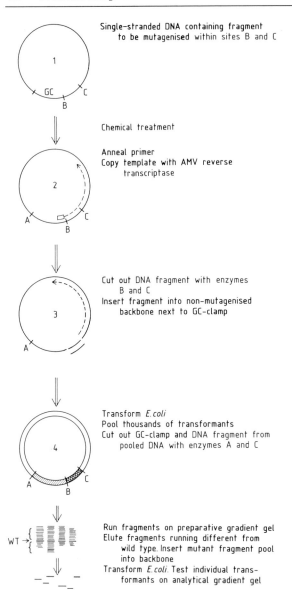

Single-stranded DNA containing fragment
    to be mutagenised within sites B and C

Chemical treatment

Anneal primer
Copy template with AMV reverse
    transcriptase

Cut out DNA fragment with enzymes
    B and C
Insert fragment into non-mutagenised
    backbone next to GC-clamp

Transform *E.coli*
Pool thousands of transformants
Cut out GC-clamp and DNA fragment from
    pooled DNA with enzymes A and C

Run fragments on preparative gradient gel
Elute fragments running different from
    wild type. Insert mutant fragment pool
    into backbone
Transform *E.coli*. Test individual trans-
    formants on analytical gradient gel

Fig. 12-20.   Schematic diagram for a generalised mutagenesis strategy.
Letters A, B and C indicate restriction enzyme sites required for cloning or removing the fragments to be
mutagenised without (B+C digestion), or with GC clamp (A+C digestion). The dashed arrows in molecules 2 and 3
represent newly synthesised DNA primed from the GC-rich primer. In molecule 4, the GC clamp is shown by a
hatched bar, the mutagenised fragment by a cross-hatched bar (Myers *et al.*, 1985a).

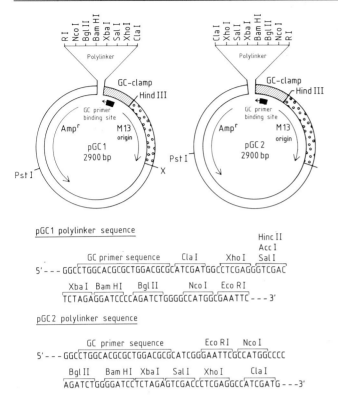

Fig. 12-21. Maps of plasmids pGC1 and pGC2.
The GC clamp plasmids pGC1 and pGC2 are composed of fours parts; (1) a 1 922 bp long fragment of pBR322 DNA (open bar) containing the ColE1 origin and the β-lactamase gene (pBR322 positions 2440 to 4363); (2) the M13 origin of replication (stippled bar; nucleotides 5330 to 6001 of M13; *cf*. Fig. 2.4-15); (3) the GC clamp (hatched bar) comprising a region between base pairs -113 to -409 from the mRNA cap site of the human α1-globin gene (Myers *et al.*, 1985b), and (4) a polylinker. The sequences show the arrangements of the polylinker with respect to a GC-rich primer oligonucleotide (black bar; 18 bases long as shown in the sequence below) used for the synthesis of the second strand DNA (Myers *et al.*, 1985a).

ually increased, the DNA will melt in domains of 50-100 bp, depending on its base composition. Domains are melted co-operatively at a certain temperature, T_m, which is specific for individual domains. A single base change in one domain may alter the T_m value. Although such temperature shifts are normally very small, they can be identified by gel electrophoresis if polyacrylamide gels with an increasing gradient of a denaturing agent, urea, for example, are used and run at approximately 60 °C. Initially the double-strand-ed DNA will enter the gel and migrate according to its molecular weight. When the position in the gel is reached at which the first domain melts, the molecules will become partially single- and parti-ally double-stranded with a concomitant marked decrease in electrophoretic mobility as compared to the truly double-stranded molecules. Mole-cules with single base changes in their melting domains will display a different melting behaviour and will thus reach this point at different positions in the gel.

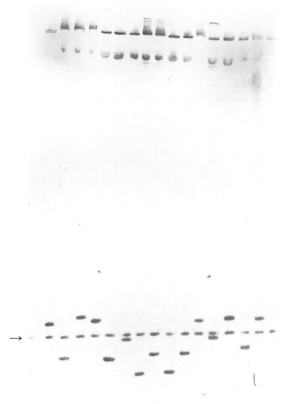

Fig. 12-22. Denaturing gradient gel pattern of mutant DNA fragments.
In this example, a 135 bp fragment carrying the mouse β-globin promoter region between positions -104 to +26 was inserted adjacent to the GC clamp and mutagenised. Electrophoresis was carried out with wild-type and mutant promoter fragments linked to the GC clamp. A mixture of wild-type and mutant fragments (always from a single clone) was loaded on each lane and run on a 6.5% polyacrylamide gel containing a linear gradient of 30 to 60% denaturing agent (100% denaturing agent = 40% formamide/7 M urea) for 5 hours at 150 V. The arrow shows the position of the wild-type fragment (Courtesy of R.M. Myers; *cf.* Myers *et al.*, 1985c).

The desired alterations in mobility are observed only for branched and partially melted DNA molecules. When the final domain melts the fragment will undergo complete strand separation. Such denatured molecules cannot be separat-

ed from each other and the resolving power of the gel is lost, since the mobility in the gel becomes independent of the sequence. In order to be able to use this method also for the detection of base substitutions in the highest temperature melting domains, and hence for the entire region to be analysed, this region must be attached to a high-melting, GC-rich DNA sequence (80% GC) called the GC clamp (Myers *et al.*, 1985b). In plasmids pGC1 and pGC2 such a GC-rich region is located next to a polylinker, which permits the analysis of DNA segments in different orientations (Fig. 12-21). In practice, molecules may be separated by up to several centimeters (Myers *et al.*, 1985c; Fig. 12-22). Mutated fragments migrating faster or slower than wild-type are then isolated from the gel, recloned, and characterised by sequencing. Transitions from GC to AT will lead to a decrease in melting temperature, and the corresponding fragments will melt earlier than wild-type and migrate slower. Transitions from AT to GC show a reverse behaviour so that the desired mutant type is easily identified. In practice, up to 95% of all possible single-base substitutions often can be resolved in this system.

In order to find suitable conditions for electrophoresis, *i.e.*, a suitable concentration gradient of the denaturing agent, it is necessary to estimate the melting temperature of the domain in question. Melting temperatures either can be calculated by suitable computer programs or ascertained experimentally by testing several different urea concentrations. This technique can also be adapted for the identification of mutations in genomic DNA, and thus for the characterisation of genetic defects (Myers *et al.*, 1985d and e).

12.4 Directed Mutagenesis by Synthetic Oligonucleotides

A universal and extremely specific way of introducing mutations is the use of synthetic oligonucleotides as mutagens. Hutchison *et al.* (1978) have

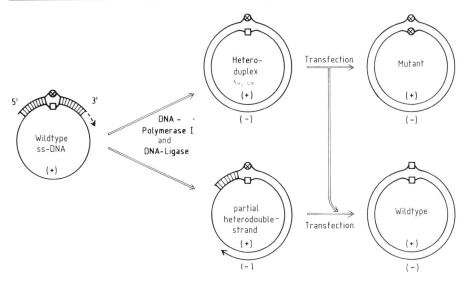

Fig. 12-23.   Directed mutagenesis with synthetic oligonucleotides.
The synthetic oligonucleotide, with a defined deviation from the wild-type sequence, hybridises with the region of single-stranded DNA to be mutagenised and then serves as a primer for *in vitro* DNA synthesis. The resulting double-stranded heteroduplex is used to transfect competent *E. coli* cells. In principle, half of the clones obtained contain wild-type plasmid DNA, the other half plasmids with the desired mutation (Kramer *et al.*, 1982). Transfection with a partial heteroduplex, which is obtained by an incomplete DNA polymerase I reaction, only yields wild-type DNA moleules.

introduced a widely used technique (Fig. 12-23) which employs chemically synthesised oligonucle-otides, the sequences of which are complementa-ry to the DNA region on the single-stranded DNA molecule to be mutagenised. Apart from one or several base changes in specific and predetermined positions, the sequence of the oligonucleotide is identical to the wild-type sequence. Under suitable conditions the mutating oligonucleotide hybridises with the single-strand-ed circular template DNA molecule and serves as a primer for the enzymatic synthesis of a comple-mentary double strand (Fig. 12-23). The resulting heteroduplex molecules are then closed by DNA ligase treatment and transfected into bacteria. The mutants must be identified in the population of wild-type and mutant clones by suitable selec-tion or screening systems.

Initially, this method was developed for bacte-riophage ΦX174 which has a life cycle involving single-stranded and double-stranded phases. Due to its icosahedral shape, the virus cannot take up excess foreign DNA; however, these difficulties can be circumvented using M13 cloning tech-niques. In addition to the vectors described in Section 2.4.2.2, which were derived directly from the filamentous phage M13 (Zinder and Boeke, 1982), newly developed single-stranded plasmid vectors are also available (*cf.* Section 4.1.5). The vector family pEMBL (*cf.* Fig. 4.1-15), for exam-ple, contains parts of pBR322 and several hun-dred base pairs of the M13 genome comprising the signals for M13 DNA replication and phage morphogenesis. pEMBL molecules therefore can be replicated and isolated as normal double-stranded plasmids. Since these plasmids contain M13-specific elements, they can also be packaged as single strands when superinfecting M13 helper viruses provide the necessary coat proteins. pEMBL phages then can be isolated from the culture medium, together with the helper virus (Dente *et al.*, 1983). These vectors offer the advantage that their size is only approximately 4 000 bp, which is considerably smaller than the M13-mp vectors (approximately 8 000 bp), and thus more DNA can be incorporated into, and replicated with, pEMBL molecules. In addition, pEMBL vectors can be amplified by chloramphenicol treatment and therefore can be obtained with high yields, while the amount of M13 replicative form which can be isolated from virus-infected cells is relatively limited. Even more useful, because of their smaller sizes, are the single-stranded miniplasmid vectors which can accommodate up to 10 kb of foreign DNA (*cf.* Fig. 4.1-17).

12.4.1 Length of the Primer

The minimum length of the primer depends on several parameters (Gillam and Smith, 1979). The oligonucleotide should hybridise only to a certain region on a circular template strand, *i.e.*, the site of mutagenesis. The length of an oligonucleotide that will hybridise only once on a chromosome can be calculated from the length of the chromosome. The probability P that a sequence with a length n, composed of four bases, will occur only once on a given chromosome is $P = 1/4^n$. A hepta- or octanucleotide would, therefore, be long enough to meet these statistical requirements and hybridise only once to adenovirus DNA, which is approximately 37 kb ($4^7 = 16 384$; $4^8 = 65 536$). In fact, however, a primer for *E. coli* DNA polymerase I should be at least nine bases long to give perfect base-pairing at 20 °C. Shorter molecules do not form stable double strands under these conditions (Goulian *et al.* 1973). If the stability of the double-stranded regions is decreased due to the presence of a mutation, this effect must be counteracted by a corresponding increase in the length of the primer. Each mismatch requires approximately two additional

bases in the oligonuleotide. A review of the literature shows that primers of 12 bases are rarely used and that oligonucleotides consisting of 16-18 bases are most frequently employed. If genomic sequence data is availabe, it is advisable to use a computer search progam to choose the best oligomer sequence to ensure maximum primer specificity. To verify this specificity a short stretch of the target DNA molecule can be sequenced by M13 primed DNA synthesis using the oligonucleotide in question as a primer.

The use of DNA polymerase I for the synthesis of the complementary strand is rather critical. The native enzyme possesses not only a polymerase activity, but also 5′ to 3′ and 3′ to 5′ exonuclease activities. The enzyme can be separated into two catalytically active fragments by proteolytic cleavage (Klenow *et al.*, 1971). The larger of the two fragments, the Klenow fragment, contains only the polymerase and the 3′ to 5′ proofreading exonuclease activity. If this Klenow fragment is used, the 5′ end of the oligonucleotide is protected from digestion. It is advisable to position the 3′ end of the primer as far away as possible from the target site of mutagenesis (*i.e.* the mismatch between wild-type template and oligonucleotide) in order to avoid the removal of misincorporated bases from the primer by the 3′ to 5′ exonuclease activity of the enzyme. This is another criterion for choosing the minimum length of a suitable primer.

12.4.2 Mutation Frequencies

The replication of M13 DNA proceeds semiconservatively, Thus the predicted ratio for marker distribution from the complementary strands is 50/50. Nevertheless, replication displays an intrinsic bias for the (−) strand markers, due to an asymmetry inherent in the rolling circle mechanism of replication (*cf.* Fig. 2.4-17). Replication of the initial parental RF I form DNA *via* the RI$_e$ intermediates yields two daughter molecules, an

RF I consisting of the parental (+) strand and a newly synthesised (−) strand and an RF II molecule with a parental (−) strand and a newly synthesised (+) strand. The latter molecule, since it is already nicked, can proceed immediately into a new round of replication while the former, a Form I molecule, has yet to be cleaved in order to initiate DNA replication. This replicative advantage of the molecules with the parental (−) strand is retained throughout a considerable part of the replication period and leads to a preponderance (66% to 75%) of this molecular species. Since it is the (−) strand which initially carries the mutated oligonucleotides this bias works in favour of the experiment. In the experiment shown in Fig. 12-23, at least two thirds of the phage plaques should contain the desired mutation. This mutation frequency, or marker yield, is defined as 100 per cent. Practical experience shows that these results are rarely, if ever, obtained. Good yields usually are around or below 10%, and the identification of the desired clone may therefore be a rather exacting task.

Assuming the mutation frequency to be F (in %), the number of clones N which must be analysed in order to achieve a 90% probability of finding the desired clones can be calculated by the formula given below:

$$N = \frac{1}{\log\left(1 - \dfrac{F}{100}\right)}$$

(Kramer *et al.*, 1982). If the marker yield is 10%, the number of clones that must be analysed is 22; for a yield of 5%, 45 clones would have to be analysed to find the desired clone with 90% probability. It is, therefore, of utmost importance to use techniques that guarantee high yields of mutants.

The low yields frequently observed can be explained in various ways. A major factor is certainly the variable efficiency of the DNA polymerase reaction. It is important to note that the single-stranded DNA of single-stranded DNA phages is also infectious. If the synthesis of complementary DNA is not quantitative, one will also obtain infectious, partially single-stranded molecules like those partial heteroduplex double strands shown in Fig. 12-23. Since it is only the wild-type strand which is intact in these molecules, the wild-type strand will replicate and therefore increase the fraction of wild-type DNA. This problem can be solved either by enriching for the desired double-stranded replicative forms in the *in vitro* reaction mix by S1 nuclease treatment of the reaction mixture, or by purifying Form I molecules by centrifugation in alkaline sucrose gradients (Zoller and Smith, 1982). Nuclease treatment destroys the interfering, partially double-stranded molecules, while centrifugation separates the desired intact RF forms (Form I), which have a high sedimentation coefficient in alkali, from the slowly sedimenting forms (see also Table 8-2).

Another factor influencing the proportion of wild-type and mutant clones and favouring the occurrence of wild-type clones is the correction of mismatches in non-base-paired regions in the RF forms before these are replicated *in vivo*. This might be due to the 3′ exonuclease activity still present in the Klenow fragment of DNA polymerase I, which theoretically could digest the primer before *in vitro* DNA synthesis has begun; in addition, even a complete heteroduplex containing mismatches (Fig. 12-23) could also be repaired by a removal of the misincorporated nucleotides. It has been postulated that such a repair system for replicational errors can distinguish between correct and incorrect nucleotides because it can recognise differences in the level of methylation of both strands. Since methylation occurs only after replication, the parental DNA would normally be methylated; the newly synthesised strand would be undermethylated and would be a good substrate for suitable correction mechanisms. Glickman and Radman (1980) have shown that undermethylation is almost certainly the guide for mismatch repair (*cf.* also Pukkila *et al.*, 1983). λ DNA heteroduplexes, for example, in which only one of the two complementary strands is methylated, will lose genetic markers only from the unmethylated strand.

In a site-directed mutagenesis experiment using oligonucleotides, initially the parental single strand carrying the wild-type sequence will be methylated and the strand synthesised *in vitro* and primed by the oligonucleotide will be unmethylated. The mismatch repair system would then correct the mismatch in the oligonucleotide strand and would therefore promote the production of wild-type DNA. This mechanism might explain why higher yields are always obtained in oligonucleotide-directed mutagenesis experiments involving ΦX174 DNA rather than M13 vectors. *Dam* methylation in *E. coli* occurs exclusively at adenine residues in 5'-GATC-3' sequences and produces N^6-methyl adenines (Grier and Modrich, 1979; Marinus and Morris, 1973). Since this GATC sequence does not occur in ΦX174 DNA, *dam* methylation, and hence repair processes based on this modification, should not be relevant to mismatches in ΦX174 DNA. In contrast, M13 and fd DNA possess three and four 5'-GATC-3' sites, respectively, and therefore are subject to such repair mechanisms.

A variety of strategies have been proposed to overcome these problems and to raise the yield of mutated molecules. Based on the correlation between GATC methylation and marker yields mentioned above, Kramer *et al.* (1982) have suggested that oligonucleotide-directed mutagenesis should be performed with unmethylated parental DNA as the single-stranded template. Such DNA can be obtained easily by using DNA methylation-defective *E. coli dam* mutants. Unfortunately, mutation frequencies in such mutants are three to four orders of magnitude higher than mutation frequencies in the wild-type; the mutants also show increased recombination frequencies and an increased sensitivity to mutagens and UV light (Bale *et al.*, 1978; Marinus and Konrad, 1976) and must, therefore, be handled with a certain care.

Kramer *et al.* (1982) have also suggested the use of partially single-stranded molecules, the so-

called "gapped duplex" approach, to avoid mismatch repair problems. As shown in Fig. 12-24 for pBR322 clones, "gapped duplex" molecules can be generated by hybridising with a vector DNA lacking the inserted sequence. According to Dalbadic-McFarland *et al.* (1982) such molecules also can be obtained by partial exonuclease III digestion, beginning at a strand break within a

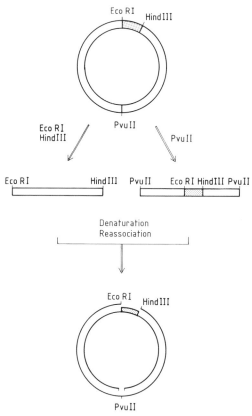

Fig. 12-24. Generation of partially single-stranded substrates for directed mutagenesis.
The principle of this approach is similar to that described in Fig. 12-15. In this example, a pBR322 clone with an insertion between the *Eco* RI and *Hind* III site is opened by cleavage at the *Pvu* II site residing opposite to the insertion, or by cleavage at the *Eco* RI and *Hind* III sites flanking the insertion. Denaturation and reassociation yield a double-stranded molecule which is single-stranded in the desired region and can interact with an oligonucleotide primer.

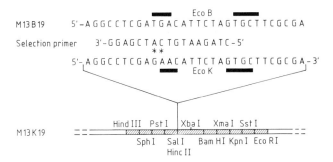

Fig. 12-25. Selection mutagenesis vectors M13K19 and M13B19.
Vector M13K19 is derived from M13mp19 (*cf.* Fig. 2.4-18) and carries a 30 bp insertion carrying the *Eco* K recognition site (*cf.* Fig. 2.1-1; black bars). The selection primer with 2 mismatches (asterisks) changes the *Eco* K into an *Eco* B site (black bars on top), resulting in the formation of vector M13B19 (Carter *et al.*, 1985).

single strand. *In vitro* DNA synthesis of the complementary strand will be simplified in either case, since it does not have to proceed along the entire single strand, but only fills-in the desired gap.

This approach could be used for example with an unmethylated (+) strand and a methylated (–) strand. The mismatch repair system would correct the (+) strand rather than the (–) strand, thereby transferring the newly introduced mutation into the template strand. Unfortunately, however, the efficiency of methyl-directed DNA mismatch repair is strongly dependent on the nature of the mismatch to be corrected. While TG and CA transition-type mismatches are readily corrected, the transversion-type mismatches GA, AG and CA are corrected with greatly reduced efficiencies. If the mismatch is not repaired, the (+) strand thus retains the wild-type sequence, thereby reducing the marker yield. In view of these difficulties direct biological selection either for the (–) strand or against the (+) strand may be desirable. (Kramer *et al.*, 1984a).

The strategy of Kramer *et al.* (1984b) uses a selection system based on α-complementation. An amber mutation is introduced into codon sixteen of the *lacZ* fragment in M13mp9. This mutation renders *lac* expression from this vector dependent on a host-encoded suppressor. A "gapped duplex" is then constructed consisting of a (+) strand carrying the *lacZ* amber mutation and a (shorter) gapped (–) strand with a wild-type codon in the *lacZ* portion. An oligonucleotide is annealed to this gap and linked to the (-) strand by gap-filling and ligation steps. Transformation into a Lac⁻ Dam⁻, non-supressing host, with selection for Lac⁺ transformants, yields mutation frequencies of up to 70%.

The strategy of Kunkel (1985) destroys the undesired (+) strand directly. A single-stranded phage vector is grown in a Dut⁻ Ung⁻ strain of *E. coli*. The lack of the *dut* gene, which codes for a dUTPase, leads to an overproduction of dUTP which in turn can be incorporated into DNA instead of dTTP. Ung⁻ strains lack an uracil-glycosylase which removes dU residues from DNA. It is therefore possible to obtain single-stranded vector DNA from Dut⁻ Ung⁻ strains carrying about 20-30 uracil residues. This DNA is used as a template (+) strand in a site-directed mutagenesis experiment as described in Fig. 12-23. Prior to transfection, the double strand is treated with uracil-glycosylase and subsequently with alkali to destroy the (+) strand. The uracil glycosylase treatment removes its U residues and leaves the abasic DNA open to attack by alkali.

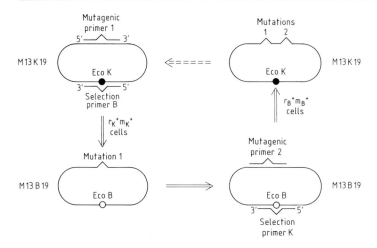

Fig. 12-26. Mutagenesis scheme for multiple rounds of selection.
The structures of M13K19 and M13B19 have been described (Fig. 12-25). Apart from the selection primer the system uses mutagenic primers to mutagenise the gene insertion in question. By alternating growth cycles in $r_K{}^+m_K{}^+$ or $r_B{}^+m_B{}^+$ strains, multiple rounds of selection can be performed with different mutagenic primers (Carter *et al.*, 1985).

Although the overall yield of plaques in this strategy is low (about 1%) the mutation frequency has been reported to exceed 80%.

An ingenious selection system based on *Eco* K and *Eco* B restriction permits multiple rounds of mutagenesis to be performed with high efficiency (Carter *et al.*, 1985). The M13 vectors M13mp18 or 19 are provided with a 30 bp insertion into the *Sal* I polylinker site to form M13K19 which retains donor ability in α-complementation tests (Fig. 12-25). The insert carries an *Eco* K recognition site (*cf.* Fig. 2.1-1). An M13K19 (+) strand is then hybridised to two different primers (Fig. 12-26); one is the so-called selection primer, which changes the *Eco* K sequence into an *Eco* B sequence, the other is the mutagenic primer which alters the target gene. The primers are extended with the Klenow fragment in the presence of T4 DNA ligase, and the double-stranded product is then transfected into an $r_K{}^+m_K{}^+$ strain. This selectively destroys all plasmids derived from the (+) strand, since these plasmids carry the *Eco* K site. The progeny carries an

Eco B site derived from the (−) strand. The selection now can be repeated with a selection primer which changes the *Eco* B site into an *Eco* K site; in this round the strand selection step is executed in an $r_B{}^+m_B{}^+$ strain. Marker yields exceeding 70% have been obtained with such strand-selection protocols.

12.4.3 Selection and Identification Techniques

Although marker yields have been improved considerably by the modifications described in the previous section, the use of additional selection and identification techniques is usually indispensable.

The first attempts to mutagenise ΦX174 DNA were aimed at the generation or reversion of *amber* mutations. In this case, a biological selection for the desired mutants is provided by suitable suppressor-free and suppressor-positive *E. coli* strains (Smith and Gillam, 1981). Another

Oligonucleotide

Wildtype-DNA
(+)strand

Gene E
Gene D

Fig. 12-27. Section of the sequence of ΦX174 DNA containing part of the coding region of gene *D*, the ribosomal binding site, and the start codon for gene *E*.
The two different reading frames are marked by parentheses (gene *E*) and brackets (gene *D*). The synthetic oligonucleotide primer leads to a G to A transition at position 557, which generates a *Hind* III cleavage site (A/AGCTT) (Smith and Gillam, 1981).

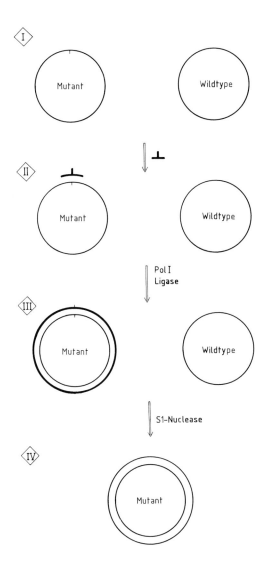

attractive selection utilises the loss or generation of a restriction enzyme cleavage site introduced by the desired mutation. A mutation in the ribosomal binding site of gene *E* of ΦX174 DNA, for example, which alters the sequence GA*G*GCTT to GA*A*GCTT, generates a *Hind* III site (A/AGCTT) without altering the amino acid sequence in gene *D* (Fig. 12-27). Since there are no other *Hind* III sites in ΦX174 DNA, mutants can be identified by a simple *Hind* III digestion of the replicative forms (Smith and Gillam, 1981). Similarly, Wasylyk *et al.* (1980) have identified a mutation in the TATA box of the conalbumin gene. In this case the transversion from TCTA*T*A to TCTA*G*A generated a new *Xba* I site (T/CTAGA).

Mutations also can be isolated, however, which do not create, or eliminate, a cleavage site. The oligonucleotide primer plays a central role: in two methods for increasing mutant yields it can be used to enrich for the desired mutant DNA in mixtures containing wild-type and mutated DNA; it also can be used as a probe for colony

Fig. 12-28. *In vitro* selection of mutant DNA by oligonucleotides.
Specific mutagenesis first leads to a mixture of wild-type and mutant DNA (I). The oligonucleotide preferentially hybridises to mutant DNA (II), and the subsequent reaction with polymerase and ligase therefore favours the formation of mutant double strands (III). Wild-type DNA is digested with S1 nuclease. This cycle is repeated several times until the proportion of mutant DNA in the mixture is satisfactory (Smith and Gillam, 1981).

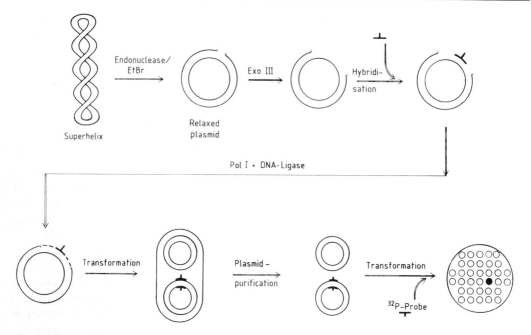

Fig. 12-29. Specific mutagenesis of plasmid DNA molecules by mutating oligonucleotides (see text; Dalbadic-McFarland *et al.*, 1982).

hybridisation. Both techniques are based on the fact that a perfect double strand between the mutated oligonucleotide and the complementary mutated DNA is much more stable than duplexes between mutated primer and wild-type DNA. In a mixture of single-stranded mutated DNA and wild-type DNA the primer will therefore react preferentially with the mutated DNA, which can then be converted much more efficiently into an S1 nuclease-resistant double strand (Fig. 12-28). S1 nuclease will preferentially digest the unprotected wild-type DNA and the remaining double-stranded forms can then be used to transfect *E. coli* spheroplasts, from which the single-stranded viral DNA can be reisolated. The isolated DNA can be subjected to another enrichment step with the primer (Smith and Gillam, 1981). Marker yields can be greatly increased by several such cycles; eventually only a few clones must be sequenced in order to identify the desired mutant.

Although this technique requires only a few pmoles of DNA, it is not unproblematic, because of the rather delicate reaction conditions of the S1 digestion.

Another technique that is widely used today was worked out by Wallace *et al.* (1981a, b). It utilises the ^{32}P-labelled primer directly as a hybridisation probe and can be applied to bacterial colonies as well as phage lysates. A similar technique was described by Dalbadic-McFarland (1982) (Fig. 12-29). Plasmid DNA is digested with a restriction enzyme in the presence of ethidium bromide, then treated with exonuclease III to convert it into a partially single-stranded form. After hybridisation with the mutated oligonucleotide, the DNA is treated with DNA polymerase I and DNA ligase, and transformed into *E. coli* spheroplasts. The resulting mixture of plasmids contains mutant and wild-type DNA. This mixed DNA is transformed back into cells to isolate the different plasmid molecules; single colonies are cultured in microtiter plates (Gergen *et al.* 1979) and transferred to paper filters. Colonies carrying the mutant plasmid are then identified by hybridisation with the ^{32}P-labelled oligonucleotide.

The same procedures can be applied to phage lysates. Single plaques are amplified in milliliter cultures and a sample of the supernatant is transferred to a nitrocellulose filter (see Winter *et al.*, 1982, for example). Several empirical parameters for the hybridisation conditions were described by Suggs *et al.* (1981). At room temperature, oligonucleotides of 16-18 bases will hybridise with mutant and wild-type DNA equally well, even if there are one or two mismatches. By carefully raising the temperature, the primer can be selectively removed from the wild-type DNA. This is accomplished by choosing a temperature which is two degrees higher than T_d, the temperature at which 50% of the primer has already been removed. T_d can be calculated as the sum of 2 °C for each AT pair plus 4 °C for each GC pair if the hybridisation is carried out in 6×SSC buffer (0.9 M NaCl; 0.09 M sodium citrate, pH 7.0) (Suggs *et al.*, 1981). For practical purposes the actual hybridisation reactions are usually performed at room temperature and the washing proceduresare then carried out by slowly raising the temperature to the T_d and slightly above. The example shown in Fig. 12-30 (Winter *et al.*, 1982) involves the use of a hexadecanucleotide. This oligonucleotide is complementary to a region around cysteine 35 of the tyrosyl-tRNA synthetase gene of *B. stearothermophilus*, with the exception of a single base change which causes a shift from cysteine (TGC) to serine (AGC). By taking into account the 9 GC pairs and the 6 AT pairs, T_d is calculated to be 48 °C. Following hybridisation at room temperature, the washing procedure is then carried out for several minutes at 50 °C (T_d + 2 °C). In the experiment described here, the Form I DNA was purified by alkaline sucrose gradients and 16 of 36 clones (44%) contained the desired mutated phage DNA.

12.4.4 Applications

There are numerous practical applications for oligonucleotide-directed mutagenesis. Such diverse problems as the analysis of eukaryotic and prokaryotic promoters and other control sequences and the study of enzyme mechanisms have been addressed by using this technique. Studies of the role of the TATA box in the transcription of the conalbumin and ovalbumin genes were described by Wasylyk *et al.* (1980) and Zaruchi-Shulz (1982), for example. The A to T and T to G transversions in positions two or three in either of these control sequences resulted in a dramatic reduction of the transcription efficiency. Miyada *et al.* (1982) constructed deletions in the *E. coli araBAD* promoter. The *araBAD* operon is a classical example of an *E. coli* operon positively regulated by the regulator proteins ara*C* and CRP. The deletions described by Miyada *et al.* (1982) located the binding sites of these two proteins in regions "-60" and "-90" bp upstream of the transcription initiation site.

The role of the consensus sequence, AG/GUAAGUA, which occurs in almost all 5' donor sites of exon-intron junctions was studied by Montell *et al.* (1982). A detailed analysis of the 12S mRNA of the early *EIa* transcription unit of adenovirus type 2 revealed that a T to G transversion in the neighbouring intron sequence com-

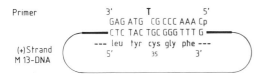

Fig. 12-30. *In vitro* mutagenesis of tyrosyl-tRNA synthetase from *B. stearothermophilus*.
The M13 clone contains a *Sal* I fragment, approximately 2.5 kb in length, which is composed of the 1,257 bp coding region of the tyrosyl-tRNA gene, approximately 1 000 bp of the 5', and 296 bp of the 3' flanking regions. Thick lines represent *B. stearothermophilus*, thin lines M13 sequences. Superscript T (bold-face) indicates the site of mutation (Winter *et al.*, 1982).

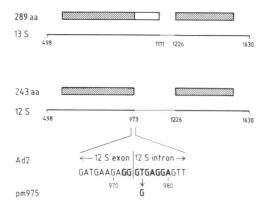

Fig. 12-31. Section of the sequence of the extreme left end of the linear adenovirus genome.
The *E1a* region codes for a 12S and a 13S mRNA which direct the synthesis of two proteins of 243 and 289 amino acids, respectively. The hatched parts of the bars, representing the proteins, indicate the regions where the sequences of the two proteins are identical. The open bar indicates the 46 amino acids only found in the larger protein. Numbers are co-ordinates of the adenovirus genome with respect to the left end. The sequence section shown comes from the region near the 5' donor site of the 12S mRNA. The T to G transversion is marked by an arrow (Montell *et al.*, 1982).

pletely suppressed the expression of this mRNA (Fig. 12-31). Inouye *et al.* (1982) investigated the influence of positively charged amino acids in the signal sequence of the prolipoprotein of the outer cellular membrane of *E. coli* on lipoprotein transport through the membrane. The influence of specific sequences in tRNA molecules on tRNA biosynthesis was studied by Kudo *et al.* (1981). Winter *et al.* (1982) and Dalbadic-McFarland *et al.* (1982) described studies employing site-directed mutagenesis for the alteration of the catalytic properties of enzymes. In the future, similar experiments may provide valuable insights into the mechanisms of enzyme action and the relationships between primary structure and biological functions of enzymes. The term "protein engineering" was coined to describe a new field that is being opened by site-directed mutagenesis.

Among the many applications of synthetic oligonucleotides, a particularly impressive one is the use of oligonucleotide probes for the antenatal diagnosis of genetic diseases. Sickle cell anaemia, for example, is a common genetic defect in certain black populations (*cf.* Table 13-1). Its cause is a single base transversion in the β-globin gene (adenine to thymine), which leads to the substitution of glutamic acid for valine at position 6 of the β-chain of the haemoglobin molecule. Conner *et al.* (1983) used two 19-base oligonucleotides which correspond to the sequence of the normal human β-globin gene (β^A) and the sickle cell allele (β^S) as probes for the sickle cell allele. Normal homozygous individuals ($\beta^A\beta^A$), heterozygous ($\beta^A\beta^S$), and homozygous sickle cell anaemia patients ($\beta^S\beta^S$) can be distinguished from each other by differential hybridisation with these probes. In principle, this technique should be applicable to the diagnosis of any other single gene defect for which the lesion has been identified on the nucleotide level.

References

Auerswald, E.A., Ludwig, G., and Schaller, H. (1981). Structural analysis of Tn5. Cold Spring Harbor Symp. Quant. Biol. 45, 107-113.

Bale, A., d'Alarco, M., and Marinus, M.G. (1978). Characterization of DNA adenine methylation mutants of E. coli K12. Mutat. Res. 59, 157-165.

Beattie, K.L., Wiegand, R., and Radding C.M. (1977). Uptake of homologous single-stranded fragments by superhelical DNA. II. Characterization of the reaction. J. Mol. Biol. 116, 783-803.

Beck, E., and Zink, B. (1981). Nucleotide sequence and genome organisation of filamentous bacteriophage f1 and fd. Gene 16, 35-58.

Beck, E., Ludwig, G., Auerswald, E.A., Reiss, B., and Schaller, H. (1982). Nucleotide sequence and exact location of the neomycin phosphotransferase gene from Transposon Tn5. Gene 19, 327-336.

Berg, D.E., Davies, J., Allet, B., and Rochaix, J.D. (1975). Transposition of R factor genes to bacteriophage λ. Proc. Natl. Acad. Sci. USA 72, 3628-3632.

Boeke, J.D. (1981). One and two codon insertion

mutants of bacteriophage f1. Mol. Gen. Genet. 181, 288-291.

Caras, J.W., MacInnes, M.A., Persing, D.H., Coffino, P., and Martin, D.W., Jr. (1982). Mechanism of 2-aminopurine mutagenesis in mouse T-lymphosarcoma cells. Mol. Cell. Biol. 2, 1096-1103.

Carbon, J., Shenk, T., and Berg, P. (1975). Biochemical procedure for production of small deletions in simian virus 40 DNA. Proc. Natl. Acad. Sci. USA 72, 1392-1396.

Carter, P., Bedouelle, H., and Winter, G. (1985). Improved oligonucleotide site-directed mutagenesis using M13 vectors. Nucleic Acids Res. 13, 4431-4443.

Conner, B.J., Reyes, A.A., Morin, C., Itakura, K., Teplitz, R.L., and Wallace, R.B. (1983). Detection of sickle cell βS-globin allele by hybridisation with synthetic oligonucleotides. Proc. Natl. Acad. Sci. USA 80, 278-282.

Dalbadic-McFarland, G., Cohen, L.W., Riggs, A.D., Morin, C., Itakura, K., and Richards, J.H. (1982). Oligonucleotide-directed mutagenesis as a general and powerful method for studies of protein function. Proc. Natl. Acad. Sci. USA 79, 6409-6413.

Dente, L., Cesarini, G., and Cortese, R. (1983). pEMBL: a new family of single-stranded plasmids. Nucleic Acids Res. 11, 1645-1655.

Ditta, G., Stanfield, S., Corbin, D., and Helinski, D.R. (1980). Broad host range DNA cloning system for gram-negative bacteria: Construction of a gene bank of *Rhizobium meliloti*. Proc. Natl. Acad. Sci. USA 77, 7347-7351.

Eckstein, F. (1985). Nucleoside phosphorothioates. Ann. Rev. Biochem. 54, 367-402.

Everett, R.D., and Chambon, P. (1982). A rapid and efficient method for region- and strand-specific mutagenesis of cloned DNA. The EMBO Journal 1, 433-437.

Fischer, G.S., and Lerman, L.S. (1983). DNA fragments differing by single base pair substitutions are separated in denaturing gels. Correspondence with melting theory. Proc. Natl. Acad. Sci. USA 80, 1579-1583.

Gergen, J.P., Stern, R.H., and Wensink, P.C. (1979). Filter replicas and permanent collections of recombinant DNA plasmids. Nucleic Acids Res. 7, 2115-2136.

Gillam, S., and Smith, M. (1979). Site-specific mutagenesis using synthetic oligodeoxyribonucleotide primers. I. Optimum conditions and minimum oligodeoxyribonucleotide length. Gene 8, 81-97.

Glickman, B.W., and Radman, M. (1980). *Escherichia coli* mutator mutants deficient in methylation-instructed DNA mismatch correction. Proc. Natl. Acad. Sci. USA 77, 1063-1067.

Goulian, M., Goulian, S.H., Lodd, E.E., and Blumenfield, A.Z. (1973). Properties of oligodeoxynucleotides that determine priming activity with *Escherichia coli* deoxyribonucleic acid polymerase I. Biochemistry 12, 2893-2901.

Green, C., and Tibbetts, C. (1980). Targeted deletions of sequences from closed circular DNA. Proc. Natl. Acad. Sci. USA 77, 2455-2459.

Grier, G.E., and Modrich, P. (1979). Recognition sequence of the dam methylase of *E. coli* K12 and mode of cleavage of *Dpn* I endonuclease. J. Biol. Chem. 254, 1408-1413.

Harris, T. (1982). *In vitro* mutagenesis. Nature 299, 298-299.

Heffron, F., So, M., and McCarthy, B.J. (1978). *In vitro* mutagenesis of a circular DNA molecule using synthetic restriction sites. Proc. Natl. Acad. Sci. USA 75, 6012-6016.

Hutchison, C.A., Phillips, S., Edgell, M.H., Gillam, S., Jahnke, D., and Smith, M. (1978). Mutagenesis at a specific position in a DNA sequence. J. Biol. Chem. 253, 6551-6560.

Inouye, S., Soberon, X., Francescini, T., Nakamura, K., Itakura, K., and Inouye, M. (1982). Role of positive charge on the amino-terminal region of the signal peptide in protein secretion across the membrane. Proc. Natl. Acad. Sci. USA 79, 3438-3441.

Kalderon, D., Oostra, B.A., Ely, B.K., and Smith, A.E. (1982). Deletion loop mutagenesis: a novel method for the construction of point mutations using deletion mutants. Nucleic Acids Res. 10, 5161-5171.

Kleckner, N. (1981). Transposable elements in prokaryotes. Ann. Rev. Genet. 15, 341-404.

Klenow, H., Overgaard-Hansen, K., and Pathar, S.A. (1971). Proteolytic cleavage of native DNA polymerase into two different catalytic fragments. Influence of assay conditions on the change of exonuclease activity and polymerase activity accompanying cleavage. Europ. J. Biochem. 22, 371-381.

Kramer, W., Schughart, K., and Fritz, H.-J. (1982). Directed mutagenesis of DNA cloned in filamentous phage: influence of hemimethylated GATC sites on marker recovery from restriction fragment. Nucleic Acids Res. 10, 6475-6485.

Kramer, B., Kramer, W., and Fritz, H.J. (1984a). Different base/base mismatches are corrected with different efficiencies by the methyl-directed DNA mismatch-repair system. Cell 38, 879-887.

Kramer, W., Drutsa, V., Jansen, H.W., Kramer, B., Pflugfelder, M., and Fritz, H.J. (1984b). The gapped duplex DNA approach to oligonucleotide-directed mutation construction. Nucleic Acids Res. 12, 9441-9456.

Kudo, J., Leineweber, M., and RayBhandary, U.L.

(1981). Site-specific mutagenesis on cloned DNAs: Generation of a mutant of *Escherichia coli* tyrosine suppressor tRNA in which the sequence G-T-T-C corresponding to the universal G-T-ψ-C sequence of tRNAs is changed to G-A-T-C. Proc. Natl. Acad. Sci. USA 78, 4753-4757.

Kunkel, T.A., Eckstein, F. Mildvan, A.S., Koplitz, R.M., and Loeb, L.A. (1981). Deoxynucleoside-1-thio triphosphates prevent proofreading during *in vitro* DNA synthesis. Proc. Natl. Acad. Sci. USA 78, 6734-6738.

Kunkel, T.A. (1985). Rapid and efficient site-specific mutagenesis without phenotypic selection. Proc. Natl. Acad. Sci. USA 82, 488-492.

Marinus, M.G., and Konrad, E.B. (1976). Hyperrecombination in *dam* mutants of *E. coli* K12. Mol. Gen. Genet. 149, 273-277.

Marinus, M.G., and Morris, N.R. (1973). Isolation of deoxyribonucleic acid methylase mutants of *E. coli* K12. J. Bacteriol. 114, 1143-1150.

McKinnon, R.D., Bacchetti, S., and Graham, F.L. (1982). Tn5 mutagenesis of the transforming genes of human adenovirus type 5. Gene 19, 33-42.

McKnight, S.L. (1982). Functional relationships between transcriptional control signals of the thymidine kinase gene of Herpes Simplex Virus. Cell 31, 355-365.

McKnight, S.L., and Kingsbury, R. (1982). Transcriptional control signals of a eukaryotic protein-coding gene. Science 217, 316-324.

Miller, J.H. (1972). Experiments in molecular genetics. Cold Spring Harbor Laboratory, Cold Spring Harbor, New York, 11734, USA.

Miyada, C.G., Soberon, X., Itakura, K., and Wilcox, G. (1982). The use of synthetic oligodeoxyribonucleotides to produce specific deletions in the *araBAD* promotor of *Escherichia coli* B/r. Gene 17, 167-177.

Montell, C., Fischer, E.F., Caruthers, M.H., and Berk, A.J. (1982). Resolving the functions of overlapping viral genes by site-specific mutagenesis at a mRNA splice site. Nature 295, 380-384.

Müller, W., Weber, H., Meyer, F., and Weissmann, C. (1978). Site-directed mutagenesis in DNA: Generation of point mutations in cloned β-globin complementary DNA at the positions corresponding to amino acids 121 to 123. J. Mol. Biol. 124, 343-358.

Myers, R.M., Lerman, L.S., and Maniatis, T. (1985a). A general method for saturation mutagenesis of cloned DNA fragments. Science 229, 242-247.

Myers, R.M., Fischer, S.G., Maniatis, T., and Lerman, L.S. (1985b). Modification of the melting properties of duplex DNA by attachment of a GC-rich DNA sequence as determined by denaturing gel electrophoresis. Nucleic Acids Res. 13, 3111-3129.

Myers, R.M., Fischer, S.G., Lerman, L.S., and Maniatis, T. (1985c). Nearly all single base substitutions in DNA fragments joined to a GC-clamp can be detected by denaturing gel electrophoresis. Nucleic Acids Res. 13, 3131-3145.

Myers, R.M., Lumelsky, N., Lerman, L.S., and Maniatis, T. (1985d). Detection of single-base substitutions in total genomic DNA. Nature 313, 495-498.

Myers, R.M., Larin, Z., and Maniatis, T. (1985e). Detection of single base substitutions by ribonuclease cleavage at mismatches in RNA:DNA duplexes. Science 230, 1242-1246.

Österlund, M., Luthman, H., Nilsson, S.V., and Magnusson, G. (1982). Ethidium-bromide inhibited restriction endonucleases cleave one strand of circular DNA. Gene 20, 121-125.

Parker, R.C., Watson, R.M., Vinograd, J. (1977). Mapping of closed circular DNAs by cleavage with restriction endonucleases and calibration by agarose gel electrophoresis. Proc. Natl. Acad. Sci. USA 74, 851-855.

Pukkila, P.J., Peterson, J., Herman, G., and Meselson, M. (1983). Effects of high levels of DNA adenine methylation on methyl-directed mismatch repair in *E. coli*. Genetics 104, 571-582.

Putney, S.D., Benkovicz, S.J., and Schimmel, P.R. (1981). A DNA fragment with an α-phosphorothioate nucleotide at one end is asymmetrically blocked from digestion by exonuclease III and can be replicated *in vivo*. Proc. Natl. Acad. Sci. USA 787, 7350-7354.

Radding, C.M. (1981). Recombination activities of *E. coli* RecA Protein. Cell 25, 3-4.

Ruvkun, G.B., and Ausubel, F.M. (1981). A general method for site-directed mutagenesis in prokaryotes. Nature 289, 85-88.

Sakonju, S., Bogenhagen, D.F., and Brown, D.D. (1980). A control region in the center of the 5S RNA gene directs specific initiation of transcription. I. The 5′ border of the region. Cell 19, 13-25.

Shenk, T.E., Rhodes, C., Rigby, P.W., and Berg, P. (1975). Biochemical method for mapping mutational alterations in DNA with S1-nuclease: The location of deletions and temperature-sensitive mutations in Simian Virus 40. Proc. Natl. Acad. Sci. USA 72, 989-993.

Shenk, T. (1977). A biochemical method for increasing the size of deletion mutations in simian virus 40 DNA. J. Mol. Biol. 113, 503-515.

Shortle, D., and Nathans, D. (1978). Local mutagenesis: A method for generating viral mutants with base substitutions in preselected regions of the viral genome. Proc. Natl. Acad. Sci. USA 75, 2170-2174.

Shortle, D., Koshland, D., Weinstock, G.M., and

Botstein, D. (1980). Segment-directed mutagenesis: Construction *in vitro* of point mutations limited to a small predetermined region of a circular DNA molecule. Proc. Natl. Acad. Sci. USA 77, 5375-5379.

Shortle, D., DiMaio, D., and Nathans, D. (1981). Directed mutagenesis. Ann. Rev. Genet. 15, 265-294.

Shortle, D., Grisafi, P., Benkovicz, S.J., and Botstein, D. (1982). Gap misrepair mutagenesis: Efficient site-directed induction of transition, transversion, and frame-shift mutations *in vitro*. Proc. Natl. Acad. Sci. USA 79, 1588-1592.

Smith, M., and Gillam, S. (1981). Constructed mutants using synthetic oligodeoxyribonucleotides as site-specific mutagens. In: Genetic Engineering, I.K. Setlow and A. Hollaender (eds.). Plenum Press, New York and London; ISBN 0-306-40729-9; pp.1-32.

Suggs, S.V., Wallace, R.B., Hirose, T., Kawashima, E.H., and Itakura, K. (1981). Use of synthetic cloned cDNA sequences for human β2-microglobulin. Proc. Natl. Acad. Sci. USA 78, 6613-6617.

Taniguchi, T., and Weissmann, C. (1978). Site-directed mutations in the initiator region of the bacteriophage Qβ coat cistron and their effect on ribosome binding. J. Mol. Biol. 118, 533-565.

Vosberg, H.P., and Eckstein, F. (1977). Incorporation of phosphorothioate groups into fd and ΦX174 DNA. Biochemistry 16, 3633-3640.

Wallace, R.B., Johnson, M.J., Hirose, T., Miyake, T., Kawashima, E.H., and Itakura, K. (1981a). The use of synthetic oliogonucleotides as hybridization probes. II. Hybridization of oligonucleotides of mixed sequence to rabbit β-globin-DNA. Nucleic Acids Res. 9, 879-894.

Wallace, R.B., Shold, M., Johnson, M.J., Dembek, P., and Itakura, K. (1981b). Oligonucleotide directed mutagenesis of the human β-globin gene: a general method for producing specific point mutations in cloned DNA. Nucleic Acids Res. 9, 3647-3656.

Warner, H.R., and Duncan, B.K. (1978). *In vivo* synthesis and properties of uracil-containing DNA. Nature 272, 32-34.

Wasylyk, B., Derbyshire, R., Guy, A., Molko, D., Roget, A., Teoule, R., and Chambon, P. (1980). Specific *in vitro* transcription of conalbumin gene is drastically decreased by a single-point mutation in T-A-T-A box homology sequence. Proc. Natl. Acad. Sci. USA 77, 7024-7028.

Weiher, H., and Schaller, H. (1982). Segment-specific mutagenesis: Extensive mutagenesis of a *lac* promotor/operator element. Proc. Natl. Acad. Sci. USA 79, 1408-1412.

Winter, G., Fersht, A.R., Wilkinson, A.J., Zoller, M., and Smith, M. (1982). Redesigning enzyme structure by site-directed mutagenesis: tyrosyl-tRNA-synthetase and ATP binding. Nature 299, 756-758.

Zaruchi-Shulz, T., Tsai, S.Y., Itakura, K., Soberon, X., Wallace, R.B., Tsai, M.J., Woo, S.L.C., and O'Mallay, B.W. (1982). Point mutagenesis of the ovalbumin gene promotor sequence and its effect on *in vitro* transcription. J. Biol. Chem. 257, 11070-11077.

Zinder, N.D., and Boeke, J.D. (1982). The filamentous phage (fd) as vector for recombinant DNA – a review. Gene 19, 1-10.

Zoller, M.J., and Smith, M. (1982). Oligonucleotide-directed mutagenesis using M13-derived vectors: an efficient and general procedure for the production of point mutations in any fragment of DNA. Nucleic Acids. Res. 10, 6487-6500.

13 Safety in Recombinant DNA Research

No significant human activity can be performed with zero risk. Those employing a technology thus have to consider the consequences of their endeavours. Many of the more beneficial applications of gene technology have been mentioned in previous chapters of this book. The *potential* risks of future applications are many; scientists themselves drew early attention to the possibility that certain problems might be caused by the application of genetic engineering. In a letter to the United States Academy of Science (July 26, 1974) a group of scientists with Paul Berg as spokesman publicly expressed concern about the "potential biohazards of recombinant DNA molecules". In particular, this group asked the scientific community for a voluntary deferment of cloning experiments involving genomes of certain animal viruses, of antibiotic resistance determinants, toxins and oncogenes. In addition, this moratorium was to be adhered to until the director of the National Institutes of Health established procedures and guidelines to be followed by investigators in this field. Finally, the letter called for an international conference to review and to discuss these issues. This conference was, in fact, held at Asilomar in February 1975 and resulted in a vote both to continue the research and to develop a useful safety code. In particular, the conference discussed the novel concept of biological containment, which requires the use of disabled vector/host systems which can only survive under special laboratory conditions.

A vivid and detailed account of the events leading to this conference, conclusions derived at, and of further developments both in the United States and worldwide is given in "The DNA story" (Watson and Tooze, 1981) which also includes facsimiles of the original documents and letters. Even before this conference, the Director of the National Institutes of Health established the Recombinant DNA Molecule Program Advisory Committee, or for short, the RAC Committee. This committee formulated a set of guidelines which were eventually issued by the United States Secretary of Health, Education and Welfare on June 23, 1976. These guidelines set up and defined categories of physical and biological containment procedures which were to be followed by researchers in the field, according to the types of vectors and the origins of the DNA to be used. The four physical containment levels ranged from standard microbiological practice (P1) to carefully designed laboratory conditions involving negative pressure, air locks, autoclaves and so forth (P4, see below). Biohazard risks dictated the establishment of three safety levels, demanding the use of either the standard laboratory strains of *E. coli* K12 (EK1) or genetically disabled derivatives which have theoretically (EK2) or experimentally (EK3) shown to be ten times less likely to survive outside the laboratory than the standard strain.

These guidelines were stricter than the Asilomar recommendations and at that time prevented many experiments due to the lack of appropriate containment facilities. While they were prepared and put into effect, the misgivings and suspicions of the general public, encouraged by the media, grew constantly. They reached a climax in Cambridge, Massachusetts, when Harvard University proposed to set up a P3 facility in its biological laboratories. Due to some resistance within the university itself, the debate spread quickly to involve the community at large and eventually the Cambridge City Council. In 1977, an Experimentation Review Board issued its recommendations, which contained even stricter requirements than the NIH guidelines. While few other cities in the United States followed suit, matters were soon taken up by the Congress of the United States. A variety of representatives and senators introduced a rash of bills which could have set up vast bureaucracies and harsh penalties for violations of the guidelines. The tide eventually turned in the summer of 1978. At that time it became obvious that thousands of experiments had been done in a multitude of laboratories with not a single accident or untoward effects on laboratory workers or the environment, in general. Furthermore, it became clear at the time, due to the efforts of many microbiologists, that *E. coli* K12 is so defective that there is no chance for it to become a pathogen which would be able to survive outside of the laboratory.

In this context, Curtiss (1976) constructed the famous strain *E. coli* χ1776, named after the year of the United States' Declaration of Independence. This strain is extremely sensitive to UV light and detergents; its growth is dependent on exogenous thymidine, biotin and diaminopimelic acid to such an extent that even its propagation in the laboratory is faced with difficulties. Due to its high efficiency of transformation, however, it is even employed today, although its use is rarely required for safety reasons.

These and other developments not only led to diminished political interest in imposing legislation, but also resulted in a careful move by the RAC to review and reconsider the NIH Guidelines. The point was particularly pressed by animal virologists who were able to point to a variety of inconsistencies within the guidelines. They argued, for example, that containment requirements for cloning of parts of a viral chromosome by far exceeded those for handling of the intact infectious virus itself. In order to address this question, *i.e.*, whether a cloned viral DNA molecule or parts thereof cloned in bacteria are infectious at all or perhaps even more infectious than the virus itself, the RAC permitted a group of scientists to perform such experiments under the highest possible levels of containment (EK2, P4) in Ft. Detrick, Maryland. Polyoma virus was chosen as the experimental system since it is both extremely infectious for mice and highly oncogenic for newborn hamsters. The viral DNA, which is similar in size and functional organisation to simian virus 40 (*cf.* Chapter 8) was cloned into pBR322 or λ vectors and transformed into *E. coli*. Introduction of purified recombinant DNA or of *E. coli* cells carrying the recombinant vectors into susceptible animals never resulted in a virus infection or tumour formation (Chan et al., 1979; Israel et al., 1979). These conclusions were confirmed by similar experiments performed in Europe (Fried et al., 1979). These findings actually are surprising because the viral DNA itself (without the vector portion of the DNA) is highly infectious, and because analogous experiments on chimpanzees with cloned hepatitis B DNA have lead to the expected outbreaks of hepatitis B (Will et al., 1982).

These and other experimental results precipitated a critical re-examination of other sections of the guidelines, finally resulting in the institution of a revised set of guidelines on January 2, 1979. As data has accumulated over the years, the guidelines have been revised and improved constantly; in fact the latest edition was published on May 7, 1986.

Meanwhile, however, the issues have changed. While the problem of laboratory safety appears to have been solved, a variety of other concerns are being voiced and many new issues are discussed both by scientists and by the public. These include:

- large-scale fermentation of organisms containing recombinant DNA;
- deliberate release of such organisms into the environment;
- the release into the environment of manipulated plants;
- the transfer of genes into embryonic mammalian cells and the generation of transgenic animals;
- gene therapy and the cloning of human beings.

Some of these points not only raise questions of safety but also of ethics, and will be discussed accordingly in the following sections.

13.1 Laboratory and Industrial Applications

The potential hazards of recombinant DNA applications have remained speculative and conjectural for over ten years and for tens of thousands of experiments. There is no evidence that a bacterium, a virus, a yeast cell or any other micro-organism carrying recombinant DNA poses any hazard beyond the hazard associated with, and known from, the components of the recombinant DNA molecule itself. On the other hand, it will always remain impossible to prove the complete safety of a particular construction under all imaginable circumstances. It thus appears useful and wise to perform recombinant DNA experiments with an added measure of caution, as is required now in all countries with facilities for modern biological research.

In the United States it is the NIH guidelines for research involving recombinant DNA molecules in their latest version (Federal Register, Vol. 51, 16958–16985, 1986; *cf.* Appendix F) which specify practices for handling and constructing recombinant DNA molecules. These guidelines are applicable to all recombinant DNA research funded by the National Institutes of Health (NIH) and thus technically do not include privately funded corporate research. Apparently, however, all private institutions have voluntarily registered with NIH and no evidence is available for noncompliance with these guidelines.

Individual institutions are responsible for ensuring that recombinant DNA activities comply with the guidelines, and within a given institution the principal investigator (PI) of a project, an Institutional Biosafety Committee (IBC), and a Biological Safety Officer (BSO) carry responsibility. The director of the NIH is responsible both for establishment of the guidelines and for their implementation, and is advised in these obligations by the RAC committee. This committee consists of 25 members, fourteen of whom are scientists working in the field of recombinant DNA research, while the other eleven include lawyers, physicians, and a housewife. Since 80% – 90% of all recombinant DNA research is currently exempt from the guidelines, the RAC committee now is involved mainly in questions of risk assessment, large-scale production, deliberate release into the environment, and gene therapy.

The guidelines recognise two means of containment: physical and biological. There are four levels of physical containment designated biosafety levels, the last providing the most stringent containment. Table 13-1 indicates some of the technical provisions required at various containment levels. Biological containment recognises two levels of protection, HV1 (or Host-Vector System 1) and HV2. A specific HV1 system which is supposed to provide only moderate containment would be the system EK1, which always comprises *E. coli* K12 and certain phages and nonconjugative plasmids. Other HV1-type systems have been described for *B. subtilis*, for *N. crassa*, and certain strains of *Streptomycetes* and *Pseudomonas putida*. HV2 systems must display high levels of inherent containment; only one in 10^8 host cells should be able to perpetuate a cloned DNA fragment under natural conditions. HV2 systems have been described for *E. coli* K12 (strain X1776 in conjuction with non-conjugative

Table 13–1. Containment safeguards.

Biological containment levels	
HV1	Host-vector systems which provide a moderate level of containment, e.g., EK1. In EK1, the host is always *E. coli* K-12 and the vectors are non-conjugative plasmids (e.g. pBR322) or bacteriophage λ derivatives.
HV2	Host-vector systems which provide a high level of containment. In EK2 systems with plasmid vectors, no more than one in 10^8 host cells should be able to perpetuate a cloned fragment in a non-laboratory environment.

Physical containment levels	
BL1	No separate laboratory; no containment equipment; personnel with general training in microbiological practice; no mouth pipetting.
BL2	Limited access to laboratory; autoclaving facilities; biological safety cabinets.
BL3	Personnel with training for handling pathogenic and potentially lethal agents; protective clothing; restricted access; protected work surfaces; biological safety cabinets with high-efficiency (HEPA) air filters; airlocks; directional air flow and negative pressure.
BL4	Separate building; no windows; exhaust air decontaminated; all liquid waste from shower room and toilets decontaminated; personnel with positive pressure suits; airtight doors etc.

from Federal Register, Part III, May 7, 1986; pp. 16958–16985

plasmids) and for *S. cerevisiae* strains rendered sterile by the *ste-VC9* mutation.

Experiments with recombinant DNA have been divided into the following four classes:

(a) Experiments requiring review by the RAC and approval by the NIH and IBC. These include, for example, experiments with cloned DNA from viruses (like foot-and-mouth-disease virus) which are forbidden entry into the United States.

(b) Experiments requiring IBC approval. These include experiments involving more than ten litres of culture and those using certain animal and human pathogens.

(c) Experiments requiring IBC notification. These include experiments involving nonpathogenic prokaryotes or lower eukaryotes.

(d) Exempt experiments. This category includes all experiments involving homologous cloning, i.e., cloning of *E. coli* DNA in *E. coli* or human DNA in human cells (with the exception of viruses). Almost 90% of all recombinant DNA experiments are currently exempt since they fall into this category.

In addition to the NIH guidelines, however, there are a variety of regulatory regimes which regulate modern biotechnology. The FDA's Office of New Drug Evaluation, for example, has ruled that drugs made by recombinant DNA technology, even if identical with currently approved drugs, are "new drugs". The "Office of Biological Research and Review – Center for Drugs and Biologicals" has released a set of "Points to Consider in the Production and Testing of new Drugs and Biologicals Produced by Recombinant DNA Technology". Many other types of regulations are discussed in "Commercial Biotechnology", Office of Technology Assessment, Congress of the USA, OTA-BA-218.

Other countries have, in general, followed the precedents set in the United States. In Japan, recombinant DNA research is monitored by the Council for Science and Technology, in the office of the prime minister, who is technically its chairman. Initially, the "Guidelines for Recombinant Experiments" from August 1979 were considerably more restrictive than the NIH model. This related particularly to an upper limit of 20 litres for large-scale fermentations. This limit, however, has been dropped in the latest revision (September 1983) of the guidelines. New guidelines for the industrial application of recombinant DNA are in preparation in the Ministry of Economics. Within its Chemical Products Deliberation Council, this ministry established a

Recombinant DNA Technology Subcommittee which will deliver a draft for new guidelines in the course of 1986. It is expected that this revision will relax the rather strict biological containment provision which, so far, has severely restricted some types of host/vector systems which might be used in either basic or applied research.

As in Japan, the guidelines in Great Britain apply not only to government-funded but also to private recombinant DNA experiments. These guidelines were originally announced by the Genetic Manipulation Advisory Group (GMAG) according to the Health and Safety at Work Act of 1974. In the meantime, the GMAG has been renamed the Health and Safety Commission Advisory Committee on Genetic Manipulation. The Health and Safety Executive (HSE) is re-onsible for enforcement of the guidelines and is directed to inspect facilities for recombinant DNA work at the two highest containment levels.

In the Federal Republic of Germany, the Central Commission for Biological Safety (Zentrale Kommission für die Biologische Sicherheit, ZKBS) within the Federal Institute of Health is responsible for all recombinant DNA activities. The present version of the recombinant DNA guidelines (dated May 1986) recognises two levels of biological (B1, B2) and four levels of physical (L1 to L4) containment. The ZKBS registers all laboratories involved in recombinant DNA research and must be notified of all experiments above, and including, levels L1B2 or L2B1. Large-scale fermentations in excess of ten litres are reviewed on a case-by-case basis, while deliberate release experiments (see below) are not permitted. The present guidelines only apply to government-funded but not to corporate recombinant DNA activites. New legislation is expected to be introduced in 1987 to cover all activities.

Both Switzerland and France have adopted the NIH guidelines almost verbatim. In France, a Control Commission (Commission de Controle) must be notified of all recombinant DNA experiments but only experiments at the highest con-tainment level have to be approved. The government of Switzerland is not involved in the control of recombinant DNA research but rather relies on self-regulation on the part of the scientists. A commission for experimental genetics (Kommission für experimentelle Genetik) within the Swiss Academy of Medical Sciences thus has been created which is responsible for compliance with the guidelines.

In general, it can be stated that the differences between the various guidelines of different countries more or less have been eliminated, such that both basic and applied research, as well as industrial activities, can be performed worldwide with no significant competitive disadvantages.

13.2 Environmental Applications

13.2.1 Micro-organisms

Most current applications of recombinant DNA technology involve the use of manipulated micro-organisms in closed surroundings, *i.e.*, the laboratory or especially designed fermentation plants. As discussed above, there has been considerable concern about the possibility that genetically engineered bacteria will escape from the laboratory and establish themselves in the natural environment; much criticism has focussed on the use of *E. coli* as a host for recombinant molecules. This organism, which has become the pet of the field, not only is a natural inhabitant of humans and all warm-blooded animals but it has been spread extensively into water and soil due to a variety of human activites. In order to minimise the risk of an inadvertant release of manipulated bacteria into the environment, debilitated strains of *E. coli* were developed to be used as host cells in cloning experiments. The example of the K12 strain X1776, with its longer generation time, its auxotrophic mutations and its sensitivity to detergents, *e.g.* bile acids, has already been discussed. The question, of course, arises whether this valuable concept of biological containment will also be useful and acceptable when microorganisms are deliberately released into the environment where, in fact, they would be asked to survive in order to fulfill a particular task, *e.g.* to degrade pollutants.

Currently at least thirteen microbial pesticides are registered with the United States Environmental Protection Agency (EPA) for use in forestry, agriculture and insect control. Among these are bacteria such as *Bacillus thuringiensis var. israeliensis* which is used against mosquitoes, and certain baculoviruses (nuclear polyhedrosis virus) with which thousands of hectares have been sprayed in the People's Republic of China, much to the dismay of the silkworm farmers (Su, 1984).

While these and other applications have led to little, if any, public concern, the advent of genetic engineering and the possibility of new widespread and extensive applications has raised a variety of novel questions. The current controversy, in particular, focuses around two species of micro-organisms, *Pseudomonas syringae* and *Erwinia herbicola*. These bacteria produce a protein which promotes the nucleation of ice at 0 °C and which thus is responsible for a substantial amount of frost damage accrued to crops just below the freezing temperature of water (Green and Warren, 1985). Deletions of the gene encoding this protein, which either occur naturally or which can be engineered using appropriate cloning techniques delays frost damage on sprayed plants until the temperature falls below –7 °C, thus augmenting the growing season by two to four weeks. In order to extend successful greenhouse tests with genetically-engineered bacteria, known as "ice-minus", into a more typical environment (see Hirano and Upper (1985) for a review) researchers from the University of California (Drs. Steven Lindow and Nicholas Panopoulos, Berkeley) had asked NIH for permis-

sion to conduct small-scale field tests on bean, tomato and potato plants. Although this permission was granted by NIH, the experiment, which had been planned for May 25, 1984, was stopped by a preliminary injunction of the United States District Court for the District of Columbia on May 16, 1984. This decision was reversed on Feb. 27, 1985 by the Federal Appeals Court with the provision that the experiment could be performed as soon as the necessary Environmental Impact Statement (EIS) required by the National Environmental Policy Act (NEPA) of 1969 had been filed. In preparation of these analyses which relate to the behaviour of genetically engineered micro-organisms in the natural environment, the NIH has released a proposal for "Points to Consider for Experiments Involving Release of Genetically Engineered Organisms". In addition, the Environmental Protection Agency (EPA) has published a notice which describes how the agency plans to deal with microbial products produced by biotechnological procedures and how it interprets certain provisions of the Toxic Substances Control Act (TSCA) and the Federal Insecticide, Fungicide and Rodenticide Act (FIFRA) with respect to new genetically engineered micro-organisms (Environmental Protection Agency, Federal Register 49, December 31, 1984, pp. 50880-50907).

Finally, there is little doubt that the United States Department of Agriculture (USDA) will become involved in these matters since, for example, the bacterium *Pseudomonas syringae* used in ice nucleation research is a plant pathogen which may induce leaf spotting and other diseases in a variety of plants. It remains to be seen whether and how these various bureaucratic and legal burdens can be overcome in order to initiate field trials with deliberately released micro-organisms.

On a more pragmatic level, the following points have to be addressed and considered by anyone planning an experiment in this realm:
– survival of the organism released;
– multiplication in the environment;

- genetic stability;
- the possibility of dissemination or transportation of the engineered organisms from the site of release to another site;
- potential deleterious effects of the engineered organism on the environment.

Prerequisites for an assessment of these parameters are adequate methods for monitoring the released micro-organisms. Given that such organisms are released in astronomical numbers, it may be difficult to determine, on the basis of population samples by standard plate count methods, whether they are actually proliferating, in particular, since cells released into a non-permissive environment may be viable but not culturable. More direct methods are thus required and are being developed, *i.e.*, fluorescent antibody staining techniques or the "superinfecting phage" technique (Colwell et al., 1985).

Survival, growth, and stability of any micro-organism within the environment depend on its genetic constitution, and in turn, on chemical, physical and biological factors of the respective habitat. The few data available on this subject have been reviewed by Stotzky and Babich (1984). Immediately after the release of a micro-organism a considerable drop in the number of viable cells is to be expected and, of course, is the more pronounced the less adapted the organism is to the particular environment. Half-lives of 68 days have been observed for *B. thuringensis* in natural soil while typical soil bacteria may reach half-lives of one year or more even under adverse conditions (West et al., 1984). The formation of spores and conidia in appropriate organisms can, of course, increase these survival times almost indefinitely.

The question of gene transfer or gene exchange between a newly released organism and the microbial flora in a natural environment is an issue of critical importance. Such gene transfer can occur by several mechanisms, including conjugation, transduction and transformation. Conjugation requires cell-to-cell contact and thus depends on large cell populations (Freter, 1984).

Its efficiency decreases by the square of the dilution, and thus conjugation occurs successfully only when the organism in question is not only able to survive but also to grow and to colonise the new habitat. Conjugation, in addition, is highly temperature-dependent and generally ceases below 15 °C. Finally, a significant factor that affects conjugation is pH. Optimal mating frequencies are observed between pH values of 6.6 to 8.5 while higher or lower values, in particular in combination with low temperature, strongly inhibit or even block conjugation.

These restraints notwithstanding, gene transfer by conjugation has been observed even in the natural environment. The phenomenon is mediated by conjugative plasmids which code for sex pili on the surface of plasmid-containing bacteria, through which plasmid as well as host chromosomal DNA can enter the recipient bacteria. This problem can be minimised by the use of nonconjugative plasmids which have lost even their ability to be mobilised by a resident conjugative plasmid (see Section 4.1). In this context, the rapid transfer of antibiotic resistance markers is a serious concern, since these markers often reside on conjugative plasmids which are efficiently transferred from environmental strains to laboratory or hospital recipients. These difficulties are of particular significance in the proximity of sewage treatment plants, or in the effluents of animal breeding stations. Little or nothing is known in this respect about the behaviour of other markers *e.g.*, those responsible for the degradation of toxic chemicals. Bacteria have been developed for example, which can live on 2,4,5-T as a sole source of carbon and which, in the space of a few weeks, can reduce the level of the toxin in heavily contaminated soil, *e.g.* from 20 000 ppm to 10 to 20 ppm. Likewise, tertiary oil recovery may require plasmid-encoded enzymes as does the production of surfactants necessary to release oil from tar sands. It is obvious that the behaviour of such markers should be observed with greatest care; it may also be wise to refrain from the use of many toxic chemicals altogether, in particular, when the residual concentrations which bacteria cannot dispose of (due to diffusion and concentration limits) are still toxic to the environment as is the case with dioxin.

The danger that mutations will transform a harmless micro-organism into a pathogenic variant is considered remote. Pathogenicity is a complex trait which involves the interaction of many genes. In view of the abundance of micro-organisms and their constant exposure to radiation and chemicals, there are ample opportunities for mutations to occur. No report, however, exists of the new and unexpected development of a pathogenic micro-organism. In this context the existing diversity of the microbial world, on the one hand, and the apparent genetic stability of bacterial strains in a given habitat, on the other, may indicate that only a few rare mutants are able to compete with an indigenous wild-type microbial population.

The scientific literature abounds with analyses of the survival, growth and stability of deliberately released organisms, though no general rules have been developed yet. In the absence of a "predictive ecology" there will be no choice other than the case-by-case study of each proposal for the deliberate release of micro-organisms.

Prior to a more final regulatory approach, the EPA thus demands notification of all small-scale field studies which involve either naturally occurring micro-organisms for use in environments where they are not native or genetically altered micro-organisms. These notifications are expected to include background information on the microbe itself (habitat description, pathogenicity etc.) as well as detailed descriptions of the test. These tests include methods for the difficult task of monitoring the released micro-organisms during the field test (Federal Register, Vol. 49, 40659-40661, 1984). It is hoped and expected that these restrictions will stimulate research on the assessment of risks incurred through the deliberate release of micro-organisms and eventually will reconcile economical with environmental interests.

13.2.2 Plants

Environmental issues and concerns are also being raised by the idea of deliberate release of genetically manipulated plants. It is argued that such plants may disturb existing ecological equilibria and may transfer their particular traits to other plants, in which they are not desired (Hauptli *et al.*, 1985). These issues are often compared with the introduction of exotic and foreign species into a novel environment. Mostly the pertinent experiments have either been harmless or even beneficial considering, for example, that many of the important crops of the United States such as wheat, soybean and rice, are not native. There are some exceptions, however, as attested to by the introduction of plant pathogens which are nonpathogenic to their native hosts but highly pathogenic to related plants. The parasitic fungus *Endothia parasitica*, for example, was introduced into the Unites States around 1900; it originated in Asia and was carried by nursery chestnut plants. While it is entirely harmless for the Asian chestnut, it had killed all American chestnuts by around 1950. In Europe, where the "chestnut blight" was introduced in the 1930's, the disease seems to spread more slowly; nevertheless it lead to a decay of the chestnut-related industry in northern Italy. A similar catastrophy was the occurrence of the Dutch elm disease, which eliminated more then 90% of all elm trees in Europe within the last 50 years. Resistant variants, however, have developed such that the disease now can be contained. There are other examples, *e.g.,* the prostrate vine kudzu (*Pueraria thunbergiana*) of China and Japan, now regarded as a pest in certain parts of the United States. In this case the issue is not the introduction of plant pathogens but the introduction of the plants themselves. In general, it can certainly be said that in comparison to the numerous cases of exotic introductions adverse examples are few, and some untoward effects are even predictable.

However, there are other concerns. These include cross contamination of genetic material to the same or related crop plants, toxicity, and gene transfer to weeds. There are some examples in the literature of biochemical changes in plants which were inadvertently selected for because they were associated with, and linked to, other more desirable traits. These changes are certainly unpredictable in most cases. Whether such changes may be toxic for humans or animals can easily be tested in each particular case. Weediness is a complex genetic trait which will not be easy to transmit by single genetic transfers. Many crop plants, however, have weedy companions to which certain traits, e.g. single-gene herbicide resistance, may be transferred by repeated backcrossing. There is no evidence that this will actually occur and will lead to weeds that are even more difficult to control. Nevertheless the problem may arise, but should become evident early in the tests required before the commercialisation of new crop plants.

A final and increasing concern relates to a possible reduction of the gene pool (at least of major crop plants) and thus to the loss of important genetic traits. This would, and does, occur due to the high degree of inbreeding required for crop improvement, due to the spread of uniform agriculture, and due to the replacement of wild-type varieties (populations) by newly introduced species.

Among the important traits to be retained and protected are those required to build disease resistance in plants (Horsfall and Cowling, 1978; Evans et al., 1981). It is estimated (Le Clerg, 1964) that crop losses caused by infectious plant diseases in the United States alone are about 15 – 25% of the potential yield. Some catastrophic epidemics are well remembered in this connection, e.g., the Irish potato blight epidemic of 1845/6 and the US maize leaf blight epidemic of 1970. In the former case, the virulent fungus, Phytophthora infestans, caused a total potato crop failure in the wet and cold summer of 1845. In a population of eight million, one million died and almost two million were forced to emigrate. The disastrous maize epidemic of 1970 was caused by the pathogenic fungus Helminthosporium maydis, which spread rapidly through a genetically uniform variety of Zea mays. It led to the destruction of 15% of the US corn crops. Fortunately, breeders were able to resort to resistant varieties and thus to supply resistant seed material within the following year. The importance of preserving a wide genetic basis, at least for food crops, is now widely recognised. Plant breeders themselves often maintain huge stocks of seed samples. The International Board of Plant Genetic Resources (IBPGR), an organisation funded by the United Nations Food and Agricultural Organisation (FAO) and the World Bank maintains a seed collection which has catalogued some 350000 food crop varieties from all over the world. Its main repository is located in Ft. Collins, Colorado, at the National Seed Storage Laboratory (NSSL), a branch of the United States Department of Agriculture. The international community will be well advised not only to maintain these and other collections throughout the world but also to significantly increase the level of support for the protection of the world's plant genetic stocks (cf. Nature 318, 96, 1985).

13.3 Reproductive Engineering

The transfer of specific genes can be accomplished not only into micro-organisms such as E. coli and yeast, or into mammalian cells in tissue culture (cf. Chapter 8) but also into intact animals. Such experiments are of considerable interest since they represent the only possible approach to the study of mammalian development (ontogeny) and an understanding of the behaviour of genes in different cell types.

A prerequisite for this type of analysis is the development of methods for the generation of healthy mammalian eggs and for their culture in vitro. In most mammals, ovulation is determined and controlled by two hormones: the follicle-stimulating hormone (FSH) which causes the final growth of the immature egg follicles, and the luteinising hormone (LH) which induces maturation and release of the eggs into the oviduct. Normally in most mammals (including humans), the time of ovulation is determined by a periodic rhythm. Unfertilised eggs, however, can also be obtained by injecting large doses of the gonadotropic hormones. The discovery, in 1970, that unfertilised eggs can be obtained in this manner from human female patients, and that they can be cultured as well as fertilised *in vitro*, opened the way to the birth of the first "test tube baby", born in England on July 25, 1978 (Edwards, 1974; Shettles, 1979; Edwards, 1981).

Mammalian eggs and embryos can be kept in tissue culture only for a short time, since they rely on the complex uterine environment for their continued development. Technical advances, however, have lengthened the permissible *in vitro* period to such an extent that the early steps of development could be studied in greater detail (Balinsky, 1981). Immediately after fertilisation and fusion of the two pronuclei, the embryo undergoes a series of rapid mitotic divisions leading to the formation of two, four, eight etc. blastomeres. The synchrony of mitosis of the different blastomeres is lost at an early stage such that an embryo may contain not only two, four or eight but also three or seven etc. blastomeres. The result of the early divisions, the morula, is a solid, compact mass of 16 to 32 cells surrounded by the *zona pellucida*. Soon after the morula is formed, a fluid-filled cavity develops between its inner cells. The resulting blastocyst is made up of two types of cells: an outer layer of tightly connected cells known as the trophoblast, and an eccentrically placed mass of cells referred to as the inner cell mass (ICM). The trophoblast is thought to provide nourishment for the embryo and eventually establishes the connection with the uterus while the inner cell mass represents the true embryo. Embryos at an early stage, *e.g.* the 8-cell stage morula, can be made to fuse and to develop into a single blastocyst. If the fusion partners are of different genetic origins, *e.g.* from different

strains of mice, or from different sexes within the same species or even from different species, the resulting blastocyst will carry cells of different genotypes. These may develop into "chimaeric" or "allophenic" animals. This technique obviously can be used to obtain chimaeric animals from parents (*e.g.* sheep and goat) which are otherwise sexually incompatible (Fehilly *et al.*, 1984; Meinecke-Tillmann and Meinecke, 1984) but it has also been employed to study the problems of immunological tolerance, "totipotency" of individual blastomers, and early cell lineages (see below).

Soon after its formation, the blastocyst escapes from the *zona pellucida* and attaches itself, *via* the trophoblast, to the endometrium. Since it is exceedingly difficult as yet to carry the development *in vitro* beyond this stage the explanted embryo must be returned to its customary environment, the uterus of a female. As the tissues of the uterus are not always ready to receive an embryo, because they undergo periodic changes which concur with changes in the ovary, the recipient female organism must be pseudo-pregnant. This state is a direct result of ovulation; it can be induced by the administration of hormones and is maintained for some time even if ovulated eggs are not fertilised. In most mammals the embryos can be reimplanted at any stage of development between the early cleavages and the blastocyst stage. Embryos of most species, including human embryos, have been brought to term in this way. This opens a variety of possibilities not only for *in vitro* fertilisation but also for specific genetic manipulations of embryos. Experimental endeavours in either of these areas are associated with a complex array of scientific, legal and ethical considerations. Only the issues raised by the targeted introduction of specific genes into embryos, will be the subject of the subsequent discussions.

13.3.1 Transgenic animals

Transgenic animals are defined (Palmiter and Brinster, 1985) as animals which, as a consequence of experimental manipulations, have incorporated foreign DNA into their germline, and are thus able to transmit this DNA to their offspring. Early experiments with the frog *Xenopus laevis* were hampered by its long generation time (2 years). In contrast, the mouse system offers several advantages, especially a generation time of only 60 days, the availability of many inbred strains, and a considerable number of mapped genetic loci. Approximately 1 200 genes of *Mus musculus* were known in August 1983, and 30% of them had been assigned a location on one of the 40 chromosomes (Roderick and Davisson, 1984).

There are several ways to produce transgenic mice. The most commonly employed method involves microinjection of isolated DNA (*e.g.* recombinant DNA) into the pronuclei of fertilised eggs which then are implanted into a recipient female (Gordon *et al.*, 1980). Other methods include the transfer of DNA into embryonal carcinoma cells followed by injection of the manipulated cells into mouse blastocysts and, finally, nuclear transplantation, *i.e.* the replacement of a nucleus from a mouse embryo with a nucleus from another embryo. In each case, transfer of DNA to the recipient embryo is necessarily followed by implantation into a foster mother.

13.3.1.1 Gene Transfer by Nuclear Injection

At present, microinjection into fertilised eggs from outbred strains of mice (inbred strains are less efficient) represents the method of choice for introducing genes into the germline.

Richard Palmiter,
Seattle

The various parameters influencing the efficiency of this procedure have been studied in detail by Brinster *et al.* (1985). In a typical experiment, one-cell embryos (8 to 12 per animal) are obtained one day after mating by flushing the oviduct of super-ovulated females. For microinjection, the eggs are held in place with a blunt-ended pipette (Fig. 13-1), while a glass capillary (2 μm outside diameter) is forced through the *zona pellucida* into the male pronucleus (the larger of the two pronuclei). Successful injection is demonstrated by the swelling of the pronucleus, which will receive approximately 2 pl of injection fluid. The optimal DNA concentration in the injection fluid should be on the order of 1 ng/μl, amounting to several hundred copies of a given DNA molecule per pronucleus. Linear molecules are approximately five times more efficient than

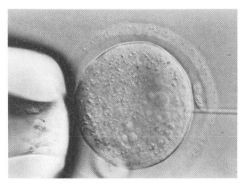

Fig. 13-1. Microinjection into mouse embryos. The injected pronucleus is about twice as big as the normal counterpart. The polar body is visible next to the needle in the perivitelline space. (Magnification ×630; Nomarski optics; Courtesy of Dr. Brem and Prof. Kräußlich, Department of Veterinary Medicine, University of Munich).

circular ones, while neither extended single-stranded ends nor the length of the DNA will influence the integration efficiency. The number of integration sites, in general, is low (usually only one), and the location of these sites is random. Multiple copies of a gene usually integrate at a single chromosomal site in a tandem head-to-tail orientation. There is no method as yet for allelic replacement, *i.e.*, for targeted introduction of genes at defined chromosomal sites. However, Smithies *et al.* (1985) have recently been able to introduce human β-globin DNA specifically into the human chromosomal β-globin locus of tissue culture cells by homologous recombination. Whether this rationale which is used successfully in yeast (*cf.* Chapter 5) can be efficiently extended to mammalian embryos remains to be seen.

The efficiency of producing transgenic mice, calculated as animals carrying the foreign gene that develop per female donor, *i.e.*, per set of 10-12 transferred embryos, averages about 25%.

Lower efficiencies (between 1% to 10%) are observed for rabbits, sheep and pigs (Hammer *et al.*, 1985). In the case of pigs, the opacity of the eggs poses a major technical obstacle, because it is difficult to see the pronuclei. A short centrifuga-

tion step, in combination with interference contrast microscopy, can be used to alleviate this problem.

The successful establishment of a transgenic animal can be verified through various methods. The presence of the foreign gene is readily identified by Southern blot hybridisation using cellular DNA obtained from tail biopsies. Expression of the transferred foreign gene is analyzed in S1 nuclease protection assays (*cf.* Section 11.2.2.2).

The final proof of the presence of a foreign gene in the germ line must be obtained, of course, in breeding experiments. The founding, or G_0 mouse, *i.e.* the parental transgenic organism, can be bred to transmit the foreign DNA into half of its offspring. It is possible to establish lines of transgenic animals which are homo- or heterozygous, and which carry the foreign DNA at different chromosomal locations by backcrosses with a founder animal and/or continued breeding with wild-type animals.

Many genes have been introduced by now into mice in order to address three types of questions. One series of studies aims at revealing the mechanisms of tissue-specific gene expression. By introducing an intact gene, the expression of which is normally regulated in a tissue-specific manner, into transgenic mice it is possible to ask whether the regulation of expression of the introduced gene at the new chromosomal locus is still tissue-specific. Experiments of this type have been performed with a mouse immunoglobulin ϰ gene (Storb *et al.*, 1984) and the rat pancreatic elastase I gene (Swift *et al.*, 1984). In both cases, tissue-specific expression was observed exclusively in the expected target organs or cells, *i.e.* B lymphocytes or the pancreas.

In addition, one may delineate the necessary *cis* elements required for a tissue-specific expression by fusing the putative control region from a highly regulated gene to the coding region of an indicator gene which is easily identified. In this connection, the human growth hormone has proved a very useful indicator gene. Its mRNA can be expressed efficiently in a variety of mouse cells and is assayed easily by hybridisation with a cDNA probe. If the mRNA is translated appropriately, growth hormone is produced and causes increased growth of the transgenic animals. Initial experiments employed fusions of the human growth hormone gene with the mouse metallothionein-I (*MT-I*) promotor, which led to the specific expression of the growth hormone in tissues that normally synthesise metallothioneins, *e.g.* liver (Palmiter *et al.*, 1982). The resulting mice grew to twice the normal size (40 g instead of 20 g). More recently, a fusion containing only 213 bp of the 5′ flanking region of the rat elastase I gene was shown to specifically direct the expession of human growth hormone in pancreatic acinar cells (Ornitz *et al.*, 1985).

Occasionally, however, the sitation is more complex. A fused gene consisting of the metallothionein promoter and the human *HGPRT* gene encoding hypoxanthine-guanine phosphoribosyl transferase for example, is not expressed preferentially in liver or kidney cells, as one might expect, but rather in tissues of the central nervous system (Stout *et al.*, 1985). Thus, in this case it is not the promoter region itself, but a sequence within the HGPRT cDNA which promotes the elevated expression of the enzyme in cells of the nervous system.

Another interesting set of marker genes are the oncogenes. Their presence is easily detected by hybridisation, and their expression causes the formation of tumours. Thus, a fusion of 5′ flanking regions of the rat insulin II gene with the coding region of SV40 large T-antigen led to the expression of the large T-antigen specifically in the β cells of the endocrine pancreas of transgenic mice (Hanahan, 1985). In addition, these fusions resulted in the development of β cell tumours in every transgenic mouse; these tumours, however, originated in only a small fraction of the islet cells, although all islet cells expressed the SV40 large T-antigen. This protein is thus necessary but may not be sufficient to produce a tumour condition in every islet cell. Another construction contained

the human *c-myc* oncogene fused to the promoter region of the mouse mammary tumour virus (MMTV) long terminal repeat (*LTR*), the expression of which is regulated by glucocorticoids (Ringold, 1983). As anticipated, transgenic female mice carrying this fusion developed mammary adenocarcinomas. These mammary tumours appeared during the second or third pregnancies of the mice, and no non-mammary tissues were affected (Stewart *et al.*, 1984).

Gene transfer experiments have also been used to explore a peculiarity of immunoglobulin (Ig) gene expression, namely a requirement for an accurate rearrangement of Ig DNA. This rearrangement, which occurs during development, accomplishes the precise juxtaposition of several specific *Ig* gene segments, and is absolutely necessary for the functional expression of *Ig* genes. Normal mice, even if they are heterozygous at the *Ig* loci, never express more than one heavy chain, or more than one light chain product. The other alleles are said to be "excluded" from participating in Ig synthesis, and it has been postulated that a rearranged *Ig* gene (or its product) effects exclusion by influencing or inhibiting the rearrangement of other alleles. This model was tested recently by the introduction of a rearranged μ heavy chain gene (Weaver *et al.*, 1985), and a rearranged ϰ light-chain gene (Ritchie *et al.*, 1984). In both cases the rearranged transgenes suppressed DNA rearrangement of their homologues, the endogenous heavy and light chain genes, respectively. An influence on the rearrangement of the heterologous chains, that is, an influence on the endogenous light chain by the transgenic heavy chain and *vice versa*, was not observed. Similar results were obtained when both a rearranged light chain and a rearranged heavy chain gene were introduced simultaneously into transgenic mice (Rusconi and Köhler, 1985). In all these cases, however, suppression of rearrangement of the endogenous alleles was never complete. Incomplete suppression may be the result of low transcriptional activity at the respective transgenic loci during the early phases of

B cell development when these rearrangements are thought to occur. It remains to be seen, however, whether the results of these experiments will fully explain the phenomenon of allelic exclusion.

A second series of experiments with transgenic mice examined the function of oncogenes. As explained previously (*cf.* Section 8.7; Table 8-4) the detailed roles of these genes, both in normal development and in tumorigenesis, remain largely unknown.

Two oncogenes have been studied in transgenic mice: the SV40 gene for large T-antigen and the oncogene *c-myc*. Transgenic mice developing from embryos injected with the SV40 early region genes (under control of their own promoter) develop tumours within the choroid plexus (Brinster *et al.*, 1984). These tumours express SV40 large T-antigen very efficiently. Although the T-antigen DNA is incorporated into the genomes of all cells, T-antigen protein and its corresponding mRNA are undetectable in other tissues. It may be that choroid plexus cells are more permissive for the activation of the T-antigen gene than are other cells. Deletion of the SV40 transcriptional enhancer region drastically reduces the incidence of tumours of the choroid plexus, indicating that this sequence element specifically directs expression of the oncogenic phenotype to the choroid plexus and that T-antigen is sufficient for oncogenesis in these cells.

The organ specificity in the appearance of tumours can be altered if the SV40 large T-antigen gene is fused with other enhancer/promoter elements. A 520 bp upstream region of the rat insulin II gene fused to the SV40 large T-antigen coding region leads to β cell insulinomas of the pancreas (Hanahan, 1985). As mentioned previously, not every β cell, however, is transformed, and this indicates that a secondary alteration may be required to induce solid β cell tumours. SV40 large T-antigen belongs to a class of oncogenes which can both immortalise primary cells and transform immortalised cells to a transformed

phenotype. In the case of pancreatic expression of T-antigen, the secondary event suspected to be required for the establishment of solid β cell tumours thus may not represent an interaction with a second oncogene but rather another event related to oncogenesis, *e.g.* the induction of angiogenesis (Hanahan, 1985).

The behaviour of the *c-myc* oncogene differs from that of SV40 T-antigen. Studies by Land *et al.* (1983) have demonstrated that the presence of an activated *c-myc* gene is a necessary prerequisite for transformation in certain malignancies but not sufficient in others. In order to study this question in an intact animal, Stewart *et al.* (1984) produced transgenic mice carrying a fusion of the human *c-myc* gene with the mouse mammary tumour virus (MMTV) *LTR* promoter. As mentioned above, this construction was expected to lead to malignant changes in the mammary glands of lactating animals.

Two out of ten transgenic founder animals indeed developed spontaneous adenocarcinomas of the breast and transmitted this phenotype to all their female offspring. However, if the expression of the *c-myc* oncogene had been sufficient for the development of these tumours all cells of the mammary epithelium of all the transgenic mice should have been affected. More recent work based on selective breeding of these founder animals indicates that a further transforming event is required for the occurrence of tumours in this system (P. Leder, personal communication). Similar conclusions were reached by Adams *et al.* (1985). In their experiments the transgenic mice produced from a fusion of the *c-myc* gene with the immunoglobulin heavy chain transcriptional enhancer developed lymphoid malignancies, as expected. The incidence of lymphomas in founder animals was high; however, the long latency periods (5 months) and the fact that many of the tumours were monoclonal, *i.e.*, derived from a single lymphocyte, seem to indicate that not only overexpression of the *c-myc* gene but also other genetic events had influenced tumour formation in this system.

Transgenic mouse technology has recently been used for the first time in an analysis of pattern formation during embryonic development. Nine day old embryos developed into chimaeric (but otherwise normal) animals after having received microinjections of cultured heterologous neural crest cells. This demonstrates that microinjected heterologous cells can participate in normal cell migration patterns (Jaenisch, 1985). Eventually, microinjection at different stages during gestation may allow dissection of cell lineage patterns.

A final application of transgenic mice exploits the mutagenic potential of recombinant DNA introduced into the germline, since it is thought that integration of foreign DNA occurs at random chromosomal sites. If insertional mutagenesis occurs in genes essential for development or morphogenesis such mutations should reveal themselves by the absence of homozygous offspring or by unusually small litter sizes. At least four transgenic lines of mice carrying recessive lethal mutations have been described from which viable and healthy homozygotes cannot be obtained. In one instance, the insertion of Moloney murine leukemia virus into the germline of mice was achieved by exposing early embryos to this retrovirus; one virus insertion isolated led to early embryonic death at day 13 of gestation (Harbers et al., 1984; Schnieke et al., 1983). A detailed analysis of the integration site showed that the proviral insertion had occured into the $\alpha 1(I)$ collagen gene.

Another transgenic line with a novel, non-lethal recessive mutation arose by systematic intercrossing of heterozygous animals carrying an inserted MMTV-c-myc fusion gene (Woychik et al., 1985). These homozygous animals display a severe defect in the pattern of limb formation which is identical to a previously known limb deformity mutation. Since the site of the mutation is tagged with the c-myc gene insertion it will be possible to isolate the inactivated gene and to study its pattern of expression during embryonic development.

In at least two other cases, the affected genes have not been identified. They involve germline insertions of the human growth hormone gene (Wagner et al., 1983) or of a metallothionein-thymidine kinase fusion (Palmiter et al., 1984). The latter case displays mosaicism in that only female founder animals are able to transmit the foreign DNA to their offspring. The male animals are fertile but never transmit the fusion gene which indicates that this insert may disrupt a gene required for sperm viability. Gene transfer into the germline thus offers the unexpected bonus of obtaining developmental mutations by insertional mutagenesis.

13.3.1.2. Gene Transfer Into Embryonic Stem Cells.

Another gene transfer system is represented by the so-called embryonal carcinoma (EC) cells, the stem cells of teratocarcinomas. Teratocarcinomas are rare tumours which originate in the gonads and which contain a wide range of differentiated cell types representing all three embryonic germ layers, the ectoderm, the mesoderm and the endoderm. The original observations derive from work by L. Stevens (1958; 1967) with strain 129/Sv of mice which displays a high frequency of spontaneous testicular teratocarcinoma (teraton (Greek) = monster). Subsequently it could be shown that teratocarcinomas can be obtained from many different strains of mice with high frequency by the implantation of embryos with an age of up to seven days into extrauterine sites such as the kidney capsule (Martin et al., 1975). Apart from differentiated cells, all teratocarcinomas contain a distinctive undifferentiated type of cells known as "embryonal carcinoma" cells. These cells are very similar to cells of the early embryo in that they are also pluripotent. The pluripotency of EC cells has been demonstrated in at least two different ways. They can give rise to a teratocarcinoma with a wide variety of differentiated tissues when implanted intraperitoneally (Kleinsmith and Pierce, 1964). In addition, when mouse blastocysts are injected with genetically distinct EC cells they develop normally and their off-spring display the expected chimaerism, both for external and enzymic markers (Papaionnou *et al.*, 1975). It is therefore possible to examine integration and expression patterns of foreign genes in isolated cell lines prior to incorporation into the intact organism. Since the founder animals are mosaic, outbreeding is required to establish pure lines that are heterozygous for the desired marker. Unfortunately, the efficiency of gene transfer into EC cells by DNA-mediated gene transfer is low, *i.e.*, about 100-fold less than in L cells. An alternative approach employs retroviruses as vectors for EC cells. Wagner *et al.* (1985) succeeded in transforming almost quantitatively an embryonic carcinoma cell line with a retroviral vector expressing the bacterial neomycin resistance gene from an internal thymidine kinase promoter. A *c-myc* gene present on the same construct under transcriptional control of the murine leukemia virus *LTR* region was not expressed in these cells. It remains to be seen whether promoters other than that of the thymidine kinase gene can be found which are functioning in embryonic carcinoma cells.

13.3.2 Gene Therapy

The term gene therapy covers two different types of applications of genetic engineering to human beings, *i.e.*, gene therapy of somatic cells, and genetic manipulation of germ cells.

The absolute prerequisite for the targeted transfer of genetic material with the expressed aim of correcting a genetic defect is an intimate knowledge of the molecular causes of the disease in question and of the afflicted gene(s). The following section will briefly describe the origin of human genetic disorders and techniques available for their diagnosis.

13.3.2.1 Human Genetic Diseases

The haploid human genome is estimated to comprise 50 000-100 000 genes, located on 23 chromosomes (22 autosomes and one of the two sex chromosomes, X and Y). At present, the chromosomal locations of approximately 400 of these genes have been determined. Almost 90 per cent of these assignments were obtained by employing the techniques of either somatic cell hybridisation or family linkage analysis. Somatic cell hybrids between cells of different species, *e.g.*, mouse and man, preferentially lose the chromosomes of one of the two fusion partners. It is therefore often possible to correlate the loss of a particular phenotype with the loss of a specific chromosome. Linkage analysis of a genetic trait usually requires a large pedigree and the knowledge of the chromosomal location of one marker adjacent to the gene under study.

Other methods for the identification of chromosomal loci include *in situ* hybridisation of cDNA or RNA probes to isolated chromosomes, or the analysis of restriction fragment length polymorphisms (RFLP) (*cf.* Section 2.1.2.6).

The complete DNA sequence of the entire haploid human genome with its estimated 3.5×10^9 bp, the knowledge of which would finally solve the problem of gene mapping, may not be available before the end of this century.

Human genetic disorders can be classified into at least three different categories, *i.e.*, chromosomal, monogenic, and multifactorial defects. Most chromosomal defects are incompatible with life. This is demonstrated by the fact that the frequency of such abnormalities may exceed 50% in first-trimester spontaneous abortions (Stanbury, 1983). Some types of chromosomal disorders which are less detrimental are characterised by numerical aberrations of autosomes or sex chromosomes. They are the result of non-disjunction occurring during meiotic division and include conditions such as Down's syndrome (trisomy 21), Klinefelter's syndrome (47, XXY), and Turner's syndrome (45, X0). Patients with an

Table 13–2. Prevalence of some common monogenic disorders among live-born infants

Disorder	*Estimated prevalence*
Autosomal dominant	
Familial hypercholesterolaemia	1 in 500
Polycystic kidney disease	1 in 1 250
Huntington's chorea	1 in 2 500
Hereditary spherocytosis	1 in 5 000
Marfan syndrome	1 in 20 000
Autosomal recessive	
Sickle-cell anaemia	1 in 625 (U.S. blacks)
Cystic fibrosis	1 in 2 000 (U.S. whites)
Tay-Sachs disease	1 in 3 000 (U.S. Jews)
Phenylketonuria	1 in 12 000
Mucopolysaccharidoses (all types together)	1 in 25 000
Glycogen storage diseases (all types together)	1 in 50 000
X-linked	
Duchenne muscular dystrophy	1 in 7 000
Haemophilia	1 in 10 000

from Stanbury *et al.* (1983); Chapter 1.

XXY karyotype are phenotypic males displaying testicular dysgenesis and infertility. Individuals with the X0 karyotype are infertile females which are deficient in secondary sexual development. Of considerable significance is the observed association between the occurrence of these (and other) non-disjunction syndromes and increasing maternal age.

Translocations, another form of chromosomal abnormalities, are observed in certain malignancies. One example is chronic myelogenous leukemia (CML), in which the long arm of chromosome 22 has become translocated to the long arm of chromosome 9. This unique structural abnormality is known as the Philadelphia chromosome.

Monogenic disorders are caused by single mutated genes. Their Mendelian patterns of inheritance either is autosomal dominant, autosomal recessive or X-linked. Autosomal dominant aberrations manifest themselves even if only one of the two alleles of the particular gene is affected. In contrast, both alleles of a gene must be affected in autosomal recessive disorders for the disease to become clinically manifest. The effects of X-linked disorders are different in females and males since the latter carry only a single X-chromosome. A female with two X-chromosomes may be homozygous or heterozygous for an X-linked mutant gene and thus expression may be dominant or recessive. Regardless of the mode of expression in females, an X-linked trait will always be expressed in a dominant mode in males (with only one X-chromosome).

According to Stanbury (1983) approximately 1 400 of all known human genetic diseases are caused by mutations in single genes. As shown in Table 13-2, their incidence is rare although, as a group, they account for five to ten per cent of all cases referred to pediatric clinics. The biochemical lesions of approximately 250 of these disorders are known; they affect mainly enzymes, but also a variety of other proteins such as blood clotting factors, transport proteins (*e.g.*, haemoglobins), peptide hormones, receptors, and even ion channels. Among the enzymic defects are those candidates most likely to be subjected to genetic therapy (see below).

Monogenic disorders may not necessarily be inherited; in fact they may be the result of a new mutation in the afflicted individual. The frequency with which such mutations occur has been estimated to be 5×10^{-6} mutations per gene per generation. Approximately one in 100 000 newborns should therefore carry a new mutation at any given genetic locus. Since many of these mutations will either be recessive or genetically silent because they do not affect the function of the protein involved, new mutations with a clinical manifestation should be extremely rare. It is assumed, for example, that all cases of Huntington's chorea (several tens of thousands) stem from a small number of afflicted individuals (presumably with independent mutations) which can be traced back to the 18th century. Reports about cases of a dominant genetic disorder suddenly appearing in the offspring of unaffected parents can be explained, more often than not, by extramarital paternity rather than the acquisition of a new mutation.

Multifactorial disorders include a variety of diseases such as cancer, diabetes mellitus, hypertension etc. The disposition to become affected is thought to be inherited. These diseases are polygenic, and since the number of genes involved is unknown, the individual risk is difficult to assess. However, due to the larger number of genes which must be affected simultaneously, this risk is regarded to be considerably lower (one to five per cent) than the risks of acquiring monogenic disorders from affected parents (50% or 25%). Only few of these predispositions can even be diagnosed at the present time. Among the few genetic loci most prominently associated with predispositions to various diseases are those of the histocompatibility or HLA (*H*uman *L*eukocyte *A*ntigen) complex. It consists of four distinct genetic loci each of which carries multiple alleles coding for distinct proteins. A particular constellation, for example, known as HLA-B27, is associated with an increased risk of contracting ankylosing spondylitis. The risk of contracting this serious disease is 90 times higher in HLA-B27-positive than in B27-negative subjects. Although this predisposition can be diagnosed, the disease itself will not be amenable to gene therapy for a long time to come since it has a multifactorial genetic basis.

Diagnosis and prevention of genetic disorders can be accomplished on very different levels. If the molecular basis of a genetic defect is not known, genetic counselling can only provide data about the statistical risk of having a child with the familial disorder. In contrast, an intimate knowledge of a disease on the molecular level in conjuction with suitable techniques of detecting the molecular defect permits positive identification of a lesion in a given individual. This can be achieved either on the level of proteins, by analysing, for example, fetal blood specimens. Cloning techniques now even allow a genetic defect to be identified on the level of the gene itself. Since these analyses can be carried out with minute amounts of tissue, obtained for example by amniocentesis and/or chorion biopsy, genetic disorders can be diagnosed at a very early stage of pregnancy; depending on the result of these tests, the pregnancy could be terminated without significant risks.

Apart from being extraordinarily beneficial for the prevention of genetic diseases, these developments raise a number of ethical, moral and legal questions. Should a positive diagnosis of Huntington's chorea, which will manifest itself only at the age of 40 to 50, be disclosed to a patient in his twenties? How long will a society accept the existence of genetically disabled individuals when the defect had been diagosed antenatally and the parents had chosen to give birth to such a child rather than to abort the pregnancy? Will employers take advantage of their knowledge of a genetic predisposition of potential employees if there is a chance that they could develop this disease in their future occupation? An advanced society may certainly accept these types of screening, for example, in connection with a heart condition of a commercial aircraft pilot; however, it may be less inclined to tolerate such infringements in less obvious cases. How will life insurance companies incorporate genetic factors and predipositions into their analyses of risk assessment and life expectancies? DNA "fingerprinting" based on the pattern of distribution of a certain class of repetitive DNA elements within the genome has recently been shown to be not only as specific and unique for a given individual (apart from identical twins) as are conventional fingerprints but also considerably more sensitive. How can we prevent this type of information, which is so useful in forensic applications (Gill *et al.*, 1985) to intrude into our personal lifes and to infringe upon our constitutional rights? There is no doubt that the potentials of genome analysis and genetic screening have not simplified our lives. An advanced society however with its abilty to perform relevant and balanced risk/benefit analyses should be able to realise this challenge and to provide appropriate solutions. These solutions must ensure confidentiality and freedom of choice; they must avoid misunderstandings and stigmatisation (Rowley, 1984). Finally, as much as genetic engineering may reduce the incidence of genetic diseases, it should maximise the options available to individuals and families at risk.

13.3.2.2 Gene Therapy of Somatic Cells

Somatic cells are all cells of an organism except the germ cells or their precursors, *i.e.*, all cells which are not involved in the transfer of genetic information to the next generation. Somatic cell gene therapy essentially is a substitution therapy. It is only applicable to recessive phenotypes in which the lack of a protein is compensated, not by the injection of this protein as, for example, in insulin therapy of diabetes mellitus patients, but by the introduction of a wild-type gene with unimpaired and normal functions. In the case of a dominant defect it is not the lack of a gene or its

product which has to be dealt with but the presence of a mutated gene product which overrides the activity of the normal gene. In this particular situation, gene therapy not only must be aimed at substituting the mutated function but also at eliminating the mutated gene or, at least, at preventing its expression. These types of corrections may have to be performed in all cells of an organism and, thus, have to involve embryos rather than a particular organ or subset of cells of a mature organism. Manipulations of this kind are referred to as genetic manipulations of germ cells and will be discussed in Section 13.3.2.3.

The experimental details and the criteria for evaluating approaches towards somatic cell gene therapy in humans have been discussed previously (Anderson, 1984 and 1985). They can be summarised in the following set of questions:

(a) Why is the particular genetic disorder a good candidate for gene therapy?

(b) Are there other and safer alternatives to treat this particular disorder?

(c) Is a cloned and functional wild-type gene available for replacement therapy?

(d) Can the new gene be inserted efficiently, and will it be stably and properly expressed?

(e) Has the therapeutic approach been tested in an appropriate animal model?

(f) How safe are the delivery systems, *i.e.*, the vector and the newly inserted gene for the manipulated organism?

The commenest and most extensively studied human gene disorders are haemoglobin abnormalities, the thalassaemias in particular. Unfortunately, however, the regulation of globin synthesis, *i.e.*, the synthesis of the protein components of haemoglobin, is extremely complex. It involves intricate switches from embryonic to fetal and to adult globin, as well as the stoichiometric control of the expression of the α and β chains. None of these questions is sufficiently understood to warrant an attempt to correct this disease by gene therapy. Unfortunately, such experiments were performed prematurely (Cline *et al.*, (1980); see also Nature 291, 369 (1981)) for a criticism of these experiments.

Similar difficulties and limitations are observed with protein hormone deficiencies such as pituitary dwarfism or diabetes mellitus, which are currently treated by administering human growth hormone and insulin, respectively. The first attempts at gene therapy will therefore have to be restricted to those types of genetic disorders in which the mere presence of the gene product of the newly introduced gene suffices and its expression does not have to be strictly controlled. Among the genes with an "always-on" type of regulation are some involved in purine metabolism, *e.g.*, hypoxanthine-guanine phosphoribosyl transferase (HGPRT; EC 2.4.2.8), adenosine deaminase (ADA), and purine nucleoside phosphorylase (PNP), as well as some others such as α1-antitrypsin, and arginino-succinate synthetase.

A complete lack of HGPRT is observed in patients with the Lesch-Nyhan (hyperuricaemia) syndrome. This X-linked recessive disorder which, thus, affects only male patients is associated with an excessive production of uric acid (25 to 143 mg per kg body weight per day as compared to an upper limit of 18 mg for normal children), and several devastating neurological features such as mental retardation and a compulsive self-destructive behaviour. The pathogenesis of the central nervous system dysfunctions is still unknown. No satisfactory treatment is available, and patients usually succumb to renal failure at the age of 20-30. Patients with only a partial deficiency of HGPRT, *i.e.*, patients which are heterozygous for HGPRT deficiency, may develop severe forms of gout but rarely show neurological symptoms. Due to the vast differences in clinical appearance and prognosis, the differential antenatal diagnosis of the two types of the syndrome is important but difficult to accomplish (Kelley and Wyngaarden, 1983).

Deficiencies in either adenosine deaminase (ADA) or purine nucleoside phosphorylase (PNP) lead to abnormalities in the metabolism of purine nucleoside. ADA and PNP patients show a predisposition to recurrent infections from which they usually die early in childhood. The reason for the selective impairment of immune functions in these patients is still unknown. It is conceivable, however that the accumulation of dATP in ADA-deficient patients leads to an inhibition of ribonucleotide reductase, which in turn decreases the levels of deoxyribonucleosides and concomitantly reduces the rate of DNA replication and cell division. The incidence of both diseases is extremely low with only 50 and 9 patients known for ADA and PNP deficiencies, respectively.

The severe combined immunodeficiency (SCID) associated with ADA deficiency can be treated by transplanting bone marrow from histocompatible donors, indicating that gene therapy may indeed be successfully employed for such patients.

The target cells envisaged for the first human gene transfer studies are the haematopoietic stem cells of bone marrow. They represent a small fraction of undifferentiated nucleated cells which still have the capacity to divide and to differentiate into either the lymphocyte or myeloid progeny cell population of the bone marrow. Efficient transformation of these cells with genes cloned in retrovirus vectors of the type described in Section 8.7 has been described by Williams *et al.* (1984) and Miller *et al.* (1984). In a recent study, Keller *et al.* (1985) demonstrated the expression of the newly introduced bacterial aminoglycoside 3'-phosphotransferase gene in differentiated myeloid and lymphoid cells obtained from the bone marrow of mice reconstituted with infected bone marrow cells. This result is only explicable if one assumes that stem cells were successfully infected and were still able to divide and differentiate after their introduction into previously irridiated mice.

A similar approach is envisaged for human gene therapy. The defective bone marrow of a human patient is first destroyed by total body irradiation and subsequently is replaced by treated bone marrow stem cells carrying the newly introduced gene. It is expected that transfected stem cells surviving this treatment will differentiate into lymphoid and myeloid cells and will then give rise to progeny, eventually populating the bone marrow. In a pilot experiment Willis *et al.* (1984) successfully and efficiently transformed a homozygous human Lesch-Nyhan lymphoblastoid cell line with a retrovirus vector containing a functional *HGPRT* gene, and obtained phenotypically HGPRT-positive cells. The *HGPRT* gene in different clones was expressed at levels varying between 4 and 23 per cent of the level observed in wild-type cells. This indicated that gene expression in these haematopoietic cells was not fully under control and that the system must still be improved considerably. More recently, Hock and Miller (1986) were able to transfect human CFU-GM cells, the progenitor cells for the myeloid lineage, with amphotropic retrovirus vectors. The efficiency of transfection however did not exceed 3-10% which is as yet too low for practical purposes.

As yet, there remain a number of unanswered questions (Robertson, 1986). It is not known, for example, how important, the chromosomal location of the transduced gene is, and whether transcriptional enhancers are required for the regulated expression of *HGPRT* and other genes in haematopoietic stem cells. Unfortunately, no animal models (including primates) for these human genetic disorders are available. Many of the open questions in this field may therefore have to be answered directly through experiments on human patients. There are still problems in obtaining high transformation rates and in expressing the genes in question in haematopoietic stem cells. Of major concern, however, is the patient's safety. At present, the optimal transducing vectors are retroviruses; other systems such as papovaviruses appear to be less efficient, and they are also less developed. The advantages of retroviral vectors have already been discussed; however, there are a number of pitfalls associated with their use. Retroviruses integrate at random chromosomal sites. The integration event may

therefore interrupt important genes and regulatory sequences. The promoter sequences in the *LTR* regions of retroviruses have been reported to activate oncogenes located downstream from the integration site. Finally, retroviruses are known to exchange genetic material with endogenous, chromosomal DNA. As transducing vectors, they are unable to spread to other cells of an organism since they lack a particular sequence required for packaging and forming infectious particles; yet, such particles may be generated by recombination with an endogenous viral sequence containing an appropriate packaging signal. Previous work with retroviral vectors has provided little or no evidence that such recombination events actually do occur. Moreover, there are certainly strategies to circumvent these problems. These possibilities notwithstanding, a gene therapy protocol should also include a search for infectious recombinant viruses in the transformed tissue, for example, in transformed bone marrow cells, before the genetically altered cells are re-introduced into the patient.

In spite of all these limitations and shortcomings, the devastating character of some genetic disorders for which there is no other treatment available may justify the use of somatic gene therapy if it were restricted to the sole purpose of correcting such a defect. The decision of when to resort to such measures should be at least as strictly controlled as those criteria generally applied to other experimental medical procedures (Anderson, 1984). According to the regulations of the United States Department of Health and Human Services, federally funded gene therapy treatment must be reviewed and approved by the Institutional Review Board of the investigator's own institution and the Recombinant DNA Advisory Committee (Fed. Regist. 49, 17846 (25 April 1984)). There is a widespread public interest in these matters; however, in contrast to germline manipulation (see below), the scientific and ethical risks of such endeavours appear to be reasonably low and acceptable (*cf.* Walters, 1986, for a critical evaluation of the issue). In addition, the public will soon realise that the number of cases is extremely limited and may decline continuously with the development and availability of ever more efficient and safe diagnostic techniques.

13.3.2.3 Genetic Manipulation of Germ Cells

Genetic material introduced into the germline allows genotypic changes to be transmitted to the offspring in a normal Mendelian fashion. In mice and other experimental animals, germline transmission has indeed been effectively achieved by microinjection of fertilised eggs (*cf.* Section 13.3.1). The reasoning behind these manipulations has been discussed in detail. In humans, the introduction of foreign genes into the germline raises overwhelming concerns, misgivings, and criticism.

The first step should be to identify those disorders and conditions which would warrant germline gene therapy. The first genetic disorders to come into mind are those dominant disorders which may require the substituion of a defective gene in every single cell of an organism. Other scenarios centre around what has been termed "enhancement" or "eugenic" engineering (Anderson, 1985; *cf.* Kevles (1984) for a detailed discussion on eugenics). Germline therapy therefore differs drastically from purely therapeutic gene therapy regimens: it may not only attempt to correct otherwise incurable and desparate medical conditions but is predestined to improve or change certain characteristics of an individual. The introduction of an additional growth hormone gene to obtain a larger body size or of a low-density lipoprotein receptor gene to lower the risk of heart attacks and strokes may seem feasable and even acceptable, at least as far as the latter case is concerned.

Even if society were to accept such objectives, there are still a number of significant technical problems which must be solved before germline gene manipulation in humans will become feasable. First of all it would require extensive experimentation on human eggs and embryos. The technique of microinjection of mammalian eggs is extremely harmful and is therefore associated with high failure rates. In mice, the yields of transgenic animals rarely exceed one to five per cent, implying that one hundred human embryos might have to be manipulated and sacrificed in order to produce a single baby with a foreign gene. As pointed out previously, (*cf.* Section 13.3) methods for obtaining human eggs are being developed and optimised in connection with such controversial procedures as *in vitro* fertilisation, embryo freezing, and surrogate motherhood (Grobstein *et al.*, 1985). This raises enormous questions, concerning, for example, the legal status of such embryos, the ownership and fate of surplus embryos, and even places genetic engineering ever closer to the tense and difficult issue of abortion.

Every researcher in this field must be aware of these problems. He or she must also realise that there is little or no indication that the public will be willing to tolerate or support this kind of research without very good reasons. These, indeed, appear to be scarce if we consider other prerequisites of, and problems associated with, germline therapy.

These problems include issues which are similar to those already discussed in connection with somatic cell therapy; yet, now the context is very different. One of these problems is the fact that, at present, the site of integration of a transduced gene cannot be predicted. In the case of a dominant disorder, however, it is not only the replacement of a defective gene which must be accomplished, but also the inactivation of the corresponding defective gene itself. Moreover, the experiments with transgenic mice have demonstrated that the incorrect integration of a gene can be deleterious in that it impairs embryonic development due to the insertional inactivation of essential genes. There are indications that the problem of allelic replacement may be solved in the distant future (Smithies *et al.*, 1985). At present, however, the risks of producing pathological conditions are overwhelming and unacceptable. Of course, the integration of a foreign gene into the genome does not necessarily have to have deleterious effects; however, even if it did not immediately precipiate a clinical condition, it could involve subtle changes affecting future generations. Although there is no such thing as a "human" gene, there certainly exists a human gene pool, a quantitative and qualitative assortment of genes which should be highly protected as long as we do not know any better.

Finally there is the difficulty of identifying the genetic status of an embryo without destroying it in the course of the analysis. Even if this problem could be solved in the future, another aspect deserves consideration. In cases where both parents are homozygous for a genetic disorder, all their offspring would inevitably be affected. These patients however are almost always infertile. In all other cases of inheritance there will always be the chance of creating healthy and unaffected embryos. A doctor, thus, would simply have to identify a defective embryo and then chose another healthy embryo which then could be carried to full term. Should antenatal diagnosis reveal a serious genetic disorder, the prospective parents might simply consider an abortion and try again without having to resort to any type of gene therapy (Williamson, 1982).

The procedure of germline gene therapy in humans thus appears not only scientifically difficult and ethically highly questionable. It is also utterly useless! It is not surprising that legislation against this type of experiment is being considered in many countries. Nevertheless, banning this kind of experimentation with human embryos altogether should be considered very carefully. While such legislation could and should certainly include germline genetic therapy as far as it concerns "eugenic" engineering in order to prevent any future misuses, the requirements for improving *in vitro* fertilisation methods, for developing effective diagnostic procedures as well as new and unexpected developments in the treatment of genetic disorders may warrant certain exemptions.

References

Adams, J.M., Harris, A.W., Pinkert, Ca.A., Corcoran, L.M., Alexander, W.S., Cory, S., Palmiter, R.D., and Brinster, R.L. (1985). The *c-myc* oncogene driven by immunoglobulin enhancers induces lymphoid malignancies in transgenic mice. Nature 318, 533-538.

Anderson, W.F., and Fletcher, J.C. (1980). Gene therapy in human beings: when is it ethical to begin. N. Engl. J. Med. 303, 1293-1297.

Anderson, W.F. (1984). Prospects for human gene therapy. Science 226, 401-409.

Anderson, W.F. (1985). Human gene therapy: Scientific and ethical considerations. J. Med. and Phil. 10, 275-291.

Balinsky, B.I. (1981). An Introduction to Embryology. Fifth Edition; Saunders College Publishing, New York.

Berg, P. (1974). Potential biohazards of recombinant DNA molecules. Proc. Natl. Acad. Sci. USA 71, 2593-2594.

Brinster, R.L., Chen, H.Y., Messing, A., van Dyke, T., Levine, A., and Palmiter, R.D. (1984). Transgenic mice harbouring SV40 T-antigen develop characteristic brain tumours. Cell 37, 367-379.

Brinster, R.L., Chen, H.Y., Trumbauer, M.E., Yagle, M.K., and Palmiter, R.D. (1985). Factors affecting the efficiency of introducing foreign DNA into mice by microinjecting eggs. Proc. Natl. Acad. Sci. USA 82, 4438-4442.

Chan, H.W., Israel, M.A., Garon, C.F., Rowe, W.P., and Martin, M.A. (1979). Molecular cloning of polyoma virus DNA in *Escherichia coli*: λ phage vector system. Science 203, 887-892.

Cline, M.J., Stang, H., Mercola, K., Morx, L., Ruprecht, R., Browne, J., and Salser, W. (1980). Gene transfer in intact animals. Nature 284, 422-425.

Colwell, R.R., Brayton, P.R., Grimes, D.J., Roszak, D.B., Hug, S.A., and Palmer, L.M. (1985). Viable but non-culturable *Vibrio Cholerae* and related pathogens in the environment: implications for release of genetically engineered microorganisms. Biotechnology 3, 817-820.

Curtiss, R., III. (1976). Genetic manipulations of microorganisms: potential benefits and biohazards. Ann. Rev. Microbiol. 30, 507-533.

Edwards, R.G. (1974). Fertilization of human eggs *in vitro*: morals, ethics and the law. Quart. Rev. Biol. 49, 3-26.

Edwards, R.G. (1981). Test-tube babies. Nature 293, 253-256.

Evans, C.G.T., Preece, T.F., and Sargeant, K. (1981).

Microbial plant pathogens: Natural Spread, and possible risks in their industrial use. Study contract No. 724-EC1 UK; Commission of the European Communities.

Fehilly, C.B., Willadsen, S.M., and Tucker, E.M. (1984). Interspecific chimaerism between sheep and goat. Nature 307, 634-636.

Freter, R. (1984). Factors affecting conjugal transfer in natural bacterial communities. In "Current perspectives in microbial ecology"; Klug, M.J., and Reddy, C.A., eds.; Amer. Soc. Microbiol.,; Washington, D.C.; pp.105-114.

Fried, M., Klein, B., Murray, K., Greenaway, P., Tooze, J., Boll, W., and Weissmann, C. (1979). Infectivity in mouse fibroblasts of polyoma DNA integrated into plasmid pBR322 or lambdoid phage DNA. Nature 279, 811-816.

Gill, P., Jeffreys, A.J., and Werzett, D.J. (1985). Forensic applications of DNA fingerprints. Nature 318, 577-579.

Gordon, J.W., Scangos, G.A., Plotkin, D.J., Barbosa, J.A., and Ruddle, F.H. (1980). Genetic transformation of mouse embryos by microinjection of purified DNA. Proc. Natl. Acad. Sci. USA 77, 7380-7384.

Green, R.L., and Warren, G.J. (1985). Physical and functional repetition in a bacterial ice nucleation gene. Nature 317, 645-648.

Grobstein, C., Flower, M., and Mendeloff, J. (1985). Frozen embryos: policy issues. N. Engl. J. Med. 312, 1584-1588.

Hammer, R.E., Pursel, V.G., Rexroad, Jr., C.E., Wall. R.J., Bolt, D.J., Ebert, K.M., Palmiter, R.D., and Brinster, R.L. (1985). Production of transgenic rabbits, sheeps and pigs by microinjection. Nature 315, 680-683.

Hanahan, D. (1985). Heritable formation of pancreatic β-cell tumours in transgenic mice expressing recombinant insulin/simian virus 40 oncogenes. Nature 315, 115-122.

Harbers, K., Kuehn, M., Delius, H., and Jaenisch, R. (1984). Insertion of retrovirus into the first intron of α1 (I) collagen gene leads to embryonic lethal mutation in mice. Proc. Natl. Acad. Sci. USA 81, 1504-1508.

Hauptli, H., Newell, N., and Goodman, R.M. (1985). Genetically engineered plants: Environmental issues. Biotechnology 3, 437-442.

Hirano, S.S., and Upper, C.D. (1985). Ecology and physiology of Pseudomonas syringae. Biotechnology 3, 1073-1078.

Hock, R.A., and Miller, A.D. (1986). Retrovirus-mediated transfer and expression of drug resistance genes in human haematopoietic progenitor cells. Nature 320, 275-277.

Horsfall, J.G., and Cowling, E.B. (1978). Some epide-mics man has known. In "Plant diseases", vol. 2; Horsfall, J.G., and Cowling,E.B., eds.; Academic Press, New York and London; pp.17-32.

Israel, M.A., Chan, H.W., Rowe, W.P., and Martin, M.A. (1979). Molecular cloning of polyoma virus DNA in Escherichia coli: plasmid vector systems. Science 203, 883-887.

Jaenisch, R. (1985). Mammalian neural crest cells participate in normal embryonic development on microinjection into post-implantation mouse embryos. Nature 318, 181-183.

Keller, G., Paige, Ch., Gilboa, E., and Wagner, E.F. (1985). Expression of a foreign gene in myeloid and lymphoid cells derived from multipotent haematopoietic precursors. Nature 318, 149-154.

Kelley, W.N., and Wyngaarden, J.B. (1983). Clinical syndromes associated with hypoxanthine-guanine phosphoribosyl transferase deficiency. In "The metabolic basis of inherited diseases". J.B. Stanbury, D.S. Frederickson, J.L. Goldstein and M.S. Brown, eds.; McGraw-Hill, New York; pp. 1115-1143.

Kevles, J.J. (1984). Annals of Eugenics: A secular faith. In "The New Yorker"; issues of Oct. 8, pp. 51 ff.; Oct. 15, pp. 52 ff.; Oct. 22, pp. 92 ff.; and Oct. 29, pp. 51 ff.

Kleinsmith, L.J., and Pierce, G.B., Jr. (1964). Multipotentiality of single embryonal carcinoma cells. Cancer Res. 24, 1544-1552.

Land, H., Parada, L.F., and Weinberg, R.A. (1983). Cellular oncogenes and multistep carcinogenesis. Science 222, 771-778.

LeClerg, E.L. (1964). Crop losses due to plant diseases in the United States. Phytophathol. 54, 1209-1313.

Martin, G.R. (1975). Teratocarcinomas as a model system for the study of embryogenesis and neoplasia. Cell 5, 229-243.

Meinicke-Tillmann, S., and Meinicke, B. (1984). Experimental chimaeras – removal of reproductive barrier between sheep and goat. Nature 307, 637-638.

Miller, A.D., Eckner, R.J., Jolly, D.J., Friedmann, T., and Verma, I.M. (1984). Expression of a retrovirus encoding human HPRT in mice. Science 225, 630-632.

Ornitz, D.M., Palmiter, R.D., Hammer, R.E., Brinster, R.L., Swift, G.H., and MacDonald, R.J. (1985). Specific expression of an elastase-human growth hormone fusion gene in pancreatic acinar cells of transgenic mice. Nature 313, 600-602.

Palmiter, R.D., Brinster, R.L., Hammer, R.E., Trumbauer, M.E., Rosenfeld, M.G., Birnberg, N.C., and Evans, R.M. (1982). Dramatic growth of mice that develop from eggs microinjected with metallothionein-growth hormone fusion genes. Nature 300, 611-615.

Palmiter, R.D., Wilkie, T.M., Chen, H.Y., and Brinster, R.L. (1984). Transmission distortion and mosaicism in an unusual transgenic mouse pedigree. Cell 36, 869-877.

Palmiter, R.D., and Brinster, R.L. (1985). Transgenic mice. Cell 41, 343-345.

Papaioannou, V.E., McBurney, and Gardner, R.L. (1975). Fate of teratocarcinoma cells injected into early mouse embryos. Nature 258, 70-73.

Ringold, G.M. (1983). Regulation of mouse mammary tumor virus gene regulation by glucocorticoid hormones. Curr. Topics Microbiol. Immunol. 106, 79-103.

Ritchie, K.A., Brinster, R.L., and Storb, U. (1984). Allelic exclusion and control of endogenous immunoglobulin gene rearrangement in kappa transgenic mice. Nature 312, 517-520.

Robertson, M. (1986). Desparate appliances. Nature 320, 213-214.

Roderick, T.H., and Davisson, M.T. (1984). Linkage map of the mouse. In "Genetic maps"; Stepen O'Brien, Ed.; Cold Spring Harbor Laboratory; Cold Spring Harbor, New York 11724; pp. 343-355.

Rowley, P.T. (1984). Genetic screening: marvel or menace. Science 225, 138-144.

Rusconi, S., and Köhler, G. (1985). Transmission and expression of a specific pair of rearranged immunoglobulin μ and ϰ genes in a transgenic mouse line. Nature 314, 330-334.

Schnieke, A., Harbers, K., and Jaenisch, R. (1983). Embryonic lethal mutation in mice induced by retrovirus insertion into the α1 (I) collagen gene. Nature 304, 315-320.

Shettles, L.B. (1979). Diploid nuclear replacement in mature human ova with cleavage. Am. J. Obstet. Gynecol., 222-225.

Smithies, O., Gregg, R.G., Boggs, S.S., Kovalewski, M.A., and Kucherlapati, R.S. (1985). Insertion of DNA sequences into the human chromosomal β-globin locus by homologous recombination. Nature 317, 230-324.

Stanbury, J.B., Wyngaarden, J.B., Frederickson, D.S., Goldstein, J.L., and Brown, M.S. (1983). The Metabolic basis of inherited diseases. McGraw-Hill, New York.

Stevens, L.C. (1958). Studies on transplantable testicular teratomas of strain 129 mice. J. Nat. Cancer Inst. 20, 1257-1276.

Stevens, L.C. (1967). The biology of teratomas. Adv. Morph. 6, 1-31.

Stewart, T.A., Pattengale, P.K., and Leder, P. (1984). Spontaneous mammary adenocarcinomas in transgenic mice that carry and express MMTV/myc fusion genes. Cell 38, 627-637.

Stotzky, G., and Babich, H. (1984). Fate of genetically-engineered microbes in natural environment. Rec. DNA Technical Bull. 7, 163-188.

Storb, U., O'Bried, R., McMullen, M., Gollahon, K., and Brinster, R.L. (1984). High expression of cloned immunoglobulin kappa gene in transgenic mice is restricted to B lymphocytes. Nature 310, 238-241.

Stout, J.T., Chen, H.Y., Brennand, J., Caskey, C.T., and Brinster, R.L. (1985). Expression of human HPRT in the central nervous system of transgenic mice. Nature 317, 250-252.

Su, De-Ming (1984). Can pathogenic microorganisms be established as conventional control agents of pests? In "Current perspectives in microbial ecology"; Klug, J.M., and Reddy, C.A., eds.; Amer. Soc. Microbiol., Washington, D.C.; pp. 383-387.

Swift, G.H., Hammer, R.E., MacDonald, R.J., and Brinster, R.L. (1984). Tissue-specific expression of the rat pancreatic elastase I gene in transgenic mice. Cell 38, 639-646.

Wagner, E.F., Covarrubias, L., Stewart, T.A., and Mintz, B. (1983). Prenatal lethalities in mice homozygous for human growth hormone gene sequences integrated in the germ line. Cell 35, 647-655.

Wagner, E.F., Vanek, M., and Vennström, B. (1985). Transfer of genes into embryonal carcinoma cells by retrovirus infection: efficient expression from an internal promoter. The EMBO J. 4, 663-666.

Walters, L. (1986). The ethics of human gene therapy. Nature 320, 225-227.

Watson, J., and Tooze, J. (1981). The DNA Story; A documentary history of gene cloning; W.H. Freeman and Co.; San Francisco.

Weaver, D., Costantini, F., Imanishi-Kari, Th., and Baltimore, D. (1985). A transgenic immunoglobulin μ gene prevents rearrangement of endogenous genes. Cell 42, 117-127.

West, A.W., Burges, H.D., Whithe, R.J., Wyborn, C.H. (1984). Persistence of Bacillus thuringiensis parasporal crystal insecticidal activity in soil. J. Invertebrate Path. 44, 128-133.

Will, H., Cattaneo, R., Hoch, H.G., Darai, G., Schaller, H., Schellekens, H., van Eerd, P.M.C.A., and Deinhardt, F. (1982). Cloned HBV DNA causes hepatitis in chimpanzees. Nature 299, 740-742.

Williamson, B. (1982). Gene therapy. Nature 298, 416-418.

Williams, D.A., Lemischka, I.R., Nathan, D.G., and Mulligan, R.C. (1984). Introduction of new genetic material into pluripotent haematopoietic stem cells of the mouse. Nature 310, 476-480.

Willis, R.C., (1984). Partial phenotypic correction of human Lesch-Nyhan (hypoxanthine-guanine phosphoribosyl transferase-deficient) lymphoblasts with a transmissable retroviral vector. J. Biol. Chem. 259, 7842-7849.

Woychik, R.P., Stewart, T.A., Davis, L.G., D'Eusta-
chio, P., and Leder, P. (1985). An inherited limb deformity created by insertional mutagenesis in a transgenic mouse. Nature 318, 36-40.

Appendix A

List of restriction endonucleases

The following data are taken from an extensive review by Kessler *et al.* (1985). Table I represents an alphabetical cross-index of the Class II enzymes mentioned in Table II. Table II begins with Class II nucleases and methylases followed by Class I and Class III enzymes. Entries are not in alphabetical order but are arranged according to homologies within the recognition sequences. Palindromic recognition sequences thus are listed as internal AT, GC, CG, and TA palindromes, in that order. The table continues with pentanucleotide sequences, palindromic sequences carrying insertions with ambiguous nucleotides and, finally, non-palindromic sequences. The complete list of Kessler *et al.* (1985) comprises 626 entries while the present one (132 entries) includes only commercially available restriction endonucleases and enzymes mentioned in the text. It thus lacks many isoschizomeric enzymes.

Table I

Alphabetical cross-index of restriction endonucleases and methylases.

38	*Aat*II	81	*Ava*II
67	*Acc*I	27	*Bal*I
28	*Acc*II	7	*Bam*HI
47	*Aha*II	48	*Ban*I
79	*Aha*III	24	*Ban*II
15	*Alu*I	44	*Bbe*I
23	*Apa*I	113	*Bbr*I
84	*Apy*I	11	*Bcl*I
98	*Asp*700	100	*Bgl*I
74	*Asp*718	6	*Bgl*II
58	*Ava*I	35	*Bsp*1286

32	*Bss*HII	49	*Mst*I
94	*Bst*EII	95	*Mst*II
85	*Bst*NI	55	*Nae*I
101	*Bst*XI	45	*Nar*I
40	*Cfo*I	87	*Nal*I
62	*Cla*I	14	*Nco*I
82	*Clm*II	71	*Nde*I
93	*Dde*I	12	*Nla*III
3	*Dpn*I	97	*Nla*IV
80	*Dra*I	26	*Not*I
1	*Eco*RI	68	*Pae*R7
86	*Eco*RII	36	*Pst*I
70	*Eco*RV	9	*Pvu*I
90	*Fnu*4HI	20	*Pvu*II
29	*Fnu*DII	72	*Rsa*I
104	*Fok*I	18	*Sac*I
46	*Hae*II	33	*Sac*II
21	*Hae*III	64	*Sal*I
50	*Hap*II	5	*Sau*3A
105	*Hga*I	89	*Sau*96I
41	*Hha*I	96	*Sau*I
43	*Hin*PII	73	*Sca*I
65	*Hinc*II	91	*Scr*FI
66	*Hind*II	108	*Sfa*NI
17	*Hind*III	102	*Sfi*I
88	*Hinf*I	83	*Sin*I
78	*Hpa*I	56	*Sma*I
51	*Hpa*II	39	*Sna*BI
106	*Hph*I	13	*Sph*I
75	*Kpn*I	19	*Sst*I
16	*M.Alu*I	34	*Sst*II
63	*M.Cla*I	22	*Stu*I
2	*M.Eco*RI	59	*Taq*I
42	*M.Hha*I	30	*Tha*I
52	*M.Hpa*II	109	*Tth*111II
54	*M.Msp*I	61	*Tth*HB8I
37	*M.Pst*I	77	*Xba*I
60	*M.Taq*I	69	*Xho*I
76	*Mae*I	8	*Xho*II
92	*Mae*III	57	*Xma*I
4	*Mbo*I	25	*Xma*III
17	*Mbo*II	99	*Xmn*I
31	*Mlu*I	10	*Xor*II
53	*Msp*I		

Numbers refer to the corresponding numbers in column (1) of Table II.

Table II
List of restriction endonucleases and methylases
(Footnotes see end of table)

1. Class II enzymes
1.1 Enzymes with palindromic recognition sequences
1.1.1 Recognition sequences of tetra-, hexa- or octanucleotides
1.1.1.1 Internal AT-palindromes

Pos. no. (1)	Recognition sequence[a,b,c] (2)	Restriction endonuclease/methylase[d,e] (3)	Number of recognition sites on[f]						Micro-organism used for isolation[g] (Source of micro-organism) (10)
			λ (4)	Ad2 (5)	SV40 (6)	φX174 (7)	M13mp7 (8)	pBR322 (9)	
1	G/A A T T C (+A, +C)	*Eco*RI(*)	5	5	1	0	2	1	*Escherichia coli* RY13 (R.N. Yoshimori)
2	G A A T T C (M)	M · *Eco*RI	5	5	1	0	2	1	*Escherichia coli* RY13 (R.N. Yoshimori)
3	G A/T C (m)	*Dpn*I	116	87	8	0	8	22	*Diplococcus pneumoniae* (S. Lacks)
4	/G A T C (+A, oC)	*Mbo*I	116	87	8	0	8	22	*Moraxella bovis* (ATCC 10900)
5	/G A T C (oA, +C)	*Sau* 3A(*)	116	87	8	0	8	22	*Staphylococcus aureus* 3A (E.E. Stobberingh)
6	A/G A T C T (oA, +C)	*Bgl*II	6	11	0	0	1	0	*Bacillus globigli* (G.A. Wilson)
7	G/G A T C C (oA, +oC)	*Bam*HI(*)	5	3	1	0	2	1	*Bacillus amyloliquefaciens* H (F.E. Young)

Table II. List of Restriction Endonucleases and Methylases   511

Table II (continued)

Pos. no. (1)	Recognition sequence[a,b,c] (2)	Restriction endonuclease/methylase[d,e] (3)	Number of recognition sites on[f]						Micro-organism used for isolation[g] (Source of micro-organism) (10)
			λ (4)	Ad2 (5)	SV40 (6)	φX174 (7)	M13mp7 (8)	pBR322 (9)	
8	(A)/G G A T C(T)/(C)	XhoII	21	22	3	0	4	8	Xanthomonas holcicola (ATCC 13461)
9	C G A T/C G	PvuI	3	7	0	0	1	1	Proteus vulgaris (ATCC 13315)
10	C G A T/C G	XorII	3	7	0	0	1	1	Xanthomonas oryzae (M. Ehrlich)
11	T/G A T C A	BclI	8	5	1	0	0	0	Bacillus caldolyticus (A. Atkinson)
12	C A T G/	NlaIII	181	183	16	22	26	26	Neisseria lactamica (NRCC 2118)
13	G C A T G/C	SphI	6	8	2	0	0	1	Streptomyces phaeochromogenes (F. Bolivar)
14	C/C A T G G	NcoI	4	20	3	0	0	0	Nocardia corallina (ATCC 19070)

1.1.1.2 Internal GC-palindromes

Pos. no.	Recognition sequence	Restriction endonuclease/methylase	λ	Ad2	SV40	φX174	M13mp7	pBR322	Micro-organism used for isolation (Source of micro-organism)
15	A G/C T	AluI	143	158	34	24	24	16	Arthrobacter luteus (ATCC 21606)
16	A G C T	M·AluI	143	158	34	24	24	16	Arthrobacter luteus (ATCC 21606)

Table II (continued)

Pos. no. (1)	Recognition sequence[a,b,c] (2)	Restriction endonuclease/ methylase[d,e] (3)	Number of recognition sites on[f]						Micro-organism used for isolation[g] (Source of micro-organism) (10)
			λ (4)	Ad2 (5)	SV40 (6)	φX174 (7)	M13mp7 (8)	pBR322 (9)	
17	A/A G C T T	HindIII(*)	6	12	6	0	0	1	*Haemophilus influenzae* R$_d$ (S.H. Goodgal)
18	G A G C T/C	SacI	2	16	0	0	0	0	*Streptomyces achromogenes* (ATCC 12767)
19	G A G C T/C	SstI(*)	2	16	0	0	0	0	*Streptomyces stanfordii* (S. Goff and A. Rambach)
20	C A G/C T G	PvuII(*)	15	24	3	0	3	1	*Proteus vulgaris* (ATCC 13315)
21	G G/C C	HaeIII(*)	149	216	18	11	15	22	*Haemophilus aegyptius* (ATCC 11116)
22	A G G/C C T	StuI	6	11	7	1	0	0	*Streptomyces tubercidicus* (A. Takahashi)
23	G G G C C/C	ApaI	1	12	1	0	0	0	*Acetobacter pasteurianus* subsp. *pasteurianus* (NCIB 7215)
24	G(A)(G) G C(C)(C) C	BanII	7	57	2	0	1	2	*Bacillus aneurinolyticus* (IAM 1077)
25	C/G G C C G	XmaIII	2	19	0	0	0	1	*Xanthomonas malvacearum* (ATCC 9924)
26	G C/G G C C G C	NotI	0	7	0	0	0	0	*Nocardia otitidis-caviarum* (ATCC 14630)
27	T G G/C C A	BalI	18	17	0	0	1	1	*Brevibacterium albidum* (ATCC 15831)
28	C G C G	AccII	157	303	0	14	18	23	*Acinetobacter calcoaceticus* (A.J. Roberts)

Table II. List of Restriction Endonucleases and Methylases   513

Table II (continued)

Pos. no. (1)	Recognition sequence[a,b,c] (2)	Restriction endonuclease/ methylase[d,e] (3)	Number of recognition sites on[f]						Micro-organism used for isolation[g] (Source of micro-organism) (10)
			λ (4)	Ad2 (5)	SV40 (6)	φX174 (7)	M13mp7 (8)	pBR322 (9)	
29	C G/C G	FnuDII [E]	157	303	0	14	18	23	Fusobacterium nucleatum D (M. Smith)
30	C G/C G	ThaI	157	303	0	14	18	23	Thermoplasma acidophilum (D. Searcy)
31	A/C G C G T	MluI	7	5	0	2	0	0	Micrococcus luteus (IFO 12992)
32	G/C G C G C	BssHII	6	52	0	1	0	0	Bacillus stearothermophilus H3 (N. Welker)
33	C C G C/G G	SacII	4	33	0	1	0	0	Streptomyces achromogenes (ATCC 12767)
34	C C G C/G G	SstII(*)	4	33	0	1	0	0	Streptomyces stanfordii (S. Goff and A. Rambach)
35	G(A/G/T)G C(T/C/A)/C	Bsp1286	38	105	4	3	4	10	Bacillus sphaericus (IAM 1286)
36	C T G C A/G	PstI(*)	28	30	2	1	1	1	Providencia stuartii 164 (J. Davies)
37	C T G C A G	M·PstI	28	30	2	1	1	1	Providencia stuartii 164 (J. Davies)

1.1.1.3 Internal CG palindromes

Pos. no. (1)	Recognition sequence[a,b,c] (2)	Restriction endonuclease/ methylase[d,e] (3)	λ (4)	Ad2 (5)	SV40 (6)	φX174 (7)	M13mp7 (8)	pBR322 (9)	Micro-organism used for isolation[g] (Source of micro-organism) (10)
38	G A C G T/C	AatII	10	3	0	1	0	1	Acetobacter aceti (IFO 3281)
39	T A C/G T A	SnaBI	1	0	0	0	1	0	Sphaerotilus natans (ATCC 15291)

Table II (continued)

Pos. no. (1)	Recognition sequence[a,b,c] (2)	Restriction endonuclease/ methylase[d,e] (3)	Number of recognition sites on[f]						Micro-organism used for isolation[g] (Source of micro-organism) (10)
			λ (4)	Ad2 (5)	SV40 (6)	φX174 (7)	M13mp7 (8)	pBR322 (9)	
40	$\overset{+}{\text{G}}$ C $\overset{+}{\text{G}}$/C	*CfoI*	215	375	2	18	25	31	*Clostridium formicoaceticum* (ATCC 23439)
41	$\overset{+}{\text{G}}$ C $\overset{+}{\text{G}}$/C	*HhaI*(*)	215	375	2	18	25	31	*Haemophilus haemolyticus* (ATCC 10014)
42	G $\overset{M}{\text{C}}$ G C	M · *HhaI*	215	375	2	18	25	31	*Haemophilus haemolyticus* (ATCC 10014)
43	G/C G C	*HinPII*	215	375	2	18	25	31	*Haemophilus influenzae* P₁ (S. Shen)
44	G G C G $\overset{\cdots}{\text{C}}$/C	*BbeI*	1	20	0	2	1	4	*Bifidobacterium breve* YIT4006 (H. Takahashi)
45	G G/C G $\overset{\cdots}{\text{C}}$ C	*NarI*	1	20	0	2	1	4	*Nocardia argentiensis* (ATCC 31306)
46	$\binom{A}{G}$ G C G C$\overset{\cdots}{\binom{T}{C}}$	*HaeII*	48	76	1	8	6	11	*Haemophilus aegyptius* (ATCC 11116)
47	G $\binom{A}{G}$/C G $\binom{C}{C}$ C	*AhaII*	40	41	0	7	1	6	*Aphanothece halophytica* (ATCC 29534)
48	G/G$\binom{C}{C}\binom{(T)(A)}{(G)}$C C	*BanI*	25	57	1	3	6	9	*Bacillus aneurinolyticus* (IAM 1077)
49	T G C/G C A	*MstI*	15	17	0	1	0	4	*Microcoleus* species (D. Comb)

Table II. List of Restriction Endonucleases and Methylases 515

Table II (continued)

Pos. no. (1)	Recognition sequence[a,b,c] (2)	Restriction endonuclease/ methylase[d,e] (3)	Number of recognition sites on[f] λ (4)	Ad2 (5)	SV40 (6)	φX174 (7)	M13mp7 (8)	pBR322 (9)	Micro-organism used for isolation[g] (Source of micro-organism) (10)
50	⁺C/C G G	HapII	328	171	1	5	19	26	*Haemophilus aphrophilus* (ATCC 19415)
51	⁺⁺C/C G G	HpaII[h]	328	171	1	5	19	26	*Haemophilus parainfluenzae* (J. Setlow)
52	ᴹC C G G	M·HpaII	328	171	1	5	19	26	*Haemophilus parainfluenzae* (J. Setlow)
53	⁺ᵒC/C G G	MspI[i]	328	171	1	5	19	26	*Moraxella species* (R.J. Roberts)
54	ᴹC C G G	M·MspI[j]	328	171	1	5	19	26	*Moraxella species* (R.J. Roberts)
55	G C C/G G C	NaeI	1	13	1	0	1	4	*Nocardia aerocolonigenes* (ATCC 23870)
56	⁺C C C/G G G	SmaI[k]	3	12	0	0	0	0	*Serratia marcescens* S_b (C. Mulder)
57	⁺ᵒC/C C G G G	XmaI	3	12	0	0	0	0	*Xanthomonas malvacearum* (ATCC 9924)
58	C/(C)C⁽ᵀ⁾⁺ G (A)(G)G	AvaI[(*)]	8	40	0	1	1	1	*Anabaena variabilis* (ATCC 27892)
59	ᵒT/C G A⁺	TaqI	121	50	1	10	14	7	*Thermus aquaticus* YTI (J.I. Harris)

Table II (continued)

Pos. no. (1)	Recognition sequence[a,b,c] (2)	Restriction endonuclease/ methylase[d,e] (3)	λ (4)	Ad2 (5)	SV40 (6)	φX174 (7)	M13mp7 (8)	pBR322 (9)	Micro-organism used for isolation[g] (Source of micro-organism) (10)
						Number of recognition sites on[f]			
60	T C G A (M)	M · TaqI	121	50	1	10	14	7	*Thermus aquaticus* YTI (J.I. Harris)
61	T/C G A (o, +)	*Tth*HB81	121	50	1	10	14	7	*Thermus thermophilus* HB8 (T. Oshima)
62	A T/C G A T (+)	ClaI	15	2	0	0	2	1	*Caryophanon latum* L (H. Mayer)
63	A T C G A T (M, +)	M · ClaI	15	2	0	0	2	1	*Caryophanon latum* L (H. Mayer)
64	G/T C G A C (+, +)	SalI(*)	2	3	0	0	2	1	*Streptomyces albus* G (J.M. Ghuysen)
65	G T (T)(A) A C / (C)(G)	HincII	35	25	7	13	2	2	*Haemophilus influenzae* R$_c$ (A. Landy and G. Leidy)
66	G T (T)(A) A C / (C)(G) (+, o)	HindII	35	25	7	13	2	2	*Haemophilus influenzae* R$_d$ (S.H. Goodgal)
67	G T/(A)(T) A C / (C)(G) (+)	AccI	9	17	1	2	2	2	*Acinetobacter calcoaceticus* (R.J. Roberts)
68	C/T C G A G (+)	PaeR7(*)	1	5	0	1	0	0	*Pseudomonas aeruginosa* (G.A. Jacoby)

Table II (continued)

Table II. List of Restriction Endonucleases and Methylases 517

Pos. no. (1)	Recognition sequence [a,b,c] (2)	Restriction endonuclease/ methylase [d,e] (3)	Number of recognition sites on [f]						Micro-organism used for isolation [g] (Source of micro-organism) (10)
			λ (4)	Ad2 (5)	SV40 (6)	φX174 (7)	M13mp7 (8)	pBR322 (9)	
69	C/T C G A G	XhoI	1	6	0	1	0	0	Xanthomonas holcicola (ATCC 13461)

1.1.1.4 Internal TA palindromes

Pos. no.	Recognition sequence	Restriction endonuclease/methylase	λ	Ad2	SV40	φX174	M13mp7	pBR322	Micro-organism
70	G A T/A T C	EcoRV(*)	21	9	1	0	0	1	Escherichia coli J62 [pLG74] (L.I. Glatman)
71	C A/T A T G	NdeI	7	2	2	0	3	1	Neisseria denitrificans (NRCC 31009)
72	G T/A C	RsaI	113	83	12	11	18	3	Rhodopseudomonas sphaeroides (S. Kaplan)
73	A G T/A C T	ScaI(*)	5	5	0	0	0	1	Streptomyces caespitosus (H. Takahashi)
74	G/G T A C C	Asp718	2	8	1	0	0	0	Achromobacter species 718 (C. Kessler)
75	G G T A C/C	KpnI(*)	2	8	1	0	0	0	Klebsiella pneumoniae OK8 (J. Davies)
76	C/T A G	MaeI	13	54	12	3	4	5	Methanococcus aeolicus PL-15/H (DSM 2835)
77	T/C T A G A	XbaI(*)	1	5	0	0	0	0	Xanthomonas badrii (ATCC 11672)
78	G T T/A A C	HpaI(*)	14	6	4	3	0	0	Haemophilus parainfluenzae (J. Setlow)

Table II (continued)

Pos. no. (1)	Recognition sequence[a,b,c] (2)	Restriction endonuclease/ methylase[d,e] (3)	λ (4)	Ad2 (5)	SV40 (6)	φX174 (7)	M13mp7 (8)	pBR322 (9)	Micro-organism used for isolation[g] (Source of micro-organism) (10)
					Number of recognition sites on[f]				
79	T T T/A A A	*Aha*III	13	12	12	2	5	3	*Aphanothece halophytica* (ATCC 29534)
80	T T T/A A A	*Dra*I	13	12	12	2	5	3	*Deinococcus radiophilus* (ATCC 27603)

1.1.2 Pentanucleotide recognition sequences

Pos. no. (1)	Recognition sequence[a,b,c] (2)	Restriction endonuclease/ methylase[d,e] (3)	λ (4)	Ad2 (5)	SV40 (6)	φX174 (7)	M13mp7 (8)	pBR322 (9)	Micro-organism used for isolation[g] (Source of micro-organism) (10)
81	G/G (A/T) C C	*Ava*II	35	73	6	1	1	8	*Anabaena variabilis* (ATCC 27892)
82	G G (A/T) C C	*Clm*II	35	73	6	1	1	8	*Caryophanon latum* (ATCC 15219)
83	G G (A/T) C C	*Sin*I	35	73	6	1	1	8	*Salmonella infantis* (A. de Waard)
84	C C/(A/T) G G	*Apy*I(*)	71	136	17	2	7	6	*Arthrobacter pyridinolis* (R. DiLauro)
85	C C/(A/T) G G	*Bst*NI	71	136	17	2	7	6	*Bacillus stearothermophilus* (D. Comb)
86	/C C (A/T) G G	*Eco*RII	71	136	17	2	7	6	*Escherichia coli* R245 (R.N. Yoshimori)

Table II (continued)

Table II. List of Restriction Endonucleases and Methylases   519

Pos. no. (1)	Recognition sequence[a,b,c] (2)	Restriction endonuclease/ methylase[d,e] (3)	Number of recognition sites on[f]						Micro-organism used for isolation[g] (Source of micro-organism) (10)
			λ (4)	Ad2 (5)	SV40 (6)	φX174 (7)	M13mp7 (8)	pBR322 (9)	
87	C C(G)(C)G G	NciI[1]	114	97	0	1	4	10	Neisseria cinerea (NRCC 31006)
88	G/A N T C	HinfI	148	72	10	21	26	10	Haemophilus influenzae R_f (C.A. Hutchison III)
89	G/G N C C	Sau96I	74	164	11	2	4	15	Staphylococcus aureus PS96 (E.E. Stobberingh)
90	G C/N G C	Fnu4HI	380	411	24	31	15	42	Fusobacterium nucleatum 4H (M. Smith)
91	C C/N G G	ScrFI	185	233	17	3	11	16	Streptococcus cremonis F (C. Daly)
92	/G T N A C	MaeIII	156	118	14	17	25	17	Methanococcus aeolicus PL-15/H (DSM 2835)
93	C/T N A G	DdeI[(*)]	104	97	20	14	29	8	Desulfovibrio desulfuricans strain Norway (H. Peck)

1.1.3 Longer Recognition sequences with internal (N)$_x$ sequences

Pos. no.	Recognition sequence	Restriction endonuclease/methylase	λ	Ad2	SV40	φX174	M13mp7	pBR322	Micro-organism used for isolation (Source of micro-organism)
94	G/G T N A C C	BstEII	13	10	0	0	0	0	Bacillus stearothermophilus ET (N. Welker)
95	C C/T N A G G	MstII	2	7	0	0	1	0	Microcoleus species (D. Comb)

Table II (continued)

Pos. no. (1)	Recognition sequence[a,b,c] (2)	Restriction endonuclease/methylase[d,e] (3)	Number of recognition sites on[f]						Micro-organism used for isolation[g] (Source of micro-organism) (10)
			λ (4)	Ad2 (5)	SV40 (6)	φX174 (7)	M13mp7 (8)	pBR322 (9)	
96	C C/T N A G G	SauI	2	7	0	0	1	0	Streptomyces aureofaciens IKA 18/4 (J. Timko)
97	G G N/N C̈ C̈	NlaIV	82	178	16	6	6	24	Neisseria lactamica (NRCC 2118)
98	G A A N N/N N T T C	Asp700	24	5	0	3	2	2	Achromobacter species 700 (C. Kessler)
99	G A A N N/N N T T C	XmnI	24	5	0	3	2	2	Xanthomonas manihotis 7AS1 (B.-C. Lin)
100	G C̈ C (N)$_4$/N G G C	BglI	29	20	1	0	1	3	Bacillus globigii (G.A. Wilson)
101	C̈ C̈ A (N)$_5$/N T G G	BstXI	13	10	1	3	0	0	Bacillus stearothermophilus XI (N. Welker)
102	G G C̈ C (N)$_4$/N G G C̈ C	SfiI	0	3	1	0	0	0	Streptomyces fimbriatus (ATCC 15051)

1.2 Enzymes with non-palindromic recognition sequences

Pos. no. (1)	Recognition sequence[a,b,c] (2)	Restriction endonuclease/methylase[d,e] (3)	λ (4)	Ad2 (5)	SV40 (6)	φX174 (7)	M13mp7 (8)	pBR322 (9)	Micro-organism used for isolation[g] (Source of micro-organism) (10)
103	G C A G C (N)$_8$ / C G T C G (N)$_{12}$	BbvI	199	179	22	14	8	21	Bacillus brevis (ATCC 9999)
104	G G A T G (N)$_9$ / C C T A C (N)$_{13}$	FokI	150	78	11	8	4	12	Flavobacterium okeanokoites (IFO 12536)
105	G A C G C (N)$_5$ / C T G C G (N)$_{10}$	HgaI	102	87	0	14	7	11	Haemophilus gallinarum (ATCC 14385)

Table II (continued)

Table II. List of Restriction Endonucleases and Methylases 521

Pos. no. (1)	Recognition sequence[a,b,c] (2)	Restriction endonuclease/ methylase[d,e] (3)	Number of recognition sites on[f]						Micro-organism used for isolation[g] (Source of micro-organism) (10)
			λ (4)	Ad2 (5)	SV40 (6)	φX174 (7)	M13mp7 (8)	pBR322 (9)	
106	G G T G A (N)$_8$ / C C A C T (N)$_7$ (with + above)	HphI	168	99	4	9	18	12	Haemophilus parahaemolyticus (C.A. Hutchison III)
107	G A A G A (N)$_8$ / C T T C T (N)$_7$ (with + below)	MboII	130	113	16	11	12	11	Moraxella bovis (ATCC 10900)
108	G C A T C (N)$_5$ / C G T A G (N)$_9$	SfaNI	169	84	6	12	7	22	Streptococcus faecalis ND547 (D. Clewell)
109	T G A C N/N N G T C / A C T G N/N/N C A G N	NTth111I(*)	1	6	0	0	0	2	Thermus thermophilus 111 (T. Oshima)

2. Class I enzymes

Pos. no. (1)	Recognition sequence[a,b,c] (2)	Restriction endonuclease/ methylase[d,e] (3)	λ (4)	Ad2 (5)	SV40 (6)	φX174 (7)	M13mp7 (8)	pBR322 (9)	Micro-organism used for isolation[g] (10)
110	G A G (N)$_7$ G T C / C T C (N)$_7$ C A G (with (+) above and + below)	EcoA	10	16	0	2	1	1	Escherichia coli 15T (T. Bickle)
111	T G A (N)$_8$ T G C T / A C T (N)$_8$ A C G A (with + above)	EcoB	9	4	1	0	1	0	Escherichia coli B (W. Arber)
112	T G A (N)$_8$ T G C T / A C T (N)$_8$ A C G A (with M above)	M·EcoB	9	4	1	0	1	0	Escherichia coli B (W. Arber)
113	A A C (N)$_6$ G T G C / T T G (N)$_6$ C A C G	EcoK	5	7	0	0	0	2	Escherichia coli K (M. Meselson)
114	A A C (N)$_6$ G T G C / T T G (N)$_6$ C A C G	M·EcoK	5	7	0	0	0	2	Escherichia coli K (M. Meselson)

Table II (continued)

Pos. no. (1)	Recognition sequence[a,b,c] (2)	Restriction endonuclease/methylase[d,e] (3)	λ (4)	Ad2 (5)	SV40 (6)	φX174 (7)	M13mp7 (8)	pBR322 (9)	Micro-organism used for isolation[g] (Source of micro-organism) (10)
115	ATCA (N)₇ ATTC / TAGT (N)₇ TAAG	EcoDXI	0	0	1	2	2	0	Escherichia coli [pDXI] (A. Piekarowicz)

3. Class III enzymes

Pos. no. (1)	Recognition sequence[a,b,c] (2)	Restriction endonuclease/methylase[d,e] (3)	λ (4)	Ad2 (5)	SV40 (6)	φX174 (7)	M13mp7 (8)	pBR322 (9)	Micro-organism used for isolation[g] (Source of micro-organism) (10)
116	(+)+ + + + AGACC / TCTGG	EcoP1	49	78	4	7	4	4	Escherichia coli phage P1 (K. Murray)
117	M AGACC / TCTGG	M·EcoP1	49	78	4	7	4	4	Escherichia coli phage P1 (K. Murray)
118	CAGCAG / GTCGTC	EcoP15	72	50	12	5	4	7	Escherichia coli [p15B] (W. Arber)
119	CAGCAG / GTCGTC	M·EcoP15	72	50	12	5	4	7	Escherichia coli [p15B] (W. Arber)
120	CGAAT / GCTTA	HineI	66	22	0	5	11	1	Haemophilus influenzae e (A. Piekarowicz)
121	CGAAT / GCTTA	HinfIII	66	22	0	5	11	1	Haemophilus influenzae Rf (C.A. Hutchison III)
122	CGAAT / GCTTA	M·HinfIII	66	22	0	5	11	1	Haemophilus influenzae Rf (C.A. Hutchison III)

a   The recognition sequences of restriction endonucleases and methylases are listed in column (2). Only one strand of the recognition sequences with a dyad symmetry (read in 5' to 3' direction) is shown, e.g., the recognition sequence GAATTC given for EcoRI represents the sequence
5'-GAATTC-3'
3'-CTTAAG-5'.
The cleavage site of restriction endonucleases is represented by a slash; e.g., for EcoRI the sequence G/AATTC indicates the following cuts:
5'-G↓AATT-C-3'
3'-C-TTAA↑G-5'.
For restriction endonucleases recognising sequences without a dyad symmetry and cleaving outside of their recognition sequences, the sequences for

Table II. List of Restriction Endonucleases and Methylases   523

both strands are listed. In these cases the sites of cleavage are identified for both strands by small numbers as subscripts after the recognition sequence. Subscript numbers refer to cleavage sites; indicated *e.g.*, for *Hga*I, the listed sequence

GACGC(N)$_5$ indicates the following cuts: 5'-GACGCGNNNNN↓NNNNN-N-3'   5'-N↓NNNNN-NNNNNGCGTC-3'   respectively.
CTGCG(N)$_{10}$ 3'-CTGCGNNNNN-NNNNN$_1$N-5'   3'-N-NNNNN↓NNNNNCGCAG-5'.
$$\phantom{CTGCG(N)_{10}\quad\quad\quad\quad\quad\quad\quad\quad} \text{M}$$

For methylases the position of methylation is marked by the symbol M; *e.g.*, for M·*Eco*RI the given sequence GAATTC indicates N^6 methylation of the two internal adenine residues, one in each strand:

$$\begin{array}{c} N^6\text{-methyl} \\ | \\ \text{5'-GAATTC-3'} \\ \text{3'-CTTAAG-5'.} \\ | \\ N^6\text{-methyl} \end{array}$$

For M·*Hpa*II the given sequence CCGG represents 5-methylation of the two internal cytosine residues, one in each strand:

$$\begin{array}{c} \text{M} \\ \text{5-methyl} \\ | \\ \text{5'-CCGG-3'} \\ \text{3'-GGCC-5'} \\ | \\ \text{5-methyl} \end{array}$$

If a restriction endonuclease or methylase recognises more than one sequence, the alternative bases are given in brackets; *e.g.*, the *Xho*II recognition sequence, (A) / GATC (T), includes all four possibilities of flanking purine and pyrimidine residues, *i.e.*
$$\text{(G)}\text{(C)}$$

5'-A↓GATC-T-3'   5'-A↓GATC-C-3'   5'-G↓GATC-T-3'   5'-G↓GATC-C-3'
3'-T-CTAG$_1$A-5'   3'-T-CTAG$_1$G-5'   3'-C-CTAG$_1$A-5'   3'-C-CTAG$_1$G-5'.

The symbol N within a recognition sequence of restriction endonucleases and methylases characterises internal ambiguous base pairs. This means that in these cases any one of the four base pair combinations of A, G, C or T may be recognised by the enzyme; *e.g.*, the *Hin*fI recognition sequence, G/ANTC, represents the following possible sequences:

5'-G↓AAT-C-3'   5'-G↓AGT-C-3'   5'-G↓ACT-C-3'   5'-G↓ATT-C-3'
3'-C-TTA$_1$G-5'   3'-C-TCA$_1$G-5'   3'-C-TGA$_1$G-5'   3'-C-TAA$_1$G-5'.

[b]   The effects of site-specific methylation of DNA on the cleavage are taken from McClelland (1983) and Roberts (1984).

+   +
A and C indicate the inhibition of a restriction endonuclease by a N^6 methyladenine or 5-methylcytosine residue within the recognition sequence.

m   m
A and C indicate that N^6 methyladenine or 5-methylcytosine residues within the recognition sequence are a prerequisite for the enzymatic activity of the restriction endonuclease.

o   o
A and C show that digestion of the DNA with the restriction endonuclease is not influenced by the presence of N^6 methyladenine or 5-methylcytosine residues within the recognition sequence.

If an adenine or a cytosine residue is not marked either by A/C, A/C or A/C, the influence of methylation on the restriction activity is still unknown.

c Dotted underlining indicates the recognition sequences for the *E. coli dam*-encoded methylase M·*Eco dam* (5'-GATC-3') or a part of this sequence.

Dotted overlining marks the overlapping recognition sequence for the *E. coli dcmI*-encoded methylase M·*Eco dcmI* (5'-CC(A/T)GG -3'), often also designated M·*Eco mec*, or a part of this sequence.

In all cases where the recognition sequence of the restriction endonuclease is only a part of the recognition sequence of M·*Eco dam* or M·*Eco dcmI*, methylation depends on the completion of the *dam*- or *dcmI*-specific recognition sequence by the flanking nucleotides.

The *Xba*I site (5'-TCTAGA-3'), for example, may be methylated by M·*Eco dam* only within the sequences 5'-TCTAGATC-3' or 5'-GATCTAGA-3'; the *Sau*96I site (5'-GGNCC-3') by M·*Eco dcmI* only within the sequences 5'-GGNCCAGG-3', 5'-GGNCCTGG-3', 5'-CCAGGNCC-3' or 5'-CCTGGNCC-3'.

Enzymes with large recognition sites containing internal (N)$_x$ sequences show the same behaviour; *e.g.* the *Sfi*I site (5'-GGCC(N)$_5$GGCC-3') is methylated by M·*Eco dcmI* at the internal cytosine residue only within the sequences 5'-GGCCAGGNNGGCC-3' and 5'-GGCCTGGNNGGCC-3'. The *dam*- or *dcmI*-specific methylation is never complete in wild-type *E. coli* cells, reflecting the limited *in vivo* activity of M·*Eco dam* or M·*Eco dcmI*. An observed or predicted inhibition of the activity of a particular restriction endonuclease by *dam*- or *dcmI*-specific methylation can be eliminated by the isolation of DNA from *E. coli dam⁻*, *dcmI⁻* or (*dam, dcmI*)⁻ mutant cells.

d The nomenclature of the enzymes, given in column (3), is in accordance with the nomenclature proposed by Smith and Nathans (1973). Methylases are specified by the symbol M· preceding the name of the enzyme.

e Relaxed specificities of restriction endonucleases (star activity) are indicated in column (3) by an asterisk given in brackets after the name of the enzyme; *e.g., Eco*RI(*).

f The numbers of recognition sites for each restriction endonuclease and methylase on the various DNAs, given in columns (4) to (9), were obtained by computer analysis of published DNA sequences. The numbers of cleavage sites on the DNA of the virus Ad2 have been determined by computer analysis of the DNA sequence stored at the EMBL nucleotide sequence data base (file AD2).
All of the sequence-derived numbers of recognition sites are given for the six DNAs independently from the size of the generated fragments which are detectable under various conditions of gel-electrophoresis. Overlapping recognition sites are also included.

g The strain listed in column (10) for a given enzyme refers to the micro-organism used as the source for the isolation of the corresponding restriction endonuclease or methylase. [ ] indicates a plasmid-carrier state.
Whenever a strain is available from a culture collection, the abbreviated strain designation is given together with the assigned strain number: ATCC, American Type Culture Collection, Rockeville (USA); CCAP, Cambridge Collection of Algae and Protozoa, Cambridge (UK); CDC, Center for Disease Control, Atlanta (USA); CMI, Commonwealth Mycological Institute, Kew (UK); DSM, Deutsche Sammlung von Mikroorganismen, Göttingen (FRG); IAM, Institute of Applied Microbiology, Tokyo (Japan); IFO, Institute for Fermentation, Osaka (Japan); KCC, Kaken Chemical Company Ltd., Tokyo (Japan); NCIB, National Collection of Industrial Bacteria, Aberdeen (UK); NRCC, National Research Council, Ottawa (Canada); PCC, Pasteur Culture Collection of Cyanobacterial Strains, Paris (France).
If a particular strain is not identified by a culture collection reference number, it may be obtained from the scientist whose name is given in brackets. Adresses are cited in Kessler *et al.*, 1985).

Table II. List of Restriction Endonucleases and Methylases 525

h Cleavage of 5′-CC̅GG-3′ sequences has been reported for *Hpa*II.

i *Msp*I is not inhibited by methylation of the internal cytosine residue within the sequence 5′-CC̅GG-3′, but is reported to be inhibited by methylation of this cytosine residue within 5′-GGCC̅GG-3′ and 5′-CC̅GGCC-3′ sequences.

j Crude extracts of M · *Msp*I methylate both cytosine residues in the recognition sequence to produce 5′-C̿C̿GG-3′. However, methylation by cloned M · *Msp*I yields 5′-CC̅GG-3′ sequences.

k *Sma*I and *Nci*I both cleave at recognition sites containing 5′-C̿C̿GG-3′ sequences. Possibly the 5′-terminal methylation site compensates the inhibitory effect of 5′-CC̅GG-3′ methylation.

References

Kessler, Ch., Neumaier, P.S., and Wolf, W. (1985). Recognition sequences of restriction endonucleases and methylases – a review. Gene 33, 1–102.

Roberts, R.J. (1984). Restriction and modification enzymes and their recognition sequences. Nucleic Acids Res. 12, r167–r204.

Smith, M.O., and Nathans, D. (1973). A suggested nomenclature for bacterial host modification and restriction systems and their enzymes. J. Mol. Biol. 81, 419–423.

McClelland, M. (1983). The effect of site specific methylation on restriction endonuclease cleavage (update). Nucleic Acids Res. 11, r169–r173.

Appendix B

DNA Sequences Of Selected Vectors

Choices were made in order to include the most commonly used genes, origins of replication and control regions. Each sequence is accompanied by a circular physical map which depicts important restriction sites as well as the various structural elements which make up the particular vector. Arrows indicate directions of transcription.

ß-lactamase and the protein responsible for tetracycline resistance, are at positions 4155/3297 (ß-lactamase) and 86/1273 (tetracycline resistance), respectively. The circular map shows characteristic restriction sites, the positions of the ß-lactamase and tetracycline genes as well as that of the origin of replication (ori).

B-1
DNA Sequence of Plasmid pBR322

The sequence of the double-stranded circular plasmid pBR322 was published by Sutcliffe (1979). The sequence shown is that of the linear single strand whose transcription proceeds in clockwise direction. Numbering begins with the first thymidine residue (position +1) within the single *Eco*RI site. The sequence includes an additional C residue at position 526 recently identified by Peden (1983). The total length of this plasmid is, therefore, 5363 bp (*cf.* Section 4.1.3). Start and stop codons of two proteins,

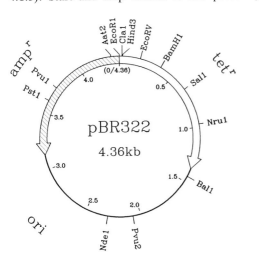

```
      1                                                    50
         TTCTCATGTTTGACAGCTTATCATCGATAAGCTTTAATGCGGTAGTTTAT
     51                                                   100
         CACAGTTAAATTGCTAACGCAGTCAGGCACCGTGTATGAAATCTAACAAT
    101                                                   150
         GCGCTCATCGTCATCCTCGGCACCGTCACCCTGGATGCTGTAGGCATAGG
    151                                                   200
         CTTGGTTATGCCGGTACTGCCGGGCCTCTTGCGGGATATCGTCCATTCCG
    201                                                   250
         ACAGCATCGCCAGTCACTATGGCGTGCTGCTAGCGCTATATGCGTTGATG
    251                                                   300
         CAATTTCTATGCGCACCCGTTCTCGGAGCACTGTCCGACCGCTTTGGCCG
    301                                                   350
         CCGCCCAGTCCTGCTCGCTTCGCTACTTGGAGCCACTATCGACTACGCGA
    351                                                   400
         TCATGGCGACCACACCCGTCCTGTGGATCCTCTACGCCGGACGCATCGTG
    401                                                   450
         GCCGGCATCACCGGCGCCACAGGTGCGGTTGCTGGCGCCTATATCGCCGA
    451                                                   500
         CATCACCGATGGGGAAGATCGGGCTCGCCACTTCGGGCTCATGAGCGCTT
    501                                                   550
         GTTTCGGCGTGGGTATGGTGGCAGGCCCCGTGGCCGGGGGACTGTTGGGC
    551                                                   600
         GCCATCTCCTTGCATGCACCATTCCTTGCGGCGGCGGTGCTCAACGGCCT
    601                                                   650
         CAACCTACTACTGGGCTGCTTCCTAATGCAGGAGTCGCATAAGGGAGAGC
    651                                                   700
         GTCGACCGATGCCCTTGAGAGCCTTCAACCCAGTCAGCTCCTTCCGGTGG
    701                                                   750
         GCGCGGGGCATGACTATCGTCGCCGCACTTATGACTGTCTTCTTTATCAT
    751                                                   800
         GCAACTCGTAGGACAGGTGCCGGCAGCGCTCTGGGTCATTTTCGGCGAGG
    801                                                   850
         ACCGCTTTCGCTGGAGCGCGACGATGATCGGCCTGTCGCTTGCGGTATTC
    851                                                   900
         GGAATCTTGCACGCCCTCGCTCAAGCCTTCGTCACTGGTCCCGCCACCAA
    901                                                   950
         ACGTTTCGGCGAGAAGCAGGCCATTATCGCCGGCATGGCGGCCGACGCGC
    951                                                  1000
         TGGGCTACGTCTTGCTGGCGTTCGCGACGCGAGGCTGGATGGCCTTCCCC
   1001                                                  1050
         ATTATGATTCTTCTCGCTTCCGGCGGCATCGGGATGCCCGCGTTGCAGGC
   1051                                                  1100
         CATGCTGTCCAGGCAGGTAGATGACGACCATCAGGGACAGCTTCAAGGAT
   1101                                                  1150
         CGCTCGCGGCTCTTACCAGCCTAACTTCGATCACTGGACCGCTGATCGTC
   1151                                                  1200
         ACGGCGATTTATGCCGCCTCGGCGAGCACATGGAACGGGTTGGCATGGAT
   1201                                                  1250
         TGTAGGCGCCGCCCTATACCTTGTCTGCCTCCCCGCGTTGCGTCGCGGTG
   1251                                                  1300
         CATGGAGCCGGGCCACCTCGACCTGAATGGAAGCCGGCGGCACCTCGCTA
   1301                                                  1350
         ACGGATTCACCACTCCAAGAATTGGAGCCAATCAATTCTTGCGGAGAACT
   1351                                                  1400
         GTGAATGCGCAAACCAACCCTTGGCAGAACATATCCATCGCGTCCGCCAT
   1401                                                  1450
         CTCCAGCAGCCGCACGCGGCGCATCTCGGGCAGCGTTGGGTCCTGGCCAC
   1451                                                  1500
         GGGTGCGCATGATCGTGCTCCTGTCGTTGAGGACCCGGCTAGGCTGGCGG
```

```
1501                 .              .              .              .         1550
      GGTTGCCTTACTGGTTAGCAGAATGAATCACCGATACGCGAGCGAACGTG
1551                                                               1600
      AAGCGACTGCTGCTGCAAAACGTCTGCGACCTGAGCAACAACATGAATGG
1601                                                               1650
      TCTTCGGTTTCCGTGTTTCGTAAAGTCTGGAAACGCGGAAGTCAGCGCCC
1651                                                               1700
      TGCACCATTATGTTCCGGATCTGCATCGCAGGATGCTGCTGGCTACCCTG
1701                                                               1750
      TGGAACACCTACATCTGTATTAACGAAGCGCTGGCATTGACCCTGAGTGA
1751                                                               1800
      TTTTTCTCTGGTCCCGCCGCATCCATACCGCCAGTTGTTTACCCTCACAA
1801                                                               1850
      CGTTCCAGTAACCGGGCATGTTCATCATCAGTAACCCGTATCGTGAGCAT
1851                                                               1900
      CCTCTCTCGTTTCATCGGTATCATTACCCCCATGAACAGAAATTCCCCCT
1901                                                               1950
      TACACGGAGGCATCAAGTGACCAAACAGGAAAAAACCGCCCTTAACATGG
1951                                                               2000
      CCCGCTTTATCAGAAGCCAGACATTAACGCTTCTGGAGAAACTCAACGAG
2001                                                               2050
      CTGGACGCGGATGAACAGGCAGACATCTGTGAATCGCTTCACGACCACGC
2051                                                               2100
      TGATGAGCTTTACCGCAGCTGCCTCGCGCGTTTCGGTGATGACGGTGAAA
2101                                                               2150
      ACCTCTGACACATGCAGCTCCCGGAGACGGTCACAGCTTGTCTGTAAGCG
2151                                                               2200
      GATGCCGGGAGCAGACAAGCCCGTCAGGGCGCGTCAGCGGGTGTTGGCGG
2201                                                               2250
      GTGTCGGGGCGCAGCCATGACCCAGTCACGTAGCGATAGCGGAGTGTATA
2251                                                               2300
      CTGGCTTAACTATGCGGCATCAGAGCAGATTGTACTGAGAGTGCACCATA
2301                                                               2350
      TGCGGTGTGAAATACCGCACAGATGCGTAAGGAGAAAATACCGCATCAGG
2351                                                               2400
      CGCTCTTCCGCTTCCTCGCTCACTGACTCGCTGCGCTCGGTCGTTCGGCT
2401                                                               2450
      GCGGCGAGCGGTATCAGCTCACTCAAAGGCGGTAATACGGTTATCCACAG
2451                                                               2500
      AATCAGGGGATAACGCAGGAAAGAACATGTGAGCAAAAGGCCAGCAAAAG
2501                                                               2550
      GCCAGGAACCGTAAAAAGGCCGCGTTGCTGGCGTTTTTCCATAGGCTCCG
2551                                                               2600
      CCCCCCTGACGAGCATCACAAAAATCGACGCTCAAGTCAGAGGTGGCGAA
2601                                                               2650
      ACCCGACAGGACTATAAAGATACCAGGCGTTTCCCCCTGGAAGCTCCCTC
2651                                                               2700
      GTGCGCTCTCCTGTTCCGACCCTGCCGCTTACCGGATACCTGTCCGCCTT
2701                                                               2750
      TCTCCCTTCGGGAAGCGTGGCGCTTTCTCAATGCTCACGCTGTAGGTATC
2751                                                               2800
      TCAGTTCGGTGTAGGTCGTTCGCTCCAAGCTGGGCTGTGTGCACGAACCC
2801                                                               2850
      CCCGTTCAGCCCGACCGCTGCGCCTTATCCGGTAACTATCGTCTTGAGTC
2851                                                               2900
      CAACCCGGTAAGACACGACTTATCGCCACTGGCAGCAGCCACTGGTAACA
2901                                                               2950
      GGATTAGCAGAGCGAGGTATGTAGGCGGTGCTACAGAGTTCTTGAAGTGG
2951                                                               3000
      TGGCCTAACTACGGCTACACTAGAAGGACAGTATTTGGTATCTGCGCTCT
3001                                                               3050
      GCTGAAGCCAGTTACCTTCGGAAAAAGAGTTGGTAGCTCTTGATCCGGCA
3051                                                               3100
      AACAAACCACCGCTGGTAGCGGTGGTTTTTTTGTTTGCAAGCAGCAGATT
3101                                                               3150
      ACGCGCAGAAAAAAAGGATCTCAAGAAGATCCTTTGATCTTTTCTACGGG
3151                                                               3200
      GTCTGACGCTCAGTGGAACGAAAACTCACGTTAAGGGATTTTGGTCATGA
3201                                                               3250
      GATTATCAAAAAGGATCTTCACCTAGATCCTTTTAAATTAAAAATGAAGT
```

```
3251                                                            33
      TTTAAATCAATCTAAAGTATATATGAGTAAACTTGGTCTGACAGTTAC
3301                                                            33
      ATGCTTAATCAGTGAGGCACCTATCTCAGCGATCTGTCTATTTCGTTC
3351                                                            34
      CCATAGTTGCCTGACTCCCCGTCGTGTAGATAACTACGATACGGGAGG
3401                                                            34
      TTACCATCTGGCCCCAGTGCTGCAATGATACCGCGAGACCCACGCTCA
3451                                                            35
      GGCTCCAGATTTATCAGCAATAAACCAGCCAGCCGGAAGGGCCGAGCG
3501                                                            35
      GAAGTGGTCCTGCAACTTTATCCGCCTCCATCCAGTCTATTAATTGTT
3551                                                            36
      CGGGAAGCTAGAGTAAGTAGTTCGCCAGTTAATAGTTTGCGCAACGTT
3601                                                            36
      TGCCATTGCTGCAGGCATCGTGGTGTCACGCTCGTCGTTTGGTATGGC
3651                                                            36
      CATTCAGCTCCGGTTCCCAACGATCAAGGCGAGTTACATGATCCCCCA
3701                                                            37
      TTGTGCAAAAAAGCGGTTAGCTCCTTCGGTCCTCCGATCGTTGTCAGA
3751                                                            38
      TAAGTTGGCCGCAGTGTTATCACTCATGGTTATGGCAGCACTGCATAA
3801                                                            38
      CTCTTACTGTCATGCCATCCGTAAGATGCTTTTCTGTGACTGGTGAGT
3851                                                            39
      TCAACCAAGTCATTCTGAGAATAGTGTATGCGGCGACCGAGTTGCTCT
3901                                                            39
      CCCGGCGTCAACACGGGATAATACCGCGCCACATAGCAGAACTTTAAA
3951                                                            40
      TGCTCATCATTGGAAAACGTTCTTCGGGGCGAAAACTCTCAAGGATCT
4001                                                            405
      CCGCTGTTGAGATCCAGTTCGATGTAACCCACTCGTGCACCCAACTGA
4051                                                            410
      TTCAGCATCTTTTACTTTCACCAGCGTTTCTGGGTGAGCAAAAACAGG
4101                                                            415
      GGCAAAATGCCGCAAAAAAGGGAATAAGGGCGACACGGAAATGTTGAA
4151                                                            420
      CTCATACTCTTCCTTTTTCAATATTATTGAAGCATTTATCAGGGTTAT
4201                                                            425
      TCTCATGAGCGGATACATATTTGAATGTATTTAGAAAAATAAACAAAT
4251                                                            430
      GGGTTCCGCGCACATTTCCCCGAAAAGTGCCACCTGACGTCTAAGAA
4301                                                            435
      ATTATTATCATGACATTAACCTATAAAAATAGGCGTATCACGAGGCCC
4351               4363
      TCGTCTTCAAGAA
```

References

Backman, K., and Boyer, H.W. (1983). Tetracycline resistance determined by pBR322 is mediated by one polypeptide. Gene 26, 197–203.

Peden, K.W.C. (1983). Revised sequence of the tetracycline-resistance gene of pBR322. Gene 22, 277–280.

Sutcliffe, J.G. (1979). Complete nucleotide sequence of the *Escherichia coli* plasmid pBR322. Cold Spring Harbor Symposia Quant. Biol. 43, 77–90.

B-2
DNA Sequence of Simian Virus 40 (SV40)

The sequence of the double-stranded circular genome of SV40 is that of Fiers *et al.* (1978) with modifications of van Heuverswyn and Fiers (1979). Numbering (in the BBB system, *cf.* also Chapter 8) begins with nucleotide 0/5243 which is the central nucleotide of the 27 bp palindromic sequence containing the single *Bgl*I site (GCCNNNN/NGCC). Nucleotide +1 is the last G residue in the recognition sequence ....GCCGAGGCGGC.... Numbering proceeds clockwise and the single strand shown corresponds to that of the late mRNA.

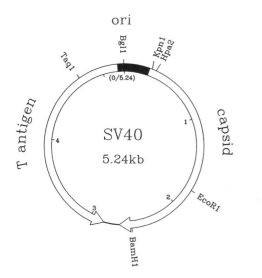

```
  1  GCCTCGGCCTCTGCATAAATAAAAAAAAATTAGTCAGCCATGGGGCGGAGA   50
 51  ATGGGCGGAACTGGGCGGAGTTAGGGGCGGGATGGGCGGAGTTAGGGGCG  100
101  GGACTATGGTTGCTGACTAATTGAGATGCATGCTTTGCATACTTCTGCCT  150
151  GCTGGGGAGCCTGGGGACTTTCCACACCTGGTTGCTGACTAATTGAGATG  200
201  CATGCTTTGCATACTTCTGCCTGCTGGGGAGCCTGGGGACTTTCCACACC  250
251  CTAACTGACACACATTCCACAGCTGGTTCTTTCCGCCTCAGAAGGTACCT  300
301  AACCAAGTTCCTCTTTCAGAGGTTATTTCAGGCCATGGTGCTGCGCCGGC  350
351  TGTCACGCCAGGCCTCCGTTAAGGTTCGTAGGTCATGGACTGAAAGTAAA  400
```

```
 401  AAAACAGCTCAACGCCTTTTTGTGTTTGTTTTAGAGCTTTTGCTGCAATT   450
 451  TTGTGAAGGGGAAGATACTGTTGACGGGAAACGCAAAAAACCAGAAAGGT   500
 501  TAACTGAAAAACCAGAAAGTTAACTGGTAAGTTTAGTCTTTTTGTCTTTT   550
 551  ATTTCAGGTCCATGGGTGCTGCTTTAACACTGTTGGGGGACCTAATTGCT   600
 601  ACTGTGTCTGAAGCTGCTGCTGCTACTGGATTTTCAGTAGCTGAAATTGC   650
 651  TGCTGGAGAGGCCGCTGCTGCAATTGAAGTGCAACTTGCATCTGTTGCTA   700
 701  CTGTTGAAGGCCTAACAACCTCTGAGGCAATTGCTGCTATAGGCCTCACT   750
 751  CCACAGGCCTATGCTGTGATATCTGGGGCTCCTGCTGCTATAGCTGGATT   800
 801  TGCAGCTTTACTGCAAACTGTGACTGGTGTGAGCGCTGTTGCTCAAGTGG   850
 851  GGTATAGATTTTTTAGTGACTGGGATCACAAAGTTTCTACTGTTGGTTTA   900
 901  TATCAACAACCAGGAATGGCTGTAGATTTGTATAGGCCAGATGATTACTA   950
 951  TGATATTTTATTTCCTGGAGTACAAACCTTTGTTCACAGTGTTCAGTATC  1000
1001  TTGACCCCAGACATTGGGGTCCAACACTTTTTAATGCCATTTCTCAAGCT  1050
1051  TTTTGGCGTGTAATACAAAATGACATTCCTAGGCTCACCTCACAGGAGCT  1100
1101  TGAAAGAAGAACCCAAAGATATTTAAGGGACAGTTTGGCAAGGTTTTTAG  1150
1151  AGGAAACTACTTGGACAGTAATTAATGCTCCTGTTAATTGGTATAACTCT  1200
1201  TTACAAGATTACTACTCTACTTTGTCTCCCATTAGGCCTACAATGGTGAG  1250
1251  ACAAGTAGCCAACAGGGAAGGGTTGCAAATATCATTTGGGCACACCTATG  1300
1301  ATAATATTGATGAAGCAGACAGTATTCAGCAAGTAACTGAGAGGTGGGAA  1350
1351  GCTCAAAGCCAAAGTCCTAATGTGCAGTCAGGTGAATTTATTGAAAAATT  1400
1401  TGAGGCTCCTGGTGGTGCAAATCAAAGAACTGCTCCTCAGTGGATGTTGC  1450
1451  CTTTACTTCTAGGCCTGTACGGAAGTGTTACTTCTGCTCTAAAAGCTTAT  1500
1501  GAAGATGGCCCCAACAAAAAGAAAAGGAAGTTGTCCAGGGGCAGCTCCCA  1550
1551  AAAAACCAAAGGAACCAGTGCAAGTGCCAAAGCTCGTCATCAAAAGGAGGA  1600
1601  ATAGAAGTTCTAGGGAGTTAAAACTGGAGTAGACAGCTTCACTGAGGTGGA  1650
1651  GTGCTTTTTAAATCCTCAAATGGGCAATCCTGATGAACATCAAAAGGAC  1700
1701  TAAGTAAAAGCTTAGCAGCTGAAAAACAGTTTACAGATGACTCTCCAGAC  1750
1751  AAAGAACAACTGCCTTGCTACAGTGTGGCTAGAATTCCTTTGCCTAATTT  1800
1801  AAATGAGGACTTAACCTGTGGAAATATTTTGATGTGGGAAGCTGTTACTG  1850
1851  TTAAAACTGAGGTTATTGGGGTAACTGCTATGTTAAACTTGCATTCAGGG  1900
1901  ACACAAAAAACTCATGAAAATGGTGCTGGAAAAACCCATTCAAGGGTCAAA  1950
1951  TTTTCATTTTTTTGCTGTTGGTGGGGAACCTTTGGAGCTGCAGGGTGTGT  2000
2001  TAGCAAACTACAGGACCAAATATCCTGCTCAAACTGTAACCCCAAAAAAT  2050
2051  GCTACAGTTGACAGTCAGCAGATGAACACTGACCACAAGGCTGTTTTGGA  2100
2101  TAAGGATAATGCTTATCCAGTGGAGTGCTGGGTTCCTGATCCAAGTAAAA  2150
2151  ATGAAAACACTAGATATTTTGGAACCTACACAGGTGGGGAAAATGTGCCT  2200
2201  CCTGTTTTGCACATTACTAACACAGCAACCACAGTGCTTCTTGATGAGCA  2250
2251  GGGTGTTGGGCCCTTGTGCAAAGCTGACAGCTTGTATGTTTCTGCTGTTG  2300
2301  ACATTTGTGGGCTGTTTACCAACACTTCTGGAACACAGCAGTGGAAGGGA  2350
2351  CTTCCCAGATATTTTAAAATTACCCTTAGAAAGCGGTCTGTGAAAAACCC  2400
2401  CTACCCAATTTCCTTTTTGTTAAGTGACCTAATTAACAGGAGGACACAGA  2450
2451  GGGTGGATGGGCAGCCTATGATTGGAATGTCCTCTCAAGTAGAGGGAGGT  2500
2501  AGGGTTTATGAGGACACAGAGGAGCTTCCTGGGGATCCAGACATGATAAG  2550
2551  ATACATTGATGAGTTTGGACAAACCACAACTAGAATGCAGTGAAAAAAAT  2600
```

```
2601                                            2650
     GCTTTATTTGTGAAATTTGTGATGCTATTGCTTTATTTGTAACCATTATA
2651                                            2700
     AGCTGCAATAAACAAGTTAACAACAACAATTGCATTCATTTTATGTTTCA
2701                                            2750
     GGTTCAGGGGGAGGTGTGGGAGGTTTTTTAAAGCAAGTAAAACCTCTACA
2751                                            2800
     AATGTGGTATGGCTGATTATGATCATGAACAGACTGTGAGGGACTGAGGGG
2801                                            2850
     CCTGAAATGAGCCTTGGGACTGTGAATCAATGCCTGTTTCATGCCCTGAG
2851                                            2900
     TCTTCCATGTTCTTCTCCCCACCATCTTCATTTTTATCAGCATTTTCCTG
2901                                            2950
     GCTGTCTTCATCATCATCATCACTGTTTCTTAGCCAATCTAAAACTCCAA
2951                                            3000
     TTCCCATAGCCACATTAAACTTCATTTTTTGATACACTGACAAACTAAAC
3001                                            3050
     TCTTTGTCCAATCTCTCTTTCCACTCCACAATTCTGCTCTGAATACTTTG
3051                                            3100
     AGCAAACTCAGCCACAGGTCTGTACCAAATTAACATAAGAAGCAAAGCAA
3101                                            3150
     TGCCACTTTGAATTATTCTCTTTTCTAACAAAAACTCACTGCGTTCCAGG
3151                                            3200
     CAATGCTTTAAATAATCTTTGGGCCTAAAATCTATTTGTTTTTACAAATCT
3201                                            3250
     GGCCTGCAGTGTTTTAGGCACACTGTACTCATTCATGGTGACTATTCCAG
3251                                            3300
     GGGGAAATATTTGAGTTCTTTTATTTAGGTGTTTCTTTTCTAAGTTTACC
3301                                            3350
     TTAACACTGCCATCCAAATAATCCCTTAAATTGTCCAGGTTATTAATTCC
3351                                            3400
     CTGACCTGAAGGCAAATCTCTGGACTCCCCTCCAGTGCCCTTTACATCCT
3401                                            3450
     CAAAAACTACTAAAAACTGGTCAATAGCTACTCCTAGCTCAAAGTTCAGC
3451                                            3500
     CTGTCCAAGGGCAAATTAACATTTAAAGCTTTCCCCCCACATAATTCAAG
3501                                            3550
     CAAAGCAGCTGCTAATGTAGTTTTACCACTATCAATTGGTCCTTTAAACA
3551                                            3600
     GCCAGTATCTTTTTTTAGGAATGTTGTACACCATGCATTTTAAAAAGTCA
3601                                            3650
     TACACCACTGAATCCATTTGGGCAACAAACAGTGTAGCCAAGCAACTCC
3651                                            3700
     AGCCATCCATTCTTCTATGTCAGCAGAGCCTGTAGAACCAAACATTATAT
3701                                            3750
     CCATCCTATCCAAAAGATCATTAAATCTGTTTGTTAACATTTGTTCTCTA
3751                                            3800
     GTTAATTGTAGGCTATCAACCCGCTTTTTAGCTAAAACAGTATCAACAGC
3801                                            3850
     CTGTTGGCATATGGTTTTTTGGTTTTTGCTGTCAGCAAATATAGCAGCAT
3851                                            3900
     TTGCATAATGCTTTTCATGGTACTTATAGTGGCTGGGCTGTTCTTTTTTA
3901                                            3950
     ATACATTTTAAACACATTTCAAAACTGTACTGAAATTCCAAGTACATCCC
3951                                            4000
     AAGCAATAACAACACATCATCACATTTTGTTTCCATTGCATACTCTGTTA
4001                                            4050
     CAAGCTTCCAGGACACTTGTTTAGTTTCCTCTGCTTCTTCTGGATTAAAA
4051                                            4100
     TCATGCTCCTTTAACCCACCTGGCAAACTTTCCTCAATAACAGAAAATGG
4101                                            4150
     ATCTCTAGTCAAGGCACTATACATCAAATATTCCTTATTAACCCCTTTAC
4151                                            4200
     AAATTAAAAAGCTAAAGGTACACAATTTTTGAGCATAGTTATTAATAGCA
4201                                            4250
     GACACTCTATGCCTGTGTGGAGTAAGAAAAAACAGTATGTTATGATTATA
```

```
4251                                            43
     ACTGTTATGCCTACTTATAAAGGTTACAGAATATTTTTCCATAATTTTC
4301                                            43
     TGTATAGCAGTGCAGCTTTTTCCTTTGTGGTGTAAATAGCAAAGCAAGC
4351                                            44
     AGAGTTCTATTACTAAACACAGCATGACTCAAAAAACTTAGCAATTCTC
4401                                            44
     AGGAAAGTCCTTGGGGTCTTCTACCTTTCTCTTCTTTTTTGGAGGAGTA
4451                                            45
     AATGTTGAGAGTCAGCAGTAGCCTCATCATCACTAGATGGCATTTCTTC
4501                                            45
     GAGCAAAACAGGTTTTCCTCATTAAAGGCATTCCACCACTGCTCCCAT
4551                                            46
     ATCAGTTCCATAGGTTGGAATCTAAAATACACAAACAATTAGAATCAGT
4601                                            46
     GTTTAACACATTATACACTTAAAAATTTTATATTTACCTTAGAGCTTT
4651                                            47
     ATCTCTGTAGGTAGTTTGTCCAATTATGTCACACCCACAGAAGTAAGGT
4701                                            47
     CTTCACAAAGATCAAGTCCAAACCACATTCTAAAGCAATCGAAGCAGTA
4751                                            48
     CAATCAACCCACACAAGTGGATCTTTCCTGTATAATTTTCTATTTTCA
4801                                            48
     CTTCATCCTCAGTAAGCACAGCAAGCATATGCAGTTAGCAGACATTTT
4851                                            49
     TTGCACACTCAGGCCATTGTTTGCAGTACATTGCATCAACACCAGGATT
4901                                            49
     AAGGAAGAAGCAAATACCTCAGTTGCATCCCAGAAGCCTCCAAAGTCAC
4951                                            50
     TTGATGAGCATATTTTACTCCATCTTCCATTTTCTTGTACAGAGTATTC
5001                                            50
     TTTTCTTCATTTTTTCTTCATCTCCTCCTTTATCAGGATGAAACTCCTT
5051                                            51
     CATTTTTTTAAATATGCCTTTCTCATCAGAGGAATATTCCCCCAGGCAC
5101                                            51
     CCTTTCAAGACCTAGAAGGTCCATTAGCTGCAAAGATTCCTCTCTGTTT
5151                                            52
     AAACTTTTATCCATCTTTGCAAAGCTTTTTGCAAAAGCCTAGGCCTCCAA
5201                                            5243
     AAAGCCTCCTCACTACTTCTGGAATAGCTCAGAGGCCGAGGCG
```

References

Fiers, W., Contreras, R., Haegerman, G., Rogiers, R., van der Voorde, A., van Heuverswyn, H., van Herreweghe, J., Volckaert, G., and Ysebaert, M. (1978). The complete nucleotide sequence of SV40 DNA. Nature 273, 113–120.

van Heuverswyn, H., and Fiers, W. (1979). Nucleotide sequence of the *Hind*-C fragment of simian virus 40 DNA. Europ. J. Biochem. 100, 51–60.

B-3
DNA Sequence of Plasmid ΠAN7

The structure of miniplasmid ΠAN7 is taken from Huang et al. (1986). The origin of pBR322 DNA replication (*cf.* Fig. 4.1-3) includes the "-35" promoter region required for RNA primer (RNAII) synthesis. A useful application of ΠAN7 is discussed in Section 9.4.5 (Fig. 9–6).

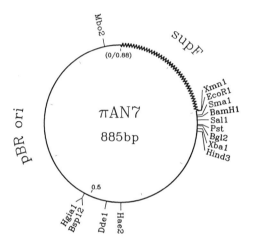

```
  1                                                50
     TTTCGGACTTTTGAAAGTGATGGTGGTGGGGGAAGGATTCGAACCTTCGA
 51                                               100
     AGTCGATGACGGCAGCTTTAGAGTCTGCTCCCTTTGGCCGCTCGGGAACC
101                                               150
     CCACCACGGGTAATGCTTTTACTGGCCTGCTCCCTTATCGGGAAGCGGGG
151                                               200
     CGCATCATATCAAATGACGCGCCGCTGTAAAGTGTTACGTTGAGAAAGAA
201                                               250
     TTCCCGGGGATCCGTCGACCTGCAGATCTCTAGAAGCTTGCGTTGCTGGC
251                                               300
     GTTTTTCCATAGGCTCCGCCCCCCTGACGAGCATCACAAAAATCGACGCT
301                                               350
     CAAGTCAGAGGTGGCGAAACCCGACAGGACTATAAAGATACCAGGCGTTT
351                                               400
     CCCCCTGGAAGCTCCCTCGTGCGCTCTCCTGTTCCGACCCTGCCGCTTAC
401                                               450
     CGGATACCTGTCCGCCTTTCTCCCTTCGGGAAGCGTGGCGCTTTCTCAAT
451                                               500
     GCTCACGCTGTAGGTATCTCAGTTCGGTGTAGGTCGTTCGCTCCAAGCTG
501                                               550
     GGCTGTGTGCACGAACCCCCCGTTCAGCCCGACCGCTGCGCCTTATCCGG
551                                               600
     TAACTATCGTCTTGAGTCCAACCCGGTAAGACACGACTTATCGCCACTGG
601                                               650
     CAGCAGCCACTGGTAACAGGATTAGCAGAGCGAGGTATGTAGGCGGTGCT
651                                               700
     ACAGAGTTCTTGAAGTGGTGGCCTAACTACGGCTACACTAGAAGGACAGT
701                                               750
     ATTTGGTATCTGCGCTCTGCTGAAGCCAGTTACCTTCGGAAAAAGAGTTG
751                                               800
     GTAGCTCTTGATCCGGCAAACAAACCACCGCTGGTAGCGGTGGTTTTTTT
801                                               850
     GTTTGCAAGCAGCAGATTACGCGCAGAAAAAAAGGATCTCAAGAAGATCC
851                                        885
     TTTGATCTTTTCTACGGGGTCTGACGCTCAAATTC
```

References

Huang, H., Little, P., and Seed, B. (1985). Suppressor miniplasmids and plasmid-phage recombination. In "Vectors: A survey of molecular cloning vectors and their applications", Rodriguez, R., ed., Butterworth Publishers, Stoneham, MA, USA.

B-4
DNA Sequence of Plasmid pUC18

The polarity of the single-stranded DNA is opposite to that of the sequence published by Yanisch-Perron *et al.* (1985) such that the ß-lactamase gene can be read conveniently as the message strand (start and stop codon at positions 201 and 1059, respectively). The sequence of the sister plasmid pUC19 has been published by Yanisch-Perron *et al.* (1985).

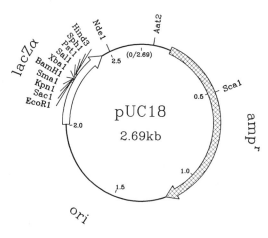

```
  1                                                   50
    GACGAAAGGGCCTCGTGATACGCCTATTTTTATAGGTTAATGTCATGATA
 51                                                  100
    ATAATGGTTTCTTAGACGTCAGGTGGCACTTTTCGGGGAAATGTGCGCGG
101                                                  150
    AACCCCTATTTGTTTATTTTTCTAAATACATTCAAATATGTATCCGCTCA
151                                                  200
    TGAGACAATAACCCTGATAAATGCTTCAATAATATTGAAAAAGGAAGAGT
201                                                  250
    ATGAGTATTCAACATTTCCGTGTCGCCCTTATTCCCTTTTTTGCGGCATT
251                                                  300
    TTGCCTTCCTGTTTTTGCTCACCCAGAAACGCTGGTGAAAGTAAAAGATG
301                                                  350
    CTGAAGATCAGTTGGGTGCACGAGTGGGTTACATCGAACTGGATCTCAAC
351                                                  400
    AGCGGTAAGATCCTTGAGAGTTTTCGCCCCGAAGAACGTTTTCCAATGAT
401                                                  450
    GAGCACTTTTAAAGTTCTGCTATGTGGCGCGGTATTATCCCGTATTGACG
451                                                  500
    CCGGGCAAGAGCAACTCGGTCGCCGCATACACTATTCTCAGAATGACTTG
501                                                  550
    GTTGAGTACTCACCAGTCACAGAAAAGCATCTTACGGATGGCATGACAGT
551                                                  600
    AAGAGAATTATGCAGTGCTGCCATAACCATGAGTGATAACACTGCGGCCA
601                                                  650
    ACTTACTTCTGACAACGATCGGAGGACCGAAGGAGCTAACCGCTTTTTTG
651                                                  700
    CACAACATGGGGGATCATGTAACTCGCCTTGATCGTTGGGAACCGGAGCT
701                                                  750
    GAATGAAGCCATACCAAACGACGAGCGTGACACCACGATGCCTGTAGCAA
751                                                  800
    TGGCAACAACGTTGCGCAAACTATTAACTGGCGAACTACTTACTCTAGCT
801                                                  850
    TCCCGGCAACAATTAATAGACTGGATGGAGGCGGATAAAGTTGCAGGACC
```

```
 851
    ACTTCTGCGCTCGGCCCTTCCGGCTGGCTGGTTTATTGCTGATAAAT
 901
    GAGCCGGTGAGCGTGGGTCTCGCGGTATCATTGCAGCACTGGGGCCA
 951                                                 1
    GGTAAGCCCTCCCGTATCGTAGTTATCTACACGACGGGGAGTCAGGC
1001                                                 1
    TATGGATGAACGAAATAGACAGATCGCTGAGATAGGTGCCTCACTGA
1051                                                 1
    AGCATTGGTAACTGTCAGACCAAGTTTACTCATATATACTTTAGATT
1101                                                 1
    TTAAAACTTCATTTTTAATTTAAAAGGATCTAGGTGAAGATCCTTTT
1151                                                 1
    TAATCTCATGACCAAAATCCCTTAACGTGAGTTTTCGTTCCACTGAG
1201                                                 1
    CAGACCCCGTAGAAAAGATCAAAGGATCTTCTTGAGATCCTTTTTTT
1251                                                 1
    CGCGTAATCTGCTGCTTGCAAACAAAAAAACCACCGCTACCAGCGGT
1301                                                 1
    TTGTTTGCCGGATCAAGAGCTACCAACTCTTTTTCCGAAGGTAACTGC
1351                                                 1
    TCAGCAGAGCGCAGATACCAAATACTGTCCTTCTAGTGTAGCCGTAGT
1401                                                 1
    GGCCACCACTTCAAGAACTCTGTAGCACCGCCTACATACCTCGCTCTC
1451                                                 1
    AATCCTGTTACCAGTGGCTGCTGCCAGTGGCGATAAGTCGTGTCTTAC
1501                                                 1
    GGTTGGACTCAAGACGATAGTTACCGGATAAGGCGCAGCGGTCGGGCT
1551                                                 1
    ACGGGGGGTTCGTGCACACAGCCCAGCTTGGAGCGAACGACCTACACC
1601                                                 1
    ACTGAGATACCTACAGCGTGAGCATTGAGAAAGCGCCACGCTTCCCGA
1651                                                 1
    GGAGAAAGGCGGACAGGTATCCGGTAAGCGGCAGGGTCGGAACAGGAC
1701                                                 1
    CGCACGAGGGAGCTTCCAGGGGGAAACGCCTGGTATCTTTATAGTCC
1751                                                 1
    CGGGTTTCGCCACCTCTGACTTGAGCGTCGATTTTTGTGATGCTCGTC
1801                                                 1
    GGGGGCGGAGCCTATGGAAAAACGCCAGCAACGCGGCCTTTTTACGG
1851                                                 1
    CTGGCCTTTTGCTGGCCTTTTGCTCACATGTTCTTTCCTGCGTTATC
1901                                                 1
    TGATTCTGTGGATAACCGTATTACCGCCTTTGAGTGAGCTGATACCG
1951                                                 2
    GCCGCAGCCGAACGACCGAGCGCAGCGAGTCAGTGAGCGAGGAAGCG
2001                                                 2
    GAGCGCCCAATACGCAAACCGCCTCTCCCCGCGCGTTGGCCGATTCA
2051                                                 2
    ATGCAGCTGGCACGACAGGTTTCCCGACTGGAAAGCGGGCAGTGAGC
2101                                                 2
    ACGCAATTAATGTGAGTTAGCTCACTCATTCGGCACCCCAGGCTTTAC
2151                                                 2
    TTTATGCTTCCGGCTCGTATGTTGTGTGGAATTGTGAGCGGATAACA
2201                                                 2
    TCACACAGGAAACAGCTATGACCATGATTACGAATTCGAGCTCGGTAC
2251                                                 2
    GGGGATCCTCTAGAGTCGACCTGCAGGCATGCAAGCTTGGCACTGGC
2301                                                 2
    CGTTTTACAACGTCGTGACTGGGAAAACCCTGGCGTTACCCAACTTA
2351                                                 2
    GCCTTGCAGCACATCCCCCTTTCGCCAGCTGGCGTAATAGCGAAGAG
2401                                                 2
    CGCACCGATCGCCCTTCCCAACAGTTGCGTAGCCTGAATGGCGAATGG
2451                                                 2
    CCTGATGCGGTATTTTCTCCTTACGCATCTGTGCGGTATTTCACACC
2501                                                 2
    TATGGTGCACTCTCAGTACAATCTGCTCTGATGCCGCATAGTTAAGC
2551                                                 2
    CCCCGACACCCGCCAACACCCGCTGACGCGCCCTGACGGGCTTGTCTC
2601                                                 2
    CCCGGCATCCGCTTACAGACAAGCTGTGACCGTCTCCGGGAGCTGCAT
2651                                            2686
    GTCAGAGGTTTTCACCGTCATCACCGAAACGCGCGA
```

References

Yanisch-Perron, C., Vieira, J., and Messing, J. (1985). Improved M13 phage cloning vectors and host strains: nucleotide sequences of the M13mp18 and pUC19 vectors. Gene 33, 103–119.

B-5
DNA Sequence of Plasmid pGcos4

The physical map of cosmid pGcos4 (*cf.* Fig. 4.2–16) was published by Gitschier *et al.* (1984). Its DNA sequence was kindly provided by William I. Wood, Genentech.

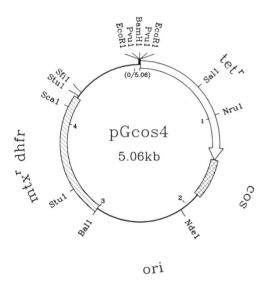

```
   1                                                        50
     GAATTCGATCGGATCCGATCGAATTCTCATGTTTGACAGCTTATCATCGA
  51                                                       100
     TAAGCTTTAATGCGGTAGTTTATCACAGTTAAATTGCTAACGCAGTCAGG
 101                                                       150
     CACCGTGTATGAAATCTAACAATGCGCTCATCGTCATCCTCGGCACCGTC
 151                                                       200
     ACCCTGGATGCTGTAGGCATAGGCTTGGTTATGCCGGTACTGCCGGGCCT
 201                                                       250
     CTTGCGGGATATCGTCCATTCCGACAGCATCGCCAGTCACTATGGCGTGC
 251                                                       300
     TGCTAGCGCTATATGCGTTGATGCAATTTCTATGCGCACCCGTTCTCGGA
 301                                                       350
     GCACTGTCCGACCGCTTTGGCCGCCGCCCAGTCCTGCTCGCTTCGCTACT
 351                                                       400
     TGGAGCCACTATCGACTACGCGATCATGGCGACCACACCCGTCCTGTGGA
 401                                                       450
     TTCTCTACGCCGGACGCATCGTGGCCGGCATCACCGGCGCCACAGGTGCG
 451                                                       500
     GTTGCTGGCGCCTATATCGCCGACATCACCGATGGGGAAGATCGGGCTCG
 501                                                       550
     CCACTTCGGGCTCATGAGCGCTTGTTTCGGCGTGGGTATGGTGGCAGGCC
 551                                                       600
     CCGTGGCCGGGGGACTGTTGGGCGCCATCTCCTTGCATGCACCATTCCTT
 601                                                       650
     GCGGCGGCGGTGCTCAACGGCCTCAACCTACTACTGGGCTGCTTCCTAAT
 651                                                       700
     GCAGGAGTCGCATAAGGGAGAGCGTCGACCGGATGCCCTTGAGAGCCTTCA
 701                                                       750
     ACCCAGTCAGCTCCTTCCGGTGGGCGCGGGGCATGACTATCGTCGCCGCA
 751                                                       800
     CTTATGACTGTCTTCTTTATCATGCAACTCGTAGGACAGGTGCCGGCAGC
 801                                                       850
     GCTCTGGGTCATTTTCGGCGAGGACCGCTTTCGCTGGAGCGCGACGATGA
```

```
 851                                                       900
     TCGGCCTGTCGCTTGCGGTATTCGGAATCTTGCACGCCCTCGCTCAAGCC
 901                                                       950
     TTCGTCACTGGTCCCGCCACCAAACGTTTCGGCGAGAAGCAGGCCATTAT
 951                                                      1000
     CGCCGGCATGGCGGCCGACGCGCTGGGCTACGTCTTGCTGGCGTTCGCGA
1001                                                      1050
     CGCGAGGCTGGATGGCCTTCCCCATTATGATTCTTCTCGCTTCCGGCGGC
1051                                                      1100
     ATCGGGATGCCCGCGTTGCAGGCCATGCTGTCCAGGCAGGTAGATGACGA
1101                                                      1150
     CCATCAGGGACAGCTTCAAGGATCGCTCGCGGCTCTTACCAGCCTAACTT
1151                                                      1200
     CGATCACTGGACCGCTGATCGTCACGGCGATTTATGCCGCCTCGGCGAGC
1201                                                      1250
     ACATGGAACGGGTTGGCATGGATTGTAGGCGCCGCCCTATACCTTGTCTG
1251                                                      1300
     CCTCCCCGCGTTGCGTCGCGGTGCATGGAGCCGGGCCACCTCGACCTGAA
1301                                                      1350
     TGGAAGCCGGCGGCACCTCGCTAACGGATTCACCACTCCAAGAATTGGAG
1351                                                      1400
     CCAATCAATTCTTGCGGAGAACTGTGAATGCGCAAACCAACCCTTGGCAG
1401                                                      1450
     AACATATCCATCGCGTCCGCCATCTCCAGCAGCCGCACGCGGCGCATCTC
1451                                                      1500
     GGGACATGAGGTTGCCCCGTATTCAGTGTCGCTGATTTGTATTGTCTGAA
1501                                                      1550
     GTTGTTTTTACGTTAAGTTGATGCAGATCAATTAATACGATACCTGCGTC
1551                                                      1600
     ATAATTGATTATTTGACGTGGTTTGATGGCCTCCACGCACGTTGTGATAT
1601                                                      1650
     GTAGATGATAATCATTATCACTTTACGGGTCCTTTCCGGTGATCCGACAG
1651                                                      1700
     GTTACGGGGCGGCGACCTCGCGGGTTTTCGCTATTTATGAAAATTTTCCG
1701                                                      1750
     GTTTAAGGCGTTTCCGTTCTTCTTCGTCATAACTTAATGTTTTTATTTAA
1751                                                      1800
     AATACCCTCTGAAAAGAAAGGAAACGACAGGTGCTGAAAGCGAGGCTTTT
1801                                                      1850
     TGGCCTCTGTCGTTTCCTTTCTCTGTTTTTGTCCGTGGAATGAACAATGG
1851                                                      1900
     AAGTCCTGCCTCGCGCGTTTCGGTGATGACGGTGAAAACCTCTGACACAT
1901                                                      1950
     GCAGCTCCCGGAGACGGTCACAGCTTGTCTGTAAGCGGATGCCGGGAGCA
1951                                                      2000
     GACAAGCCCGTCAGGGCGCGTCAGCGGGTGTTGGCGGGTGTCGGGGCGCA
2001                                                      2050
     GCCATGACCCAGTCACGTAGCGATAGCGGAGTGTATACTGGCTTAACTAT
2051                                                      2100
     GCGGCATCAGAGCAGATTGTACTGAGAGTGCACCATATGCGGTGTGAAAT
2101                                                      2150
     ACCGCACAGATGCGTAAGGAGAAAATACCGCATCAGGCGCTCTTCCGCTT
2151                                                      2200
     CCTCGCTCACTGACTCGCTGCGCTCGGTCGTTCGGCTGCGGCGAGCGGTA
2201                                                      2250
     TCAGCTCACTCAAAGGCGGTAATACGGTTATCCACAGAATCAGGGGATAA
2251                                                      2300
     CGCAGGAAAGAACATGTGAGCAAAAGGCCAGCAAAAGGCCAGGAACCGTA
2301                                                      2350
     AAAAGGCCGCGTTGCTGGCGTTTTTCCATAGGCTCCGCCCCCCTGACGAG
2351                                                      2400
     CATCACAAAAATCGACGCTCAAGTCAGAGGTGGCGAAACCCGACAGGACT
2401                                                      2450
     ATAAAGATACCAGGCGTTTCCCCCTGGAAGCTCCCTCGTGCGCTCTCCTG
2451                                                      2500
     TTCCGACCCTGCCGCTTACCGGATACCTGTCCGCCTTTCTCCCTTCGGGA
2501                                                      2550
     AGCGTGGCGCTTTCTCAATGCTCACGCTGTAGGTATCTCAGTTCGGTGTA
2551                                                      2600
     GGTCGTTCGCTCCAAGCTGGGCTGTGTGCACGAACCCCCCGTTCAGCCCG
2601                                                      2650
     ACCGCTGCGCCTTATCCGGTAACTATCGTCTTGAGTCCAACCCGGTAAGA
2651                                                      2700
     CACGACTTATCGCCACTGGCAGCAGCCACTGGTAACAGGATTAGCAGAGC
2701                                                      2750
     GAGGTATGTAGGCGGTGCTACAGAGTTCTTGAAGTGGTGGCCTAACTACG
2751                                                      2800
     GCTACACTAGAAGGACAGTATTTGGTATCTGCGCTCTGCTGAAGCCAGTT
2801                                                      2850
     ACCTTCGGAAAAAGAGTTGGTAGCTCTTGATCCGGCAAACAAACCACCGC
2851                                                      2900
     TGGTAGCGGTGGTTTTTTTGTTTGCAAGCAGCAGATTACGCGCAGAAAAA
2901                                                      2950
     AAGGATCTCAAGAAGATCCTTTGATCTTTTCTACGGGGTCTGACGCTCAG
2951                                                      3000
     TGGAACGAAAACTCACGTTAAGGGATTTTGGTCATGAGATTATCAAAAAG
3001                                                      3050
     GATCTTCACCTAGATCCTTTTGGCCACGATGCGTCCGGCGTAGAGGATCT
```

```
3051                                                  3100
     CTGACGGAAGGAAAGAAGTCAGAAGGCAAAAACGAGAGTAACTCCACAGT
3101                                                  3150
     AGCTCCAAATTCTTTATAAGGGTCAATGTCCATGCCCCAAAGCCACCCAA
3151                                                  3200
     GGCACAGCTTGGAGGCTTGAACAGTGGGACATGTACAAGAGATGATTAGG
3201                                                  3250
     CAGAGGTGAAAAAGTTGCATGGTGCTGGTGCGCAGACCAATTTGTGCCTA
3251                                                  3300
     CAGCCTCCTAATACAAAGACCTTTAACCTAATCTCCTCCCCCAGCTCCTC
3301                                                  3350
     CCAGTCCTTAAACACACAGTCTTTGAAGTAGGCCTCAAGGTCGGTCGTTG
3351                                                  3400
     ACATTGCTGGGAGTCCAAGAGTCCTCTTATGTAAGACCTTGGGCAGGATC
3401                                                  3450
     TGATGGGCGTTCACGGTGGTCTCCATGCAACGTGCAGAGGTGAAGCGAAG
3451                                                  3500
     TGCACACGGACCGGCAGATGAGAAGGCACAGACGGGGAGACCGCGTAAAG
3501                                                  3550
     AGAGGTGCGCCCCGTGGTCGGCTGGAACGGCAGACGGAGAAGGGGACGAG
3551                                                  3600
     AGAGTCCCAAGCGGCCCCGAGAGGGGTCGTCCGCGGGATTCAGCGCCGAC
3601                                                  3650
     GGGACGTAAACAAAGGACGTCCCGCGAAGGATCTAAAGCCAGCAAAAGTC
3651                                                  3700
     CCATGGTCTTATAAAAATGCATAGCTTTAGGAGGGGAGCAGAGAACTTGA
3701                                                  3750
     AAGCATCTTCCTGTTAGTCTTTCTTCTCGTAGATTTCAAACTTATACTTG
3751                                                  3800
     ATCCCTTTTTCCTCCTGGACCTCAGAGAGGACGCCTGGGTATTCTGGGAG
3801                                                  3850
     AAGTTTATATTTCCCCAAATCAATTTCTGGGAAAAACGTGTCACTTTCAA
3851                                                  3900
     ATTCCTGCATGATCCTTGTCACAAAGAGTCTGAGGTGGCCTGGTTGATTC
3901                                                  3950
     ATGGCTTCCTGGTAAACAGAACTGCCTCCGACTATCCAAACCATGTCTAC
3951                                                  4000
     TTTACTTGCCAATTCCGGTTGTTCAATAAGTCTTAAGGCATCATCCAAAC
4001                                                  4050
     TTTTGGCAAGAAAATGAGCTCCTCGTGGTGGTTCTTTGAGTTCTCTACTG
4051                                                  4100
     AGAACTATATTAATTCTGTCCTTTAAAGGTCGATTCTTCTCAGGAATGGA
4101                                                  4150
     GAACCAGGTTTTCCTACCCATAATCACCAGATTCTGTTTACCTTCCACTG
4151                                                  4200
     AAGAGGTTGTGGTCATTCTTTGGAAGTACTTGAACTCGTTCCTGAGCGGA
4201                                                  4250
     GGCCAGGGTCGGTCTCCGTTCTTGCCAATCCCCATATTTTGGGACACGGC
```

```
4251                                                  43
     GACGATGCAGTTCAATGGTCGAACCATGATGGCAAATTCTAGAATCGA
4301                                                  43
     AGCTTTTTGCAAAAGCCTAGGCCTCCAAAAAAGCCTCCTCACTACTTC
4351                                                  4
     GAATAGCTCAGAGGCCGAGGCGGCCTCGGCCTCTGCATAAATAAAAAA
4401                                                  44
     TTAGTCAGCCATGGGGCGGAGAATGGGCGGAACTGGGCGGAGTTAGGG
4451                                                  45
     GGGATGGGCGGAGTTAGGGGCGGGACTATGGTTGCTGACTAATTGAGA
4501                                                  45
     CATGCTTTGCATACTTCTGCCTGCTGGGGAGCCTGGGGACTTTCCACA
4551                                                  45
     TGGTTGCTGACTAATTGAGATGCATGCTTTGCATACTTCTGCCTGCTC
4601                                                  46
     GAGCCTGGGGACTTTCCACACCCTAACTGACACACATTCCACAGAAAA
4651                                                  47
     GCTCATCATTGGAAAACGTTCTTCGGGGCGAAAACTCTCAAGGATCTT
4701                                                  47
     CGCTGTTGAGATCCAGTTCGATGTAACCCACTCGTGCACCCAACTGAT
4751                                                  48
     TCAGCATCTTTTACTTTCACCAGCGTTTCTGGGTGAGCAAAAACAGGA
4801                                                  48
     GCAAAATGCCGCAAAAAAGGGAATAAGGGCGACACGGAAATGTTGAAT
4851                                                  49
     TCATACTCTTCCTTTTTCAATATTATTGAAGCATTTATCAGGGTTATT
4901                                                  49
     CTCATGAGCGGATACATATTTGAATGTATTTAGAAAAATAAACAAATA
4951                                                  5
     GGTTCCGCGCACATTTCCCCGAAAAGTGCCACCTGACGTCTAAGAAAC
5001                                                  5
     TTATTATCATGACATTAACCTATAAAAATAGGCGTATCACGAGGCCC
5051      5059
     CGTCTTCAA
```

References

Gitschier, J., Wood, W.I., Goralka, T.M., Wion, K.L., Chen, E.Y., Eaton, D.H., Vehar, G.A., Capon, D.J., and Lawn, R.M. (1984). Characterisation of the human factor VIII gene. Nature 312, 326–330.

B-6
DNA Sequence of Plasmid pMLS12

The construction of the single-stranded vector pMLS12 has been described by Levinson *et al.* (1984). The sequence was kindly provided by Brian Seed, Mass. General Hospital. It is identical to that of plasmid pSDL12 (*cf.* Fig. 4.1–17) except for the presence of an *Nhe*I site (G/CTAGC) instead of the second *Hin*dIII site at the transition between the M13 ori and the ColE1 ori sequences.

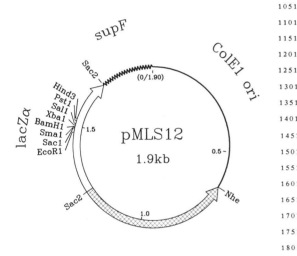

```
401                                            450
    GACCTACACCGAACTGAGATACCTACAGCGTGAGCATTGAGAAAGCGCCA
451                                            500
    CGCTTCCCGAAGGGAGAAAGGCGGACAGGTATCCGGTAAGCGGCAGGGTC
501                                            550
    GGAACAGGAGAGCGCACGAGGGAGCTTCCAGGGGGAAACGCCTGGTATCT
551                                            600
    TTATAGTCCTGTCGGGTTTCGCCACCTCTGACTTGAGCGTCGATTTTTGT
601                                            650
    GATGCTCGTCAGGGGGGCGGAGCCTATGGAAAAACGCCAGCAACGCAAGC
651                                            700
    TAGCTTCTAGCTAGAAATTGTAAACGTTAATATTTTGTTAAAATTCGCGT
701                                            750
    TAAATTTTTGTTAAATCAGCTCATTTTTTAACCAATAGGCCGAAATCGGC
751                                            800
    AAAATCCCTTATAAATCAAAAGAATAGCCCGAGATAGGGTTGAGTGTTGT
801                                            850
    TCCAGTTTGGAACAAGAGTCCACTATTAAAGAACGTGGACTCCAACGTCA
851                                            900
    AAGGCGAAAAACCGTCTATCAGGGCGATGGCCGCCCACTACGTGACCCA
901                                            950
    TCACCCAAATCAAGTTTTTTGGGGTCGAGGTGCCGTAAAGCACTAAATCG
951                                            1000
    GAACCCTAAAGGGAGCCCCCGATTTAGAGCTTGACGGGGAAAGCCGGCGA
1001                                           1050
    ACGTGGCGAGAAAGGAAGGGAAGAAAGCGAAAGGAGCGGGCGCTAGGGCG
1051                                           1100
    CTGGCAAGTGTAGCGGTCACGCTGCGCGTAACCACCACACCCGCCGCGCT
1101                                           1150
    TAATGCGCCGCTACAGGGCGCGTACTATGGTTGCTTTGACGAGCACGTAT
1151                                           1200
    AACGTGCTTTCCTCGTTGGAATCAGAGCGGGAGCTAAACAGGAGGCCGAT
1201                                           1250
    TAAAGGGATTTTAGACAGGAACGGTACGCCAGAATCCTGGATCACCGCGG
1251                                           1300
    TCCCAATACGCAAACCGCCTCTCCCCGCGCGTTGGCCGATTCATTAATGC
1301                                           1350
    AGCTGGCACGACAGGTTTCCCGACTGGAAAGCGGGCAGTGAGCGCAACGC
1351                                           1400
    AATTAATGTGAGTTAGCTCACTCATTCGGCACCCCAGGCTTTACACTTTA
1401                                           1450
    TGCTTCCGGCTCGTATGTTGTGTGGAATTGTGAGCGGATAACAATTTCAC
1451                                           1500
    ACAGGAAACAGCTATGACCATGATTACGAATTCGAGCTCGCCCGGGGATC
1501                                           1550
    CTCTAGAGTCGACCTGCAGCCCAAGCTTGGCACTGGCCGTCGTTTTACAA
1551                                           1600
    CGTCGTGACTGGGAAAAACCCTGGCGTTACCCAACTTAATCGCCTTGCAGC
1601                                           1650
    ACATCCCCCTTTCGCCAGCTGGCGTAATAGCGAAGAGGCCGCCCGCACCG
1651                                           1700
    ATCGCCCTTCCCAACAGTTGCGTAGCCTGAATGGCGAATGACCGCGGTAA
1701                                           1750
    TTCTTTCTCAACGTAACACTTTACAGCGGCGCGTCATTTGATATGATGCG
1751                                           1800
    CCCCGCTTCCCGATAAGGGAGCAGGCCAGTAAAAGCATTACCCGTGGTGG
1801                                           1850
    GGTTCCCGAGCGGCCAAAGGGAGCAGACTCTAAAGCTGCCGTCATCGACT
1851                                           1900
    TCGAAGGTTCGAATCCTTCCCCCACCACCATCACTTTCAAAAGTCCGAAA
```

```
1                                              50
    GAATTTGAGCGTCAGACCCCGTAGAAAAGATCAAAGGATCTTCTTGAGAT
51                                             100
    CCTTTTTTTCTGCGCGTAATCTGCTGCTTGCAAACAAAAAAACCACCGCT
01                                             150
    ACCAGCGGTGGTTTGTTTGCCGGATCAAGAGCTACCAACTCTTTTTCCGA
51                                             200
    AGGTAACTGGCTTCAGCAGAGCGCAGATACCAAATACTGTCCTTCTAGTG
01                                             250
    TAGCCGTAGTTAGGCCACCACTTCAAGAACTCTGTAGCACCGCCTACATA
51                                             300
    CCTCGCTCTGCTAATCCTGTTACCAGTGGCTGCTGCCAGTGGCGATAAGT
01                                             350
    CGTGTCTTACCGGGTTGGACTCAAGACGATAGTTACCGGATAAGGCGCAG
51                                             400
    CGGTCGGGCTGAACGGGGGGTTCGTGCACACAGCCCAGCTTGGAGCGAAC
```

References

Levinson, A., Silver, D., and Seed, B. (1984). Minimal size plasmids containing an M13 origin for the production of single-stranded transducing particles. J. Mol. Appl. Gen. 2, 507–517.

Appendix C

Glossary

Adapter A short chemically synthesised DNA double strand which can be used to link the ends of two DNA molecules.

Allele One of different variants of a gene.

Alu sequences A family of sequence-related elements about 300 bp in length approximately 500 000 copies of which are scattered along the human genome.

Antibody Soluble serum proteins (immunoglobulins) which are synthesised by a special set of cells of the immune system, the B-lymphocytes, in response to antigens.

Antigen Specific chemical structure, for example a particular sequence of a protein, which induces the production of antibodies in an organism and specifically binds to antibodies.

Attenuator A special form of a transcriptional terminator whose strength can be controlled.

Auxotrophy Inability of micro-organisms (usually mutants of bacteria or yeasts) to synthesise certain compounds, amino acids, for example, from simple precursors. In contrast to corresponding wild-type strains auxotrophic variants do not grow on so-called minimal media. Auxotrophic strains only grow on complete media which contain all components required for growth. Micro-organisms may be auxotrophic for more than one essential nutrient (Latin: auxilium = help). Wild-type cells which grow on minimal media are called prototrophic (Greek: protos = the first).

B-DNA Conformation of the Watson-Crick double helix in which the two strands form a right-handed double helix.

Chromatin Complex of DNA and proteins (histones, in particular) in the nuclei of eukaryotic cells.

Chromosome Threadlike particles of chromatin which contain the entire genetic material of an organism or parts thereof. A haploid human cell contains 23 chromosomes, a diploid cell 46 chromosomes.

cis/trans action A *cis*-acting genetic element, such as an enhancer, influences the biological activity of DNA sequences lying on the same DNA molecule as the element itself. *Trans*-acting elements are usually proteins and do not only interact with DNA sequences on the same molecule but also other DNA molecules carrying a suitable recognition sequence.

Clone Colony of genetically uniform cells derived from a single parent cell; also used for unique populations of recombinant DNA molecules (phage clones; plasmid clones etc.).

Codon Sequence of three nucleotides on a DNA or mRNA molecule, which directs the incorporation of a certain amino acid into proteins. There are 64 possible condons. 20 amino acids are coded for by 61 different codons, and three codons are used as stop signals for translation.

Concatemer Tandem arrays of monomeric DNA molecules with complementary ends. *Intra*molecular reassociation of such molecules leads to circularisation while *inter*molecular reactions produce concatemers.

Concatenates Interlocked circular molecules.

Conjugation Physical contact with the establishment of plasma bridges between two different bacterical cells allowing directed transfer of DNA.

Cosmid Plasmid which does not only carry the elements required for replication but, in addition, the so-called *cos* site of bacteriophage λ. The presence of this sequence element, which is approximately 400 bp in length and which includes the overhanging molecular ends of the λ genome, allows cosmids to be packaged into λ heads *in vitro*.

Dalton (Da)   Unit of atomic mass. One Dalton corresponds to the mass of a hydrogen atom = 3.32×10^{-24} g.

Diploid cell   Cell with a double set of chromosomes as found, for example, in so-called somatic cells.

Enhancer   *Cis*-acting genetic element, the presence of which stimulates eukaryotic transcription irrespective of its location or orientation within a DNA molecule. Enhancers are usually 70–80 bp in length and are found, for example, in viral DNA molecules.

Episome   Extrachromosomal genetic element. Generally used synonymously for plasmid.

Exon   Coding region of a eukaryotic gene which is represented in mRNA molecules. Different exons are usually interrupted by so-called introns and are linked to each other during splicing of the primary transcript when mRNA is generated.

Gene   Region on a chromosome which codes for a protein. Apart from coding regions a gene also comprises other elements such as promoters, introns, and terminators.

Gene library   Collection of cloned DNA fragments comprising the entire genome of an organism.

Genome   Entire genetic material of an organism. Also used for the genetic material of viruses and cell organelles.

Genotype   Set of genes and genetic elements which is characteristic for an organism.

Haploid cell   Cells with half of the normal set of chromosomes such as are found in reproductive (germ) cells.

Heterozygote   Individual with a pair of different alleles of a certain gene. Different alleles can be generated by mutation.

Homozygote   Individual with a pair of identical alleles of a gene.

Intron   Region within a eukaryotic gene which interrupts coding sequences (exons).

Leader sequence   a) *N*-terminal presequence of

secretory proteins such as peptide hormones and membrane proteins; b) the untranslated sequences at the 5'-ends of mRNA molecules.

Linker   Short chemically synthesised double stranded oligonucleotide which contains the recognition site for a restriction endonuclease.

Liposome   Artificial vesicle surrounded by a synthetic membrane of phospholipids.

Lysogen   Bacterium which contains a prophage in its genome. The dormant prophage can be activated by a variety of treatments (greek: lysogen = lysis inducing).

Lytic infection cycle   Interaction of a virus with a cell which leads to cell death and the generation of virus (phage) particles.

Minicells   Bacterial cells with only 10 % of the normal cell volume, which do not contain chromosomal DNA. Minicells are generated by certain mutants defective in cell division.

Modification   Process which is responsible for the protection of the organism's own DNA against digestion by restriction endonucleases. Protection is achieved by methylation of the DNA in those regions which are usually attacked by these endonucleases.

Mutation   Alteration of a gene which may be caused by base exchanges but also by loss (deletion) or addition (insertion) of base sequences.

Nucleosome   Unit particle into which chromosomal DNA is packaged. A nucleosome consists of approximately 200 bp of DNA complexed with two copies of each of the four histones H2A, H2B, H3, and H4.

Oncogene   Class of genes whose presence and expression within a cell may lead to a tumour phenotype. In normal cells oncogenes occur as so-called c-*onc* genes, in retroviruses as v-*onc* genes.

Operator   Short DNA sequence which interacts with repressor proteins.

Operon   Set of neighbouring prokaryotic genes whose transcription is simultaneously controlled; mRNA transcribed off an operon is polycistronic.

Palindrome   Sequence which can be read backwards and forwards to give the same sense.

(ABLE WAS I ERE I SAW ELBA). In double-stranded DNA this term is used for DNA sequences which read alike on both strands if read in the same (*e.g.* 5' to 3') direction.

Phage   Bacteriophages. In genetic engineering, single-stranded DNA phages, such as M13 or fd, and double-stranded DNA phages, such as phage λ are widely used.

Phenotype   Observable traits of an organism which are caused by its genetic make-up.

Plasmid   Circular DNA molecule which carries genetic elements permitting its autonomous replication in bacteria or yeast cells. Conjugative plasmids contain a set of genes which allow such plasmids and also other plasmids to be transferred to plasmid-free host cells. Cryptic plasmids are naturally occurring plasmids whose genotype and biological importance for the host cells are unknown.

Polycistronic mRNA   mRNA which does not only carry coding regions for several proteins but also corresponding control sequences which allow several proteins to be translated from a single mRNA molecule.

Polynucleotide   Linear polymer of ribonucleotides (RNA) or deoxyribonucleotides (DNA).

Polypeptide   Linear polymer of amino acids.

Polysome   Complex of mRNA and several ribosomes.

Primary transcript   First intermediate in the biosynthesis of mRNA. In eukaryotes this transcript which is an exact copy of the DNA is processed by splicing away non-coding regions to produce functional translatable mRNA.

Promoter   Sequence region on a DNA molecule which is important for transcription of a gene. Promoters are recognised by RNA polymerases and/or other DNA binding proteins.

Prophage   Viral genome integrated into a bacterial chromosome.

Protoplast   Cell without cell wall but with intact cell membranes.

Pseudogene   Eukaryotic gene without biological activity, usually lacking introns and promoter regions. It is generally assumed that pseudogenes are DNA copies of mRNA. Pseudogenes are known in the globin system but also in many other multigene families.

Recombinant DNA   DNA molecule consisting of portions of different DNA molecules which have been ligated *in vitro*.

Recombination   Process which involves the exchange of genetic material between two chromosomes. General recombination takes place between two homologous DNA sequences while illegitimate recombination involves the exchange of material between two non-homologous DNA molecules.

Restriction   Phenomenon describing the degradation of foreign DNA in bacterial cells by restriction endonucleases. Cells protect their own DNA against restriction by modification.

Repressor   Proteins which interact with short DNA sequences, so-called operator sequences.

Retrovirus   Single-stranded RNA virus which is replicated and expressed *via* a double-stranded DNA intermediate. Virus coats contain a special RNA-dependent DNA polymerase, so-called reverse transcriptase for this purpose.

R factor   Resistance Factor. Naturally occurring plasmid encoding antibiotic resistance genes.

Ribosome   Subcellular particle which is used for the translation of proteins. Ribosomes contain ribosomal RNA and approximately 60 different proteins.

SDS   Sodium dodecyl hydrogensulfate. Ionic detergent.

Shuttle vector   Vector which replicates in different host systems, for example, in *E. coli* and in *B. subtilis*. For this purpose shuttle vectors carry different origins of DNA replication which are characteristic for the desired host systems.

Sigma factor   Subunit of bacterial RNA polymerase which controls the correct initiation of transcription.

snRNP   (small nuclear ribonucleoprotein particles; pronounced snurp) Complexes of small RNA molecules (so-called U-RNAs) with 7-10 proteins. snRNPs have been identified by antibodies of patients with systemic Lupus erythematosus and are engaged in splicing of heterogenous nuclear RNA to mRNA.

Somatic cells All cells of a higher organism which are not reproductive cells (eggs or spermatozoa).

Splicing Process of conversion of primary RNA transcripts to mature and translatable mRNA. Since many genes are split, *i.e.* since they contain introns and exons, primary transcripts also contain sequence regions which do not code for proteins. Such sequences are removed from the primary transcript while the ends of the coding regions are correctly fused (spliced).

Superhelix Tertiary structure of double-stranded circular DNA molecules which is characterised by supertwists introduced into these molecules.

Suppressor tRNA molecule which directs the incorporation of an amino acid at the position of a stop codon. Suppressors have been found in *E. coli* and *Saccharomyces cerevisiae.*

Temperate phage Class of bacterial viruses which do not only undergo a lytic growth cycle but are also capable of integrating a copy of their genome into the host cell. Host cells which contain an integrated bacteriophage are called lysogens.

Terminator Sequence region on a DNA molecule which signals the termination of transcription.

Transcript RNA copy of a DNA region.

Transcription Process of RNA synthesis from a double-stranded DNA as template.

Transcription unit DNA region which comprises exons, introns and all other control elements required for gene expression.

Transfection (*cf.* also transformation) Both terms describe the transfer of pure DNA into intact viable cells. Transfection is usually used for manipulations involving higher cells. Transformation in higher cells usually denotes the transition from a normal to a tumour phenotype.

Transformation Transfer of purified DNA, recombinant DNA, for example, to bacteria or higher cells.

Transient expression Short-lived expression of a gene from vectors which are introduced into a non-permissive cell. Since such vectors cannot replicate, the copy numbers gradually decrease upon cell division with a concomitant reduction in gene expression.

Translation Formation of protein molecules with mRNA as template.

Transposon Movable genetic element which is capable of jumping from one chromosome to another chromosome. Structurally transposons are genes which are flanked by identical nucleotide sequences in opposite orientation.

Tumour antigen Protein which is specifically found in tumour cells or transformed cells and which can be recognised by immune reactions with sera from tumour-bearing animals.

Vector DNA molecule which serves as a recipient or carrier for foreign DNA. Vectors are usually plasmid or phage DNA molecules which carry an origin of DNA replication and genetic markers which allow them to be detected in host cells.

Z-DNA Alternative left-handed form of the DNA double helix which retains Watson-Crick-type base pairing.

Zygote Single cell obtained by fusion of an egg cell with a spermatozoon. The genes of a zygote represent a random assortment of genes from both parent cells.

Appendix D

Important Lambda Genes

Since its discovery by Lederberg (1951) bacteriophage λ has played an important role for the elucidation of a variety of biological phenomena. λ also plays a prominent role in gene cloning. It is not the structural genes of this bacteriophage but rather regulatory genes and their products which are of major importance.

The genome of λ has a molecular weight of 32×10^6 Dalton and comprises 48 502 bp which have been completely sequenced (Sanger *et al.*, 1982). The genome can be divided into three parts, genes *Nu1-J* which correspond to the left arm (0 to ca. 40 %), the *b* region or the so-called central region, and genes *int-Rz* on the right arm (57–100 %). The sequence of genes is *m-Nu1-J-b-att-int-Rz-m'*.

Some of the genes and coding regions lying on the λ genome will be described below. The glossary is in alphabetical order and does not reflect the physical arrangement of genes on the λ genome (*cf.* also section 4.2). An almost complete discussion can be found in Szybalski and Szybalski (1979) and in the book Lambda II by Hendrix *et al.* (1983).

A: positions 711–2633; 641 amino acids; $M_r = 73 300$. The A-protein binds to DNA concatemers and cleaves viral DNA at the *cos* site during packaging. This cleavage generates the overhanging single-stranded ends *m* and *m'* (*cf.* also Fig. 4.2–4).

att, aa' or att POP': Site at which phage and bacterial DNA sequences are linked by specific recombination. The so-called core sequence which is common to viral and bacterial DNA lies between positions 27723 and 27742 on the phage genome.

b-region: Non-essential region extending from 40–57 % on the λ genome (positions 19000–27700). It is flanked by gene *J* and the *att* site.

Deletion *b2* is one of the most frequently employed *b*-deletions.

cI: positions 37940–37230; 237 amino acids; $M_r = 26 200$. Gene for λ cI repressor.

cII: positions 38360–38650; 97 amino acids; $M_r = 11 100$. The complex of proteins cII and cIII plays an important role in influencing the decision between lytic or lysogenic cycles of infection.

cIII: positions 33463–33302; 54 amino acids; $M_r = 6 000$. The product of gene *cIII* stabilises repressor cII.

cos: A region on prophage or circular λ DNA which comprises the fused regions *m* and *m'* and neighbouring sequences. *Cos* sites are required for packaging of λ DNA into phage heads. The sequence is found as a 402 bp insert in plasmid pGcos4 at positions 1454–1855 (Appendix B-5) with the overhanging termini at positions 1657–1671.

cro: positions 38041–38238; 66 amino acids; $M_r = 7 400$. Gene product is essential for the lytic cycle, a so-called anti-repressor.

D: positions 5747–6076; 110 amino acids; $M_r = 11 600$. One of the major structural proteins. λD⁻ mutants are viable if the viral DNA is shorter than 82 % of its normal length.

E: positions 6135–7157; 341 amino acids; $M_r = 38 200$. Major structural protein.

exo or redA: positions 32028–31351; 226 amino acids; $M_r = 25 900$. Gene for λ exonuclease. The protein consists of two subunits and specifically digests double-stranded DNA from the 5'ends. The reaction products are 5'mononucleotides.

Fec⁺ phenotype: Ability of λ to grow on *recA*-negative strains of *E. coli* due to the combined action of the products of genes *exo*, *redB*, and *gam*. Fec-negative mutants of λ grow only on *recA*-negative strains of *E. coli* if λ DNA carries a *cis*-dominant *chi* mutation (chi⁺) (*cf.* also Section 4.2.2.4).

gam: positions 33232–32819; 138 amino acids; M_r = 16300. The *gam* gene product inhibits a nuclease of *E. coli* host cells encoded by genes *rec*BC. In the absence of *gam*, late DNA replication of λ DNA becomes dependent on host or phage recombination systems.

ice: inceptor, *cis*-element essential for DNA replication. *Ice* is positioned 600 bp to the left of the origin of DNA replication within gene *cII*.

***imm* region**: immunity region covers the region of genes *cI*, *cro*, and the corresponding control regions. If this region is present in a λ lysogen, lysogenic bacteria are immune and cannot be superinfected with phages carrying the same immunity region. However, if the superinfecting phage carries an unrelated immunity region, such as *imm434* of bacteriophage 434, superinfection is possible.

int*, *xis: positions 28882–27815; 356 amino acids; M_r = 40 300, and positions 29078–28863; 72 amino acids; M_r = 8 600. The gene product of the *int* gene catalyses integration, that of gene *xis* the excision of the prophage at the *att* site.

m* and *m': The overlapping single-stranded ends of λ DNA as found in mature phage particles. The sequence at the 5'-end of the l-strand is 5'-GGGCGGCGACCT-3'. The sequence of the r-strand terminus is complementary (*m'*) (*cf.* Fig. 4.2-1).

N: positions 35360–35040; 107 amino acids; M_r=12 300. The product of gene *N* abolishes the rho-dependent termination of transcription at the so-called t-termination sites. Protein N serves as an antiterminator.

***nut*L**: Recognition site for N protein at position 35528 which serves for antitermination of the leftward early transcription.

***nut*R**: positions 38265–38280. Recognition site for N protein where it effects antitermination of rightward early transcription.

O,P: positions 38686–39582; 299 amino acids; M_r = 33 900, and positions 39582–40280; 233 amino acids; M_r = 26 500. Both gene products are required for the initiation of DNA replication.

***P*$_I$**: positions 29065–29105. Promoter for *int*-

mRNA which is activated by the products of genes *cII* and *cIII*.

***P*$_{RM}$**: positions 37941–37980. Promoter for leftward transcription of gene *cI*.

***P*$_L O_L$**: positions 35580–35660. Left early promoter/operator region.

***P*$_R O_R$**: positions 37941–38040. Right early promoter/operator region (*cf.* Figure 4.2-5).

Q: positions 43886–44506; 207 amino acids; M_r = 22500. Antiterminator for late transcription.

R: positions 45493–45966; 158 amino acids; M_r = 17800. Gene coding for λ endolysin, a murein transglycosidase activity.

Rz: positions 45966–46423; 153 amino acids; M_r = 17 200; in conjunction with *S* and *R*, the product of gene *Rz*, is important for the lysis of bacterial cell walls. Presumably Rz is an endopeptidase which cleaves the bonds between diaminopimelic acid and D-alanine residues in murein.

red function: Combined function involving genes *red*A and *red*B. Both genes are essential for general genetic recombination. The analogous gene in *E. coli* is *rec*A.

***red*B**: positions 32810–32028; 261 amino acids; M_r = 29700. Gene coding for the so-called β protein required for, and catalysing general genetic recombination.

S: positions 45186–45506; 107 amino acids; M_r = 11 500. The product of gene *S* is important for the lysis of infected bacterial cells.

***Spi*$^+$ phenotype:** Inability of λ to grow on P2-lysogens. The products of λ genes *exo*, *red*B, and *gam* are responsible for this phenotype. Red⁻gam⁻ mutants of λ can grow on P2-lysogens if they are chi⁺ (*cf.* also Section 4.2.2.4). λ derivative λ1059 is such a mutant.

References

Hendrix, R.W., Roberts, J.W., Stahl, F.W., and Weisberg, R.A. (1983). Lambda II. Cold Spring Harbor Laboratory, Cold Spring Harbor, New York 11724.

Lederberg, E.M. (1951) Lysogenicity in *E. coli* K–12. Genetics 36, 560–564.

Sanger, F., Coulson, A.R., Hong, G.F., Hill, D.F., and Petersen, G.B. (1982). Nucleotide sequence of bacteriophage Lambda DNA. J. Mol. Biol. 162, 29–773.

Szybalski, E.H., and Szybalski, W. (1969). A comprehensive molecular map of bacteriophage Lambda. Gene 7, 217–270.

Appendix E

Properties of Commonly Used Strains of E. coli

Although the rules proposed by Demerec *et al.* (1966) are generally followed, there is no consistent genetic nomenclature for bacteria in the literature. Genotypic as well as phenotypic notations are used. The former describe the existence of certain genes such as *gal*K, for example. This abbreviation stands for the gene coding for galactokinase (E.C.2.7.1.6) of *E. coli*. An addition, such as *gal*K2, denotes a specific mutation, mutation 2, in the *gal*K gene. When the genotype of a particular *E. coli* strain is described, only those alleles are normally mentioned which carry mutations and can therefore be distinguished from wild-type. This convention avoids the use of hyphens (–) for characterising mutated alleles.

If a certain mutation, for example *gal*K, is introduced into the genotype, the result is a specific phenotype which in this case in known as Gal⁻. Gal⁻ describes the inability of the particular strain to grow on galactose as sole source of carbon.

In genetic engineering it is predominantly functions involved in the metabolism and modification of DNA which are of interest. Some relevant genes are listed in Table III (according to Bachmann and Low, 1980).

The presence or absence of an F-factor, *i.e.* a conjugative plasmid (*cf.* Table 4.1–1), may be of special significance. F^+ signifies the presence of the so-called sex factor, F^- its absence. F' stands

Table III Selection of genetic symbols from the linkage map of *E. coli*

Gene symbol	Phenotypic defect or phenotype
ara	utilisation of arabinose
end	endonuclease I
*gal*K	galactokinase (EC 2.7.1.6)
*hsd*M	host-specific modification
*hsd*R	host-specific restriction
lac	utilisation of lactose
*leu*B	2-isopropylmalate dehydrogenase (EC 1.1.1.85)
mtl	utilisation of D-mannitol
min	formation of minicells
pro	utilisation of proline
rec	general genetic recombination
*rps*L	ribosomal protein L12
*str*A	streptomycin resistance
thi	utilisation of thiamine
thy	utilisation of thymidine
*ton*A	resistance towards bacteriophage T1 (T-one)
xyl	utilisation of xylose

A complete list can be found in Bachmann and Low (1980).

for F-factors which carry parts of the bacterial chromosome. F' *proA*$^+$B$^+$, for example, signifies a transducing F-factor which carries the *proA*$^+$B$^+$ region of the bacterial chromosome, and the (+) signs indicate the wild-type genotype. Mutated alleles of chromosomal DNA residing on the F-factor are normally written without a superscript (–) sign. F' *lacZ*△M15, therefore, describes a transducing F-factor which carries deletion M15 in the *lacZ* region. Of course, an F-factor may also carry mutations in its own genes, which are indicated as usual. Mutation *traD36*, for example, is a mutation in gene *traD* which is responsible for conjugational transfer of DNA.

Quite frequently it is very difficult to trace back the origins of *E. coli* strains and their genotypes. The reader is referred here to two sources, namely Bachmann (1972) and Bachmann and Low (1980).

The properties of some bacterial strains which are used in cloning experiments are summarised below:

(1) χ**1776** (Curtiss-Clark and Curtiss III, 1983).
F', *tonA53*, *dapD8*, *minA1*, *supE42*, △40(*gal-uvrB*), λ$^-$, *minB2*, *rfb-2*, *nalA25*, *oms-2*, *thyA57*, *metC65*, *oms-1*, △29(*bioH-asd*), *cycB2*, *hsdR2*.
This strain is used as an EK2 host. It requires diaminopimelic acid (100 µg/ml), L-threonine (40 µg/ml), L-methionine (10 µg/ml), biotin (1 µg/ml), and thymine or thymidine (10–40 µg/ml) for growth.
The strain is also sensitive towards UV light, elevated temperatures (37 °C), and traces of detergents (SDS or sarcosyl). In addition, χ1776 shows an increased sensitivity towards chloramphenicol, tetracycline, kanamycin, and rifampicin but it is resistent against nalidixic acid (up to 50 µg/ml), cycloserine (up to 10 µg/ml) and trimethoprim.
There are certain difficulties in handling this strain. Nevertheless, χ1776 is often used since transformation frequencies in this strain are particularly high.

(2) **HB101** (Boyer and Roulland-Dussoix, 1969).
F$^-$, *hsdS20* (r$_B^-$m$_B^-$), *supE44*, *ara-14*, *galK2*, *lacY1*, *proA2*, *rpsL20* (*str*^{R}), *xyl-5*, *mtl-1*, λ$^-$, *recA13*.
This frequently employed strain was obtained by crossing *E. coli* K12 and *E. coli* B. It lacks B-specific (but not K-specific) restriction and modification systems. In addition, the strain is deficient in general recombination (recA$^-$).

(3) **RR1** (Bolivar et al., 1977).
RR1 and HB101 are so-called isogenic recA$^+$/recA$^-$ strains, the genotypes of which only differ in the presence or absence of the *recA* function. RR1, therefore, possesses the same genotype as HB101 apart from the *recA13* allele. Its phenotype thus is RecA$^+$. It is frequently used for cDNA cloning.

(4) **MM294** (Meselson and Yuan, 1968; Hanahan, 1983).
F$^-$, *endA1*, *hsdR17* (r$_K^-$,m$_K^+$), *supE44*, *thi-1*, *gyrA96*, λ$^-$,

(5) **DH1** (Hanahan, 1983).
This strain is isogenic with MM294 apart from the mutated alleles *recA1* and *gyrA96*. DH1 and also MM294 show particular high transformation frequencies with pBR322 under the conditions described by Hanahan (1983).

(6) **C600 or CR34** (Appelyard, 1954).
F$^-$, *supE44*, *thi-1*, *thr-1*, *leuB6*, *lacY1*, *tonA21*, λ$^-$.
This *E. coli* K12 strain is used as a so-called wild-type strain although apparently it carries some auxotrophic mutations. The strain codes for an active *supE* amber suppressor and is sensitive towards infection with bacteriophage λ.

(7) **JM103** (Messing et al., 1981).
endA, △[*lac,pro*], *thi-1*, *strA*, *sbcB15*, *hsdR4*, *supE*, λ$^-$, [F'*traD36*, *proA*$^+$B$^+$, *lacI*^{q}Z△M15].
This strain is well suited for the replication of M13 phages and its corresponding mp

derivatives. The chromosomal wild-type allele *pro*A$^+$B$^+$, which is carried on the F'-factor, complements the corresponding chromosomal deletion. In the absence of proline (proline prototophy) it thus allows a selection for the presence of F'-factors which are required for M13 propagation. The *lac*I^q allele of the *lac*I gene on the F'-factor is responsible for an overproduction of *lac* repressor. α-complementation can, therefore, only occur in the presence of a suitable inducer such as IPTG.

(8) **JM83** (Vieira and Messing, 1982).
ara, Δ[*lac*, *pro*], *str*A, *thi*-1, (Φ80d*lac*ZΔM15).
This strain is a Φ80 lysogen. It lacks the F'-factor and is therefore not suited for M13 replication. However, this strain can be used for the replication of pUC plasmids. Due to the lack of the *lac*I^q allele, this strain permits α-complementation in the absence of IPTG. JM83 clones obtained by transformation with pUC plasmids, therefore, form blue colonies in the presence of Xgal even in the absence of IPTG (see also cover photograph).

(9) **BHB2688** (Hohn, 1979).
N205*rec*A$^-$, [λimm^{434}, *c*Its*857*, *b*2, *red*3, *Eam*4, *Sam*7].
BHB2690. This lysogen is isogenic with BHB2688 apart from the fact that it carries the *Dam*15 instead of the *Eam*4 mutation. Temperature-induced extracts of BHB2690 provide empty phage heads, extracts from BHB2688 contain proteins required for *in vitro* packaging of DNA and maturation of phage heads.

(10) **BNN102** (Young and Davis, 1983a).
*E. coli hsd*R$^-$, *hsd*M$^+$, *sup*E, *thr*, *leu*, *thi*, *lac*Y1, *ton*A21, *hfl*A150 (chr::Tn10).
This strain is required for cloning in λ gt10 (Huynh *et al.*, 1985).

(11) **Y1088** (Young and Davis, 1983b).
E. coli Δ*lac*U169, *sup*E, *sup*F, *hsd*R-,

*hsd*M+, *met*B, *trp*R, *ton*A21, *pro*C::Tn5 (pMC9 = pBR322-*lac*I^q).
Y1090. *E. coli* Δ*c*U169, *pro*A$^+$, Δ*lon*, *ara*D139,*str*A,*sup*F(*trp*C22::Tn10)(pMC9). These strains are required for cloning in λgt11 (Huynh et al., 1985).

References

Achtman, M., Willets, N., and Clark, A.J. (1971). Beginning of a genetic analysis of conjugational transfer determined by the F-factor in *Escherichia coli* by isolation and characterisation of transfer-deficient mutants. J. Bacteriol. 106, 529–538.

Appleyard, R.K. (1954). Segregation of new lysogenic types during growth of a doubly lysogenic strain derived from *Escherichia coli*. Genetics 39, 440–452.

Bachmann, B.J. (1972). Pedigrees of some mutant strains of *Escherichia coli* K12. Bacteriological Reviews 36, 525–557.

Bachmann, B.J., and Low, K.B. (1980). Linkage map of *Escherichia coli* K12, Edition 6. Microbiological Reviews 44, 1–56.

Bolivar, F., Rodroguez, R.L., Greene, P.J., Betlach, M.C., Heyneker, H.L., and Boyer, H.W. (1977). Construction and characterisation of new cloning vectors. II. A multipurpose cloning system. Gene 2, 95–113.

Boyer, H.W., and Roulland-Dussoix, D. (1969). A complementation analysis of the restriction and modification of DNA in *Escherichia coli*. J. Mol. Biol. 41, 459–472.

Curtiss-Clark, J.E. and Curtiss III, R. (1983). Analysis of recombinant DNA using *Escherichia coli* minicells. Methods in Enzymology 101, 347–362.

Demerec, M., Adelberg, E.A., Clark, A.J., and Hartmann, P.A. (1966). A proposal for a uniform nomenclature in bacterial genetics. Genetics 54, 61–76.

Hanahan, D. (1983). Studies on transformation of *Escherichia coli* with plasmids. J. Mol. Biol. 166, 557–580.

Hohn, B. (1979) *In vitro* packaging of λ and cosmid DNA. Methods in Enzymology 68, 299–309.

Huynh, T.V., Young, R.A., and Davis, R.W. (1985). Construction and screening cDNA libraries in λ gt10 and λ gt11. In "DNA cloning techniques: A practical approach"; D. Glover, ed.; IRL Press, Oxford.

Meselson, M., and Yuan, R. (1968). DNA restriction enzyme from *E. coli*. Nature 217, 1110–1114.

Messing, J., Crea, R., and Seeburg, P.H. (1981). A system for shotgun DNA sequencing. Nucleic Acids Res. 9, 309–321.

Vieira, J., and Messing, J. (1982). The pUC plasmid, an M13mp7-derived system for insertion mutagenesis and sequencing with synthetic primers. Gene 19, 259–268.

Young, R.A., and Davis, R.W. (1983a). Efficient isolation of genes by using antibody probes. Proc. Natl. Acad. Sci. USA 80, 1194–1198.

Young, R.A., and Davis, R.W. (1983b). Yeast RNA polymerase II genes: isolation with antibody probes. Science 222, 778–782.

Appendix F

Guidelines for Research Involving Recombinant DNA Molecules

Department of Health and Human Services

May 1986.

These NIH Guidelines supersede earlier versions and will be in effect until further notice.

Table of Contents

I. Scope of the Guidelines

I - A – Purpose

The purpose of these Guidelines is to specify practices for constructing and handling (i) recombinant DNA molecules and (ii) organisms and viruses containing recombinant DNA molecules.

I - B – Definition of Recombinant DNA Molecules

In the context of these Guidelines, recombinant DNA molecules are defined as either (i) molecules which are constructed outside living cells by joining natural or synthetic DNA segments to DNA molecules that can replicate in a living cell, or (ii) DNA molecules that result from the replication of those described in (i) above.

Synthetic DNA segments likely to yield a potentially harmful polynucleotide or polypeptide (e.g., a toxin or a pharmocologically active agent) shall be considered as equivalent to their natural DNA counterpart. If the synthetic DNA segment is not expressed *in vivo* as a biologically active polynucleotide or polypeptide product, it is exempt from the Guidelines.

I - C – General Applicability

The Guidelines are applicable to all recombinant DNA research within the United States or its territories which is conducted at or sponsored by an institution that receives any support for recombinant DNA research from the National Institutes of Health (NIH). This includes research performed by NIH directly.

An individual receiving support for research involving recombinant DNA must be associated with or sponsored by an institution that can and does assume the responsibilities assigned in these Guidelines.

The Guidelines are also applicable to projects done abroad if they are supported by NIH funds. If the host country, however, has established rules for the conduct of recombinant DNA projects, then a certificate of compliance with those rules may be submitted to NIH in lieu of compliance with the NIH Guidelines. The NIH reserves the right to withhold funding if the safety practices to be employed abroad are not reasonably consistent with the NIH Guidelines.

I - D – General Definitions

The following terms, which are used throughout the Guidelines, are defined as follows:

I - D - 1. "Institution" means any public or private entity (including Federal, State, and local government agencies).

I - D - 2. "Institutional Biosafety Committee" or "IBC" means a committee that (i) meets the requirements for membership specified in Section IV - B - 2, and (ii) reviews, approves, and oversees projects in accordance with the responsibilities defined in Sections IV - B - 2 and IV - B - 3.

I - D - 3. "NIH Office of Recombinant DNA Activities" or "ORDA" means the office within NIH with responsibility for (i) reviewing and coordinating all activities of NIH related to the Guidelines, and (ii) performing other duties as defined in Section IV - C - 3.

I - D - 4. "Recombinant DNA Advisory Committee" of "RAC" means the public advisory committee that advises the Secretary, the Assistant Secretary for Health, and the Director, NIH, concerning recombinant DNA research. The RAC shall be constituted as specified in Section IV - C - 2.

I - D - 5. "Director, NIH" or "Director" means the Director, NIH, or any other officer or employee of NIH to whom authority has been delegated.

II. Containment

Effective biological safety programs have been operative in a variety of laboratories for many years.

Considerable information, therefore, already exists for the design of physical containment facilities and the selection of laboratory procedures applicable to organisms carrying recombinant DNAs 3 – 16. The existing programs rely upon mechanisms that, for convenience, can be divided into two categories: (i) A set of standard practices that are generally used in microbiological laboratories; and (ii) special procedures, equipment, and laboratory installations that provide physical barriers which are applied in varying degrees according to the estimated biohazard. Four biosafety levels (BL) are described in Appendix G. These biosafety levels consist of combinations of laboratory practices and techniques, safety equipment, and laboratory facilities appropriate for the operations performed and the hazard posed by agents and for the laboratory function and activity. Biosafety level 4 (BL4) provides the most stringent containment conditions, BL1 the least stringent.

Experiments on recombinant DNAs by their very nature lend themselves to a third containment mechanism – namely, the application of highly specific biological barriers. In fact, natural barriers do exist which limit either (i) the infectivity of a *vector* or *vehicle* (plasmid or virus) for specific hosts, or (ii) its dissemination and survival in the environment. The vectors that provide the means for replication of the recombinant DNAs and/or the host cells in which they replicate can be genetically designed to decrease by many orders of magnitude the probability of dissemination of recombinant DNAs outside the laboratory. Further details on biological containment may be found in Appendix I.

As these three means of containment are complementary, different levels of containment appropriate for experiments with different recombinants can be established by applying various combinations of the physical and biological barriers along with a constant use of the standard practices. We consider these categories of containment separately in order that such combinations can be conveniently expressed in the Guidelines.

In constructing these Guidelines, it was necessary to define boundary conditions for the different levels of physical and biological containment and for the classes of experiments to which they apply. We recognize that these definitions do not take into account all existing and anticipated information on special procedures that will allow particular experiments to be carried out under different conditions than indicated here without affecting risk. Indeed, we urge that individual investigators devise simple and more effective containment procedures, and that investigators and IBCs recommend changes in the Guidelines to permit their use.

III. Guidelines for Covered Experiments

Part III discusses experiments involving recombinant DNA. These experiments have been divided into four classes:

III - A. Experiments which require specific RAC review and NIH and IBC approval before initiation of the experiment;

III - B. Experiments which require IBC approval before initiation of the experiment;

III - C. Experiments which require IBC notification at the time of initiation of the experiment;

III - D. Experiments which are exempt from the procedures of the Guidelines.

IF AN EXPERIMENT FALLS INTO BOTH CLASS III - A AND ONE OF THE OTHER CLASSES, THE RULES PERTAINING TO CLASS III - A MUST BE FOLLOWED. If an experiment falls into Class III - D and into either Class III - B or III - C as well, it can be considered exempt from the requirements of the Guidelines.

Changes in containment levels from those specified here may not be instituted without the express approval of the Director, NIH (see Sections IV - C - 1 - b - (1), IV - C - 1 - b - (2), and subsections).

III - A – Experiments That Require RAC Review and NIH and IBC Approval Before Initiation

Experiments in this category cannot be initiated without submission of relevant information on the proposed experiment to NIH, the publication of the proposal in the Federal Register for thirty days of comment, review by the RAC, and specific approval by NIH. The containment conditions for such experiments will be recommended by RAC and set by NIH at the time of approval. Such experiments also require the approval of the IBC before initiation. Specific experiments already approved in this section and the appropriate containment conditions are listed in Appendices D and F. If an experiment is similar to those listed in Appendices D and F, ORDA may determine appropriate containment conditions according to case precedents under Section IV - C - 1 - b - (3) - (g).

If the experiments in this category are submitted for review to another Federal agency, the submitter shall notify ORDA, ORDA may then determine that such review serves the same purpose, and based on that determination, notify the submitter that no RAC review will take place, no NIH approval is necessary, and the experiment may proceed upon approval from the other Federal agency.

III - A - 1. Deliberate formation of recombinant DNAs containing genes for the biosynthesis of toxic molecules lethal for vertebrates at an LD_{50} of less than 100 nanograms per kilogram body weight (e.g., microbial toxins such as the botulinum toxins, tetanus toxin, diphtheria toxin, *Shigella dysenteriae* neurotoxin). Specific approval has been given for the cloning in *E. coli* K - 12 of DNAs containing genes coding for the biosynthesis of toxic molecules which are lethal to vertebrates at 100 nanograms to 100 micrograms per kilogram body weight. Containment levels for these experiments are specified in Appendix F.

III - A - 2. Deliberate release into the environment of any organism containing recombinant DNA, except certain plants as described in Appendix L.

III - A - 3. Deliberate transfer of a drug resistance trait to microorganisms that are not known to acquire it naturally [2] if such acquisition could compromise the use of the drug to control disease agents in human or veterinary medicine or agriculture.

III - A - 4. Deliberate transfer of recombinant DNA or DNA or RNA derived from recombinant DNA into human subjects [21]. The requirement for RAC review should not be considered to preempt any other required review of experiments with human subjects. Institutional Review Board (IRB) review of the proposal should be completed before submission to NIH.

III - B – Experiments That Require IBC Approval Before Initiation

Investigators performing experiments in this category must submit to their IBC, prior to initiation of the experiments, a registration document that contains a description of: (i) The source(s) of DNA; (ii) the nature of the inserted DNA sequences; (iii) the hosts and vectors to be used; (iv) whether a deliberate attempt will be made to obtain expression of a foreign gene, and, if so, what protein will be produced; and (v) the containment conditions specified in these Guidelines. This registration document must be dated and signed by the investigator and filed only with the local IBC. The IBC shall review all such proposals prior to initiation of the experiments. Requests for lowering of containment for experiments in this category will be considered by NIH (see Section IV - C - 1 - b - (3)).

III - B - 1 – Experiments Using Human or Animal Pathogens (Class 2, Class 3, Class 4, or Class 5 Agents [1] as Host-Vector Systems

III - B - 1 - a. Experiments involving the introduction of recombinant DNA into Class 2 agents can be carried out at BL2 containment.

III - B - 1 - b. Experiments involving the introduction of recombinant DNA into Class 3 agents can be carried out at BL3 containment.

III - B - 1 - c. Experiments involving the introduction of recombinant DNA into Class 4 agents can be carried out at BL4 containment.

III - B - 1 - d. Containment conditions for experiments involving the introduction of recombinant DNA into Class 5 agents will be set on a case-by-case basis following ORDA review. A U.S. Department of Agriculture (USDA) permit is required for work with Class 5 agents [18, 20].

III - B - 2 - Experiments in Which DNA From Human or Animal Pathogens (Class 2, Class 3, Class 4, or Class 5 Agents [1] is Cloned in Nonpathogenic Prokaryotic or Lower Eukaryotic Host-Vector Systems

III - B - 2 - a. Recombinant DNA experiments in which DNA from Class 2 or Class 3 agents [1] is transferred into nonpathogenic prokaryotes or lower eukaryotes may be performed under BL2 containment. Recombinant DNA experiments in which DNA from Class 4 agents is transferred into nonpathogenic prokaryotes or lower eukaryotes can be performed at BL2 containment after demonstration that only a totally and irreversibly defective fraction of the agent's genome is present in a given recombinant. In the absence of such a demonstration, BL4 containment should be used. Specific lowering of containment to BL1 for particular experiments can be approved by the IBC. Many experiments in this category will be exempt from the Guidelines (see Sections III - D - 4 and III - D - 5). Experiments involving the formation of recombinant DNAs for certain genes coding for molecules toxic for vertebrates require RAC review and NIH approval (see Section III - A - 1) or must be carried out under NIH specified conditions as described in Appendix F.

III - B - 2 - b. Containment conditions for experiments in which DNA from Class 5 agents is transferred into nonpathogenic prokaryotes or lower eukaryotes will be determined by ORDA following a case-by-case review. A USDA permit is required for work with Class 5 agents [18, 20].

III - B - 3 - Experiments Involving the Use of Infectious Animal or Plant DNA or RNA Viruses or Defective Animal or Plant DNA or RNA Viruses in the Presence of Helper Virus in Tissue Culture Systems

Caution: Special care should be used in the evaluation of containment levels for experiments which are likely to either enhance the pathogenicity (e.g., insertion of a host oncogene) or to extend the host range (e.g., introduction of novel control elements) of viral vectors under conditions which permit a productive infection. In such cases, serious consideration should be given to raising the physical containment by at least one level.

Note. – Recombinant DNA molecules or RNA molecules derived therefrom, which contain less than two-thirds of the genome of any eukaryotic virus (all virus from a single Family [17] being considered identical [19] may be considered defective and can be used in the absence of helper under the conditions specified in Section III - C.

III - B - 3 - a. Experiments involving the use of infectious Class 2 animal viruses 1 or defective Class 2 animal viruses in the presence of helper virus can be performed at BL2 containment.

III - B - 3 - b. Experiments involving the use of infectious Class 3 animal viruses 1 of defective Class 3 animal viruses in the presence of helper virus can be carried out at BL3 containment.

III - B - 3 - c. Experiments involving the use of infectious Class 4 viruses [1] or defective Class 4 viruses in the presence of helper virus may be carried out under BL4 containment.

III - B - 3 - d. Experiments involving the use of infectious Class 5 [1] viruses or defective Class 5 viruses in the presence of helper virus will be determined on a case-by-case basis following ORDA review. A USDA permit is required for work with Class 5 pathogens [18, 20].

III - B - 3 - e. Experiments involving the use of infectious animal or plant viruses or defective animal or plant viruses in the presence of helper virus not covered by Sections III - B - 3 - a, III - B - 3 - b, III - B - 3 - c, or III - B - 3 - d may be carried out under BL1 containment.

III - B - 4 – Recombinant DNA Experiments Involving Whole Animals or Plants

III - B - 4 - a. Recombinant DNA, or RNA molecules derived therefrom, from any source except for greater than two-thirds of a eukaryotic viral genome may be transferred to any non-human vertebrate organism and propagated under conditions of physical containment comparable to BL1 and appropriate to the organism under study [2]. It is important that the investigator demonstrate that the fraction of the viral genome being utilized does not lead to productive infection. A USDA permit is required for work with Class 5 agents [18, 20].

III - B - 4 - b. For all experiments involving whole animals and plants and not covered by Section III - B - 4 - a, the appropriate containment will be determined by the IBC [22].

III - B - 5 – Experiments Involving More Than 10 Liters of Culture

The appropriate containment will be decided by the IBC. Where appropriate, Appendix K, *Physical Containment for Large-Scale Uses of Organisms Containing Recombinant DNA Molecules,* should be used.

III – C. Experiments That Require IBC Notice Simultaneously With Initiation of Experiments

Experiments not included in Sections III - A, III - B, III - D, and subsections of these sections are to be considered in Section III - C. All such experiments can be carried out at BL1 containment. For experiments in this category, a registration document as described in Section III - B must be dated and signed by the investigator and filed with the local IBC at the time of initiation of the experiment. The IBC shall review all such proposals, but IBC review prior to initiation of the experiment is not required. (The reader should refer to the policy statement in the first two paragraphs of Section IV - A).

For example, experiments in which all components derive from non-pathogenic lower eukaryotes and non-pathogenic lower eukaryotes fall under Section III - C and can be carried out at BL1 containment.

CAUTION: Experiments Involving Formation of Recombinant DNA Molecules Containing no more than Two-Thirds of the Genome of any Eukaryotic Virus. Recombinant DNA molecules containing no more than two-thirds of the genome of any eukaryotic virus (all viruses from a single Family [17] being considered identical [19] may be propagated and maintained in cells in tissue culture using BL1 containment. For such experiments, it must be shown that the cells lack helper virus for the specific Families of defective viruses being used. If helper virus is present, procedures specified under Section III - B - 3 should be used. The DNA may contain fragments of the genome of viruses from more than one Family but each fragment must be less than two-thirds of a genome.

III - D – Exempt Experiments

The following recombinant DNA molecules are exempt from these Guidelines and no registration with the IBC is necessary:

III - D - 1. Those that are not in organisms or viruses.

III - D - 2. Those that consist entirely of DNA segments from a single nonchromosomal or viral DNA source though one or more of the segments may be a synthetic equivalent.

III - D - 3. Those that consist entirely of DNA from a prokaryotic host including its indigenous plasmids or viruses when propagated only in that host (or a closely related strain of the same species) or when transferred to another host by well established physiological means; also, those that consist entirely of DNA from an eukaryotic host including its chloroplasts, mitochondria, or plasmids (but excluding viruses) when propagated only in that host (or a closely related strain of the same species).

III - D - 4. Certain specified recombinant DNA molecules that consist entirely of DNA segments from different species that exchange DNA by known physiological processes though one or more of the segments may be a synthetic equivalent. A list of such exchangers will be prepared and periodically revised by the Director. NIH, with advice of the RAC after appropriate notice and opportunity for public comment (see Section IV - C - 1 - b - (1) - (c)). Certain classes are exempt as of publication of these revised Guidelines. This list is in Appendix A. An updated list may be obtained from the Office of Recombinant DNA Activities, National Institutes of Health, Building 31, Room 3B10, Bethesda, Maryland 20892.

III - D - 5. Other classes of recombinant DNA molecules – if the Director, NIH, with advice of the RAC, after appropriate notice and opportunity for public comment, finds that they do not present a significant risk to health or the environment (see Section IV - C - 1 - b - (1) - (c)). Certain classes are exempt as of publication of these revised Guidelines. The list is in Appendix C. An updated list may be obtained from the Office of Recombinant DNA Activities, National Institutes of Health, Building 31, Room 3B10, Bethesda, Maryland 20892.

IV. Roles and Responsibilities

IV - A – Policy

Safety in activities involving recombinant DNA depends on the individual conducting them. The Guidelines cannot anticipate every possible situation. Motivation and good judgment are the key essentials to protection of health and the environment.

The Guidelines are intended to help the institution, Institutional Biosafety Committee (IBC), Biological Safety Officer (BSO), and Principal Investigator (PI) determine the safeguards that should be implemented. These Guidelines will never be complete or final, since all conceivable experiments involving recombinant DNA cannot be foreseen. Therefore, *it is the responsibility of the institution and those associated with it to adhere to the intent of the Guidelines as well as to their specifics.*

Each institution (and the IBC acting on its behalf) is responsible for ensuring that recombinant DNA activities comply with the Guidelines. General recognition of institutional authority and responsibility properly establishes accountability for safe conduct of the research at the local level.

The following roles and responsibilities constitute an administrative framework in which safety is an essential and integral part of research involving recombinant DNA molecules. Further clarifications and interpretations of roles and responsibilities will be issued by NIH as necessary.

IV - B – Responsibility of the Institution

IV - B - 1. General Information. Each institution conducting or sponsoring recombinant DNA research covered by these Guidelines is responsible for ensuring that the research is carried out in full conformity with the provisions of the Guidelines. In order to fulfill this responsibility, the institution shall:

IV - B - 1 - a. Establish and implement policies that provide for the safe conduct of recombinant DNA research and that ensure compliance with the Guidelines. The institution as part of its general responsibilities for implementing the Guidelines may establish additional procedures as deemed necessary to govern the institution and its components in the discharge of its responsibilities under the Guidelines. This may incluse: (i) Statements formulated by the institution for general implementation of the Guidelines, and (ii) whatever additional precautionary steps the institution may deem appropriate.

IV - B - 1 - b. Establish an IBC that meets the requirements set forth in Section IV - B - 2 and carries out the functions detailed in Section IV - B - 3.

IV - B - 1 - c. If the institution is engaged in recombinant DNA research at the BL3 or BL4 containment level, appoint a BSO, who shall be a member of the IBC and carry out the duties specified in Section IV - B - 4.

IV - B - 1 - d. Require that investigators responsible for research covered by these Guidelines comply with the provisions of Section IV - B - 5 and assist investigators to do so.

IV - B - 1 - e. Ensure appropriate training for the IBC chairperson and members, the BSO, PIs, and laboratory staff regarding the Guidelines, their implementation, and laboratory safety. Responsibility for training IBC members may be carried out through the IBC chairperson. Responsibility for training laboratory staff may be carried out through the PI. The institution is responsible for seeing that the PI has sufficient training but may delegate this responsibility to the IBC.

IV - B - 1 - f. Determine the necessity in connection with each project for health surveillance of recombinant DNA research personnel, and conduct, if found appropriate, a health surveillance program for the project. [The "Laboratory Safety Monograph" (LSM) discusses various possible components of such a program – for example, records of agents handled, active investigation of relevant illnesses, and the maintenance of serial serum samples for monitoring serologic changes that may result from the employees' work experience. Certain medical conditions may place a laboratory worker at increased risk in any endeavor where infectious agents are handled. Examples given in the LSM include gastrointestinal disorders and treatment with steroids, immunosuppressive drugs, or antibiotics. Workers with such disorders or treatment should be evaluated to determine whether they should be engaged in research with potentially hazardous organisms during their treatment or illness. Copies of the LSM are available from ORDA].

IV - B - 1 - g. Report within 30 days to ORDA any significant problems with and violations of the Guidelines and significant research-related accidents and illnesses, unless the institution determines that the PI or IBC has done so.

IV - B - 2. Membership and Procedures of the IBC. The institution shall establish an IBC whose responsibilities need not be restricted to recombinant DNA. The committee shall meet the following requirements:

IV - B - 2 - a. The IBC shall comprise no fewer than five members so selected that they collectively have experience and expertise in recombinant DNA technology and the capability to assess the safety of recombinant DNA research experiments and any potential risk to public health or the environment. At least two members shall not be affiliated with the institution (apart from their membership on the IBC) and shall represent the interest of the surrounding community with respect to health and protection of the environment. Members meet this requirement if, for example, they are officials of State or local public health or environmental protection agencies, members of other local governmental bodies, or persons active in medical, occupational health, or environmental concerns in the community. The BSO, mandatory when research is being conducted at the BL3 and BL4 levels, shall be a member (see Section IV - B - 4).

IV - B - 2 - b. In order to ensure the competence necessary to review recombinant DNA activities, it is recommended that: (i) The IBC include persons with expertise in recombinant DNA technology, biological safety, and physical containment; (ii) the IBC include, or have available as consultants, persons knowledgeable in institutional commitments and policies, applicable law, standards of professional conduct and practice, community attitudes, and the environment; and (iii) at least one member be from the laboratory technical staff.

IV - B - 2 - c. The institution shall identify the committee members by name in a report to ORDA and shall include relevant background information on each member in such form and at such times as ORDA may require.

IV - B - 2 - d. No member of an IBC may be involved (except to provide information requested by the IBC) in the review or approval of a project in which he or she has been or expects to be engaged or has a direct financial interest.

IV - B - 2 - e. The institution, who is ultimately responsible for the effectiveness of the IBC, may establish procedures that the IBC will follow in its initial and continuing review of applications, proposals, and activities. (IBC review procedures are specified in Section IV - B - 3 - a).

IV - B - 2 - f. Institutions are encouraged to open IBC meetings to public whenever possible, consistent with protection of privacy and proprietary interests.

IV - B - 2 - g. Upon request, the institution shall make available to the public all minutes of IBC meetings and any documents submitted to or received from funding agencies which the latter are required to make available to the public. If comments are made by members of the public on IBC actions, the institution shall forward to NIH both the comments and the ICB's response.

IV - B - 3. Functions of the IBC. On behalf of the institution, the IBC is responsible for:

IV - B - 3 - a. Reviewing for compliance with the NIH Guidelines recombinant DNA research as specified in Part III conducted at or sponsored by the institution, and approving those research projects that it finds are in conformity with the Guidelines. This review shall include:

IV - B - 3 - a - (1). An independent assessment of the containment levels required by these Guidelines for the proposed research, and

IV - B - 3 - a - (2). An assessment of the facilities, procedures, and practices, and of the training and expertise of recombinant DNA personnel.

IV - B - 3 - b. Notifying the PI of the results of their review.

IV - B - 3 - c. Lowering containment levels for certain experiments as specified in Sections III - B - 2.

IV - B - 3 - d. Setting containment levels as specified in Section III - B - 4 - b and III - B - 5.

IV - B - 3 - e. Reviewing periodically recombinant DNA research being conducted at the institution to ensure that the requirements of the Guidelines are being fulfilled.

IV - B - 3 - f. Adopting emergency plans covering accidental spills and personnel contamination resulting from such research.

Note. – Basic elements in developing specific procedures for dealing with major spills of potentially hazardous materials in the laboratory are detailed in the LSM. Included are information and references on decontamination and emergency plans. The NIH and the Centers for Disease Control are available to provide consultation and direct assistance, if necessary, as posted in the LSM. The institution shall cooperate with the State and local public health departments reporting any significant research-related illness or accident that appears to be a hazard to the public health.

IV - B - 3 - g. Reporting within 30 days to the appropriate institutional official and to ORDA any significant problems with or violations of the Guidelines and any significant research-related accidents or illnesses unless the IBC determines that the PI has done so.

IV - B - 3 - h. The IBC may not authorize initiation of experiments not explicitly covered by the Guidelines until NIH (with the advice of the RAC when required) establishes the containment requirement.

IV - B - 3 - i. Performing such other functions as may be delegated to the IBC under Section IV - B - 1.

IV - B - 4. Biological Safety Officer. The institution shall appoint a BSO if it engages in recombinant DNA research at the BL3 or BL4 containment level. The officer shall be a member of the IBC, and his or her duties shall include (but need not be limited to):

IV - B - 4 - a. Ensuring through periodic inspections that laboratory standards are rigorously followed;

IV - B - 4 - b. Reporting to the IBC and the institution all significant problems with and violations of the Guidelines and all significant research-related accidents and illnesses of which the BSO becomes aware unless the BSO determines that the PI has done so;

IV - B - 4 - c. Developing emergency plans for dealing with accidental spills and personnel contamination and investigating recombinant DNA research laboratory accidents,

IV - B - 4 - d. Providing advice on laboratory security;

IV - B - 4 - e. Providing technical advice to the PI and the IBC on research safety procedures.

Note. – See the LSM for additional information on the duties of the BSO.

IV - B - 5. Principal Investigator (PI). On behalf of the institution, the PI is responsible for complying fully with the Guidelines in conducting any recombinant DNA research.

IV - B - 5. PI – General. As part of this general responsibility, the PI shall:

IV - B - 5 - a - (1). Initiate or modify no recombinant DNA research requiring approval by the IBC prior to initiation (see Sections III - A and III - B) until that research or the proposed modification thereof has been approved by the IBC and has met all other requirements of the Guidelines;

IV - B - 5 - a - (2). Determine whether experiments are covered by Section III - C and follow the appropriate procedures;

IV - B - 5 - a - (3). Report within 30 days to the IBC and NIH (ORDA) all significant problems with and violations of the Guidelines and all significant research-related accidents and illnesses;

IV - B - 5 - a - (4). Report to the IBC and to NIH (ORDA) new information bearing on the Guidelines;

IV - B - 5 - a - (5). Be adequately trained in good microbiological techniques;

IV - B - 5 - a - (6). Adhere to IBC-approved emergency plans for dealing with accidental spills and personnel contamination; and

IV - B - 5 - a - (7). Comply with shipping requirements for recombinant DNA molecules. (See Appendix H for shipping requirements and the LSM for technical recommendations.)

IV - B - 5 - b. Submissions by the PI to NIH. The PI shall:

IV - B - 5 - b - (1). Submit information to NIH (ORDA) in order to have new host-vector systems certified;

IV - B - 5 - b - (2). Petition NIH with notice to the IBC for exemptions to these Guidelines;

IV - B - 5 - b - (3). Petition NIH with concurrence of the IBC for approval to conduct experiments specified in Section III - A of the Guidelines;

IV - B - 5 - b - (4). Petition NIH for determination of containment for experiments requiring case-by-case review;

IV - B - 5 - b - (5). Petition NIH for determination of containment for experiments not covered by the Guidelines.

IV - B - 5 - c. Submissions by the PI to the IBC. The PI shall:

IV - B - 5 - c - (1). Make the initial determination of the required levels of physical and biological containment in accordance with the Guidelines;

IV - B - 5 - c - (2). Select appropriate microbiological practices and laboratory techniques to be used in the research;

IV - B - 5 - c - (3). Submit the initial research protocol if covered under Guidelines Section III - A, III - B, or III - C (and also subsequent changes – e.g., changes in the source of DNA or host-vector system) to the IBC for review and approval or disapproval; and

IV - B - 5 - c - (4). Remain in communication with the IBC throughout the conduct of the project.

IV - B - 5 - d. PI Responsibilities Prior to Initiating Research. The PI is responsible for:

IV - B - 5 - d - (1). Making available to the laboratory staff copies of the protocols that describe the potential biohazards and the precautions to be taken;

IV - B - 5 - d - (2). Instructing and training staff in the practices and techniques required to ensure safety and in the procedures for dealing with accidents; and

IV - B - 5 - d - (3). Informing the staff of the reasons and provisions for any precautionary medical practices advised or requested, such as vaccinations or serum collection.

IV - B - 5 - e. PI Responsibilities During the Conduct of the Research. The PI is responsible for:

IV - B - 5 - e - (1). Supervising the safety performance of the staff to ensure that the required safety practices and techniques are employed;

IV - B - 5 - e - (2). Investigating and reporting in writing to ORDA, the BSO (where applicable), and the IBC any significant problems pertaining to the operation and implementation of containment practices and procedures;

IV - B - 5 - e - (3). Correcting work errors and conditions that may result in the release of recombinant DNA materials;

IV - B - 5 - e - (4). Ensuring the integrity of the physical containment (e.g., biological safety cabinets) and the biological containment (e.g., purity and genotypic and phenotypic characteristics).

IV - C – Responsibilities of NIH

IV - C - 1. Director. The Director, NIH, is responsible for (i) establishing the NIH Guidelines for Research Involving Recombinant DNA Molecules, (ii) overseeing their implementation, and (iii) their final interpretation.

The Director has responsibilities under the Guidelines that involve ORDA and RAC. The ORDA's responsibilities under the Guidelines are administrative. Advice from the RAC is primarily scientific and technical. In certain circumstances, there is specific opportunity for public comment with published response before final action.

IV - C - 1 - a. General Responsibilities of the Director. NIH. The responsibilities of the director shall include the following:

IV - C - 1 - a - (1). Promulgating requirements as necessary to implement the Guidelines;

IV - C - 1 - a - (2). Establishing and maintaining the RAC to carry out the responsibilities set forth in Section IV - C - 2. The RAC's membership is specified in its charter and in Section IV - C - 2;

IV - C - 1 - a - (3). Establishing and maintaining ORDA to carry out the responsibilities defined in Section IV - C - 3.

IV - C - 1 - b. Specific Responsibilities of the Director, NIH. In carrying out the responsibilities set forth in this section, the director or a designee shall weigh each proposed action through appropriate analysis and consultation to determine that it complies with the Guidelines and presents no significant risk to health or the environment.

IV - C - 1 - b - (1). Major Actions. To execute major actions the director must seek the advice of the RAC and provide an opportunity for public and Federal agency comment. Specifically, the agenda of the RAC meeting citing the major actions will be published in the Federal Register at least 30 days before the meeting, and the director will also publish the proposed actions in the Federal Register for comment as least 30 days before the meeting. In addition, the director's proposed decision, at his discretion, may be published in the Federal Register for 30 days of comment before final action is taken. The director's final decision, along with response to the comments, will be published in the Federal Register and the *Recombinant DNA Technical Bulletin.* The RAC and IBC chairpersons will be notified of this decision:

IV - C - 1 - b - (1) - (a). Changing containment levels for types of experiments that are specified in the Guidelines when a major action is involved;

IV - C - 1 - b - (1) - (b). Assigning containment levels for types of experiments that are not explicitly considered in the Guidelines when a major action is involved;

IV - C - 1 - b - (1) - (c). Promulgating and amending a list of classes of recombinant DNA molecules to be exempt from these Guidelines because they consist entirely of DNA segments from species that exchange DNA by known physiological processes or otherwise do not present a significant risk to health or the environment;

IV - C - 1 - b - (1) - (d). Permitting experiments specified by Section III - A of the Guidelines;

IV - C - 1 - b - (1) - (e). Certifying new host-vector systems with the exception of minor modifications of already certified systems (the standards and procedures for certification are described in Appendix I - II - A. Minor modifications constitute, for example, those of minimal or no consequence to the properties relevant to containment); and

IV - C - 1 - b - (1) - (f). Adopting other changes in the Guidelines.

IV - C - 1 - b - (2). Lesser Actions. To execute lesser actions, the director must seek the advice of the RAC. The director's decision will be transmitted to the RAC and IBC chairpersons and published in the *Recombinant DNA Technical Bulletin.*

IV - C - 1 - b - (2) - (a). Interpreting and determining containment levels upon request by ORDA;

IV - C - 1 - b - (2) - (b). Changing containment levels for experiments that are specified in the Guidelines (see Section III);

IV - C - 1 - b - (2) - (c). Assigning containment levels for experiments not explicitly considered in the Guidelines;

IV - C - 1 - b - (2) - (d). Revising the "Classification of Etiologic Agents" for the purpose of these Guidelines [1].

IV - C - 1 - b - (3). Other Actions. The director's decision will be transmitted to the RAC and IBC chairpersons and published in the *Recombinant DNA Technical Bulletin:*

IV - C - 1 - b - (3) - (a). Interpreting the Guidelines for experiments to which the Guidelines specifically assign containment levels;

IV - C - 1 - b - (3) - (b). Setting containment under Section III - B - 1 - d and Section III - B - 3 - d;

IV - C - 1 - b - (3) - (c). Approving minor modifications of already certified host-vector systems (the standards and procedures for such modifications are described in Appendix I - II);

IV - C - 1 - b - (3) - (d). Decertifying already certified host-vector systems;

IV - C - 1 - b - (3) - (e). Adding new entries to the list of molecules toxic for vertebrates (see Appendix F);

IV - C - 1 - b - (3) - (f). Approving the cloning of toxin genes in host-vector systems other than *E. coli* K - 12 (see Appendix F); and

IV - C - 1 - b - (3) - (g). Determining appropriate containment conditions for experiments according to case precedents developed under Section IV - C - 1 - b - (2) - (c).

IV - C - 1 - b - (4). The director shall conduct, support, and assist training programs in laboratory safety for IBC members, BSOs, PIs, and laboratory staff.

IV - C - 2. Recombinant DNA Advisory Committee. The Recombinant DNA Advisory Committee (RAC) is responsible for carrying out specified functions cited below as well as others assigned under its charter or by the Secretary, HHS, the Assistent Secretary for Health, and the Director, NIH.

The committee shall consist of 25 members including the chair, appointed by the Secretary or his or her designee, at least fourteen of whom shall be selected from authorities knowledgeable in the fields of molecular biology or recombinant DNA research or in scientific fields other than molecular biology or recombinant DNA research or in scientific fields other than molecular biology or recombinant DNA research, and at least six of whom shall be persons knowledgeable in applicable law, standards of professional conduct and practice, public attitudes, the environment, public health, occupational health, or related fields. Representatives from Federal agencies shall serve as non-voting members. Nominations for the RAC may be submitted to the Office of Recombinant DNA Activities, National Institutes of Health, Building 31, Room 3B10, Bethesda, MD 20892.

All meetings of the RAC will be announced in the Federal Register, including tentative agenda items, 30 days in advance of the meeting with final agendas (if modified) available at least 72 hours before the meeting. No item defined as a major action under Section IV - C - 1 - b - (1) may be added to an agenda after it appears in the Federal Register.

The RAC shall be responsible for advising the Director, NIH, on the actions listed in Section IV - C - 1 - b - (1) and IV - C - 1 - b - (2).

IV - C - 3. The Office of Recombinant DNA Activities. The ORDA shall serve as a focal point for information on recombinant DNA activities and provide advice to all within and outside NIH including Institutions, BSOs, PIs, Federal agencies, State and local governments and institutions in the private sector. The ORDA shall carry out such other functions as may be delegated to it by the Director, NIH, including those authorities described in Section IV - C - 1 - b - (3). In addition, ORDA shall be responsible for the following:

IV - C - 3 - a. Reviewing and approving IBC membership;

IV - C - 3 - b. Publishing in the Federal Register:

IV - C - 3 - b - (1). Announcements of RAC meetings and agendas at least 30 days in advance;

Note. – If the agenda for an RAC meeting is modified, ORDA shall make the revised agenda available to anyone upon request at least 72 hours in advance of the meeting.

IV - C - 3 - b - (2). Proposed major actions of the type falling under Section IV - C - 1 - b - (1) at least 30 days prior to the RAC meeting at which they will be considered; and

IV - C - 3 - b - (3). The NIH director's final decision on recommendations made by the RAC.

IV - C - 3 - c. Publishing the *Recombinant DNA Technical Bulletin;* and

IV - C - 3 - d. Serving as executive secretary of the RAC.

IV - C - 4. Other NIH Components. Other NIH components shall be responsible for certifying maximum containment (BL4) facilities, inspecting them periodically, and inspecting other recombinant DNA facilities as deemed necessary.

IV - D – Compliance

As a condition for NIH funding of recombinant DNA research, institutions must ensure that such research conducted at or sponsored by the institution, irrespective of the source of funding, shall comply with these Guidelines. The policies on noncompliance are as follows:

IV - D - 1. All NIH-funded projects involving recombinant DNA techniques must comply with the NIH Guidelines. Noncompliance may result in (i) suspension, limitation, or termination of financial assistance for such projects and a requirement for prior NIH approval of any or all recombinant DNA research at the institution, or (ii) a requirement for prior NIH approval of any or all recombinant DNA projects at the Institution.

IV - D - 2. All non-NIH funded projects involving recombinant DNA techniques conducted at or sponsored by an institution that receives NIH funds for projects involving such techniques must comply with the NIH Guidelines. Noncompliance may result in (i) Suspension, limitation, or termination of NIH funds for recombinant DNA research at the institution, or (ii) a requirement for prior NIH approval of any or all recombinant DNA projects at the institution.

IV - D - 3. Information concerning noncompliance with the Guidelines may be brought forward by any person. It should be delivered to both NIH (ORDA) and the relevant Institution. The institution, generally through the IBC, shall take appropriate action. The institution shall forward a complete report of the incident to ORDA, recommending any further action.

IV - D - 4. In cases where NIH proposes to suspend, limit, or terminate financial assistance because of noncompliance with the Guidelines, applicable DHHS and Public Health Service procedures shall govern.

IV - D - 5. Voluntary Compliance. Any individual, corporation, or institution that is not otherwise covered by the Guidelines is encouraged to conduct recombinant DNA research activities in accordance with the Guidelines through the procedures set forth in Part VI.

V. Footnotes and References of Sections I - IV

1. The original reference to organisms as Class 1, 2, 3, 4, or 5 refers to the classification in the publication *Classification of Etiologic Agents of the Basis of Hazard*, 4th Edition, July 1974; U.S. Department of Health, Education, and Welfare, Public Health Service, Centers for Disease Control, Office of Biosafety, Atlanta, Georgia 30333.

The Director, NIH, with advice of the Recombinant DNA Advisory Committee, may revise the classification for the purposes of these Guidelines (see Section IV - C - 1 - b - (2) - (d)). The revised list of organisms in each class is reprinted in Appendix B to these Guidelines.

2. In Part III of the Guidelines, there are a number of places where judgments are to be made. In all these cases the principal investigator is to make the judgment on these matters as part of his responsibility to "make the initial determination of the required levels of physical and biological containment in accordance with the Guidelines" (Section IV - B - 5 - c - (1)). In the cases falling under Sections III - A, - B or - C, this judgment is to be reviewed and approved by the IBC as part of its responsibility to make "an independent assessment of the containment levels required by these Guidelines for the proposed research" (Section IV - B - 3 - a - (1)). If the IBC wishes, any specific cases may be referred to ORDA as part of ORDA's functions to "provide advice to all within and outside NIH" (Section IV - C - 3), and ORDA may request advice from the RAC as part of the RAC's responsibility for "interpreting and determining containment levels upon request by ORDA" (Section IV - C - 1 - b - (2) - (a)).

3. *Laboratory Safety at the Center for Disease Control* (Sept. 1974). U.S. Department of Health, Education and Welfare Publication No. CDC 75 - 8118.

4. *Classification of Etiologic Agents on the Basis of Hazard* (4th Edition, July 1974). U.S. Department of Health, Education and Welfare. Public Health Service. Centers for Disease Control, Office of Biosafety, Atlanta, Georgia 30333.

5. *National Cancer Institute Safety Standards for Research Involving Oncogenic Viruses* (Oct. 1974). U.S. Department of Health, Education and Welfare Publication No. (NIH) 75 - 790.

6. *National Institutes of Health Biohazards Safety Guide* (1974). U.S. Department of Health, Education and Welfare, Public Health Service, National Institutes of Health. U.S. Government Printing Office, Stock No. 1740 - 00383.

7. *Biohazards in Biological Research* (1973). A. Hellmann, M.N. Oxman, and R. Pollack (ed.) Cold Spring Harbor Laboratory.

8. *Handbook of Laboratory Safety* (1971). 2nd Edition. N.V. Steere (ed.). The Chemical Rubber Co., Cleveland.

9. Bodily, J.L. (1970). *General Administration of the Laboratory*, H.L. Bodily, E.L. Updyke, and J.O. Mason (eds.), Diagnostic Procedures for Bacterial, Mycotic and Parasitic Infections. American Public Health Association, New York, pp. 11 - 28.

10. Darlow, H.M. (1969). *Safety in the Microbiological Laboratory*. In J.R. Norris and D.W. Robbins (ed.), Methods in Microbiology, Academic Press, Inc., New York, pp. 169 - 204.

11. *The Prevention of Laboratory Acquired Infection* (1974). C.H. Collins, E.G. Hartley, and R. Pilsworth. Public Health Laboratory Service, Monograph Series No. 6.

12. Chatigny, M.A. (1961). *Protection Against Infection in the Microbiological Laboratory: Devices and Procedures*. In W.W. Umbreit (ed.): Advances in Applied Microbiology. Academic Press, New York, N.Y. 3:131 - 192.

13. *Design Criteria for Viral Oncology Research Facilities* (1975). U.S. Department of Health, Education and Welfare, Public Health Service, National Institutes of Health, DHEW Publication No. (NIH) 75 - 891.

14. Kuehne, R.W. (1973). *Biological Containment*

Facility for Studying Infectious Disease. Appl. Microbiol. 26 - 239 - 243.

15. Runkle, R.S., and G.B. Phillips (1969). *Microbial Containment Control Facilities*. Van Nostrand Reinhold, New York.

16. Chatigny, M.A., and D.I. Clinger (1969). *Contamination Control in Aerobiology*. In R.L. Dimmick and A.B. Akers (eds.). An Introduction to Experimental Aerobiology. John Wiley & Sons, New York, pp. 194 - 263.

17. As classified in the Third Report of the International Committee on Taxonomy of Viruses: Classification and Nomenclature of Viruses, R.E.F. Matthews, Ed. Intervirology 12 (129 - 296) 1979.

18. A USDA permit, required for import and interstate transport of pathogens, may be obtained from the Animal and Plant Health Inspection Service, USDA, Federal Building, Hyattsville, MD 20782.

19. i.e., the total of all genomes within a Family shall not exceed two-thirds of the genome.

20. All activities, including storage of variola and whitepox, are restricted to the single national facility (World Health Organization (WHO) Collaborating Center for Smallpox Research, Centers for Disease Control, in Atlanta).

21. Section III - A - 4 covers only those experiments in which the intent is to modify stably the genome of cells of a human subject. Other experiments involving recombinant DNA in human subjects such as feeding of bacteria containing recombinant DNA or the administration of vaccines containing recombinant DNA are not covered in Section III - A - 4 of the Guidelines.

22. For recombinant DNA experiments in which the intent is to modify stably the genome of cells of a human subject, see Section III - A - 4.

VI. Voluntary Compliance

VI - A. – Basic Policy

Individuals, corporations, and institutions not otherwise covered by the Guidelines are encouraged to do so by following the standards and procedures set forth in Parts I - IV of the Guidelines. In order to simplify discussion, references hereafter to "institutions" are intended to encompass corporations, and individuals who have no organizational affiliation. For purposes of complying with the Guidelines, an individual intending to carry out research involving recombinant DNA is encouraged to affiliate with an institution that has an IBC approved under the Guidelines.

Since commercial organizations have special concerns, such as protection of proprietary data, some modifications and explanations of the procedures in Parts I - IV are provided below, in order to address these concerns.

VI - B - IBC Approval

The ORDA will review the membership of an institution's IBC, and where it finds the IBC meets the requirements set forth in Section IV - B - 2 will give its approval to the IBC membership.

It should be emphasized that employment of an IBC member solely for purposes of membership on the IBC does not itself make the member an institutionally affiliated member for purposes of Section IV - B - 2 - a.

Except for the unaffiliated members, a member of an IBC for an institution not otherwise covered by the Guidelines may participate in the review and approval of a project in which the member has a direct financial interest so long as the member has not been, and does not expect to be, engaged in the project. Section IV - B - 2 - d is modified to that extent for purposes of these institutions.

VI - C – Certification of Host-Vector Systems

A host-vector system may be proposed for certification by the Director, NIH, in accordance with the procedures set forth in Appendix I - II - A.

In order to ensure protection for proprietary data, any public notice regarding a host-vector system which is designated by the institution as proprietary under Section VI - E - 1 will be issued only after consultation with the institution as to the content of the notice.

VI - D – Requests for Exemptions and Approvals

Requests for exemptions or other approvals required by the Guidelines should be requested by following the procedures set forth in the appropriate sections in Parts I - IV of the Guidelines.

In order to ensure protection for proprietary data, any public notice regarding a request for an exemption or other approval which is designated by the institution as proprietary under Section VI - E - 1 will be issued only after consultation with the institution as to the content of the notice.

VI - E – Protection of Proprietary Data

In general, the Freedom of Information Act requires Federal agencies to make their records available to the public upon request. However, this requirement does not apply to, among other things, "trade secrets and commercial and financial information obtained from a person and privileged or confidential." 18 U.S.C. 1905, in turn makes it a crime for an officer or employee of the United States for any Federal department or agency to publish, divulge, disclose, or make known "in any manner or to any extent not authorized by law any information coming to him in the course of his employment for official duties or by reason of any examination or investigation made by, or return, report or record made to or filed with, such department or agency or officer or employee thereof, which information concerns or relates to the trade secrets, [or] processes ... of any person, firm, partnership, corporation, or association." This provision applies to all employees of the Federal Government, including special Government employees. Members of the Recombinant DNA Advisory Committee are "special Government employees."

VI - E - 1. In submitting to NIH for purposes of complying voluntarily with the Guidelines, an institution may designate those items of information which the institution believes constitute trade secrets, privileged, confidential commercial, or financial information.

VI - E - 2. If NIH receives a request under the Freedom of Information Act for information so designated, NIH will promptly contact the institution to secure its views as to whether the information (or some portion) should be released.

VI - E - 3. If the NIH decides to release this information (or some portion) in response to a Freedom of Information request or otherwise, the institution will be advised; and the actual release will not be made until the expiration of 15 days after the institution is so advised except to the extent that earlier release in the judgment of the Director, NIH, is necessary to protect against an imminent hazard to the public or the environment.

VI - E - 4. Presubmission Review.

VI - E - 4 - a. Any institution not otherwise covered by the Guidelines, which is considering submission of data or information voluntarily to NIH, may request presubmission review of the records involved to determine whether if the records are submitted NIH will or will not make part or all of the records available upon request under the Freedom of Information Act.

VI - E - 4 - b. A request for presubmission review should be submitted to ORDA along with the records involved. These records must be clearly marked as being the property of the institution on loan to NIH solely for the purpose of making a determination under the Freedom of Information Act. The ORDA will then seek a determination from the HHS Freedom of Information Officer, the responsible official under HHS regulations (45 CFR Part 5) as to whether the records involved (or some portion) are or are not available to members of the Public under the Freedom of Information Act. Pending such a determination the records will be kept separate from ORDA files, will be considered records of the institution and not ORDA, and will not be received as part of ORDA files. No copies will be made of the records.

VI - E - 4 - c. The ORDA will inform the institution of the HHS Freedom of Information Officer's determination and follow the institution's instructions as to whether some or all of the records involved are to be returned to the institution or to become a part of ORDA files. If the institution instructs ORDA to return the records, no copies or summaries of the records will be made or retained by HHS, NIH, or ORDA.

VI - E - 4 - d. The HHS Freedom of Information Officer's determination will represent that official's judgement at the time of the determination as to whether the records involved (or some portion) would be exempt from disclosure under the Freedom of Information Act if at the time of the determination the records were in ORDA files at a request were received for them under the Act.

Appendix A – Exemptions Under Section III - D - 4

Section III - D - 4 states that exempt from these Guidelines are "certain specified recombinant DNA molecules that consist entirely of DNA segments from different species that exchange DNA by known physiological processes though one or more of the segments may be a synthetic equivalent. A list of such exchangers will be prepared and periodically revised by the Director, NIH, with advice of the RAC after appropriate notice and opportunity for public comment (see Section IV - C - 1 - b - (1) - (c)). Certain classes are exempt as of publication of these revised Guidelines. The list is in Appendix A."

Under Section III - D - 4 of these Guidelines are recombinant DNA molecules that are: (1) Composed entirely of DNA segments from one or more of the organisms within a sublist and (2) to be propagated in any of the organisms within a sublist.

(Classification of *Bergey's Manual of Determinative Bacteriology*, 8th edition. R.E. Buchanan and N.E. Gibbons, editors. Williams and Wilkins Company: Baltimore, 1974).

Although these experiments are exempt, it is recommended that they be performed at the appropriate biosafety level for the host or recombinant organism (for biosafety levels see *Biosafety in Microbiological and Biomedical Laboratories*, 1st Edition (March 1984), U.S. Department of Health and Human Services, Public Health Service, Centers for Disease Control, Atlanta, Georgia 30333, and National Institutes of Health, Bethesda, Maryland 20892).

Sublist A

1. Genus *Escherichia*
2. Genus *Shigella*
3. Genus *Salmonella"* (including *Arizona*)
4. Genus *Enterobacter*
5. Genus *Citrobacter* (including *Levinea*)
6. Genus *Klebsiella*
7. Genus *Erwinia*
8. *Pseudomonas aeruginosa, Pseudomonas Putida* and *Pseudomonos fluorescens*
9. *Serratia marcescens*
10. *Yersinia enterocolitica*

Sublist B

1. *Bacillus subtilis*
2. *Bacillus licheniformis*
3. *Bacillus pumilus*

4. *Bacillus globigii*
5. *Bacillus niger*
6. *Bacillus nato*
7. *Bacillus amyloliquefaciens*
8. *Bacillus aterrimus*

Sublist C

1. *Streptomyces aureofaciens*
2. *Streptomyces rimosus*
3. *Streptomyces coelicolor*

Sublist D

1. *Streptomyces griseus*
2. *Streptomyces cyaneus*
3. *Streptomyces venezuelae*

Sublist E

1. One way transfer of *Streptococcus mutans* or *Streptococcus lactis* DNA into *Streptococcus sanguis*.

Sublist F

1. *Streptococcus sanguis*
2. *Streptococcus pneumoniae*
3. *Streptococcus faecalis*
4. *Streptococcus pyogenes*
5. *Streptococcus mutans*

Appendix B – Classification of Microorganisms on the Basis of Hazard

Appendix B - I – Classification of Etiologic Agents

The original reference for this classification was the publication *Classification of Etiological Agents on the Basis of Hazard*, 4th edition, July 1974, U.S. Department of Health, Education, and Welfare, Public Health Service, Center for Disease Control, Office of Biosafety, Atlanta, Georgia 30333. For the purposes of these Guidelines, this list has been revised by the NIH [1].

Appendix B - I - A, Class 1 Agents. All bacterial, parasitic, fungal, viral, rickettsial, and chlamydial agents not included in higher classes.

Appendix B - I - B. Class 2 Agents.
Appendix B - I - B - 1. Bacterial Agents.

Acinetobacter calcoaceticus
Actinobacillus-all species
Aeromonas hydrophila
Arizona hinshawii-all serotypes
Bacillus anthracis
Bordetella-all species
Borrelia recurrentis, B. vincenti
Campylobacter fetus
Campylobacter jejuni
Chlamydia psittaci
Chlamydia trachomatis
Clostridium botulinum,
  Cl. chauvoei, Cl. haemolyticum,
  Cl. histolyticum, Cl. novyi,
  Cl. septicum, Cl. tetani
Corynebacterium diphtheriae,
  C. equi, C. haemolyticum,
  C. pseudotuberculosis,
  C. pyogenes, C. renale
Edwardsiella tarda
Erysipelothrix insidiosa

Escherichia coli-all enteropathogenic, enterotoxigenic, enteroinvasive and strains bearing K1 antigen
Haemophilus ducreyi, H. influenzae
Klebsiella-all species and all serotypes
Legionella pneumophila
Leptospira interrogans-all serotypes
Listeria-all species
Moraxella-all species
Mycobacteria-all species except those listed in Class 3
Mycoplasma-all species except *Mycoplasma mycoides* and *Mycoplasma agalactiae*, which are in Class 5
Neisseria gonorrhoeae, N. meningitidis
Pasteurella-all species except those listed in Class 3
Salmonella-all species and all serotypes
Shigella-all species and all serotypes
Sphaerophorus necrophorus
Staphylococcus aureus
Streptobacillus moniliformis
Streptococcus pneumoniae
Streptococcus pyogenes
Treponema carateum, T. pallidum, and T. pertenue
Vibrio cholerae
Vibrio parahemolyticus
Yersinia enterocolitica

Appendix B - I - B - 2. Fungal Agents.

Actinomycetes (including *Nocardia* species, *Actinomyces* species, and *Arachnia propionica*) [2]
Blastomyces dermatitidis
Cryptococcus neoformans
Paracoccidioides braziliensis

Appendix B - I - B - 3. Parasitic Agents.

Endamoeba histolytica
Leishmania sp.
Naegleria gruberi
Schistosoma mansoni
Toxoplasma gondii
Toxocara canis
Trichinella spiralis
Trypanosoma cruzi

Appendix B - I - B - 4. Viral, Rickettsial, and Chlamydial Agents.

Adenoviruses – human – all types
Cache Valley virus
Coxsackie A and B viruses
Cytomegaloviruses
Echoviruses – all types
Encephalomyocarditis virus (EMC)
Flanders virus
Hart Park virus
Hepatitus – associated antigen material
Herpes viruses-except *Herpesvirus simiae* (Monkey B virus) which is in Class 4
Corona viruses
Influenza viruses – all types except A/PR8/34, which is in Class 1
Langat virus
Lymphogranuloma venereum agent
Measles virus
Mumps virus
Parainfluenza virus – all types except Parainfluenza virus 3, SF4 strain, which is in Class 1
Polioviruses – all types, wild and attenuated
Poxviruses – all types except *Alastrim, Smallpox,* and *Whitepox* which are Class 5 and *Monkey pox* which depending on experiments is in Class 3 or Class 4
Rabies virus – all strains except *Rabies street virus* which should be classified in Class 3
Reoviruses – all types
Respiratory syncytial virus
Rhinoviruses – all types
Rubella virus

Simian viruses – all types except *Herpesvirus simiae (Monkey B virus)* and *Marburg virus* which are in Class 4
Sindbis virus
Tensaw virus
Turlock virus
Vaccinia virus
Varicella virus
Vesicular stomatitis virus [3]
Vole rickettsia
Yellow fever virus, 17D vaccine strain

Appendix B - I - C. Class 3 Agents.
Appendix B - I - C - 1. Bacterial Agents.

Bartonella – all species
Brucella – all species
Francisella tularensis
Mycobacterium avium, M. bovis, M. tuberculosis
Pasteurella multocide type *B* ("buffalo" and other foreign virulent strains) [3]
Pseudomonas mallei [3]
Pseudomonas pseudomallei [3]
Yersinia pestis

Appendix B - I - C - 2. Fungal Agents.

Coccidioides immitis
Histoplasma capsulatum
Histoplasma capsulatum var. *duboisii*

Appendix B - I - C - 3. Parasitic Agents.

None.

Appendix B - I - C - 4. Viral, Rickettsial, and Chlamydial Agents.

Monkey pox, when used *in vitro* [4]
    Arboviruses-all strains except those in Class 2 and 4 (*Arboviruses indigenous to the United States are in Class 3 except those listed in Class 2. West Nile and Semliki Forest viruses may be classified up or down depending on the conditions of use and geographical location of the laboratory.)*
Dengue virus, when used for transmission or animal inoculation experiments
Lymphocytic choriomeningitis virus (LCM)
Rickettsia – all species except *Vole rickettsia* when used for transmission or animal inoculation experiments
Yellow fever virus – wild, when used *in vitro*

Appendix B - I - D. Class 4 Agents.
Appendix B - I - D - 1. Bacterial Agents.

None.

Appendix B - I - D - 2. Fungal Agents.

None.

Appendix B - I - D - 3. Parasitic Agents.

None.

Appendix B - I - D - 4. Viral, Rickettsial, and Chlamydial Agents.

Ebola fever virus
Monkey pox, when used for transmission or animal inoculation experiments [4]
Hemorrhagic fever agents, including *Crimean hemorrhagic fever, (Congo), Junin,* and *Machupo* viruses, and others as yet undefined

Herpesvirus simiae (Monkey B virus)
Lassa virus
Marburg virus
Tick-borne encephalitis virus complex, including *Russian spring-summer encephalitis, Kyasanur forest disease, Omsk hemorrhagic fever,* and *Central European encephalitis viruses*
Venezuelan equine encephalitis virus, epidemic strains, when used for transmission or animal inoculation experiments
Yellow fever virus – wild, when used for transmission or animal inoculation experiments

Appendix B - II – Classification of Oncogenic Viruses on the Basis of Potential Hazard [5]
Appendix B - II - A. Low-Risk Oncogenic Viruses.

Rous sarcoma
SV - 40
CELO
Ad7 - SV40
Polyoma
Bovine papilloma
Rat mammary tumor
Avian leukosis
Murine leukemia
Murine sarcoma
Mouse mammary tumor
Rat leukemia
Hamster leukemia
Bovine leukemia
Dog sarcoma
Mason-Pfizer monkey virus
Marek's
Guinea pig herpes
Lucke (frog)
Adenovirus
Shope fibroma
Shope papilloma

Appendix B - II - B. Moderate-Risk Oncogenic Viruses.

Ad2 - SV40
FeLV
HV Saimiri
EBV
SSV - 1
GaLV
HV ateles
Yaba
FeSV

Appendix B - III – Class 5 Agents
Appendix B - III - A. Animal Disease Organisms Which are Forbidden Entry into the United States by Law.

Foot and mouth disease virus.

Appendix B - III - B. Animal Disease Organisms and Vectors Which are Forbidden Entry into the United States by USDA Policy.

African horse sickness virus
African swine fever virus
Besnoitia besnoiti
Borna disease virus
Bovine infectious petechial fever
Camel pox virus
Ephemeral fever virus
Fowl plague virus
Goat pox virus
Hog cholera virus
Louping ill virus
Lumpy skin disease virus
Nairobi sheep disease virus
Newcastle disease virus (Asiatic strains)
Mycoplasma mycoides (contagious bovine pleuropneumonia)

Mycoplasma agalactiae (contagious agalactia of sheep)
Rickettsia ruminatium (heart water)
Rift valley fever virus
Rinderpest virus
Sheep pox virus
Swine vesicular disease virus
Teschen disease virus
Trypanosoma vivax (Nagana)
Trypanosoma evansi
Theileria parva (East Coast fever)
Theileria annulata
Theileria lawrencei
Theileria bovis
Theileria hirci
Vesicular exanthema virus
Wesselsbron disease virus
Zyonema

Appendix B - III - C. Organisms Which may not be Studied in the United States Except at Specified Facilities.

Small pox [4]
Alastrim [4]
White pox [4]

Appendix B - IV – Footnotes and References of Appendix B.

1. The original reference for this classification was the publication *Classification of Etiologic Agents on the Basis of Hazard,* 4th edition, July 1974, U.S. Department of Health Service, Center for Disease Control, Office of Biosafety, Atlanta, Georgia 30333. For the purposes of these Guidelines, this list has been revised by the NIH.
2. Since the publication of the classification in 1974 [1], the *Actinomycetes* have been reclassified as bacterial rather than fungal agents.
3. A USDA permit, required for import and interstate transport of pathogens, may be obtained from the Animal and Plant Health Inspection Service, USDA, Federal Building, Hyattsville, MD 20782.
4. All activities, including storage of variola and whitepox, are restricted to the single national facility [World Health Organization (WHO) Collaborating Center for Smallpox Research, Centers for Disease Control, in Atlanta].
5. *National Cancer Institute Safety Standards for Research Involving Oncogenic Viruses* (October 1974). U.S. Department of Health, Education, and Welfare Publication No. (NIH) 75 - 790.
6. U.S. Department of Agriculture, Animal and Plant Health Inspection Service.

Appendix C – Exemptions Under Section III - D - 5

Section III - D - 5 states that exempt from these Guidelines are "Other classes of recombinant DNA molecules if the Director, NIH, with advice of the RAC, after appropriate notice and opportunity for public comment finds that they do not present a significant risk to health or the environment (see Section IV - C - 1 - b - (1) - (c)). Certain classes are exempt as of publication of these revised Guidelines."
The following classes of experiments are exempt under Section III - D - 5 of the Guidelines:

Appendix C - I – Recombinant DNAs in Tissue Culture.

Recombinant DNA molecules containing less than one-half of any eukaryotic genome (all viruses from a single Family [4] being considered identical [5]) that are propagated and maintained in cells in tissue culture are exempt from these Guidelines with the exceptions listed below.

Exceptions. Experiments described in Section III - A which require specific RAC review and NIH approval before initiation of the experiment.

Experiments involving DNA from Class 3, 4, or 5 organisms [1] or cells known to be infected with these agents.

Experiments involving the deliberate introduction of genes coding for the biosynthesis of molecules toxic for vertebrates (see Appendix F).

Appendix C - II – Experiments Involving E. coli K - 12 Host-Vector Systems

Experiments which use *E. coli* K - 12 host-vector systems, with the exception of those experiments listed below, are exempt from these Guidelines provided that: (i) the *E. coli* host shall not contain conjugation proficient plasmids or generalized transducing phages; and (ii) lambda or lambdoid or Ff bacteriophages or nonconjugative plasmids [2] shall be used as vectors. However, experiments involving the insertion into *E. coli* K - 12 of DNA from prokaryotes that exchange genetic information [3] with *E. coli* may be performed with any *E. coli* K - 12 vector (e.g., conjugative plasmid). When a nonconjugative vector is used, the *E. coli* K - 12 host may contain conjugation-proficient plasmids either autonomous or integrated, or generalized transducing phages.

For these exempt laboratory experiments, BL1 physical containment conditions are recommended.

For large-scale (LS) fermentation experiments BL1-LS physical containment conditions are recommended. However, following review by the IBC of appropriate data for a particular host-vector system, some latitude in the application of BL1 - LS requirements as outlined in Appendix K - II - A through K - II - F is permitted.

Exceptions. Experiments described in Section III - A which require specific RAC review and NIH approval before initiation of the experiment.

Experiments involving DNA from Class 3, 4, or 5 organisms [1] or from cells known to be infected with these agents may be conducted under containment conditions specified in Section III - B - 2 with prior IBC review and approval.

Large-scale experiments (e.g., more than 10 liters of culture) require prior IBC review and approval (see Section III - B - 5).

Experiments involving the deliberate cloning of genes coding for the biosynthesis of molecules toxic for vertebrates (see Appendix F).

Appendix C - III – Experiments Involving Saccharomyces Host-Vector Systems

Experiments which use *Saccharomyces cerevisiae* host-vector systems, with the exception of experiments listed, below, are exempt from these Guidelines.

Experiments which use *Saccharomyces uvarum* host-vector systems, with the exception of experiments listed below, are exempt from these Guidelines.

For these exempt laboratory experiments, BL1 physical containment conditions are recommended.

For large-scale fermentation experiments BL1 - LS physical containment conditions are recommended. However, following review by the IBC of appropriate data for a particular host-vector system some latitude in the application of BL1 - LS requirements as outlined in Appendix K - II - A through K - II - F is permitted.

Exceptions. Experiments described in Section III - A which require specific RAC review and NIH approval before initiation of the experiment.

Experiments involving Class 3, 4, or 5 organisms [1] or cells knowns to be infected with these agents may be conducted under containment conditions specified in Section III - B - 2 with prior IBC review and approval.

Large-scale experiments (e.g., more than 10 liters of culture) require prior IBC review and approval (see Section III - B - 5).

Experiments involving the deliberate cloning of genes coding for the biosynthesis of molecules toxic for vertebrates (see Appendix F).

Appendix C - IV – Experiments Involving Bacillus subtilis Host-Vector Systems

Any asporogenic *Bacillus subtilis* strain which does not revert to a sporeformer with a frequency greater than 10^{-7} can be used for cloning DNA with the exception of those experiments listed below.

For these exempt laboratory experiments, BL1 physical containment conditions are recommended.

For large-scale fermentation experiments BL1 - LS physical containment conditions are recommended. However, following review by the IBC of appropriate data for a particular host-vector system, some latitude in the application of BL1 - LS requirements as outlined in Appendix K - II - A through K - III - F is permitted.

Exceptions. Experiments described in Section III - A which require specific RAC review and NIH approval before initiation of the experiment.

Experiments involving Class 3, 4, or 5 organisms [1] or cells known to be infected with these agents may be conducted under containment conditions specified by Section III - B - 2 with prior IBC review and approval.

Large-scale experiments (e.g., more than 10 liters of culture) require prior IBC review and approval (see Section III - B - 5).

Experiments involving the deliberate cloning of genes coding for the biosynthesis of molecules toxic for vertebrates (see Appendix F).

Appendix C - V – Extrachromosomal Elements of Gram Positive Organisms

Recombinant DNA molecules derived entirely from extrachromosomal elements of the organisms listed below (including shuttle vectors constructed from vectors described in Appendix C), propagated and maintained in organisms listed below are exempt from these Guidelines.

Bacillus subtilis
Bacillus pumilus
Bacillus licheniformis
Bacillus thuringiensis
Bacillus cereus
Bacillus amyloliquefaciens
Bacillus brevis
Bacillus natto
Bacillus niger
Bacillus aterrimus
Bacillus amylosacchariticus
Bacillus anthracis
Bacillus globigii
Bacillus megaterium
Staphylococcus aureus
Staphylococcus epidermidis
Staphylococcus carnosus
Clostridium acetobutylicum
Pediococcus damnosus
Pediococcus pentosaceus
Pediococcus acidilactici
Lactobacillus casei
Listeria grayi
Listeria murrayi
Listeria monocytogenes
Streptococcus pyogenes
Streptococcus agalactiae
Streptococcus sanguis
Streptococcus salivarious
Streptococcus cremoris
Streptococcus pneumoniae
Streptococcus avium
Streptococcus faecalis
Streptococcus anginosus
Streptococcus sobrinus
Streptococcus lactis
Streptococcus mutans
Streptococcus equisimilis
Streptococcus thermophylus
Streptococcus milleri
Streptococcus durans
Streptococcus mitior
Streptococcus ferus

Exceptions. Experiments described in Section III - A which require specific RAC review and NIH approval before initiation of the experiment.

Large-scale experiments (e.g., more than 10 liters of culture) require prior IBC review and approval (see Section III - B - 5).

Experiments involving the deliberate cloning of genes coding for the biosynthesis of molecules toxic for vertebrates (see Appendix F).

Appendix C - VI – Footnotes and References of Appendix C

1. The original reference to organisms as Class 1, 2, 3, 4, or 5 refers to the classification in the publication *Classification of Etiologic Agents on the Basis of Hazard,* 4th Edition, July 1974; U.S. Department of Health, Education and Welfare, Public Health Service, Centers for Disease Control, Office of Biosafety, Atlanta, Georgia 30333.

The Director, NIH, with advice of the Recombinant DNA Advisory Committee, may revise the classification for the purposes of these Guidelines (see Section IV - C - 1 - b - (2) - (d)). The revised list of organisms in each class is reprinted in Appendix B to these Guidelines.

2. A subset of non-conjugative plasmid vectors are also poorly mobilizable (e.g., pBR322, pBR313). Where practical, these vectors should be employed.

3. Defined as observable under optimal laboratory conditions by transformation, transduction, phage infection, and/or conjugation with transfer of phage, plasmid, and/or chromosomal genetic information. Note that this definition of exchange may be less stringent than that applied to exempt organisms under Section III - D - 4.

4. As classified in the Third Report of the International Committee on Taxonomy of Viruses: Classification and Nomenclature of Viruses, R.E.F. Matthews, ed. Intervirology 12 (129 - 296) 1979.

5. i.e., the total of all genomes within a Family shall not exceed one-half of the genome.

Appendix D – Actions Taken Under the Guidelines

As noted in the subsections of Section IV - C - 1 - b - (1), the Director, NIH, may take certain actions with regard to the Guidelines after the issues have been considered by the RAC. Some of the actions taken to date include the following:

Appendix D - I

Permission is granted to clone foot and mouth disease virus in the EK1 host-vector system consisting of *E. coli* K - 12 and the vector pBR322, all work to be done at the Plum Island Animal Disease Center.

Appendix D - II

Certain specified clones derived from segments of the foot and mouth disease virus may be transferred from Plum Island Animal Disease Center to the facilities of Genentech, Inc., of South San Francisco, California. Further development of the clones at Genentech has been approved under BL1 + EK1 conditions.

Appendix D - III

The Rd strain of *Hemophilus influenzae* can be used as a host for the propagation of the cloned Tn 10 tet R gene derived from *E. coli* K - 12 employing the non-conjugative *Hemophilus* plasmid, pRSF0885, under BL1 conditions.

Appendix D - IV

Permissions is granted to clone certain subgenomic segments of foot and mouth disease virus in HV1 *Bacillus subtilis* and *Saccharomyces cerevisiae* host-vector systems under BL1 conditions at Genentech, Inc., South San Francisco, California.

Appendix D - V

Permission is granted to Dr. Ronald Davis of Stanford University to field test corn plants modified by recombinant DNA techniques under specified containment conditions.

Appendix D - VI

Permission is granted to clone in *E. coli* K - 12 under BL1 physical containment conditions subgenomic segments of rift valley fever virus subject to conditions which have been set forth by the RAC.

Appendix D - VII

Attenuated laboratory strains of *Salmonella typhimurium* may be used under BL1 physical containment conditions to screen for the *Saccharomyces cerevisiae* pseudouridine synthetase gene. The plasmid YEp13 will be employed as the vector.

Appendix D - VIII

Permission is granted to transfer certain clones of subgenomic segments of foot and mouth disease virus from Plum Island Animal Disease Center to the laboratories of Molecular Genetics, Inc., Minnetonka, Minnesota, and to work with these clones under BL1 containment conditions. Approval is contingent upon review of data on infectivity testing of the clones by a working group of the RAC.

Appendix D - IX

Permission is granted to Dr. John Sanford of Cornell University to field test tomato and tobacco plants transformed with bacterial (*E. coli* K - 12) and yeast DNA using pollen as a vector.

Appendix D - X

Permission is granted to Drs. Steven Lindow and Nicholas Panopoulos of the University of California, Berkeley, to release under specified conditions *Pseudomonas syringae* pv. *syringae* and *Erwinia herbicola* carrying *in vitro* generated deletions of all or part of the genes involved in ice nucleation.

Appendix D - XI

Agracetus of Middleton, Wisconsin, may field test under specified conditions disease resistant tobacco conditions prepared by recombinant DNA techniques.

Appendix E – Certified Host-Vector Systems
(See also Appendix I)

While many experiments using *E. coli* K - 12, *Saccharomyces cerevisiae* and *Bacillus subtilis* are currently exempt from the Guidelines under Section III - D - 5, some derivatives of these host-vector systems were previously classified as HV1 or HV2. A listing of those systems follows:

Appendix E - 1 – Bacillus subtilis

HV1. The following plasmids are accepted as the vector components of certified *B. subtilis* HV1 systems: pUB110, pC194, pS194, pSA2100, pE194, pT127, pUB112, pC221, pC223, and pAB124. *B. subtilis* strains RUB 331 and BGSC 1S53 have been certified as the host component of HV1 systems based on these plasmids.
HV2. The asporogenic mutant derivative of *Bacillus subtilis*, ASB 298, with the following plasmids as the vector component: pUB110, pC194, pS194, pSA2100, pE194, pT127, pUB112, pC221, pC223, and pAB 124.

Appendix E - II – Saccharomyces cerevisiae

HV2. The following sterile strains of *Saccharomyces cerevisiae*, all of which have the ste - VC9 mutation, SHY1, SHY2, SHY3, and SHY4. The following plasmids are certified for use: YIp1, YEp2, YEp4, YIp5, YEp6, YRp7, YEp20, YEp21, YEp24, YIp25, YIp26, YIp27, YIp28, YIp29, YIp30, YIp32, and YIp33.

Appendix E - III – Escherichia coli

EK2 Plasmid Systems. The *E. coli* K - 12 strain chi-1776. The following plasmids are certified for use; pSC101, pMB9, pBR313, pBR322, pDH24, pBR325, pBR327, pGL101, and pHB1. The following *E. coli/S. cerevisiae* hybrid plasmids are certified as EK2 vectors when used in *E. coli* chi - 1776 or in the sterile yeast strains, SHY1, SHY2, SHY3, and SHY4; YIp1, YEp2, YEp4, YEp5, YEp6, YRp7, YEp20, YEp21, YEp24, YIp25, YIp26, YIp27, YIp28, YIp29, YEp30, YIp31, YIp32, and YIp33.
EK2 Bacteriophage Systems. The following are certified EK2 systems based on bacteriophage lambda:

Vector	Host
λgt*WES*λB'	DP50*supF*
λgt*WES*λB★	DP50*supF*
λgt*ZJvir*λB'	*E. coli* K - 12
λgtALOλB	DP50*supF*
Charon 3A	DP50 or DP50*supF*
Charon 4A	DP50 or DP50*supF*
Charon 16A	DP50 or DP50*supF*
Charon 21A	DP50*supF*
Charon 23A	DP50 or DP50*supF*
Charon 24A	DP50 or DP50*supF*

E. coli K - 12 strains chi - 2447 and chi2281 are certified for use with lambda vectors that are certified for use with strain DP50 or DP50*supF* provided that the su⁻ strain not be used as a propagation host.

Appendix E IV – Neurospora crassa

HV1. The following specified strains of *Neurospora crassa* which have been modified to prevent aerial dispersion:
Inl (inositolless) strains 37102, 37401, 46316, 64001, and 89601.
Csp - 1 strain UCLA37 and csp - 2 strains FS 590, UCLA 101 (these are conidial separation mutants).
Eas strain UCLA191 (an "easily wettable" mutant).

Appendix E - V – Streptomyces

HV1. The following *Streptomyces* species: *Streptomyces coelicolor*, *S. lividans*, *S. parvulus*, and *S. griseus*. The following are accepted as vector components of certified *Streptomyces* HV1 systems: *Streptomyces* plasmids SCP2, SLP1.2, pIJ101, actionophage phi C31, and their derivatives.

Appendix E - VI – Pseudomonas putida

HV1. *Pseudomonas putida* strains KT2440 with plasmid vectors pKT262, pKT263, and pKT264.

Appendix F – Containment Conditions for Cloning of Genes Coding for the Biosynthesis of Molecules Toxic for Vertebrates

Appendix F - 1 – General Information.

Appendix F specifies the containment to be used for the deliberate cloning of genes coding for the biosynthesis of molecules toxic for vertebrates. The cloning of genes coding for molecules toxic for vertebrates that have an LD_{50} of less than 100 nanograms per kilogram body weight (e.g., microbial toxins such as the botulinum toxins, tetanus toxin, diphtheria toxin, *Shigella dysenteriae* neurotoxin) is covered under Section III - A - 1 of the Guidelines and requires RAC review and NIH and IBC approval before initiation. No specific restrictions shall apply to the cloning of genes if the protein specified by the gene has an LD_{50} of 100 micrograms or more per kilogram of body weight. Experiments involving genes coding for toxic molecules with an LD_{50} of 100 micrograms or less per kilogram body weight shall be registered with ORDA prior to initiating the experiments. A list of toxic molecules classified as to LD_{50} is available from ORDA. Testing precedures for determining toxicity of toxic molecules not on the list are available from ORDA. The results of such tests shall be forwarded to ORDA which will consult with the RAC Working Group on Toxins prior to inclusion of the molecules on the list (see Section IV - C - 1 - b - (2) - (e)).

Appendix F - II – Containment Conditions for Cloning of Toxic Molecule Genes in E. coli K - 12

Appendix F - II - A. Cloning of genes coding for molecules toxic for vertebrates that have an LD_{50} in the range of 100 nanograms to 1000 nanograms per kilogram body weight (e.g., abrin, *Clostridium perfringens* epsilon toxin) may proceed under BL2 + EK2 or BL3 + EK1 containment conditions.
Appendix F - II - B. Cloning of genes for the biosynthesis of molecules toxic for vertebrates with an LD_{50} in the range of 1 microgram to 100 micrograms per kilogram body weight may proceed under BL2 + EK1 containment conditions (e.g., *Staphylococcus* aureus alpha toxin, *Staphylocossus aureus* beta toxin, ricin, *Pseudomonas aeruginosa* exotoxin A, *Bordatella pertussis* toxin, the lethal factor of *Bacillus anthracis*, the *Pasteurella pestis* murine toxins, the oxygen-labile hemolysins such as streptolysin O, and certain neurotoxins present in snake venoms and other venoms).
Appendix F - II - C. Some enterotoxins are substantially more toxic when administered enterally than parenterally. The following enterotoxins shall be subject to BL1 + EK1 containment conditions: cholera toxin, the heat labile toxins of *E. coli*, *Klebsiella*, and other related proteins that may be identified by neutralization with an antiserum monospecific for cholera toxin, and the heat stable toxins of *E. coli* and of *Yersinia enterocolitica*.

Appendix F-III – Containment Conditions for Cloning of Toxic Molecule Genes in Organisms Other Than E. coli K-12

Requests involving the cloning of genes coding for molecules toxic for vertebrates in host-vector systems other than *E. coli* K-12 will be evaluated by ORDA which will consult with the Working Group on Toxins (see Section IV-C-1-b-(3)-(f)).

Appendix F-IV – Specific Approvals

Appendix F-IV-A. Permission is granted to clone the Exotoxin A gene of *Pseudomonas aeruginosa* under BL1 conditions in *Pseudomonas aeruginosa* and in *Pseudomonas putida.*

Appendix F-IV-B. The pyrogenic exotoxin type A (Tox A) gene of *Staphylococcus aureus* may be cloned in an HV2 *Bacillus subtilis* host-vector system under BL3 containment conditions.

Appendix F-IV-C. Restriction fragments of *Corynephage Beta* carrying the structural gene for diphtheria toxin may be safely cloned in *e. coli* K-12 in high containment Building 550 at the Frederick Cancer Research Facility. Laboratory practices and containment equipment are to be specified by the IBC. If the investigators wish to proceed with the experiments, a prior review will be conducted to advise NIH whether the proposal has sufficient scientific merit to justify the use of the NIH BL4 facility.

Appendix F-IV-D. The genes coding for the *Staphylococcus aureus* determinants, A, B, and F, which may be implicated in toxic shock syndrome may be cloned in *E. coli* K-12 under BL2 + EK1 conditions. The *Staphylococcus aureus* strain used as the donor is to be alpha toxin minus. It is suggested that, if possible, the donor *Staphylococcus aureus* strain should lack oder toxins with LD$_{50}$s in the range of one microgram per kilogram body weight such as the exfoliative toxin.

Appendix F-IV-E. Fragments F-1, F-2, and F-3 of the diphtheria toxin gene (tox) may be cloned in *E. coli* K-12 under BL1 + EK1 containment conditions and may be cloned in *Bacillus Subtilis* host-vector systems under BL1 containment conditions. Fragment F-1 and fragment F-2 both contain: (i) Some or all of the transcriptional control elements of *tox*; (ii) the signal peptide; and (iii) fragment A (the center responsible for ADP-ribosylation of elongation factor 2). Fragment F-3 codes for most of the non-toxic fragment B of the toxin and contains no sequences coding for any portion of the enzymatically active fragment A moiety.

Appendix F-IV-F. The gene(s) coding for a toxin (designated LT-like) isolated from *E. coli* which is similar to the *E. coli* heat labile enterotoxin (LT) with respect to its activities and mode of action but is not neutralized by antibodies against cholera enterotoxin or against LT from human or porcine *E. coli* strains, and sequences homologous to the *E. coli* LT-like toxin gene may be cloned under BL1 + EK1 conditions.

Appendix F-IV-G. Genes from *Vibrio fluvialis*, *Vibrio mimicus*, and non 0-1 *Vibrio cholerae*, specifying virulence factors for animals, may be cloned under BL1 + EK1 conditions. The virulence factors to be cloned will be selected by testing fluid induction in suckling mice and Y-1 mouse adrenal cells.

Appendix F-IV-H. The intact structural gene(s) of the Shiga-like toxin from bacterial species classified in the families *Enterobacteriaceae* or *Vibrionaceae* including *Campylobacter* species may be cloned in *E. coli* K-12 under BL3 + EK1 containment conditions.

E. coli host-vector systems expressing the Shiga-like toxin gene product may be moved from BL3 + EK1 to BL2 + EK1 containment conditions provided that: (1) The amount of toxin produced by the modified host-vector systems be no greater than that produced by the positive control strain *Shigella dysenteriae* 60R, grown and measured under optimal conditions; and (2) the cloning vehicle is to be an EK1 vector preferably belonging to the class of poorly mobilizable plasmids such as pBR322, pBR328, and pBR325.

Nontoxinogenic fragments of the Shiga-like toxin structural gene(s) may be moved from BL3 + EK1 to BL2 + EK1 containment conditions or such nontoxic fragments may be directly cloned in *E. coli* K-12 under BL2 + EK1 conditions provided that the *E. coli* host-vector systems containing the fragments do not contain overlapping fragments which together would encompass the Shigalike toxin structural gene(s).

Appendix F-IV-I. A hybrid gene in which to gene coding for the melanocyte stimulating hormone (MSH) is joined to a segment of the gene encoding diphtheria toxin may be safely propagated in *E. coli* K-12 under BL4 containment in high containment building 550 at the Frederick Cancer Research Facility. If the investigators wish to proceed with the experiment, a prior review will be conducted to advise NIH whether the proposal has sufficient scientific merit to justify the use of the NIH BL4 facility. Before any of the strains may be removed from the BL4 facility, data on their safety shall be evaluated by the Working Group in Toxins and the working group recommendation shall be acted upon by NIH.

Appendix F-IV-J. The gene segment encoding the A subunit of cholera toxin of *Vibrio cholerae* may be joined to the transposons Tn5 and Tn5-131 and the A-subunit: Tn5-131 hybrid gene cloned in *E. coli* K-12 and *V. cholerae* under BL1 containment conditions.

Appendix F-IV-K. A hybrid gene in which the gene coding for interleukin 2 (IL-2) is joined to a specific segment of the gene encoding diphtheria toxin may be propagated in *E. coli* K-12 host-vector systems under BL2 containment plus BL3 practices, with the use of poorly mobilizable plasmid vectors such as EK2 certified plasmids.

Appendix G – Physical Containment

Appendix G-I – Standard Practices and Training

The first principle of containment is a strict adherence to good microbiological practices [1-10]. Consequently, all personnel directly or indirectly involved in experiments on recombinant DNAs must receive adequate instruction (see Sections IV-B-1-e and IV-B-5-d). This shall, as a minimum, include instructions in aseptic techniques and in the biology of the organisms used in the experiments so that the potential biohazards can be understood and appreciated.

Any research group working with agents with a known or potential biohazard shall have an emergency plan which describes the procedures to be followed if an accident contaminates personnel or the environment. The PI must ensure that everyone in the laboratory is familiar with both the potential hazards of the work and the emergency plan (see Sections IV-B-3-d and IV-B-5-e). If a research group is working with a known pathogen for which there is an effective vaccine, the vaccine should be made available to all workers. Where serological monitoring is clearly appropriate, it shall be provided (see Section IV-B-1-f).

The "Laboratory Safety Monograph" and *Biosafety in Microbiological and Biomedical Laboratories* [2] booklets describe practices, equipment, and facilities in detail.

Appendix G-II – Physical Containment Levels

The objective of physical containment is to confine organisms containing recombinant DNA molecules and thus to reduce the potential for exposure of the laboratory worker, persons outside of the laboratory, and the environment to organisms containing recombinant DNA molecules. Physical containment is achieved through the use of laboratory practices, containment equipment, and special laboratory design. Emphasis is placed on primary means of physical containment which are provided by laboratory practices and containment equipment. Special laboratory design provides a sec-

ondary means of protection against the accidental release of organisms outside the laboratory or to the environment. Special laboratory design is used primarily in facilities in which experiments of moderate to high potential hazards are performed.

Combinations of laboratory practices, containment equipment, and special laboratory design can be made to achieve different levels of physical containment. Four levels of physical containment, which are designated as BL1, BL2, BL3, and BL4, are described. It should be emphasized that the descriptions and assigments of physical containment detailed below are based on existing approaches to containment of pathogenic organisms [2]. The National Cancer institute describes three levels for research on oncogenic viruses which roughly correspond to our BL2, BL3, and BL4 levels [3].

It is recognized that several different combinations of laboratory practices, containment equipment, and special laboratory design may be appropriate for containment of specific research activities. The Guidelines, therefore, allow alternative selections of primary containment equipment within facilities that have been designed to provide BL3 and BL4 levels of physical containment. The selection of alternative methods of primary containment is dependent, however, on the level of biological containment provided by the host-vector system used in the experiment.

Consideration will also be given by the Director, NIH, with the advice of the RAC to other combinations which achieve an equivalent level of containment (see Section IV-C-1-b-(2)-(b)).

Appendix G-II-A – Biosafety Level 1 (BL1) [13]

Appendix G-II-A-1. Standard Microbiological Practices.

Appendix G-II-A-1-a. Access to the laboratory is limited or restricted at the discretion of the laboratory director when experiments are in progress.

Appendix G-II-A-1-b. Work surfaces are decontaminated once a day and after any spill of viable material.

Appendix G-II-A-1-c. All contaminated liquid or solid wastes are decontaminated before disposal.

Appendix G-II-A-1-d. Mechanical pipetting devices are used; mouth pipetting is prohibited.

Appendix G-II-A-1-e. Eating, drinking, smoking, and applying cosmetics are not permitted in the work area. Food may be stored in cabinets or refrigerators designated and used for this purpose only.

Appendix G-II-A-1-f. Persons wash their hands after they handle materials involving organisms containing recombinant DNA molecules, and animals, and before leaving the laboratory.

Appendix G-II-A-1-g. All procedures are performed carefully to minimize the creation of aerosols.

Appendix G-II-A-1-h. It is recommended that laboratory coats, gowns, or uniforms be worn to prevent contamination or soiling of street clothes.

Appendix G-II-A-2 – Special Practices

Appendix G-II-A-2-a. Contaminated materials that are to be decontaminated at a site away from the laboratory are placed in a durable leakproof container which is closed before being removed from the laboratory.

Appendix G-II-A-2-b. An insect and rodent control program is in effect.

Appendix G-II-A-3 – Containment Equipment

Appendix G-II-A-3-a. Special containment equipment is generally not required for manipulations of agents assigned to Biosafety Level 1.

Appendix G - II - A - 4 – Laboratory Facilities

Appendix G - II - A - 4 - a. The laboratory is designed so that it can be easily cleaned.

Appendix G - II - A - 4 - b. Bench tops are impervious to water and resistant to acids, alkalis, organic solvents, and moderate heat.

Appendix G - II - A - 4 - c. Laboratory furniture is sturdy. Spaces between benches, cabinets, and equipment are accessible for cleaning.

Appendix G - II - A - 4 - d. Each laboratory contains a sink for hand-washing.

Appendix G - II - A - 4 - e. If the laboratory has windows that open, they are fitted with fly screens.

Appendix G - II - B – Biosafety Level 2 (BL2) [14]

Appendix G - II - B - 1. Standard Microbiological Practices.

Appendix G - II - B - 1 - a. Access to the laboratory is limited or restricted by the laboratory director when work with organisms containing recombinant DNA molecules is in progress.

Appendix G - II - B - 1 - b. Work surfaces are decontaminated at least once a day and after any spill of viable material.

Appendix G - II - B - 1 - c. All contaminated liquid or solid wastes are decontaminated before disposal.

Appendix G - II - B - 1 - d. Mechanical pipetting devices are used; mouth pipetting is prohibited.

Appendix G - II - B - 1 - e. Eating, drinking, smoking, and applying cosmetics are not permitted in the work area. Food may be stored in cabinets or refrigerators designated and used for this purpose only.

Appendix G - II - B - 1 - f. Persons wash their hands after handling materials involving organisms containing recombinant DNA molecules, and animals, and when they leave the laboratory.

Appendix G - II - B - 1 - g. All procedures are performed carefully to minimize the creation of aerosols.

Appendix G - II - B - 1 - h. Experiments of lesser biohazard potential can be carried out concurrently in carefully demarcated areas of the same laboratory.

Appendix G - II - B - 2 – Special Practices

Appendix G - II - B - 2 - a. Contaminated materials that are to be decontaminated at a site away from the laboratory are placed in a durable leakproof container which is closed before being removed from the laboratory.

Appendix G - II - B - 2 - b. The laboratory director limits access to the laboratory. The director has the final responsibility for assessing each circumstance and determining who may enter or work in the laboratory.

Appendix G - II - B - 2 - c. The laboratory director establishes policies and procedures whereby only persons who have been advised of the potential hazard and meet any specific entry requirements (e.g., immunization) enter the laboratory or animal rooms.

Appendix G - II - B - 2 - d. When the organisms containing recombinant DNA molecules in use in the laboratory require special provisions for entry (e.g., vaccination), a hazard warning sign incorporating the universal biohazard symbol is posted on the access door to the laboratory work area. The hazard warning sign identifies the agent, lists the name and telephone number of the laboratory director or other responsible person(s), and indicates the special requirement(s) for entering the laboratory.

Appendix G - II - B - 2 - e. An insect and rodent control program is in effect.

Appendix G - II - B - 2 - f. Laboratory coats, gowns, smocks, or uniforms are worn while in the laboratory. Before leaving the laboratory for nonlaboratory areas (e.g., cafeteria, library, administrative offices), this protective clothing is removed and left in the laboratory or covered with a clean coat not used in the laboratory.

Appendix G - II - B - 2 - g. Animals not involved in the work being performed are not permitted in the laboratory.

Appendix G - II - B - 2 - h. Special care is taken to avoid skin contamination with organisms containing recombinant DNA molecules; gloves should be worn when handling experimental animals and when skin contact with the agent is unavoidable.

Appendix G - II - B - 2 - i. All wastes from laboratories and animal rooms are appropriately decontaminated before disposal.

Appendix G - II - B - 2 - j. Hypodermic needles and syringes are used only for parenteral injection and aspiration of fluids from laboratory animals and diaphragm bottles. Only needle-locking syringes or disposable syringe-needle units (i.e., needle is integral to the syringe) are used for the injection or aspiration of fluids containing organisms that contain recombinant DNA molecules. Extreme caution should be used when handling needles and syringes to avoid autoinoculation and the generation of aerosols during use and disposal. Needles should not be bent, sheared, replaced in the needle sheath or guard, or removed from the syringe should be promptly placed in a puncture-resistant container and decontaminated, preferably by autoclaving, before discard or reuse.

Appendix G - II - B - 2 - k. Spills and accidents which result in overt exposures to organisms containing recombinant DNA molecules are immediately reported to the laboratory director. Medical evaluation, surveillance, and treatment are provided as appropriate and written records are maintained.

Appendix G - II - B - 2 - l. When appropriate, considering the agent(s) handled, baseline serum samples for laboratory and other at-risk personnel are collected and stores. Additional serum specimens may be collected periodically depending on the agents handled or the function of the facility.

Appendix G - II - B - 2 - m. A biosafety manual is prepared or adopted. Personnel are advised of special hazards and are required to read instructions on practices and procedures and to follow them.

Appendix G - II - B - 3 – Containment Equipment

Appendix G - II - B - 3 - a. Biological safety cabinets (Class I or II) (see Appendix G - III - 12) or other appropriate personal protective or physical containment devices are used whenever:

Appendix G - II - B - 3 - a - (1). Procedures with a high potential for creating aerosols are conducted [15]. These may include centrifuging, grinding, blending, vigorous shaking or mixing, sonic disruption, opening containers of materials whose internal pressures may be different from ambient pressures, inoculating animals intranasally, and harvesting infected tissues from animals or eggs.

Appendix G - II - B - 3 - a - (2). High concentrations or large volumes of organisms containing recombinant DNA molecules are used. Such materials may be centrifuged in the open laboratory if sealed heads or centrifuge safety cups are used and if they are opened only in a biological safety cabinet.

Appendix G - II - B - 4. – Laboratory Facilities

Appendix G - II - B - 4 - a. The laboratory is designed so that it can be easily cleaned.

Appendix G - II - B - 4 - b. Bench tops are impervious to water and resistant to acids, alkalis, organic solvents, and moderate heat.

Appendix G - II - B - 4 - c. Laboratory furniture is sturdy and spaces accessible between benches, cabinets, and equipment are accessible for cleaning.

Appendix G - II - B - 4 - d. Each laboratory contains a sink for handwashing.

Appendix G - II - B - 4 - e. If the laboratory has windows that open, they are fietted with fly screens.

Appendix G - II - B - 4 - f. An autoclave for decontaminating laboratory wastes is available.

Appendix G - II - C. – Biosafety Level 3 (BL3) [16]

Appendix G - II - C - 1. Standard Microbiological Practices.

Appendix G - II - C - 1 - a. Work surfaces are decontaminated at least once a day and after any spill of viable material.

Appendix G - II - C - 1 - b. All contaminated liquid or solid wastes are decontaminated before disposal.

Appendix G - II - C - 1 - c. Mechanical pipetting devices are used; mouth pipetting is prohibited.

Appendix G - II - C - 1 - d. Eating, drinking, smoking, storing food, and applying cosmetics are not permitted in the work area.

Appendix G - II - C - 1 - e. Persons wash their hands after handling materials involving organisms containing recombinant DNA molecules and animals, and when they leave the laboratory.

Appendix G - II - C - 1 - f. All procedures are performed carefully to minimize the creation of aerosols.

Appendix G - II - C - 1 - g. Persons under 16 years of age shall not enter the laboratory.

Appendix G - II - C - 1 - h. If experiments involving other organisms which require lower levels of containment are to be conducted in the same laboratory concurrently with experiments requiring BL3 level physical containment, they shall be conducted in accordance with all BL3 level laboratory practices.

Appendix G - II - C - 2 – Special Pracites

Appendix G - II - C - 2 - a. Laboratory doors are kept closed when experiments are in progress.

Appendix G - II - C - 2 - b. Contaminated materials that are to be decontaminated at a site away from the laboratory are placed in a durable leakproof container which is closed before being removed from the laboratory.

Appendix G - II - C - 2 - c. The laboratory director controls access to the laboratory and restricts access to persons whose presence is required for program or support purposes. The director has the final responsibility for assessing each circumstance and determining who may enter or work in the laboratory.

Appendix G - II - C - 2 - d. The laboratory director establishes policies and procedures whereby only persons who have been advised of the potential biohazard, who meet any specific entry requirements (e.g., immunization), and who comply with all entry and exit procedures enter the laboratory or animals rooms.

Appendix G - II - C - 2 - e. When organisms containing recombinant DNA molecules or experimental animals are present in the laboratory or containment module, a hazard warning sign incorporation the universal biohazard symbol is posted on all laboratory and animal room access doors. The hazard warning sign identifies the agent, lists the name and telephone number of the laboratory director or other responsible person(s), and indicates any special requirements for entering the laboratory, such as the need for immunizations, respirators, or other personal protective measures.

Appendix G - II - C - 2 - f. All activities involving organisms containing recombinant DNA molecules are conducted in biological safety cabinets or other physical containment devices within the containment module. No work in open vessels is conducted on the open bench.

Appendix G - II - C - 2 - g. The work surfaces of biological safety cabinets and other containment equipment are decontaminated when work with organisms containing recombinant DNA molecules is finished. Plastic-backed paper toweling used on nonperforated work surfaces within biological safety cabinets facilitates clean-up.

Appendix G - II - C - 2 - h. An insect and rodent program is in effect.

Appendix G - II - C - 2 - i. Laboratory clothing that protects street clothing (e.g., solid front or wrap-around gowns, scrub suits, coveralls) is worn in the laboratory. Laboratory clothing is not worn outside the laboratory, and it is decontaminated before being laundered.

Appendix G - II - C - 2 - j. Special care is taken to avoid skin contamination with contaminated materials; gloves should be worn when handling infected animals and when skin contact with infectious materials is unavoidable.

Appendix G - II - C - 2 - k. Molded surgical masks or respirators are worn in rooms containing experimental animals.

Appendix G - II - C - 2 - l. Animals and plants not related to the work being conducted are not permitted in the laboratory.

Appendix G - II - C - 2 - m. Laboratory animals held in a BL3 area shall be housed in partial-containment caging systems, such as Horsfall units [11], open cages placed in ventilated enclosures, solid-wall and -bottom cages covered by filter bonnets, or solid-wall and -bottom cages placed on holding racks equipped with ultraviolet in radiation lamps and reflectors.

Note. – Conventional caging systems may be used provided that all personnel wear appropriate personal protective devices. These shall include at a minimum wrap-around gowns, head covers, gloves, shoe covers, and respirators. All personnel shall shower on exit from areas where these devices are required.

Appendix G - II - C - 2 - n. All wastes from laboratories and animal rooms are appropriately decontaminated before disposal.

Appendix G - II - C - 2 - o. Vacuum lines are protected with high efficiency particulate air (HEPA) filters and liquid disinfectant traps.

Appendix G - II - C - 2 - p. Hypodermic needles and syringes are used only for parenteral injection and aspiration of fluids from laboratory animals and diaphragm bottles. Only needle-locking syringes or disposable syringe-needle units (i.e., needle is integral to the syringe) are used for the injection or aspiration of fluids containing organisms that contain recombinant DNA molecules. Extreme caution should be used when handling needles and syringes to avoid autoinoculation and the generation of aerosols during use and disposal. Needles should not be bent, sheared, replaced in the needle sheath or guard or removed from the syringe following use. The needle and syringe should be promptly placed in a puncture-resistant container and autoclaving, before discard or reuse.

Appendix G - II - C - 2 - q. Spills and accidents which result in overt or potential exposures to organisms containing recombinant DNA molecules are immediately reported to the laboratory director. Appropriate medical evaluation, surveillance, and treatment are provided and written records are maintained.

Appendix G - II - C - 2 - r. Baseline serum samples for all laboratory and other at-risk personnel should be collected and stored. Additional serum specimens may be collected periodically depending on the agents handled or the function of the laboratory.

Appendix G - II - C - 2 - s. A biosafety manual is prepared or adopted. Personnel are advised of special hazards and are required to read instructions on practices and procedures and to follow them.

Appendix G - II - C - 2 - t. Alternative Selection of Containment Equipment. Experimental procedures involving a host-vector system that provides a one-step higher level of biological containment than that specified can be conducted in the BL3 laboratory using containment equipment specified for the BL2 level of physical containment. Experimental procedures involving a host-vector system that provides a one-step lower level of biological containment than that specified can be conducted in the BL3 laboratory using containment equipment specified for the BL4 level of physical containment. Alternative combination of containment safeguards are shown in Table 1.

Appendix G - II - C - 3 – Containment Equipment

Appendix G - II - C - 3 - a. Biological safety cabinets (Class I, II, or III) (see Appendix G - III - 12) or other appropriate combinations of personal protective or physical containment devices (e.g., special protective clothing, masks, gloves, respirators, centrifuge safety cups, sealed centrifuge rotors, and containment caging for animals) are used for all activities with organisms containing recombinant DNA molecules which pose a threat of aerosol exposure. These include: manipulation of cultures and of those clinical or environmental materials which may be a source of aerosols; the aerosol challenge of experimental animals; and harvesting infected tissues or fluids from experimental animals and ambryonate eggs; and necropsy of experimental animals.

Appendix G - II - C - 4 – Laboratory Facilities

Appendix G - II - C - 4 - a. The laboratory is separated from areas which are open to unrestricted traffic flow within the building. Passage through two sets of doors is the basic requirement for entry into the laboratory from access corridors or other contiguous areas. Physical separation of the high containment laboratory from access corridors or other laboratories or activities may also be provided by a double-doored clothes change room (showers may be included), airlock, or other access facility which requires passage through two sets of doors before entering the laboratory.

Appendix G - II - C - 4 - b. The interior surfaces of walls, floors, and ceilings are water resistant so that they can be easily cleaned. Penetrations in these surfaces are sealed or capable of being sealed to facilitate decontaminating the area.

Appendix G - II - C - 4 - c. Bench tops are impervious to water and resistant to acids, alkalis, organic solvents, and moderate heat.

Appendix G - II - C - 4 - d. Laboratory furniture is sturdy and spaces between benches, cabinets, and equipment are accessible for cleaning.

Appendix G - II - C - 4 - e. Each laboratory contains a sink for hand-washing. The sink is foot, elbow, or automatically operated and is located near the laboratory exit door.

Appendix G - II - C - 4 - f. Windows in the laboratory are closed and sealed.

Appendix G - II - C - 4 - g. Access doors to the laboratory or containment module are self-closing.

Appendix G - II - C - 4 - h. An autoclave for decontaminating laboratory wastes is available preferably within the laboratory.

Appendix G - II - C - 4 - i. A ducted exhaust air ventilation system is provided. This system creates directional airflow that draws air into the laboratory through the entry area. The exhaust air is not recirculated to any other area of the building, is discharged to the outside, and is dispersed away from the occupied areas and air intakes. Personnel must verify that the direction of the airflow (into the laboratory) is proper. The exhaust air from the laboratory room can be discharged to the outside without being filtered or otherwise treated.

Appendix G - II - C - 4 - j. The HEPA-filtered exhaust air from Class I or Class II biological safety cabinets is discharged directly to the outside or through the building exhaust system. Exhaust air from Class I or II biological safety cabinets may be recirculated within the laboratory if the cabinet is tested and certified at least every twelve months. If the HEPA-filtered exhaust air from Class I or II biological safety cabinets is to be discharged to the outside through the Building exhaust air system, it is connected to this system in a manner (e.g., thimble unit connection [12]) that avoids any interference with the air balance of the cabinets or building exhaust system.

Appendix G - II - D – Biosafety Level 4 (BL4)

Appendix G - II - D - 1. Standard Microbiological Practices.

Appendix G - II - D - 1 - a. Work surfaces are decontaminated at least once a day and immediately after any spill of viable material.

Appendix G - II - D - 1 - b. Only mechanical pipetting devices are used.

Appendix G - II - D - 1 - c. Eating, drinking, smoking, storing food, and applying cosmetics are not permitted in the laboratory.

Appendix G - II - D - 1 - d. All procedures are performed carefully to minimize the creation of aerosols.

Appendix G - II - D - 2 – Special Practices

Appendix G - II - D - 2 - a. Biological materials to be removed from the Class III cabinets or from the maximum containment laboratory in a viable or intact state are transferred to a nonbreakable, sealed primary container and then enclosed in a nonbreakable sealed secondary container which is removed from the facility through a disinfectant dunk tank, fumigation chamber, or an airlock designed for this pupose.

Appendix G - II - D - 2 - b. No materials, except for biological materials that are to remain in a viable or intact state, are removed from the maximum containment laboratory unless they have been autoclaved or decontaminated before they leave the facility. Equipment or material which might be damaged by high temperatures or steam is decontaminated by gaseous or vapor methods in an airlock or chamber designed for this purpose.

Appendix G - II - D - 2 - c. Only persons whose presence in the facility or individual laboratory rooms is required for program or support purposes are authorized to enter. The supervisor has the final responsibility for assessing each circumstance and determining who may enter or work in the laboratory. Access to the facility is limited by means of secure, locked doors; accessibility is managed by the laboratory director, biohazards control officer, or other person responsible for the physical security of the facility. Before entering, persons are advised of the potential biohazards and instructed as to appropriate safeguards for ensuring their safety. Authorized persons comply with the instructions and all other applicable entry and exit procedures. A logbook signed by all personnel indicates the date and time of each entry and exit. Practical and effective protocols for emergency situations are established.

Appendix G - II - D - 2 - d. Personnel enter and leave the facility only through the clothing change and shower rooms. Personnel shower each time they leave the facility. Personnel use the airlocks to enter or leave the laboratory only in an emergency.

Appendix G - II - D - 2 - e. Street clothing is removed in the outer clothing change room and kept there. Complete laboratory clothing, including undergarments, pants and shirts or jumpsuits, shoes, and gloves, is provided and used by all personnel entering the facility. Head covers are provided for personnel who do not wash their hair during the exit shower. When leaving the laboratory and before proceeding into the shower area, personnel remove their laboratory clothing and store it in a locker or hamper in the inner change room.

Appendix G - II - D - 2 - f. When materials that contain organisms containing recombinant DNA molecules or experimental animals are present in the laboratory or animal rooms, a hazard warning sign incorporating the universal biohazard symbol is posted on all access doors. The sign identifies the agent, lists the name of the laboratory director or other responsible person(s), and indicates any special requirements for entering the area (e.g., the need for immunizations or respirators).

Appendix G - II - D - 2 - g. Supplies and materials needed in the facility are brought in by way of the double-doored autoclave, fumigation chamber, or airlock which is appropriately decontaminated between each use. After securing the outer doors, personnel within the facility retrieve the materials by opening the interior doors of the autoclave, fumigation chamber, or airlock. These doors are secured after materials are brought into the facility.

Appendix G - II - D - 2 - h. An insect and rodent control program is in effect.

Appendix G - II - D - 2 - i. Materials (e.g., plants, animals, and clothing) not related to the experiment being conducted are not permitted in the facility.

Appendix G - II - D - 2 - j. Hypodermic needles and syringes are used only for parenteral injection and aspiration of fluids from laboratory animals and diaphragm bottles. Only needle-locking syringes or disposable syringe-needle units (i.e., needle is integral part of unit) are used for the injection or aspiration of fluids containing organisms that contain recombinant DNA molecules. Needles should not be bent, sheared, replaced in the needle sheath or guard, or removed from the syringe following use. The needle and syringe should be placed in a puncture-resistant container and decontaminated, preferably by autoclaving before discard or reuse. Whenever possible, cannulas are used instead of sharp needles (e.g., gavage).

Appendix G - II - D - 2 - k. A system is set up for reporting laboratory accidents and exposures and employee absenteeism and for the medical surveillance of potential laboratory-associated illnesses. Written records are prepared and maintained. An essential adjunct to such a reporting-surveillance system is the availability of a facility for quarantine, isolation, and medical care of personnel with potential or known laboratory associated illnesses.

Appendix G - II - D - 2 - l. Laboratory animals involved in experiments requiring BL4 level physical containment shall be housed either in cages contained in Class III cabinets or in partial containment caging systems (such as Horsfall units 11, open cages placed in ventilated enclosures, or solid-wall and -bottom cages placed on holding racks equipped with ultraviolet irradiation lamps and reflectors that are located in a specially designed area in which all personnel are required to wear one-piece positive pressure suits.

Appendix G - II - D - 2 - m. Alternative Selection of Containment Equipment. Experimental procedures involving a host-vector system that provides a one-step higher level of biological containment than that specified for the BL4 facility using containment equipment requirements specified for the BL3 level of physical containment. Alternative combinations of containment safeguards are shown in Table I.

Appendix G - II - D - 3. – Containment Equipment

Appendix G - II - D - 3 - a. All procedures within the facility with agents assigned to Biosafety Level 4 are conducted in the Class III biological safety cabinet or in Class I or II biological safety cabinets used in conjunction with one-piece positive pressure personnel suits ventilated by a life-support system.

Appendix G - II - D - 4. – Laboratory Facilities

Appendix G - II - D - 4 - a. The maximum containment facility consists of either a separate building or a clearly demarcated and isolated zone within a building. Outer and inner change rooms separated by a shower are provided for personnel entering and leaving the facility.

A double-doored autoclave, fumigation chamber, or ventilated airlock is provided for passage of those materials, supplies, or equipment which are not brought into the facility through the change room.

Appendix G - II - D - 4 - b. Walls, floors, and ceilings of the facility are constructed to form a sealed internal shell which facilitates fumigation and is animal and and insect proof. The internal surfaces of this shell are resistant to liquids and chemicals, thus facilitating cleaning and decontamination of the area. All penetrations in these structures and surfaces are sealed. Any drains in the floors contain traps filled with a chemical disinfectant of demonstrated efficacy against the target agent, and they are connected directly to the liquid waste decontamination system. Sewer and other ventilation lines contain HEPA filters.

Appendix G - II - D - 4 - c. Internal facility appurtenances, such as light fixtures, air ducts, and utility pipes, are arranged to minimize the horizontal surface area on which dust can settle.

Appendix G - II - D - 4 - d. Bench tops have seamless surfaces which are impervious to water and resistant to acids, alkalis, organic solvents, and moderate heat.

Appendix G - II - D - 4 - e. Laboratory furniture is of simple and sturdy construction, and spaces between benches, cabinets, and equipment are accessible for cleaning.

Appendix G - II - D - 4 - f. A foot, elbow, or automatically operated hand-washing sink is provided near the door of each laboratory room in the facility.

Appendix G - II - D - 4 - g. If there is a central vacuum system, it does not serve areas outside the facility. In-line HEPA filters are placed as near as practicable to each use point or service cock. Filters are installed to permit in-place decontamination and replacement. Other liquid and gas services to the facility are protected by devices that prevent backflow.

Appendix G - II - D - 4 - h. If water fountains are provided, they are foot operated and are located in the facility corridors outside the laboratory. The water service to the fountain is not connected to the backflow-protected distribution system supplying water to the laboratory areas.

Appendix G - II - D - 4 - i. Access doors to the laboratory are self-closing and lockable.

Appendix G - II - D - 4 - j. Any windows are breakage resistant.

Appendix G - II - D - 4 - k. A double-doored autoclave is provided for decontaminating materials passing out of the facility. The autoclave door which opens to the area external to the facility is sealed to the outer wall and automatically controlled so that the outside door can only be opened after the autoclave "sterilization" cycle has been completed.

Appendix G - II - D - 4 - l. A pass-through dunk tank, fumigation chamber, or an equivalent decontamination method is provided so that materials and equipment that cannot be decontaminated in the autoclave can be safely removed from the facility.

Appendix G - II - D - 4 - m. Liquid effluents from laboratory sinks, biological safety cabinets, floors, and autoclave chambers are decontaminated by heat treatment before being released from the maximum containment facility. Liquid wastes from shower rooms and toilets may be decontaminated with chemical disinfectants or by heat in the liquid waste decontamination system. The procedure used for heat decontamination of liquid wastes is evaluated mechanically and biologically by using a recording thermometer and an indicator microorganism with a defined heat susceptibility pattern. If liquid wastes from the shower room are decontaminated with chemical desinfectants, the chemical used is of demonstrated efficacy against the target or indicator microorganisms.

Appendix G - II - D - 4 - n. An individual supply and exhaust air ventilation system is provided. The system maintains pressure differentials and directional airflow as required to assure flows inward from areas outside of the facility toward areas of highest potential risk within the facility. Manometers are used to sense pressure differentials between adjacent areas maintained at different pressure levels. If a system malfunctions, the manometers sound an alarm. The supply and exhaust airflow is interlocked to assure inward (or zero) airflow at all times.

Appendix G - II - D - 4 - o. The exhaust air from the facility is filtered through HEPA filters and discharged to the outside so that it is dispersed away from occupied buildings and air intakes. Within the facility, the filters are located as near the laboratories as practicable in order to reduce the length of potentially contaminated air ducts. The filter chambers are designed to allow in *situ* decontamination before filters are removed and to facilitate certification testing after they are replaced. Coarse filters and HEPA filters are provided to treat air supplied to the facility in order to increase the lifetime of the exhaust HEPA filters and to protect the supply and exhaust HEPA filters and to protect the supply air system should air pressures become unbalanced in the laboratory.

Appendix G - II - D - 4 - p. The treated exhaust air from Class I and II biological safety cabinets can be discharged into the laboratory room environment or the outside through the facility air exhaust system. If exhaust air from Class I or II biological safety cabinets is discharged into the laboratory the cabinets are tested and certified at 6-month intervals. *The exhaust air from Class III biological safety cabinets is discharged, without recirculation through two sets of HEPA filters in series, via the facility exhaust air system.* If the treated exhaust air from any of these cabinets is discharged to the outside through the facility exhaust air system, it is connected to this system in a manner (e.g., thimble unit connection [12]) that avoids any interference with the air balance of the cabinets or the facility exhaust air system.

Appendix G - II - D - 4 - q. A specially designed suit area may be provided in the facility. Personnel who enter this area wear a one-piece positive pressure suit that is ventilated by a life-support system. The life-support system includes alarms and emergency backup breathing air tanks. Entry to this area is through an airlock fitted with airtight doors. A chemical shower is provided to decontaminate the surface of the suit before the worker leaves the area. The exhaust air from the suit area is filtered by two sets of HEPA filters installed in series. A duplicate filtration unit, exhaust fan, and an automatically starting emergency power source are provided. The air pressure within the suit area is lower than that of any adjacent area. Emergency lighting and communication systems are provided. All penetrations into the internal shell of the suit area are sealed. A double-doored autoclave is provided for decontaminating waste materials to be removed from the suit area.

Appendix G - III – Footnotes and References of Appendix G

1. *Laboratory Safety at the Center for Disease Control* (Sept. 1974). U.S. Department of Health Education and Welfare Publication No. CDC 75 - 8118.
2. *Biosefety in Microbiological and Biomedical Laboratories.* 1st Edition (March 1984). U.S. Department of Health and Human Services, Public Health Service, Centers for Disease Control, Atlanta, Georgia 30333,

TABLE I. – POSSIBLE ALTERNATE COMBINATIONS OF PHYSICAL AND BIOLOGICAL CONTAINMENT SAFEGUARDS

Classification of physical and biological containment	Alternate physical containment			Alternate biological containment
	Laboratory facilities	Laboratory practices	Containment equipment	
BL3/HV2.........	BL3	BL3	BL3	HV2
	BL3	BL3	BL4	HV1
BL3/HV1.........	BL3	BL3	BL3	HV1
	BL3	BL3	BL2	HV2
BL4/HV1.........	BL4	BL4	BL4	HV1
	BL4	BL4	BL3	HV2

and National Institutes of Health, Bethesda, Maryland 20205.

3. *National Cancer Institute Safety Standards for Research Involving Oncogenic Viruses* (Oct. 1974). U.S. Department of Health, Education and Welfare Publication No. (NIH) 75 - 790.

4. *National Institutes of Health Biohazards Safety Guide* (1974). U.S. Department of Health, Education, and Welfare, Public Health Service, National Institutes of Health, U.S. Government Printing Office, Stock No. 1740 - 00383.

5. *Biohazards in Biological Research* (1973). A. Hellmann, M.N. Oxman, and R. Pollack (cd.) Cold Spring Harbor Laboratory.

6. *Handbook of Laboratory Safety* (1971). 2nd. Edition. N.V. Steere (ed.). The Chemical Rubber Co., Cleveland.

7. Bodily, J.L. (1970). *General Administration of the Laboratory*, H.L. Bodily, E.L. Updyke, and J.O. Mason (eds.), Diagnostic Procedures for Bacterial, Mycotic and Parasitic Infections. American Public Health Association, New York, pp. 11 - 28.

8. Darlow, H.M. (1969). *Safety in the Microbiological Laboratory*. In J.R. Norris and D.W. Robbins (ed.), Methods in Microbiological Laboratory. *In J.R. Norris and D.W. Robbins (ed.), Methods in Microbiology, Academic Press, Incl., New York*, pp. 169 - 204.

9. The Prevention of Laboratory Acquired Infection (1974). *C.H. Collins, E.G. Hartley, and R. Pilsworth, Public Health Laboratory Service, Monograph Series No. 6.*

10. Chatigny, M.A. (1961). *Protection Against Infection in the Microbiological laboratory: Devices and Procedures.* In W.W. Umbreit (ed.). Advances in Applied Microbiology, Academic Press, New York, N.Y. 3:131 - 192.

11. Horsfall, F.L., Jr., and J.H. Baner (1940). *Individual Isolation of Infected Animals in a Single Room.* J. Bact. *40*, 569 - 580.

12. Biological safety cabinets referred to in this section are classified as *Class I, Class II,* or *Class III* cabinets. A *Class I* is a ventilated cabinet for personnel protection having an inward flow of air away from the operator. The exhaust air from this cabinet is filtered through a high-efficiency particulate air (HEPA) filter. This cabinet is used in three operational modes: (1) with a full-width open front, (2) with an installed front closure panel (having four 8-inch diameter openings) without gloves, and (3) with an installed front closure panel equipped with arm-length rubber gloves. The face velocity of the inward flow of air through the full-width open front is 75 feet per minute or greater.

A *Class II* cabinet is a ventilated cabinet for personnel and product protection having an open front with inward air flow for personnel protection, and HEPA filtered mass recirculated air flow for product protection. The cabinet exhaust air is filtered through a HEPA filter. The face velocity of the inward flow of air through the full-width open front is 75 feet per minute or greater. Design and performance specifications for *Class II* cabinets have been adopted by the National Sanitation Foundation. Ann Arbor, Michigan. A *Class III* cabinet is a closed-front ventilated cabinet of gas-tight construction which provides the highest level of personnel protection of all biohazard safety cabinets. The interior of the cabinet is protected from contaminants exterior to the cabinet. The cabinet is fitted with arm-length rubber gloves and is operated under a negative pressure of at least 0.5 inches water gauge. All supply air is filtered through HEPA filters. Exhaust air is filtered through two HEPA filters or one HEPA filter and incinerator before being discharged to the outside environment. National Sanitation Foundation Standard 49. 1976. Class II (Laminar Flow) Biohazard Cabinetry. Ann Arbor, Michigan.

13. Biosafety Level 1 is suitable for work involving agents of no known or minimal potential hazard to laboratory personnel and the environment. The laboratory is not separated from the general traffic patterns in the building. Work is generally conducted on open bench tops. Special containment equipment is not required or generally used. Laboratory personnel have specific training in the procedures conducted in the laboratory and are supervised by a scientist with general training in microbiology or a related science (see Appendix G - III - 2).

14. Biosafety Level 2 is similar to Level 1 and is suitable for work involving agents of moderate potential hazard to personnel and the environment. It differs in that: (1) laboratory personnel have specific training in handling pathogenic agents and are directed by competent scientists; (2) access to the laboratory is limited when work is being conducted; and (3) certain procedures in which infectious aerosols are created are conducted in biological safety cabinets or other physical containment equipment (see Appendix G - III - 2).

15. Office of Reseach Safety, National Cancer Institute, and the Special Committee of Safety and Health Experts, 1978. "Laboratory Safety Monograph: A Supplement to the NIH Guidelines for Recombinant DNA Research." Bethesda, Maryland, Natinal Institutes of Health.

16. Biosafety Level 3 is applicable to clinical, diagnostic, teaching, research, or production facilities in which work is done with indigenous or exotic agents which may cause serious or potentially lethal disease as a result of exposure by the inhalation route. Laboratory personnel have specific training in handling pathogenic and potentially lethal agents and are supervised by competent scientists who are experienced in working with these agents. All procedures involving the manipulation of infectious material are conducted within biological safety cabinets or other physical containment devices or by personnel wearing appropriate personal protective clothing and devices. The laboratory has special engineering and design features. It is recognized, however, that many existing facilities may not have all the facility safeguards recommended for Biosafety Level 3 (e.g., access zone, sealed penetrations, and directional airflow, etc.). In these circumstances, acceptable safety may be achieved for routine or repetitive operations (e.g., diagnostic procedures involving the propagation of an agent for identification, typing, and susceptibility testing) in laboratories where facility features satisfy Biosafety Level 2 recommendations provided the recommended "Standard Microbiological Practices," "Special Practices," and "Containment Equipment" for Biosafety Level 3 are rigorously followed. The decision to implement this modification of Biosafety Level 3 recommendations should be made only by the laboratory director (see Appendix G - III - 2).

Appendix H – Shipment

Recombinant DNA molecules contained in an organism or virus shall be shipped only as an etiologic agent under requirements of the U.S. Public Health Service, and the U.S. Department of Transportation (§ 72.3, Part 72, Title 42, and §§ 173.386 - .388, Part 173, Title 49, U.S. Code of Federal Regulations (CFR) as specified below:

Appendix H - I

Recombinant DNA molecules contained in an organism or virus requiring BL1, BL2, or BL3 physical containment, when offered for transportation or transported, are subject to all requirements of §§ 72.3(a) - (e), Part 72, Titel 42 CFR, and §§ 173.386 - .388, Part 173, Title 49 CFR.

Appendix H - II

Recombinant DNA molecules contained in an organism or virus requiring BL4 physical containment, when offered for transportation or transported, are subject to the requirements listed above under Appendix H - I and are also subject to § 72.3(f), Part 72, Title 42 CFR.

Appendix H - III

Information on packaging and labeling of etiologic agents is shown in Figures 1, 2, and 3. Additional information on packaging and shipment is given in the "Laboratory Safety Monograph - A Supplement to the NIH Guidelines for Recombinant DNA Research," available from the Office of Recombinant DNA Activities and in *Biosafety in Microbiological and Biomedical Laboratories* (see Appendix G - III - 2).

Appendix I – Biological Containment

(See also Appendix E)

Appendix I - I – Levels of Biological Containment.

In consideration of biological containment, the vector (plasmid, organelle, or virus) for the recombinant DNA and the host (bacterial, plant, or animal cell) in which the vector is propagated in the laboratory will be considered together. Any combination of vector and host which is to provide biological containment must be chosen or constructed so that the following types of "escape" are minimized: (i) Survival of the vector in its host outside the laboratory, and (ii) transmission of the vector from the propagation host to other nonlaboratory hosts.

The following levels of biological containment (HV, or *Host-Vector*, systems) for prokaryotes will be establishted; specific criteria will depend on the organisms to be used.

Appendix I - I - A. HV1. A host-vector system which provides a moderate level of containment. *Specific systems are:*

Appendix I - I - A - 1. EK1. The host is always *E. coli* K - 12 or a derivative thereof, and the vectors include nonconjugative plasmids (e.g., pSC101, ColEl, or derivatives thereof [1 - 7] and variants of bacteriophage, such as lambda [8 - 15]. The *E. coli* K - 12 hosts shall not contain conjugation-proficient plasmids, whether autonomous or integrated, or generalized transducing phages.

Appendix I - I - A - 2. Other HV1. Hosts and vectors shall be, at a minimum, comparable in containment to *E. coli* K - 12 with a non conjugative plasmid or bacteriophage vector. The data to be considered and a mechanism for approval of such HV1 systems are described below (Appendix I - II).

Appendix I - I - B HV2. These are host-vector systems shown on provide a high level of biological containment as demonstrated by data from suitable tests performed in the laboratory. Escape of the recombinant DNA either via survival of the organisms or via transmission of recombinant DNA to other organisms should be less than $1/10^8$ under specified conditions. Specific systems are:

Appendix I - I - B - 1. For EK2 host-vector systems in which the vector is a plasmid, no more than one in 10^8 host cells should be able to perpetuate a cloned DNA fragment under the specified nonpermissive laboratory conditions designed to represent the natural environment, either by survival of the original host or as a consequences of transmission of the cloned DNA fragment.

Appendix I - I - B - 2. For EK2 host-vector systems in which the vector is a phage, no more than one in 10^8 phage particles should be able to perpetuate a cloned DNA fragment under the specified nonpermissive laboratory conditions designed to represent the natural environment either: (i) as a prophage (in the inserted or plasmid form) in the laboratory host used for phage propagation or (ii) by surviving in natural environments and transferring a cloned DNA fragment to other hosts (or their resident prophages).

Appendix I - II – Certification of Host-Vector Systems

Appendix I - II - A. Responsibility. HV1 systems other than *E. coli* K - 12 and HV2 host-vector systems may not be designated as such until they have been certified by the Director, NIH. Application for certification of a host-vector system is made by written application to the Office of Recombinant DNA Activities, National Institutes of Health, Building 31, Room 3B10, Bethesda, Maryland 20892.

Host-vector systems that are proposed for certification will be reviewed by the RAC (see Section IV - C - 1 - b - (1) - (e)). This will first involve review of the data on construction, properties, and testing of the proposed host-vector system by a working group composed of one or more members of the RAC and other persons chosen because of their expertise in evaluating such data. The committee will then evaluate the report of the working group and any other available information at a regular review meeting. The Director, NIH, is responsible for certification after receiving the advice of the RAC. Minor modifications of existing certified host-vector systems where the modifications are of minimal or no consequence to the properties relevant to containment may be certified by the Director, NIH, without review by the RAC (see Section IV - C - 1 - b - (3) - (c)).

When new host-vector systems are certified, notice of the certification will be sent by ORDA to the applicant and to all IBCs and will be published in the *Recombinant DNA Technical Bulletin*. Copies of a list of all currently certified host-vector systems may be obtained from ORDA at any time.

The Director, NIH, may at any time rescind the certification of any host-vector system (see Section IV - C - 1 - b - (3) - (d)). If certification of a host-vector system is rescinded, NIH will instruct investigators to transfer cloned DNA into a different system or use the clones at a higher physical containment level unless NIH determines that the already constructed clones incorporate adequate biological containment.

Certification of a given system does not extend to modifications of either the host or vector component of that system. Such modified systems must be independently certified by the Director, NIH. If modifications are minor, it may only be necessary for the investigator to submit data showing that the modifications have either improved or not impaired the major phenotypic traits on which the containment of the system depends. Substantial modifications of a certified system require the submission of complete testing data.

Appendix I - II - B. Data to be Submitted for Certification.

Appendix I - II - B - 1. HV1 Systems Other than E. coli K - 12. The following types of data shall be submitted, modified as appropriate for the particular system under consideration: (i) A description of the organism and vector; the strain's natural habitat and growth requirements; its physiological properties, particularly those related to its reproduction and survival and the mechanisms by which it exchanges genetic information; the range of organisms with which this organism normally exchanges genetic information and what sort of information is exchanged; and any relevant information on its pathogenicity or toxicity; (ii) a description of the history of the particular strains and vectors to be used, including data on any mutations which render this organism less able to survive or transmit genetic information; and (iii) a general description of the range of experiments contemplated with emphasis on the need for developing such an HV1 system.

Appendix I - II - B - 2. HV2 Systems. Investigators planning to request HV2 certification for host-vector system can obtain instructions from ORDA concerning data to be submitted [14 - 15]. In general, the following types of data are required: (i) Description of construction steps with indication of source, properties, and manner of introduction of genetic traits; (ii) quantitative data on the stability of genetic traits that contribute to the containment of the system, (iii) data on the survival of the host-vector system under nonpermissive laboratory conditions designed to represent the relevant natural environment; (iv) data on transmissibility of the vector and/or a cloned DNA fragment under both permissive

and nonpermissive conditions; (v) data on all other properties of the system which affect containment and utility, including information on yields oh phage or plasmid molecules, ease of DNA isolation, and ease of transfection or transformation; and (vi) in some cases, the investigator may be asked to submit data on survival and vector transmissibility from experiments in which the host-vector is fed to laboratory animals and human subjects. Such *in vivo* data may be required to confirm the validity of predicting in vivo *survival on the basis of in vitro* experiments.

Data must be submitted in writing to ORDA. Ten to twelve weeks are normally required for review and circulation of the data prior to the meeting at which such data can be considered by the RAC. Investigators are encouraged to publish their data on the construction, properties, and testing of proposed HV2 systems prior to consideration of the system by the RAC and its subcommittee. More specific instructions concerning the type of data to be submitted to NIH for proposed EK2 systems involving either plasmids or bacteriophage in *E. coli* K - 12 are available from ORDA.

Appendix I - III – Footnotes and References of Appendix I

1. Hersfield, V., H.W. Boyer, C. Yanofsky, M.A. Lovett, and D.R. Helinski (1974). *Plasmid ColEl as a Molecular Vehicle for Cloning and Amplification of DNA.* Proc. Nat. Acad. Sci. USA *71*, 3455 - 3459.
2. Wensink, P.C., D.J. Finnegan, J.E. Donelson, and D.S. Hogness (1974). *A System for Mapping DNA Sequences in the Chromosomes of Drosophila Melanogaster.* Cell *3*, 315 - 335.
3. Tanaka, T., and B. Weisblum (1975). *Construction of a Colicin El-R Factor Composite Plasmid In Vitro: Means for Amplifications of Deoxyribonucleic Acid.* J. Bacteriol. *121*, 354 - 362.
4. Armstrong, K.A., V. Hershfield, and D.R. Helinski (1977). *Gene Cloning and Containment Properties of Plasmid Col El and Its Derivatives*, Science *196*, 172 - 174.
5. Bolivar, F., R.L. Rodriguez, M.C. Betlach, and H.W. Boyer (1977). *Construction and Characterization of New Cloning Vehicles: I. Ampicillin-Resistant Derivative of pMB9.* Gene *2*, 75 - 93.
6. Cohen, S.N., A.C.W. Chang, H. Boyer, and R. Helling (1973). *Construction of Biologically Functional Bacterial Plasmids in Vitro*, Proc. Natl. Acad, Sci. USA *70*, 3240 - 3244.
7. Bolivar, F., R.L. Rodriguez, R.J. Greene, M.C. Batlach, H.L. Reyneker, H.W. Boyer, J.H. Crosa, and S. Falkow (1977). *Construction and Characterization of New Cloning Vehicles: II. A Multi-Purpose Cloning System.* Gene *2*, 95 - 113.
8. Thomas, M., J.R. Cameron, and R.W. Davis (1974). *Viable Molecular Hybrids of Bacteriophage Lambda and Eukaryotic DNA.* Proc. Nat. Aca. Sci. USA *71*, 4579 - 1583.
9. Murray, N.E., and K. Murray (1974). *Manipulation of Restriction Targets in Phage Lambda to Form Receptor Chromosomes for DNA Fragments.* Nature *251*, 476 - 481.
10. Ramback. A., and P. Tiollais (1974). *Bacteriophage Having EcoRI Endonuclease Sites Only in the Non-Essential Region of the Genome.* Proce. Nat. Acad. Sci., USA *71*, 3927 - 3930.
11. Blattner, F.R., B.G. Williams, A.E. Bleche, K. Denniston-Thompson, H.E. Faber, L.A. Furlong, D.J. Gunwald, D.O. Kiefer, D.D. Moore, J.W. Shumm, E.L. Sheldon, and O. Smithies (1977). *Charon Phages: Safer Derivatives of Bacteriophage Lambda for DNA Cloning.* Science *196*, 163 - 169.
12. Donoghoue, D.J., and P.A. Sharp (1977). *An Improved Lambda Vector: Construction of Model Recombinants Coding for Kanamycin Resistance*, Gene *1*, 209 - 227.
13. Leder, P., D. Tiemeier, and L. Enquist (1977). *EK2 Derivatives of Bacteriophage Lambda Useful in the Cloning of DNA from Higher Organisms: The λgt WES System.* Science *196*, 175 - 177.
14. Skalka, A. (1978). *Current Status of Coliphage λ EK2 Vectors.* Gene *3*, 29 - 35.
15. Szybalski, W., A. Skalka, S. Gottesman, A. Campbell, and D. Botstein (1978). *Standardized Laboratory Tests for EK2 Certification.* Gene *3*, 36 - 38.

Appendix J – Biotechnology Science Coordinating Committee

The following excerpts from its charter (signed October 30, 1985) describe the Biotechnology Science Coordinating Committee:

Purpose

The Domestic Policy Working Group on Biotechnology has determined that in the area of biotechnology with its rapid growth of scientific discovery, scientific issues of interagency concern will arise frequently and need to be communicated among the various agencies involved with reviews of biotechnology applications. The Federal Coordinating Council for Science. Engineering, and Technology (FCCSET) established by 42 U.S.C. 6651 is an interagency science committee chaired by the Director of the Office of Science and Technology Policy with the mission of coordinating science activities affecting more than one agency. Committees may be established under FCCSET for addressing particular science issues. Thus, the Biotechnology Science Coordinating Committee (BSCC) is established to provide formally an opportunity for interagency science policy coordination and guidance and for the exchange of information regarding the scientific aspects of biotechnology applications submitted to federal research and regulatory agencies for approval.

Functions

The BSCC will coordinate interagency review of scientific issues related to the assessments and approval of biotechnology research applications and biotechnology product applications and postmarketing surveillance when they involve the use of recombinant RNA, recombinant DNA, cell fusion or similar techniques. The BSCC will:
(a) Serve as a coordinating forum for addressing scientific problems, sharing information, and developing consensus;
(b) Promote consistency in the development of Federal agencies' review procedures and assessments;
(c) Facilitate continuing cooperation among Federal agencies on emerging scientific issues; and
(d) Identify gaps in scientific knowledge.

Authority

To accomplish these functions the BSCC is authorized to:
(a) Receive documentation from agencies necessary for the performance of its function;
(b) Conduct analyses of broad scientific issues that extend beyond those of any one agency;
(c) Develop generic scientific recommendations that can be applied to similar, recurring applications;
(d) Convene workshops, symposia, and generic research projects related to scientific issues in biotechnology; and
(e) Hold periodic public meetings.

Members and Chairman

The BSCC includes the following initial members:
Department of Agriculture
  Assistant Secretary for Marketing and Inspection Services

Assistant Secretary for Science and Education
Department of Health and Human Services
  Commissioner, Food and Drug Administration
  Director, National Institutes of Health
Environmental Protection Agency
  Assistant Administrator for Pesticides and Toxic Substances
  Assistant Administrator for Research and Development
National Science Foundation
  Assistant Director of Biological, Behavorial & Social Sciences
The BSCC is chaired by the Assistant Director for Biological, Behavioral and Social Sciences of the National Science Foundation and the Director of the National Institutes of Health on a rotating basis.

Administrative Provisions

(a) The BSCC will report to the FCCSET through the Chair.
(b) Meetings of the BSCC shall be held periodically. Some public meetings will be held.
(c) Confidential business information and proprietary information shall be protected under the confidentiality requirements of each member agency.
(d) Subcommittees and working groups, with participation not restricted to BSCC members or full-time Federal employees, may be formed to assist the BSCC in its work.
(e) All BSCC members will be fulltime Federal employees whose compensation, reimbursement for travel expenses and other costs shall be borne by their respective agencies.
(f) Each member of the BSCC shall provide such agency support and resources as may be available and necessary for the operation of the BSCC including undertaking special studies as come within the functions assigned herein.
(g) An Office of Science and Technology Policy staff member will serve as BSCC Executive Secretary.

Appendix K – Physical Containment for Large-Scale Uses of Organisms Containing Recombinant DNA Molecules

This part of the Guidelines specifices physical containment guidelines for large-scale (greater than 10 liters of culture) research or production involving viable organisms containing recombinant DNA molecules. It shall apply to large-scale research or production activities as specified in Section III - B - 5 of the Guidelines.
All provisions of the Guidelines shall apply to large-scale research or production activities with the following modifications:
• Appendix K shall replace Appendix G when quantities in excess of 10 liters of culture are involved in research or production.
• The institutions shall appoint a Biological Safety Officer (BSO) if it engages in large-scale research or production activities involving viable organis..as containing recombinant DNA molecules. The duties of the BSO shall include those specified in Section IV - B - 4 of the Guidelines.
• The institution shall establish and maintain a health surveillance program for personnel engaged in large-scale research or production activities involving viable organisms containing recombinant DNA molecules which require BL3 containment at the laboratory scale. The program shall include: preassignment and periodic physical and medical examinations; collection, maintenance and analysis of serum specimens for monitoring serologic changes that may result from the employee's work experience; and provisions for the investigation of any serious, unusual or extended illnesses of employees to determine possible occupational origin.

Appendix K - I. – Selection of Physical Containment Levels.

The selection of the physical containment level required for recombinant DNA research or production involving more than 10 liters of culture is based on the containment guidelines established in Part III of the Guidelines. For purposes of large-scale research or production, three physical containment levels are established. These are referred to as BL1 - LS, BL2 - LS, and BL3 - LS. The BL - LS level of physical containments is required for large-scale research or production of viable organisms containing recombinant DNA molecules which require BLI containment at the laboratory scale. (The BL1 - LS level of physical containment is recommended for large-scale research or production of viable organisms for which BLI is recommended at the laboratory scale such as those described in Appendix C). The BL2 - LS level of physical containment is required for large-scale research or production of viable organisms containing recombinant DNA molecules which require BL2 containment at the laboratory scale. The BL3 - LS level of physical containment is required for large-scale research or production of viable organisms containing recombinant DNA molecules which require BL3 containment at the laboratory scale. No provisions are made for large-scale research or production of viable organisms containing recombinant DNA molecules which require BL4 containment at the laboratory scale. If necessary, these requirements will be established by NIH on an individual basis.

Appendix K - II – BL1 - LS Level

Appendix K - II - A. Cultures of viable organisms containing recombinant DNA molecules shall be handled in a closed system (e.g., closed vessel used for the propagation and growth of cultures) or other primary containment equipment (e.g., biological safety cabinet containing a centrifuge used to process culture fluids) which is designed to reduce the potential for escape of viable organisms. Volumes less than 10 liters may be handled outside of a closed system or other primary containment equipment provided all physical containment requirements specified in Appendix G - II - A of the Guidelines are met.
Appendix K - II - B. Culture fluids (except as allowed in Appendix K - II - C) shall not be removed from a closed system or other primary containment equipment unless the viable organisms containing recombinant DNA molecules have been inactivated by a validated inactivation procedure. A validated inactivation procedure is one which has been demonstrated to be effective using the organism that will serve as the host for propagating the recombinant DNA molecules.
Appendix K - II - C. Sample collection from a closed system, the addition of materials to a closed system, and the transfer of culture fluids from one closed system to another shall be done in a manner which minimizes the release of aerosols or contamination of exposed surfaces.
Appendix K - II - D. Exhaust gases removed from a closed system or other primary containment equipment shall be treated by filters which have efficiencies equivalent to HEPA filters or by other equivalent procedures (e.g., incineration) to minimize the release of viable organisms containing recombinant DNA molecules to the environment.
Appendix K - II - E. A closed system or other primary containment equipment that has contained viable organisms containing recombinant DNA molecules shall not be opened for maintenance or other purposes unless it has been sterilized by a validated sterilization procedure. A validated sterilization procedure is one which has been demonstrated to be effective using the organism that will serve as the host for propagating the recombinant DNA molecules.
Appendix K - II - F. Emergency plans required by Section IV - B - 3 - f shall include methods and procedures for handling large losses of culture on an emergency basis.

Appendix K - III - BL2 - LS Level

Appendix K - III - A. Cultures of viable organisms containing recombinant DNA molecules shall be handled in a closed system (e.g., closed vessel used for the propagation and growth of cultures) or other primary containment equipment (e.g., Class III biological safety cabinet containing a centrifuge used to process culture fluids) which is designed to prevent the escape of viable organisms. Volumes less than 10 liters may be handled outside of a closed system or other primary containment equipment provided all physical containment requirements specified in Appendix G - II - B of the Guidelines are met.
Appendix K - III - B. Culture fluids (except as allowed in Appendix K - III - C) shall not be removed from a closed system or other primary containment equipment unless the viable organisms containing recombinant DNA molecules have been inactivated by a validated inactivation procedure. A validated inactivation proceudre is one which has been demonstrated to be effective using the organism that will serve as the host for propagating the recombinant DNA molecules.
Appendix K - III - C. Sample collection from a closed system, the addition of materials to a closed system, and the transfer of cultures fluids from one closed system to another shall be done in a manner which prevents the release of aerosols or contamination of exposed surfaces.
Appendix K - III - D. Exhaust gases removed from a closed system or other primary containment equipment shall be treated by filters which have efficiencies equivalent to HEPA filters or by other equivalent procedures (e.g., incineration) to prevent the release of viable organisms containing recombinant DNA molecules to the environment.
Appendix K - III - E. A closed system or other primary containment equipment that has contained viable organisms containing recombinant DNA molecules shall not be opened for maintenance or other purposes unless it has been sterilized by a validated sterilization procedure. A validated sterilization procedure is one which has been demonstrated to be effective using the organisms that will serve as the host for propagating the recombinant DNA molecules.
Appendix K - III - F. Rotating seals and other mechanical devices directly associated with a closed system used for the propagation and growth of viable organisms containing recombinant DNA molecules shall be designed to prevent leakage or shall be fully enclosed in ventilated housings that are exhausted through filters which have efficiencies equivalent to HEPA filters or through other equivalent treatment devices.
Appendix K - III - G. A closed system used for the propagation and growth of viable organisms containing recombinant DNA molecules and other primary containment equipment used to contain operations involving viable organisms containing recombinant DNA molecules shall include monitoring or sensing devices that monitor the integrity of containment during operations.
Appendix K - III - H. A closed system used for the propagation and growth of viable organisms containing the recombinant DNA molecules shall be tested for integrity of the containment features using the organism that will serve as the host for propagating recombinant DNA molecules. Testing shall be accomplished prior to the introduction of viable organisms containing recombinant DNA molecules and following modification or replacement of essential containment features. Procedures and methods used in the testing shall be appropriate for the equipment design and for recovery and demonstration of the test organism. Records of tests and results shall be maintained on file.
Appendix K - III - I. A closed system used for the propagation and growth of viable organisms containing recombinant DNA molecules shall be permanently identified. This identification shall be used in all records reflecting testing, operation, and maintenance and in all documentation relating to use of this equipment for research or production activities involving viable organisms containing recombinant DNA molecules.

Appendix K-III-J. The universal biohazard sign shall be posted on each closed system and primary containment equipment when used to contain viable organisms containing recombinant DNA molecules.

Appendix K-III-K. Emergency plans required by Section IV-B-3-f shall include methods and procedures for handling large losses of culture on an emergency basis.

Appendix K-IV–BL3-LS Level

Appendix K-IV-A. Cultures of viable organisms containing recombinant DNA molecules shall be handled in a closed system (e.g., closed vessels used for the propagation and growth of cultures) or other primary containment equipment (e.g., Class III biological safety cabinet containing a centrifuge used to process culture fluids) which is designed to prevent the escape of viable organisms. Volumes less than 10 liters may be handled outside of a closed system provided all physical containment requirements specified in Appendix G-II-C of the Guidelines are met.

Appendix K-IV-B. Culture fluids (except as allowed in Appendix K-IV-C) shall not be removed from a closed system or other primary containment equipment unless the viable organisms containing recombinant DNA molecules have been inactivated by a validated inactivation procedure. A validated inactivation procedure is one which has been demonstrated to be effective using the organisms that will serve as the host for propagating the recombinant DNA molecules.

Appendix K-IV-C. Sample collection from a closed system, the addition of materials to a closed system, and the transfer of culture fluids from one closed system to another shall be done in a manner which prevents the release of aerosols or contamination of exposed surfaces.

Appendix K-IV-D. Exhaust gases removed from a closed system or other primary containment equipment shall be treated by filters which have efficiencies equivalent to HEPA filters or by other equivalent procedures (e.g., incineration) to prevent the release of viable organisms containing recombinant DNA molecules to the environment.

Appendix K-IV-E. A closed system or other primary containment equipment that has contained viable organisms containing recombinant DNA molecules shall not be opened for maintenance or other purposes unless it has been sterilized by a validated sterilization procedure. A validated sterilization procedure is one which has been demonstrated to be effective using the organims that will serve as the host for propagating the recombinant DNA molecules.

Appendix K-IV-F. A closed system used for the propagation and growth of viable organisms containing recombinant DNA molecules shall be operated so that the space above the culture level will be maintained at a pressure as low as possible, consistent with equipment design, in order to maintain the integrity of containment features.

Appendix K-IV-G. Rotating seals and other mechanical devices directly associated with a closed system used to contain viable organisms containing recombinant DNA molecules shall be designed to prevent leakage or shall be fully enclosed in ventilated housings that are exhausted through filters which have efficiencies equivalent to HEPA filters or through other equivalent treatment devices.

Appendix K-IV-H. A closed system used for the propagation and growth of viable organisms containing recombinant DNA molecules and other primary containment equipment used to contain operations involving viable organisms containing recombinant DNA molecules shall include monitoring or sensing devices that monitor the integrity of containment during operations.

Appendix K-IV-I. A closed system used for the propagation and growth of viable organisms containing recombinant DNA molecules shall be tested for integrity of the containment features using the organisms that will serve as the host for propagating the recombinant DNA molecules. Testing shall be accomplished prior to the introduction of viable organisms containing recombinant DNA molecules and following modification or replacement of essential containment features. Procedures and methods used in the testing shall be appropriate for the equipment design and for recovery and demonstration of the test organism. Records of tests and results shall be maintained on file.

Appendix K-IV-J. A closed system used for the propagation and growth of viable organisms containing recombinant DNA molecules shall be permanently identified. This identification shall be used in all records reflecting testing, operation, and maintenance and in all documentation relating to the use of this equipment for research production activities involving viable organisms containing recombinant DNA molecules.

Appendix K-IV-K. The universal biohazard sign shall be posted on each closed system and primary containment equipment when used to contain viable organisms containing recombinant DNA molecules.

Appendix K-IV-L. Emergency plans required by Section IV-B-3-f shall include methods and procedures for handling large losses of culture on an emergency basis.

Appendix K-IV-M. Closed systems and other primary containment equipment used in handling cultures of viable organisms containing recombinant DNA molecules shall be located within a controlled area which meets the folloging requirements:

Appendix K-IV-M-1. The controlled area shall have a separate entry area. The entry area shall be a double-doored space such as an air lock, anteroom, or change room that separates the controlled area from the balance of the facility.

Appendix K-IV-M-2. The surfaces of walls, ceilings, and floors in the controlled area shall be such as to permit ready cleaning and decontamination.

Appendix K-IV-M-3. Penetrations into the controlled area shall be sealed to permit liquid or vapor phase space decontamination.

Appendix K-IV-M-4. All utilities and service or process piping and wiring entering the controlled area shall be protected against contamination.

Appendix K-IV-M-5. Hand-washing facilities equipped with foot, elbow, or automatically operated valves shall be located at each major work area and near each primary exit.

Appendix K-IV-M-6. A shower facility shall be provided. This facility shall be located in close proximity to the controlled area.

Appendix K-IV-M-7. The controlled area shall be designed to preclude release of culture fluids outside the controlled area in the event of an accidental spill or release from the closed systems or other primary containment equipment.

Appendix K-IV-M-8. The controlled area shall have a ventilation system that is capable of controlling air movement. The movement of air shall be from areas of lower contamination potential to areas of higher contamination potential. If the ventilation system provides positive pressure supply air, the system shall operate in a manner that prevents the reversal of the direction of air movement or shall be equipped with an alarm that would be actuated in the event that reversal in the direction of air movement were to occur. The exhaust air from the controlled area shall not be recirculated to other areas of the facility. The exhaust air from the controlled area may be discharged to the outdoors without filtration or other means for effectively reducing an accidental aerosol burden provided that it can be dispersed clear or occupied buildings and air intakes.

Appendix K-IV-N. The following personnel and operational practices shall be required:

Appendix K-IV-N-1. Personnel entry into the controlled area shall be through the entry area specifed in Appendix K-IV-M-1.

Appendix K-IV-N-2. Persons entering the controlled area shall exchange or cover their personal clothing with work garments such as jumpsuits, laboratory coats, pants and shirts, head cover, and shoes or shoe covers. On exit from the controlled area the work clothing may be stores in a locker separate from that used for personal clothing or discarded for laundering. Clothing shall be decontaminated before laundering.

Appendix K-IV-N-3. Entry into the controlled area during periods when work is in progress shall be restricted to those persons required to meet program or support needs. Prior to entry all persons shall be informed of the operating practices, emergency procedures, and the nature of the work conducted.

Appendix K-IV-N-4. Persons under 18 years of age shall not be permitted to enter the controlled area.

Appendix K-IV-N-5. The universal biohazard sign shall be posted on entry doors to the controlled area and all internal doors when any work involving the organism is in progress. This includes periods when decontamination procedures are in progress. The sign posted on the entry doors to the controlled area shall include a statement of agents in use and personnel authorized to enter the controlled area.

Appendix K-IV-N-6. The controlled area shall be kept neat and clean.

Appendix K-IV-N-7. Eating, drinking, smoking, and storage of food are prohibited in the controlled area.

Appendix K-IV-N-8. Animals and plants shall be excluded from the controlled area.

Appendix K-IV-N-9. An effective insect and rodent control program shall be maintained.

Appendix K-IV-N-10. Access doors to the controlled area shall be kept closed, except as necessary for access, while work is in progress. Serve doors leading directly outdoors shall be sealed and locked while work is in progress.

Appendix K-IV-N-11. Persons shall wash their hands when leaving the controlled area.

Appendix K-IV-N-12. Persons working in the controlled area shall be trained in emergency procedures.

Appendix K-IV-N-13. Equipment and materials required for the management of accidents involving viable organisms containing recombinant DNA molecules shall be available in the controlled area.

Appendix K-IV-N-14. The controlled area shall be decontaminated in accordance with established procedures following spills or other accidental release of viable organisms containing recombinant DNA molecules.

Appendix L – Release Into the Environment of Certain Plants

Appendix L-I– General Information

Appendix L specifies conditions under which certain plants as specified below, may be approved for release into the environment. Experiments in this category cannot be initiated without submission of relevant information on the proposed experiment to NIH, review by the RAC Plant Working Group, and specific approval by NIH. Such experiments also require the approval of the IBC before initiation. Information on specific experiments which have been approved will be available in ORDA and will be listed in Appendix L-III when the Guidelines are republished.

Experiments which do not meet the specifications of Appendix L-II fall under Section III-A and require RAC review and NIH and IBC approval before initiation.

Appendix L-II–Criteria Allowing Review by the RAC Plant Working Group Without the Requirement for Full RAC Review

Approval may be granted by ORDA in consultation with the Plant Working Group without the requirement for full RAC review (IBC review is also necessary) for growing plants containing recombinant DNA in the field under the following conditions:

Appendix L-II-A. The plant species is a cultivated crop of a genus that has no species known to be a noxious weed.

Appendix L-II-B. The introduced DNA consists of well-characterized genes containing no sequences harmful to humans, animals, or plants.

Appendix L-II-C. The vector consists of DNA: (i) From exempt host-vector systems (Appendix C); (ii) from plants of the same or closely related species; (iii) from nonpathogenic prokaryotes or nonpathogenic lower eukaryotic plants; (iv) from plant pathogens only if sequences resulting in production of disease symptoms have been deleted; or (v) chimeric vectors constructed from sequences defined in (i) to (iv) above. The DNA may be introduced by any suitable method. If sequences resulting in production of disease symptoms are retained for purposes of introducing the DNA into the plant, greenhouse-grown plants must be shown to be free of such sequences before such plants, derivatives, or seed from them can be used in field tests.

Appendix L-II-D. Plants are grown in controlled access fields under specified conditions appropriate for the plant under study and the geographical location. Such conditions should include provisions for using good cultural and pest control practices, for physical isolation from plants of the same species outside of the experimental plot in accordance with pollination charac-

teristics of the species, and for further preventing plants containing recombinant DNA from becoming established in the environment. Review by the IBC should include an appraisal by scientists knowledgeable of the crop, its production practices, and the local geographical conditions. Procedures for assessing alterations in and the spread of organisms containing recombinant DNA must be developed. The results of the outlined tests must be submitted for review by the IBC. Copies must also be submitted to the Plant Working Group of the RAC.

Appendix L-III – Specific Approvals

As of publication of the revised Guidelines, no specific proposals have been approved. An updated list may be obtained from the Office of Recombinant DNA Activities, National Institutes of Health, Building 31, Room 3B10, Bethesda, Maryland 20892.

(OMB's "Mandatory Information Requirements for Federal Assistance Program Announcements" (45 FR 39592) requires a statement concerning the official government programs contained in the *Catalog of*

Federal Domestic Assistance. Normally NIH lists in its announcements the number of title of affected individual programs for the guidance of the public. Because the guidance in this notice covers not only virtually every NIH program but also essentially every Federal researchprogram in which DNA recombinant molecule techniques could be used, it has been determined to be not cost effective or in the public interest to attempt to list these programs. Such a list would likely require several additional pages. In addition, NIH could not be certain that every federal program would be included as many Federal agencies, as well as private organizations, both national and international, have elected to follow the NIH Guidelines. In lieu of the individual program listing, NIH invites readers to direct questions to the information address above whether individual programs listed in the *Catalog of Federal Domestic Assistance* are affected.)

Dated: April 18, 1986.

Thomas E. Malone,

Acting Director, National Institutes of Health.
(FR Doc. 89-10120)

Index

A

A DNA
- DNA structure 71

AATAAA sequence
- transcription termination 194

Abelson murine leukaemia virus
- structure 363, 365

ABM paper
 - hybridisation techniques 442

acetolactate synthetase
- inhibition by chlorsulfuron 410, 411

acetyl CoA
- substrate for chloramphenicol acetyl transferase 327

acetyl dimethoxyphenol
- virulence region 405

acetylcholine receptor
- cDNA cloning 46

acid phosphatase
- *S. cerevisiae*
-- expression vectors 202, 203

actin gene
- S. cerevisiae
-- introns 182

actinorhodin
- biosynthesis
-- gene cluster 226

activated cellulose 442, 443

ADA
- gene therapy 502

adaptor
- definition 537

adaptors 113, 114

adenine methylation
- absence in higher cells 18
- assay for DNA replication 18, 19
- mechanism 5

adenine phosphoribosyl transferase
- selectable marker 324

adenosine deaminase
- gene therapy 502

adenosyl methionine, S-   5

adenovirus
- DNA replication *in vitro* 19
- expression in oocytes of type 2 DNA binding protein 355
- late promoter, strength 370
- oligonucleotide-directed mutagenesis of E1A region 478
- transposon mutagenesis of type 5 transforming gene 456

ADH1 gene
- *S. cerevisiae*
-- promoter region 198

ADH1 promoter
- expression of leukocyte interferon 199
- sequence 198

aerial hyphae
- *Streptomycetes* 223

affinity chromatography
- mRNA purification 34

agarose gels
- dissolution in chaotropic salts 14
- low-melting, for recovery of DNA 12
- migration of D loops 459
- preparation 10
- separation of DNA fragments 10

bacteriophage SP6
- DNA sequence of promoter 435
- RNA polymerase 433-436
- SP6 vectors 435
- use of SP6 promoter for labelling RNA 433-436

bacteriophage T4
- DNA ligase 103-109
- DNA polymerase 41, 42, 78, 111
- *E. coli* promoters 216
- polynucleotide kinase 19, 78

bacteriophage T5
- promoter P25
-- DNA sequence 241

bacteriophage T7
- genetic map 268
- φ10 promoter sequence 268
- promoter A2
-- DNA sequence 241
- regulation of transcription 267
- RNA polymerase 267
- size of genome 267

baculoviruses
- deliberate release 488

Bal31 nuclease
- digestion of double-stranded DNA 19
- generation of deletion mutants 453

Baltimore, David 38

base composition
- effects of hybridisation 426

base-specific cleavage
- Maxam-Gilbert sequencing 80-81

BBB nomenclature
- SV40 DNA sequence 330

B DNA
- structure 71

Benton and Davis procedure
- plaque hybridisation 424, 425
- sensitivity 425, 426

Berg, Paul 338
- letter to US Academy of Science 483

Berk and Sharp method
- mapping of RNA 436

bifunctional agents
- in oligonucleotide synthesis 58

bioassay, of translation products
- in cell-free systems 35
- in *Xenopus* oocytes 36, 441

biohazards, see biological safety

biological containment 485, 486
- deliberate release of micro-organisms 488
- host-vector systems 485

biological safety
- Berg letter 483
- biological containment 485, 486
- cDNA synthesis 42
- cloning in M13 vectors 98
- cloning in microrganisms
- *E. coli* χ1776 484, 546
- F factor mutants 98
- gene therapy 501
- guidelines 483, 549
- λ DNA
-- *in vitro* packaging 169
- λ vectors 161
- physical containment 486
- Polyoma virus 484
- risk assessment 585, 586
- separation of transfer and mob functions 140, 141

biological safety officer
- NIH guidelines 485

biotin
- affinity towards avidin 433
- labelling of DNA 433, 434

biotinylated DNA
- detection with avidin 434

biotinylated dUTP
- structure 434

bipartite genome
- Gemini viruses 413

Birnstiel, Max 20

bisulfite mutagenesis
- mechanism 458-461
- mutagenesis in deletion loops 461
- reaction scheme 459

BK virus
- enhancer 358

bladder carcinoma
- oncogene expression with cosmids 173

blastocyst 494

blastomere
- mammalian development 492

bld genes
- *Streptomycetes*
-- isolation via mutational cloning 231

C

H

M

Q

R

S